CHRONICA BOTANICA
● NEW SERIES OF PLANT SCIENCE BOOKS ●
edited by Frans Verdoorn
Volume XXVIII

HAROLD NORMAN MOLDENKE *was born March 11, 1909, at Watchung, New Jersey, and educated at Wardlaw School, Plainfield, N. J. (1919-20), Susquehanna Academy, Selinsgrove, Pa. (1921-25), Susquehanna University (1925-29), and Columbia University (1920-34). He was graduated with first honors and the B.S. degree from Susquehanna and received the degrees of M.A. and Ph.D. from Columbia. He came to the New York Botanical Garden on a fellowship to study algae under the late M. A.* HOWE *in 1929, was appointed part-time assistant (1929-30), then Assistant Curator (1932-37), Associate Curator (1937-48), and finally Curator & Administrator of the Herbarium (1949-52). He is at present Director of the Trailside Museum, Watchung Reservation, and Supervisor of Nature Activities for the Union County, New Jersey, Park Commission. He has been a member of the graduate faculty at Columbia, department of botany (1936-42, 1946-52), Hunter College, extension session, since 1948, and State Teachers College, Newark, N. J., since 1951, and a member of the Board of Directors of the John Burroughs Memorial Association since 1938 and the Alice Rich Northrop Memorial since 1949. In 1935-36 he held a National Research Council Fellowship and studied in most of the major botanical institutions of Europe. In 1949 he was a delegate to the Second South American Botanical Congress at Tucumán and in 1950 represented 17 botanical institutions and organizations at the Seventh International Botanical Congress at Stockholm. He has traveled and botanized in 47 states of the U.S.A., the District of Columbia, southern Canada, Mexico, Guatemala, Honduras, El Salvador, Nicaragua, Costa Rica, Panama, Cuba, Dominican Republic, Puerto Rico, Curacao, Trinidad, 10 countries of South America and 19 countries of Europe. He is a member of Sigma Xi and Pi Gamma Mu, a Fellow of the A.A.A.S., N.Y. Academy of Science, and American Geographical Society, and member or former member of 21 scientific organizations in the U.S.A., England, Mexico, Cuba, Argentina, Ecuador, and Brazil. His publications embrace 661 titles, covering over 9,400 printed pages, chiefly on the nomenclature and taxonomy of the Verbenaceae, Avicenniaceae, Stilbaceae, Symphoremaceae, Eriocaulaceae, Menispermaceae, and Amaryllidaceae, the flora of extratropical and tropical South America, plants strategic to America's war effort, North American wild flowers, tautonyms, shamrock, silphium, vernacular plant names, and state floras. His interest in the plants of the Bible is understandable since his father, Dr.* CHARLES E. MOLDENKE, *was a well-known Egyptologist and student of the antiquities, and there are no less than nine members of the clergy in his immediate family.*

ALMA LANCE MOLDENKE *(neé* ERICSON*), wife of the senior author, was born in New York City on April 29, 1908 and received her training in the New York City schools and Hunter College where she was elected to Phi Sigma (B.A. 1931). She studied at the Woods Hole Marine Biological Station in 1931 on a scholarship. Her graduate work in botany, zoology, psychology and education was done at Columbia University. She has been a teacher of biology in the Model School of Hunter College (1931-1934) and in the Evander Childs High School in New York City (1934-1952). She has also been active in social service, religious, educational, first aid training, scientific, camping, and nature study groups and organizations, such as the Northrop Memorial Camp where she was in charge of the nature study program and later served as director. With her husband she was a delegate to both the Second South American Botanical Congress held at Tucumán and the Seventh International Botanical Congress held at Stockholm. She has traveled and done some botanizing throughout most of the states of this country, the Central and South American countries, the Scandinavian peninsula, and the British Isles. Mrs.* MOLDENKE *has written a few articles independently on camping and educational subjects, and she has collaborated with her husband in a few of his more popular scientific works.*

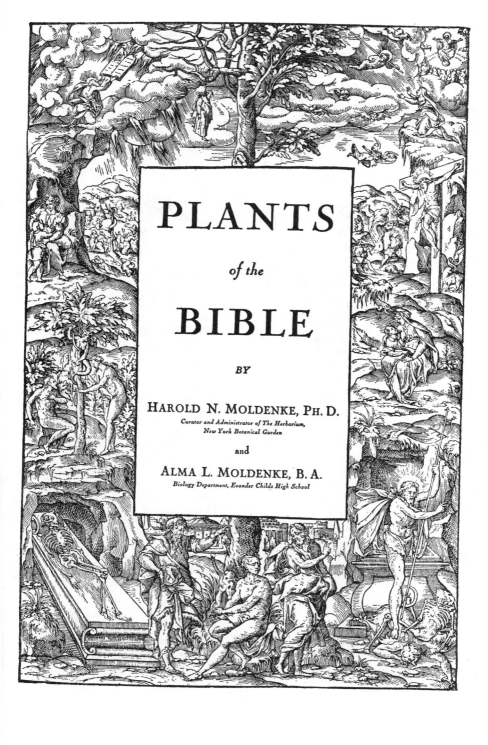

PLANTS

of the

BIBLE

BY

HAROLD N. MOLDENKE, PH. D.
*Curator and Administrator of The Herbarium,
New York Botanical Garden*

and

ALMA L. MOLDENKE, B. A.
Biology Department, Evander Childs High School

FIGURE 1.—*Cedrus libani*. A grove of giant Cedars-of-Lebanon on white limestone in Lebanon, with the snow-capped peaks of Dar el Kodīb in the background. Thus did this Prince of Trees grow in Biblical times before man brought destruction and near-extinction to the species! (Wood engraving from C. W. Wilson's Picturesque Palestine, 1883; see also p. 64 and fig. 52).

PLANTS
of the
BIBLE

BY
HAROLD N. MOLDENKE, Ph.D.
Curator and Administrator of The Herbarium,
New York Botanical Garden

and

ALMA L. MOLDENKE, B.A.
Biology Department, Evander Childs High School

Dover Publications, Inc.
New York

Copyright © 1952 by The Ronald Press Company.
All rights reserved under Pan American and International Copyright Conventions.

Published in Canada by General Publishing Company, Ltd., 30 Lesmill Road, Don Mills, Toronto, Ontario.
Published in the United Kingdom by Constable and Company, Ltd.

This Dover edition, first published in 1986, is an unabridged and unaltered republication of the work first published by the Chronica Botanica Company, Waltham, Massachusetts in 1952.

Manufactured in the United States of America
Dover Publications, Inc., 31 East 2nd Street, Mineola, N.Y. 11501

Library of Congress Cataloging-in-Publication Data

Moldenke, Harold N. (Harold Norman), 1909–
  Plants of the Bible.

  Reprint. Originally published: Waltham, Massachusetts : Chronica Botanica Co., 1952.
  Bibliography: p.
  Includes indexes.
  1. Plants in the Bible. I. Moldenke, Alma L. (Alma Lance), 1908–   II. Title.
BS665.M6 1986   220.8′582   85-27374
ISBN 0-486-25069-5

Preface

Twelve years have elapsed since we began work on this résumé of present day knowledge of Biblical botany. During the course of the work much assistance was rendered us by many friends and colleagues, and by the staffs of various libraries. Especially do we wish publicly to acknowledge our deep appreciation of the help given us by various members of the staff of the American Museum of Natural History, the American Geographical Society, the College of Physicians and Surgeons, Columbia University Library, the Jewish Theological Seminary, the New York Academy of Medicine, the New York Botanical Garden, the New York Public Library (especially the Jewish Reading Room), and Union Theological Seminary, all in New York City, and of the Arnold Arboretum at Jamaica Plain, Massachusetts, and the Hebrew University at Jerusalem. We are also most grateful to Mr. T. K. BARRIE, of Coudersport, Pa., for information graciously supplied about the lentisk tree; to Mrs. GRACE ESTERNAUX, of the Biology Department of Evander Childs High School, New York City, for checking all the Greek words of the text in the Septuagint and in various concordances, Greek dictionaries, and lexicons; to Dr. MICHAEL EVENARI, of the Botany Department of the Hebrew University at Jerusalem, for his gift of illustrations and for his reading of and valued criticisms of the whole text; to the late Dr. J. M. GREENMAN, of the Missouri Botanical Garden, St. Louis, for his checking of certain bibliographic materials; to Dr. ROBERT F. GRIGGS, of George Washington University, Washington, for information relative to the banana-grape controversy; to Miss ELIZABETH C. HALL, librarian of the New York Botanical Garden, for her perennial assistance, so cheerfully rendered, in many bibliographic problems; to Rev. ALBERT K. HERLING, then of Tufts College and pastor of the Unity Community Church, Natick, Mass., for his careful checking of references from the Apocrypha and for other assistance; to Rev. Dr. JOHN HAYNES HOLMES, distinguished pastor emeritus of the Community Church, New York City, for his reading of certain chapters and especially for his helpful advice and enthusiastic encouragement of the whole project; to Miss BARBARA HOSKINS, then of the New York Botanical Garden library staff, for her copious and very efficient help in bibliographic matters; to Dr. and Mrs. MOSES BARRAS, and Mrs. SYLVIA KAPLAN, of the Hebrew Department of Evander Childs High School, New York City, for their invaluable aid in checking all the Hebrew characters and transliterations; to Miss ROSLYN NORTON, erstwhile instructor in Hebrew at the Schiff Jewish Cultural Center, New York City, and Mr. MARTIN NORTON, Hebrew student, for checking the Hebrew words and Old Testament references in ancient and modern Hebrew Bibles and for providing some and checking all their English transliterations; to the late Professor ALFRED REHDER, of Arnold Arboretum, for valuable information regarding the geographic distribution of certain species of trees; to Mr. W. T. STEARN, of the Royal Horticultural Society, London, for information about lilies, onions, papyrus, etc.; to Mr. JASON R. SWALLEN, of the United States National Museum, Washington, for information relative to grasses; to Dr. LOUIS C. WHEELER, of the University of Southern California, Los Angeles, for information about Biblical dyes and their sources; to Miss CAROL WOODWARD, then editorial assistant at the New York Botanical Garden, for many helpful suggestions during the course of the work; to Mr. E. R. VAN LIEW, of Givaudan-Delawanna Company, New York, for information on labdanum; to Dr. JOHN C. TREVOR, Director of the Department of English Bible, National Council of Churches of Christ in the U.S.A., Chicago; and to Dr. FRANS VERDOORN, the editor of Chronica Botanica, for his invaluable cooperation and painstaking help so generously extended throughout the preparation of this book.

We are grateful to the following publishers and copyright owners for permission to quote from the books listed in full in our bibliography and indicated here by the numbers in parentheses: John Murphy Company, Baltimore, Maryland (1); Hebrew Publishing Company, 632-634 Broadway, New York City (6); Jewish Press Publishing Company (9); University of Chicago Press, Chicago, Ill. (5)*; A. J. Holman Company, Philadelphia, Pa. (8); Harper & Brothers, Publishers, New York City (10, 11, 12); Thomas Nelson & Sons, New York City (14); Oxford University Press, New York City (3); E. P. Dutton & Company, New York City (7); The Confraternity of Christian Doctrine, St. Anthony Guild Press, Paterson, N. J. (13)**; The International Council of Religious Education, Thomas Nelson & Sons, New York City (14); The Pilgrim Press, Boston, Mass. (15); The Jewish Publication Society of America, Philadelphia, Pa. (16); S. S. Scranton & Company, Hartford, Conn. (306); Hardwicke & Bogue, 192 Piccadilly West, London, England (299); Methuen & Company, 36 Essex Street, London, W. C., England (95); Lippincott, Philadelphia, Pa. (298); American University Press (Macmillan Company, 60 Fifth Avenue, New York City) (265, 266, 267); Virtue & Yorsten, New York City (119); and John B. Alden, New York City (128a).

We wish also to express here our gratitude for the copious assistance given us in semantic and historic matters by Mr. B. D'ARLON, head of the foreign language department, Evander Childs High School, Dr. J. BLOCH, of the New York Public Library, Rabbi A. OPHER and Mr. H. WESLER, both of Paterson, New Jersey, Rabbis A. ROSENBERG and D. SHOHET, both of Yonkers, New York, Rev. Dr. V. E. BECK, of the Gustavus Adolphus Church in New York City, and the library staff of the American Bible Society, New York City. We are greatly indebted to Dr. M. ZOHARY, Dr. H. R. OPPENHEIMER, and Dr. BANDMANN, all of Israel, for their great kindness in supplying us with illustrations, to Miss W. WALKER and Mrs. M. KOTELES for their drawings for vignettes and tailpieces, and to Miss F. W. SMITH, Methodist missionary, and our son, A. R. MOLDENKE, for much time-saving assistance in the preparation of the index materials.

It should be explained here that in the preparation of the transliterations into English characters of the Hebrew words referred to in most of the chapters, we came upon many variations in spelling and accentuation in the different authorities listed in the bibliography. Probably only a very few of these variations can be credited to inaccuracies in copying, typographic errors, or similar mistakes. Most of them are due to differences in Hebrew pronunciation, such as the Ashkenazi (used so widely in Europe) or Sephardic (which is the classical Palestinian) and the various intermediate dialects, and to the lack of exact equivalents of Hebrew letter-sounds in our language. In cases where we have given several transliterations for the same Hebrew word, these are the forms common in the quoted literature. They should, in most cases, be easily recognizable to Hebrew students with either a classical or an Ashkenazi background. Dr. J. C. TREVOR, in several letters to us, takes exception to many of these transliterations, stating that they are obsolete forms not generally accepted by Hebrew scholars today. He states, for instance, that today "b", "g", "d", "k", "p", and "t" are used in place of the "bh", "gh", "dh", "kh", "ph", and "th" of older writers quoted by us for the softened letters (used when they follow a vowel). Also "ṣ" is used more frequently today than "sh". The "sheᵂa" or shortest Hebrew vowel is now indicated by raising the letter (according to its color, *i.e.*, a, e, or o) slightly above the line; "aleph" is indicated by an apostrophe opened to the left; and "'ayin" is indicated by an apostrophe opened to the right, as in their letter names. Dr. TREVOR's corrections have come to us too late to be incorporated in the general text of the book, but are given in part in the Supplementary Notes.

In various places in the text we have had occasion to assign approximate dates to Biblical occurrences. In this we have followed the so-called Archbishop USSHER chronology. We realize that modern Biblical scholars do not place complete reliance on this old chronology, but in view of the fact that it is still widely accepted by the general Bible-reading public, we have decided to follow it here.

THE AUTHORS

---

*These publishers prefer to have what we and others call the "Goodspeed Version" referred to as "The University of Chicago Version".
**These publishers prefer to have what we and others call the "O'Hara Version" referred to as "The Confraternity Edition".

To the Memory of
John Hendley Barnhart
Botanical Bibliographer and Biographer
An unfailing Font of Knowledge
Always cheerfully shared, a thorough Scientist
Abundantly imbued with Humanity
Genial Adviser and Friend
this Book is gratefully dedicated

Contents

The Authors	ii
Preface	vii
Contents	xi
List of Illustrations	xv
Historical Sketch	1
Description of the Land	13
Helps to Users of this Work	17
1. *Acacia nilotica* (L.) Forsk.	23
2. *Loranthus acaciae* Zucc.	23
3. *Acacia seyal* Delile	24
4. *Acacia tortilis* Hayne	24
5. *Acanthus syriacus* Boiss.	26
6. *Acetobacter acetigenum* (Henneberg) Bergey	27
7. *Acetobacter acetum* (Thomsen) H.B.H.H.	27
8. *Acetobacter plicatum* Fuhrmann	27
9. *Acetobacter xylinum* (Brown) Bergey	27
10. *Aegilops variabilis* Eig	28
11. *Alopecurus anthoxanthoides* Boiss.	28
12. *Avena sterilis* L.	28
13. *Eragrostis megastachya* (Koel.) Link	28
14. *Nardurus orientalis* Boiss.	28
15. *Polypogon monspeliensis* (L.) Desf.	28
16. *Agrostemma githago* L.	29
17. *Solanum incanum* L.	29
18. *Alhagi camelorum* var. *turcorum* (Boiss.) Boiss.	30
19. *Alhagi maurorum* Medic.	31
20. *Fraxinus ornus* L.	31
21. *Tamarix mannifera* (Ehrenb.) Bunge	31
22. *Allium ascalonicum* L.	32
23. *Allium sativum* L.	32
24. *Allium cepa* L.	33
25. *Allium porrum* L.	34
26. *Trigonella foenum-graecum* L.	34
27. *Aloë succotrina* Lam.	35
28. *Amygdalus communis* L.	35
29. *Anastatica hierochuntica* L.	38
30. *Gundelia tournefortii* L.	38
31. *Andropogon aromaticus* Roxb.	39
32. *Anemone coronaria* L.	41
33. *Anthemis palaestina* Reut.	41
34. *Anethum graveolens* L.	46
35. *Apinus pinea* (L.) Neck.	46
36. *Aquilaria agallocha* Roxb.	47
37. *Santalum album* L.	47
38. *Artemisia herba-alba* Asso	48
39. *Artemisia judaica* L.	48
40. *Arundo donax* L.	50
41. *Astragalus gummifer* Labill.	51
42. *Astragalus tragacantha* L.	51
43. *Atriplex dimorphostegia* Kar. & Kir.	53
44. *Atriplex halimus* L.	53
45. *Atriplex rosea* L.	53
46. *Atriplex tatarica* L.	53
47. *Balanites aegyptiaca* (L.) Delile	55
48. *Boswellia carterii* Birdw.	56
49. *Boswellia papyrifera* Hochst.	56
50. *Boswellia thurifera* Roxb.	56
51. *Brassica nigra* (L.) W. Koch	59
52. *Butomus umbellatus* L.	62
53. *Buxus longifolia* Boiss.	62
54. *Capparis sicula* Duham.	65
55. *Cedrus libani* Loud.	66
56. *Centaurea calcitrapa* L.	70
57. *Centaurea iberica* Trev.	70
58. *Centaurea verutum* L.	70
59. *Silybum marianum* (L.) Gaertn.	70
60. *Ceratonia siliqua* L.	72
61. *Cercis siliquastrum* L.	73
62. *Cichorium endivia* L.	74
63. *Cichorium intybus* L.	74
64. *Lactuca sativa* L.	74
65. *Nasturtium officinale* R. Br.	74
66. *Rumex acetosella* var. *multifidus* (L.) P. DC. & Lam.	74
67. *Taraxacum officinale* Weber	74
68. *Cinnamomum cassia* Blume	75
69. *Cinnamomum zeylanicum* Nees	76
70. *Cistus creticus* L.	77
71. *Cistus salvifolius* L.	77
72. *Cistus villosus* L.	77
73. *Citrullus colocynthis* (L.) Schrad.	78
74. *Citrullus vulgaris* Schrad.	80

75. *Cucumis melo* L. 80
76. *Commiphora africana* (Arn.) Engl. 81
77. *Commiphora kataf* (Forsk.) Engl. 82
78. *Commiphora myrrha* (Nees) Engl. 82
79. *Commiphora opobalsamum* (L.) Engl. 84
80. *Coriandrum sativum* L. 86
81. *Crocus cancellatus* var. *damascenus* (Herb.) G. Maw 86
82. *Crocus hyemalis* Boiss. & Bl. 86
83. *Crocus vitellinus* Wahlenb. 86
84. *Crocus zonatus* J. Gay 86
85. *Crocus sativus* L. 87
86. *Cucumis chate* L. 88
87. *Cucumis sativus* L. 88
88. *Cuminum cyminum* L. 89
89. *Cupressus sempervirens* var. *horizontalis* (Mill.) Gord. 89
90. *Cynomorium coccineum* L. 91
91. *Cyperus papyrus* L. 92
92. *Diospyros ebenaster* Retz. 95
93. *Diospyros ebenum* König 95
94. *Diospyros melanoxylon* Roxb. 95
95. *Eberthella typhi* (Schröt.) Buchanan 97
96. *Rickettsia prowazeki* da Rocha-Lima 97
97. *Streptococcus erysipelatis* Fahleisen 97
98. *Elaeagnus angustifolia* L. 97
99. *Epidermophyton rubrum* Castellani 98
100. *Favotrichophyton violaceum* (Sabouraud) Dodge 98
101. *Trichophyton rosaceum* Sabouraud 98
102. *Faba vulgaris* Moench 101
103. *Ferula galbaniflua* Boiss. & Buhse. 102
104. *Ficus carica* L. 103
104a. *Ficus carica* var. *silvestris* Nees.. 103
105. *Ficus sycomorus* L. 106
106. *Gossypium herbaceum* L. 109
107. *Hedera helix* L. 111
108. *Hordeum distichon* L. 111
109. *Hordeum hexastichon* L. 111
110. *Hordeum vulgare* L. 111
111. *Hyacinthus orientalis* L. 114
112. *Lilium candidum* L. 114
113. *Iris pseudacorus* L. 117
114. *Juglans regia* L. 119
115. *Juncus effusus* L. 120
116. *Juncus maritimus* Lam. 120
117. *Scirpus holoschoenus* var. *linnaei* (Reichenb.) Asch. & Graebn. ... 120
118. *Scirpus lacustris* L. 120
119. *Scirpus maritimus* L. 120
120. *Juniperus oxycedrus* L. 121
121. *Laurus nobilis* L. 123
122. *Lawsonia inermis* L. 124
123. *Lecanora affinis* Eversm. 125
124. *Lecanora esculenta* (Pall.) Eversm. 125
125. *Lecanora fruticulosa* Eversm. 125
126. *Nostoc* spp. 125
127. *Lens esculenta* Moench 128
128. *Lilium chalcedonicum* L. 129
129. *Linum usitatissimum* L. 129
130. *Lolium temulentum* L. 133
131. *Lycium europaeum* L. 134

132. *Mandragora officinarum* L. 137
133. *Mentha longifolia* (L.) Huds. 139
134. *Morus nigra* L. 140
135. *Mucor mucedo* L. 141
136. *Mycobacterium leprae* (A. Hansen) Lehm. & Neum. 142
137. *Mycobacterium tuberculosis* var. *hominis* (Koch) Lehm. & Neum. 143
138. *Myrtus communis* L. 143
139. *Narcissus tazetta* L. 147
140. *Nardostachys jatamansi* (Wall.) P. DC. 148
141. *Neisseria gonorrhoeae* Trevis. 149
142. *Treponema pallidum* (Schaud. & Hoffm.) Schaud. ... 149
143. *Nerium oleander* L. 151
144. *Nigella sativa* L. 152
145. *Notobasis syriaca* (L.) Cass. 153
146. *Scolymus maculatus* L. 153
147. *Nymphaea alba* L. 154
148. *Nymphaea caerulea* Sav. 154
149. *Nymphaea lotus* L. 154
150. *Olea europaea* L. 157
151. *Origanum maru* L. 160
152. *Origanum maru* var. *aegyptiacum* (L.) Dinsm. 160
153. *Ornithogalum umbellatum* L. 162
154. *Paliurus spina-christi* Mill. 165
155. *Panicum miliaceum* L. 166
156. *Pasteurella pestis* (Lehm. & Neum.) Bergey 167
157. *Phoenix dactylifera* L. 169
158. *Phragmites communis* Trin. 172
159. *Pinus brutia* Tenore 173
160. *Pinus halepensis* Mill. 175
161. *Pistacia lentiscus* L. 177
162. *Pistacia terebinthus* var. *palaestina* (Boiss.) Post 178
163. *Pistacia vera* L. 179
164. *Platanus orientalis* L. 180
165. *Populus alba* L. 181
166. *Populus euphratica* Oliv. 183
167. *Prunus armeniaca* L. 184
168. *Pterocarpus santalinus* L. f. 188
169. *Punica granatum* L. 189
170. *Quercus aegilops* L. 193
171. *Quercus coccifera* L. 193
172. *Quercus coccifera* var. *pseudococcifera* (Desf.) Boiss. 193
173. *Quercus ilex* L. 193
174. *Quercus lusitanica* Lam. 193
175. *Retama raetam* (Forsk.) Webb. & Berth. 201
176. *Rhamnus palaestina* Boiss. 202
177. *Ricinus communis* L. 203
178. *Roccella tinctoria* Lam. & P. DC... 204
179. *Rosa phoenicia* Boiss. 205
180. *Rubus sanctus* Schreb. 206
181. *Rubus ulmifolius* Schott 206
182. *Ruscus aculeatus* L. 207
183. *Ruta chalepensis* var. *latifolia* (Salisb.) Fiori 208
184. *Ruta graveolens* L. 208
185. *Sabina excelsa* (Bieb.) Antoine.... 209
186. *Sabina phoenicia* (L.) Antoine.... 209
187. *Saccharomyces cerevisiae* Meyen.. 210

188. *Saccharomyces ellipsoideus* Reess. 212
189. *Saccharum officinarum* L. 214
190. *Salicornia fruticosa* (L.) L. 215
191. *Salicornia herbacea* (L.) L. 215
192. *Salsola inermis* Forsk. 215
193. *Salsola kali* L. 215
194. *Salix acmophylla* Boiss. 216
195. *Salix alba* L. 216
196. *Salix fragilis* L. 216
197. *Salix safsaf* Forsk. 216
198. *Salvia judaica* Boiss. 218
199. *Saussurea lappa* (Decaisne)
  C. B. Clarke 218
200. *Shigella ambigua* (Andrews)
  Weldin 219
201. *Shigella dysenteriae* (Shiga)
  Castel. & Chalm. 219
202. *Shigella paradysenteriae* (Collins)
  Weldin 219
203. *Sinapis arvensis* L. 220
204. *Solanum incanum* L. 221
205. *Sorghum vulgare* var. *durra*
  (Forsk.) Dinsm. 222
206. *Staphylococcus albus* Rosenb. 223
207. *Staphylococcus aureus* Rosenb. ... 223
208. *Streptococcus pyogenes* Rosenb. .. 223
209. *Styrax benzoin* Dryand. 223
210. *Styrax officinalis* L. 224

211. *Tamarix articulata* Vahl 227
212. *Tamarix pentandra* Pall. 227
213. *Tamarix tetragyna* Ehrenb. 227
214. *Tetraclinis articulata* (Vahl)
  Masters 228
215. *Triticum aestivum* L. 228
216. *Triticum compositum* L. 228
217. *Triticum aestivum* var. *spelta* (L.)
  L. H. Bailey 233
218. *Tulipa montana* Lindl. 234
219. *Tulipa sharonensis* Dinsm. 234
220. *Typha angustata* Bory & Chaub... 235
221. *Urtica caudata* Vahl 237
222. *Urtica dioica* L. 237
223. *Urtica pilulifera* L. 237
224. *Urtica urens* L. 237
225. *Vitis orientalis* (Lam.) Boiss. 239
226. *Vitis vinifera* L. 240
227. *Xanthium spinosum* L. 245
228. *Zizyphus lotus* (L.) Lam. 247
229. *Zizyphus spina-christi* (L.) Willd.. 248
230. *Zostera marina* L. 249
UNIDENTIFIED PLANT REFERENCES 251
BIBLIOGRAPHY 259
SUPPLEMENTARY NOTES 275
INDEX TO BIBLE VERSES 293
GENERAL INDEX 301

Figures 2-4. — Renaissance Bibles and Biblical commentaries were often illustrated with decorative and symbolic engravings. On this page are reproduced three charming woodcuts from an old German, emblematic commentary on the Psalms, Hohberg's "Psalter Davids" (1680), with illustrations of plants at that time considered as very typical Bible plants. In more recent times, it has been shown that two of the three species illustrated here are not actually Bible plants. *Cercis siliquastrum*, long considered to be the tree in which Judas hanged himself, is not mentioned in the Scriptures. *Passiflora incarnata*, one of the passionflowers, is an American plant and was unknown in Biblical times, though it has been popularly supposed to be a Bible plant since 1610, because of a fancied symbolism of the Crucifixion seen in its remarkable flowers. The "Adam's Apple" (labeled "Malus assyrica" here) is not a true apple (*cf.* figs. 91, 94 and 95), nor is it the plant referred to as "apple" in the Old Testament—it appears, rather, to be the sycamore fig, *Ficus sycomorus*. (Courtesy Arnold Arboretum of Harvard University).

List of Illustrations

VIGNETTE by Professor OLLE HJORTZBERG: Moses smiting the rock in Horeb and obtaining water therefrom, as recounted in the book of Exodus. Though the rod appears somewhat generalized, it is probable that an almond branch was intended. (Swedish Bible, 1927, cf. sub fig. 9) i

HALF TITLE: In this wood engraving, GERARD JANSSEN, of Kampen, attempted to crowd, into a single picture, the leading events of the Old and New Testaments, including (on the left) MOSES receiving the Commandments, HANNAH and the infant SAMUEL, the raising of the golden serpent, the temptation of ADAM and EVE in the Garden of Eden, with the ultimate end of the Old Dispensation being death. On the right, or New Dispensation, side of the engraving are depicted the Ascension into Heaven, the angels heralding the good news of JESUS' birth, the Annunciation to MARY, the Crucifixion, ELIZABETH and the infant JOHN, and JESUS' triumph over death. In the center is the Tree of Life, green on one side, but withered and dead on the side where JUDAS has hanged himself. At its base are ISAIAH and JOHN THE BAPTIST pointing Mankind from the Old to the New Dispensation. (Frontispiece to the Flemish Bible, published by Plantin in 1566) iii

FIG. 1: *Cedrus libani*, Cedar - of - Lebanon (see also p. 64 and fig. 52) iv

HEADPIECE for the book of Genesis: The artist attempts to symbolize the Garden of Eden, with the intertwined stems of the fig, pomegranate, and the vine, and prancing deer indicating the fruitfulness of the earth. At the lower left is a symbolic representation of the all-seeing eye of God watching over the earth. (Swedish Bible, 1927, cf. sub fig. 9) vii

HEADPIECE: The Good Shepherd, bearing the Lost Sheep on his shoulders and carrying the symbolic shepherd's staff. With him are four figures representing the four seasons, symbolic of the shortness of life. (After a painting in the crypt of St. Calixtus, Roman catacombs), this headpiece, as well as the vignettes on p. 11, 15, DDD, and at the end of the book, have been drawn by Mrs. MARTHA KOTELES from tracings reproduced in LOUISE TWINING, Symbols and Emblems of Early and Mediaeval Christian Art, 1852 xi

TAILPIECE: Emblematic representation from HOHBERG'S Psalter Davids (cf. p. xiv) xiii

FIGS. 2-4: *Cercis siliquastrum*, Judas Tree; *Passiflora*; "Adam's Apple" xiv

HEADPIECE for the book of Judges: The artist has depicted the burning of the grain of the Philistines by SAMSON after he had been tricked out of a Philistine wife. The "corn" of the Bible was chiefly barley, spelt, millet, and a poor grade of wheat. (Swedish Bible, 1927, cf. sub fig. 9) xv

FIGS. 5-8: Four representations of the Temptation in the Garden of Eden. xvi

FIG. 9: Title page for the book of Deuteronomy, depicting the aged MOSES describing the Promised Land, with date palms *Phoenix dactylifera* (see also fig. 28, fig. 70, and fig. 92) xviii

FIGS. 10-12: Three pages from LEMNIUS'S Herbarium ... in Bibliis (1566), the first book ever published on the plants of the Holy Scriptures xx

INITIAL used by the Plantin Press 1

TAILPIECE: An artist's attempt to portray the "good" and the "corrupt" trees described in MATTHEW. The good tree is distinguished by its flourishing leaves and by the lamps, signifying good deeds, suspended from its branches. The corrupt tree has bare branches and the feller's ax already is buried in its trunk. (After sculpture in the Cathedral of Amiens) 11

FIG. 13: Vegetation Map of Modern Palestine. (Courtesy Dr. ZOHARY) 12

INITIAL used by the Plantin Press 13

FIGURES 5-8. — Three representations of the Temptation in the Garden of Eden, a favorite subject through the ages. The Tree of Knowledge has sometimes been depicted as a fig tree, sometimes as a citron, sometimes as a mountain-ash, but most often as an apple (as in figs. 91, 94 and 95). Fig. 5 is from the typographically famous "Methodus Primum Olimpiade" (published in Basel, 1504), in fig. 6, a Renaissance woodcut, the skeleton tree bears fruits which look like citrons or pomelos. In fig. 7 the central tree is obviously a fig, to the right of which is a grape vine and to the left apples (from a modern advertisement, designed by JOHN FARLEIGH, for the British Brewers' Society).—Fig. 8, a fine, contemporary drawing by JOSEPH SCHARL for a series of interpretations of the Old and New Testament, symbolizes King SOLOMON and his Beloved enjoying the manifold blessings of an Eden-like paradise, possibly his own famous garden.—See also p. 329, fig. 91 and fig. 95.

TAILPIECE: Two doves holding olive branches over a burial urn, apparently an allusion to conjugal affection lasting even after death. (From a tomb in the Roman catacombs) 15
FIG. 14: Bethlehem from the southwest. 16
INITIAL used by the Plantin Press (*cf.* figs. 64 and 65) 17
TAILPIECE: *Amygdalus communis*, the almond. The twelve "rods" represent the twelve tribes of Israel, laid up by MOSES in the Tabernacle. The "rod" of AARON is said to have sprouted and "brought forth buds, and bloomed blossoms, and yielded almonds", see also fig. 37 and p. 274. (Swedish Bible, 1927, cf. sub fig. 9) 21
FIG. 15: *Acacia tortilis*, the "shittah tree" 22
FIG. 16: *Astragalus tragacantha*, the gum-tragacanth 54
Cedrus libani, the majestic Cedar-of-Lebanon, see also fig. 1 and fig. 52. (Tailpiece by WINIFRED WALKER) 64
FIGS. 17-18: Seeds of *Cuminum cyminum* (cummin) and *Coriandrum sativum* (coriander, see also fig. 57) 94
FIG. 19: *Cupressus sempervirens* var. *stricta*, the columnar cypress 96
Commiphora myrrha, the myrrh. (Vignette by WINIFRED WALKER) 99
FIG. 20: *Lawsonia inermis*, the henna-plant 100
FIG. 21: *Gundelia tournefortii* 108
Acacia seyal, source of "shittim wood". (Vignette by WINIFRED WALKER) .. 110
FIG. 22: *Anastatica hierochuntica*, the "rolling thing" of ISAIAH 118
Cistus salvifolius, a Palestinian rockrose, the source of ladanum used in perfumery, see also fig 53. (Vignette by WINIFRED WALKER) 122
Linum usitatissimum, the flax, source of the Biblical linen. (Vignette by WINIFRED WALKER) 135
FIG. 23: *Morus nigra*, the red mulberry 136
Myrtus communis, the true myrtle, see also fig. 68. (Vignette by WINIFRED WALKER) 145
FIG. 24: *Narcissus tazetta*, the polyanthus narcissus 146
Nymphaea caerulea, the blue lotus. (Tailpiece by WINIFRED WALKER) 155
FIG. 25: *Olea europaea*, olive trees in the Garden of Gethsemane (see also p. 15 and fig. 26) 156
FIG. 26: NOAH shown with olive leaves and dove (see also p. 15 and fig. 25).... 163
FIG. 27: *Lecanora esculenta*, the manna lichen (see also fig. 93); *Paliurus spina-christi*; *Commiphora kataf* 164
FIG. 28: *Phoenix dactylifera*, the date palm in a Haifa garden (see also fig. 9, fig. 70, and fig. 92) 192
Punica granatum, the pomegranate, see also fig. 38 and fig. 76. (Vignette by WINIFRED WALKER) 199
FIG. 29: Abraham's Oak (*Quercus coccifera* var. *pseudococcifera*) near Hebron, in the 1870's (see also fig. 84) 200
Styrax officinalis, the storax-tree, source of the Biblical "stacte" and "sweet storax". (Vignette by WINIFRED WALKER) .. 225
FIG. 30: *Tamarix* sp., tamarisks in the Sinai Desert 226
Apinus pinea, the stone pine; referred to as "green fir tree" by HOSEA. (Tailpiece by WINIFRED WALKER) 236
FIGS. 31-32: *Vitis vinifera*, the common grape-vine of the Old World (see also fig. 71) 238
FIG. 33: *Saussurea lappa*, the Indian orris 246
FIG. 34: *Nardostachys jatamansi*, the Himalayan spikenard 246
Nigella sativa, the nutmeg-flower; referred to by ISAIAH as "fitches". (Vignette by WINIFRED WALKER) 249
FIG. 35: *Tabernaemontana alternifolia*, the divi-ladner 250
Cinnamomum zeylanicum, the cinnamon, famed for its fragrant bark. (Vignette by WINIFRED WALKER) 257
FIGS. 36-39: *Anastatica hierochuntica*, the Palestinian tumbleweed or "the rose plant in Jericho" (see also fig. 22); *Amygdalus communis*, the almond (see also p. 21 and p. 274); *Punica granatum*, the pomegranate (see also p. 199 and fig. 76); *Ricinus communis*, the castor-bean 258
Amygdalus communis, the almond; also referred to as "hazel" and by the proper name "Luz", see also p. 21 and fig 37. (Vignette by WINIFRED WALKER)... 274
FIG. 40: The Garden of Eden as represented in an incunabulum 292
HEADPIECE for the book of Ecclesiastes: The artist contrasts the worldly man who spends his time in gaiety and the amassing of wealth, yet must leave all this behind, with the spiritual man who rises above earthly things to the lasting things of the soul. Note the use of the *Acanthus* leaf to sheathe the vine parts. (Swedish Bible, 1927, cf. sub fig. 9) 293
FIG. 41: The Parable of the Tares as recounted in the 13th chapter of Matthew (with *Lolium temulentum*, darnel grass, see also fig. 72, and *Triticum aestivum*, wheat) 300
HEADPIECE for the book of Wisdom of Solomon, in the Apocrypha: Here the artist depicts the contrasting fate of the wicked and ungodly, who are uprooted and destroyed by the serpents of falsehood which they served, and that of the wise and righteous who flourish like a date palm entwined by a fruitful vine and lifting her head to the bright sunshine of

FIGURE 9. — The large Swedish Family and Pulpit Bible, published in 1927 by A. B. Nordisk Familjeboks Bibelförlag in Stockholm, Sweden, an unusually fine typographical accomplishment, is also unique in being one of the very few modern Bibles illustrated with, in general, quite accurate as well as highly ornamental and artistic drawings of various Bible plants by Professor OLLE HJORTZBERG. We reproduce this title page for the book of Deuteronomy, wherein the distinguished artist has depicted the aged MOSES describing the Promised Land, rich in blessings, into which he himself could not take his people, but which he commanded them to "go up and possess it". The date palms (see also fig. 28, fig. 70, and fig. 92) in the drawing are heavy with fruit symbolizing the fruitfulness of the land.

God's favor. (Swedish Bible, 1927, cf. sub fig. 9) 301
Another representation of the Temptation of ADAM and EVE (see also figs. 5-8, fig. 91 and fig. 95). Evil spirits appear, in bodily form, as tempters to the act of disobedience. The fruit seems plainly intended to be an apple. (After a 14th Century manuscript in the British Museum) 329

— Plates (after page 329) —

FIG. 42: *Alhagi camelorum* var. *turcorum*, the camel's thorn.
FIG. 43: *Artemisia herba-alba*, wormwood.
FIG. 44: *Aquilaria agallocha*, the eaglewood.
FIG. 45: *Loranthus acaciae*, the acacia strapflower.
FIG. 46: *Anemone coronaria*, the Palestine anemony.
FIG. 47: *Laurus nobilis*, the bay-tree.
FIG. 48: *Acanthus spinosus*.
FIG. 49: *Aloë succotrina*.
FIG. 50: *Anthemis palaestina*, the Palestine chamomile.
FIG. 51: *Capparis sicula*, the Palestine caper.
FIG. 52: *Cedrus libani*, the Cedar-of-Lebanon (see also fig. 1 and p. 64).
FIG. 53: *Cistus salvifolius* (see also p. 122).
FIG. 54: *Citrullus colocynthis*, the colocynth.
FIG. 55: *Arundo donax*, the giant reed, on the banks of the Crocodile River.
FIG. 56: *Ceratonia siliqua*, the carob-tree.
FIG. 57: *Coriandrum sativum*, the coriander (see also fig. 18).
FIG. 58: *Crocus sativus*, the saffron crocus.
FIG. 59: *Cyperus papyrus*, on the banks of the Nile (see also fig. 63).
FIG. 60: *Cynomorium coccineum*.
FIG. 61: *Ferula galbaniflua*, source of the Biblical "galbanum".
FIG. 62: *Ficus sycomorus*, the "sycomore" of the Bible.
FIG. 63: *Cyperus papyrus*, the "bulrush" or "paper reed" of the Bible (see also fig. 59).
FIG. 64: *Lilium candidum*, the Madonna Lily (see also the initial on p. 17 and fig. 89).
FIG. 65: The Annunciation, with the angel GABRIEL bearing a Madonna lily.
FIGS. 66-67: *Mandragora officinarum*, the mandrake.
FIG. 68: *Myrtus communis*, the true myrtle (see also p. 145).
FIG. 69: *Nerium oleander*, the oleander.
FIG. 70: *Phoenix dactylifera*, the date palm, a view near Haifa (see also fig. 9, fig. 28 and fig. 92).
FIG. 71: Watchtower in vineyard (*Vitis vinifera*), see also fig. 31 and fig. 32.
FIG. 72: *Lolium temulentum*, the bearded darnel-grass (see also fig. 41).
FIG. 73: *Pinus halepensis*, the Aleppo pine.
FIG. 74: *Pistacia lentiscus*, the lentisk.
FIG. 75: *Pistacia terebinthus*, the common terebinth.
FIG. 76: *Punica granatum*, the pomegranate, fruits (see also p. 199 and fig. 38).
FIG. 77: *Prunus armeniaca*, the apricot.
FIG. 78: *Retama raetam*, the white broom, in the S. Sahara.
FIG. 79: *Solanum incanum*, the Palestine nightshade.
FIG. 80: *Salsola kali*, the prickly saltwort.
FIG. 81: *Salvia judaica*, the Judean sage.
FIG. 82: *Sorghum vulgare* var. *durra*, the dhura.
FIG. 83: *Tulipa sharonensis*, the Sharon tulip.
FIG. 84: View of the Sea of Galilee, with *Quercus coccifera* var. *pseudococcifera* (see also fig. 29).
FIG. 85: Plants (hyssop?) on walls in Jerusalem.
FIG. 86: Scene in the "wilderness" of Judea.
FIG. 87: Mount Tabor in northern Palestine, with *Silybum marianum*, etc. (see also fig. 92).
FIG. 88: View of the valley of Nazareth, with *Opuntia*.
FIG. 89: The parable of the "lilies of the field" (*Lilium candidum*), see also the initial on p. 17 and figs. 64-65.
FIG. 90: Esau eating the "red pottage" (*Lens esculenta*, the lentil plant).
FIG. 91: The Temptation in the Garden of Eden (from one of Père DAVID's Emblem Books).
FIG. 92: Flight of JOSEPH and MARY and the infant JESUS into Egypt, with *Phoenix dactylifera* and *Silybum marianum*. (Engraving by MARTIN SCHONGAUER).
FIG. 93: The Gathering of the Manna, see also fig. 27. (Painting by TINTORETTO).
FIG. 94: RUBENS's famous painting of the Garden of Eden.
FIG. 95: The Temptation in the Garden of Eden. (Painting by LUCAS CRANACH the Elder).
VIGNETTE: The cross of Christ represented quite literally as a leafy tree. (From a 14th Century religious poem in Old English, in the British Museum).

HERBARVM

ATQVE ARBORVM QVÆ
IN BIBLIIS PASSIM OBVIÆ
sunt, & ex quibus sacri vates similitudines desumunt, ac collationes rebus accommodant, dilucida explicatio: in qua narratione singula loca explanantur, quibus Prophetæ obseruata stirpium natura, conciones suas illustrant, diuinaq; oracula fulciunt, LIVINO LEMNIO Sacrarum Literarum studioso autore.

ANTVERPIÆ
Apud Gulielmum Simonem
sub scuto Basileensi.
1566.
Cum Priuilegio.

*EXIMIO SVMMAEQVE VIRTVTIS
AC DIGNITATIS VIRO
D. THOMAE THIELDIO DIVI
Bernardi in confinijs Antuerpianis
ad Scaldam, Antistiti cumprimis
reuerendo Liuinus Lemnius
Medicus Zirizæus sacrarum literarum studiosus.*
S. P. D.

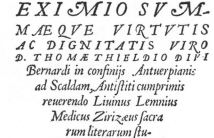

VVM mihi nuper amplissimè deprædicaret virtutis tuæ decus, & mentis præstantiam, Antistes spectatissime, Iacobus Susius vir, præter antiquæ nobilitatis insignia, omni disciplinarum genere excultissimus: cœpit ilico animus noster in tui nominis amorem ac venerationem incitari. Quis enim erga illum non afficiatur? qui in pacificandis optimatum dissidijs, in dirimendis sopiendisq, controuer-
A 2 *sijs*

FIGURES 10-12. — Facsimile reproductions of three pages from the first book ever published on the plants of the Holy Scriptures. This was a 161-page work, published in Antwerp in 1566, by LEVINUS LEMMENS (1505-1568), a Dutch physician of Zierikzee, a friend of DODOENS and GESSNER. Shown are the title-page, the author's portrait given on the back of the title-page, and the first page of preface. Early works on the subject, like this one, depended almost entirely on the evidence furnished by comparative philology, etymology, the opinions of the "Church Fathers", and the "internal evidence" of the Scriptures themselves. Not until 1757 was botanical evidence brought to bear on the subject—and this was long after some of the Bible versions now in most general use were formulated. (Courtesy Arnold Arboretum of Harvard University).

Historical Sketch

POPULAR INTEREST IN THE BOTANY OF THE BIBLE DATES FROM VERY EARLY TIMES. PLANTS AND PLANT PRODUCTS ARE REFERRED TO IN SO MANY HUNDREDS OF VERSES OF BOTH THE OLD AND NEW TESTAMENTS, AS WELL AS IN THE BOOKS OF THE APOCRYPHA, THAT IT IS NOT AT ALL SURPRISING TO FIND EARLY THEOLOGIANS, DIVINES, AND SCHOLARS EXCEEDINGLY INTERESTED IN THEM. THEOLOGICAL LITERATURE, FROM ITS VERY BEGINNINGS, IS FILLED WITH REFERENCES TO AND COMMENTS UPON THE PLANTS OF THE BIBLE AND OF BIBLICAL LANDS. THERE WAS, OF COURSE, A LONG PERIOD OF TIME WHEN NO ONE DARED TO CHALLENGE ANY OF THE TRANSLATIONS OR INTERPRETATIONS OF BIBLICAL PASSAGES BY THE LEADERS OF THE CHURCH THEN IN AUTHORITY. THE TIME CAME, HOWEVER, WHEN QUESTIONS WERE ASKED AND various sects or denominations appeared—at first persecuted and suppressed as "heretics". Translations were questioned and even the canonicity of some of the chapters and of entire books was challenged. There naturally followed heated and often bitter discussions and arguments among scholars and theologians concerning the translation and interpretation of certain Hebrew or Greek words or phrases in the various passages referring to, or thought to refer to, plants or plant products. No attempt has been made by us to review the huge theological literature on this subject, or to list it in our bibliography, primarily because these writers were not botanically trained and their arguments, while in many cases interesting to read, are largely metaphysical, philosophical, moralistic, or philological in nature.

Passing over the incidental—though none the less important—contributions to the subject made by ARISTOTLE, PLATO, PLINY, DIOSCORIDES, HERODOTUS, THEOPHRASTUS, and even PLUTARCH and JOSEPHUS, we find that the first book (of which we have a record) that dealt entirely with the plants mentioned in the Scriptures was that of LEVINUS LEMMENS in 1566 (206). This was a 161-page work with the imposing title of "Herbarum atque arborum quae in Bibliis passim obviae sunt et ex quibus sacri vates similitudines desumunt, ac collationes rebus accommodant, dilucida explicatio; in qua narratione singula loca explanantur quibus Prophetae observata stirpium natura, conciones suas illustrant, divina oracula fulciunt." It was reissued, in 1568, with another title, as "Similitudinum ac parabolarum quae in Bibliis ex herbis atque arboribus desumuntur dilucida explicatio . . ." Then followed THOMAS NEWTON's "An herbal for the Bible" in 1587, with 287 pages, which was actually only a translation, albeit with alterations, of LEMMENS' work. The only other noteworthy contribution to the subject in the 16th century was F. VALLES' "De iis, quae scripta sunt physice in Libris Sacris; sive, de sacre philosophia"—a 978-page work first issued in the year 1588.

The 17th century was ushered in by L. RUMETIUS' "Sacrorum Bibliorum arboretum morale", with 118 pages, published in 1606, but later (in 1626) expanded into a 3-volume and 901-page work, and by CLEMENS ANOMOEUS' "Sacrorum arborum, fruticum et herbarum", in 1609, with 223 pages. These were followed by a Portuguese work of 582 pages in 1622 by F. I. DE BARREIRA, by JOHANNES MEURS' "Arboretum sacrum" in 1642, and by A. COCQUIUS' two books—"Historia ac contemplatio sacra plantarum arborum & herbarum quarum fit mentio in Sacra Scriptura" in 1664 and his "Observationes et exercitationes philologico-physiologicae" in 1671. E. CASTELLI produced in 1667 a book on the plants mentioned in the Bible, and there were smaller tracts on various single plants by DU PAS (lily of the field), MUNDELSTRUP (apple of Sodom), RAVIUS (mandrake), and MEYER (sycomore). In 1663 the first of the many editions of J. H. URSINUS' extremely popular 638-page

"Arboretum biblicum" was printed, to be followed, in 1694, by the "Scripture herbal" of WILLIAM WESTMACOTT.

In the 18th century there appeared the numerous and very important works of CELSIUS, HILLER, SCHEUCHZER, SHAW, OEDMAN, HARRIS, and TAYLOR. In the year 1740 the first edition of J. B. ROHR's "Phytotheologia" was published, and, in 1745 and 1747, the two volumes of CELSIUS' "Hierobotanicon". The latter are usually regarded as being among the foundation-stones of this study.

There is no other branch of botany in which so very many persons have spoken, argued, and written without ever bothering to investigate the controversial matters by direct observation,—in other words, without applying the scientific method! It seemed entirely unnecessary—and even irrelevant— to these older writers, chiefly theologians, divines, and classical scholars, to inquire as to what plants actually were growing in Biblical lands in their day. Even scholars with a fair knowledge of natural history were slow to realize that the plants of one region might differ considerably from those of another region. For centuries scholars and even herbalists attempted to identify in Germany, France, England, or Scandinavia the plants described by THEOPHRASTUS or DIOSCORIDES from Greece. That a plant described from Greece might not be also native to Sweden or England or South Africa apparently never occurred to these writers. Conversely, for centuries herbalists searched the writings of the ancient Greeks, Romans, and Arabs for descriptions of the wayside plants of Great Britain and Holland, rarely, if ever, realizing the futility of their task! So it is not surprising that most of the early writers on the plants of the Bible should have based the greater portion or even all of their arguments on such lines of evidence as furnished by comparative philology, etymology, the opinions of the "Church Fathers", and the "internal evidence" of the Scriptures themselves. That they should have arrived by these means at various often extremely divergent and in not a few cases completely erroneous conclusions was inevitable. KITTO has quite aptly summed up the situation in the following words: "The Natural Histories of the Bible form a class by themselves, having less connexion than any other with the science of nature. They are rather works of criticism than of natural history—rather the production of philologists than of natural historians. Whatever learning could do on such subjects has been done; and whatever might be done by science, observation, and well directed research has been left undone. The process usually taken in works of this class has been to exhaust the resources of philology and conjecture in the attempt to discover the meaning of the Hebrew name and the object denoted by it. From the very nature of the thing, the conclusion arrived at is often unsatisfactory or uncertain. But a conclusion being taken, the ancient writers of Greece and Rome are ransacked to supply this history and description of the object, and in particular to furnish such intimations as may coincide with or illustrate those of the sacred writers. All this was very proper; but the value of the information thus collected as *contributory* to a Natural History of Palestine might have been very greatly enhanced had corroborations and elucidations been sought in the *actual* condition of the country, and the character of its products in the various departments of nature" (202).

In 1757 the "Immortal Swede", CAROLUS LINNAEUS, edited and published the "Iter palaestinum" (160) of his lamented student and devoted follower, F. HASSELQUIST. This 619-page work was immediately translated, by command of the Queen of Sweden, into German, English, and French. The English edition is entitled "Voyages and travels in the Levant in the years 1749, 50, 51, 52: containing observations in natural history, physick, agriculture, and commerce: particularly in the Holy Land and the natural history of the Scriptures". This book marks the beginning of an entirely new epoch in the study of the plants of the Bible. LINNAEUS himself in the introduction to the book says: "In one of my botanical lectures in 1747 I enumerated the countries of which we knew the natural history and those of which we were ignorant. Among the latter was Palestine: with this we were less acquainted than with the remotest parts of India; and although the natural history of this remarkable country was most necessary for divines and writers on the Scriptures, who have used their greatest endeavours to know the animals therein mentioned, yet they could not with any degree of certainty determine which they were before someone had been in the country and informed himself of its natural history" (162).

Impressed by this lament of LINNAEUS in 1747, his student, HASSELQUIST, though weak and delicate in health, undertook two years later a journey to Egypt and Palestine. He succeeded in exploring a large part of the Holy Land and in making extensive notes and collections for shipment back to Sweden, but the excessive heat of Palestine proved too much for him and he died in Smyrna at the age of 31—"wasting away", as his great tutor and biographer laments, "like a lamp whose oil is spent." But the results of HASSELQUIST's investigations—the first to be made by a qualified naturalist in that area—were given to the world by LINNAEUS in 1757. Imperfect though this work is because of the short time that HASSELQUIST had in Palestine before he succumbed, it is still one of the most valuable books ever written on the subject. It marks the beginning of a new era because now for the first time in history a writer on the natural history of the Bible had actually visited Biblical lands and had there studied firsthand the natural features of the region. The writings of all previous workers, including LEMMENS, URSINUS, SHAW, and even CELSIUS—friend and patron of LINNAEUS—were based on hearsay and suppositions, often quite erroneous. Beginning with HASSELQUIST at least some of the writers on Biblical natural history based their arguments and commentaries on actual observations rather than on mere suppositions, assumptions, and linguistic maneuverings.

HASSELQUIST was followed in the exploration of Biblical lands by another ill-starred naturalist, PEHR FORSKÅL, who traveled in and explored Egypt, Arabia, and the regions about the Red Sea. Misfortune and hard luck dogged his footsteps. His body wracked by diseases, attacked repeatedly by robbers and bandits who once stripped him of all his worldly possessions save only his pressed plant specimens which they spurned as "worthless", his misfortunes came to a climax when his guides and carriers deserted him in the midst of the Arabian desert and he was left to die of hunger and thirst. But in 1775 and 1776, after his untimely death, the results of his brave pioneering were published by C. NIEBUHR. One 164-page volume described the mammals, birds, fish, insects, amphibians, worms, and materia medica of the region (121); another volume, of 377 pages, described the plants (122). So, thanks to the self-sacrificial labors of HASSELQUIST and FORSKÅL, both of whom paid with their life's blood, the world was finally given a firsthand description of at least some of the plants and animals of Biblical lands.

It is astonishing to note how slow botanists were to follow up the work of these two pioneers, and to explore intensively the Holy Land. In spite of the unnaturally truncated work of HASSELQUIST and FORSKÅL, the lament of LINNAEUS held true for an unbelievably long time, during which far more was learned and was known of vastly more distant and isolated lands than was known of Palestine and its surrounding countries. In 1863 a writer in the "London Quarterly Review" (301) tells of visiting the great British Museum of natural history in South Kensington and of finding there specimens of plants and animals of all the rest of the world in great abundance, but scarcely any at all from Palestine! Even today this singular dearth of material from the Holy Land in our leading museums and herbaria is most noticeable, and, to the student of the plants of the Bible, most deplorable.

The books of URSINUS are typical of that class of publications on the natural history of the Bible described so fitly by KITTO. URSINUS gives diagrammatic illustrations, often laughable in their inaccuracy, of the more important plants, and brief descriptions, culled mostly from the old Greek and Roman natural history writers, but goes deeply into the linguistic origins of their names, considering the original Hebrew and Greek texts. In 1793 the first edition of T. M. HARRIS' 297-page "The natural history of the Bible" appeared. The great botanist, C. P. THUNBERG, in 1828, published nine dissertations on the plants of the Bible, entitled "Afhandling om de wäxter, som i Bibelon omtales" (321). KITTO's splendid work on "Palestine: the physical geography and natural history of the Holy Land" was printed in 1841. Very frequently cited is MARIE CALLCOTT's famous "Scripture herbal", on whose 568 pages, in 1842, was brought together most of the pertinent information assembled by all the previous authors on this subject. PRATT's "Plants and trees of Scripture", which made its appearance, unsigned, in 1851, is well worthy of mention, as is also J. H. BALFOUR's "The plants of the Bible" of 1857. Canon TRISTRAM's "Natural history of the Bible"—a 524-page work whose first edition appeared in 1867—is very frequently quoted especially in Biblical commentaries and theological works. We have personally found the volume by the distinguished Kew botanist, JOHN SMITH, entitled "Bible plants, their history", of 265 pages, published in 1878, extremely useful, and it will be cited hundreds of times on the following pages. In the decade of 1883 to 1893 the first edition of G. E. POST's monumental "Flora of Syria, Palestine, and Sinai " was published. In 1932—1933 a second edition of this work appeared. It is still the only comprehensive and authoritative flora that we have for this most important and fascinating part of the earth's surface. The work of W. H. GROSER, entitled "Scriptural natural history. I. Trees and plants mentioned in the Bible", first issued in 1888, is a fine and handy book. In 1928 and 1929 an interesting series of articles by A. BORAH on the trees of the Bible appeared in "American Forests & Forest Life". Several authors have written on Biblical botany in the "Gardeners' Chronicle", while as late as 1934 a stimulating paper on "Flowers of the Bible" appeared in "Nature Magazine".

Thus we see that interest in the subject has remained keen right down to the present day. In fact, there is at present a professor of Biblical botany at the Hebrew University of Jerusalem, where intensive research is being carried on in this subject. It is confidently expected that much valuable light will be shed on the plants of the Bible by Dr. HA-REUBENI, his students, and his successors. Further evidence of the continued popular appeal of this branch of botany is seen in the fact that so many of the better books on this subject were reprinted in 2, 3, 4, or more editions, and some translated into several languages. The edition of our own previous work on the subject in 1940 (233) was quickly exhausted. This led to the publication by Miss E. A. KING and Miss C. H. WOODWARD of their résumé in the "Journal of the New York Botanical Garden" (192, 347) and to Miss KING's volume on "Bible plants for American gardens" in 1941 (187). These works, too, were soon out of print. The two former were reprinted twice and another edition is now under way. Realizing the popularity of the topic, the New York Botanical Garden made it the subject of its annual entry at the International Flower Show held at Grand Central Palace, New York City, in March, 1941, where it proved to be the cynosure of all eyes and where it won a first prize gold medal award. Since then the Garden has many times reconstructed the scene of the Nativity at Bethlehem in one of its display greenhouses during the Christmas season, with only authentic Bible plants used in the scenery. That these exhibits have captured the public imagination and interest is attested by the huge crowds which invariably visit them and by the long series of newspaper, magazine, and church paper notices cited in our bibliography (318, 348-372), which, in fact, is only a very partial listing of the more important notices which happen to have come to our attention. Letters have poured into the Garden from every section of the United States and Canada, and from several foreign countries, on this subject. At present there are at least two Biblical gardens established by churches in the West, and a well-known architect plans to offer special designs for gardens of Biblical plants for churches, synagogues, and other religious institutions such as monasteries, convents, and theological seminaries. Several persons have written to us of their intentions to establish Biblical gardens in their towns, similar to the Shakespeare gardens which have been so popular.

One of the most frequent errors into which the non-botanically trained preachers and theological writers have fallen in the past—and, surprisingly, continue to fall even at the present time—is that of identifying the plants of the Scriptures with plants growing naturally in the regions in which these men or women happen to be living—France, Germany, England, Scotland, the United States, Canada, etc. This error was understandable and forgivable in the distant past, before the days of LINNAEUS, HASSELQUIST, and FORSKAL, because in those days few persons except soldiers traveled extensively and hardly anyone realized the great differences that exist in the floras of various lands. To LINNAEUS, among many other things, goes the credit for proving and emphasizing the fact that floras and faunas differ from place to place, depending on climate, soils, natural barriers, latitude and longitude, etc. Even he had not realized this until after his trip to Lapland. Yet this is now such a common piece of general knowledge that it seems unbelievable that preachers should continue to identify the "elm", "sycamore", "lily", "rose", "vine", etc., of the Scriptures with plants bearing these names in our own American woods and fields! And yet this has been done very extensively and is still being done today, much to the amusement of the botanist. Another common error of the non-botanist who visits the Holy Land today is the supposition that all the plants which he sees there now were naturally there in ancient Biblical days, or that he must find there now the plants referred to in the Bible as having been there then. Both of these suppositions are fallacious, since they fail to take into account that fact that floras change, especially in regions like Palestine and Egypt where man, notorious for his aptitude in upsetting the delicately adjusted balances in nature, has been most active, in one way or another, for at least 6000 years. Many plants now extremely common in the Holy Land were definitely *not* there in Biblical days. Many examples might be mentioned, but outstanding are the American locust, *Robinia pseudo-acacia* L., which has been introduced into Palestine only rather recently, but is thriving and spreading there now as well as it does in its native land, and the American pricklypear cactus, *Opuntia ficus-indica* (L.) Mill., which was introduced some time ago and has found Palestine so favorable to its growth that it is now found "everywhere" (266) and is one of the conspicuous and distinctive features of the landscape. So characteristic is it now that artists who make Biblical pictures have time and again fallen into the ridiculous error of painting some episode in the life of JESUS or of one of the Old Testament characters, showing the surrounding landscape filled with cactus plants in flower or fruit! Every botanist, of course, knows that no *Opuntia* cacti were known anywhere in the Old World until after the discovery and exploration of America and their subsequent introduction from America. Similarly, the opopanax, *Vachellia farnesiana* (L.) Wight & Arn., is now found all through the Levant in hedges and along watercourses, but is an American plant. Again, the white mulberry, *Morus alba* L., is now cultivated "everywhere" (267) in the Holy Land and one might be led to assume that it was the mulberry of the Bible. However, *M. alba* is a native of China, and it is fairly certain that trade had not yet been established with China in Biblical days, so it could not have been in the Holy Land in those days. The same is true of the weeping willow, *Salix babylonica* L., now widely cultivated and escaped in Palestine, but actually a native of Japan or China. It could not have been known to the Biblical children of Israel. Similarly, the badly-named rose-of-Sharon (*Hibiscus syriacus* L.), the soapwort (*Saponaria officinalis* L.), and many species of *Citrus* are now common in the Holy Land due to recent introduction. Hundreds of other exotic plants exist at present in Syria, Lebanon, and Palestine which were certainly not there in Biblical days, and many of these are now escaping into the countryside and grow like natives. There is even a legend that the blue vervain, *Verbena hastata* L., grew at the foot of the cross when JESUS was crucified, but this, again, is impossible since that species is confined to portions of the United States and Canada. EIG lists 115 plant species as endemic to Palestine and has computed that they comprise only about one quarter of all plants there today (119). How different the plant life of Biblical Palestine must have been from that of the present time!

The complications which this introduction of exotic species into Palestine has brought about is well illustrated by a passage in Dr. J. C. GEIKIE's "Life and Words of Christ" where the present mixture of native and exotic species in the flora as observed through the eyes of a non-botanist is dramatically pictured: "Within the extent of a single landscape, there is every climate, from the cold of Northern Europe to the heat of India. The oak, the pine, the walnut, the maple, the juniper, the alder, the poplar, the willow, the ash, the ivy, and the hawthorn, grow luxuriantly on the heights of Hermon, Bashan, and Galilee. Hence the traveller from the more northerly temperate lands finds himself, in some parts, surrounded by the trees and vegetation of his own country. He sees the apple, the pear, and the plum, and rejoices to meet the familiar wheat and barley, the peas, potatoes, cabbage, carrots, lettuce, endive, and mustard. The Englishman is delighted to find himself surrounded by many of the flowers of his native land; for out of the 2,000 or 2,500 flowers of Palestine, perhaps 500 are British. It looks like home to see the anemony, ranunculus, yellow water-lily, tulip, crocus, and hyacinth, the mignonette, geraniums, mallows, the common bramble, the dog-rose, the daisy, the well-known groundsel, the dandelion,—sage, thyme, and sweet marjoram, blue and white pimpernel, cyclamens, vervain, mint, horehound, road-way nettles, and thistles; and ponds with the wonted water-cress, duckweed, and rushes.

"The traveller from more southern countries is no less at home; for from whatever part he comes, be it sunny Spain or Western India, he will recognize well-known forms in one or the other of such a list as the carob, the oleander and willow skirting the streams and water-courses; the sycamore, the fig, the olive, the date-palm, the pride of India, the pistachio, the tamarisk, the acacia, and the tall tropical grasses and reeds; or in such fruits as the date, the pomegranate, the vine, the orange, the shaddock, the lime, the banana, the almond, and the prickly-pear. The sight of fields of cotton, millet, rice, sugar-cane, maize, or even of Indian indigo, and of patches of melons, gourds, pumpkins, tobacco, yam, sweet potato, and other southern or tropical field or garden crops, will carry him back in thought to his home (128a)."

There is, however, another side to the picture. Many plants which grew in abundance in the Holy Land or surrounding countries in Biblical days are now no longer found there or else grow in far smaller numbers. Some have been driven out by foreign, more weedy invaders, much as the native vegetation in many parts of America is being driven out by European or Asiatic weeds. Others of the native Palestinian plants have been exterminated or almost exterminated by the over-cultivation of the land, the destruction of the forests, and the resulting changes in climatic and other environmental conditions. If we stop a moment to recall the vast changes which have occurred in the American Middle West due to the cultivation of the land over the past 75 years, and then recall that the Palestine area has been under cultivation continuously for the past 3000 or 4000 years, we may obtain a fair idea of the changes that must have taken place there. At one time Palestine was a land of palm trees, with the date-palm as abundant and characteristic there as it was in Egypt. This we know not only from the testimony of early historians, but also from the abundant fossil remains of palm trunks buried in the shifting sands. Now the date-palm must be carefully cultivated there. In Biblical days the towering cedars clothed the slopes of the Lebanon and other mountain ranges. Now the few remaining specimens must be carefully fenced in and protected by the government, lest the species be completely exterminated in its own homeland.

Anyone delving even very superficially into the literature of Bible plants will be impressed at once by the amazing discrepancies, contradictions, palpable mis-identifications, erroneous statements, and general confusion which exist there. The reasons for this lamentable confusion are several. First, it must be remembered that the exact science of botany, as such, is a very recent development in human knowledge. It is true that men were always concerned with and interested in the plants about them, but this interest could hardly be termed as anything approaching an exact science until quite recently in man's history. Certainly during the days when the books of the Bible, especially those of the Old Testament, were composed and later recorded, there was no such thing as an exact science of botany or botanical nomenclature, hence exactness and accuracy of terminology were impossible. Secondly, the writers of the books of the Bible were not botanists or even "natural historians" (with, perhaps, one possible exception). Mostly they were very plain men, with a limited vocabulary. Nor was the purpose of their writing botanical. The Bible was not intended by its writers to serve as a textbook of natural history. To the writers of the Scriptures the botanical aspects were strictly secondary and subservient to the moral, ethical, theological, and historical aspects, which were to them of far greater importance and significance. Thus it would be more surprising to *find* accuracy in botanical matters than *not* to find it. Thirdly, one must keep in mind the history of the Scriptures. It is generally agreed by students of the subject (102) that much of the material of the Old Testament originated in the form of poetry—songs and ballads, or what we now call folksongs—handed down from generation to generation of bards and singers for many hundreds of years before ever they were written down; indeed, in some cases, before ever written language was invented. It is thought that the so-called "Books of the Law" (Pentateuch, Torah), the events recorded in which are purported to have covered the period of 4004-1451 B.C. in the Ussher chronology, were not actually completed in written form until about 444 B.C.; the "Prophets" (Nebi'im) at about 250 B.C.; and the remaining books of the Old Testament or "Sacred Writings" (Ketubim) and poetical books at about 150-100 B.C. The Hexateuch (Genesis to Joshua) was probably written in part, by many different scribes, with more constantly added, in the 10th or 9th centuries B.C. In such a situation as this, with much of the material existing for hundreds of years merely as folksongs or ballads, then written down in part by various writers, at various dates, and mostly long after the events described, it is not surprising that confusion of terminology should result, along with probable mis-identifications, the use of different words for the same plant and of the same word for different plants, even in the "original" texts.

Fourthly, one must remember that even after the books of the Bible were once recorded in written form, they were still subject to much possible change. The Hebrew Bible was first translated into the Greek between the 3rd and 1st centuries B. C., while it was the Reformers, much later, who separated the books which had no Hebrew original into the Apocrypha. The oldest Hebrew text of the Bible now known to us is the Masoretic text, which may have been written between the 6th and 8th centuries A. D., although no less an authority than Goodspeed and others of similar merit say not until the 9th century (5). It is probable, although not at all certain, that it became fixed at about the 2nd century A. D. After that time it was copied and recopied by lonely monks in the solitude of their cloistered monastery cells. This copying was undoubtedly done very faithfully and religiously, by persons to whom every "jot and tittle" was sacred, and so it can be rather safely assumed that no more than a very few purely accidental changes crept in during that time (8). This leaves, however, a tremendous period of years between the time when it was first written down (from the 10th century B. C.) to the time when it became fixed (in the 2nd century A. D.), during which errors, corruptions, and accidental or intentional additions or subtractions, and various other modifications could have and probably did creep into the texts (10, 11, 12). And then there was before that the vastly much longer period of time during which the material was handed down by singers and bards, by word of mouth, from generation to generation, much as the folklore of any nation or race. The changes that must have occurred then can only be imagined.

The Masoretic text has also been subject to many "editions" and translations, among the more important of which are the Samaritan (written in the ancient Hebrew or Ibri, the so-called Samaritan character, in the 1st or 2nd century A. D.), those of Aquila, Symmachus, Theodotion, Origen, Hesychius, and Lucian, all thought to date back to about the 2nd century, the various uncial editions of the Septuagint of the 4th to 7th centuries and the later cursive editions of the 10th to 14th centuries, the Vulgate or Latin translation of Jerome, thought to date to the 2nd century, and the Chaldee versions (or Targums) of about the 2nd to 4th centuries A. D. (306). There have also been various Egyptian versions (of the 3rd and 4th centuries, including the so-called Memphitic or Coptic, Thebaic, and Bashmuric or Ammonian), as well as Ethiopian and Syriac (of the 4th century), Armenian (5th century), Gothic (6th century), Slavonic (9th century), and Arabic (10th century) versions. According to the American Bible Society, the Bible has at present been translated into 1118 different languages and dialects! It must be remembered in this connection that it is humanly impossible ever to make a translation from one language to another and still express *exactly* the same meaning. There are shades of meaning and connotations in every language which it is impossible to translate accurately into any other language. Thus, every time a translation into a new language is made, something of the original is lost, or something is added. And if translations are made one from the other in a linear series, instead of in each case from the "original" texts, these differences in meaning become ever greater. And the "original" texts are not always available to translators. In fact, it is relevant to point out here that many of the older versions and translations were made without recourse to "original" texts now available. On the other hand, many ancient texts then available have since been lost. All of this helps to explain some of the confusion which exists. In this connection a statement from the preface of the Jewish Publication Society's revision of 1946 is worth quoting: "The historic necessity for translation was repeated with all the great changes in Israel's career. It is enough to point to the Septuagint, or the Greek translation of the Scriptures, the product of Israel's contact with the Hellenistic civilization dominating the world at that time; to the Arabic translation by the Gaon Saadya, when the great majority of the Jewish people came under the sceptre of Mohammedan rulers; and the German translation by MENDELSSOHN and his school, at the dawn of a new epoch, which brought the Jews in Europe, most of whom spoke a German dialect, into closer contact with their neighbours. These translations are all historical products intimately connected with Israel's wanderings among the nations and with the great events of mankind in general.

"Ancient and continuous as this task of translation was, it would be a mistake to think that there were no misgivings about it. At least it is certain that opinions were divided as to the desirability of such undertakings. While PHILO and his Alexandrian coreligionists looked upon the translation of the Seventy as a work of inspired men, the Palestinian Rabbis subsequently considered the day on which the Septuagint was completed as one of the most unfortunate in Israel's history, seeing that the Torah could never be adequately translated. And there are indications enough that the consequences of such translations were not all of a desirable nature" (16).

Careful comparison of these many editions and versions reveals numerous differences. This holds for the botanical aspects as well as for any other, as will be abundantly illustrated on the following pages. Each time that it was translated the text was subjected to editorial criticism—"editing", as it were—and the opinions of the various editors have frequently changed the meaning of the original texts. The early vernacular translations were mostly translations, not of the original texts, but of 4th century Latin versions, which, for all their merits, were in turn based on totally inadequate knowledge of the original sources (10). In 1611 the English version known as the Authorized Version of King James I was completed, and to this, again, with all its merits, we owe much of the confusion regarding the plants of the Bible which has misled writers from that day to this. It was the Authorized Version which perpetuated, and, indeed, in some cases, originated, the mis-identification of Biblical plants with common English ones. It is most unfortunate that botanical knowledge was not in a more advanced stage at the time that this version was undertaken, for in this almost universally-used version among Protestants for the past 336 years aspens are called "mulberries", mulberries are called "sycamine", a species of fig is called "sycomore", the eaglewood is called "aloes", the acanthus is referred to as a "nettle", the almond becomes a "hazel", the juniper is called a "heath", the dill is called "anise", the apricot becomes an "apple", the box is called "ivory", the cypress is called "box", the saltwort is referred to as "mallows", the terebinth becomes an "elm", and the planetree becomes a "chestnut"! Similarly, in the field of zoology, the antelope is called a "wild ox", the bull a "unicorn", the hippopotamus a "behemoth", hyenas are "doleful creatures", the jackal is called "fox" and "dragon", the pelican is a "cormorant", the goose is referred to as a "fatted fowl", the buzzard is called "glede", the lämmergeier is called "ossifrage", ostriches are called "owls", the crane becomes a "swallow", the crocodile is called a "whale", "leviathan", and "dragon", locusts are called "palmerworms", and sea-cows become "badgers"!

The King James Version, however, was not the only "authorized" English version. There were others by WYCLIFFE (14th century), TYNDALE (1525), COVERDALE (1535), MATTHEW (1537), TAVERNER (1539), and CRANMER (1539), as well as the so-called Geneva Bible of WHITTINGHAM (1557 and 1560)—the Bible adopted by the Puritans and the first to omit the entire Apocrypha. Later appeared the Bishops' Bible of 1568 and 1572. LUTHER's Bible appeared first as the "Septemberbibel", which was a New Testament translation dated September 1522 at Wittemberg. The title page was without any date or publisher's name. The Zurich Bible appeared in 1530, a German version, which consisted of LUTHER's translation and the addition of the Prophets and the Apocrypha. The Berleburg Bible (1726-42) was a German translation based on the two preceding ones and illustrating mystical tendencies. In 1735 the WERTHEIM Bible, a German rationalistic version, was published by WERTHEIM. The Probe-Bibel or Proof Bible (1892) is the revision of LUTHER's Bible as ordered by the Eisenach German Protestant Church Congress.

GREGORY MARTIN's Rheims and Douay Bible of 1582 and 1609, became the official English Catholic version. Both the Geneva and the Douay versions added to the texts notes (political as well as theological) of highly controversial nature (306). The Douay version was revised by Bishop CHALLONER in the middle of the 18th century.

The MAZARIN Bible, GUTENBERG Bible, or Bible of Forty-two Lines, an edition of the Vulgate printed at Metz about 1450-55 by GUTENBERG and others, was the first Bible, and probably the first complete book, printed with movable type. It derives one of its names from being found about 1760 in the library of MAZARIN. The Bible of Thirty-six Lines, a folio edition of the Biblia Latina, is so-called because of the page format of two columns having thirty-six lines. It is also called the Bamberg Bible because that was probably its place of printing in 1460.

The Leopolita Bible (1561) was a Polish translation by JOHN OF LEMBERG from the Vulgate. It was published at Cracow. It was intended for Roman Catholic use, but was not sanctioned by the Pope.

The St. Wenceslaus or Wenzel Bible was published at Prague for Bohemian Roman Catholics by the Jesuits (1677-1715).

WUYECK's or WUJEK's Bible is the Authorized Polish Roman Catholic version made by the Jesuit WUJEK and published at Cracow—New Testament 1593—Old Testament 1599.

The Bible of Ferrara (1553) was the first edition of the Old Testament in Spanish by DUARTE PIVEL, and was made directly from the Hebrew expressly for the Jews. Another edition appeared in the same year for Christians.

PFISTER's Bible was so-called because it was printed by ALBERT PFISTER. SCHELHORN's Bible was so-called because it was described by SCHELHORN in 1760.

The Kralitz or Brothers' Bible is the most important Bohemian version published by the United Brethren at Kralitz in Moravia (1579-93).

The Ostrog Bible is the first complete Bible printed in Slavonic. It was printed in Ostrog in 1581.

Of the many other editions there are those whose chief importance is derived from some peculiarity of or mistake in printing or translation such as:

Chained Bible (1539) for the Great Bible of CRANMER since it was often chained in churches for public reading.

Bug Bible (1551) reading in Psalm 91:5 "So that thou shalt not need to be afraid for any bugges

[terrors] by night", as it is also rendered in the COVERDALE and TAVERNER works.
Breeches or Geneva Bible (1560) reading in Genesis 3:7 "They sewed fig leaves together and made themselves breeches" [aprons].
Placemaker's Bible (about 1562) reading in Matthew 5:9 "Blessed are the placemakers" [peacemakers].
Bible of the Bear or Biblia del Oso, being the first published Spanish translation of the whole Bible, so-called from the animal which appeared as the frontispiece, and dated 1567-9 at Basel.
Treacle Bible or Bishops' Bible (1568) reading in Jeremiah 8:22 "Is there no tryacle [treacle] in Gilead?"
Thumb Bible (Aberdeen about 1607) about the size of a thumb, one inch square, and a half inch thick.
Rosin Bible, the Douay Bible of 1609, having in Jeremiah 8:22 the wording "Is there no rosin [balm] in Galaad [Gilead]?"
He Bible, the first edition (1611) of the King James Version, reading in Ruth 3:15 "He [she] went into the city", as does the Revised Version.
She Bible, the second edition of the King James, appearing in the same year, and correcting this error.
Adulterous or *Wicked Bible* (1631) having the seventh commandment printed "Thou shalt commit adultery" in Exodus 20:14.
Printer's Bible (about 1702) reading in Psalm 119:161 "Printers [princes] have persecuted me without a cause."
Vinegar Bible (Clarendon Press, Oxford, 1717) having the title "Parable of the Vinegar [vineyard]" over the twentieth chapter of Luke.
Murderer's Bible (about 1801) reading in Jude 16 "These are murderers [murmurers]."
To-remain Bible (Bible Society of Cambridge 1805) having the words "to remain" inserted in Galatians 4:29 instead of a comma.
Standing Fishes' Bible (1806) reading in Ezekiel 47:10 "The fishes [fishers] shall stand upon it".
Discharge Bible (1806) reading in Timothy 5:21 "I discharge [charge] thee before God."
Wifehater Bible (1810) reading in Luke 14:26 "If any man come to me and hate not his own father ... Yea, and his own wife [life] also".
Ears-to-ear Bible (1810) reading in Matthew 13:43 "Who hath ears to ear [hear], let him hear".
Rebekah's camels Bible (1823) reading in Genesis 24:61 "And Rebekah arose, and her camels [damsels]". (158)

In very recent years there have appeared several excellent new Bible translations. Among these must be mentioned the translation of the New Testament into everyday English by R. F. WEYMOUTH in 1947 (15), the revision of the Challoner-Rheims version prepared under the direction of Bishop EDWIN V. O'HARA, completed in 1941, and the Revised Standard Version published by the International Council of Religious Education under the direction of L. A. WEIGLE in 1946. Another of the important recent contributions is the English translation in 1940 of the New Testament according to the Eastern text from the Peshitta or original Aramaic sources, by G. M. LAMSA. The rendition of the New Testament into Basic English in 1941 is of considerable interest, although not of any particular value in our study. The attempt to render the entire New Testament into a language of 1000 words naturally does not lend itself to scientific accuracy (7).

Most important of all, however, from our point of view are the splendid translations of Dr. JAMES MOFFATT in 1922-1925 (10, 11, 12) and of Dr. EDGAR J. GOODSPEED and his associates in 1939 (5). These two translations are almost revolutionary in form. Not only have the authors gone back to all the original sources, including many not available to previous translators, but they have studied and compared all the previous versions and have couched their final rendition in everyday English as it is spoken in America today. In scientific matters it is very obvious that they have studied the works of botanical and zoological writers, including the various works discussed by us in this historical sketch. They have not been afraid to adopt the conclusions of modern scientific research wherever these apply. The same high commendation holds for the LEESSER and JASTROW versions of the Old Testament prepared for Jewish-English readers. It is a decided joy to read their renditions of many passages whose meaning was effectively obscured by the antiquated and medieval English of previous versions. Many errors have been corrected. Unfortunately, in our opinion there are still instances of mis-identification in even these splendid versions, and even a few new ambiguities and ill-advised renditions have been initiated. The invention of the name "wake-tree" for the almond is one of these.

As a single typical example of the differences which one finds in rendition of the same passage in different English versions alone may be cited the case of Nahum 2: 3. This is rendered by the King James Version—"and the *fir trees* shall be terribly shaken", by the JASTROW version—"and the *cypress spears* are made to quiver", by the GOODSPEED version—"and the *chargers* will prance", by MOFFATT—"and their *horses* prance at the muster", and by the Douay version—"and the *drivers* are stupefied." Here the same word runs the gamut of interpretation from trees to spears to horses to men!

In addition to inaccuracies and ambiguities in the translations themselves, commentators have published many utterly fantastic claims. One, for instance, has stated that the bright tint produced with henna dye on the soles of the feet and the palms of the hands of Oriental women needed renewing once in about two weeks, while the dye on their nails "is permanent for years" (146). It would appear from this that fingernails do not grow as rapidly in the Orient as they do in the Occident—an assumption not at all borne out by the facts! Again, a commentator has said that "blood oranges are produced from a branch (of orange) grafted on a pomegranate stem" (184). These are typical examples of hundreds of such erroneous statements which crowd the literature on Bible plants.

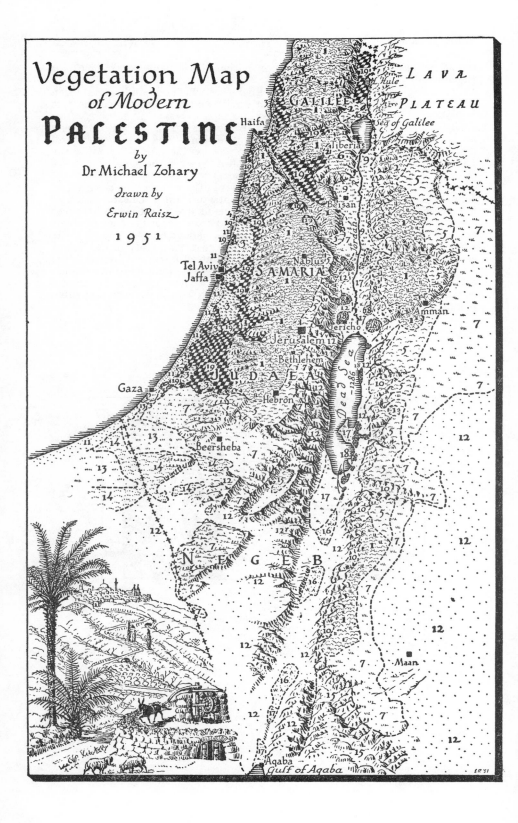

Description of the Land

LOOKING OVER A LIST OF THE ANIMALS AND PLANTS TO WHICH MENTION IS MADE IN THE BIBLE, IT BECOMES EVIDENT AT ONCE THAT THEY ARE PRINCIPALLY THOSE OF EGYPT AND PALESTINE (301), ALTHOUGH, OF COURSE, THERE ARE SOME REFERENCES TO THOSE THAT OCCUR IN ARABIA, THE SINAI PENINSULA, AND SYRIA, AS WELL AS VARIOUS ARTICLES OF MERCHANDISE CONSISTING OF ANIMAL OR VEGETABLE PRODUCTS IMPORTED FROM YEMEN, SOMALILAND, INDIA, AND CEYLON. IT HAS BEEN POINTED OUT (184) THAT THE FAUNA AND FLORA OF THE BIBLE ACCORD WELL WITH THE KNOWN FACTS OF THE GEOGRAPHY AND CHRONOLOGY INVOLVED. THE DOMESTIC ANIMALS AND CEREALS BELONG MOSTLY TO MESOPOTAMIA, PERSIA, AND OTHER LANDS TO THE EAST OF PALESTINE, an area thought by many anthropologists to represent the cradle of the human race; those of the Pentateuch are in general such as had their origin, or, at least, were very prevalent, in Egypt and the Sinai Peninsula; those of the historical books and most of the Prophets belong more particularly to Palestine; while those of Job are more typically Assyrian than they are Egyptian, Arabian, or Palestinian. The marked paucity of marine references is in accord with the historical fact that the seacoast was never held by the Hebrews.

A few words regarding the present climate and phytogeography of Bible lands are perhaps appropriate here (146). The climate of the Lebanon-Syria-Palestine-Transjordania-Egypt region is extremely varied. Dr. POST has well stated (266) that this area "is unequalled by any of the same size on the globe, not only for the thrilling and important events of human history of which it has been the theatre, but for its unique geological structure, its great diversity of surface and climate, and its remarkable fauna and flora. It is the meeting point of three continents, since Asia Minor must be regarded, from the standpoint of its Natural History, as belonging to Europe rather than Asia, and as such, a link of connection between them all."

On the snow-capped peaks of the Lebanon mountains, which rise to 10,200 feet, and in parts of the Antilebanon chain (4000 to 8700 feet altitude) the climate is almost arctic in severity, while in the lower portions of the Ghor or Jordan Valley it is decidedly tropical. Between these two extremes of temperature, and the extremes in elevation of Lebanon and the Dead Sea (1292 feet below sea level), we find almost every intermediate type of environment. There is the distinct western coastal region with its sand dunes and its lush meadows, the region of inland plains and foothills, the higher uplands, and the lofty tablelands beyond the Jordan. Then there are the surrounding deserts of Syria and, to the south and southwest, of Arabia, the Sinai Peninsula, and Egypt. In Egypt the most outstanding features are the extremely fertile Nile River delta and valley, the extensive dry deserts, and the scattered wadies with their oases, which are also characteristic of the Arabian deserts. The diverse climatic and physiographic features have brought about a corresponding variety in wild life. Tropical bats, Indian owls, and Ethiopian sunbirds are

---

FIGURE 13. — *Vegetation Map of Modern Palestine* (prepared by Dr. MICHAEL ZOHARY, 1951). — (1) *Mediterranean evergreen forests and Maquis* (Jerusalem pine, Kermes oak, Palestine Terebinth, Lentisk, Carob). — (2) *Mediterranean deciduous forests and Maquis* (Tabor oak, Storax, Atlantic pistachio). — (3) *Light soils belt vegetation* (Eragrostis, Tabor oak, Carob, Lentisk). — (4) *Carob-Lentisk forests on sandy soils*. — (5) *Mediterranean dwarf shrub formation* (Ononis, Salvia, Echinops, Carlina, Ballota). — (6) *Mediterranean dwarf shrub formation as above accompanied by Zizyphus lotus*. — (7) *Irano-Turanian dwarf shrub steppes* (Artemisia herba-alba, Noëa, Haloxylon). — (8) *Irano-Turanian shrub steppes* (Retama duriaei). — (9) *Irano-Turanian shrub steppes* (Zizyphus lotus). — (10) *Irano-Turanian steppes* (no communities identified as yet). — (11) *Vegetation of mobile and semi-stable coastal sands* (Artemisia, Lotus). — (12) *Saharo-Sindian desert vegetation on Hammadas and gypseous soils* (Anabasis, Salsola, Zygophyllum, Suaeda). — (13) *Saharo-Sindian desert vegetation on sandy-loess soils* (Artemisia monosperma, steppes of annual grasses). — (14) *Saharo-Sindian dune vegetation* (Aristida, Retama) *in W. Negev*. — (15) *Sandy desert vegetation of E. Negev and Edom* (Retama, Anabasis, Haloxylon, Zilla). — (16) *Sandy desert vegetation of Wadi Araba* (Haloxylon persicum, Haloxylon salicornicum).— (17) *Saline desert vegetation* (Tamarix, Arthrocnemum, Suaeda, Atriplex, Salsola, Nitraria). — (18) *Sudano-Deccanian oases vegetation* (Zizyphus, Balanites, Acacia). — (19) *Weeds and hydrophilous vegetation of alluvial plains* (Prosopis, Phragmites, Populus euphratica, Potamogeton).

found in Palestine, along with European robins, skylarks, finches, and wrens. The plants of the coastal plain and southern highlands are mostly the same as, or very similar to, those common in the Mediterranean region as a whole east of Gibraltar. Here one finds the Aleppo pine, myrtle, holm, olive, arbutus, carob, orange, citron, grapevine, fig, and pomegranate, "and the streams are overhung with the roseate blossoms of the oleander" (146).

The remainder of the tablelands which make up the greater part of Palestine east of the Jordan and of Transjordania west of that river, have a flora of a more widely diffused character. There are plants of Europe and central Asia, including even some of the same species as are found naturally in England (although not as many as some writers would have us believe—the relationships in most cases are generic, not specific). Among these plants of the tablelands are pines, junipers, terebinths, almond, apricot, peach, hawthorn, mountain-ash, ivy, honeysuckle, walnut, mulberry, oaks, poplars, willows, "the majestic cedars of Lebanon, the melancholy cypress, and the plane-tree with its wide-spreading shade" (146).

The vegetation of the Jordan valley is of a type most closely related to that of northern Africa, with some western Indian and also some European species interspersed. Here the date-palm once flourished in great numbers, although now only a very few remain. Here grow the acacias and the retam of the Sinai desert and many less well-known plants common to northern Africa and Arabia.

The mountain chains of Lebanon and Antilebanon, in spite of their height and cold, are completely lacking in any truly Arctic species of plants, due, it is thought, to a very hot geological period which preceded the present and which doubtless exterminated them (266). Endemic plants, however, are there in large numbers. The chain of the Antilebanons more or less parallels the higher Lebanons and terminates in the massive Mount Hermon (9400 feet altitude) which dominates northern and eastern Palestine. Between Hermon and the mountains of Gilead is a great lava plateau 40 miles long, formed by the ancient eruptions of the numerous volcanic cones which still dot its surface. This lava outpouring is largely responsible for the fertile wheat fields of Bashan. Rainfall in the westernmost range of mountains varies from 36 to 50 inches due to the fact that these mountains impede the eastward flow of the heavy moisture-laden winds from the Mediterranean Sea. The rainfall on the second range of mountains is about half that of the Lebanons, while that of the eastern plateau is not more than 10 to 12 inches. Although the entire Lebanon-Syria-Palestine-Transjordania-Sinai region is only about the size of New York state or of England, it has a vascular flora of at least 126 families, 850 genera, and 3500 species (266). If the Biblical portion of Egypt is included, the entire area of Bible lands would be about equivalent to that of New York and Maryland, or less than that of England and Scotland combined. And over all this area, except the most remote, are now the introduced weeds from Europe, Asia, Africa, and even America, and, in cultivation, exotics of hundreds of kinds from every corner of the globe.

The barrenness and desolation of so much of Palestine, Syria, and Transjordania today are laid to two chief causes: first, the cutting down of the natural timbers by the natives and by successive hordes of invaders from the time of SENNACHERIB (705-681 B. C.) to that of TITUS (70 A. D.), leaving the exposed land to be eroded and the fertile topsoil to be carried off, much as in parts of our own American Middle West; second, the neglect of the ancient terrace-type cultivation, which was the precursor of our modern contour-plowing. "Originally Palestine was emphatically 'a good land, a land of brooks of water, of fountains and depths that spring out of valleys and hills'; a land whose inhabitants could 'eat bread without scarceness' and 'not lack anything in it'. It was 'a pleasant portion' and 'a delightsome land' (Deut. 8: 7-9; Jerem. 12: 10; Mal. 3: 12). But centuries of misrule and neglect have combined with natural agencies to make desolate this once favored heritage. The winter rains have swept the thin soil from the hill-sides, the sword of the conqueror and the ax of the peasant have demolished both forest and fruit-tree; many a spring has thus run dry, and many a stream now feeds only a pestilential marsh; the 'soil mourneth and languisheth' and the ancient prediction is fulfilled by the operation of natural but unerring laws" (146)—"When Lebanon withers in shame, and Sharon sinks to a desert, and Bashan and Karmel are leafless" (Isaiah 33: 9, MOFFATT translation). And the modern botanist confirms the equally complete and literal accuracy of the ancient prophetic denunciation: "Upon the land of my people shall come up thorns and briers; yea, upon all the houses of joy in the joyous city . . . The thorn and the thistle shall come upon their altars" (Isaiah 32: 13; Hosea 10: 8).

The year in Palestine is naturally divided into two seasons beginning, according to DALMAN (97) with the winter "rains" from December or a trifle earlier and lasting for about two hundred days, after which the winds reverse their course and the east or desert winds complete the rest of the calendar year with a "summer" dryness. It seems that even today Palestinians refer very infrequently to time and events as measured by formal calendar dates, but rather to them in terms of a natural calendar which is marked off by the time of sowing, pruning, and harvesting for each of the main crops. Throughout the Bible there are found no formal calendar references. The Hebrews set their natural calendars according to the moon. Their gentile neighbors had similar natural calendar time reckoning, but it was set according to the sun (97).

FONDA of the United States Soil Conservation Service writes: "Even in the most material sense the psalmist who scribed the 24th psalm, 'The earth is the Lord's and the fullness thereof', knew what he was writing. Throughout the ages, in one way or another, man has disputed this stewardship of the land, claiming its use for personal and often selfish benefits. Parts of the old Biblical land are among the countries which offer some of the world's most stark testimonials of soil erosion and land exploitation. Dr. W. C. LOWDERMILK, an eminent student of land use in the Biblical regions as well as other areas, states: 'Out of the land of Mesopotamia came the stories of the "Flood" and of NOAH and the Ark; JONAH and Nineveh and of the "Tower of Babel". Here also was the traditional site of the Garden of Eden, where ADAM and EVE were food gatherers. Today it is as barren of vegetation as though shaved with a razor.' Fabulous old cities, now covered deep with shifting sands, broken and weathered olive presses and other relics of human endeavor and life offer mute testimony to the bountiful prosperity man once enjoyed from the then rich soil and flourishing agriculture. 'The land which we passed through to spy it out is an exceeding good land—a land which floweth with milk and honey.' This was the minority report submitted by JOSHUA and CALEB to MOSES, who had sent these two and ten other men to spy out the land of Canaan. Dr. LOWDERMILK writes that 'The Promised Land', which three thousand years ago was 'flowing with milk and honey', has been so devastated by soil erosion that the soils have been swept off fully half the area of the hill lands. A survey of ancient village sites abandoned and now occupied discloses how the lands of Palestine have been depopulated since the seventh century" (120).

FIGURE 14. — A view of Bethlehem from the southwest. Neglect of the ancient terrace type cultivation on steep hillsides, like this, soon completed the total destruction of the land begun originally by the ruthless deforestation and carried on by the thoughtless exploitation of man over a history of 6,000 or more years. The leafy trees are olives. Ruins such as are seen in the foreground are now often covered by caper and prickly-pear plants. (Wood engraving from C. W. Wilson's Picturesque Palestine, 1883).

Helps to Users of this Work

SCRIPTURAL CITATIONS AS GIVEN AT THE BEGINNING OF THE CHAPTERS ARE TAKEN IN EVERY CASE FROM THE AUTHORIZED VERSION OF KING JAMES, SINCE THAT IS THE MOST GENERALLY USED VERSION AMONG ENGLISH-SPEAKING PEOPLE. TO CONSERVE SPACE ONLY THE MOST IMPORTANT OR INTERESTING VERSES ARE QUOTED IN EACH CHAPTER, AND IN CASES WHERE THE REFERENCES ARE LENGTHY, ONLY THE MOST RELEVANT PARTS OF EACH VERSE. AT THE END OF EACH CHAPTER WE HAVE PLACED THE NAMES OF THE PLANTS WHICH WE REGARD AS THE ONES REFERRED TO IN THE CITED VERSES OF SCRIPTURE IMMEDIATELY BELOW THEM. THESE NAMES AND AUTHORITIES ARE, IN GENERAL, IN ACCORDANCE WITH POST (266, 267), OR, IN THE CASE OF BACTERIA AND RELATED SCHIZOPHYTES, IN ACCORDANCE WITH THE LATEST EDITION OF BERGEY'S MANUAL (52). WHERE VERSES OR PARTS OF VERSES ARE OMITTED THIS fact is indicated by a series of dots—the length of this series of dots bears no definite relation to the number of verses or words omitted. Where a quotation ends at the conclusion of a verse a period is placed regardless of the punctuation mark which may occur there in the original (except for interrogation and exclamation points, which are naturally retained). In all other cases the spelling, capitalization, and punctuation of the original are retained. The italicization of supplied words as given in the King James Version is, however, not retained. Instead, we are italicizing in each quotation the word or words chiefly involved in that particular chapter. All personal names are printed in small capitals as per the rules of the Cambridge and Oxford University Presses, unless they are employed in such a way as not actually to refer to persons.

In the body of each chapter we refer to the 1611 Authorized Version of King James I either as the "Authorized Version" or the "King James Version" (2, 3, 17); the 1914 edition of the Challoner-Rheims-Douay version as the "Douay version" (1); the 1941 revision of the Challoner-Rheims-Douay version edited under the patronage of the Episcopal Committee of the Confraternity of Christian Doctrine as the "O'HARA version" (13); Dr. MOFFATT'S three volumes as the "MOFFATT version" (10, 11, 12); Dr. LAMSA's translation of the New Testament as the "LAMSA version" (8); the 5th edition (1943) of Dr. WEYMOUTH's translation into everyday English as the "WEYMOUTH version" (15); the Jewish Publication Society of America's 1946 edition as the "JASTROW version" (16); the Revised Standard Version of 1946 as the "WEIGLE version" (14); the 1939 "Complete Bible: an American translation" by E. J. GOODSPEED, T. J. MEEK, L. WATERMAN, J. M. P. SMITH, and A. R. GORDON as the "GOODSPEED version" (5); and the "Twenty-four books of the Holy Scriptures carefully translated after the best Jewish authorities" by I. LEESSER in 1913 as the "LEESSER version" (9).

In parentheses throughout the work will be found the numbers referring to items in our bibliography on pages 259 to 270 from which the statements quoted have been taken or which are authority for statements there made. No attempt has been made thus to refer to every place in the bibliography where a statement is made, for very many of the statements occur in numerous of the works cited. Only the most striking or controversial statements are thus indicated. As a matter of general popular interest the procedure of previous authors on this subject has been followed, and interesting myths, superstitions, and legends about the plants in question, or their close relatives, are given in very brief form in many of the chapters. No attempt has been made to make this discussion of the mythology and folklore of the plants or their use in medicine or present day arts complete. Such material of this character as is given is taken largely from R. A. COTES' "Bible flowers" (95), H. FRIEND'S "Flowers and flower lore" (125, 126), SKINNER'S "Myths and legends of flowers, trees, fruits, and plants in all ages and in all climes" (298), previous works on Scriptural natural history, and such standard sources as WEBSTER'S dictionary (158) and the Twentieth Century Encyclopedia (373). For more information on this subject the reader is referred to these books. Our sincere thanks are hereby rendered to the publishers and copyright owners of these books, the various editions and translations of the Bible or its several parts, and the many other publications listed in our bibliography and cited in various parts of this work, for their kind permission to quote from their works.

Some of the versions of the Bible do not employ the same names for the various component books, nor are they always included in the same sequence. The names and sequence adopted by us in this work are those of the King James Version. Similarly, the chapter and verse numbers used by us are those of that version. Readers desirous of looking up a given reference in another version must bear in mind the discrepancy that often occurs in the numbering of the chapters and verses. To conserve space we refer to the Song of Solomon simply as "Song" and the Acts of the Apostles as "Acts". The following table indicates the different names under which the various books of the Bible will be found in the leading Protestant, Catholic, and Jewish Bibles in use today. The number preceding each name indicates the numerical position of that book in the sequence of that particular version:

Authorized Version:	*Douay Version*:	*Jastrow Version*:
1. GENESIS	1. GENESIS	1. GENESIS
2. EXODUS	2. EXODUS	2. EXODUS
3. LEVITICUS	3. LEVITICUS	3. LEVITICUS
4. NUMBERS	4. NUMBERS	4. NUMBERS
5. DEUTERONOMY	5. DEUTERONOMY	5. DEUTERONOMY
6. JOSHUA	6. JOSUE	6. JOSHUA
7. JUDGES	7. JUDGES	7. JUDGES
8. RUTH	8. RUTH	31. RUTH
9. I SAMUEL	9. I KINGS	8. I SAMUEL
10. II SAMUEL	10. II KINGS	9. II SAMUEL
11. I KINGS	11. III KINGS	10. I KINGS
12. II KINGS	12. IV KINGS	11. II KINGS
13. I CHRONICLES	13. I PARALIPOMENON	38. I CHRONICLES
14. II CHRONICLES	14. II PARALIPOMENON	39. II CHRONICLES
15. EZRA	15. I ESDRAS	36. EZRA
16. NEHEMIAH	16. II ESDRAS	37. NEHEMIAH
17. ESTHER	19. ESTHER	34. ESTHER
18. JOB	20. JOB	29. JOB
19. PSALMS	21. PSALMS	27. PSALMS
20. PROVERBS	22. PROVERBS	28. PROVERBS
21. ECCLESIASTES	23. ECCLESIASTES	33. ECCLESIASTES
22. SONG OF SOLOMON	24. CANTICLE OF CANTICLES	30. SONG OF SONGS
23. ISAIAH	27. ISAIAS	12. ISAIAH
24. JEREMIAH	28. JEREMIAS	13. JEREMIAH
25. LAMENTATIONS	29. LAMENTATIONS	32. LAMENTATIONS
26. EZEKIEL	31. EZECHIEL	14. EZEKIEL
27. DANIEL	32. DANIEL	35. DANIEL
28. HOSEA	33. OSEE	15. HOSEA
29. JOEL	34. JOEL	16. JOEL
30. AMOS	35. AMOS	17. AMOS
31. OBADIAH	36. ABDIAS	18. OBADIAH
32. JONAH	37. JONAS	19. JONAH
33. MICAH	38. MICHAEAS	20. MICAH
34. NAHUM	39. NAHUM	21. NAHUM
35. HABAKKUK	40. HABACUC	22. HABAKKUK
36. ZEPHANIAH	41. SOPHONIAS	23. ZEPHANIAH
37. HAGGAI	42. AGGEUS	24. HAGGAI
38. ZECHARIAH	43. ZACHARIAS	25. ZECHARIAH
39. MALACHI	44. MALACHIAS	26. MALACHI
40. I ESDRAS		
41. II ESDRAS		
42. TOBIT	17. TOBIAS	
43. JUDITH	18. JUDITH	
44. ADDITIONS TO ESTHER		

Authorized Version	Douay Version	Jastrow Version
45. WISDOM OF SOLOMON	25. WISDOM	
46. ECCLESIASTICUS (WISDOM OF SIRACH)	26. ECCLESIASTICUS	
47. BARUCH	30. BARUCH	
48. SUSANNAH		
49. SONG OF THE THREE CHILDREN		
50. BEL AND THE DRAGON		
51. MANASSEH		
52. I MACCABEES	45. I MACHABEES	
53. II MACCABEES	46. II MACHABEES	
54. MATTHEW	47. ST. MATTHEW	
55. MARK	48. ST. MARK	
56. LUKE	49. ST. LUKE	
57. JOHN	50. ST. JOHN	
58. ACTS OF THE APOSTLES	51. ACTS OF THE APOSTLES	
59. ROMANS	52. ST. PAUL TO THE ROMANS	
60. I CORINTHIANS	53. I CORINTHIANS	
61. II CORINTHIANS	54. II CORINTHIANS	
62. GALATIANS	55. GALATIANS	
63. EPHESIANS	56. EPHESIANS	
64. PHILIPPIANS	57. PHILIPPIANS	
65. COLOSSIANS	58. COLOSSIANS	
66. I THESSALONIANS	59. I THESSALONIANS	
67. II THESSALONIANS	60. II THESSALONIANS	
68. I TIMOTHY	61. I TIMOTHY	
69. II TIMOTHY	62. II TIMOTHY	
70. TITUS	63. TITUS	
71. PHILEMON	64. PHILEMON	
72. HEBREWS	65. TO THE HEBREWS	
73. JAMES	66. ST. JAMES	
74. I PETER	67. I ST. PETER	
75. II PETER	68. II ST. PETER	
76. I JOHN	69. I ST. JOHN	
77. II JOHN	70. II ST. JOHN	
78. III JOHN	71. III ST. JOHN	
79. JUDE	72. ST. JUDE	
80 REVELATION	73. APOCALYPSE	

The following table indicates some of the instances among plant references where the three leading versions of the Bible differ in the numbering of the chapters and verses:

Authorized Version	Douay Version	Jastrow Version
LEVITICUS 6: 15	LEVITICUS 6: 15	LEVITICUS 6: 8
LEVITICUS 6: 21	LEVITICUS 6: 21	LEVITICUS 6: 14
NUMBERS 16:46-50	NUMBERS 16: 46-50	NUMBERS 17: 11-15
NUMBERS 17: 1-8	NUMBERS 17: 1-8	NUMBERS 17: 16-23
DEUTERONOMY 29: 18	DEUTERONOMY 29: 18	DEUTERONOMY 29: 17
I KINGS 4: 25	III KINGS 4: 25	I KINGS 5: 5
I KINGS 4: 28	III KINGS 4: 28	I KINGS 5: 8
I KINGS 4: 33	III KINGS 4: 33	I KINGS 5: 13
I KINGS 5: 8	III KINGS 5: 8	I KINGS 5: 22
I KINGS 5: 10	III KINGS 5: 10	I KINGS 5: 24
I KINGS 5: 11	III KINGS 5: 11	I KINGS 5: 25
I CHRONICLES 12: 40	I PARALIPOMENON 12: 40	I CHRONICLES 12:41
II CHRONICLES 2: 7-8	II PARALIPOMENON 2: 7-8	II CHRONICLES 2: 6-7
II CHRONICLES 2: 10	II PARALIPOMENON 2: 10	II CHRONICLES 2: 9
II CHRONICLES 2: 14-15	II PARALIPOMENON 2: 14-15	II CHRONICLES 2: 13-14
JOB 40: 21-22	JOB 40: 16-17	JOB 40: 21-22
JOB 41: 2	JOB 40: 21	JOB 40: 26
PSALMS 4: 7	PSALMS 4: 8	PSALMS 4: 8
PSALMS 37: 36	PSALMS 36: 35	PSALMS 37: 35
PSALMS 38: 11	PSALMS 37: 12	PSALMS 38: 12
PSALMS 45	PSALMS 44	PSALMS 45
PSALMS 45: 8	PSALMS 44: 9	PSALMS 45: 9
PSALMS 51: 7	PSALMS 50: 9	PSALMS 51: 9
PSALMS 52: 8	PSALMS 51: 10	PSALMS 52: 10
PSALMS 58: 9	PSALMS 57: 10	PSALMS 58: 10
PSALMS 60	PSALMS 59	PSALMS 60
PSALMS 65: 9	PSALMS 64: 10	PSALMS 65: 10
PSALMS 65: 13	PSALMS 64: 12	PSALMS 65: 14
PSALMS 69: 21	PSALMS 68: 22	PSALMS 69: 22
PSALMS 72: 6	PSALMS 71: 6	PSALMS 72: 6
PSALMS 72: 16	PSALMS 71: 16	PSALMS 72: 16
PSALMS 75: 8	PSALMS 74: 9	PSALMS 75: 9

Authorized Version	Douay Version	Jastrow Version
PSALMS 78: 24	PSALMS 77: 24	PSALMS 78: 24
PSALMS 78: 47	PSALMS 77: 47	PSALMS 78: 47
PSALMS 80: 8-16	PSALMS 79: 9-17	PSALMS 80: 9-17
PSALMS 81: 16	PSALMS 80: 17	PSALMS 81: 17
PSALMS 83: 13	PSALMS 82: 14	PSALMS 83: 14
PSALMS 84: 6	PSALMS 83: 6-7	PSALMS 84: 7
PSALMS 92: 12	PSALMS 91: 13	PSALMS 92: 13
PSALMS 102: 4	PSALMS 101: 4	PSALMS 102: 5
PSALMS 102: 11	PSALMS 101: 12	PSALMS 102: 12
PSALMS 103: 15	PSALMS 102: 15	PSALMS 103: 15
PSALMS 120: 4	PSALMS 119: 4	PSALMS 120: 4
PSALMS 128: 3	PSALMS 127: 3	PSALMS 128: 3
PSALMS 137: 2	PSALMS 136: 2	PSALMS 137: 2
SONG OF SOLOMON 6: 2-3	CANTICLE OF CANTICLES 6: 1-2	SONG OF SONGS 6: 2-3
SONG OF SOLOMON 6: 7 & 11	CANTICLE OF CANTICLES 6: 6 & 10	SONG OF SONGS 6: 7 & 11
SONG OF SOLOMON 7: 8	CANTICLE OF CANTICLES 7: 8	SONG OF SONGS 7: 9
SONG OF SOLOMON 7: 12-13	CANTICLE OF CANTICLES 7: 12-13	SONG OF SONGS 7: 13-14
ISAIAH 9: 10	ISAIAS 9: 10	ISAIAH 9: 9
JEREMIAH 9: 15	JEREMIAS 9: 15	JEREMIAH 9: 14
DANIEL 4: 10-12	DANIEL 4: 7-9	DANIEL 4: 7-9
HOSEA 2: 5-6	OSEE 2: 5-6	HOSEA 2: 7-8
HOSEA 2: 8-9	OSEE 2: 8-9	HOSEA 2: 10-11
HOSEA 2: 12	OSEE 2: 12	HOSEA 2: 14
HOSEA 12: 1	OSEE 12: 1	HOSEA 12: 2
HOSEA 14: 7-8	OSEE 14: 8-9	HOSEA 14: 8-9
JOEL 3: 18	JOEL 3: 18	JOEL 4: 18
AMOS 6: 14	AMOS 6: 15	AMOS 6: 14
JONAH 2: 5	JONAS 2: 6	JONAH 2: 6
NAHUM 2: 3	NAHUM 2: 3	NAHUM 2: 4
ECCLESIASTICUS 24: 15	ECCLESIASTICUS 24: 21	
MATTHEW 17: 20	ST. MATTHEW 17: 19	

* * *

No work on the plants of the Bible could really begin without quoting first of all the beautiful legend of the creation of the plant world and the equally beautiful legend of the Garden of Eden. These are given herewith in the Authorized Version and in the MOFFATT translation:

GENESIS 1: 11-13; 2: 8-9 & 15-17; and 3: 1-6—And God said, Let the earth bring forth grass, and herb yielding seed, and the fruit tree yielding fruit after his kind, whose seed is in itself, upon the earth: and it was so. And the earth brought forth grass, and herb yielding seed after his kind, and the tree yielding fruit, whose seed was in itself, after his kind: and God saw that it was good. And the evening and the morning were the third day ... And the Lord God planted a garden eastward in Eden; and there he put the man whom he had formed. And out of the ground made the Lord God to grow every tree that is pleasant to the sight, and good for food; the tree of life also in the midst of the garden, and the tree of knowledge of good and evil ... And the Lord God took the man, and put him into the garden of Eden to dress it and to keep it. And the Lord God commanded the man, saying, Of every tree of the garden thou mayest freely eat: But of the tree of the knowledge of good and evil, thou shalt not eat of it: for in the day that thou eatest thereof thou shalt surely die ... Now the serpent was more subtil than any beast of the field which the Lord God had made. And he said unto the woman, Yea, hath God said, Ye shall not eat of every tree of the garden? And the woman said unto the serpent, We may eat of the fruit of the trees of the garden: But of the fruit of the tree which is in the midst of the garden, God hath said, Ye shall not eat of it, neither shall ye touch it, lest ye die. And the serpent said unto the woman, Ye shall not surely die: For God doth know that in the day ye eat thereof, then your eyes shall be opened, and ye shall be as gods, knowing good and evil. And when the woman saw that the tree was good for food, and that it was pleasant to the eyes, and a tree to be desired to make one wise, she took of the fruit thereof, and did eat, and gave also unto her husband with her; and he did eat.

MOFFATT translation: And God said, "Let the earth put out verdure, plants that bear seed and trees yielding fruit of every kind, fruit with seed in it." And so it was; the earth brought forth verdure, plants bearing seed of every kind and trees yielding fruit of every kind, fruit with seed in it. God saw that it was good. Evening came and morning came, making the third day ... In the land of Eden, to the far east, God the Eternal then planted a park, where he put the man whom he had moulded. And from the ground God the Eternal made all sorts of trees to grow that were delightful to see and good to eat, with the tree of life and the tree that yields knowledge of good and evil in the center of the park ... God the Eternal took man and put him in the park of Eden to till it and to guard it. And God the Eternal laid a command upon the man: "You are free to eat from any tree in the park," he said, "but you must not eat from the tree that yields knowledge of good and evil, for on the day you eat from that tree you shall die." . . . Now the serpent was cunning, more cunning than any creature that God the Eternal had made; he said to the woman, "And so God has said you are not to eat fruit from any tree in the park?" The woman said to the serpent, "We can eat fruit from the trees in the park, but, as for the tree in the center of the park, God has said, 'You must not eat from it, you must not touch it, lest you die'." "No," said the serpent to the woman, "you shall not die; God knows that on the day you eat it your eyes will be opened and you will be like gods, knowing good and evil." So, when the woman saw that the tree was good to eat and delightful to see, desirable to look upon, she took some of the fruit and ate it; she also gave some to her husband, and he ate.

FIGURE 15. — *Acacia tortilis*, the "shittah tree", a source of the "shittim wood" of the Old Testament. This is the largest and commonest tree in the Sinai desert. Its wood was used to build the Ark of the Tabernacle. (Wood engraving from C. W. Wilson's Picturesque Palestine, 1883).

1. Acacia nilotica (L.) Forsk.   2. Loranthus acaciae Zucc.
(FIGURE 45)

EXODUS 3: 2-4—And the angel of the Lord appeared unto him in a *flame of fire* out of the midst of a *bush*: and he looked, and, behold, the *bush* burned with fire, and the *bush* was not consumed. And MOSES said, I will now turn aside, and see this great sight, why the *bush* is not burnt. And ... God called unto him out of the midst of the *bush*.

There is much diversity of opinion about these verses. There are, of course, those who feel that the event described here was supernatural and therefore a true miracle. Among the commentators who feel that a natural explanation may be found are some who think that the phenomenon of the bush that "burned with fire" and yet "was not consumed" may have been caused by its being a particularly vigorous plant of the gasplant or fraxinella, *Dictamnus albus* L.[1] This is a strong-growing herb to three feet tall, with pinnately compound leaves of 5 to 8 pairs of leaflets and panicles of purple flowers (266). The entire plant is covered with tiny oil glands, the oil of which is so volatile that it escapes continually into the air about the plant and the close approach of an uncovered light will cause a flash of flame to envelop the plant. Other authorities maintain that the bush involved was *Acacia seyal* (which see). The Hebrew word used is "seneh" or "s'neh" (Greek, βάτος, meaning any prickly bush) and TRISTRAM and others feel (27, 146) that the bush must have been the thorny acacia or Egyptian mimosa, *Acacia nilotica*[2], which is common throughout the Arabian peninsula and Egypt, locally called "sunt" (146, 184). TRISTRAM reports it from the shores of the Dead Sea (326). It is even thought that the mountains on which it grows derived their name of Sinai from it and that this region is the "wilderness of Sin" or "seneh" (184).

The most logical explanation seems to be that of SMITH who suggests (299) that the "flame of fire" may have been the crimson-flowered mistletoe known as the acacia strap-flower, *Loranthus acaciae*, which grows in great profusion on various thorry *Acacia* shrubs and trees in the Holy Land and Sinai, including *A. nilotica* and *A. seyal*. This mistletoe, when in full bloom, imparts to the shrub or tree the appearance of being ablaze with fire because of its brilliant flame-colored blossoms, which stand out most conspicuously against the green foliage and the yellow flowers of the host plants. SMITH calls attention to the almost breath-taking appearance of this mistletoe on the yellow-flowered opoponax, *Vachellia farnesiana* (Willd.) Wight & Arn.[3], now so common in hedges and along watercourses in Lebanon and Palestine. This, however, is an American plant, only recently introduced into the Holy Land, and could not possibly be involved in the story of the "burning bush". Neither could the shrubs now called "burning-bush", *Euonymus americanus* L. and *E. atropurpureus* Jacq., have been involved, as some have conjectured, because these, too, are American plants, unknown in the Holy Land or Sinai (158).

The JASTROW and Douay renditions of the verses involved are virtually identical with that of the King James Version, and the GOODSPEED translation also continues to make use of the non-committal word "bush". MOFFATT, however, is more explicit—and, we feel, rightly so. His version is: "The angel of the Eternal appeared to him in a *flame of fire* rising out of a *thorn-bush*. When he looked, there was the *thorn-bush* ablaze with fire, yet not consumed! 'I will step aside,' said MOSES, 'and see this marvel, why the *thorn-bush* is not yet burnt up.' ... God called to him out of the *thorn-bush* ... "

---
[1] Also known as *Dictamnus fraxinella* Pers.
[2] Also called *Mimosa nilotica* L. and *Acacia arabica* var. *nilotica* (L.) Asch. & Schweinf.
[3] Also called *Acacia farnesiana* (L.) Willd., *Mimosa farnesiana* L., and *M. scorpioides* Forsk.

3. Acacia seyal Delile   4. Acacia tortilis Hayne
(PAGE 110 — FIGURE 15)

EXODUS 25: 5, 10, 13, 23, & 28—And rams' skins dyed red, and badgers' skins, and *shittim wood* . . . And they shall make an ark of *shittim wood*: two cubits and a half shall be the length thereof, and a cubit and a half the breadth thereof, and a cubit and a half the height thereof . . . And thou shalt make staves of *shittim wood*, and overlay them with gold . . . Thou shalt also make a table of *shittim wood* . . . And thou shalt make the staves of *shittim wood*, and overlay them with gold, that the table may be borne with them.

EXODUS 26: 15-16, 26, 32, & 37—And thou shalt make boards for the tabernacle of *shittim wood* standing up . . . And thou shalt make bars of *shittim wood* . . . And thou shalt hang it upon four pillars of *shittim wood* . . . And thou shalt make for the hanging five pillars of *shittim wood*.

EXODUS 27: 1 & 6—And thou shalt make an altar of *shittim wood*, five cubits long, and five cubits broad . . . And thou shalt make staves for the altar, staves of *shittim wood*, and overlay them with brass.

EXODUS 30: 1 & 5—And thou shalt make an altar to burn incense upon: of *shittim wood* shalt thou make it . . . And thou shalt make the staves of *shittim wood* . . .

EXODUS 35: 7 & 24—And rams' skins dyed red, and badgers' skins, and *shittim wood* . . . and every man, with whom was found *shittim wood* for any work of the service, brought it.

EXODUS 36: 20, 31, & 36—And he made boards for the tabernacle of *shittim wood*, standing up . . . And he made bars of *shittim wood* . . . And he made thereunto four pillars of *shittim wood*.

EXODUS 37: 1, 4, 10, 15, 25, & 28—And BEZALEEL made the ark of *shittim wood* . . . and he made staves of *shittim wood* and overlaid them with gold . . . And he made the table of *shittim wood* . . . And he made the staves of *shittim wood* . . . And he made the incense altar of *shittim wood* . . . And he made the staves of *shittim wood*.

EXODUS 38: 1 & 6—And he made the altar of burnt offering of *shittim wood* . . . And he made the staves of *shittim wood*, and overlaid them with brass.

NUMBERS 25: 1—And Israel abode in *Shittim*, and the people began to commit whoredom with the daughters of Moab.

NUMBERS 33: 49—And they pitched by Jordan, from Beth-jesimoth even unto *Abel-shittim* in the plains of Moab.

DEUTERONOMY 10: 3—And I made an ark of *shittim wood*.

JOSHUA 2: 1—And JOSHUA the son of Nun sent out of *Shittim* two men to spy secretly.

JOSHUA 3: 1—And JOSHUA rose early in the morning; and they removed from *Shittim*, and came to Jordan.

ISAIAH 41: 19—I will plant in the wilderness the cedar, the *shittah tree*, and the myrtle, and the oil-tree; I will set in the desert the fir tree, and the pine, and the box tree together.

JOEL 3: 18—And it shall come to pass in that day, that the mountains shall drop down new wine, and the hills shall flow with milk, and all the rivers of Judah shall flow with waters, and a fountain shall come forth of the house of the Lord, and shall water the valley of *Shittim*.

MICAH 6: 5—O my people, remember now what BALAK king of Moab consulted, and what BALAAM the son of BEOR answered him from *Shittim* unto Gilgal.

There is no doubt about the identity of the plant referred to by the Hebrew word "shittah" (singular) or "shittim" (plural). In the Authorized Version of King James the name "shittah tree" occurs but once (Isaiah 41: 19), while "shittim wood" occurs twenty-six times, always in connection with the ark of the tabernacle, which, along with its altar and table, was ordered to be made of this wood. The Douay version uses "setim wood." In the Revised Version the terms employed are "acacia tree" and "acacia wood", and with this rendition practically all the modern versions, including those of MOFFATT, JASTROW, and GOODSPEED, agree. The "shittah tree" is without doubt a species of the genus *Acacia*, of which there are three or four in Bible lands. Most authorities are of the opinion that *A. seyal* and *A. tortilis* are the most probable species involved in these references since they are the only timber trees of any considerable size on the Arabian desert. They are essentially trees of barren regions, seemingly able to flourish in very dry lands where no other tree is able to find subsistence. *A. tortilis* is by far (268) the largest and commonest tree on the deserts of Arabia where the Israelites wandered for forty years. It is especially conspicuous on Mount Sinai. The present-day nomadic Arabs of the desert gather its wood, which they burn for fuel, and collect its foliage and flowers to serve as food for their cattle. Its wood is very hard, close-grained, and durable, orange-brown in color, very splendid for and still highly valued in cabinet-work (299). It was, thus, admirably suited for employment in the construction of the tabernacle. Mummy-coffins of sycomore were clamped shut with acacia wood by the ancient Egyptians. In favorable localities the tree may attain to a considerable size, reaching a height of 20 or even 25 feet, but in the desert it is usually more shrubby, often small, twisted, gnarled, and windblown. Its branches are armed with strong slender white spines, about 1½ inches long, borne in pairs. The leaves are bicompound and the flowers are borne in numerous, small, yellow heads. The bark is used for tanning leather. The fruit is a slender, curved, leguminous pod. The Greek for "shittah tree" or "shittim wood" is "ξύλον ἄσηπτον", meaning wood not liable to rot.

The Egyptian name for the acacia tree is "sont", "sant", or "santh". JABLONSKI, CELSIUS, and many other authors state that the Hebrew name is derived from this Egyptian word. STANLEY and W. SMITH (306) are of the opinion that the predominant use of the plural form of the word in the Scriptures—"shittim" rather than "shittah"—may be traced to the fact that these trees seldom grow singly, but usually form tangled thickets.

Many authors feel that *A. seyal* is more probably the tree referred to (178) while *A. nilotica* (L.) Forsk. and *A. arabica* Willd. may also be involved. *A. seyal* is very common in some parts of the Sinai peninsula. The wood of *A. seyal* and *A. arabica* yields the well-known brownish gum known as gum arabic. It is not known if the ancient Hebrews were acquainted with this substance and its uses. The almost complete absence of references to the shittah tree in the later books of the Bible implies that it was not a native of northern Palestine (146). *A. seyal* meets this requirement perfectly, for it is quite common in the region of Sinai, but is not found in Palestine save where it has straggled up the valley of the Jordan. There is a valley on the west side of the Dead Sea, the Wady Seyal, which is said to derive its name from the presence of a few acacias there. From the Dead Sea southward the acacias increase in abundance. Their favorite haunts are the wadies or ravines far to the south. The specific name of *A. seyal* is derived from the Arabic word "seyal", meaning a torrent, and this acacia is sometimes referred to as the "torrent tree" in allusion to its habitat in the wadies through which fast-moving streams flow in the rainy season. POST is of the opinion that these trees were more numerous in ancient times "filling most of the desert valleys, and growing in clefts of the rock and on the now bare mountain sides."

In the books of Numbers, Joel, Joshua, and Micah the word "shittim" is used as a place name, according to the Authorized Version, probably because of the abundance of acacias at those places at that time. The "Abel-shittim"—or "Abel has-Shittim" according to SMITH (306)—of Numbers 33:49 literally means "the meadow (or moist place) of the acacias" and is regarded as probably being the same place referred to simply as "Shittim" in the other references except that of Joel. The "Nachal-Shittim", or Wady Sunt as it would now be called, of Joel can hardly have been the same spot, but, according to SMITH, there is nothing to give a clue to its exact location. Since the acacia is not a tree of northern Palestine, it is not at all strange that the "valley of Shittim" and the last camping site of Israel between the conquest of the trans-Jordanic highlands and the passage of the Jordan, the plains of "Shittim", where it did grow, should have been named for it. The Douay version translates "the valley of Shittim" in Joel 3:18 as "the torrent of thorns", although in the Joshua and Micah references it continues to call the place "Setim."

TRISTRAM is of the opinion that the "burning bush" of MOSES (Exodus 3:2) was *A. seyal* or perhaps the Egyptian *A. nilotica*, known locally as "sunt", and that it represents the Hebrew word "seneh" or "senna" wherever this occurs in the Bible. J. SMITH (299) dismisses this theory as not very plausible, but many authorities are now of the opinion that the story of MOSES and the "burning bush" may, indeed, be an allegory referring to the flame-like appearance of the parasitic mistletoe, *Loranthus acaciae*, among the branches of an acacia (see under *Acacia nilotica*).

While there is no doubt about the identity of the "shittim" of the Bible, some doubt has been expressed about the "shittah" of ISAIAH. ISAIAH predicted that myrtles, olives, firs, planetrees, and cypresses—and the "shittah tree"—would be planted in what were then desert places. The general meaning is obvious enough, *viz.*, that trees of rich and fertile soils should be made to grow in the dry and sterile waste-lands of the desert. But the acacia is *normally* a tree of the deserts, so its inclusion here detracts from the force of the figure. It was probably from a realization of this incongruity that the Septuagint translates the word "box" in this first clause of the verse (and not in the final clause). Numerous writers are also of this opinion, namely, that ISAIAH's "shittah tree" is *Buxus longifolia*. MOFFATT, however, still regards it as the acacia, although he renders the "box tree" of the latter part of the same verse as "cypress". His rendition of the verse is: "I will plant cedars in the desert, *acacias*, myrtles, olive-trees; I will put fir-trees in the wilderness, and planes, and cypresses." The Douay version is: "I will plant in the wilderness the cedar, and the *thorn*, and the myrtle, and the olive tree: I will set in the desert the fir tree, the elm, and the box tree together." The use of the word "thorn" by the Douay version is open to the same objection as is advanced against the acacia in this passage. The GOODSPEED version is: "I will plant in the wilderness the cedar, the *acacia*, the myrtle, and the olive; I will set in the desert the cypress, the plane, and the larch as well."

GROSER points out (146) that linguistically "shittah" and "shittim" appear to be forms of the same word, but that the context of the ISAIAH passage seems to demand a tree which is normally unknown in the desert. Elsewhere in the book of Isaiah the word employed for box tree is an entirely different word. It is very possible that the original recorders or one of the later copiers of the material now comprising the book made an error in this clause, substituting the word meaning acacia for the one meaning box.

Dr. EVENARI writes us that the acacias, "*A. Seyal, A. tortilis* and *A. spirocarpa*, which are the three species involved, grow only in desert wadis—places which are sometimes during the year filled with water and from whence the salt has been washed out. They are never to be found in the plain desert. So ISAIAH could really mean the acacias because he is speaking about real desert."

The tree possesses still another economic virtue, for, besides yielding the famous gum arabic for the trade, charcoal burners of the present day cut the acacia extensively for burning. Its charcoal is said to be the best that can be obtained in Palestine.

Numerous non-botanical writers and travelers in the Holy Land have advanced the opinion that the acacia of the Bible is *Robinia pseudo-acacia* L., the common black locust of eastern North America. This exotic tree was introduced into Palestine at the end of the 17th or beginning of the 18th century and is now thriving there very well. It is spreading and becoming naturalized (71), so it is not at all strange for travelers who are not acquainted with its history to mistake it for a native resident. However, it was not present in the Holy Land—or anywhere else outside of eastern North America—in Biblical times! Nor does the "shittimwood" of the southern United States (*Bumelia*) have anything to do with the Biblical plant of that name!

5. Acanthus syriacus Boiss.
(*Cf.* FIGURE 48)

JOB 30: 7—Among the bushes they brayed; under the *nettles* they were gathered together.
ZEPHANIAH 2: 9—Surely Moab shall be as Sodom, and the children of Ammon as Gomorrah, even the breeding of *nettles*, and salt pits, and a perpetual desolation.

Many writers, including GROSER (146), regard the above passages as referring to true nettles of the genus *Urtica*. TRISTRAM, however, maintains that the "nettles" of Job, at least, were not true nettles, for it is most difficult to picture any creatures "gathered together" ("huddle" according to GOODSPEED, "coupling" according to MOFFATT, "counted it delightful" according to Douay!) in the fashion described under such plants, which in Palestine are even more virulently stinging than they are by us. Also, the fact that in the Job reference the Hebrew word "chârûl" is used instead of the word "kimmosh" which appears in the Isaiah and Hosea references to true nettles, seems to indicate that a different plant is intended. TRISTRAM believed (299) that the "nettles" of Job were acanthus, spiny-leaved, strong-growing, perennial plants common as weeds in all eastern countries and used since time immemorial as models for the leaf or scroll decorations in art. Most authors regard *Acanthus spinosus* L. as the species involved, but POST (265) identifies the Holy Land variety as *A. syriacus*.

MOFFATT and GOODSPEED both translate the word "nettles" of the Job passage as "scrub". The Douay version renders the verse "They pleased themselves among these kind of things, and counted it delightful to be under the *briers*." The JASTROW rendition is the same here as that of the Authorized Version.

The word used in the Zephaniah passage is also "chârûl" in the original text. It is rendered "nettles" in the Authorized and JASTROW versions, "weeds" by MOFFATT and GOODSPEED, and "thorns" in Douay. Thus we find the very same Hebrew word translated as "nettles", "scrub", "briers", "thorns", and "weeds". It is possible that the acanthus plant is actually intended in each case. CELSIUS and SMITH (306) maintained that the word "chârûl" applied to the shrub, *Paliurus aculeatus*, the Christ-thorn, and ROYLE has thought that it was the charlock, *Sinapis arvensis*. The latter interpretation is here accepted for Proverbs 24: 31, but not for the Job and Zephaniah references cited at the head of this chapter. DALMAN suggests for the reference either the tall growing thistle, *Cynara syriaca* Boiss., which often surpasses the six-foot mark, or *Gymnarrhena micrantha* Desf. which grows tall enough for people to sit under it (98).

6. Acetobacter acetigenum (Henneberg) Bergey
7. Acetobacter acetum (Thomsen) H.B.H.H.
8. Acetobacter plicatum Fuhrmann
9. Acetobacter xylinum (Brown) Bergey

RUTH 2: 14—And BOAZ said unto her, At mealtime come thou hither, and eat of the bread, and dip thy morsel in the *vinegar*.

PSALMS 69: 21— . . . and in my thirst they gave me *vinegar* to drink.

PROVERBS 10: 26—As *vinegar* to the teeth, and as smoke to the eyes, so is the sluggard to them that send him.

PROVERBS 25: 20—As he that taketh away a garment in cold weather, and as *vinegar* upon nitre, so is he that singeth songs to an heavy heart.

MATTHEW 27: 34 & 48—They gave him *vinegar* to drink mingled with gall . . . And straightway one of them ran, and took a spunge, and filled it with *vinegar*.

MARK 15: 36—And one ran and filled a spunge full of *vinegar*.

LUKE 23: 36—And the soldiers also mocked him, coming to him, and offering him *vinegar*.

JOHN 19: 29-30—Now there was set a vessel full of *vinegar*: and they filled a spunge with *vinegar*, and put it upon hyssop, and put it to his mouth . . . When JESUS therefore had received the *vinegar*, he said, It is finished.

The word translated "vinegar" in the Old Testament passages is the Hebrew word "chômets". This liquid generally consisted of wine or some other strong drink which had turned sour. It was sometimes made artificially by the addition of barley to wine, thus rendering it liable to fermentation. That it was quite acid is indicated by the proverbs cited above, and that it was a nauseous drink is indicated by the passage from the Psalms. It was, however, used as a drink by laborers, as is attested by the verse in the book of Ruth.

Quite similar to the "chômets" or "ch'metz" of the Jews was the "acetum" or "posca" of the Romans. This was a thin sour wine, popular among the soldiers, and it was this that was offered to JESUS on the cross in an attempt to keep life in him a little while longer.

In both cases the "vinegar" was probably made from wine through the fermenting action of the four bacteria listed at the head of this chapter. These bacteria are acetic acid-forming and comprise the bulk of what is popularly known as "mother-of-vinegar".

It is interesting to note that the MOFFATT and LAMSA versions continue to use the word "vinegar" in both the Old and New Testament passages quoted above, while GOODSPEED uses it only in the Old, substituting "wine", "sour wine", and "common wine" in the New. The Douay version uses "vinegar" in all the references except the first part of the Matthew passage, where "wine" is used. The Basic English version uses "bitter drink" and "bitter wine"; the O'HARA version uses "gall" and "common wine"; while WEYMOUTH uses "gall", "sour wine", and "wine". The JASTROW version uses "vinegar".

In the Vinegar Bible of the Clarendon Press at Oxford (1717) the parable of the vineyard is mislabeled "Parable of the Vinegar" (158).

10. **Aegilops variabilis** Eig   11. **Alopecurus anthoxanthoides** Boiss.
12. **Avena sterilis** L.   13. **Eragrostis megastachya** (Koel.) Link
14. **Nardurus orientalis** Boiss.   15. **Polypogon monspeliensis** (L.) Desf.

GENESIS 1: 11-12—And God said, Let the earth bring forth *grass* . . . And the earth brought forth *grass* . . .
NUMBERS 22: 4— . . . Now shall this company lick up all that are round about us, as the ox licketh up the *grass* of the field . . .
DEUTERONOMY 11: 15—And I will send *grass* in thy fields for thy cattle . . .
DEUTERONOMY 32: 2—My doctrine shall drop as the rain . . . and as the showers upon the *grass*.
I KINGS 18: 5— . . . Go into the land, unto all fountains of water, and unto all brooks: peradventure we may find *grass* to save the horses and mules alive, that we lose not all the beasts.
II KINGS 19: 26— . . . they were as the *grass* of the field, and as the green herb, as the *grass* on the house tops, and as corn blasted before it is grown up.
JOB 40: 15— . . . he eateth *grass* as an ox.
PSALMS 37: 2—For they shall be cut down like the *grass*, and wither as the green herb.
PSALMS 72: 6—He shall come down like rain upon the mown *grass* . . .
PSALMS 90: 5-6— . . . they are like *grass* which groweth up. In the morning it flourisheth, and groweth up; in the evening it is cut down, and withereth.
PSALMS 102: 4 & 11—My heart is smitten, and withered like *grass* . . . I am withered like *grass*.
PSALMS 103: 15—As for man, his days are as *grass*: as a flower of the field, so he flourisheth.
PSALMS 104: 14—He causeth the *grass* to grow for the cattle, and herb for the service of man . . .
PSALMS 129: 6—Let them be as the *grass* upon the housetops, which withereth afore it groweth up.
PROVERBS 27: 25—The *hay* appeareth, and the tender *grass* sheweth itself . . .
ISAIAH 15: 6—For the waters of Nimrim shall be desolate: for the *hay* is withered away, the *grass* faileth, there is no green thing.
ISAIAH 35: 7— . . . in the habitation of dragons, where each lay, shall be *grass* with reeds and rushes.
ISAIAH 40: 6-8— . . . All flesh is *grass*, and all the goodliness thereof is as the flower of the field: the *grass* withereth, the flower fadeth . . . surely the people is *grass*. The *grass* withereth, the flower fadeth . . .
JEREMIAH 14:6—And the wild asses did stand in the high places . . . their eyes did fail because there was no *grass*.
MATTHEW 6: 30—Wherefore, if God so clothe the *grass* of the field, which today is, and tomorrow is cast into the oven, shall he not much more clothe you, O ye of little faith.
LUKE 12: 28—If then God clothe the *grass*, which is to day in the field, and to morrow is cast into the oven . . .
JOHN 6: 10— . . . Now there was much *grass* in the place . . .
JAMES 1: 10— . . . because as the flower of the *grass* he shall pass away.
I PETER 1: 24—For all flesh is as *grass*, and all the glory of man as the flower of *grass*. The *grass* withereth, and the flower thereof falleth away.
REVELATION 8: 7— . . . and all green *grass* was burnt up.
REVELATION 9: 4—And it was commanded them that they should not hurt the *grass* of the earth . . .

The word "grass" occurs 48 times in the Bible, but in many cases its use is very obviously strictly figurative. Because its green herbage rapidly fades and withers under the parching heat of the Palestinian sun, it has afforded to Biblical writers a ready symbol of the fleeting nature of human fortunes, fame, and reputation (306). It has become symbolic of early withering away, decay, and death (Psalms 37: 2 and 90: 5, Isaiah 40: 6—8). In many such cases the word probably referred to all tender green herbaceous plants, rather than to grasses in particular, some of which are quite tough and hardy. Indeed, one of the Hebrew words employed—"yarok" or "yêrek" (Greek, τα χλωρά,χλω)—means "green" and seems to be used here as a general term for all green herbage growing in fields and meadows and fit for food of cattle. It is the word used in Numbers 22: 4. The "grass" of the field mentioned in Matthew and Luke seems obviously to refer back to the "lily" of the field, herein discussed under *Anemone coronaria* (which see). The "flower of the field" in Isaiah 40: 6-8 probably also refers to this anemony.

About 460 different kinds of grasses are recorded from the Holy Land area by Post (267); some of these are discussed separately by us (see *Arundo, Lolium, Phragmites, Saccharum, Triticum, Hordeum, Sorghum*). Many others are species that have been introduced since Biblical times, others are rare or local in distribution and do not fit the context of the cited verses. In our opinion, the six grasses listed at the head of this chapter are among the most probable for involvement in the Biblical passages quoted. They are natives, abundant and widespread, and were doubtless known familiarly by both Old and New Testament characters.

The Hebrew word used in Genesis 1: 11 is "déshe" or "dehsheh" (Greek, βοτάνη) and is perhaps the nearest equivalent of our "grass" as distinguished from "herbs". In Psalms 72: 6 the Hebrew word "gēz" or "gayz" is used for the mown grass or mown field of other translations. In Isaiah 15:6 and 35:7, I Kings 18: 5, Job 40: 15, and Psalms 104: 14 a third word is used—"châtzîr" (Greek χόρτος, πόα or πόια)—and this more closely resembles our word "fodder" in meaning, that is, dry food for cattle. The same word in Proverbs 27: 25 is translated "hay" and in Numbers 11: 5 is rendered "leeks" (see under *Allium porrum*). It probably refers, in most cases, at least, to the stems of tall grasses, for there is no such thing as true hay, in our sense of the word, in Palestine (184). In fact, there are scarcely any pastures or grassy meadows in Palestine, at least, not of the lush type as are found in Europe and eastern North America, except on the maritime plains, where there is tall and luxuriant meadow grass somewhat resembling ours (184). In the Jordan valley there is rank, rapidly growing, prairie-like herbage, including grasses, but in the other parts of the land the grasses are mostly scattered, often growing only near watercourses (*cfr.* I Kings 18: 5). Some species are harsh, tough, desert plants.

The "grass" of Jeremiah 1:11, Authorized Version, is an error for "corn" (see under *Triticum aestivum*). The word "'eseb" signifies herbs for human food in Genesis 1: 30 and 2: 5, Exodus 9: 22, and Isaiah 42: 15, but also signifies fodder for cattle, as in Psalms 104: 14, Deuteronomy 11: 15, and Jeremiah 14: 6.

MOFFATT translates the "grass" of Isaiah 35: 7 as "pasture": "the jackals' and hyenas' lair shall turn to *pasture* for your flocks, and reeds and rushes shall be flourishing." The JASTROW version closely parallels the Authorized for all the references discussed in this chapter, with the following exceptions. In Isaiah 15: 6 JASTROW substitutes "grass" and "herbage" for the Authorized Version's "hay" and "grass", and in Jeremiah 14: 6 he uses "herbage" in place of "grass". His rendition of Isaiah 35: 7 is quite different: "in the habitation of the jackals, herds shall lie down, It shall be an *enclosure* for reeds and rushes."

16. Agrostemma githago L.   17. Solanum incanum L.
(FIGURE 79)

JOB 31: 40—Let thistles grow instead of wheat, and *cockle* instead of barley.

There has been much discussion and argument over the identity of the "cockle" in this reference. The Hebrew word employed is "caoshah" or "coash" (299) or "boshâh" (306); the Greek word is βάτος, meaning any prickly plant. Marginal Bibles give the alternative translation of "noisesome weeds". The plural form of the same word is rendered "wild grapes" in Isaiah 5: 2-4 according to the Bible Encyclopedia (179). Since the original meaning of the Hebrew word actually is "stink" (299) TRISTRAM is of the opinion that it is meant in the Job passage to include noxious weeds in general. This is also the conclusion of MOFFATT who translates it "foul weeds", GOODSPEED who says "weeds", and the JASTROW version which says "noisome weeds". The Douay version employs the word "thorns". WILLIAM SMITH (306) says "we are inclined to believe that the *boshâh* denotes any bad weeds or fruit, and may in Job signify bad or smutted barley. Or it may mean some of the useless grasses which have somewhat the appearance of barley, such as *Hordeum murinum*, &c."

Some writers have suggested species of *Aconitum* and of the *Araceae*, but it does not seem that these are as probable as the ordinary corn-cockle, *Agrostemma githago*[4], for these plants are not usually the grainfield pests that the corn-cockle is. Some commentators (268) have suggested blackberry bushes, poppies, dwarf elder, and the white aconite. CELSIUS has argued in favor of *Aconitum napellus* L., but this is predominantly a plant of mountainous woods, never of grainfields. PRATT, HASSELQUIST, and ROYLE all incline toward the hoary nightshade, *Solanum incanum*. This nightshade is said by some to be a common weed in Palestine and Egypt and its berries resemble grapes in form, although they are bitterly narcotic and poisonous. Possibly this species would fit both the Job and the Isaiah references. Dr. EVENARI claims that it is uncommon, appearing now only in the Jordan valley and in Wadi Arabah. The corn-cockle plant, on the other hand, is said by POST (265) to be common in grainfields throughout the area. It is a strong-growing and very troublesome weed in grainfields, growing one to three feet tall, covered with a dense whitish pubescense, and producing showy purple, red, or white campion-like flowers. It would seem to fit the context of the Job passage very well.

18. Alhagi camelorum var. turcorum (Boiss.) Boiss.
(FIGURE 42)

ECCLESIASTICUS 24: 15—I gave a sweet smell like cinnamon and *aspalathus*, and I yielded a pleasant odour like the best myrrh, as galbanum, and onyx, and sweet storax, and as the fume of frankincense in the tabernacle.

There has been much uncertainty concerning the "aspalathus" of this passage from the Apocrypha (158). This is the only place in the Bible where the word occurs and it seems obviously to be the Greek name of some sweet-smelling plant. THEOPHRASTUS mentions an "aspalathus", along with cinnamon and cassia, as a plant of Indian origin. His plant seems to be *Myrica sapida* Wall., a shrub or small tree of Nepal, and is, of course, a possibility for the Scriptural plant, especially since it is mentioned there also in connection with Indian cinnamon. DIOSCORIDES states that aspalathus was used in his day for thickening ointments.

GERARDE speaks of an "aspalathus" which was called "lignum Rhodium" or "lignum Rhodianum". Modern botanists are of the opinion that this was the wood of two small, erect, branching, shrubby species of morning-glory relatives, *Convolvulus floridus* L. f. and *C. scoparius* L. f., native to the Canary Islands. The wood is sweet-scented and yields an oil called "oil of Rhodium". The leaves are small and silky and the flowers white and pink (299). Although these plants are often mentioned in connection with the "aspalathus" of the Bible (306), it does not seem probable that they were known in those ancient times. SMITH thinks that "some allied species, native of the south of Europe, and possessing the same qualities" may be involved (299). PLINY writes of an "aspalathus" that grows in Spain and on the island of Cyprus, is a white thorny shrub about the height of a medium-sized tree, and is employed as an ingredient of perfumes and ointments (299).

GOODSPEED substitutes "camel's thorn" in the Ecclesiasticus reference and it would appear to us that this is the most likely of all the suggested identifications. The camel's thorn, *Alhagi camelorum* var. *turcorum*, is a many-stemmed, much-branched shrub, thickly beset with sharp axillary spines arising from aborted peduncles. The simple leaves are oblong-obovate and entire. It is reported by POST (266) from Lebanon, Damascus, and Palestine. The even more widely distributed *A. maurorum* Medic. (which see) is also sometimes called "camel's-thorn", but is probably not the plant here referred to.

LINNAEUS adopted the name *Aspalathus* for a genus of small, handsome, leguminous shrubs with silky or heath-like leaves, native to South Africa. These plants, however, have absolutely no connection with the "aspalathus" of the Greeks and Romans, nor that of the Hebrews (299).

---
[4] Sometimes discussed under the synonymous names *Githago segetum* Link and *Lychnis githago* (L.) Scop.

19. Alhagi maurorum Medic.   20. Fraxinus ornus L.
21. Tamarix mannifera (Ehrenb.) Bunge

BARUCH 1: 10—And they said, Behold, we have sent you money to buy burnt-offerings, and sin-offerings, and incense, and prepare ye *manna*, and offer upon the altar of the Lord our God.

This verse from the Apocryphal book of Baruch refers to something which happened long after the manna that fell in the "wilderness" had ceased being used as bread by the Israelites. Since the "manna" here spoken of apparently could be and was purchased with money, there is little doubt that it was a resinous gum derived from trees native to the Levant. Such gummy manna is the solidified sap of the stems or gummy exudations from the leaves of the prickly alhagi or Sinai-manna, *Alhagi maurorum*[5], of the manna tamarisk, *Tamarix mannifera*[6], and of the flowering ash, *Fraxinus ornus*.

The prickly alhagi belongs to the complex *Astragalus* section of the pea family and is a low, scrubby, many-stemmed, much-branched shrub to about 3 feet tall, with slender somewhat hairy twigs thickly beset with slender axillary spines. The simple leaves are obovate-oblong and entire-margined. The flowers are pea-like, borne in few-flowered axillary racemes. It grows in waste places from Syria and Lebanon, through Palestine, to Arabia Petraea and Sinai. During the heat of the day a sweet gummy substance exudes from the leaves and stems. This hardens upon contact with the air, and is then collected by shaking the bushes over a spread-out cloth. Professor DON was so convinced that this was the "manna" of the Israelites that he proposed changing the scientific name of the plant to *Manna hebraica* (299)! It is closely related to the camel-thorn, *A. camelorum* var. *turcorum* (which see), regarded by GOODSPEED as the "aspalathus" of Ecclesiasticus 24: 15.

The manna tamarisk is a much-branched shrub or small tree 9 to 15 or more feet tall, with rather rigid but deciduous branches and minute scale-like half-clasping leaves. The tiny pink flowers are borne in rather short, dense, spreading, terminal panicled racemes, which appear after the leaves. It is found on deserts from Palestine to Arabia Petraea and Sinai. According to HUME, it is found "everywhere" in the Sinai country (266). At certain seasons of the year its tender stems are punctured by the proboscis of a small scale-insect (*Coccus manniparus*), and from these punctures a honey-like liquid exudes (184, 299). This liquid quickly hardens and drops from the tree. It is collected even today by the Bedouin Arabs, who preserve it like honey or make it into cakes, regarding it as a great delicacy (101). WOLSTEAD attributes the sudden disappearance of the manna after the Israelites had crossed the Jordan to the fact that although the tamarisk continues to grow there, the climate there is not favorable to the life and wellbeing of the insect that does the puncturing (299). Thus, after entering the Promised Land, the Hebrews had to secure manna through purchase and trade. A recent author (27a, 27b) has stated that "it was established beyond doubt that the appearance of manna is a phenomenon well known in other countries under the name of 'honey-dew', which is a sweet excretion of plant-lice and scale-insects." This is a confusion of terminology. Honeydew is a term usually applied in America to the very sweet excretion of the insects mentioned above, but not to the exuded sap of the tree which has not actually passed through the insects' digestive tracts. Honeydew, in our sense, occurs in very minute quantities and is used as food by ants, but cannot conveniently be harvested for food for human consumption.

The flowering or manna ash is a tree 15 to 50 feet tall, with odd-pinnate leaves composed of 3 or 4 pairs of lanceolate to round-ovate denticulate leaflets. Unlike our ordinary American ashes, the flowers are produced at the same time or later than the leaves, and are borne in racemes which form a panicle in the upper and terminal leaf-axils. The fruits are winged samaras very similar to those produced by our species. This European tree grows also in Syria and

---
[5]Also known as *Hedysarum alhagi* L. and *Alhagi mannifera* Desv.
[6]Also known as *Tamarix gallica* var. *mannifera* Ehrenb. and *T. nilotica* var. *mannifera* (Ehrenb.) Schweinf.

Lebanon. The manna obtained from it is a sweetish exudate secured either in the form of flakes ("flake manna") or fragments ("common manna") or as a viscid mass ("fat manna"). Its chief chemical constituent is known as mannin. It is used medicinally at present as a gentle laxative, demulcent, and expectorant (158).

The Greek term for the type of "manna" which is discussed in the present chapter is μάννα, while the Hebrew term employed is "mân".

Some commentators have thought that all of the "manna" of the Bible, including that of Exodus 16: 13-15, came from these three species of trees and shrubs, but this is not now generally believed possible (see under *Lecanora affinis*). Other commentators have suggested that the "heath" of Jeremiah 17: 6 and 48: 6 was *Tamarix mannifera*, which has heath-like leaves and branches, but the "heath" is by us regarded as *Juniperus oxycedrus* (which see). The flowering ash has been suggested for the so-called "ash" of Isaiah 44: 14, but the latter is herein discussed under *Pinus halepensis* (which see).

22. Allium ascalonicum L.   23. Allium sativum L.

NUMBERS 11: 5—We remember the fish, which we did eat in Egypt freely; the cucumbers, and the melons, and the leeks, and the onions, and the *garlick*.

The common garlic, *Allium sativum*, is a hardy bulbous perennial plant, which is cultivated in Europe, western Asia, and Egypt (126), as it doubtless also was at the time of Moses (98, 299). It is well-known as a culinary stimulant and is extremely popular with the peoples of the Mediterranean region even today. It is much used in cooking and is often eaten raw on slices of bread. Its bulb has a very characteristic strong scent and pungent flavor and is composed of a number of smaller bulblets, called "cloves", closely crowded together.

There are about 67 kinds of onion and garlic recorded by POST (265, 267) from the Holy Land region, so it is not at all strange that the Hebrews should have developed a liking for these plants. JOHN SMITH (299) thinks that the shallot, *A. ascalonicum*[7], is probably the species referred to by the Hebrew word "shoomim" in the above verse and uniformly translated "garlic". The Greek word for "garlic" here is τά σκόρδα or κόροδοτ. It is milder in taste than *A. sativum* and occurs wild, as well as cultivated, about the once-famous city of Escalom in Palestine—hence its specific name. WILLIAM SMITH, however, is emphatic (306) in his assertion that *A. sativum* is the species involved as it "abounds in Egypt." W. T. STEARN, who has done considerable research on the genus *Allium*, informs us that *A. sativum* "is not a European species; it is a very old cultivated type not known with absolute certainty as a wild plant but which doubtlessly originated in central Asia. The shallot (*A. ascalonicum*) is likewise a cultivated plant unknown in a wild state but is obviously a variety of *A. cepa* L., of which the nearest wild forms occur in central Asia. The epithet Ascalonia applied by the Roman writers Columella and Pliny to a variety of onion merely indicates that it was cultivated about Escalon. Both onion and garlic were well known to the Romans."

The Talmud directs (57) that many kinds of food are regularly to be seasoned with garlic, and it is still a favorite with the Hebrews, whose customs in all lands today still retain so many traces of their Oriental origin. Garlic is said to have been introduced into western Europe by the Crusaders. In medicine it is employed as a digestive stimulant, diuretic, and antispasmodic. There is said to be a tradition in the Orient that when Satan stepped out of the garden of Eden after the "fall" of man, onions sprang up from the spot where he placed his right foot and garlic from the spot where he placed his left foot. This legend alludes to the magic powers (125, 126) once attributed to these vegetables and is not meant to be an aspersion on them because of their odor, for their powerful odor is not in the least objectionable to Oriental peoples. Garlic is fed to dogs, cocks, and ganders in Bohemia in the supposition that this will make them fearless and strong.

---

[7]Inaccurately referred to by JOHN SMITH as *"Allium Escallonicum"* (261).

24. Allium cepa L.

NUMBERS 11:5—We remember the fish, which we did eat in Egypt freely; the cucumbers, and the melons, and the leeks, and the *onions*, and the garlick.

There is no doubt that the "onions" referred to here—"betsâlîm" (306) or "belsal" (299) in the Hebrew and κρόμμυα or κρόμυον in the Greek—are *Allium cepa*, probably in its variety known to us now as the Egyptian onion, in which the compact coated bulb, formed of layers consisting of the broad and fleshy bases of closely overlapping leaves, produces numerous offsets. The leaves are slender, terete, and hollow. The entire plant has a characteristic pungent (alliaceous) taste and odor. While its native country is not definitely known (299), it has been cultivated in Egypt since time immemorial (306). POST says (265) that the onion is cultivated everywhere in the region. Doubtless it was used extensively as food in the days of MOSES as it is today in both Egypt and Palestine (98).

HERODOTUS records an ancient inscription in the Great Pyramid (102) of Cheops (or Khufu) stating that the sum of 1600 talents of silver had been paid to supply the workers with onions, garlic, and radishes while the pyramid was being built. Since the estimated value (158) of the Hebrew talent of silver is about $2176 in American money, this means that about $3,481,600 were spent for that purpose. This gives one a fair idea of the amount of these vegetables consumed in the building of this one pyramid at about 3700 B.C.

HASSELQUIST says (268): "whoever has tasted onions in Egypt, must allow that none can be had better in any part of the universe. Here, they are sweet; in other countries they are nauseous and strong. Here, they are soft; whereas in the northern and other parts they are hard, and their coats are so compact that they are difficult of digestion. Hence they cannot in any place be eaten with less prejudice and more satisfaction than in Egypt." Onion soup is made of them, or else they are cut into four quarters, baked and eaten. Some of the poorest people in the region are said to live almost entirely on them (126), while the richer people eat them cooked with roasted meat. The ancient Egyptians swore by the onion, and have even been accused of worshipping it as a god (126). In it they saw symbolized the universe, since in their cosmogeny the various spheres of heaven, earth, and hell were concentric, like the onion bulb's layers (298). Therefore Egyptian priests were forbidden to eat it (*cfr.*, PLUTARCH, De Isis et Osiris 2: 253).

In view of the glowing accounts of the excellence of Egyptian onions it is not difficult to understand why the Israelites, in a moment of rebellious dissatisfaction in the desert just before the terrible plague at Kibroth-hattaavah, remembered them and even longed to return to Egypt for them! In this time of privation and hunger their freedom, the vines and olives of the Promised Land, and even the happy and independent homes which they were afterwards to enjoy under the fig and myrtle trees of Palestine "counted as nothing against the onions, leeks, and garlic of captivity" (95).

In more recent times the onion has been regarded as a plant of ill-omen and misfortune (95). To dream of onions was indicative of domestic strife and a portent of impending sickness. Other simple folks have endowed the onion with magic properties (125, 126) and believe that if a bunch of onions is hung in rooms where people congregate, the bulbs will draw to themselves the diseases that might otherwise afflict the people (298). SKINNER states that the onion is regarded as sacred to SAINT THOMAS and at Christmas time is frequently a rival to mistletoe in popularity. "At the old holiday sports, a merry fellow who represented the saint would dance into the firelight when the Yule logs blazed, and give to the girls in the company an onion which they were to cut into quarters, each whispering to it the name of the young man from whom she awaited an offer of marriage, waving it over her head, and reciting this spell: 'Good Saint Thomas do me right, and send my true love come tonight, That I may see him in the face, and him in my kind arms embrace.' The damsel will be in her bed by the stroke of twelve, and if the fates are kind she will have a comforting vision of the wedding" (298).

25. Allium porrum L.   26. Trigonella foenum-graecum L.

NUMBERS 11: 5—We remember the fish, which we did eat in Egypt freely; the cucumbers, and the melons, and the *leeks*, and the onions, and the garlick.

There is some question whether the "leeks", here referred to by the Hebrew word "châtsîr" or "chatzir" and by the Greek words τὰ πράσα or πράσος, are the true leek, *Allium porrum,* or an entirely unrelated leguminous plant known as fenugreek, *Trigonella foenum-graecum.* Both have been known in Egypt and the Holy Land since time immemorial and both are popular as food there.

The bulb of the leek differs from that of the onion and garlic in being slender, long-cylindric, and six or more inches in length. The leaves are succulent, but flat and broadly linear. The flavor resembles that of the onion, but is more pungent. The leaves are eaten as a relish, especially in Europe, or are cooked in soups (158). The bulbs are cut into small pieces and employed as a seasoning for meat.

The Hebrew word "châtsîr" is in other parts of the Bible translated otherwise. It occurs twenty times in the Hebrew text. In I Kings 18: 15, Job 40: 15, Psalms 37: 2, 90: 15, 103: 15, 104: 14, 129: 6, and 147: 8, and Isaiah 37: 27, 40: 6-8, 44: 4, and 51: 12 it is rendered "grass" in the Authorized Version; in Job 8: 12 it appears as "herb"; in Proverbs 27: 25 and Isaiah 15: 6 it is translated "hay"; and in Isaiah 34: 13 it is rendered "court" (249). Actually this word literally denotes "grass" (306) and is derived from a linguistic root signifying "to be green". Various commentators, like LUDOLF and MAILLET, have on this account maintained that the reference in this passage is to any green food like lettuce or endive. It would thus have a meaning similar to the present term "greens". However, since the word is used in Numbers 11: 5 in connection with words that unquestionably mean onion and garlic, since it applies to a plant that was very common in Egypt, and since all the most ancient as well as modern translations have uniformly translated it as "leek", it is our opinion that in this verse it is *Allium porrum* to which reference is made (98).

HENGSTENBERG, KITTO, and other writers (299, 306) are of the opinion that the fenugreek is the more likely plant for which the children of Israel longed in the desert. The fenugreek is a three-leaved, clover-like, annual plant with small, axillary, yellow flowers. The seeds are mucilaginous and strongly aromatic, and are eaten boiled or raw, mixed with honey. They were formerly employed in medicine and are still used by veterinarians to give flavor to horse medicines. They are also used to scent damaged hay. Great quantities of this plant in the young state are eaten with considerable relish (*cfr.,* MAYER, Reise nach Aegypten, p. 226) as a salad by poorer Egyptians and are grown in gardens and sold on the streets of Cairo even today (122). SONNINI says (307): "In this fertile country, the Egyptians themselves eat the fenu-grec so largely, that it may be properly called the food of man. In the month of November they cry 'green halbeh for sale!' in the streets of the town; it is tied up in large bunches, which the inhabitants purchase at a low price, and which they eat with incredible greediness without any kind of seasoning." FORSKÅL includes it in his materia medica of Egypt (121).

The true leek, on the other hand (268), is still a favorite article of food of the Hebrews and is as common an article in the Egyptian diet as it probably was 2000 years ago. Both the onion and the leek are said to grow wild on the deserts about Cairo. That the Hebrews use it so abundantly today, rather than the fenugreek, is a potent argument against the latter plant, since the Hebrews are recognized as being noteworthy for the tenacity and persistence of their customs through the ages. Onions and leeks are regarded by the Turks as delicacies fit for paradise. GROSER (146) suggests that since the Hebrew word "châtsîr" literally means "herb", the leek must have been regarded by the Semites as *"the* herb" par excellence. STEARN questions the possibility of people living almost entirely upon onions in view of the low carbohydrate and vitamin content of the onion, "which is essentially a flavouring agent making other and more nutritious but possibly dull foods more palatable." He also expresses doubt about leeks and onions growing wild in the deserts about Cairo, saying that "The leek of Egypt called *A. kurrat* is reputed to be that found in the Egyptian tombs" (203a).

Because the leek has always been the food of the poor (126) in the Orient, it has come to be regarded as a symbol of humility. PLINY records that the emperor NERO was very fond of leeks and thus raised them to the level of respectability among the Romans, but after his death he was often referred to derisively as "Porrophagus", or, "the leek-eater" (81). The leek has more recently become the national emblem of Wales (125) because the old Cymric colors were white and green—white for the snow of their mountain fastnesses, and green for the luxuriant verdure of their lowlands. In their first struggles for independence some Welsh soldier plucked a white-stemmed green-leaved leek and fastened it to his cap as a reminder to his companions of their ancestral colors. Since this time the Welshman regards the leek much as the Irishman does the shamrock or the Scotsman the thistle, and religiously wears one on Saint David's Day (March 1) (125, 126).

27. Aloë succotrina Lam.
(FIGURE 49)

JOHN 19: 39— . . . and brought a mixture of myrrh and *aloes*.

The "aloes" of the Old Testament were probably either *Aquilaria agallocha* or *Santalum album*, which see, but those brought by NICODEMUS to embalm the body of JESUS were doubtless the true aloes, *Aloë succotrina*[8]. The inspissated juice of these aloes forms the purgative and emmenagogue drug of that name. This drug was known to the ancients and was used by the Egyptians in their highly perfected art of embalming. Its smell, however, is not very agreeable, and its taste is very bitter (125). It is now used principally by veterinarians as a horse medicine (299).

In spite of the fact that most recent authors agree that the aloes of NICODEMUS were true aloes, GROSER (146) still regards them as having been the same as the Old Testament "aloes", while SHAW (293) does not attempt to identify any of the plants referred to in the Bible under this name. The Hebrew words used in the Old Testament are "ahâlim" and "a'haloth" or "a'halot", while the Greek word used in the NICODEMUS story is ἀλόη.

Aloë succotrina is a succulent plant with stiff, thick, fleshy leaves borne in a dense rosette like those of the familiar centuryplant, and tubular red flowers, tipped with green, borne in a single dense terminal spike, blooming every year. The drug is manufactured principally from the pulp of the fleshy leaves, although every part of the plant is purgative. It is a native of the island of Socotra, off the east coast of Africa at the mouth of the Red Sea, and its name is taken from the Arabic name of the plant, "alloeh". That this drug and myrrh, also imported, were expensive in Palestine goes without saying, and since NICODEMUS is said to have brought about a hundred pounds of these substances, it may safely be assumed that he was quite a wealthy man.

The plant here under discussion must not be confused with the American aloe, *Agave americana*, more properly called centuryplant.

28. Amygdalus communis L.
(PAGE 21; FIGURE 37; PAGE 274)

GENESIS 28: 19—And he called the name of that place Bethel: but the name of that city was called *Luz* at the first.
GENESIS 30: 37—And JACOB took him rods of green poplar, and of the *hazel* and chestnut tree.
GENESIS 35: 6—So JACOB came to *Luz*, which is in the land of Canaan.
GENESIS 43: 11— . . . carry down the man a present, a little balm, and a little honey, spices, and myrrh, nuts, and *almonds*.

---
[8]Sometimes inaccurately referred to as "*Aloe socotrina*."

EXODUS 25: 33-36—Three bowls made like unto *almonds*, with a knop and a flower in one branch; and three bowls made like *almonds* in the other branch, with a knop and a flower: so in the six branches that come out of the candlestick. And in the candlestick shall be four bowls made like unto *almonds*, with their knops and their flowers. And there shall be a knop under two branches of the same, and a knop under two branches of the same, and a knop under two branches of the same, according to the six branches that proceed out of the candlestick. Their knops and their branches shall be of the same.
EXODUS 37: 19-20—Three bowls made after the fashion of *almonds* in one branch, a knop and a flower; and three bowls made like unto *almonds* in another branch, a knop and a flower: so throughout the six branches going out of the candlestick. And in the candlestick were four bowls made like *almonds*, his knops, and his flowers.
NUMBERS 17: 1-8—And the Lord spake unto MOSES, saying, Speak unto the children of Israel, and take of every one of them a rod according to the house of their fathers, of all their princes according to the house of their fathers twelve rods: write thou every man's name upon his rod. And thou shalt write AARON's name upon the rod of LEVI: for one rod shall be for the head of the house of their fathers. And thou shalt lay them up in the tabernacle of the congregation before the testimony, where I will meet with you. And it shall come to pass, that the man's rod, whom I shall choose, shall blossom: and I will make to cease from me the murmurings of the children of Israel, whereby they murmur against you. And MOSES spake unto the children of Israel, and every one of the princes gave him a rod apiece, for each prince one, according to their father's houses, even twelve rods: and the rod of AARON was among their rods. And MOSES laid up the rods before the Lord in the tabernacle of witness. And it came to pass, that on the morrow MOSES went into the tabernacle to witness; and, behold, the rod of AARON for the house of LEVI was budded, and brought forth buds, and bloomed blossoms, and yielded *almonds*.
JOSHUA 16: 2 & 18: 13—And goeth out from Beth-el to *Luz* . . . And the border went over from thence toward *Luz*, to the side of *Luz*, which is Beth-el, southward.
ECCLESIASTES 12: 5—Also when they shall be afraid of that which is high, and fears shall be in the way, and the *almond tree* shall flourish, and the grasshopper shall be a burden, and desire shall fail.
JEREMIAH 1: 11—Moreover the word of the Lord came unto me, saying, JEREMIAH, what seest thou? And I said, I see a rod of an *almond tree*.

The almond, *Amygdalus communis*, is supposed to be a native of western temperate India and Persia (287a, 293, 299) and to have spread westward in very early times. It was apparently common in Palestine at the time when JACOB sent for corn from Egypt, about 1707 B.C. His act of sending a present of almonds, however, implies that the almond did not yet grow in Egypt. It was undoubtedly introduced into Egypt soon thereafter and became common there during the 200 years' residence of the Israelites in Egypt. Thus they must have become thoroughly acquainted with its flowers and fruits. This must have been the case because in 1491 B.C., while they were encamped in the barren desert of Sinai where no almonds grow, they adopted them as models for ornamenting the cups of the golden lampstands. Unless they had been well acquainted with them from Egypt they could not have done so. To this day the glass drops or pieces of rock crystal used for ornamenting branched candlesticks are called "almonds" by English craftsmen and lepidaries (178). The rod or stick of AARON, since it was from an almond tree, must have been brought from Egypt during the exodus. Whether the other eleven rods were also almond wood is not stated. Possibly they were not, thus accounting for their failure to blossom. Almond branches are famous for the speed and ease with which they can be forced into premature bloom when placed in a glass of water in a warm place. KITTO remarks that these rods or staves were doubtless official ensigns of the authority with which the heads (or "princes") of the clans or tribes were invested. Hence the Bible often uses the word "rod" as equivalent to "sceptre". It is worth noting that the awkward expression "a knop and a flower" used in the Authorized Version in the description of the candlesticks is far better translated "a calyx and a flower" by MOFFATT and "calyx and petals" by GOODSPEED, and, similarly, "almond" is far better translated "almond blossoms" in the JASTROW version.

MOFFATT's translation of the Jeremiah passage cited at the head of this chapter is: "This word from the Eternal came to me: 'JEREMIAH, what do you see?' I said, 'The shoot of a *wake-tree*'." The Hebrew word translated as "almond" in the Authorized Version references cited above is "shâkêd." Literally it means "to watch for", "to be wakeful", or "to hasten". The almond is a tree which blooms very early in the spring. Its beautiful pink flowers symbolize the awakening of spring. To the Jews it was a welcome harbinger of spring, a reminder that the winter was passing away, that the flowers would soon appear on the earth again, that the time of the singing of birds had arrived, and that the voice of the turtle-dove would soon be heard again through the land (306). It is therefore a "wakeful tree" or "wake-tree". MOFFATT's use of this term is merely an effort on his part to show the play on words in the original Hebrew. This play on words is completed in the next sentence, which reads, in the MOFFATT translation: "You have seen right; for I am wakeful over my word, to carry it out." Some German translations of this passage render "shâkêd" as "Wacholder", meaning "juniper". This "juniper" of the Authorized Version, however, has nothing whatever to do with the true evergreen juniper as we know it now, but is a legume, *Retama raetam* (which see), whose white pea-like flowers also bloom on leafless branches very early in the year. Dr. EVENARI states that another reason for applying the word "shâkêd" to the almond tree is to be found in the fact that this word also signifies "to try hard". The connection is that this tree has the outstanding ability of penetrating deeply into rock crevices in the mountainous parts of Palestine— apparently trying hard or struggling hard to live.

The Greek καρύα is rendered "almond" in the Revised Version and ἀμύγδαλη is rendered "almond" in the King James Version.

The almond spread early through southern and middle Europe and reached England some 350 years ago, where it now produces one of the most beautiful sights of spring. It is commonly cultivated there and is quite hardy, attaining a height of 12 to 14 feet, blooming in very early spring (even as early as January 9) with a great abundance of light-pink peach-like blossoms usually borne in pairs and appearing before the leaves. Some writers claim that it is native to southern Europe (158) or northern Africa (306), but the evidence does not bear out either claim. Its leaves are long and ovate, with a serrate margin and an acute apex. Its fruit is like that of the peach, but is oval in shape, with a downy succulent covering enclosing a hard shell in which is borne the edible kernel. It was much valued in the Orient (268) because it furnished a very pleasant oil. Any plant producing an oil would naturally be highly prized in lands where the use of oil is so general as it is in the Orient. It is still cultivated in Syria and grows wild on the northern and eastern hills (146), but it was doubtless much more abundant in ancient times, when the Holy Land was a much more fertile land than it is now.

GOODSPEED's rendition of the Ecclesiastes verse cited above is quite different from that of the Authorized Version and equally obscure: "Also, he is afraid of a height, And terrors are on the road; And he rejects the *almond*, And the locust is burdensome, And the caper-berry is ineffectual." The Douay version is different again: "And they shall fear high things, and they shall be afraid in the way, the *almond tree* shall flourish, the locust shall be made fat, and the caper tree shall be destroyed." Of all the versions it seems that only that of MOFFATT catches and transmits the true meaning of the passage in understandable form: "when old age fears a height, and even a walk has its terrors, when his hair is *almond white*, and he drags his limbs along, and the spirit flags and fades." This reference to the almond whiteness of aged peoples' hair alludes to the fact that although their flowers are actually light-pink, when viewed from a distance almond trees in full bloom have a decidedly snowy-white aspect.

It is now generally agreed (179) that the "hazel" of Genesis 30: 37 in the King James Version was actually the almond and not the true hazel, *Corylus avellana* L., as was supposed by earlier writers. The Hebrew word here employed is "luz", and if this word really also stands for the almond tree, as MOFFATT, JASTROW, and GOODSPEED agree that it does (MOFFATT translates the Genesis verse: "But JACOB took fresh boughs of poplar, almond, and plane"), then the proper place name, Luz, in Genesis 28: 19 and 35: 6, Joshua 16: 2 and 18: 13, and Judges 1: 23, probably refers to the abundance of almond trees in that locality at that time. The JASTROW version for Joshua 18: 13 speaks of "Beth-el-luz" (Hebrew, "luza"). Evidence pointing to this conclusion includes the fact that the identical Arabic word "luz" denotes the almond. Among people as fond of poetic imagery as the Jews, it is not strange that the word "shâkêd", indicating a wakeful or watchful tree, hastening to put forth its blossoms before those of other trees, should have come to be used as a synonym for "luz", the almond tree.

The almond has always been regarded with reverence by the Jews, and even to this day the modern English Jews carry branches of flowering almonds into their synagogues on their spring feast days, much as the Hebrews in olden times were wont to carry palm "branches" to the temple (306). In Tuscany branches of almond are employed as divining-rods to locate hidden treasure. Catholics ascribe the tree to the Virgin MARY and Mohammedans see in it a symbol of the hope of heaven.

There are many legends concerning the "miraculous" budding of almond branches. WAGNER has familiarized the world with one of them in his famous opera "Tannhäuser", whose lesson is that God's judgments are far gentler and more merciful than ours. TURPIN's history of CHARLEMAGNE relates the legendary tale of the spears of the great emperor's troops, which had been thrust into the ground when he made camp and which sprouted during the night and shaded the tents the next day. VIRGIL relates a story of a blossoming staff.

29. Anastatica hierochuntica L.   30. Gundelia tournefortii L.
(FIGURE 22; FIGURE 36 — FIGURE 21)

PSALMS 83: 13—O my God, make them like a *wheel*; as the stubble before the wind.
ISAIAH 17: 13—... they shall flee far off, and shall be chased as the chaff of the mountains before the wind, and like a *rolling thing* before the whirlwind.
ECCLESIASTICUS 24: 14—... as the *rose plant* in Jericho.

MOFFATT translates the Psalms passage cited above: "My God, whirl them away like *dust*, like straw before the wind", but in spite of this use of the word "dust" by an authority who is usually so reliably exact, it is most probable that the Hebrew word "gulgal" translated in the Authorized and Douay versions as "wheel" and "a rolling thing" refers to a plant, and a very particular one at that. Marginal Bibles give the alternative translation "thistle-down", but this identification is not probable, as we shall point out below. "Whirling dust" is given in both references by the JASTROW version. The GOODSPEED rendition is the only one, in this case, which appears to be botanically correct: "My God, make them like a *tumble-weed*, Like chaff before the wind."

The plant referred to in these passages may be the Palestinian tumbleweed, *Anastatica hierochuntica*[9] (97). This is a typical tumbleweed and resurrection-plant. It is a member of the mustard family and has been reported as growing abundantly about Jericho and throughout Syria and the Mediterranean region (299), although POST (265) and EVENARI give its range now as Wady Arabah and Sinai. HART and others report it from Wady Arabah and Ghores-Sofi. AARONSOHN made two collections of it around the Dead Sea. NABELEK found one station of it there, too (18, 110, 257). TRISTRAM also reports it from the shores of the Dead Sea (326). It is an annual, dichotomous, prostrate plant with a heavy tap-root at the apex of which numerous slender branches are produced. These branches form a circular disk about a foot in diameter, which lies at first almost flat upon the ground. The branches bear small, obovate, repand-margined leaves and small, white, axillary flowers. When the seeds have matured, the stems become dry, hardened, and incurved, the apices meeting on top, forming a globe or skeletonized hollow ball. This ball, in time, through the force of the wind pressing against it, breaks off at ground level and is blown about, rolling over and over like a "wheel" or "rolling thing", shedding its seeds as it rolls. Dr. THOMSON says: "When ripe and dry in autumn, these branches become rigid and light as a feather, the parent stem breaks off at the ground, and the wind carries these vegetable globes wheresoever it pleaseth. At the proper season, thousands of them come scudding over the plain, rolling, leaping, bounding with vast racket, to the dismay both of the horse and his rider. Once, on the plain north of Hamath, my horse became quite unmanageable among them... If this is not the 'wheel' of DAVID and the

---
[9] Referred to by some authors as "*Anastatica hiero-chuntina* L." (159, 376).

'rolling thing' of Isaiah, I have seen nothing in the country to suggest the comparison" (319).

Dr. M. Evenari, of the Hebrew University at Jerusalem, states that the Palestinian *Gundelia tournefortii*—a member of the thistle family—is the "gulgal". It is a prickly herb, commonly called "gundelia", has milky juice and the aspect of an *Eryngium*. Dalman also considers this plant even a better choice than the *Anastatica* for the "rolling thing". It also rolls well over the land and gathers in tremendous heaps in hollows. It is the more common of the two today. In November, 1906, Dalman found it on the plains of Moab where he burned it in brewing his tea (97). He further suggests that this plant may be the one with the "thorns that cracked under the pot" (98).

The translation "thistle-down", suggested by Marginal Bibles, is not at all apt, since thistle-down does not roll over and over like a "wheel", but, like Moffatt's "dust", rises in the air with the wind. The Goodspeed version falls into the same error in the Isaiah passage which it renders: "And will be chased like chaff of the mountains before the wind, Or like *whirling dust* before the hurricane." Other thistle-like plants suggested for the "gulgal" or "galgal" are *Gypsophila, Eryngium, Centaurea microcephala,* and the artichoke.

On being moistened after landing in a wet spot, the dry globe of the tumbleweed again expands and thus resembles the so-called "resurrection-plant" of desert places in America. This fact has quite naturally caused *Anastatica hierochuntica* to be held in superstitious veneration by many of the natives of the region where it abounds. It is called the "rose of Jericho" or "Mary's flower" by the monks (125). The latter name is applied because of a legend which affirms that all of the plants of this species expanded, became green, and blossomed again at the birth of Jesus, and still do so in commemoration of this event. The vernacular name "rose of Jericho"[9a] is interesting to note because the translators of the apocryphal book of Ecclesiasticus so rendered the Hebrew word "gulgal" there. Goodspeed, however, translates the verse: "Or like the *rosebushes* in Jericho", apparently implying that true roses were intended.

It has been suggested to the writers by some travelers in the Holy Land that an entirely different plant, *Allium schuberti* Zucc., may be the "rolling thing" of these references. These travelers state that the mature umbels of this wild onion attain a diameter of up to a foot, break off at the apex of the peduncle, and roll before the wind like the balls of the tumbleweed. Post (267) makes no mention of the heads breaking off in this fashion.

It is interesting to note that the same Hebrew word "gulgal" or "galgal" is translated "heaven" in Psalms 73: 18 and "wheel" in Ezekiel 10: 13 in the Authorized Version, and by Gesenius is rendered "whirlwind"—a concept borrowed by the Douay version which renders the Isaiah passage: "as a whirlwind before the tempest."

31. Andropogon aromaticus Roxb.

Exodus 30: 23-24—Take thou also unto thee principal spices, of pure myrrh five hundred shekels, and of sweet cinnamon half so much, even two hundred and fifty shekels, and of *sweet calamus* two hundred and fifty shekels, And of cassia five hundred shekels, after the shekel of the sanctuary, and of oil olive an hin.

I Kings 10: 10—And she gave the king an hundred and twenty talents of gold, and of *spices* very great store, and precious stones: there came no more such abundance of *spices* as these which the queen of Sheba gave to king Solomon.

Song 4: 14—Spikenard and saffron; *calamus* and cinnamon, with all trees of frankincense.

Jeremiah 6: 20—To what purpose cometh there to me incense from Sheba, and the *sweet cane* from a far country?

Ezekiel 27: 19—Dan also and Javan going to and fro occupied thy fairs: bright iron, cassia, and *calamus*, were in thy markets.

[9a] A reviewer of Skinner's book on plant legends and myths notes that on page 19 of that work the author confuses the "rose-of-Sharon" with the "rose-of-Jericho" when he claims that the former is "held to be a symbol of resurrection, for when its blossoms fall away they are borne by the wind to a distant place, there to root and bloom anew" (27e).

Most commentators agree that the "calamus", "sweet calamus", and "sweet cane" of the references cited above were the ginger-grass, *Andropogon aromaticus*. The Hebrew word involved is "keneh" (Ezekiel 27:19; Song 4:14) or, more fully, "keneh bosem", meaning "spiced or sweet cane" (Exodus 30:23) or "keneh hattob" or "v'kaneh hatov", meaning "and the good cane" (Jeremiah 6:20). The Greek words used are κάλαμος εὐώδης, meaning "a sweet-smelling calamus, reed or cane". This is a common grass of northwestern and central (268) India—a "far country" indeed!—the leaves of which are highly odoriferous when bruised and taste strongly like ginger. Cattle are very fond of it (299), but it has the property of scenting their flesh, milk, and butter. It yields an oil known as "ginger-grass oil". It is related to *A. schoenanthus* L., the well-known "lemon-grass", "sweet rush", or "camel's hay" of India and Arabia, with which some early commentators erroneously identified the Biblical plant. TRISTRAM found the lemon-grass in Palestine at Gennesaret (265, 267), and, of course, it is extensively cultivated for its oil in Ceylon. Another relative, *A. muricatus* Retz., also from India, has sweet-scented fibrous roots which are woven into screens for windows and verandas in that country. All three of these closely related aromatic grasses form tufts, with leaves 3 to 5 feet long. In age these tufts become more or less elevated through the gradual elongation of their thick stem-like rootstocks. Often great numbers of these plants grow close together, in the usual fashion of grasses, forming extensive fields.

In the 1860 edition of SMITH's "Dictionary of the Bible" (300) the "calamus" of the Old Testament is identified as the "sweet-flag" or "sweet sedge" (306), *Acorus calamus* L., of Europe and eastern North America, and STANTON and STURTEVANT (309) also identify it thus. The sweet-flag is a plant of wet places and streamsides, highly aromatic throughout, with flat iris-like leaves. It is said to be common in many countries of temperate Europe and Asia, but POST (265, 267) does not record it from the Holy Land and it seems far less probable that this species should have been imported in those ancient days than the ginger-grass from northwestern India. In his 1876 edition (306) SMITH abandons this theory in favor of *Andropogon*. SHAW, in 1884, still regarded (293) the Ezekiel, Jeremiah, Song of Solomon, and even Isaiah 43:24 (see below) passages as applying to *Acorus calamus*, but HORNER (178) truly points out that it is extremely doubtful that this plant was even known in Biblical times. She identifies the Biblical plant as *"Calamus aromaticus"* (*Andropogon aromaticus*). Sir GILBERT BLANE believed it to be the spikenard, *Nardostachys jatamansi* (which see), but this theory, too, is now rejected.

Because some species of *Cyperus* are sweet-scented and because *C. pertenuis* Roxb. is still used by the women of India for scenting their hair, it has been supposed by some writers that this plant may be included here. This, however, does not seem very probable as there is no evidence that this plant was an article of trade in Old Testament days.

The GOODSPEED and JASTROW versions use the words "calamus" and "sweet cane" as in the King James version, but MOFFATT renders the "calamus" of Ezekiel as "sweet cane" and that of the Song of Solomon as "cassia". The Exodus reference he translates "eight pounds of scented cane" and the Jeremiah verse: "What care I for incense that you bring from Sheba, or for *perfume* fetched from lands afar?" The Douay version uses "sweet cane" for "calamus" in Song 4:14. There is, however, an entirely different plant which is called "sweet cane" in the Bible. The "sweet cane" of Isaiah 43:24, for example, even though also represented by the Hebrew word "kănêh" in the original text, is now usually regarded as sugar-cane, *Saccharum officinarum* (which see), although many early commentators regarded the expressions "sweet cane" and "sweet calamus" wherever they occur in the Old Testament as names for one and the same plant. The botanist, JOHN SMITH, was apparently the first to point out (299) that the Isaiah plant is described as sweet-*tasting*, while all the other references are to a plant that is sweet-*smelling*. The Douay version commendably renders Jeremiah 6:20: "To what purpose do you bring me frankincense from Saba, and the sweet smelling cane from a far country?"

For the benefit of those readers who may be surprised at the identification of Bible plants with plants native to India, it may be pointed out that historians believe (268) that India and Egypt had commercial relations as early as the time of the Pharaohs. If this is so, Indian products may very easily have been known to the Israelites, although they would certainly always have been very costly products brought from "a far country". Besides the ginger-grass, other Biblical plants native to India and adjacent regions are the eaglewood, sandalwood, frankincense, cassia, cinnamon, ebony, cotton, spikenard, sugar-cane, orris, and perhaps benzoin.

It is of interest to note that the "κάλαμος ἀρωματικός" of DIOSCORIDES is identified as *Andropogon aromaticus* by ROYLE (306). This is doubtless the basis for Miss HORNER's identification mentioned above. According to DIOSCORIDES this plant was well-known in trade at his time.

32. Anemone coronaria L.   33. Anthemis palaestina Reut.

(FIGURE 46 — FIGURE 50)

NEHEMIAH 1: 1— . . . as I was in *Shushan* the palace.
ESTHER 1: 2— . . . when the king AHASUERUS sat on the throne of his kingdom, which was in *Shushan* the palace.
PSALMS 45, title—To the chief musician upon *Shoshannim*, for the sons of Korah, Maschil, A Song of loves.
PSALMS 60, title—To the chief musician upon *Shushan*-eduth, Michtam of DAVID to teach.
ADDITIONS TO ESTHER 11: 3—Who was a Jew, and dwelt in the city of *Susa* . . .
ADDITIONS TO ESTHER 16: 18—For he, that was the worker of these things, is hanged at the gates of *Susa* with all his family . . .
MATTHEW 6: 28-30—And why take ye thought for raiment? Consider the *lilies* of the field, how they grow; they toil not, neither do they spin: And yet I say unto you, That even SOLOMON in all his glory was not arrayed like one of these.
LUKE 12: 27-28—Consider the *lilies* how they grow: they toil not, they spin not; and yet I say unto you, that SOLOMON in all his glory was not arrayed like one of these. If then God so clothe the grass, which is to day in the field, and tomorrow is cast into the oven; how much more will he clothe you, O ye of little faith?

The "lily" is probably the most famous of all the plants of the Bible. It is likewise the one concerning which there has been the most difference of opinion and the most argument. It seems most probable that several different kinds of plants—perhaps five or six—are referred to under this name in the Authorized Version of King James. Many authors have even claimed that the "rose" of Isaiah 35: 1—"the desert shall rejoice, and blossom as the *rose*"—is the same plant as is referred to under the name of "lily" elsewhere, but MOFFATT now believes the former is the polyanthus narcissus, *Narcissus tazetta* (which see), and in this view we concur.

In the Song of Solomon there are numerous references to "lilies" and to a "lily of the valleys". These are now regarded as referring to the hyacinth, *Hyacinthus orientalis* (which see).

SHAW (293) expresses the general opinion of non-botanical Bible readers when he opines that all the "lilies" of both the Old and New Testaments were some species of *Lilium*, but he does not venture to say which species. DONEY (101) emphasizes that the Madonna lily, *L. candidum* L., is usually considered in the popular mind to be the "lily" of the Bible, doubtless because of its former general use at Easter time (now mostly replaced in America by the Japanese *L. longiflorum* var. *eximium* Nichols.) He is inclined to believe that all the "lilies" of the Bible were *L. chalcedonicum*, which, he says, grows abundantly about Galilee. PRATT (268) calls attention to the innumerable paintings and carvings of the Madonna and of the Annunciation, Resurrection, and Ascension in which representations of the Madonna lily appear. He thinks, however, that in spite of this strong popular conviction, all the Bible references are to *Sternbergia lutea* (Herb.) Ker[10]—a species, by the way, that is not known from Palestine according to Post, who calls the Holy Land species *S. aurantiaca* Dinsm. (267). SMITH states (299) that the only true lily in Palestine is the scarlet martagon lily, *Lilium chalcedonicum*, while Post (265, 267) records

---

[10]Referred to by some authors as *Oporanthus luteus* Herb.

only *L. candidum*, which SMITH admits is abundantly cultivated and escaped, but is doubtfully native. NAFTOLSKY's collection of the latter in Palestine in 1925 settles this point. These two lilies seem to be the only ones present in Bible lands in Biblical times (341). Dr. WARBURG believes that some definite plant is meant in the New Testament "lily" references since other definite plants and animals are also mentioned in the same or neighboring verses. He further believes that abstraction of thought was not well developed yet in ancient Bible times (341).

Most authorities now regard the Palestine anemony or windflower, *Anemone coronaria*, as the famous "lily of the fields" which surpassed "SOLOMON in all his glory". With its various brilliant colors it is certainly the most conspicuous plant of the sort thought by most writers to be referred to here. TRISTRAM has this to say of the Palestine anemone "In every part of the country in profusion, almost invariably the red variety, yellow, blue, and purple occurring very rarely. The most gorgeously painted, the most conspicuous in spring, the most universally spread of all the floral treasures of the Holy Land, if any one plant can claim pre-eminence among the wondrous richness of bloom which clothes the land of Israel in spring, it is the anemone, and therefore it is on this we fix as the most probable 'lily of the field' of our Lord's discourse" (326). The Greek words rendered "lilies of the field" are κρίνον ἄγριον, literally meaning "a lily of any kind living in the fields". The Hebrew words used for "lily" in the Old Testament are "shûsan" and "shûshân"; and HORNER calls attention (178) to the fact that "shûsan" is now the name applied by the Arabs to any brilliantly colored flower which to them resembles a lily. "Susan" is also the Persian, Syriac, and Coptic word with the same meaning (306). Curiously, the Chaldee Targums (written from the second to tenth centuries A.D.), as well as MAIMONIDES and other rabbinical writers, translate the Hebrew words "shûsan" and "shôsannâh" as "rose", probably because of the similarity in these words. The Hebrew singular and plural words used for roses are "shoshanot" and "shoshanah".

Neither of the two lilies native to Palestine, *Lilium candidum* and *L. chalcedonicum* (which see), is common there. *Anemone coronaria*, on the other hand, in one or more of its forms is found almost everywhere, in all kinds of soils and in all manner of environmental situations. It is abundant on the Mount of Olives today, as it doubtless also was in JESUS' day. It is abundant on the plains, and is especially luxuriant on the shores of the Lake of Galilee. It would, thus, be quite familiar to JESUS' listeners—something which cannot be said for most of the other plants advanced by various writers. It has a tuberous root. The showy flowers, each with 5-7 colored petal-like sepals, may attain a diameter of 2 ¾ inches and are scarlet, yellow, blue, purple, rose, or white in color, depending on the variety, the typical and commonest forms being scarlet or yellow.

BLANCHAN (56a) states that *Lilium chalcedonicum* is not uncommon in Palestine, but still she believes that the Palestine anemony is the "lily of the field". She points out that "Opinions differ as to the Lily of Scripture. Eastern peoples use the same word interchangeably for the tulip, anemone, ranunculus, iris, the water-lilies, and those of the field" and yet "whoever has seen the large anemones there is inclined to believe that JESUS, who always chose the most familiar object in the daily life of his simple listeners to illustrate His teachings, rested His eyes on the slopes about Him glowing with anemones in all their matchless loveliness."

Some authorities claim that the "lily" of the Old Testament was *Nymphaea lotus* and that of the New Testament was either *Tulipa montana* or *Lilium chalcedonicum* (100). It is our opinion that the "lilies" of II Chronicles 4: 5 and I Kings 7: 19, 22, & 26, used as models for the carving or "lily-work" in the ornamentation of SOLOMON's temple, were, indeed, *Nymphaea* (which see), but only those of one passage, Song 5: 13, *Lilium chalcedonicum* (which see). *Tulipa montana* we regard as the "rose of Sharon" and the "rose in the spring of the year" of Song 2: 1 and Ecclesiasticus 50: 8.

Dr. THOMSON describes a showy flowering plant which he calls the "Hülch lily", said to be very abundant in many places in Palestine, which seems to be an iris. *Iris pseudacorus* (which see) is, in fact, regarded by us as the "lily" of Ecclesiasticus 1: 8 and Hosea 14: 5. POST quotes Professor DAY, of Beirut, to the effect that "sûsân" is the present Arabic name for members of the genus *Iris*, of which there are 51 species and varieties in and about the Holy Land (267).

Some writers are of the opinion that the alpine-violet or cyclamen, *Cyclamen persicum* Mill., is one of the "lilies" of the Bible. This species is common among rocks and on walls throughout the Holy Land and two related species grow in the subalpine regions. According to GERARDE (95) this plant is one of decidedly good reputation. If placed in a room it will guard the inmate from all evil influences, as is attested by this old couplet: "St. John's Wort and fresh Cyclamen she in his chamber kept, From the power of evil angels to guard him while he slept." SOUCIET endeavored to prove (306) that the crown-imperial, *Fritillaria imperialis* L., was the Biblical "lily" (170), but there is no proof that this plant was ever common in Palestine or even known there at all, although other quite dissimilar species of that genus do occur there. DIOSCORIDES praises the "lilies" of Syria and Pisidia (an ancient country and Roman province in southern Asia Minor), from which the best perfumes were made in his day. JONES (184) suggests *Ranunculus asiaticus* L. and *Adonis palestina* Boiss., but POST (265) says for the former "notwithstanding its brilliant flowers, not to be regarded as the 'lily of the field'." LUNDGRUN suggests an *Asphodelus*, FISCHER and BENSON some red lily ("Feuerlilien"), and DALMAN a *Gladiolus* (100).

Commentators are of the opinion that "Shushan", and its variants, "Susa", "Shushan-eduth", and "Shoshannim", used as place names in the books of Esther and Nehemiah, being the same word as that used for "lily", refer to this plant, if not because of an abundance of these plants there, then perhaps in a figurative way. MOFFATT has also discovered it in the titles of two of the Psalms, which he translates, respectively: "From the choirmaster's collection of Korahite songs. To the tune of 'The Lilies'. An ode or love-song" (Psalm 45) and "From the choirmaster's collection. To the tune of 'Lily of the Law'. A golden ode for recitation, sung by DAVID" (Psalm 60). The female names, Susan and Susanna, are derived from the same source.

PAULINE ROSENBERG (276) reports that Dr. EPHRAIM HA-REUBENI, professor of Biblical botany at the Hebrew University in Jerusalem, after years of study, feels convinced that the "lilies of the field" were really the Palestine chamomile, *Anthemis palaestina*, a common white daisy-like plant of the Holy Land. He feels that JESUS' words "And yet I say unto you . . ." indicate plainly that the plant to which he referred was not a showy one, as everyone previous to him seems to have assumed, but, rather, was one whose innate beauty had to be carefully pointed out, much as a hand-lens and microscope now reveal to us hidden beauties in the lowliest objects of Nature, beauties of form and figure and pattern never before suspected. He feels also that the words "If God so clothe the grass of the field, which today is, and tomorrow is cast into the oven . . ." apply best to this chamomile, for this, he says, when it has dried up "is gathered with the dried grass and cast into the furnace." The anemonies of the Holy Land bloom from January to April, and soon after blooming mature their seeds and wither away. The chamomile, however, comes out later and lasts longer; it would therefore be more likely to be present at the time of hay gathering.

It is interesting to note that while O'HARA, MOFFATT, and the Douay versions all translate the "lilies of the field" and "lilies" passages in Matthew and Luke the same as does the King James version, the GOODSPEED version differs in rendering the former "See how the *wild flowers* grow . . ." and the latter "See how the *lilies* grow . . ." The Basic English version translates the two passages: "And why are you troubled about clothing? See the *flowers* of the field, how they come up . . ." and "Give thought to the *flowers*: they do no work, they make no thread; and still I say to you, Even SOLOMON, in all his glory, was not clothed like one of these. But if God gives such clothing to the grass in the fields which today is living, and tomorrow will be burned in the oven, how much more will he give clothing to you, O men of little faith?" WEYMOUTH says "Observe well the wild lilies . . ." and "Look at the lilies, how they grow . . .", but the LAMSA translation also makes use of "wild flowers" in the Matthew and "flowers" in the Luke verse. We can see, thus, that there is a definite tendency among modern translators to broaden the concept of "lilies", at least in the Matthew and Luke references, to embrace wild flowers in general. This doubtless approaches much closer to the actual meaning of JESUS.

Speaking of the firm hold that *Lilium candidum* has taken on the popular fancy, ROSEMARY COTES has well said (95) that no artist seems to have lived in the Middle Ages who has not "painted a picture of the Blessed Virgin with a white lily,—LIPPO-LIPPI, FRA ANGELICO, TITIAN, MURILLO, BOTTICELLI, CORREGGIO, and a host of followers" have all painted the white Madonna lily; and no poet seems to have lived since the tenth century who has not sung of its purity and grace. In the year 1618 a special papal edict laid down stringent rules "as to the proper treatment of certain sacred subjects in art, and the necessity of the introduction of the white lily into pictures treating of the Immaculate Conception. In these rules—besides the stringent ones as to the colour of the Virgin's hair, the expression of her 'grave, sweet eyes', her robe of spotless white with a scarf of blue, and the twelve stars to crown her head—it was commanded that roses, palms, and lilies should be employed as the flowers suitable to be scattered by angels, or as the decorations for the picture. MURILLO occasionally departed a little from the letter of the law, as for example when he painted the Virgin's hair brown instead of golden, but in the main he, like all the other artists of the day, seemed to have desired to show his orthodoxy by a strict observance of the papal rules" (95).

The lily has always been considered, both in the Orient and Occident, the traditional emblem of purity and grace. The Greeks and Romans crowned their brides and grooms with wreaths of lilies and wheat, symbolizing, respectively, a pure and fertile life. Also among other nations it symbolized virginity and innocence, and we still use it today in this same sense when the bride, robed in a white wedding gown and veil, carries a bouquet of white lilies if she has not been married before. White lilies placed on Christian altars formerly always had their stamens and pistils removed "so that they might remain virgin." That the lily was also regarded as able to bring good luck to the wearer is indicated by the story of the Jewish heroine, JUDITH, in the Apocryphal book of that name, who wore a lily on the night that she killed OLOFERNES, general of King NEBUCHADNEZZAR, during the siege Bethulia (125, 126).

According to a well-known legend, ST. THOMAS, not believing the reports which he had heard about the resurrection of the Blessed Virgin, had her tomb opened. Inside, instead of her body, he found the tomb to be filled with lilies and roses (95, 298). In the symbolism of the Catholic church the lily is the "attribute" of ST. ANTHONY, ST. BERNARD, ST. CLARA, St. DOMINICK, ST. FRANCIS, ST. JOSEPH, ST. KATHARINE OF SIENA, ST. LOUIS DE GONZAGUE, and the angel GABRIEL (298). The Greek goddesses DIANA and JUNO, as well as LILITH (the first wife of ADAM), are usually represented in art as carrying a lily as an emblem.

An ancient legend states that lilies were all white, with erect flowers, before the time of JESUS' agony in the garden of Gethsemane. There, while all the other flowers drooped their heads in sympathetic grief over his suffering, the white lily in her conceit at having been officially proclaimed more beautiful than SOLOMON's royal robes, proudly held up her head for him to admire. But when JESUS' eye fell upon the flower the unspoken rebuke in his glance caused the haughty lily to contrast her conceit with his humility, and she blushed with shame. The red flush that covered her face then tinges it still, and the flowers of our red lilies today still droop in shame for their former arrogance.

In old Spain it was believed by the ignorant that the lily could restore human form to any unfortunate men or women who had been transformed into beasts by witchcraft. Tradition tells of a Spanish garden where in 1048 the image of MARY appeared in a white lily flower at the exact moment when the dying king suddenly recovered his health and strength. In commemoration of this miracle the order of the Knights of St. Mary of the Lily was founded. Three centuries later a similar religious order was established by King LOUIS IX of France (298).

Lilies are employed by girls in the Caucasus Mountains to determine the faithfulness or faithlessness of their lovers. The color of the lily flower there is supposed to be determined during and immediately after rain showers. If a chosen bud opens yellow after such a shower, the girl believes that her lover is faithless, but if it opens red she knows that he is true to her (298). An eleventh century legend from the same region tells of a beautiful maiden who was transformed into a lily in order to escape marriage to a man whom she did not love. Her true lover, on finding her in the form of a drooping lily, weeps bitterly and his falling teardrops refresh and comfort her. He is subsequently transformed into a rain cloud by the sympathetic gods so that he can continually refresh his loved one. Even today in times of drought inhabitants will scatter lilies over their parched fields in the hope that rain clouds will appear to water the desiccated land.

In the folklore of Normandy is a charming tale which explains the origin of snow and winter. It tells of a beautiful maiden who, on Christmas Eve, faded away in her husband's arms. He was left holding a white lily in her place. And the earth, which before that had always been bright and warm and sunny, suddenly became cold and bleak and covered with tiny fragments of the white lily tepals, which fell from heaven as the first snow.

But the anemony, too, has its store of legend and tradition. ROSEMARY COTES writes (95): "The anemone of Palestine is a flower of ancient romance. It was supposed to have sprung from the tears of VENUS sorrowing for her beloved ADONIS. The Greek poet BION (125) says—'Where streams his blood, there blushing springs the rose; And where a tear has dropped, the windflower blows.' RAPIN adopts the same legend for the origin of the anemone, and speaking of VENUS' grief—'Showers of tears on the pale body shed, Lovely anemones in order rose, And veiled, with purple palls, the cause of all her woes.' PLINY declared that anemones never come into bloom except when the wind is blowing, and that the magicians gathered them in his day as a remedy against disease, tying the flowers round the neck or arm of the patient as a charm to cure an illness (126). The ancient Egyptians also held the same tradition with regard to the anemone, and it has been handed down even to our own day, as shown in the following lines from an old English ballad—'The first spring-blown anemone she in his doublet wove, To keep him safe from pestilence wherever he should rove' " (95). This latter, of course, is the European wood anemony, *Anemone nemorosa*, not the Palestinian species. Other traditions state that the anemony is called "windflower" because it first appears when the first gentle winds of spring blow.

Another legend states that before the time of the Crucifixion all the Palestine anemonies were white or blue, but some blood dropped from the cross on Golgotha and stained the anemonies growing at its foot. Since then most of the anemonies of Palestine have borne flowers of a vivid crimson.

As implied by the reference to PLINY above, the Romans ascribed medical virtues to the anemony, especially as a cure for malaria. They also used anemonies to decorate the altars of VENUS (125) and to cover the faces of their dead. The Chinese regard anemonies as a symbol of death and so associate them with suffering and grief. In Greek mythology ANEMONE was a lovely maid in "the court of CHLORIS, where she was seen and loved by ZEPHYRUS, the god who caused flowers and fruits to spring from the earth by breathing on it. CHLORIS fancied that the wind god was about to sue for her hand; hence, on discovering his passion for ANEMONE, she drove that nymph from her presence in anger. Finding her broken-hearted, and hence dismal company, and having also to make his peace with CHLORIS, ZEPHYRUS abandoned the poor creature, but in taking his leave changed her into the flower that bears her name" (298).

34. Anethum graveolens L.

MATTHEW 23: 23—Woe unto you, scribes and Pharisees, hypocrites! for you pay tithe of mint and *anise* and cummin, and have omitted the weightier matters of the law . . .

It is very certain that the plant to which the Greek word ἄνηθον here refers is not anise, but dill, *Anethum graveolens*[11] (98, 158, 184). Dill is a weedy annual umbellifer resembling parsley and fennel, 12 to 20 inches tall, with yellow flowers (268, 299). The error of considering this plant to be the anise is one that the Authorized Version has caused to be perpetuated for over two centuries (272).

The dill plant is widely cultivated for its seeds, which are aromatic and carminative, similar to those of caraway. It is also used in cookery for flavoring dishes, especially pickles, and in medicine. Dill-water is obtained from the seeds by distillation, and it contains an ethereal oil used in medicine. A native of Europe, it is a common weed in the grain fields of the southern part of that continent and Egypt, and is found both wild and cultivated in Palestine. TRISTRAM found it growing on the plain of Sharon.

The true anise, *Pimpinella anisum* L., is now also cultivated and escaped in the Holy Land, but almost all authorities are agreed that this was not the plant to which JESUS referred. MOFFATT, WEYMOUTH, GOODSPEED, and LAMSA all use the word "dill" in the passage cited above, although the Douay and O'HARA versions still retain "anise". The Basic English translation is interesting: "A curse on you, scribes and Pharisees, false ones! for you make men give a tenth of all sorts of sweet-smelling plants, but you give no thought to the more important things of the law." Dill is much more characteristically a plant of Oriental cultivation than is anise (306).

35. Apinus pinea (L.) Neck.
(PAGE 236)

HOSEA 14: 8— . . . I am like a *green fir tree*. From me is thy fruit found.

There is no unanimity of opinion regarding the "green fir tree" of Hosea— "berosh raanan" in the Hebrew. Most of the "fir tree" references in the Bible are now regarded as applying to the Aleppo pine, *Pinus halepensis*, but the one in Hosea differs from all the others in containing a reference to a fruit, presumably an edible sort. In all other cases the "fir" was valued for its wood. The only conifer of the region with an edible nut-like seed is the stone pine, *Apinus pinea*[12]. This tree attains a height of 30 to 60 or more feet, with a hemispheric top, the lower branches horizontal. The leaves are borne in bundles of two, are 4 to 5 inches long, stiff, ascending, and serrulate-margined. The cones are solitary or paired, nearly sessile, practically terminal on the branches, and horizontal or somewhat reflexed (267). The seeds are practically wingless, sweet, and nut-like. Somewhat resembling almonds (158), they are extensively eaten where the tree is native, and are even today used in confectionery. The tree is extensively cultivated in the Holy Land, usually pruned to a wide-spreading umbrella-like top over a naked trunk. It is probably originally native to southern Europe (158) and Asia Minor (267), and was doubtless well known at the time of Hosea, about 725 B.C., even though EVENARI doubts whether it was known in Biblical times.

The Douay and GOODSPEED versions agree with the King James in identifying the Hosea plant as a "green fir tree", but MOFFATT calls it a "cypress evergreen" and the JASTROW rendition is "leafy cypress-tree". Some writers have suggested the stone pine for others of the various "fir" references discussed by us under *Pinus halepensis* (179).

---

[11]Also called *Peucedanum graveolens* (L.) Benth. & Hook. f.
[12]Also referred to as *Pinus pinea* L.

36. Aquilaria agallocha Roxb. 37. Santalum album L.
(FIGURE 44)

PSALMS 45: 8—All thy garments smell of myrrh, and *aloes*, and cassia.
PROVERBS 7: 17—I have perfumed my bed with myrrh, *aloes*, and cinnamon.
SONG 4: 14— . . . myrrh and *aloes*, with all the chief spices.

The Hebrew words translated as "aloes" in the Authorized Version are "ahâlîm" and "ahâlôth". The "aloes" of the Old Testament are thought by most authorities to be different from those of the New Testament. Those of John 19: 39 are considered now to be the true or bitter aloes, *Aloë succotrina* (which see), which is still esteemed in the Orient for its perfume and its cordial properties, and which has been widely used as a medicine in treating rheumatic complaints. It was probably mixed with myrrh when wrapped among the burying clothes in embalming. The immense quantities of these substances brought by NICODEMUS show how profuse was the quantity of fragrant essences used at ancient Jewish burials. Or, more pointedly, we are forced to realize from these texts how necessary the use of the pungent and fragrant substances was in this warm-weather country for the benefit of the people who took part in such ceremonials near the deceased bodies.

The Old Testament "aloes", on the other hand, were doubtless not the true aloes, but probably were the eaglewood, *Aquilaria agallocha*[13]. The eaglewood is a lofty tree, 100 to 120 feet tall and to 12 feet in trunk circumference, native to Cochin-china, Malaya, and northern India. It is still often called "aloes wood" or "agallochum". It is also known as "lign aloes" from the Greek στακτή or σκηναί. The wood, especially the darker part, is fragrant, particularly when in a state of partial decay, and is then highly valued in perfumery (178), as incense, and for fumigation (306). SHAW, curiously enough, does not attempt (293) to identify either the "aloes" of the Old or New Testaments, while PRATT (268) regards all of the "aloes" of the Bible, both those of the Old and those of the New Testaments, as either *Aquilaria agallocha* or *A. ovata* Cav. JONES (184) claims that the "aloes" used as a perfume in connection with myrrh, cassia, and cinnamon *and* as a spice for embalming the dead are one and the same, the gum of what he calls the "eagle-tree". He says "It has no connexion with our 'bitter aloes' ". As has been indicated above, we do not concur in this statement, but regard the ones used for embalming as true *Aloë succotrina* and only those used for perfume as *Aquilaria agallocha*. JONES, however, rightly points out that the "lign-aloes" used by BALAAM, with the cedars of Lebanon, as symbolic of the noble position of Israel, planted in a choice land (Numbers 24: 6) could not have been the eaglewood, which is not native to the Holy Land. Although the GOODSPEED version still champions "aloes" in this passage—"Like gardens beside a river, Like *aloes* planted of the Lord . . ."— the Douay version is "As watered gardens near the rivers, as *tabernacles* which the Lord hath pitched . . ." and MOFFATT says "Like gardens by a river, like *oaks* planted by the Eternal . . ." MOFFATT's use of the word "oaks" for the "lign-aloes" of this passage is doubtless the most correct because the context of the verse plainly indicates a lofty and sturdy tree that was native to the region, and not either the eaglewood or sandalwood, nor the true aloes, which were known to the Israelites only as imported products.

It has been suggested by some commentators that perhaps the sandalwood, *Santalum album*, may represent the Old Testament "aloes". This tree is said to be native to the East Indies (158) and peninsular India (300), and may have been more accessible to the people of Palestine than the eaglewood, although this is by no means certain. It is also more generally used by all Oriental nations for its compact, close-grained, fragrant, yellowish wood, which is much valued in ornamental carving and cabinet work. Its odor is repellent to insects and so it is highly esteemed for the manufacture of chests and storage boxes.

It is interesting to note that while the GOODSPEED, JASTROW, and Douay versions still continue to use the word "aloes" for the Psalms, Proverbs, and Song of Solomon passages, MOFFATT correctly substitutes the word "eagle-

---
[13] Also called *Aloëxylum agallochum* (Roxb.) Lour.

wood" in the Proverbs and Song of Solomon verses, although he still translates the same Hebrew word in Psalms 45: 8 as "aloes". Whether he believes that the "aloes" of Psalms 45: 8 were the true aloes of the New Testament, which is not at all probable, is not clear. HENSLOW (172a) says "Aloes (*Aloë vulgaris*) is not the Lign-aloes, but was probably used, like galbanum, for fixing the more evanescent scents."

38. Artemisia herba-alba Asso   39. Artemisia judaica L.
(FIGURE 43)

DEUTERONOMY 29: 18— . . . lest there should be among you a root that beareth gall and *wormwood*.
JOB 30: 4—Who cut up mallows by the *bushes*, and juniper roots for their meat.
PROVERBS 5: 4—But her end is bitter as *wormwood*, sharp as a two-edged sword.
JEREMIAH 9: 15— . . . Behold, I will feed them, even this people, with *wormwood*, and give them water of gall to drink.
JEREMIAH 23: 15— . . . Behold, I will feed them with *wormwood*, and make them drink the water of gall.
LAMENTATIONS 3: 15 & 19—He hath filled me with bitterness, he hath made me drunken with *wormwood* . . . Remembering mine affliction and my misery, the *wormwood* and the gall.
HOSEA 10: 4—They have spoken words, swearing falsely in making a covenant: thus judgment springeth up as *hemlock* in the furrows of the field.
AMOS 5: 7—Ye who turn judgment to *wormwood*, and leave off righteousness in the earth.
AMOS 6: 12— . . . for ye have turned judgment into gall, and the fruit of righteousness into *hemlock*.
REVELATION 8: 10-11—And the third angel sounded, and there fell a great star from heaven, burning as it were a lamp, and it fell upon a third part of the rivers, and upon the fountains of waters; And the name of the star is called *wormwood*: and the third part of the waters became *wormwood*; and many men died of the waters, because they were made bitter.

Wormwood, of which there are many species, is the name given to a group of annual, biennial, or perennial, often somewhat woody plants with a hoary aspect and a strong aromatic odor, related to the common mugwort, which is one of the plants causing hayfever among us. Related, also, are the sagebrush plants of the West. Wormwood plants have a strong bitter taste, their young shoots and branch-tips furnishing the "wormwood" of commerce. It seems rather certain that the Hebrew word "la'anah" (also meaning "hemlock") and the Greek word ἄψϑος or ἀψίνϑιον of these texts refer to either *Artemisia judaica* or *A. herba-alba*. Its very bitter taste readily accounts for wormwood being spoken of with gall as symbolic of bitter calamity and sorrow (299). Oriental people usually typify sorrow, cruelty, and calamity of any kind by plants of a poisonous nature. Since the Hebrews considered all bitter-tasting plants to be poisonous, the "root of wormwood" and the "wormwood and the gall" would offer to them a most emphatic and unmistakable metaphor (268). EVENARI maintains that *A. judaica* occurs only in Sinai now, and that *A. herba-alba* is the commonest species in Palestine today, that it is strongly aromatic, bitter, and camphor-bearing. Its common name is "herba-alba". TRISTRAM mentions *A. judaica* from the southern desert, and records *A. herba-alba*, in its many varieties, from bare dry places through all the warm parts of Palestine, the entire Holy Land area, northern Africa, and the Canary Islands (326).

A drink called "absinthe" is made from species of this group, of which thousands of gallons are consumed annually, especially in Paris. It first produces activity and pleasant sensations and inspires the mind with grandiose ideas—well illustrating the Biblical phrase "he hath made me drunken with wormwood". The habitual use of it, however, brings on a stupor and a gradual diminution of the intellectual faculties, ending in delirium and even death. The flowers have been used in some countries in the brewing industry to impart an inebriating quality to ale. Wormwood was also formerly used as a tonic and vermifuge, and to protect garments from moths.

The related species, *A. arborescens* L.—rare according to EVENARI—, may also be involved in the Biblical references quoted above, but the *A. absinthium* L., *A. arenaria* P. DC.[14], and *A. camphorata* Vill.[15], suggested by some authors

---
[14]Referred to as *Artemisia fruticosa* Willd. by WILLIAM SMITH (306).
[15]Referred to as *Artemisia cinerea* P. DC. by WILLIAM SMITH (306).

(184, 306) are not regarded by POST as native to the Holy Land. *A. absinthium* is not even listed in the first edition of his "Flora" (265), but is listed in his second (267). It is probably a recent introduction. *Cotula anthemoides* L.[16], suggested by SMITH and JONES, is not very probable because it is not common in the region and may even not be native.

Some authorities claim that the "hemlock" of Amos and Hosea was not a plant or plant product at all. Certainly it had nothing whatever to do with our Canadian hemlock tree, *Tsuga canadensis* (L.) Carr., nor our American water-hemlock, *Cicuta maculata* L., nor is it very probable that it was the European poison-hemlock, *Conium maculatum* L., the juice of which was given to SOCRATES to drink (268). According to JONES (184) the Hebrew word "rôsh" is also sometimes rendered "hemlock". The usual translation for this word is "gall" (see under *Citrullus colocynthis*).

It is interesting to compare the more modern translations of some of the passages cited above. For instance, in Lamentations 3: 15 instead of "he hath made me drunken with wormwood", the GOODSPEED version says "he hath sated me with anguish". In Lamentations 3: 19 in place of "the wormwood and the gall", MOFFATT says merely "bitterness" while the GOODSPEED version is "anguish and bitterness". Deuteronomy 29: 18 is rendered by MOFFATT "never may there be any root within your soil that bears such bitter poison", the GOODSPEED version is "lest there should be among you a root bearing poison and wormwood", while the Douay translation is ". . . gall and bitterness". In the Proverbs passage the GOODSPEED and Douay versions continue to use "wormwood", while MOFFATT replaces it with "poison". In the two Jeremiah passages the word "wormwood" of the Authorized Version is retained by MOFFATT and the GOODSPEED and Douay versions, but in Amos 5: 7 "wormwood" is rendered "a bitter thing" by MOFFATT and "gall" by the GOODSPEED version. In Amos 6: 12 the "hemlock" is changed to "a bitter, deadly thing" by MOFFATT and "wormwood" by the GOODSPEED and Douay versions. The "hemlock" of Hosea is rendered merely "weeds" by the GOODSPEED version, "poisonous weeds" by MOFFATT, and "bitterness" by the Douay version. All the translators continue to use "wormwood" in the Revelation reference, but GOODSPEED renders the passage for more lucid: "Then the third angel blew his trumpet, and there fell from the sky a great star blazing like a torch, and it fell upon one third of the streams of water. The star is called *Absinthus*, that is, *Wormwood*. Then one third of the waters turned to *wormwood*, and numbers of people died of the waters, for they had turned bitter." The JASTROW version uses the same interpretations as the Authorized Version except that "wormwood" is used instead of "hemlock" in Amos 6: 12.

WEBSTER's dictionary (158) gives as the second definition of "wormwood" "anything very bitter or grievous; bitterness." In the Job reference the King James, MOFFATT, and GOODSPEED versions all say "bushes", the Douay version says "bark of trees", while JASTROW uses "wormwood". It is very probable that JASTROW is correct in this interpretation because wormwood bushes regularly grow in the same plant association with saltworts.

SKINNER (298) records numerous legends and superstitions concerning wormwood. "If it be rubbed over a child's hands before he is twelve weeks old, wormwood will keep moths out of his hair, and he will never suffer from heat or cold." Added to wines it was supposed to "counteract their alcohol". Its actual effect was probably exactly opposite. It was also thought to afford protection against hemlock poisoning, shrews, mice, and sea-dragons! *A. vulgaris* L. was used in treating female disorders, and by magicians and witches in secret incantations to bring departed spirits back from the underworld (125). If some is eaten in the month of May it is supposed to protect the eater from consumption, poisoning, physical exhaustion, wild beasts, and the necessity of paying too many bills! "Made into a cross and put on the roof, mugwort will be blessed by Christ Himself, hence it must not be taken down for a year." A Russian name for it is "herb of forgetfulness" because of a legend that

---

[16] Referred to as *Artemisia nilotica* L. by various authors (184, 306).

mankind lost the power of understanding the speech and uses of plants because of a peasant's failure to comply with an injunction never to mention the name of mugwort (105).

40. Arundo donax L.
(FIGURE 55)

II KINGS 18: 21—Now, behold, thou trusteth upon the staff of this bruised *reed*, even upon Egypt, on which if a man lean, it will go into his hand and pierce it.
JOB 8: 11—Can the *rush* grow up without mire? can the flag grow without water?
JOB 40: 21—He lieth under the shady trees, in the covert of the *reed*, and fens.
ISAIAH 19: 6—And they shall turn the rivers far away; and the brooks of defense shall be emptied and dried up: the *reeds* and the flags shall wither.
ISAIAH 35: 7—And the parched ground shall become a pool, and the thirsty land springs of water: in the habitation of dragons, where each lay, shall be grass with *reeds* and rushes.
ISAIAH 42: 3—A bruised *reed* shall he not break, and the smoking flax shall he not quench . . .
JEREMIAH 51: 32—The passages are stopped, and the *reeds* they have burned with fire, and the men of war are affrighted.
EZEKIEL 29: 6-7—And all the inhabitants of Egypt shall know that I am the Lord, because they have been a staff of *reed* to the house of Israel. When they took hold of thee by thy hand, thou didst break, and rend all their shoulder: and when they leaned upon thee, thou brakest, and madest all their loins to be at a stand.
EZEKIEL 40: 3— . . . there was a man, whose appearance was like the appearance of brass, with a line of flax in his hand, and a measuring *reed* . . .
MATTHEW 11: 7—What went ye out into the wilderness to see? A *reed* shaken with the wind?

The giant reed, *Arundo donax*, also known as "Persian reed" and "bulrush", is common throughout Palestine, Syria, and the Sinai peninsula (267, 299). Along the margins of the Dead Sea, in the Jordan Valley, and elsewhere it forms almost impenetrable thickets which may almost be compared with the bamboo jungles of India. It is a gigantic grass, growing from 8 to 18 feet tall, with a stem-diameter of 2 or 3 inches at the base, terminated by a beautiful and massive, narrow-cylindric plume of white flowers, similar to that of the sugar-cane and pampas-grass. The leaves are alternate, 1 to 3 feet long, and glaucous-green in color. It is thought that most of the "reeds" of the Bible represented by the Hebrew words "agmôn", "agam" and "agamim", refer to this species. This plant was used for many purposes by the ancients—for walking-sticks, fishing-rods, measuring-rods, and musical pipes, and is still so employed today (306). It is therefore quite possible that the "reed" of Matthew 27: 48 and Mark 15: 36 was a carpenter's reed or measuring-rod and not a stem of *Sorghum vulgare* (which see) as is commonly supposed.

Many writers identify all the New Testament "reeds" with *Typha latifolia* L., our common cattail or reedmace, which, with *T. angustifolia* L. and *T. angustata* Bory & Chaub., also occurs in the Holy Land. Many old paintings of the mock-trial of JESUS depict him with the inflorescence stalk of a species of *Typha* in his hand as a mock-sceptre (125). Apparently for this reason many writers think that *Typha latifolia* (which see) is the plant referred to in Matthew 27: 29.

In connection with the II Kings reference to the stem of a broken reed piercing a man's hand if he leans on it, it is a well-known fact that the culms of all the stout bamboo-like grasses break into many thin, sharp, pointed slivers which can be most dangerous. During their tribal wars natives in tropical Asia and Oceanica often use sharpened bamboo stems as spears, and even during World War II bamboo stakes were driven into the ground in the East Indies as pickets to impale landing parachutists. It is interesting to recall, also, in this connection that one of the tortures inflicted upon the early Christians, and revived again during the Inquisition, was to have slivers of reed stems driven under the finger- and toe-nails (268).

CRUDEN in his "Bible Concordance" defines "reed" as "a Jewish measure of six cubits three inches", and he defines "cubit" as "the distance from the elbow to the extremity of the middle finger, this is called a common cubit or the cubit of a man, containing a foot and a half." Mrs. BARNEVELD adds: "There is also a sacred cubit which is a full yard" (see also the definition under *Hordeum*). Ezekiel's reed was 6 cubits long (158).

The Isaiah 42: 3 reference is often said to refer to *Typha latifolia*, but *Arundo donax* seems more likely from the context. Some writers suppose that "Kanah" (Hebrew" kaneh"), used as the name of a brook in Joshua 16: 8, refers to the abundance of reeds there, but it seems more likely to us that the plant involved was the sugar-cane, *Saccharum officinarum* (which see).

MOFFATT uses "reed-grass" in place of the Authorized Version's "meadow" in Genesis 41: 2, but this "reed" upon which cattle grazed is obviously not *Arundo donax*. We are discussing it herein under *Scirpus lacustris* (which see). In place of "measuring reed" in Ezekiel 40: 3, MOFFATT and the GOODSPEED version say "measuring rod", but as has been mentioned above, measuring rods were mostly made of *Arundo* stems. In Jeremiah 51: 32 MOFFATT replaces "reeds" with the non-botanical term "bastions", JASTROW substitutes the word "castles", and the GOODSPEED version uses "outworks", while the Douay version sticks closer to the original with "marshes". MOFFATT'S translation of Isaiah 35: 7 is noteworthy for its much-needed lucidity: "Parched land becomes a pool, dry ground gushes with water; the jackals' and hyenas' lair shall turn to pasture for your flocks, and *reeds* and rushes shall be flourishing where once the ostrich quartered."

POST (267) identifies the "reed shaken by the wind" of Matthew 11: 7 as *Arundo donax*, but says that Professor DAY of Beirut believes that *Phragmites communis* (which see) was the plant referred to. The Basic English version of Matthew 11: 7 is: " 'What went you out into the waste land to see? *grasses* moving in the wind?' " The lofty stems of the giant reed are very flexible and are well known to bend almost to the ground before a high wind, rising again readily when the wind has abated. This plant, therefore, fits the context of this verse perfectly.

The Hebrew words "agmôn", "agam", or "agamim" usually regarded as referring to the giant reed (paraphrased in the Septuagint), are rendered "hook" in Job 41: 2, "caldron" in Job 41: 20, "rush" in Isaiah 9: 14 & 19: 15, and "bulrush" in Isaiah 58: 5. The Isaiah references are often included by commentators among those applying to *Arundo*, but we are herein discussing the first two under *Scirpus* (which see) and the last under *Cyperus* (which see). CELSIUS was of the opinion that "agam", "agmôn", and "agamim" referred in all cases to the plant which we now call *Phragmites communis* (306).

Arundo donax is common along the banks of the Nile, and it is now generally believed to have been the "staff of the bruised reed" to which SENNACHERIB compared the power of Egypt in II Kings 18: 21 and Ezekiel 29: 6-7, upon which it was so dangerous to lean. As mentioned above, the thick strong stems of this reed were extensively employed by the ancients—as they are yet today—for walking-sticks or canes.

41. Astragalus gummifer Labill. 42. Astragalus tragacantha L.
(FIGURE 16)

GENESIS 37: 25—. . . Ishmeelites came from Gilead with their camels bearing *spicery* and balm and myrrh, going to carry it down to Egypt.
GENESIS 43: 11—. . . carry down the man a present, a little balm, and a little honey, *spices*, and myrrh, nuts, and almonds.
II KINGS 20: 13—And HEZEKIAH hearkened unto them, and shewed them all the house of his *precious things*, the silver, and the gold, and the spices.
SONG 4: 10 & 14—. . . how much better is thy love than wine! and the smell of thine ointments than all *spices*! . . . myrrh and aloes, with all the *chief spices*.
SONG 5: 1 & 13—. . . I have gathered my myrrh with my *spice* . . . His cheeks are as a bed of *spices* . . .
SONG 6: 2—My beloved is gone down into his garden, to the beds of *spices* . . .
SONG 8: 2 & 14—I would cause thee to drink of *spiced* wine of the juice of my pomegranate . . . Make haste, my beloved, and be thou like a roe or a young hart upon the mountains of *spices*.
ISAIAH 39: 2—And HEZEKIAH was glad of them, and shewed them the house of his *precious things*, the silver, and the gold, and the *spices* . . .

The Hebrew words here rendered "spicery", "spices", and "precious things" are "nechôth", "nâcôth", "nĕcôth", and "n'chot" (Greek, θυμίαμα, meaning "that which is burned as incense", or θυώματα, meaning "spice"), except in Song 5: 1 where the words used are "roshay besamim", meaning "chief of spices", in Song 8:14 "harehkâch", meaning "spiced", and in II Kings 20:13 "bosem" or "basam", meaning "spices or precious things". There is considerable doubt as to the exact identity of the plant or plants involved. Possibly more than one kind of plant is included in these references. MOFFATT uses "resin" in the Genesis passages, "balsam" in the Song of Solomon 5:1 and 6:2 verses, and "balsam-flower" in Song of Solomon 5:13. PRATT seems to regard all the references cited above as applying to *Commiphora opobalsamum*, while SHAW (293) the references in Song of Solomon 4:10 & 14, 6:2, and 8:14 as applying to *Pimenta dioica* (L.) Merr.[17] It does not seem possible to determine definitely what kind of spices are referred to in Song 4:10 & 14, for the term seems to be a collective one including all the principal spices then known. However, since the other passages in this book in which the word "něcôth" is used seem to apply to the gum-tragacanth, it is assumed that the same plant is referred to in these verses. That the allspice, *Pimenta dioica*, was intended, or even included, is not possible, for this tree is native only to tropical America and, of course, was unknown in Biblical times.

Some commentators think that the "spicery" of Genesis was the balsamic gum known now as "storax", yielded by *Styrax officinalis* L. (which see), a native of the Levant, but this does not seem probable (184). It was definitely not a general term, but, as TRISTRAM has pointed out (184), applied to the product of a particular kind of native plant.

Because the Hebrew word here rendered "spicery" and "spices" is so similar to the Arabic word, "neca'at", now used for the gum-tragacanth, SMITH (299) is inclined to identify all the references cited at the head of this chapter with *Astragalus tragacantha* and *A. gummifer*, although he is apparently not so sure about the Song of Solomon references where the context indicates the use of the word in a collective sense. The use of the expression "beds of spices" in Song 5:13 and 6:2 plainly eliminates arborescent spices like cinnamon, cassia, eaglewood, sandalwood, etc., and points to herbaceous or, at least, low-growing plants which would be cultivated in beds in a garden. The gum-tragacanth fits this context well. The expression "upon the mountains of spices" in Song 8:14 implies a kind of spice-producing plant which grows naturally in mountainous areas of the Holy Land. *Astragalus gummifer* is a native of "dry subalpine regions", according to POST (267) and so, again, fits the context perfectly. It is a dwarf shrubby plant, one or two feet tall, with woolly twigs armed with long, stiff, yellow spines, pinnately compound leaves, and white or yellow pealike flowers borne in the leaf-axils or in the loose spike-like heads. The gum is a natural exudation from the stem and branches. It has no scent, and flakes without making a powder (172a).

SMITH (299) suggests that it is possible that SOLOMON "had in his gardens at Etham all kinds of sweet-smelling plants common to Palestine, as also those natives of South Europe, such as lavender, rosemary, sage, thyme, savory, marjoram, etc."

The GOODSPEED version substitutes "gum" for "spicery" and "spices" in the two Genesis references, doubtless meaning the gum-tragacanth, but retains "spices" in the Song of Solomon passages. The Douay version uses "spices" in Genesis 37:25 and "storax" in 43:11, "aromatical spices" in Song 4:10, 5:1 & 13, 6:2, and 8:14, and "chief perfumes" in Song 4:14. Although the Authorized Version uses "precious things" in II Kings 20:13 and Isaiah 39:2, MOFFATT says "treasures", and the JASTROW and GOODSPEED versions have "treasure-house", but marginal Bibles give the alternative translation "spicery". JASTROW uses the word "spices" in the same way as the Authorized Version.

The "spices" of Song 5:1 & 13 and 6:2 are regarded by SMITH (306) as referring to spices in general, of which the chief was derived from *Commiphora opobalsamum*. In this view, as has been stated above, we do not concur.

---

[17]Referred to as *Eugenia Pimenta* P. DC. by SHAW (293); also known as *Pimenta officinalis* Lindl.

43. **Atriplex dimorphostegia** Kar. & Kir.   44. **Atriplex halimus** L.
45. **Atriplex rosea** L.   46. **Atriplex tatarica** L.

JOB 30: 4—Who cut up *mallows* by the bushes, and juniper roots for their meat.

The Hebrew word here translated "mallows" in the Authorized Version is "malluach" (Greek, ἁλιμόν, related to ἁλμάς, meaning "salted", or 'αλιαρός meaning "of the sea"). Some commentators are of the opinion that this term actually refers to true species of the genus *Malva*, which, being soft and mucilaginous and not unwholesome, might well have been gathered and eaten in times of privation. CARRUTHERS (184), for instance, suggests *Malva rotundifolia* L. (dwarf mallow), *M. sylvestris* L. (marsh mallow), *Hibiscus syriacus* L.[18] (tree-mallow, rose-of-Sharon), only the first two of which are listed as occurring wild in the Holy Land by POST (266). *Hibiscus syriacus*, in spite of its scientific and common names, is a Chinese plant, now cultivated in the Holy Land, but not native there, and certainly not known there in Biblical days.

Because the Hebrew word "malluach" implies saltiness, either of taste or location (184), most modern authorities believe that species of saltwort or orach, *Atriplex*, are the plants referred to here. Twenty-one kinds of saltwort occur in the Holy Land today, almost all of which are common and could well comply with the requirements of the text. *A. halimus* is the species usually suggested (184, 306). This species, known as "sea purslane", "Spanish sea purslane", or "shrubby orach", is a strong-growing bushy shrub related to the spinach, one to three feet tall, with simple, somewhat hoary, ovate, obtuse, cuneate leaves and small inconspicuous flowers. It is abundant on the shores of the Mediterranean and in the region about the Dead Sea, where it is said to attain a height of 5 to 10 feet (299). Some commentators have doubted that *A. halimus* could be the plant because it is a strictly littoral plant of salty maritime regions, and they have supposed that the "land of Uz", where JOB lived, was situated in eastern Syria, near the Euphrates River. However, modern opinion seems to place the "land of Uz" much farther south, just east of the Sinai Peninsula and northeast of the Gulf of Akaba, in what is now Transjordania (306). According to POST all the species of *Atriplex* listed at the head of this chapter are common in that area.

Some writers (299) are of the opinion that the word "mallows" is a general term covering any and all plants with spinach-like properties, including *Atriplex* and also *Chenopodium*, of which there are several species in the region, mostly annual weedy plants with soft mucilaginous leaves.

CELSIUS, SPRENGEL, and other early writers thought that *Corchorus olitorius* L. might be one of the plants known as "mallows", but this is very improbable, although POST (265, 266) states that the species exists now farther to the north, in Palestine and Lebanon, in fields and along roadsides, as an escape from cultivation. It is an Indian plant which furnishes the valuable fiber known as "jute", which is soft, silky, and easily spun, used in the manufacture of carpets and some kinds of cloth, but not suitable for cordage because it softens in water. It is used extensively in surgery as a cheap drainage material. Its young shoots are eaten like asparagus, and in Syria and Egypt are now known by the name of "Jews' mallows". It is, however, very questionable whether this plant was cultivated in Transjordania at the time of JOB (about 1520 B.C.), or, if so, if it would be the type of plant to which the very abject poor and miserable people described by JOB would have access. A native weed, normally ignored, like the saltwort, would fit the context far better.

SMITH in his "Dictionary of the Bible" (300) recounts that in the year 1600 WILLIAM BIDDULPH was traveling from Aleppo to Jerusalem and "saw many poor people gathering mallows and three-leaved grass, and asked them what they did with it, and they answered that it was all their food and they did eate it." The "three-leaved grass" of this quotation was undoubtedly the

---
[18]Referred to as *Althaea frutex* Mill. by CARRUTHERS (184).

fenugreek, *Trigonella foenum-graecum*, a clover-like plant eaten extensively in Egypt and Palestine even to the present day.

PRATT suggests (268) species of *Mesembryanthemum* for the "mallows" of JOB but, although several species of this genus occur in JOB'S country, the identification seems highly improbable. The saltiness implied in the word "malluach", in our estimation, points too strongly to *Atriplex* to admit of any other identification (306).

LEVI, LUTHER, and the old Swedish and Danish translations all regarded the plants as "nettles". MOFFATT, JASTROW, and the GOODSPEED versions all use "salt-wort" in place of "mallows"—"They pluck *salt-wort* with wormwood; And the roots of the broom are their food", according to JASTROW. The Douay version is decidedly different: "And they ate *grass*, and barks of trees, and the root of junipers was their food."

FIGURE 16. — *Astragalus tragacantha*, the gum-tragacanth. This was probably the source of the "spicery" of the Ishmeelites and one of the "spices", "chief spices", and "precious things" of Hezekiah and Solomon. Being native to the region, it did not have to be imported as did so many of the other incense spices used in the service of the Temple. (Tristram, Natural History of the Bible, 1867).

47. Balanites aegyptiaca (L.) Delile

GENESIS 37: 25— . . . and, behold, a company of Ishmeelites came from Gilead with their camels bearing spicery and *balm* and myrrh . . .
JEREMIAH 8: 22—Is there no *balm* in Gilead; is there no physician there? . . .
JEREMIAH 46: 11—Go up into Gilead, and take *balm*, O virgin, the daughter of Egypt . . .
JEREMIAH 51: 8—Babylon is suddenly fallen and destroyed: howl for her; take *balm* for her pain, if so be she may be healed.

The above references are given by SMITH and others (299), who regard the Hebrew word here used, "tzŏri", tzĕrî", or "tzari" (Greek, $ρητίνη$, meaning "resin or gum"), as referring either to the gum of *Balanites aegyptiaca* or of *Pistacia lentiscus* L. Both these species are common in Palestine, the latter abundant in the rocky country of Gilead.

The Jericho balsam, *Balanites aegyptiaca*, abounds throughout Egypt and northern Africa, Palestine, the plains of Jericho, and the hot plains bordering the Dead Sea. It is a truly desert-loving plant and is held in veneration by the Mohammedans in western India, where it is also found. It is a small tree, 9 to 15 feet tall, with slender thorny branches, leathery, obovate-elliptic, woolly 2-foliolate leaves, and small axillary clusters of green flowers. Its fruits are pounded and boiled to extract the oil which is said to possess medicinal and healing properties. An intoxicating drink is also made from the fruit by fermentation. Because of its reputed healing properties it is generally thought that this is the "balm" of the three Jeremiah references cited above. Indeed, while the Douay, JASTROW, and GOODSPEED versions still use the word "balm", MOFFATT has substituted "balsam" in all of the four passages.

It is interesting to note that the Douay Bible of 1609 renders the twenty-second verse of chapter eight of Jeremiah as "Is there no rosin in Galaad?", and this has caused that edition to be known as the Rosin Bible. The Bishops' Bible of 1568 reads for this verse: "Is there no tryacle in Gilead?", and it has therefore acquired the name of Treacle Bible.

HASSELQUIST, the ill-starred botanist, whose untimely death was so lamented by the great LINNAEUS because of the pioneer work which he was doing on Holy Land plants and animals, has given (306) a description of the gum of this plant. It "is of a yellow color, and pellucid. It has a most fragrant smell, which is resinous, balsamic, and very agreeable. It is very tenacious or glutinous, sticking to the fingers, and may be drawn into long threads . . . I have seen it at a Turkish surgeon's, who had it immediately from Mecca, described it, and was informed of its virtues; which are, first, that it is the best stomachic they know, if taken to three grains, to strengthen a weak stomach; second, that it is a most excellent and capital remedy for curing wounds, for if a few drops are applied to the fresh wound, it cures it in a very short time."

Since the context of the Genesis text does not imply any healing properties to the "balm" there mentioned, it is possible that it may there apply to *Pistacia lentiscus* (which see), as suggested by some authors. However, MOFFATT believes that the Jericho balsam is the plant also referred to there, and in this view we concur.

48. Boswellia carterii Birdw.   49. Boswellia papyrifera Hochst.
50. Boswellia thurifera Roxb.

EXODUS 30: 1, 7-9, & 34—And thou shalt make an altar to burn *incense* upon: of shittim wood shalt thou make it . . . And AARON shall burn thereon *sweet incense* every morning: when he dresseth the lamps, he shall burn *incense* upon it. And when AARON lighteth the lamps at eve, he shall burn *incense* upon it, a perpetual *incense* before the Lord throughout your generations. Ye shall offer no strange *incense* thereon . . . And the Lord said unto MOSES, Take unto thee sweet spices, stacte and onycha, and galbanum; these sweet spices with pure *frankincense* . . .
LEVITICUS 2: 1-2 & 15-16—And when any will offer a meat offering unto the Lord, his offering shall be of fine flour; and he shall pour oil upon it, and put *frankincense* thereon. And he shall take thereout . . . all the *frankincense* thereof . . . And thou shalt put oil upon it, and lay *frankincense* thereon: it is a meat offering. And the priest shall burn the memorial of it, part of the beaten corn thereof, and part of the oil thereof, with all the *frankincense* thereof . . .
LEVITICUS 5: 11— . . . he shall put no oil upon it, neither shall he put any *frankincense* thereon . . .
LEVITICUS 6: 15—And he shall take of it his handful, of the flour of the meat offering, and of the oil thereof, and all the *frankincense* which is upon the meat offering, and shall burn it upon the altar for a sweet savour, even the memorial of it, unto the Lord.
LEVITICUS 10: 1—And NADAB and ABIHU, the sons of AARON, took either of them his censer, and put fire therein, and put *incense* thereon . . .
LEVITICUS 16: 12-13—And he shall take a censer full of burning coals of fire from off the altar before the Lord, and his hands full of *sweet incense* beaten small . . . And he shall put the *incense* upon the fire . . . that the cloud of the *incense* may cover the mercy seat . . .
LEVITICUS 24: 7—And thou shalt put pure *frankincense* upon each row.
NUMBERS 5: 15— . . . he shall pour no oil upon it, nor put *frankincense* thereon . . .
NUMBERS 7: 14, 20, 26, 32, 38, 44, 50, 56, 68, 74, 80, & 86—One spoon of ten shekels of gold, full of *incense* . . . One spoon of gold . . . full of *incense* . . . The golden spoons were twelve, full of *incense*, weighing ten shekels apiece . . .
NUMBERS 16: 46— . . . Take a censer, and put fire therein from off the altar, and put on *incense* . . .
II KINGS 17: 10-11—And they set them up images and groves in every high hill, and under every green tree: And there they burned *incense* in all the high places, as did the heathen . . .
II KINGS 18: 4—He removed the high places . . . and brake in pieces the brasen serpent that MOSES had made: for unto those days the children of Israel did burn *incense* to it . . .
II KINGS 23: 5—And he put down the idolatrous priests, whom the kings of Judah had ordained to burn *incense* in the high places in the cities of Judah, and in the places round about Jerusalem; them also that burned *incense* unto Baal, to the sun, and to the moon, and to the planets, and to all the host of heaven.
I CHRONICLES 9: 29—Some of them also were appointed to oversee . . . the *frankincense*.
II CHRONICLES 28: 4—He sacrificed also and burnt *incense* in the high places . . .
II CHRONICLES 34: 25—Because they have forsaken me, and have burned *incense* unto other gods . . .
NEHEMIAH 13: 5 & 9—And he had prepared for him a great chamber, where aforetime they laid the . . . *frankincense* . . . and thither brought I again the . . . *frankincense*.
SONG 3: 6—Who is this that cometh out of the wilderness like pillars of smoke, perfumed with myrrh and *frankincense*, with all powders of the merchant?
SONG 4: 6 & 14—Until the day break, and the shadows flee away, I will get me to the mountain of myrrh, and to the hill of *frankincense* . . . Spikenard and saffron; calamus and cinnamon, with all trees of *frankincense*; myrrh and aloes, with all the chief spices.
ISAIAH 43: 23— . . . I have not caused thee to serve with an offering, nor wearied thee with *incense*.
ISAIAH 60: 6— . . . they shall bring gold and *incense*; and they shall shew forth the praises of the Lord.
ISAIAH 65: 3—A people that provoketh me to anger continually to my face; that sacrificeth in gardens, and burneth *incense* upon altars of brick.
JEREMIAH 6: 20—To what purpose cometh there to me *incense* from Sheba, and the sweet cane from a far country? . . .
JEREMIAH 11: 12 & 17—Then shall the . . . inhabitants of Jerusalem go, and cry unto the gods unto whom they gave *incense* . . . to provoke me to anger in offering *incense* to Baal.
JEREMIAH 17: 26—And they shall come from the cities of Judah, and from the places about Jerusalem, and from the land of Benjamin, and from the plain, and from the mountains, and from the south, bringing burnt offerings, and sacrifices, and meat offerings, and *incense*, and bringing sacrifices of praise, unto the house of the Lord.
JEREMIAH 41: 5—That there came certain from Shechem, from Shiloh, and from Samaria, even fourscore men, having their beards shaven, and their clothes rent, and having cut themselves, with offerings and *incense* in their hand to bring them to the house of the Lord.
JEREMIAH 48: 35—Moreover I will cause to cease in Moab, saith the Lord, him that offereth in the high places, and him that burneth *incense* to his gods.
MATTHEW 2: 11—And when they were come into the house, they saw the young child with MARY his mother, and fell down, and worshipped him: and when they had opened their

treasures, they presented unto him gifts; gold, and *frankincense*, and myrrh.
LUKE 1: 9-10—His lot was to burn *incense* when he went into the temple of the Lord. And the whole multitude of people were praying without at the time of *incense*.
REVELATION 5: 8—... having every one of them harps, and golden vials full of *odours*, which are the prayers of saints.
REVELATION 8: 3-4—... and there was given unto him much *incense* ... and the smoke of the *incense* ... ascended up before God.
REVELATION 18: 13—And cinnamon, and odours, and ointments, and *frankincense*, and wine.

The Hebrew word used in these references and translated "frankincense" or (in some places) "incense" is "lebonah" or "levônâh" (Greek, λιβανος, meaning "the frankincense tree"), while the words rendered "incense" (in most places) are "miktar", "kitter" or "koter", and "kitteroth" or "ketoret" (184). There is little doubt that the plant product here referred to is the true frankincense, obtained from three species of the genus *Boswellia*, chiefly *B. carterii*, *B. papyrifera*, and *B. thurifera*[19] (267). These and related species grow in southern Arabia, Abyssinia, Somaliland, India, and the East Indies (158), whence they must have been procured by the Hebrews through trade (299, 306). They are trees of large size, related to the turpentine or terebinth tree and to those yielding balsam and myrrh (146). The flowers are star-shaped, pure white or green tipped with rose (95, 184); the leaves are compound, composed of 7 to 9 glossy serrate leaflets resembling those of the mountain-ash. The gum exudes in the form of brittle, glittering, roundish or oblong drops and is of a white, yellow, or pale red color. It has a bitter taste and gives off a strong balsamic odor in the form of a volatile oil when warmed or burned. It is mentioned 22 times in the Bible—16 times in relation to its use in religious worship, twice as a tribute of honor, once as an article of merchandise, and 3 times as a product of the royal garden of SOLOMON. It was probably employed almost exclusively in the sacrificial service of the Tabernacle and Temple until the time of SOLOMON's reign. Frankincense gum is obtained by making successive incisions in the bark of the trunk and branches of living trees. The first such incision yields the purest and whitest gum, while that obtained from later incisions is spotted with yellow or red, and, upon aging, loses its whiteness altogether (306).

SMITH is of the opinion (299) that the "incense" of Exodus is probably not the same as that of the other references, and that it may be, instead, a mixture of the resinous gum of *Pinus halepensis* Mill. and the fragrant wood of *Sabina phoenicia* (L.) Antoine[20] and *Juniperus oxycedrus* L. (which see), natives of Lebanon. The source of this material, Lebanon, perhaps accounts for the form employed in the Hebrew word "lebonah". These two cedar woods, while they yield no gum, or but very little, are quite fragrant when burned. GROSER, however, thinks that the Authorized Version is in error in translating "lebonah" in some places as "incense" (implying a mixture of substances) and in other places as "frankincense" (implying a pure substance). He is of the opinion that it was burned pure and unadulterated at all times. Incense today in Chinese temples is made of the sawdust of sandalwood, *Santalum album*, mixed with swine dung. On being lighted, a fragrant smoke is given off, which, as it curls heavenward, is thought by the worshiper to carry his prayers to the Eternal. SMITH remarks that the "frankincense" used in Europe today is a mixture of the gummy exudation of the European spruce-fir or Norway spruce (*Picea abies* (L.) Karst.) and the American loblolly pine (*Pinus taeda* L.) and arbor-vitae (*Thuja occidentalis* L.), none of which, naturally, had anything whatever to do with the frankincense of Biblical times (299). He suggests that the wood of *Aquilaria agallocha* and *Santalum album* may also have figured in the "incense" of some of the references cited at the head of this chapter, if this was not always pure frankincense, especially after the time of SOLOMON (975 B.C.).

It is of interest to note that where the Authorized and GOODSPEED versions say "frankincense" in Leviticus 2: 1-2 & 15-16 and 6: 15, MOFFATT says "incense", but, on the other hand, where the Authorized Version says "incense"

---
[19]Sometimes referred to as *B. serrata* Roxb.
[20]Sometimes referred to as *Juniperus lycia* L. or *J. phoenicia* L.

in Isaiah 43: 23 & 60: 6 and Jeremiah 6: 20, 17: 26, & 41: 5, the GOODSPEED version says "frankincense", and so does the Douay version in Isaiah 60: 6, Jeremiah 6: 20 and 17: 26. In Nehemiah 13: 5 & 9 MOFFATT continues to use "frankincense", while the GOODSPEED version is "incense". Where the King James Version uses "frankincense" in I Chronicles 9: 29, both the GOODSPEED and MOFFATT translations use "incense". It is, thus, obvious that the two terms are inextricably connected. In some cases, however, some modern translations depart completely from the use of either of the controversial words and give an entirely different translation. The "incense" of Isaiah 65: 3, for instance, is rendered "sacrifice" by the Douay version and the "incense" of II Kings 17: 10-11 is translated "sacrificing" by MOFFATT and both it and that of II Kings 23: 5 are rendered "sacrifices" in the GOODSPEED version. MOFFATT's rendition of Song 4: 6 & 14 is most revealing: "Yes, till the cool of the dawn, till the shadows depart, I will hie me to your *scented slopes*, your fragrant charms . . . and spikenard and saffron, with cassia and cinnamon, *all sorts of frankincense*, with myrrh and with eaglewood, all the best spices." The Douay version replaces the phrase "all trees of frankincense" of the Authorized Version with "all the trees of Libanus" in the latter part of this quotation, although it still continues to interpret the first verse in the botanical rather than anatomical sense.

The writings of THEOPHRASTUS and other ancient authors combine with Biblical evidence to point to the conclusion that the Hebrews imported all their frankincense from Arabia, especially from the region about Sheba (Saba). At the present time, however, the frankincense produced in Arabia (from *B. carterii*) is of a very inferior sort. The finest gum imported to Turkey and other lands of Asia Minor comes not *from* Arabia, but from India and the islands of the Indian Archipelago (from *B. thurifera*) *through* Arabia (306). Probably this was also true in Biblical days.

LAMARCK has suggested *Commiphora opobalsamum* (L.) Engl.[21] for the Biblical "frankincense". This may, indeed, be involved where the incense used was a mixture and it is most probable that this was the native "frankincense tree" of Ecclesiasticus 50: 8 where the context seems plainly to require a native tree whose habit of growth would be quite familiar to the Israelites.

Contrary to the opinion of GROSER, mentioned above, modern authorities maintain that the "incense" used in the service of the Tabernacle was a mixture, in definite proportions, of stacte, onycha, galbanum, and frankincense, and that the use of any incense not composed of these four ingredients in the proper proportions was strictly and explicitly forbidden. RASHI (306) enumerates seven other ingredients; JOSEPHUS lists 13. Among these other ingredients, according to MAIMONIDES, were myrrh, cassia, spikenard, saffron, costus, cinnamon, and "sweet bark", to which was added "salt of Sodom, with amber of Jordan, and an herb called 'the smoke-raiser', known only to the cunning in such matters, to whom the secret descended by tradition." Almost two pounds of incense were used per day in the Temple, or 700 pounds a year! Probably it would have had to be a mixture of several or many substances because of this huge amount required annually, but that it always contained all of the ingredients listed above, in definite proportions, seems hardly likely.

Frankincense was highly valued by the Egyptians and other ancient peoples for embalming and fumigating. It was always, and still is, the most important incense resin in the world. Frankincense gum is known as "gogul" or "olibanum" in Bombay and the tree is called "salai" by the natives of India (95). BIRDWOOD and POST claim that only *Boswellia carterii* was the true frankincense tree of the ancient Hebrews, it being native to Arabia, especially around Sheba, along the coast of Hadramaut, and on the Somali coast of east Africa, and "never grown in Syria, or in India" (184, 266). It seems more likely, however, that the Hebrews purchased any and all the frankincense that was available, including that of all three species listed at the head of this chapter.

The Basic English version renders Matthew 2: 11 ". . . gold, *perfume*, and

---

[21] As *Amyris gileadensis* L.

spices" and Luke 1: 9-10 "he had to go into the temple to see to the burning of *perfumes*. And all the people were offering prayers outside, at the time of the burning of *perfumes*." On the other hand, the O'HARA, LAMSA, and WEYMOUTH versions all use "incense" instead of "odours" in Revelation 5: 8, and, indeed, this translation is also suggested in marginal Bibles. The Basic English version, of course, again uses "perfumes" in this passage. Modern versions exhibit great diversity in the translation of Revelation 18: 13—the "odours" in the first part of this verse according to the King James Bible are rendered "perfumes" by LAMSA, "incense" by WEYMOUTH, and "amomum" by O'HARA. The Basic English version, with its mania for a small vocabulary at the cost of accuracy of meaning, renders the whole series of items enumerated as merely "and sweet-smelling plants, and perfumes."

In the JASTROW version there are several changes from the King James version which ought to be noted under this topic. In Exodus 30: 7, for instance, JASTROW uses "incense of sweet spices" instead of "sweet incense". In many passages he omits the terms "frankincense" or "incense" and merely says "offerings", viz., Numbers 16: 46, II Kings 17: 10-11 & 25: 5, II Chronicles 34: 25, Jeremiah 11: 12 & 17 and 48: 35. In certain other passages "frankincense" is substituted for "incense", viz., Isaiah 43: 23 & 60: 6 and Jeremiah 17: 26 & 41: 5.

51. Brassica nigra (L.) W. Koch

MATTHEW 13: 31-32—Another parable put he forth unto them, saying, The kingdom of heaven is like to a *grain of mustard seed*, which a man took, and sowed in his field: Which indeed is the least of all seeds: but when it is grown, it is the greatest among herbs, and becometh a tree, so that the birds of the air come and lodge in the branches thereof.
MATTHEW 17: 20— . . . If ye have faith as a *grain of mustard seed*, ye shall say unto this mountain, Remove hence to yonder place . . .
MARK 4: 31-32—It is like a *grain of mustard seed*, which, when it is sown in the earth, is less than all the seeds that be in the earth: But when it is sown, it groweth up, and becometh greater than all herbs, and shooteth out great branches; so that the fowls of the air may lodge under the shadow of it.
LUKE 13: 19—It is like a *grain of mustard seed*, which a man took, and cast into his garden; and it grew, and waxed a great tree; and the fowls of the air lodged in the branches of it
LUKE 17: 6—And the Lord said, If ye had faith as a *grain of mustard seed*, ye might say unto this sycamine tree, Be thou plucked up by the root, and be thou planted in the sea; and it should obey you.

Here, again, there has been much discussion and argument as to what the "mustard" plant of JESUS' parables really was. The Greek word in the original texts is "sinapi" (σίναπι). Most modern commentators agree that it was the ordinary black mustard, *Brassica nigra*[22] (228). This plant is extensively cultivated for its seeds, which are not only ground up to produce the mustard of commerce, but also yield a useful oil similar to colza oil. It was for the oil that this plant was probably cultivated in the fields of Palestine in JESUS' day, for, as has been pointed out previously, the Hebrews and other Oriental peoples are great users of oils. POST states (265) that it grows "along roadsides and waste places; everywhere"; and the same applies to the closely related white mustard, *Sinapis alba* L.[23]. While these mustards do not usually grow more than 3 or 4 feet tall, plants have been found to 10 and even 15 feet tall, with a main stem as thick as a man's arm. Although they are only annual plants, their stems and branches in autumn become hard and rigid and of quite sufficient strength to bear the weight of small birds that are attracted by the edible seeds. PRATT (268) thinks that *Sinapis orientalis* L.[24] is more likely than any other species. ALONZO DE AVALLO tells of having traveled many miles through mustard groves where the plants were taller than man and horse, looking like trees, with stems fully as big as a man's arm, and in which small birds actually built their nests. Lord CLAUDE HAMILTON saw one plant in upper Egypt taller than he could reach, with a stem as thick as his arm.

[22]Also referred to as *Sinapis nigra* L.
[23]Also referred to as *Brassica alba* (L.) Rabenh.
[24]Also referred to as *S. arvensis* var. *orientalis* (L.) Koch & Ziz.

DALMAN has seen wild mustards, both *Brassica* and *Sinapis*, growing commonly over three feet tall in the Holy Land and almost 10 feet tall near the Sea of Tiberias. He identifies the "mustard" of the Gospels as either *Sinapis alba* or *S. arvensis* (98). L. H. BAILEY describes the black mustard (34) as a much-branched, green, more or less hispid-hairy (sometimes glabrous or even slightly glaucous) annual "3 to 10 or more feet tall" (35). The leaves are lyrate, pinnatifid, or dentate-lobed and long-petiolate, the uppermost lanceolate and entire. The naked twig-like racemes bear many bright yellow flowers of typical mustard shape. IRBY and MANGLES have reported that in the Jordan valley they crossed a small plain on which the mustard plants reached fully as high as their horses' heads. THOMSON is quoted by GROSER (145) as reporting that on the rich plains of Akkar (in Phoenicia) the black mustard grows as tall as horse and rider (47, 306).

Many commentators, especially ROYLE (283, 306), have suggested that the "mustard" of JESUS was actually an entirely different plant, *Salvadora persica* L., which is found in thickets around the Dead Sea, especially in the hot valleys at its southern end. It is a shrub or small soft-wooded tree, 3 to 9 feet tall [or 25 feet tall, according to SMITH (299)], with ample ovate-oblong or oblong-linear leaves, obtuse at the apex and tapering to the base. It bears somewhat fleshy berries in bunches resembling currants, but with the color of plums. The plant has a pleasant, though strong, aromatic taste resembling that of mustard, and if taken in any considerable quantity will produce a similar irritation of the nose and eyes. PRATT (268) states that *Salvadora* grows so plentifully by the shores of the Sea of Galilee that JESUS and his audience must have been looking right at it when the parable was spoken. The present Arabic name of the tree is "khardal", meaning "mustard-tree", but this, according to POST (267) is a false application of the word based on an erroneous assumption. That this tree was the plant referred to by JESUS is highly improbable. Most authorities now agree that, contrary to what PRATT has stated, it does not grow as far north as Galilee, where the climate is far too cold for it. The Dead Sea is about 100 miles from where JESUS spoke his parables and it is almost certain that *Salvadora* would be totally unknown to at least the vast majority of the multitude gathered at his feet on the Mount. Furthermore, its fruits are rather large one-seeded drupes; its good-sized stony seeds would hardly fit the description of the parable. PRATT's statement that the seeds of *Salvadora* are tiny is entirely wrong. Nor would a farmer "cast into his garden" the seeds of *Salvadora* as he would those of mustard. The famous English botanist, Sir JOSEPH HOOKER, and others have, it seems to us, quite effectively eliminated *Salvadora* from possible consideration as the "mustard" of the Scriptures (178), and yet as late as 1884 we find SHAW so considering it (293). BLANCHAN says (56a) "Inasmuch as the mustard which is systematically planted for fodder by Old World farmers grows with the greatest luxuriance in Palestine, and the comparison between the size of the seed and the plant's great height was already proverbial in the East when JESUS used it, evidence strongly favors this wayside weed" over the rarer shrub-like tree, *Salvadora persica*.

FROST, in 1827, advanced the theory that the pokeberry, *Phytolacca decandra* L.[25], was the "mustard" tree (77, 127, 128). This species, however, is now usually regarded as an American plant, so its claims to having been in Palestine in JESUS' time are certainly far-fetched, although POST (265, 267) records it now from that region along with other American plants recently introduced. Of FROST's claim the Kew botanist, JOHN SMITH, says (299) that it is founded "upon no positive evidence of his own acquiring . . . his opinion is not worth a moment's consideration; he was a vain charlatan." These are harsh words and may surprise some of our readers today. Yet they are, in the main, typical of the bitterness with which these matters of the identification of Bible plants have been argued in the past. The claims and counter-claims, charges and counter-charges, leading rapidly to bitter recriminations and personalities and finally mutual damnation, so characteristic of many theologi-

---

[25]Now usually known as *P. americana* L.

cal discussions through the ages, have found their way—lamentably—even into this realm of Biblical botany which ROHR (274, 275) called "phytotheologia".

Other writers have suggested *Phytolacca dodecandra* L'Hér., an Abyssinian species of pokeweed, but POST does not record the species from anywhere in Syria, Palestine, Lebanon, or Sinai, even as an introduction, so it can safely be eliminated along with its American cousin.

It is perhaps worthy of note that although the seeds of *Brassica* are small, and were probably the smallest seeds known to the common country folk comprising JESUS' audience in Galilee, yet they are far from being "the least of all seeds" (or, as GOODSPEED says: "the smallest of all seeds"). The seeds of orchids are now usually regarded as the smallest in the world, being actually as fine as powder. Such statements as that concerning the size of the mustard seed must always be judged in the light of the knowledge of the time and of the people involved. Although there are 70 kinds of orchids native to the Holy Land, it is not probable that JESUS' listeners were at all acquainted with their dust-like seeds. These apparent dust motes they probably would not even have recognized as seeds. Of the seeds with which they were most apt to be familiar, such as the seeds of the grains and vegetables which they planted in their fields, that of the mustard was certainly one of the smallest. According to SMITH (306) the "mustard" is described in Matthew and Mark by a word rendered "herb" in all translations, but actually signifying a *garden* herb. Mustard was probably a commonly cultivated garden herb in JESUS' day, as it still is in the Orient, for its oil. Its seeds would, therefore, be very familiar to country folk. Similarly, the statement in Mark 4: 32 that the mustard plant, when full grown, "becometh greater than all herbs" (or as MOFFATT says: "larger than any plant", and GOODSPEED: "grows to be the largest of all the plants") cannot be taken literally. If we may believe the testimony of travelers, however, *Brassica nigra* did, and perhaps yet does in favorable spots, become one of the largest of *annual herbs* in the region.

Users of the Douay and O'HARA versions will find the verse numbered Matthew 17: 20 in the King James version under "17: 19" in their versions, but translated essentially the same. In fact, all the modern translations, including MOFFATT, GOODSPEED, O'HARA, Douay, LAMSA, WEYMOUTH, and even the Basic English versions, render these passages concerning the "mustard" seed essentially the same and all continue to use the word "mustard" employed by the King James version. This unanimity is noteworthy because it is so unusual!

Dr. ROYLE's argument (283) against *Brassica* for the "mustard" of Scripture is based almost wholly on the Biblical statements concerning its becoming "a tree", or, as Luke has it, "a great tree; and the fowls of the air lodged in the branches of it." These expressions, like so many others in the Bible, were to a large extent figurative. In an Oriental proverbial simile not much absolutely literal accuracy is to be looked for. ROYLE was of the opinion that the passage implies that birds built their nests in the branches of the "mustard". The Greek word employed has no such connotation; it merely means "to settle or rest upon" (306). In spite of the testimony of ALONZO DE AVALLO, quoted above, it seems most improbable that any birds ever actually nest in the branches of *Brassica*. Nor is it justified to suppose, as some advocates of *Salvadora* have done, that the expression "fowls of the air" denotes large and heavy chicken-like or hawk-like birds. It seems most probable that the word was used to denote the common insessorial (perching) birds of the region, like linnets and finches. HILLER's explanation is probably the correct one, *viz*., that these small sparrow-like birds perched temporarily on the branches of the mature mustard plants in order to eat the seeds in the ripening pods. Birds even today are very fond of mustard seed.

Both the Babylonian and Jerusalem Talmud tales include stories involving huge mustard plants (262). Interesting to recall in this connection is the fact that GUATAMA BUDDHA (died about 480 B.C.) also told a parable of the mustard, only his story was intended to call attention to the human selfishness which motivates much of our grief on the death of a loved one. In India the mustard is the symbol of reproductive generation (298).

52. **Butomus umbellatus** L.

GENESIS 41: 2—And, behold, there came up out of the river seven well favoured kine and fat fleshed; and they fed in a *meadow*.
JOB 8: 11—Can the rush grow up without mire? can the *flag* grow without water?

There is considerable uncertainty over the identification of the word "achu" (Greek, ἄχι βούτομον). It is an Egyptian, not Hebrew, word (184). It is translated as "flag" in the Job reference by the Authorized, "sedge-bush" by the Douay, "rush" by the GOODSPEED, "reed-grass" by the JASTROW, "meadow-grass" by LEESSER, and "reed" by the MOFFATT version. In the Genesis passage the Authorized Version uses "meadow" for it, the Douay version says "marshy places", the GOODSPEED version says "sedge", and MOFFATT, JASTROW, and the Revised Version use "reed-grass". It would seem from the use of the word in the Job passage, where it is mentioned along with the papyrus, that it refers to a specific kind of plant, rather than to a general ecologic formation like a meadow. From the context of being a plant on which PHARAOH's cattle might feed along the banks of the Nile and yet not the papyrus, it seems that it may well refer to the flowering-rush or water gladiole, *Butomus umbellatus*, which flourishes both in Egypt and in Palestine, growing along with and among the papyrus (184). The *Cyperus esculentus* L. also suggested by some commentators (184) seems out of the question for it does not meet the requirements of the texts. Although it grows abundantly both in Egypt and in Palestine and is edible, it inhabits sandy places and fields (267), not marshy river banks.

The flowering-rush is a showy aquatic or bog plant growing from horizontal rootstocks, producing linear triquetrous leaves and a cylindric scape overtopping the leaves. The scape bears a terminal umbel of many pink or rose-colored flowers, subtended by an involucre of three inflated leaf-like bracts (267). The plant grows luxuriantly in wet ditches and still water along the margins of lakes and streams and spreads rapidly when it once gets established.

53. **Buxus longifolia** Boiss.

ISAIAH 41: 19—I will plant in the wilderness the cedar, the *shittah tree*, and the myrtle, and the oil tree; I will set in the desert the fir tree, and the pine, and the *box tree* together.
ISAIAH 60: 13—The glory of Lebanon shall come unto thee, the fir tree, the pine tree, and the *box* together . . .
EZEKIEL 27: 6— . . . the company of *the Ashurites* have made thy benches of ivory, brought out of the isles of Chittim.
II ESDRAS 14: 24—But look thou prepare thee many *boxtrees* . . .

The Hebrew word translated "box" in the Authorized Version is "teasshûr" or "t'ashur" (Greek, πύξος). EZEKIEL makes use of a proper name "Asshur" or "Ashur"—rendered "the Ashurites" in the Authorized Version. Some authorities (178) maintain that this expression of the prophet in describing the commerce with Tyre is actually a contraction of "teasshûr" and therefore probably has a similar meaning. These authorities would replace the Authorized Version's "the company of *the Ashurites* have made thy benches of ivory" with "the benches of the rowers have they made of *box-wood*, inlaid with ivory." JASTROW says "Of the oaks of Bashan Have they made thine oars; Thy deck have they made of ivory inlaid in *larch*, From the isles of Kittites". Certainly it seems that "Assur" here does not refer to Assyria (Asshur) or to Asshurim, a tribe descended from DEDAN, the grandson of ABRAHAM, supposed to be residents of southern Arabia (301). MOFFATT and GOODSPEED disagree with the "box-wood" rendition in this verse. The former says: "with ivory inlaid in *larch* from Cyprus for your deck" and the latter says: "your deck they made of *larch* from the coast-lands of Cyprus." Just what tree JASTROW, MOFFATT, and GOODSPEED had in mind when they speak of *"larch"* is not clear, for, as far as we are aware, there is no species of larch native to Cyprus or anywhere in the Mediterranean region. Larches are trees of the northern and colder regions of the earth; the European species, *Larix decidua* Mill., is found only in northern and central Europe north of 44° N. latitude, and there is no larch native to northern Africa or western Asia. Box-wood seems a far more likely timber for this context. The Douay version neatly passes over the controversy by saying: "they have made thee benches of Indian ivory and cabins with *things* brought from the islands of Italy."

SMITH (299) considers the references to "box" in Isaiah to apply to the long-leaved box, *Buxus longifolia*, a hardy evergreen tree found in the mountainous regions of the northern part of Palestine, the Galilean hills, and more particularly in Lebanon. It grows to about 20 feet tall, with a stem, always comparatively slender, seldom more than 6 or 8 inches in diameter (268). The wood is very hard and takes a fine polish. It is valued for all purposes where hardness is required, such as wood carving, turnery, and the manufacture of combs, spoons, and mathematical instruments. The blocks for wood-engraving used to be made wholly of box-wood. Writing tablets were also made of this wood (306), and GOODSPEED now renders the "boxtrees" of II Esdras 14:24 as "writing tablets", saying "But you must prepare for yourselves many *writing tablets*." It was cultivated by the Romans for its hard wood, which they inlaid with ivory for cabinet work and jewel caskets. This is all the more reason for supposing that it, and nothing else, was the wood inlaid with ivory to which EZEKIEL referred. In the Revised Version it is translated "ashur wood" and in the Vulgate "buxus". Dr. EVENARI writes us that this tree is not now found in Palestine or Galilee. Since it has been used so extensively for thousands of years in religious festivities, it seems safe to assume that it has actually been exterminated from many places in this land where once it certainly grew.

SHAW (293) considered both the Isaiah references to apply to the closely related European *Buxus sempervirens* L., which by some botanists is regarded as a polymorphic species including the western Asiatic and north African forms of box. POST, however, considers *B. longifolia* as a distinct species and the only one in the Holy Land region, and in this opinion we concur.

Both the JASTROW and GOODSPEED versions are consistent in rendering "teasshûr" as "larch" also in the Isaiah references, but MOFFATT renders it "cypresses" in those passages. It would appear that these renditions of "larch" and "cypress" are based on W. SMITH's "Dictionary of the Bible" (300) where it was maintained that the tree in question "is properly a species of cedar called 'scherbin' (or 'sherbin'), to be recognized by the small size of its cones, and the inward tendency of its branches." This plant, according to TRISTRAM (184), however, is the Phoenician cedar, *Sabina phoenicia* (which see). The Talmudic and other Jewish writers generally are of the opinion that the box-tree is the tree referred to in Isaiah and Ezekiel, while the Syriac writers and the Arabic version of SAADIAS incline toward the cedar. WILLIAM SMITH, in the 1876 edition of his "Dictionary", has summed up the matter quite well when he says: "Although the claim which the box-tree has to represent the *teasshûr* of Isaiah and Ezekiel is far from being satisfactorily established, yet the evidence rests on a better foundation than that which supports the claims of the *sherbin*" (306).

In regard to the first part of Isaiah 41:19 there is also some argument. Here the word "shittah", usually regarded as referring to *Acacia seyal* (which see) and so translated by MOFFATT, JASTROW, and GOODSPEED, is thought by some to refer, instead, to the box tree. The verse states that myrtles, olives, firs, planetrees, and cypresses—and the "shittah" tree—will be made to grow in desert places. The meaning seems clear enough, that trees of rich and fertile soils would be made to grow in what were then dry and sterile wastelands. But the acacia is *normally* a tree of the deserts, so its inclusion here detracts from the force of the prophecy. Probably sensing this incongruity, the writers of the Septuagint translated the word "shittah" as "box". This seems logical, although Moffatt has eliminated the box entirely from the verse, retaining "acacia" for "shittah" and saying "cypress" for "teasshûr".

Box leaves somewhat resemble those of the myrtle and bay and because of this fact the box was regarded with considerable apprehension by the ancient Greeks and Romans, for they feared that if box leaves were used by mistake for those of myrtle or bay in the ceremonial rites of VENUS, "that goddess would revenge herself by destroying their virility" (298). Leafy branches of the box were used by the Jews at their thanksgiving Feast of the Tabernacles "and to this practice of symbolizing or conventionalizing the lodges in the wilderness with a green bough may be due that of masking English fireplaces at Whitsuntide with foliage" (298) and American mantels at Thanksgiving time. In many lands even today the box is commonly planted in cemeteries (126), and until rather recently it was the custom in England to cast a sprig of box into the open grave at the time of burial, much as the bereaved relatives now cast a flower on the coffin in America.

54. Capparis sicula Duham.
(FIGURE 51)

ECCLESIASTES 12: 5—Also when they shall be afraid of that which is high, and fears shall be in the way, and the almond tree shall flourish, and the grasshopper shall be a burden, and *desire* shall fail: because man goeth to his long home, and the mourners go about the streets.

This verse is not very intelligible as rendered by the King James version. In MOFFATT's translation the meaning becomes more plain: "when old age fears a height, and even a walk has its terrors, when his hair is almond white, and he drags his limbs along, and the *spirit* flags and fades. So man goes to his long, long home, and mourners pass along the street." The Septuagint and Vulgate both translate the word "desire" of the Authorized Version ("spirit" of MOFFATT) as "caper", the GOODSPEED version is "and the *caper-berry* is ineffectual", the JASTROW version is "and the *caper-berry* shall fail", while the Douay renders it "and the *caper tree* shall be destroyed." The Hebrew word used is "tapher". The explanation for this seemingly amazing series of translations may be summed up as follows.

The common caper or caper-berry, *Capparis sicula* [26], grows in profusion in Syria, Lebanon, and Palestine, on the hills around Jerusalem (268), in the wilderness of Judea, at Nazareth, and in the mountain valleys of Sinai (266). Its flowers are large and white, beautifully set off by large stamens with purple filaments. The plant may sometimes grow upright, but more generally spreads itself weakly over the ground like a vine, with decumbent or pendulous branches. It covers rocks, ruins, and old walls like ivy. The young flower-buds, pickled in vinegar, were used by the ancients, as they are by us, as a condiment for meat. Its berries are also prized in cookery. The peculiar suitability of this plant to a description of man's old age is owing both to the structure of the caper-fruit and to its stimulating nature, exciting, as it does, both hunger and thirst and thus strengthening the appetite which in old age has a tendency to become sluggish. Therefore commentators feel that the passage rendered "and the caper shall fail" by the Revised Version is intended to indicate that even the stimulating effects of the caper are unable to excite the appetite of the old man any longer. The fruit hangs on long stalks and the over-ripe berry is thought to be suggestive also to the Oriental mind, with its fondness for symbolism, of the drooping head of the old veteran bowed low by the weight of many years on his shoulders, who has just about reached the end of his days and must momentarily expect to fall into his grave.

The caper plant has also been advanced by a great many authorities for a large series of Biblical references, including Exodus 12: 22, Leviticus 14: 6, Numbers 19: 18, I Kings 4: 33, Psalms 51: 7, John 19: 29, and Hebrews 9: 19. In all of these references it is the "hyssop" of the Authorized Version which is thought to be the caper. SHAW (293) definitely regards the caper as the "hyssop" of Scriptures, and even HORNER (178), as late as 1931, thinks that all the requirements of the numerous Biblical references to "hyssop" are filled by *Capparis*. There have, however, been considerable arguments on this subject. It is doubtful whether there is any plant in the Bible over which there has been more argument and discussion and difference of opinion than the "hyssop". CELSIUS devotes 42 pages to the subject—and still arrives at no conclusion! TRISTRAM (326), JONES, and PRATT (268) favor the caper as the "hyssop" because it is found in lower Egypt, where MOSES commanded the

---

[26] Mostly referred to as *C. spinosa* L. by early writers, but this species, according to POST does not occur in the Holy Land. EVENARI regards it as a variety of *C. spinosa* for which he adopts the name *C. spinosa* var. *canescens* Coss. He believes that the connection between the caper and old age is based on the supposed aphrodisiac effect of the plant.

children of Israel to use it for sprinkling the blood of the sacrificial lamb onto the lintel and door-posts of their houses so that the destroying angel would pass them by. It is also found on the desert of Sinai, where it was to be used in the ceremonial cleansing of the leper. It is found abundantly on the ruined walls of Jerusalem where SOLOMON spoke of seeing the "hyssop that springeth out of the wall". It has always been considered to possess the implied cleansing properties, and has long been used in the cure of a disease closely allied to leprosy. The bush in warm climates would furnish a stick long enough to support the sponge offered to JESUS on the cross; "while the fact that the caper buds, and indeed every part of the caper plant, were in ancient times preserved in vinegar, would explain the presence on the spot of a vessel filled with vinegar." This makes a very strong case for the caper plant, it must be admitted, but the subject is discussed further by us under *Origanum maru* (which see).

The Greek word κάππαρις is translated in dictionaries both as "caper" and "desire", doubtless, as EVENARI has pointed out, because of the supposed aphrodisiac effect of this plant, and this explains the interchange of these translations in the various Biblical texts.

55. Cedrus libani Loud.
(FIGURE 1; PAGE 64; FIGURE 52)

NUMBERS 24: 6— . . . as *cedar trees* beside the waters.
JUDGES 9: 15— . . . let fire come out of the bramble, and devour the *cedars* of Lebanon.
II SAMUEL 5: 11—And HIRAM king of Tyre sent messengers to DAVID, and *cedar trees*, and carpenters, and masons: and they built DAVID an house.
II SAMUEL 7: 2— . . . See now, I dwell in an house of *cedar* . . .
I KINGS 4: 33—And he spake of trees from the *cedar tree* that is in Lebanon even unto the hyssop that springeth out of the wall: he spake also of beasts, and of fowl, and of creeping things, and of fishes.
I KINGS 5: 6-10—Now therefore command thou that they hew me *cedar trees* out of Lebanon; and my servants shall be with thy servants: and unto thee will I give hire for thy servants according to all that thou shalt appoint: for thou knowest that there is not among us any that can skill to hew timber like unto the Sidonians . . . And HIRAM sent to SOLOMON, saying, I have considered the things which thou sentest to me for: and I will do all thy desire concerning timber of *cedar*, and concerning timber of fir. My servants shall bring them down from Lebanon unto the sea: and I will convey them by sea in floats unto the place that thou shalt appoint me, and I will cause them to be discharged there, and thou shalt receive them: and thou shalt accomplish my desire, in giving food for my household. So HIRAM gave SOLOMON *cedar trees* and fir trees according to all his desire.
I KINGS 6: 9, 15-16, 18, & 36—So he built the house and finished it; and covered the house with beams and boards of *cedar*. And then he built chambers against all the house, five cubits high: and they rested on the house with timber of *cedar* . . . And he built the walls of the house within with boards of *cedar*, both the floor of the house, and the walls of the cieling: and he covered them on the inside with wood, and covered the floor of the house with planks of fir. And he built twenty cubits on the sides of the house, both the floor and the walls with boards of *cedar* . . . And the *cedar* of the house within was carved with knops and open flowers: all was *cedar*; there was no stone seen . . . And he built the inner court with three rows of hewed stone, and a row of *cedar* beams.
I KINGS 7: 2-3, 7, & 11-12—He built also the house of the forest of Lebanon; the length thereof was an hundred cubits, and the breadth thereof fifty cubits, and the height thereof thirty cubits, upon four rows of *cedar* pillars, with *cedar* beams upon the pillars. And it was covered with *cedar* above upon the beams, that lay on forty five pillars, fifteen in a row . . . Then he made a porch for the throne where he might judge, even the porch of judgment: and it was covered with *cedar* from one side of the floor to the other . . . And above were costly stones, after the measures of hewed stones, and *cedars*. And the great court round about was with three rows of hewed stones, and a row of *cedar* beams, both for the inner court of the house of the Lord, and for the porch of the house.
I KINGS 9: 11—Now HIRAM the king of Tyre had furnished SOLOMON with *cedar* trees . . .
II KINGS 10: 27—And the king made silver to be in Jerusalem as stones, and *cedars* made he to be as the sycomore trees that are in the vale, for abundance.
II KINGS 14: 9— . . . The thistle that was in Lebanon sent to the *cedar* that was in Lebanon . . .
II KINGS 19: 23— . . . With the multitude of my chariots I am come up to the height of the mountains, to the sides of Lebanon, and will cut down the tall *cedar trees* thereof . . .
I CHRONICLES 14: 1—Now HIRAM king of Tyre sent messengers to DAVID, and timber of *cedars*, with masons and carpenters, to build him an house.
I CHRONICLES 17: 1— . . . Lo, I dwell in an house of *cedars*.
I CHRONICLES 22: 3-4—And DAVID prepared iron in abundance for the nails for the doors of the gates, and for the joinings; and brass in abundance without weight; Also *cedar trees* in abundance: for the Zidonians and they of Tyre brought much *cedar wood* to DAVID.

II CHRONICLES 1: 15— . . . and *cedar trees* made he as the sycomore trees that are in the vale for abundance.
II CHRONICLES 2: 3 & 8—And SOLOMON sent to HURAM the king of Tyre, saying, As thou didst deal with DAVID my father, and didst send him *cedars* to build him an house to dwell therein, even so deal with me . . . Send me also *cedar trees,* fir trees, and algum trees, out of Lebanon . . .
II CHRONICLES 9: 27—And the king made silver in Jerusalem as stones, and *cedar trees* made he as the sycomore trees that are in the low plains in abundance.
II CHRONICLES 25: 18— . . . The thistle that was in Lebanon sent to the *cedar* that was in Lebanon . . .
EZRA 3: 7—They gave money also unto . . . them of Tyre, to bring *cedar trees* from Lebanon to the sea of Joppa, according to the grant that they had of CYRUS king of Persia.
EZRA 6: 4—With three rows of great stones, and a row of new *timber* . . .
NEHEMIAH 2: 8—And a letter unto ASAPH the keeper of the king's *forest,* that he may give me *timber* to make *beams* for the gates of the palace . . .
JOB 40: 17—He moveth his tail like a *cedar* . . .
PSALMS 29: 5—The voice of the Lord breaketh the *cedars*; yea, the Lord breaketh the *cedars* of Lebanon.
PSALMS 80:10— . . . and the boughs thereof were like the goodly *cedars.*
PSALMS 92: 12—The righteous shall flourish like the palm tree: he shall grow like a *cedar* of Lebanon.
PSALMS 104: 16—The trees of the Lord are full of sap; the *cedars* of Lebanon, which he hath planted.
PSALMS 148: 9—Mountains, and all hills; fruitful trees, and all *cedars.*
SONG 1: 17—The beams of our house are *cedar,* and our rafters of fir.
SONG 3: 9—King SOLOMON made himself a chariot of the *wood of Lebanon.*
SONG 5: 15— . . . His countenance is as Lebanon, excellent as the *cedars.*
SONG 8: 9— . . . and if she be a door, we will inclose her with boards of *cedar.*
ISAIAH 2: 13—And upon all the *cedars* of Lebanon . . .
ISAIAH 9: 10— . . . the sycomores are cut down, but we will change them into *cedars.*
ISAIAH 14: 8—Yea, the fir trees rejoice at thee, and the *cedars* of Lebanon, saying, Since thou art laid down, no feller is come up against us.
ISAIAH 37: 24— . . . By the multitude of my chariots am I come up to the height of the mountains, to the sides of Lebanon; and I will cut down the tall *cedars* thereof, and the choice fir trees thereof: and I will enter into the height of his border, and the forest of his Carmel.
ISAIAH 41: 19—I will plant in the wilderness the *cedar,* the shittah tree, and the myrtle, and the oil-tree . . .
ISAIAH 44: 14—He heweth him down *cedars* . . .
JEREMIAH 22: 7, 14-15, & 23— . . . and they shall cut down thy choice *cedars,* and cast them into the fire . . . That saith, I will build me a wide house and large chambers, and cutteth him out windows; and it is cieled with *cedar,* and painted with vermilion. Shalt thou reign, because thou closest thyself in *cedar?* . . . O inhabitants of Lebanon, that makest thy nest in the *cedars* . . .
EZEKIEL 17: 3 & 22-24— . . . A great eagle with great wings, long-winged, full of feathers, which had divers colours, came unto Lebanon, and took the highest branch of the *cedar*: He cropped off the top of his young twigs, and carried it into a land of traffick; he set it in a city of merchants . . . Thus saith the Lord God: I will also take of the highest branch of the high *cedar,* and will set it; I will crop off from the top of his young twigs a tender one, and will plant it upon an high mountain and eminent: In the mountain of the height of Israel will I plant it: and it shall bring forth boughs, and bear fruit, and be a goodly *cedar*: and under it shall dwell all fowl of every wing; in the shadow of the branches thereof shall they dwell. And all the trees of the field shall know that I the Lord have brought down the high tree, have exalted the low tree, have dried up the green tree, and have made the dry tree to flourish: I the Lord have spoken and have done it.
EZEKIEL 27: 24—These were thy merchants in all sorts of things . . . made of *cedar* . . .
EZEKIEL 31: 3-18—Behold the Assyrian was a *cedar* in Lebanon with fair branches, and with a shadowing shroud, and of an high stature; and his top was among the thick boughs. The waters made him great, the deep set him up on high with her rivers running round about his plants, and sent out her little rivers unto all the trees of the field. Therefore his height was exalted above all the trees of the field, and his boughs were multiplied, and his branches became long because of the multitude of waters, when he shot forth. All the fowls of heaven made their nests in his boughs, and under his branches did all the beasts of the field bring forth their young, and under his shadow dwelt all great nations. Thus was he fair in his greatness, in the length of his branches: for his root was by great waters. The cedars in the garden of God could not hide him: the fir trees were not like his boughs, and the chestnut trees were not like his branches; nor any tree in the garden of God was like unto him in his beauty. I have made him fair by the multitude of his branches: so that all the trees of Eden, that were in the garden of God, envied him. Therefore thus saith the Lord God; Because thou hast lifted up thyself in height, and he hath shot up his top among the thick boughs, and his heart is lifted up in his height; I have therefore delivered him into the hand of the mighty one of the heathen; he shall surely deal with him: I have driven him out for his wickedness. And strangers, the terrible of the nations, have cut him and off, have left him: upon the mountains and in all the valleys his branches are fallen, and his boughs are broken by all the rivers of the land; and all the people of the earth are gone down from his shadow,

and have left him. Upon his ruin shall all the fowls of the heaven remain, and all the beasts of the field shall be upon his branches: To the end that none of all the trees by the waters exalt themselves for their height, neither shoot up their top among the thick boughs, neither their trees stand up in their height, all that drink water: for they are all delivered unto death, to the nether parts of the earth, in the midst of the children of men, with them that go down to the pit. Thus saith the Lord God; In the day when he went down to the grave I caused a mourning: I covered the deep for him, and I restrained the floods thereof, and the great waters were stayed: and I caused Lebanon to mourn for him, and all the trees of the field fainted for him. I made the nations to shake at the sound of his fall, when I cast him down to hell with them that descend into the pit: and all the trees of Eden, the choice and best of Lebanon, all that drink water, shall be comforted in the nether parts of the earth. They also went down into hell with him unto them that be slain with the sword; and they that were his arm, that dwelt under his shadow in the midst of the heathen. To whom art thou thus like in glory and in greatness among the trees of Eden? yet shalt thou be brought down with the trees of Eden unto the nether parts of the earth.

AMOS 2: 9—Yet destroyed I the Amorite before them, whose height was like the height of the *cedars*, and he was strong as the oaks.

ZEPHANIAH 2: 14— . . . for he shall uncover the *cedar* work.

ZECHARIAH 11: 1-2—Open thy doors, O Lebanon, that the fire may devour thy *cedars*. Howl, fir trees; for the *cedar* is fallen; because the mighty are spoiled: howl, O ye oaks of Bashan; for the forest of the vintage is come down.

With only the few exceptions noted below, chiefly from the Pentateuch, there is no doubt about the identity of the Hebrew word "erez" or "ahrahzim" (Greek, κέδρος), which is rendered "cedar" in the Authorized and all other versions of the Bible. The word is derived from an old Arabic root signifying a "firmly rooted and strong tree" (306). This tree is the famous cedar-of-Lebanon, *Cedrus libani*[27], although some doubt has been expressed about the "cedar" of Numbers 24: 6, where the word may be a corruption and actually refer to some other tree, as is the case in Leviticus 14: 4, 6-8, & 49-52 and Numbers 19: 6 (see under *Sabina phoenicia*) and Ezekiel 27: 5 & 31: 8 (see under *Pinus halepensis*).

The cedars-of-Lebanon are noble trees, the noblest, tallest, and most massive with which the Israelites were acquainted (101, 146, 178, 268). They grow quite rapidly, attain a height of up to 120 feet and a trunk diameter of up to 8 feet, filling all who see them now with awe and reverence even as in Biblical days. When still young the trees are more or less pyramidal in shape, but with age the top flattens out and the horizontal branches become large and wide-spreading, imparting a most distinctive and beautiful appearance. In SOLOMON's day these trees were obviously abundant on the mountains of Lebanon, but now, because of excessive lumbering (even as so dramatically described in about 594-588 B.C. by the prophet EZEKIEL) they are very rare. In fact, according to SMITH (306) they are now limited to one valley of the Lebanon range, that of the Kedisha River, which flows from the highest point in the range westward to Tripoli on the Mediterranean. "The grove is at the very upper part of the valley, about 15 miles from the sea, 6,500 feet above that level, and their position is moreover above that of all other arboreous vegetation. The valley here is very broad, open, and shallow, and the grove forms a mere speck on its flat floor. On nearer inspection, the cedars are found to be confined to a small portion of a range of low stony hills of rounded outlines, and perhaps 60 or 100 ft. above the plain, which sweep across the valley. These hills are believed by Dr. HOOKER to be old moraines, deposited by glaciers that once debouched on to the plain from the surrounding tops of Lebanon."

The tree was held in high esteem not only for its vigor, beauty, and age, but also for the fragrance and the remarkable lasting qualities of the wood. It was employed to symbolize grandeur, might, majesty, dignity, lofty stature, and wide expansion (126). "It was the prince of trees, to the plant world what the lion was to the animal world." The two lengthy Ezekiel references illustrate beautifully how these lofty kings of the forest were used by prophet-orators to symbolize and typify worldly strength, power, and glory. Thus one obtains a fair idea of the crowning insolence of SENNACHERIB, the invader, when he boasted in 700 B.C.: "I am come up to the height of the mountains, to the sides of Lebanon; and I will cut down the tall *cedars* thereof."

---

[27]Also referred to as *Pinus cedrus* L., *P. libanotica* Link, and *Cedrus libanotica* Link.

TRISTRAM says (95) of the cedar: "from it extended downwards the botanical knowledge of Solomon... Of all monstrous presumption the most outrageous was the proposal of the thistle to ally itself with the cedar [II Kings 14: 9] ... Every one who has seen these noble trees recognises the force of the majestic imagery of the Prophets. With their gnarled and contorted stems, and scaly bark, with their massive branches,—spreading their foliage rather in layers than in flakes,— with their dark green leaves shot with silver in the sunlight, as they stand a lovely group in the stupendous mountain amphitheatre—they assert their title to be the monarchs of the forest." WARBURTON, in his "Crescent and the Cross", tells of a cedar tree whose trunk was 45 feet in circumference. BURCKHARDT tells of a spot where he saw twelve very ancient trees, called "The Saints" by the Arabs, which had 4, 5, and even 7 gigantic trunks "springing from the same base", bearing, like our American sequoias, leaves only at their very tops.

The many lengthy quotations from the books of Chronicles, Kings, and Samuel show clearly how the wood of the cedar-of-Lebanon was preferred above all other wood for building purposes. This was because of its tremendous size, its remarkable durability, and its delightful fragrance. AMOS compared the mighty sons of ANAK, the Amorite, to cedars-of-Lebanon. King SOLOMON, HIRAM, CYRUS, the kings of Assyria, and the rulers of all neighboring countries of those ancient days plundered the mountain heights of Lebanon for this wood to embellish their palaces. Planks and perhaps also masts for the great fleet maintained at Tyre were made of this wood, as were also SOLOMON'S palanquins or sedans ("chariots"). The magnificent Temple of Solomon, begun by DAVID and completed by SOLOMON, was constructed of this wood. It required seven years of back-breaking toil to complete this temple and 13 more years to build SOLOMON'S private house, the house of the Forest of Lebanon, the great Porch of Judgment, and the house for Pharoah's daughter, SOLOMON's favorite wife. SOLOMON conscripted 30,000 Israelites, who were sent to Lebanon in shifts, to aid King HIRAM and his slaves in the shameful destruction. He also sent 150,000 laborers—captive slaves from past wars—and 3,300 officers—a veritable army, indeed, going out to war against the trees of God! King HIRAM, of Tyre, even had to supply thousands of his own men to assist in the felling of these trees. And from the days of King SOLOMON even to the present day— some 3,000 years—unscrupulous, unthinking, unappreciative, ungrateful, selfish men have waged war against the earth's green forests, seemingly never content to leave these magnificent examples of the handiwork of a beneficent Nature, seemingly ever eager to "clear" more land for "civilization", to fell more trees, to reduce more acres to pitiful ruins of their former glory—all to satisfy their insatiable greed. But Nature takes her revenge, even as foretold by the prophets of old—revenge and retribution in the form of vanishing wild life and game, an upset balance of Nature leading to all manner of unexpected complications, erosion that cuts up a slope into Martian landscapes, gullies that carry off the valuable soil that has been "cleared", floods that wreak havoc and destruction in their wake, dust storms that lift off the precious topsoil, dust-bowls that drive out presumptuous man in mass migrations, and finally deserts wherein none can live. This, to a large extent, has been the history of the Holy Land, which has been transformed from a land of palms, "flowing with milk and honey" to its present inhospitable desert condition. How ironical it is that this plundering of the forests was begun by men who wanted to build a temple to God, a place where men might worship their Creator, when, as WILLIAM CULLEN BRYANT has so beautifully put it:

> "The groves were God's first temples,
> Ere man learned
> To hew the shaft, and lay the architrave,
> And spread the roof above them—ere he framed
> The lofty vault, to gather and roll back
> The sound of anthem; in the darkling wood
> Amid the cool and silence, he knelt down,
> And offered to the Mightiest solemn thanks
> And supplication. For his simple heart

> Might not resist the sacred influences
> Which, from the stilly twilight of the place,
> And from the gray old trunks that high in heaven
> Mingled their mossy boughs, and from the sound
> Of the invisible breath that swayed at once
> All their green tops, stole over him, and bowed
> His spirit with the thought of boundless power
> And inaccessible majesty. *Ah, why
> Should we, in the world's riper years, neglect
> God's ancient sanctuaries, and adore
> Only among the crowd, and under roofs
> That our frail hands have raised?"*

The "timber" referred to in Ezra 6:4 and II Chronicles 22:14 is often cited as referring to cedar-of-Lebanon, and, from the context, it would appear that it probably does, at least in part. The "Lebanon" of Hosea 14:9, however, also often cited as referring to cedar, is now regarded as applying to *Populus alba* (which see). The "cedars" of Ezekiel 31:8 probably were not the cedars-of-Lebanon, either, as we can tell from the context. This passage will be discussed under *Pinus halepensis* (which see).

According to legend, when ADAM was dying he sent his son, SETH, to the garden of Eden to beg the angel there on guard for a little of the precious juice of the "tree of life". Instead, the angel gave him a small slip from the tree which was later planted on ADAM's grave. This slip eventually grew up into a tree with three branches, one being cypress, one cedar, and one olive. Another version of the legend has it that three seeds were given to SETH, from which a cypress, a cedar, and an olive tree grew. From this tree, or trees, the cross was made 5000 years later on which JESUS was crucified—the upright beam of cedar, the cross arm of cypress, and the title of olive (95, 336a).

Another legend tells us that when the Queen of Sheba visited SOLOMON she was asked to cross a marshy piece of ground by means of a bridge built of cedar wood. As she was about to step on the bridge she suddenly refused to do so, stating that she would not tread on the wood which she had seen in a dream would someday bear the crucified body of the Christ. The Jews became very indignant at this suggestion of a degrading death in store for their Messiah and had to be restrained from doing the queen harm because of her words (95).

Still another legend tells of an angel who took refuge under a massive cedar tree during a fearful storm. After the storm had abated he prayed to God that this tree whose wood was so fragrant and whose shade so refreshing might also in the future bear some fruit of benefit to the human race. This fruit was the sacred body of JESUS (95). Being regarded as a tree of good fortune, its wood has always been much used in the making of sacred icons. Because of its lasting qualities the ancient Egyptians employed it for making mummy-cases. It was also used to keep insects out of the ancient tombs, and to this use of it may be traced the ancient name, literally translated "life from the dead", which it bore then. Figures carved from this wood, still in excellent condition, have been unearthed in ancient tombs.

Cedars in China are called "the trees of the faithful loves" because of a legend about a king who cast a good man into prison in order that the man's beautiful wife might be available to the king. The imprisoned husband died of grief and his wife killed herself, but, although their bodies were buried far apart from each other at the king's express command, cedar trees grew from each grave, attained a vast height, and lovingly interlaced their branches and roots (105).

56. Centaurea calcitrapa L.   57. Centaurea iberica Trev.
58. Centaurea verutum L.   59. Silybum marianum (L.) Gaertn.
(FIGURE 87; FIGURE 92)

GENESIS 3:17-18—And unto ADAM he said, Because thou hast hearkened unto the voice of thy wife, and hast eaten of the tree, of which I commanded thee, saying, Thou shalt not eat of it: cursed is the ground for thy sake; in sorrow shalt thou eat of it all the days of thy life; Thorns also and *thistles* shall it bring forth to thee . . .

II KINGS 14: 9— ... The *thistle* that was in Lebanon sent unto the cedar that was in Lebanon, saying, Give thy daughter to my son to wife: and there passed by a wild beast that was in Lebanon, and trode down the *thistle*.
II CHRONICLES 25: 18— ... The *thistle* that was in Lebanon sent to the cedar that was in Lebanon, saying, Give thy daughter to my son to wife: and there passed by a wild beast that was in Lebanon, and trode down the *thistle*.
HOSEA 10: 8— ... the thorn and the *thistle* shall come up on their altars; and they shall say to the mountains, Cover us; and to the hills, Fall on us.
MATTHEW 7: 16—Ye shall know them by their fruits. Do men gather grapes of thorns, or figs of *thistles*?
MATTHEW 13: 7—And some fell among *thorns*; and the *thorns* sprung up, and choked them.
HEBREWS 6: 8—But that which beareth *thorns* and briers is rejected, and is nigh unto cursing ...

It is most probable that both the Hebrew word "dardar"[28] used in the Old Testament passages cited above, and the Greek word τρίβολος, meaning "a prickly (water) plant", used in the New Testament, refer to thistles, as, indeed, has been generally supposed by translators. Some 125 kinds of thistles now grow in the Holy Land (267) [29], lending point to the prophecies in Genesis and Hosea. Of these the commonest and the ones usually cited for these Biblical passages are the true star-thistle, *Centaurea calcitrapa*[30], dwarf centaury, *C. verutum*[31], Iberian centaury, *C. iberica*, and lady's-thistle, *Silybum marianum*[32]. Dr. EVENARI agrees that these references are generic in character, to thistles in general, whose most common species in Palestine is *C. iberica*. This plant is also recorded from this general area by HART (159) and AARONSOHN (18). *C. calcitrapa* is similarly reported by AARONSOHN (18), DINSMORE (100), and PAINE. As alternate possibilities Dr. EVENARI lists *Silybum marianum* and *Notobasis syriaca*. Some writers, apparently misled by the Greek word employed, have supposed that the small caltrop, Malta-cross, or bur-nut, *Tribulus terrestris* L., was the plant referred to. The use of the word "tribulus" in the Septuagint was merely an effort to translate the Hebrew word "dardar" and means literally only "to tear". It should, therefore, not be thought to refer in any way to the plants which were much later given the generic name *Tribulus* by TOURNEFORT (299). Caltrops have spiny fruits, but totally unarmed stems and foliage. The three species known from the Holy Land are common along roadsides and in fields throughout the area, forming especially a very distinctive feature along the shores of the Mediterranean and the Dead Sea. They do not, however, fit the context of our passages nearly as well as true thistles do, and may well be eliminated.

Thistles are now very abundant in Palestine (268), some even attaining a height of 5 or 6 feet. Between Nazareth and Tiberias CLARKE was amazed to find the earth covered with immense tracts of these plants. LINDSAY reports that the thistles of Palestine rival those of his native Scotland. CARRUTHERS states that "thistles of gigantic size, overtopping the horse and his rider, abound in the rich plains of Gennesaret, Sharon, Esdraëlon, and Jericho" (184).

In the famous II Kings 14: 9 passage LEESSER substitutes "thornbush" for "thistle" (9). In Job 31: 40 the Authorized Version says: "Let *thistles* grow instead of wheat, and cockle instead of barley", the Douay version agrees with this rendition, and the LEESSER uses "thorns". The Hebrew word there employed, however, is an entirely different one, "choach", although also rendered "tribulus" in the Greek versions. It is now supposed that the "thistles" of Job were not the ones listed at the head of this chapter, but were *Notobasis syriaca* (which see), more common in JOB's country. DALMAN is of the opinion that the plant mentioned in Genesis 3: 18 and Hosea 10: 8 was probably *Carthamus glaucus* Bieb. (98).

According to Greek mythology the earth goddess, TELLUS MATER, created the thistle as an expression of her grief at the loss of DAPHNIS, the pain inflicted by the spines being symbolic of the pain in her heart at his loss. In Teutonic

---

[28]This term should not be confused with the similar Arabic word which is the name for *Fraxinus oxycarpa* Willd., nor with "durdar" which is the name for *Centaurea pallescens* Del. (98).
[29]Today *Calycotome villosa* (Poir.) Link is the most common Palestinian thorny plant (98).
[30]Referred to as "the genus *Calcitropa*" by CARRUTHERS (184, page 87).
[31]Also referred to as "*Centaurea venustum*" by SMITH (299, page 89).
[32]Also known as *Carduus marianus* L.

legendry it was called the "lightning plant" and was associated with THOR, the thunder god, who protected all who wore it (126). The Emperor CHARLEMAGNE once found his army threatened with destruction by an epidemic of bubonic plague. He was informed by an angel that if he would shoot an arrow from his crossbow into the air, it would fall upon a plant that would cure the disease. The arrow fell on a thistle of the species since known as blessed thistle, holy thistle, or Our Lady's thistle, *Cnicus benedictus* L. This plant became the emblem of the Order of the Thistle, founded in France in the 14th century, and, according to at least some historians, of an Order of the Thistle created by ARCHIUS, King of the Scots, in the 10th century (298).

The thistle became the national emblem of Scotland when, during the Danish wars, an invading Danish army attempted to surprise a Scottish encampment by advancing upon it barefoot during the night. A Danish soldier, stepping accidentally upon a thistle, let out such a howl of pain that it wakened the Scotsmen who promptly repelled the attack. The "guardian thistle" has since been Scotland's symbol, with the appropriate motto "Nemo me impune lacessit" (126).

Herbalists have ascribed many virtues to the thistle, including the cure of ague and jaundice (125). Soaked in wine, it "expels superfluous melancholy out of the body and makes a man merry as a cricket" (298). After removal of its thorns it is said to make a nutritious vegetable, pot herb, and blood purifier.

Although the King James and LAMSA versions use the expression "thorns and briers" in Hebrews 6: 8, MOFFATT, GOODSPEED, JASTROW, O'HARA, and WEYMOUTH all say "thorns and thistles". The Basic English rendition is "thorns and evil plants". All the modern translations continue to use the word "thistles" in the other references quoted.

60. Ceratonia siliqua L.
(FIGURE 56)

LUKE 15: 16—And he would fain have filled his belly with the *husks* that the swine did eat: and no man gave unto him.

There is no doubt whatever that the pods of the carob-tree or locust-tree, *Ceratonia siliqua*, were the "husks" of JESUS' parable of the prodigal son. The Greek word, κεράτια or κερατέα, is rendered "pods" by MOFFATT, O'HARA, WEYMOUTH, and GOODSPEED. The Basic English translation is "he would have been glad to take the pigs' *food*".

The carob is a fine, evergreen, leguminous tree, with a hemispheric top, reaching thirty feet in height and an approximately equal spread of branches. The trunk becomes a foot or more in diameter. The leaves are pinnately compound and similar to those of our ash, but the individual leaflets are coriaceous and glistening. The flowers are small, apetalous, and yellow or red—the staminate ones with a disagreeable odor—borne solitary or clustered along the branches of the previous year. The tree is very common throughout the Holy Land, Syria, and Egypt. The pods are most abundant in April and May, are flat, narrow, horn-shaped, 6 to 10 inches long, 1 to 1½ inches wide, containing numerous pea-like seeds embedded in an agreeably flavored, mucilaginous, saccharine pulp. This pulp is ground up extensively, with or without the pods, and used for making candies. The pods are also abundantly used now, as in Biblical days, for feeding cattle, horses, and pigs. In time of scarcity they are used for human consumption (125, 299) and perhaps even regularly so by the very poorest of the people (184). The pods are extremely sour when unripe, but when completely ripe are full of a sweet, dark-colored, very palatable, honey-like syrup (268). The Arabs eat them today with considerable relish and consider their flavor to be like that of manna. Early Greek writers mention them as being a source of food. HORACE and JUVENAL speak of the fruit of this tree as being the food of the poorest and most miserable of the natives (178). The Mussulmen mix carob pulp with licorice root, dried grapes, and other fruit, and make sherbets. The poor peasants of civil war-wracked Spain are happy to be able to eat carob "husks". As "St. John's bread", "carob bean", "algaroba", "algaroba bean", or "locust pod", the ripe pods may be purchased in some New York City markets.

The carob is frequently mentioned in the Talmud as a source of good food for domestic animals. The tree is grown quite extensively in almost all Mediterranean lands, and in Malta is cultivated in large numbers for its pods, which are exported to England to serve there as fodder for horses. The seeds of the carob were formerly employed as standards of weight, and are the source of the term "carat" (127, 158).

Many commentators believe that the "locusts" eaten by JOHN the Baptist (*cf.* Matthew 3: 4) were not actually the insects, locusts, but were the fruits of the carob. HENSLOW, for instance, says (172a): " 'Husks' were the carob, as the Greek word implies but also the fruit that the Baptist ate; the error appears to have arisen from a transcriber substituting the Hebrew G for the R in *cherev*, which thus turns the word from 'carob' to locust."

It is perhaps not generally known that our own familiar legend of RIP VAN WINKLE, immortalized by WASHINGTON IRVING, has its origin in a Talmudic tale and finds its counterpart in the folklore of most of the world, wherever Hebrew people have penetrated. In the Talmudic story a young rabbi once came upon an old man planting a carob seed along a roadside. Laughed at by the rabbi for planting a seed which would require at least thirty years to grow into a fruit-producing tree, by which time the planter would most surely be dead, the old man answered "I am not planting for myself. I have eaten carobs that other men have planted, so why may not I do the like for other men? The sons of my sons will eat of this and thank me." Soon afterwards the rabbi became tired and lay down in a woods to rest. He awoke 70 years later to find the carob tree matured and bearing fruit, and himself an aged man totally unfamiliar to the people about him (298).

61. Cercis siliquastrum L.
(FIGURE 2)

MATTHEW 27: 5—*And he cast down the pieces of silver in the temple, and departed, and went and hanged himself.*

Although there is no actual basis for its inclusion, no discussion of the plants of the Bible could be complete without at least a mention of the famous "Judas-tree", *Cercis siliquastrum*. This tree has long been associated with Bible legends and traditions. In the Matthew verse quoted above it is not stated that JUDAS hanged himself on a tree, or, if so, on what kind of a tree, but according to well-established legends it was on a tree of this species (101, 125, 299). The Judas-tree grows in southern Europe and western Asia and is regarded as a native of Palestine. It is a medium-sized tree to about 30 feet tall, with round or heart-shaped leaves and small clusters of purplish flowers which are borne on the old branches as well as on the young ones and may even appear on the trunk itself. Just how it was that tradition selected this tree of all Palestinian trees as the one on which JUDAS committed suicide, is not clear, but it was probably because of the color of the flowers and shape of the leaves. It has been known as "Judas-tree" for well over two hundred years. Tradition claims that the color of the flowers indicates how the tree burned with shame when JUDAS selected it.

Other legends state that it was a fig tree (*Ficus carica*), the poplar (*Populus euphratica*), the terebinth (*Pistacia terebinthus* var. *palaestina*), or even the black elder (*Sambucus nigra* L.) (125), which was used by JUDAS. The elder, however, although growing to be a tree to 18 feet tall, is not native to Palestine, occurring there only in cultivation and as a very recent escape. It is worthy of note in this connection that in Acts 1: 25 it is definitely implied that JUDAS did not die by hanging, the Matthew passage to the contrary notwithstanding.

62. Cichorium endivia L.   63. Cichorium intybus L.
64. Lactuca sativa L.   65. Nasturtium officinale R. Br.
66. Rumex acetosella var. multifidus (L.) P. DC. & Lam.
67. Taraxacum officinale Weber.

EXODUS 12: 8—And they shall eat the flesh in that night, roast with fire, and unleavened bread; and with *bitter herbs* they shall eat it.
NUMBERS 9: 11— . . . eat it with unleavened bread and *bitter herbs*.

Several commentators, including SMITH (299), are of the opinion that the "bitter herbs" of these two references—"merôrîm" or "m'rorim" in the original Hebrew (Greek, πικρίδες)—were not mints, as most other writers have assumed, but were plants like lettuce, endive, the young leaves of chicory, dandelion, and sorrel, eaten as a salad. Most of these are weedy plants of wide distribution, common in Egypt and western Asia, and so it is quite reasonable to suppose that one or more of them might have constituted the "bitter herbs" eaten at the time of the Passover. At the present day these and many other bitter-tasting herbs are eaten by the Arabs. PRATT (268) feels that the "bitter herbs" of these two references consisted only of endive, *Cichorium endivia*. However, most botanical authorities regard endive as native to India, and it is doubtful whether it had been introduced into Egypt and Palestine at the time of MOSES (about 1491 B.C.) or that trade in endive could have been established. While numerous products were undoubtedly imported from India in MOSES' day, they were products which would keep readily during the long journey from their homeland. Endive was not such a product. Yet ROSENMÜLLER observes that endive has all the oldest authorities in its favor. Most of the ancient Alexandrian Greek translations say "endives" in place of "bitter herbs". Dr. GEDDES, who definitely regards the endive as the "bitter herbs" of MOSES, remarks that the Jews of Alexandria, who translated the Pentateuch, could not have been ignorant of what herbs were eaten with the Paschal lamb in their day, at least. POST states that the endive is only rarely cultivated in the Holy Land today (267). It must, however, be remembered that the term "endive" is rather loosely applied today to several green herbs and may well have been similarly applied in those days to both the young leaves of the common chicory, *Cichorium intybus*, and those of the true Indian endive. Dr. EVENARI is of the opinion that neither the chicory nor the endive was known from Palestine. He says that *C. pumilum* Jacq. is the common species there now. Löw claims that the endive of the Passover feast was *C. intybus* (197). DALMAN regards both *C. endivia* and *C. intybus* as likely. He states that the latter is used as a salad and as a cooked vegetable; the source of supply being both cultivated and wild plants. He also suggests another wild chicory (*C. pumilum?*), which, he says, may prove to be conspecific with *C. intybus* (98).

The leaves of the ordinary garden lettuce, *Lactuca sativa*[33], as well as those of chicory and endive, are intensely bitter when unbleached. Mrs. BARNEVELD says (47) that the "Standard Bible Dictionary" identifies the "bitter herbs" of the Passover as endive, chicory, lettuce, and water-cress (*Nasturtium officinale*[34]). In certain European countries today horseradish, *Armoracia lapathifolia* Gilib., replaces the traditional herbs at the feast of the Passover. ABEN EZRA states that some bitter herbs were always eaten by the Jews with their food and that often in the East they still eat some with every mouthful of bread or meat, just as bitter gourds are now so constantly eaten with food in India. The ancient Egyptians used to place various kinds of green herbs upon the table, mixed with mustard, and then dipped morsels of their bread into this mixture while they ate. It is probable that the children of Israel derived their custom of eating bitter herbs with their meat and bread from the Egyptians. It seems most probable to us that all these herbs, including also the common dandelion,

---

[33] Also referred to as *Lactuca scariola* var. *sativa* (L.) Boiss.
[34] Also referred to as *Cardamine fontanum* Lam., *Sisymbrium nasturtium-aquaticum* L., *Radicula nasturtium* Druce, *Rorippa nasturtium-aquaticum* (L.) Schinz & Thell., and "*Nasturtium fontanum* (Lam.) Aschers."

Taraxacum officinale[35], and the sorrel, *Rumex acetosella* var. *multifidus*[36], were so employed, but not including the horseradish which could not have been known to the Hebrews at the time of MOSES. JONES maintains categorically that "Five bitter herbs were eaten with the Paschal lamb, *viz.*, lettuce, endive, chicory, mint, and one other not identified" (184).

The chicory has for centuries been regarded as an aphrodisiac, the seed of which if given by a lover to his mistress, or *vice versa*, would act as a love potion and increase his or her ardor.

A beautiful Algonquin legend concerning the dandelion tells of SHAWON-DESEE, the fat lazy south wind, who was resting on the greensward beneath some live oaks and magnolias when he observed a beautiful golden-haired maiden on the meadow near him. But he was too lazy to pursue her and after a few days in her place he saw a bent old woman with grizzled white hair. In his disappointment he heaved a tremendous sigh and was amazed to see her white hair become detached and fly away on the breeze. Other maidens like her come and grow old, but in the spring of the year the south wind still sighs for the lost beauty with the golden hair who might have been his had he but exerted himself a little more (298).

68. Cinnamomum cassia Blume

EXODUS 30: 23-24—Take thou also unto thee principal spices, of pure myrrh five hundred shekels, and of sweet cinnamon half so much, even two hundred and fifty shekels, and of sweet calamus two hundred and fifty shekels, and of *cassia* five hundred shekels.

EZEKIEL 27: 19—Dan also and Javan going to and fro occupied in thy fairs: bright iron, *cassia*, and calamus were in thy market.

There is little doubt that the "cassia" of Exodus and Ezekiel is the cassia-bark tree, *Cinnamomum cassia*[37]. The Hebrew word here employed is "kiddah" or "kiddad" (Greek, ιρις; literally "iris" and therefore a likely error in the translation into Greek from the original Hebrew; κασια is the Greek term for "cassia"). In Psalms 45: 8 the Authorized Version says "All thy garments smell of myrrh, and aloes, and *cassia*", but this is now almost uniformly regarded as a mis-translation. The Hebrew word employed in the Psalms passage is "ketziah", "ketzioth", or "k'tziot" (Greek, κασια) and it is herein discussed under *Saussurea lappa* (which see).

The cassia and various other Indian spices were apparently secured by MOSES and SOLOMON through trade, the cassia-bark probably from Ceylon. Cassia and cinnamon, although natives of Ceylon, the Malabar coast, Malaya, and the East Indies (not Arabia!), have been known in Palestine and probably also in Egypt since a very remote date (146). They are mentioned by HERODOTUS (about 484-425 B.C.) and DIOSCORIDES (about the 1st century A.D.) as Arabian products, but this was an error which many of the Greek writers made concerning plant products which merely reached them *via* the Arabian ports. GALEN (about 130-200 A.D.) reported that there was a kind of cassia in his day known as "mosyletis" or "mosyllos", imported from the ancient city and promontory of Mosyllon on the coast of Africa and the sea of Bab-el-mandeb. This throws light on Ezekiel 27: 19 which the Revised Version renders "Dan and Javan and Meuzal traded in thy markets with *cassia*, calamus". It seems probable that cassia from India and Ceylon was brought first to Meuzal or Mosyllon and thence exported to Tyre and other countries under the name of "meuzalites", "mosyletis", or "Meuzal cassia".

HERODOTUS describes an absolutely absurd method of collecting these spices (see under cinnamon). Actually the spices are secured from the inner bark of the tree and this is obtained by making longitudinal incisions in the branches. The bark then dries and peels off, rolling itself into tubes varying in diameter with the size of the branch and in quality with the thickness of

---

[35] Also called *Leontodon taraxacum* L., *Hedypnois taraxacum* (L.) Scop., *Leontodon vulgaris* Lam., *Taraxacum vulgare* Schrank, and *T. dens-leonis* Desf.

[36] Also referred to as *Rumex multifidus* L. and *R. acetoselloides* Bal.

[37] Also referred to as *Laurus cassia* (Blume) Nees, "*Kinnamomun cassia*" (268, page 43), and *Cassia cinnamomum*" (47).

the bark (299). Cassia bark has always been considered inferior to that of true cinnamon, being coarser and more pungent. Cassia was one of the ingredients of the holy oil, sold on the markets of Tyre. The coarser types of cassia bark are known today in the trade as "cassia lignea" and are often used to adulterate cinnamon. It is also known as "Chinese cinnamon" and "bastard cinnamon" (158). It has nothing whatever to do with the true laurel, *Laurus nobilis* (which see), or with the genus *Cassia* or the American *Sassafras* which SMITH mentions in this connection.

It is of interest to note that the Septuagint interpreted the "cassia" to be a flag, like the violet-scented European orris-root, which is obtained from some species of *Iris*, an utterly absurd interpretation. The Revised Version regarded it as the aromatic root of the "koost" or "costus" of Arabia, *Saussurea lappa* (which see). Modern opinion is that only the "cassia" of Psalms 45: 8 represents this Indian or "Arabian" orris or costus. The Douay version translates the "cassia" of Ezekiel 27: 19 as "stacte", but Douay and GOODSPEED both retain "cassia" in Psalms 45: 8. The JASTROW version uses "cassia" in both the Exodus and Ezekiel references, just as the King James does.

69. Cinnamomum zeylanicum Nees
(PAGE 257)

EXODUS 30: 23—Take thou also unto thee principal spices, of pure myrrh five hundred shekels, and of *sweet cinnamon* half so much, even two hundred and fifty shekels . . .
PROVERBS 7: 17—I have perfumed my bed with myrrh, aloes, and *cinnamon*.
SONG 4: 14—Spikenard and saffron; calamus and *cinnamon*, with all trees of frankincense . . .
REVELATION 18: 13—And *cinnamon*, and odours, and ointments, and frankincense . . .

There is no doubt about the identity of the "cinnamon" of the Bible. The Hebrew word here used is "kinamôn" or "kinnemôn" (Greek, κιννάμωμον) and unquestionably refers to *Cinnamomum zeylanicum*[38]. The tree is a rather low growing one, growing no more than 30 feet tall, with a smooth ash-colored bark and wide-spreading branches and white flowers. Its shiny, evergreen, lanceolate, beautifully veined leaves are 8 to 9 inches long and about 2 inches wide, bright-green above and white beneath. Its young shoots are crimson, with the bark often speckled with dark-green and orange spots. The cinnamon of commerce is the inner bark, which is peeled from trees 4 or 5 years old (in low valleys). In hilly country the trees must grow a few years longer before they can be thus stripped. The finest grade of cinnamon comes from the younger branches (146, 268, 293). Incisions are made lengthwise with a sharp knife on both sides of the branch. The bark is then removed in the form of a hollow cylinder, and these cylinder-sticks or "quills" are tied into bundles of about a pound each. Cinnamon is, thus, gathered in approximately the same way as cassia bark, but is far superior to it in quality. It is at present secured from various parts of India, Ceylon, Malaya, China, and the East Indies, the very best quality coming from the southwestern part of Ceylon. Oil of cinnamon is procured from the ripe fruit or else by soaking small broken pieces of the bark in sea water and then distilling them (184). The Hebrews always regarded cinnamon as a deliciously fragrant substance and valued it highly as a spice and as a perfume. It was one of the principal ingredients used in the manufacture of the precious ointments or "holy oil" which MOSES was commanded to use in the Tabernacle for anointing the sacred vessels and the persons of the officiating priests. It was undoubtedly very costly and precious (178).

The Basic English version of Revelation 18: 13 is "And sweet-smelling plants and perfumes", but all other translations use the word "cinnamon" in all the references in which the King James version uses it.

The cinnamon tree is called "korunda-guahah" in Ceylon. In Biblical times it was imported into Judea by the Phoenicians or by the Arabians, or both. Sir E. TENNANT believed that it "first reached India and Phoenicia overland by way of Persia from China, and that at a later period the cassia

---

[38]Also referred to as *Laurus cinnamomum* L., "*Laurus kinnamomum*" (254), and "*laurus cinnamonum*" (298, page 87).

of the Troglodytic coast supplanted the cinnamon of the Far East" (306).

Cinnamon leaves, like those of laurel, were woven into wreaths for decorating Roman temples. Greek writers state that in "Arabia" (where it actually does not grow!) only priests were allowed to collect it and they were required to place the first bundle of it on their altars for the sun god to ignite with a spark of divine fire; also they claimed that it grew regularly in valleys infested by venomous serpents so that the collectors had to gather it with hands and arms and legs well covered. This story (298) seems to be a sad hodgepodge of misinformation including material about Indian cinnamon, Arabian traders, and Persian Zoroastrian priests, and is fairly typical of the type of misinformation that has been recorded in the past concerning Bible plants.

70. Cistus creticus L.   71. Cistus salvifolius L.   72. Cistus villosus L.
(PAGE 122; FIGURE 53)

GENESIS 37: 25— ... A company of Ishmeelites came from Gilead with their camels, bearing spicery and balm and *myrrh*, going to carry it down to Egypt.
GENESIS 43: 11— ... Carry down the man a present, a little balm, and a little honey, spices, and *myrrh*, nuts, and almonds.

In these two references the Hebrew word used, "lôt" (Greek, στακτή), is wrongly translated "myrrh" in the Authorized Version. The word for the true myrrh is "môr", as is herein pointed out under *Commiphora myrrha* (which see). Myrrh is not indigenous to Gilead or Palestine as the context of these passages demands. For this reason MOFFATT renders "lôt" as "fragrant gum" in the two Genesis passages cited above, but GOODSPEED is the more accurate in saying "laudanum", a translation which the Revised Version suggests in a marginal note (as "ladanum") but does not actually employ. The JASTROW version is most accurate in using "ladanum" for both passages. The LEESSER translation recognizes the error of the myrrh rendition and substitutes "lotus" (9). The material referred to is now generally regarded to have been ladanum (Arabic, "ladan")—formerly written "labdanum" or "laudanum"—a product of a beautiful-flowered rockrose, *Cistus ladaniferus* L. and related species, represented in Palestine chiefly by *C. creticus*[39], *C. salvifolius*[40], and *C. villosus*. Dr. EVENARI questions whether *C. creticus* is really Palestinian, but it is listed from Moab and Gilead by DINSMORE (100) and from "Palestine" by POST on the authority of TRISTRAM (267). The ladanum of commerce is a soft, dark-brown or black, viscid, gummy exudation or oleoresin from the stems and leaves of these bushy little plants. It is collected during the heat of the day by means of drawing a bunch of leather thongs or some woven material to which the gum adheres, over the bushes. According to SMITH (299), it also adheres copiously to the beards of goats which browse among the bushes, and at eventide, when the goats return home, it is combed out of their beards and thus collected. The gum has a fragrant odor and bitter taste. It was at one time used extensively in medicine, but now is employed only in perfumery and plasters. Commentators are of the opinion that this was the "myrrh" of the Ishmeelites and of JACOB.

Cistus ladaniferus, known in England as "gum cistus", is sometimes suggested as the source of the Scriptural ladanum (299), but this species is indigenous only to southern Europe and the islands of the Mediterranean. It is not within the realm of possibility that it was known to the Jews at the time of JACOB (1707 B.C.).

PRATT is of the opinion (268) that the "rose of Sharon" of Song 2: 1 may have been a species of *Cistus*. He seems to base this contention chiefly on the use of the word "rose" because he points out that the shape and general appearance of these flowers, popularly known as "rockroses", are very much like the shape and appearance of the single wild brair-rose. The "rose of Sharon", however, is herein discussed under *Tulipa montana* (which see). Two of the species of rockrose listed at the head of this chapter are common practically throughout the Holy Land, especially on and about Mount Carmel.

---
[39]Also referred to as *C. villosus* var. *creticus* (L.) Boiss.
[40]Also referred to as "*Cistus salviaefolius*" (184, page 83).

73. Citrullus colocynthis (L.) Schrad.
(FIGURE 54)

DEUTERONOMY 29: 18—... lest there should be among you a root that beareth *gall* and *wormwood*.
DEUTERONOMY 32: 32—For their vine is the *vine of Sodom*, and of the fields of Gomorrah: their grapes are *grapes of gall*, their clusters are bitter.
I KINGS 6: 18—And the cedar of the house within was carved with *knops* and open flowers ...
I KINGS 7: 24—And under the brim of it round about there were *knops* compassing it, ten in a cubit, compassing the sea round about: the *knops* were cast in two rows, when it was cast.
II KINGS 4: 39-40—And one went out into the field to gather herbs, and found a *wild vine*, and gathered thereof *wild gourds* his lap full, and came and shred them into the pot of pottage: for they knew them not. So they poured out for the men to eat. And it came to pass, as they were eating of the pottage, that they cried out, and said, O thou man of God, there is death in the pot. And they could not eat thereof.
PSALMS 69: 21—They gave me also *gall* for my meat ...
JEREMIAH 8: 14—... the Lord our God hath put us to silence, and given us water of *gall* to drink, because we have sinned against the Lord.
JEREMIAH 9: 15—... Behold, I will feed them, even this people, with wormwood, and give them water of *gall* to drink.
JEREMIAH 23: 15—... I will feed them with wormwood, and make them drink the water of *gall* ...
LAMENTATIONS 3: 5 & 19—He hath builded against me, and compassed me with *gall* and travel ... Remembering mine affliction and my misery, the wormwood and the *gall*.
AMOS 6: 12—... for ye have turned judgment into *gall*, and the fruit of righteousness into hemlock.
MATTHEW 27: 34—They gave him vinegar to drink mingled with *gall* ...
ACTS 8: 23—For I perceive that thou art in the *gall of bitterness* ...

There has been considerable divergence of opinion concerning the interpretation of the words employed in the above references. The Hebrew word used in the quotation from II Kings, where the Authorized Version has said "wild gourds", is "k'la'at pkaim" or "pakknoth-sadeh" (Greek, τολύπη, meaning "a round kind of gourd", ἀγρία, meaning "wild"). The word rendered "gall" is "rôsh"[41] (Greek, χολή). The Hebrew original for the "wild vine" is "gefen sadeh". In all these instances it is now believed that the plant referred to was the colocynth, *Citrullus colocynthis*[42]. The colocynth is a vine which either trails on the ground or climbs over shrubs or fences by means of many branched tendrils (299) in the fashion of the cucumber. The leaves are triangular, deeply 3-7-lobed or -parted, with rounded sinuses, each lobe again lobed or incised. The fruit is globular, about the size and color of an orange, mottled with green and yellow, with a smooth hard rind. It contains a soft spongy pulp which is intensely bitter and poisonous, being a drastic cathartic. It is used medicinally as a purgative. The colocynth is common in western Asia and the Mediterranean region in general. In Palestine it covers dry sandy flats in the regions around the Mediterranean, Red, and Dead Seas, the plains of Engedi, Beersheba, Jericho, Jaffa, Sidon, and elsewhere, producing a great abundance of its orange-like fruits, very tempting to those unacquainted with their nature. KITTO states that "in the desert parts of Syria, Egypt, and Arabia, and on the banks of the river Tigris and Euphrates, its tendrils run over vast tracts of ground, offering a prodigious number of gourds, which are crushed under foot by camels, horses, and man. In winter we have seen the extent of many miles covered with the connecting tendrils and dry gourds of the preceding season." The powdered gourds, besides being a powerful medicinal drug, are also used now for keeping moths out of woolen clothing.

PRATT points out that the globe-cucumber, *Cucumis prophetarum* L., was long considered to be the plant of ELISHA's miracle (268), and, indeed, this is attested by the specific name given it by LINNAEUS. The fruits of this vine, however, are only one or two centimeters in diameter, are covered with prickles, and, though bitter, are not poisonous. It therefore does not fit the context of the ELISHA story nearly as well as does the colocynth. PRATT is of the opinion that ELISHA's plant was either the colocynth or the squirting-cucumber, *Ecballium elaterium* (L.) A. Rich.[43]. The latter, it is true, is also common throughout the region, but, like the globe-cucumber, its fruits are covered with

---

[41]LEESSER translates "rôsh" as "poison-water".
[42]Also known as *Cucumis colocynthis* L. and "*Cucumis colocynthus* L." (233).
[43]Also known as *Momordica elaterium* L. and "*Ecbalium elaterium*" (306, page 300).

prickles and spines and so would hardly be collected for food in the manner described in the ELISHA story. Furthermore, as Dr. EVENARI points out, when fully ripe it bursts open on being touched, and one would hardly be likely to collect such fruit for food. The squirting-cucumber has a large stout root and thick, rough, trailing stem. The leaves are borne on long stalks and the fruit is the size of a cucumber. When ripe, the fruit breaks off from the stem and ejects its seeds and a milky juice with considerable force. The etymology of the Hebrew word from "pâkâ", "to split or burst open", has been thought to allude to this dehiscence of the fruits by those who, like CELSIUS, ROSENMÜLLER, WINER, and GESENIUS, advocate this plant as the "wild gourd" of ELISHA. The dry fruit of the colocynth, however, when stepped upon, also bursts "with a crashing noise" and so the allusion may as well be to it. CARRUTHERS has well stated the case when he said: "Many kinds of wild gourd are found in Palestine, but only the bitter fruit of the colocynth would be likely to be mistaken from its appearance [for a wholesome melon], and yet reveal itself by taste. It grows wild in profusion about Gilgal" (184).

There has also been considerable difference of opinion as to the identity of the "vine of Sodom" of Deuteronomy 32: 32. Some writers think that it was *Solanum sodomeum* L. or *Calotropis procera* (Willd.) R. Br., but it seems far more probable that this was also the colocynth. TRISTRAM says "Observation of the relative abundance, and of the geographic distribution of plants of the neighbourhood of the Dead Sea, would lead me to the conclusion that the simile of the Vine of Sodom is taken from the fruit of the colocynth (*Citrullus Colocynthis*), which has long, straggling tendrils or runners like the vine, with a fruit fair to look at, but nauseous beyond description to the taste, and, when fully ripe, merely a quantity of dusty powder with the seeds, inside its beautiful orange rind" (184).

POST is of the opinion—and this view seems very logical—that the colocynth was also referred to when Old Testament writers spoke of "gall", along with wormwood, as symbolic of bitter calamity (95), although MOFFATT translates this term "poisonous drugs" in Psalms 69: 21 and Jeremiah 9: 15 & 23: 15. In Deuteronomy 32: 32 he renders the passage "their grapes are grapes of *gall*" as "*poisonous* are their grapes." Deuteronomy 29: 18 he renders "Never may there be any root within your soil that bears such *bitter poison*." In Jeremiah 8: 14 he changes the phrase "given us water of *gall* to drink" to read "has drugged us with *poison*." In Lamentations 3: 19 he substitutes for "the wormwood and the *gall*" merely the word "bitterness", and in Matthew 27: 34 he changes "vinegar ... mingled with *gall*" to "wine mixed with *bitters*" (O' HARA says "wine mixed with *bitter drink*"). The "gall of bitterness" of Acts 8: 23 he changes to "a bitter poison" (as also does the Douay version). The "gall" of Amos 6: 12 is rendered "poison" both by MOFFATT and the Douay version. The Hebrew word "rôsh" is twice translated "poison" in the Authorized Version and once as "hemlock", which fact is probably the basis for so many modern writers regarding it as applying to almost any bitter and at the same time poisonous herb.

Marginal Bibles suggest, and the GOODSPEED version adopts, the translation of "gourds" for the mysterious "knops" of I Kings 6: 18, while both MOFFATT and GOODSPEED believe that gourds were meant by the "knops" of I Kings 7: 24. The latter verse MOFFATT renders: "Under its brim on the outside ran a double row of *gourds* all around, cast in one piece with the tank itself" and which the GOODSPEED version renders: "And under its brim were *gourds* encircling it for ten cubits, completely surrounding the sea. The *gourds* were in ten rows, cast when it was cast." The Douay version uses the strange expression "chambered sculptures" for "knops". GOODSPEED translates as "poisoned water" all the "water of gall" passages in Jeremiah, and for "they gave me also *gall* for my meat" in Psalms he says "Yea, they put *poison* in my food". The "wormwood and *gall*" of Lamentations he renders "bitterness and *hardship*." The "vine of Sodom" is rendered "vineyard of Sodom" by the Douay version (thus definitely implying *Vitis vinifera*, which see) and the "stock of Sodom" by MOFFATT and GOODSPEED. The JASTROW version uses the terms "knops", "wild vine", "wild gourd", and "gall" in all of these references just as the Authorized Version does except for Psalms 69: 21, where it says "Yea, they put poison into my food." Of all the translations available only that of LEESSER gives "colocynth" or "wild colocynth" in the passages here cited. The LEESSER translation even renders "knops", mentioned above, as "colocynth-shaped knobs" (9). The Hebrew words involved here, "k'la'at p'kaim", mean literally "an expelling blossom", in allusion to the explosive debiscence of the fruit.

It is interesting to note that in the ELISHA story where the King James, MOFFATT, JASTROW, and GOODSPEED versions all unite in saying "found a wild vine" or "finding a wild vine", the Douay version adds a very meaningful qualifying phrase: "he found something like a wild vine."

Acts 8: 23—"For I perceive that thou art in the *gall of bitterness*"—is rendered "For I see your heart is as bitter as *gall*" by LAMSA and "for I see that you are prisoned in *bitter envy*" by the Basic English version, while MOFFATT and GOODSPEED both agree on "bitter poison" for the "gall". In Job 20: 14 the word "gall" (Hebrew, "m'reeroot" or "mĕrôrâh") plainly indicates a venom of venomous serpents, rather than any plant product; the same being true of the Hebrew word "rôsh" used in Job 20: 16 and translated "poison" in the Authorized Version. Elsewhere, as we have seen, the word "rôsh" usually refers to a plant product. Where "m'reeroot" or "mĕrôrâh" occur elsewhere in the Scriptures they refer mostly to the "gall" which is the intensely bitter bile of animal and human bodies (Job 13: 26, 16: 13, and 20: 25). LEESSER translates these words as "bitterness".

GESENIUS was of the opinion that the Hebrew word "rôsh", meaning literally "a head", referred to the capsules of poppies, which are often still spoken of as "poppy heads". The various species of poppies grow rapidly in grain fields and their juice is extremely bitter. A steeped solution of poppy "heads" has been thought by some to be the "water of gall" of Jeremiah.

In regard to the Matthew passage the following quotation from SMITH (306) is relevant: "The passages in the Gospels which relate the circumstance of the Roman soldiers offering our Lord, just before his crucifixion, 'vinegar mingled with gall', according to St. Matthew (xxvii.34), and 'wine mingled with myrrh', according to St. Mark's account (xv.23), require some consideration. 'MATTHEW, in his usual way,' as HENGSTENBERG remarks, 'designates the drink theologically: always keeping his eye on the prophecies of the Old Testament, he speaks of gall and vinegar for the purpose of rendering the Psalms more manifest. MARK again (xv.23), according to *his* way, looks rather at the *outward* quality of the drink.' 'Gall' is not to be understood in any other sense than expressing the bitter nature of the draught. Notwithstanding the almost concurrent opinion of ancient and modern commentators that the 'wine mingled with myrrh' was offered . . . as an anodyne, we cannot readily come to the same conclusion. Had the soldiers intended a mitigation of suffering, they would doubtless have offered a draught drugged with some substance having narcotic properties. The drink in question was probably a mere ordinary beverage of the Romans."

74. Citrullus vulgaris Schrad.   75. Cucumis melo L.

NUMBERS 11:5—We remember the fish, which we did eat in Egypt freely; the cucumbers, and the *melons*, and the leeks, and the onions, and the garlick.

The Hebrew words here translated "melons" are "avatiach", "abattichim", or "avatichim" (Greek, πέπονες or πεπων). There is some doubt as to whether it was the muskmelon, *Cucumis melo*, or the watermelon, *Citrullus vulgaris*[44], for which the children of Israel longed in the desert of Sinai. Their longing, in either case, is quite understandable under the extremely trying conditions of the barren desert. Most commentators are inclined to favor the muskmelon

---

[44] Also known as *Cucurbita citrullus* L., *Cucurbita bettich* Forsk., *Citrullus citrullus* (L.) Karst., and *Cucumis citrullus* Ser.

as the "melon" of this passage, but SMITH is of the opinion (299) that it was the watermelon. Probably both fruits were meant, as, indeed, is the contention of the "Bible Encyclopedia and Concordance" (179, 184). Both are at present extensively cultivated in Egypt and Palestine (98). All the Bible versions regularly cited in this work employ the term "melons" in this verse.

The true melon or muskmelon is considered by PRATT (268) to be a native of Egypt and the Levant; POST states that it came from India (266); and BAILEY gives its place of origin as probably central Asia. At any rate, it has been cultivated in Egypt since time immemorial. Its fruit is said to grow to a large size in hot countries, and is very cooling. There are many minor varieties. It was introduced into England at about the middle of the sixteenth century (299).

The watermelon, considered to be a native of tropical central Africa (266), has also been cultivated in Egypt since before the dawn of recorded history. It is now likewise extensively cultivated in Palestine, especially on the plain between Haifa and Jaffa (266). It serves the Egyptians for food, drink, and medicine. It is the only means of relief that the poorer people have from fevers. For this purpose the fruit is taken when it is so over-ripe and soft as to be almost in a state of decay. The juice is pressed out, mixed with sugar, and imbibed freely. HASSELQUIST reports that the watermelon is eaten in abundance by the richer inhabitants, while the poorer people eat "scarcely anything else" and consider the melon season the best season of the year, since they are "obliged to put up with a less pleasant diet during the other seasons." The seeds are also roasted and salted, forming a popular side-dish. TRISTRAM, SMITH, and others report that the watermelons of Egypt and the Holy Land often attain a weight of twenty or thirty pounds (184, 299). In many places there is a steady succession of crops of these fruits from May through November.

SMITH states that the Arabic word "batêkh", which corresponds to the Hebrew "abattichim", is used for both the muskmelon and the watermelon. POST says that the former is now called "battîkh-asfar" and the latter "battîkh-akhdar" (266).

According to a popular legend in the Holy Land, the prophet ELISHA became so incensed at all gourd-like fruits because of his experience with the colocynth (see under *Citrullus colocynthis*) that he cursed them vehemently. On the top of Mount Carmel is a field of small roundish stones which are said to have been melons, but which were petrified by ELISHA's wrath (298).

76. Commiphora africana (Arn.) Engl.

GENESIS 2: 12—And the gold of that land is good: there is *bdellium* and the onyx stone.
NUMBERS 11: 7—And the manna was as coriander seed, and the colour thereof as the colour of *bdellium*.

The Hebrew word translated "bdellium" in the King James, JASTROW, and Douay versions is "bedôloch" or "b'dolach" (Greek, ἄνθραξ). Although some authorities (184) have identified this plant with *Borassus flabellifer* L.[45], a palm growing in tropical Asia (especially India), Africa, and the Sunda Islands (158), this identification seems extremely unlikely. Much more cogent is the argument of SMITH (299), who is of the opinion that the term applied either to what is now called "African bdellium" or to "Indian bdellium". The former is a gum resin obtained from *Commiphora africana*[46] of northwestern Africa and Arabia. The latter is a product of the closely related *C. roxburghii* (Stocks) Engl.[47], a native of northwestern India and Baluchistan, known there as "mukul" or "gugul". Which of these two species is regarded as the more probable depends on one's interpretation of the garden of Eden story and on the geographic location of Eden as described in the second chapter of Genesis.

---

[45] Also referred to as *Borassus flabelliformis* Murr.
[46] Also known as *Balsamea africana* Baill. and *Balsamodendron africanum* Arn.
[47] Also known as *Balsamodendron roxburghii* Stocks, *B. mukul* Hook. f., *Balsamea mukul* Baill., and *Commiphora mukul* (Hook. f.) Engl.

There is hardly a subject on which Biblical scholars are more divided in opinion than the location of Eden (306). If the land "Havilah", which is the land described in the verse from Genesis at the head of this chapter, is the same as the "Havilah" mentioned in Genesis 10: 29 and 25: 18 or I Samuel 15: 7, then it must have been located in the eastern or southern part of Arabia, probably not far from the Persian Gulf (306) or possibly in what is now Yemen. Others have supposed that the "Pison" river which is described as watering the land of Havilah was actually the Ganges: then the land watered by the Pison would be India. For those holding to the latter theory the Indian bdellium would be the more probable plant involved; according to the former theory, the African bdellium would be the plant. It would appear to us that the region referred to was more probably Arabia and the plant spoken of as "bdellium" was *Commiphora africana*.

MOFFATT, however, denies that any plant product is referred to at all. His translation of the Genesis passage is: "where there is gold—fine gold in that land! and *pearls* and beryls." In the Numbers verse he says: "The manna was like grains of coriander seed, resembling *pearls*." The GOODSPEED version continues to use "bdellium" in Genesis, but in Numbers says " . . . its color was like that of *resinous gum*." The JASTROW version abandons the use of the word "colour" in describing the bdellium and substitutes "appearance", which is an interesting change because of its reference to the manna with which this bdellium was being compared. The LEESSER translation is "the colour of bdellium".

77. **Commiphora kataf** (Forsk.) Engl.
78. **Commiphora myrrha** (Nees) Engl.
(FIGURE 27 — PAGE 99)

EXODUS 30: 23—Take thou also unto thee principal spices, of pure *myrrh* five hundred shekels, and of sweet cinnamon half so much, even two hundred and fifty shekels, and of sweet calamus two hundred and fifty shekels.
ESTHER 2: 12— . . . for so were the days of their purification accomplished, to wit, six months with oil of *myrrh*, and six months with sweet odours, and with other things for the purifying of the women.
PSALMS 45: 8—And thy garments smell of *myrrh*, and aloes, and cassia . . .
PROVERBS 7: 17—I have perfumed my bed with *myrrh*, aloes, and cinnamon.
SONG 1: 13—A bundle of *myrrh* is my well-beloved unto me . . .
SONG 3: 6—Who is this that cometh out of the wilderness like pillars of smoke, perfumed with *myrrh* and frankincense . . . ?
SONG 5: 5 & 13—I rose up to open to my beloved: and my hands dropped with *myrrh*, and my fingers with sweet smelling *myrrh*, upon the handles of the lock . . . His cheeks are as a bed of spices, as sweet flowers: his lips like lilies, dropping sweet smelling *myrrh*.
MATTHEW 2: 11— . . . and when they had opened their treasures, they presented unto him gifts; gold, and frankincense, and *myrrh*.
MARK 15: 23—And they gave him to drink wine mingled with *myrrh* . . .
JOHN 19: 39— . . . and brought a mixture of *myrrh* and aloes . . .
REVELATION 18: 13—And cinnamon, and odours, and *ointments*, and frankincense, and wine, and oil, and fine flour, and wheat . . .

The Hebrew word translated "myrrh" in these references is "môr" (Greek, σμύρνα). There is no doubt about its referring, at least in the majority of cases, to *Commiphora myrrha*[48], although the closely related *C. kataf*[49] may also be involved because it grows in the same region and is very similar in all respects. These two trees are native to Arabia, Abyssinia, and the Somali Coast of eastern Africa (89, 106) and yield a gummy exudation which constitutes most of the myrrh of commerce. JOHN SMITH states (299) that present day commercial myrrh is also derived from *C. opobalsamum* (which see), but POST (266) insists that it is derived only from *C. abyssinica* (Berg.) Engl. HORNER points out (178) that the name for this well-known spice is practically the same in all languages—Arabic, "murr"; Hebrew, "môr"; Greek, μύρρα; Latin, "myrrha", "murra"; French, "myrrhe"; Old French, "mirre"; Middle

---
[48] Also known as *Balsamodendrum* (or *Balsamodendron*) *myrrha* Nees and *Balsamea myrrha* (Nees) Baill.
[49] Also known as *Balsamodendrum* (or *Balsamodendron*) *kataf* (Forsk.) Kunth and *Amyris kataf* Forsk.

English "mirre"—and so there can be no doubt about its identity. The Arabic name literally signifies "bitter", and the Hebrew word for "bitter" is very similar: "mar" (158). Contrary to the assertions of some writers that myrrh grows only along the Somali coast of Africa, there are abundant statements by HERODOTUS, DIOSCORIDES, THEOPHRASTUS, DIODORUS SICULUS, STRABO, and PLINY to the effect that myrrh-producing trees grow in Arabia. The botanical explorer FORSKÅL mentions two myrrh-producing species occurring in Arabia Felix (the southeastern part of ancient Arabia), one of which was *C. kataf*. EHRENBERG and HEMPRICH also found *C. myrrha* in Arabia Felix (306).

The two species of myrrh here listed are low, scrubby, thick- and stiff-branched, thorny shrubs or small trees, growing in rocky places and especially on limestone hills. The leaves are trifoliolate and the fruit is oval and plum-like. The wood and bark are strongly odorous (266) and a gum exudes naturally from the stems and branches. A more abundant flow of this gum is obtained from artificial incisions. As it first exudes from the tree it is a soft, clear, sticky, white or yellowish-brown resin (268). It is at first somewhat oily, but quickly solidifies when it drops to the stones beneath the branches. It is bitter and slightly pungent to the taste and not at all palatable to Occidentals generally, but it was for a time esteemed in medicine as an astringent tonic, and, externally, as a cleansing agent. In Oriental countries it is highly regarded as an aromatic substance (125), perfume, and medicine. The ancient Egyptians burned it in their temples and embalmed their dead with it. The classic Greeks and Romans made much use of it, and its employment by the Hebrews in embalming is attested by the story of NICODEMUS. According to WEBSTER, the term "myrrhophore" is applied to any of the women, especially the MARYS, who bore spices to the sepulcher of JESUS. They are usually depicted as carrying vases filled with myrrh (158). The Hebrews also held it in high regard as a perfume, for DAVID sings of its fragrance and SOLOMON delighted in it. It was an ingredient of the holy oil and of a domestic perfume with aloes, cassia, and cinnamon (184).

The resin obtained from *C. kataf* is now known as "bisabol", "bissabol", or "opopanax", but is very similar to and often mixed with true myrrh (158). Many authorities hold that the "myrrh" of the Bible was never a pure substance, but was a mixture of myrrh and ladanum (see under *Cistus*) (158). The "myrrh" of Genesis 37: 25 and 43: 11, represented by the Hebrew word "lôt", is now generally regarded as having been only ladanum, since true myrrh was probably not known in Palestine at the time of these references (1729-1707 B.C.).

There is some reasonable doubt about the "myrrh" of Song 1: 13 being the same as that of all the other references cited. This doubt is based on the fact that a "bundle" or "bunch" of "myrrh" is mentioned in this passage, instead of a "lump" as one would expect for such resinous gums. SMITH suggests (299) that SOLOMON may have had the European myrrh, *Myrrhis odorata* Scop., in his gardens. This plant is a common European pot-herb, a member of the carrot family—a hirsute-pubescent, erect, branching, perennial herb, 2 to 3 feet tall, with thin, soft, 2 or 3 times pinnately compound leaves, and umbels of small whitish flowers. It is pleasantly fragrant throughout and would probably be gathered in bunches. The general consensus among botanists, however, is against the possibility of any strictly European plants having been cultivated in SOLOMON's gardens.

It is asserted by many authorities that the "wine mingled with *myrrh*" of ST. MARK is not to be taken as referring to any actual use of the myrrh plant. MATTHEW's description of the same event uses the expression "vinegar . . . mingled with gall" (Matthew 27: 34). It is probable that both reporters merely meant to indicate a strongly bitter drink, possibly only the ordinary "acetum" of the Roman soldiery. GOODSPEED substitutes the expression "*drugged wine*," but whether any additional substance actually was put into the wine is still a moot point among theologians.

Of considerable interest is the GOODSPEED version's substitution of "beautifying" for "purifications" in Esther 2: 12, for this implies that myrrh was also employed by Hebrew women as a cosmetic. The word "ointments" of Revelation 18: 13 is rendered "myrrh" by MOFFATT. The Basic English version uses merely "spices" in Matthew 2: 11, although, curiously, it continues to use the word "myrrh" in Mark 15: 23 and John 19: 39, the implication plainly being that the gifts of the Wise Men included more kinds of spices than frankincense and myrrh.

STEUER has recently published (311a) a botanical and philological study of the myrrh of early times, in which he concludes that it may have been derived from *Commiphora molmol* (Engl.) Engl.[49a] This species is native to Somaliland. Comparative studies of the resins of living plants of the various species of this genus might serve to confirm or contradict this conclusion (276).

An ancient legend states that MYRRHA, daughter of the king of Cyprus, became obsessed with an unnatural love for her father and was exiled by him to the barren deserts of Arabia, where the gods transformed her into the myrrh tree, "in which guise she remains, weeping tears perfumed of repentance" (298). Myrrh was the incense burned on the altars of the sun god at Heliopolis at noon daily. Persian kings wore it in their regal crowns. As lately as the reign of King GEORGE III of England (1760-1820) myrrh, along with frankincense, was one of the spices burned ceremonially in the royal chapels.

79. Commiphora opobalsamum (L.) Engl.

I KINGS 10: 10—And she gave the king an hundred and twenty talents of gold, and of *spices* very great store, and precious stones: there came no more such abundance of *spices* as these which the queen of Sheba gave to king SOLOMON.
II KINGS 20: 13—And HEZEKIAH hearkened unto them, and shewed them all the house of his *precious things,* the silver, and the gold, and the *spices,* and the precious ointment ...
SONG 3: 6—Who is this that cometh out of the wilderness like pillars of smoke, perfumed with myrrh and frankincense, with all *powders* of the merchant?
ISAIAH 39: 2—And HEZEKIAH was glad of them, and shewed them the house of his *precious things,* the silver, and the gold, and the *spices,* and the precious ointment ...
EZEKIEL 27: 17—. . . they traded in thy market wheat of Minnith, and Pannag, and honey, and oil, and *balm.*
ECCLESIASTICUS 50: 8—And as the flower of roses in the spring of the year, as lilies by the rivers of waters, and as the branches of the *frankincense tree* in the time of summer.

The "balm" of the Ezekiel reference cited above is the same Hebrew word "tzo'ri" or "tzari" (Greek, ῥητίνη) as occurs in Genesis 43: 11, and is translated "balsam" by MOFFATT in both places, just as in the Genesis and Jeremiah references cited herein under *Balanites aegyptiaca* (which see). SMITH, however, is of the opinion (299) that the plant referred to in Genesis 43: 11 is the lentisk, *Pistacia lentiscus,* while the "balm" of Ezekiel and the "powders" (Hebrew, "avkat") of Song 3: 6 are the balm-of-Gilead, *Commiphora opobalsamum*[50]. In spite of the specific name *"gileadense"* often applied to this plant and its English names of "balm-of-Gilead" and "balsam-of-Gilead" (266), it is not a native of Gilead, nor even of Palestine, but is indigenous to Arabia, especially the mountainous regions of Yemen. Thus, it could not have provided the "balm" sent by JACOB to Egypt in 1707 B.C. It was, however, according to JOSEPHUS (who speaks of it under the name of "myrobalanum") (299), cultivated in Palestine at the time of SOLOMON, particularly about the city of Jericho. It is supposed by JOSEPHUS and many other authors, with apparently good reason, that these Arabian trees were grown from seeds brought by the Queen of Sheba and which were part of her gift of "spices".

Contrary to SMITH's statement that "it is . . . curious that no mention is made of these balsam trees from the time of SOLOMON to that of JOSEPHUS, a period of about 1,000 years", they are actually mentioned, or, rather, alluded to, according to reliable modern opinion, at least four times in the interim, the

---

[49a] Also known as *C. myrrha* var. *molmol* Engl. and *C. molmol* Engl. It is not certain that this plant is specifically distinct from *C. playfairii* (Hook. f.) Engl.
[50] Also referred to as *Amyris gileadensis* L., *A. opobalsamum* L., *Balsamodendron gileadense* (L.) Kunth, and *B. opobalsamum* (L.) Engl.

approximate dates assigned to the first five references cited at the head of this chapter being, respectively, 992, 906, 1014, 713, and 588 B.C. The trees were still in existence on the plain of Jericho at the time of the Roman conquest and were highly valued by the Roman conquerors (299), who carried branches from them to Rome as trophies of their victories over the Jews. After the final subjugation of the Jews by TITUS VESPASIAN in 70 A.D. an imperial guard was placed over the orchards to prevent their destruction by vandals. However, after the Holy Land came under the dominion of the Turks the trees were neglected and no trace of them remained at the time of the Crusades in 1099-1244 A.D. (299). POST lists the balm-of-Gilead for the Holy Land, but says "It has long since vanished" (266).

The balm-of-Gilead is a small, stiff-branched, evergreen tree, seldom more than 15 feet tall, with straggling branches and scanty trifoliate leaves. The individual leaflets are small, obovate, and entire. The white flowers are borne in clusters of 3's and appear at the same time as the leaves. The "balm" or "balsam" is a gum which is obtained by making incisions in the stem and branches of the trees. The exuding sap soon hardens into small irregular nodules, which are collected and at present are shipped to Bombay where all types of "balsam" appear as articles of commerce. The gum is procured also from the green and the ripe fruit (306).

PRATT is of the opinion that all the "balm", "balsam", and "spices" of the Bible (268) are to be referred to *Commiphora opobalsamum*, but this is far too much of a simplification of the situation. Other commentators believe that the "spicery" of Genesis 37: 25 was also this species, but this is herein discussed under *Astragalus gummifer* (which see). It is generally believed that the "spices" presented to SOLOMON by the Queen of Sheba included also the sweet-calamus, *Andropogon aromaticus* (which see).

The use of the Hebrew word "bâsâm" or "bosem" in Song 5: 1 leads W. SMITH (306) to believe that the "spice" of this passage refers principally to the balm-of-Gilead, whose Arabic name is "basham" or "balasân" even today (266, 306). The identity of the Hebrew and Arabic words, he says, "leaves no reason to doubt that the substances are identical." Yet the "spice" of Song 5: 1 is by us regarded as *Astragalus* (which see). The variant forms of the same word, "besem" and "bôsem", which occur much more frequently in the Old Testament, he thinks, may represent a more generalized concept of "spices" or "sweet odours" in accordance with the translations of the Septuagint and Vulgate versions.

As has been pointed out in the chapter on the frankincense tree (see under *Boswellia thurifera*), this was a tree never known in Palestine except in the form of the commercial incense. The context of Ecclesiasticus 50: 8 demands a tree with whose growth and appearance the Jews were thoroughly familiar and which was still a rare spice. It seems most probable, therefore, that the translation "frankincense tree" in that passage is an error. The tree referred to was probably the balm-of-Gilead.

Marginal Bibles suggest, and the Douay version uses, "rosin" for the "balm" of Ezekiel 27: 17, but MOFFATT and the GOODSPEED version both employ the word "balsam", while JASTROW follows the Authorized. The final phrase (Hebrew, "avkat rochel") of Song 3: 6 is rendered "all the *powders* of the perfumer" by the Douay and "all kinds of merchants' *spices*" by the GOODSPEED version, while MOFFATT translates it "every *scent* to be bought."

Marginal Bibles suggest the translation "spicery" for the word "nâcôth" or "n'coth" used in II Kings 20: 13 and Isaiah 39: 2, translated "precious things" in the Authorized and "treasure house" in the GOODSPEED and JASTROW versions. The Douay version says "aromatical spices" in both cases. It seems most likely that the balm-of-Gilead was at least included among these spices, if, indeed, it did not comprise the bulk of them.

Some authorities think (27) that the "balm" of Jeremiah 8: 22 was also this plant, but because of the implication of healing properties connected with this text it is herein discussed under *Balanites aegyptiaca* (which see). The criticism that the gum of *Balanites* has only a very faint odor is not a pertinent one, because in the references herein cited for that species the context demands not an especially fragrant gum, but one with supposed healing properties.

Three other plants known today as "balm" are sometimes cited by non-botanical writers—*Melissa officinalis* L., *Dracocephalum canariense* L., and *Populus candicans* Ait. (the latter two usually called "balm-of-Gilead" for no good reason)—but have absolutely nothing to do with the "balm" of the Scriptures.

80. Coriandrum sativum L.
(PAGE 94; FIGURE 57)

EXODUS 16: 31—... and it was like *coriander* seed, white; and the taste of it was like wafers made with honey.
NUMBERS 11: 7—And the manna was as *coriander* seed, and the colour thereof as the colour of bdellium.

The Hebrew word "gad" (Greek, κόριον), rendered "coriander" in all translations, seems very definitely to refer to the common coriander plant, *Coriandrum sativum* (98). This is an annual white- or reddish-flowered herb, 16 to 20 inches tall, with slender round dichotomous stems, deeply cut leaves, and a heavy odor. It differs from caraway and dill in its globular, grayish, minutely striated, pearl-like seeds, which are quite aromatic and are now employed chiefly in confectionery. They are also used in medicine as a stomachic and carminative (158). The coriander is reported by POST (265) to be quite common among grain in cultivated fields throughout the Holy Land. It grows wild also in Egypt (184) and was used by the ancients both as a condiment and as a medicine (268) and is frequently mentioned in the Talmud. The leaves are also very aromatic and are used in soups and for flavoring puddings, curries, and wines. In ancient times a favorite drink was made by steeping the plants in wine; and the seeds, afterwards dried and thus rendered milder, were eaten with various dishes. The coriander is still used today as a spice by the Arabs and is much relished in Egypt, Persia, and India. In the Bible it is mentioned only in connection with the "manna", which was said to resemble coriander seeds in size, shape, and color (see under *Lecanora*).

81. Crocus cancellatus var. damascenus (Herb.) G. Maw
82. Crocus hyemalis Boiss. & Bl.   83. Crocus vitellinus Wahlenb.
84. Crocus zonatus J. Gay.

SONG 6: 4—Thou art beautiful, O my love, as *Tirzah,* comely as Jerusalem ...

The GOODSPEED and JASTROW versions translate this verse essentially the same as the Authorized, and the Douay version renders it: "Thou art beautiful, O my love, *sweet* and comely as Jerusalem", but MOFFATT is of the opinion that the words translated "Tirzah"[51] and "Jerusalem" are intended to signify flowers and not cities. Since this verse is the beginning of a speech of a young man to a maiden, MOFFATT's view is certainly plausible, since one usually compares the beauty of a girl to that of flowers rather than that of cities. MOFFATT's rendition is: "You are fair as a *crocus*, my dear, lovely as a lily of the valley."

Fifteen or more kinds of crocus are known from the Holy Land. Apparently the reference, in MOFFATT's opinion, at least, is not to the saffron crocus, *Crocus sativus* (which see), for which the Hebrew word used in Song 4: 14 is "karkôm". Of the other species we may eliminate the ones not found in Lebanon, since it is thought by modern scholars that if the Song of Solomon was actually composed by SOLOMON it was done at "a hunting-seat somewhere on the slopes of Lebanon" where "the influence of the scenery and the language of the surrounding peasantry" may well have inspired the Song (306). The four species listed at the head of this chapter are common in alpine and subalpine regions and rocky places in the mountains of Lebanon, as well as elsewhere

---
[51]The Hebrew word means "pleasantness" by derivation.

in the Holy Land (267), and it is very probable that it was to one or more of these that reference was made by SOLOMON. In the Damascus crocus, *C. cancellatus* var. *damascenus*, the flowers are grayish-blue; in the Syrian crocus, *C. vitellinus*, they are orange-yellow; in the ringed crocus, *C. zonatus*, they are pale-lilac; and in the winter crocus, *C. hyemalis*, they are whitish, with lilac bands within, sometimes spotted on the outer surface, the throat orange.

According to Greek mythology, the first crocus sprang from the ground on a grassy bank on Mount Ida, where it had been warmed by the body of JOVE who had lain there with JUNO (125). Another legend, part of the story of the Golden Fleece, states that the crocus sprang up from the ground where some drops fell of an elixir of life that MEDEA, the enchantress, was preparing to restore the aged AESON, father of JASON, to youth (298).

85. Crocus sativus L.
(FIGURE 58)

SONG 4: 14—Spikenard and *saffron*, calamus and cinnamon, with all trees of frankincense . . .

Saffron is the product of several species of *Crocus*, especially of the blue-flowered saffron crocus, *C. sativus*, native to Greece and Asia Minor[51a]. The commercial product consists of the stigma and upper portion of the style, which are collected as or shortly after the flower opens. It requires at least 4,000 such stigmas to make an ounce of saffron, so it is not difficult to picture the terrific inroads made on the unfortunate plant to supply the commercial demand for saffron. After being gathered, the stigmas are dried in the sun, pounded, and made into small cakes (299). Saffron is used principally as a yellow dye and also for coloring curries and stews. In spring many parts of Palestine are brilliant with the white, pink, purple, lilac, blue, or orange-yellow flowers of 14 or 15 different kinds of crocus, several of which yield saffron (184). Although POST does not record *C. sativus* as definitely wild in the Holy Land, L. H. BAILEY asserts that it is native to Asia Minor and some commentators (27) say that it "abounds in Palestine", so it is very probable that this is the species referred to in the Song of Solomon. This plant was common enough and important enough for JOSHUA to enact laws concerning it (98, 218). The Hebrew word employed is "karkôm" or "carcôm" (Greek, κρόκος). The Arabic name of the plant is "kurkum" (27) and of the commercial substance is "za'farān" or "zafran", meaning "yellow", and from this we have derived our English word "saffron" (158).

The ancients are said to have scattered saffron on the floors of their theaters, mixed with wine, and used it extensively during wedding ceremonies (268). It was also used in the Orient as a perfume and was sprinkled over the clothes of guests as they entered a house. It is still employed to lend color (126) to confectionery, liquors, and varnishes. Formerly it was much used in medicine as a stimulant, antispasmodic, and emmenagogue (158). In olden times, according to ROSENMÜLLER, it was employed for the same purpose as the old-fashioned pot-pourri or mixture of flower petals and spices used to scent a room (306).

Another and entirely different kind of plant, *Carthamus tinctorius* L., called "carthamine", "bastard saffron", or "safflower", is a member of the thistle family, and its red florets yield a dye used extensively for coloring silk, in cooking, and for adulterating genuine saffron. It is an annual, spiny, thistle-like plant, 3 to 4½ feet tall, native to Syria and Egypt, as well as intervening and neighboring lands. POST states that it is now "cultivated everywhere for its flowerlets, which are used as a dye in cooking" (265). In Egypt the grave-clothes of mummies were dyed with this material (267). It is very possible that this plant may also be involved in the "saffron" of the Bible. KITTO admits that the safflower is at present widely cultivated in Syria for the dye, but feels certain that the true saffron was the actual plant to which reference is made in the Song of Solomon (306).

[51a] W. T. STEARN comments that *"Crocus sativus* is an old cultivated type, apparently one sterile clone of uncertain origin. It is doubtful if saffron has ever been prepared from the style and stigmas of any crocus but this." MAW suggests (223a) that at the time of SOLOMON saffron may have been a drug imported into Palestine.

86. Cucumis chate L.⁵²   87. Cucumis sativus L.

NUMBERS 11: 5—We remember the fish, which we did eat in Egypt freely; the *cucumbers*, and the melons, and the leeks, and the onions, and the garlick.
ISAIAH 1: 8—And the daughter of Zion is left as a cottage in a vineyard, as a lodge in a garden of *cucumbers*, as a besieged city.

The cucumber is an annual climbing or trailing vine too well known to need description. Its country of origin is unknown, as it has been cultivated in all the warm countries of the Old World since prehistoric times, but is thought to be somewhere in southern Asia (299), probably India (266). The Hebrew words⁵³ rendered "cucumbers" in the Bible are "kishuim" (Greek, σίκνοι or σικνός, meaning "the common cucumber") in the book of Numbers and "mikshah", meaning "a cucumber field" (Greek, σικνήρατον, meaning "a forcing bed of cucumbers") in Isaiah.

In Egypt and Palestine hundreds of acres of moist level lands, or such as can be irrigated, are planted in cucumbers, which form an important item in the food of the poorer classes of natives in the summer. The cucumbers are usually eaten raw—a cucumber and a barley cake or some other kind of bread often constituting a meal. The 6- to 9-inch size of the fruit as grown there is small due to the lack of proper tillage (184, 299). The Scriptural expression of "a lodge in a garden of *cucumbers*" refers to the small house or lodge often built in Palestinian cucumber fields and vineyards. It is usually very crudely and poorly constructed and soon falls to ruin. In it an attendant is placed, whose duty it is to tend and watch over the fields. Some commentators are of the opinion that the proper name of the city "Dilean"—"Delean" in the Douay version, "Dilan" in MOFFATT and GOODSPEED—mentioned in Joshua 15: 38, refers to an abundance of cucumber plants there. Cucumber plantations by the acre still attract the traveler's eye as he wanders over the plains of Palestine—each with its elevated cottage from which the wary watchman can observe and make certain that none stop and steal the fruit (299).

The cucumber was very extensively cultivated in Egypt in ancient times, and, along with fish, melons, leeks, onions, and garlic, formed the common diet of the land, on which the Israelites, although slaves, subsisted even as did the Egyptian masses. They apparently became quite attached to this diet during their 400-years' sojourn in Egypt, since they longed for it so plaintively after they had left. KITTO relates that in 1218 when Damietta was besieged, many of the more delicate Egyptians died for want of these staples of their diet, even though they had plenty of grain and other food to eat. In Egypt the hairy cucumber or "round-leaved Egyptian melon" (306), *C. chate*, is also widely cultivated, especially in the Nile river floodplain around Cairo, and is usually considered the finest of all the melons and melon-like fruits. The entire plant, including the curved young fruit, is covered with soft, white, transparent hairs. Its flesh is melon-like and more watery than that of the common cucumber. HASSELQUIST calls it "the Queen of the Cucumbers".

The "wild gourds" of II Kings 4: 39 are herein discussed under *Citrullus colocynthis* (which see), and the "gourd" of JONAH under *Ricinus communis* (which see).

Gherkins and pickles are, in part, at least, the immature fruits of cucumbers, although the fruit of the related *Cucumis anguria*, the true or West Indian gherkin, and other fruits and plant parts are also so prepared.

A Buddhist legend tells that "of the sixty thousand offspring of SAGARA'S wife ... the first was a cucumber, whose descendant climbed to heaven on his own vine" (298). In England cucumbers were avoided for centuries in the superstitious belief that their natural coldness was indicative of death and would bring death to any person foolish enough to eat them. In phallic symbolism the cucumber was regarded as the symbol of fecundity (298).

---

⁵²Also referred to as *Cucumis sativa* var. *chaete* Dalman (98).
⁵³SMITH's Bible plants: their history, page 53, gives "trispium" as the Hebrew term for "*cucumber*", but this term is not Hebrew and does not occur in the original text.

88. Cuminum cyminum L.
(FIGURE 17)

ISAIAH 28: 25 & 27—... doth he not cast abroad the fitches, and scatter the *cummin*?... For the fitches are not threshed with a threshing instrument, neither is a cart wheel turned about upon the *cummin*; but the fitches are beaten out with a staff, and the *cummin* with a rod.
MATTHEW 23: 23—... for ye pay tithe of mint and anise and *cummin* ...

The Hebrew word here translated "cummin" is "cammoin" or "kammon" (Greek, κύμινον). There is no doubt at all about the identity of the plant. Cummin or cumin, *Cuminum cyminum*[54], is a common annual plant of the carrot family, said by some to be native to Egypt (299) and the region of the eastern Mediterranean, notably Syria (158), and by others considered to have come originally from Turkestan (266). It grows from one to two feet tall, the slender stem being abundantly branched from the base, bearing very finely divided (biternately dissected) leaves and few-flowered umbels of minute white or rose-colored flowers. It has long been cultivated in southern Europe (299) for its powerfully aromatic and pungent seeds, which are similar to caraway seeds but larger. They do not have as agreeable a taste as caraway seeds, but nevertheless are extensively used in Palestine as a flavor or spice and are even mixed with flour in making bread. Cummin grows wild throughout Syria (146, 268), but POST states that at present it is only rarely cultivated in Palestine and only sometimes found as an escape (266), although *C. cyminum* var. *hirsutum* Boiss. occurs wild there.

In ancient times Ethiopian cummin was considered to be the very best and that from Egypt the second best. It was used as a condiment with fish and meats, especially in stews, as a stimulant to the appetite, and as a medicine (184). Egyptian cooks sprinkled the seeds on bread and cakes, much as we do caraway seeds. In Malta the people still grow cummin and thresh it in the manner described by ISAIAH (306).

Cummin was only included by inference in the Mosaic law covering tithing (126), hence the charge of JESUS that the Scribes and Pharisees punctiliously tithed mint, dill, and cummin, but overlooked or ignored more important matters, such as justice, mercy, and trustworthiness. The Basic English version substitutes "all sorts of sweet-smelling plants" for "mint, and anise, and *cummin*" in the Matthew reference. All the other English versions agree on "cummin" in both Isaiah and Matthew, although they disagree widely on the identification of the "fitches" in the same Isaiah reference—MOFFATT calling it "fennel", GOODSPEED "dill", JASTROW "black cummin", and the Douay version "gith". SPENSER states that cummin is "good for the eyes" (158).

89. Cupressus sempervirens var. horizontalis (Mill.) Gord.
(*Cf.* FIGURE 19)

GENESIS 6: 14—Make thee an ark of *gopher wood* ...
ISAIAH 41: 19—... I will set in the desert the *fir tree*, and the pine, and the box tree together.
ISAIAH 60: 13—The glory of Lebanon shall come unto thee, the *fir tree*, the pine tree, and the box tree together, to beautify the place of my sanctuary ...
ECCLESIASTICUS 24: 13—I was exalted like a cedar in Lebanon, and as the *cypress tree* upon the mountains of Hermon.
ECCLESIASTICUS 50: 10—... and as a *cypress tree* which groweth up to the clouds.

The evergreen cypress, *Cupressus sempervirens* var. *horizontalis*, is a massive, tall-growing evergreen which is widely distributed in the mountainous regions of the Bible lands. It grows in company with the cedar and the oak on Mount Lebanon and Mount Hermon as is required by the passages cited above. The Authorized and Revised versions do not mention the cypress as having been used by SOLOMON in the building of the temple, but according to HERODOTUS, King HIRAM told SOLOMON: "I will give order to cut down, and to export such quantities of the fairest cedars and cypress trees as you shall have occasion for." It seems most unlikely that the woodcutters would have spared the magnificent cypresses of Lebanon. Therefore MOFFATT's rendition of the "algum trees" of II Chronicles 2: 8 as "cypress logs" seems most plausible, and GOODSPEED's

---

[54]Referred to as "*Cuminum sativum*" in numerous works (47, 98, 184, 299), a name apparently overlooked by the editors of the "Index Kewensis".

use of "sandalwood", as will be pointed out below, quite implausible. The JASTROW version uses "gopher wood" in Genesis 6:14, "cedar" in Isaiah 41:19, and "cypress" in Isaiah 60:13. For the King James' "cedar trees, fir trees, and algum trees" JASTROW substitutes "cedar-trees, cypress-trees, and sandalwood". The cypress is said by POST (265) to be found in the middle mountain zones throughout the area. It grows to 80 feet tall, but its usual height in Palestine is 50 or 60 feet. Indeed, one specimen, which is said to be the oldest tree in Europe and which grows in Lombardy, is reported as being 120 feet tall, with a trunk circumference of 8 or 10 feet (299). It was a good-sized tree at the time of JULIUS CAESAR. NAPOLEON, the conqueror, was so impressed with it that he spared it when he built his road through the Simplon pass (298). Cypress wood is very hard and durable, and was much employed by the ancients in the manufacture of idols. It is said to have been used extensively in shipbuilding by the Phoenicians, Cretans, and Greeks. ALEXANDER THE GREAT constructed his galleys out of cypress wood. A pyramidal form of the tree, variety *pyramidalis* (Targ.-Rozz.) Nym., is very extensively planted in Mohammedan and Armenian cemeteries.

Several Hebrew words are rendered "cypress" in the various Scriptural translations. The only passage in which the word occurs in the Authorized version is Isaiah 44:14 and there the original Hebrew word is "tirzah" or "tizza". This passage has been the source of much discussion. The Greek Septuagint omits it altogether (184) and the Revised and Douay versions render it "holm tree", but the true holm or holly, *Ilex aquifolium* L., does not grow in the region indicated. Others have suggested the holm oak, *Quercus ilex*, while GROSER is convinced that it is the Valonia oak, *Q. aegilops* (146). One writer says that it is *Q. aegilops* "without any doubt." JASTROW calls it "ilex", which might be interpreted as meaning either *Ilex aquifolium* or *Quercus ilex*, but more probably the latter. MOFFATT and GOODSPEED both regard it as the planetree, *Platanus orientalis* (which see), and in this identification we concur.

In Song 1:14 and 4:13 marginal Bibles suggest "cypress" for the word "camphire" of the Authorized Version, but these two references are now generally regarded as applying to the henna, *Lawsonia inermis* (which see).

While the Authorized and Revised Versions' "algum trees" and "almug trees" (I Kings 10:11-12 and II Chronicles 9:10-11) are often considered to have been the red sandalwood, *Pterocarpus santalinus* (which see), MOFFATT believes that the "algum trees" of II Chronicles 2:8 were not sandalwood, but were cypress. This seems plausible because the sandalwood is a native of India and does not grow with the cedars and firs of Lebanon as is required by this passage. The cypress, on the other hand, is a common tree of mountainous parts of Bible lands, including Lebanon. The Douay version regards them as "pine trees". By us the "algum" is regarded as *Sabina excelsa* (which see).

The "gopher wood" of Genesis—Hebrew "gopher" (Greek, ξύλα τετράγωνα or τετραγωνία)—of which NOAH's ark was made, presents another problem. It is often left untranslated as in the Authorized and JASTROW Versions, while some versions of the Bible render it "pine" and some others "cedar". The GOODSPEED version calls it "oleander wood", but the small size of the oleander shrub and its thin stems definitely eliminate this as a possibility. The weight of opinion now seems to be with MOFFATT in favor of the cypress. That its wood was the "gopher wood" of which NOAH's ark was constructed is rendered even more plausible by the fact that it is so very durable. The ancient Egyptians used it for making coffins and the modern Greeks still employ it thus. The doors of St. Peter's in Rome are made of cypress and after 1200 years still show no signs of decay. It is a common timber tree in parts of the Levant and is found throughout Syria and Palestine. CELSIUS records the Hebrew name "gofer" for this tree (267), obviously the same word as used in Genesis, yet CARRUTHERS dismisses the whole subject with the categorical statement: "cedar, pine, and cypress have been conjectured, for no valid reason" (184). The Douay version sidesteps the issue by translating the verse

"Make thee an ark of *timber planks* . . ." ST. JEROME points out that the etymology of the word "gopher" signifies a resinous wood. The word even resembles the Greek word for the cypress "kupros, cupar, or cuper", according to FOLKROD (95).

The "box" of Isaiah 41: 19 and 60: 13 is regarded by MOFFATT as also applying to the cypress, but other authorities hold with us in considering it to be a true box, *Buxus longifolia* (which see).

The cypress occurs abundantly in legendry (298). The Greek god APOLLO was very fond of a mortal boy named CYPARISSOS, who, in turn, was greatly attached to a tame stag that grazed on sacred Ceos. One day CYPARISSOS accidentally killed his playmate stag, and, overcome with grief, he begged the gods to let him mourn forever. Thereupon APOLLO transformed him into a cypress tree, in which guise he still mourns. VENUS, mourning ADONIS, wore a wreath of cypress, as also did the tragic Muse, MELPOMENE. A cypress tree growing at the tomb of King CYRUS, the Persian, was said to drip blood every Mohammedan sabbath (our Friday). Cypress wood was used for roofing temples. The Persian fire-worshippers planted cypresses at their temples because the shape of its cones suggested fire to them; and ZOROASTER lived under a cypress tree. The island of Cyprus was named for this tree and on it the cypress was worshipped. In classical mythology the earth goddess, CERES, plugged the crater of Mount Etna with a huge cypress trunk, and thereby imprisoned forever the fire god, VULCAN, at his forges beneath the mountain (298). Eruptions of Etna were regarded as attempts of VULCAN to escape his imprisonment.

The handsome tree of the southeastern United States, *Cladrastis lutea* (Michx. f.) Koch, often called "gopher-wood", has nothing whatever to do with the Biblical tree of that name.

90. Cynomorium coccineum L.
(FIGURE 60)

JOB 30: 4—Who cut up mallows by the bushes, and *juniper roots* for their meat.

The roots of the "juniper" of the Holy Land—actually a species of broom, *Retama raetam* (which see)—are extremely nauseous, even somewhat poisonous (299), so it is difficult to accept the Authorized Version's translation of this passage from Job. The Douay version is similar: "And they ate grass, and barks of trees, and the *root of junipers* was their food". The GOODSPEED rendition is: "They pluck salt-wort by the bushes, And their food is the *root of the broom*." The JASTROW rendition is "They pluck salt-wort with wormwood; And the *roots of the broom* are their food." MOFFATT, apparently realizing the incongruity of stating that a nauseous-poisonous plant was eaten as food, changes the verse to read: "gathering salt-wort under bushes, using *broom-roots* for fuel." That the roots of the broom could be used as fuel is easily believable, but not that they were used as food. Yet, LEESSER translates this verse "Who crop off mallows by the bushes, and have *broom-bush roots* as their bread." The Hebrew words involved here are "r'tamim sho'resh", meaning literally "roots of the broom-bush."

Still it is not at all impossible that the people referred to might have collected something which they supposed to be "juniper" roots and used this for food. There is a parasitic plant known as the scarlet cynomorium, *Cynomorium coccineum*, which grows in salt marshes and maritime sands (267), the favorite habitat of saltwort. The broom also grows abundantly in sandy places and wadies and has often been observed with great masses of the parasitic cynomorium attached to its roots, especially in the region about the Dead Sea. The parasite is cylindric and fleshy, about a foot tall, covered with imbricate deciduous scales in place of leaves, and bearing at its apex a thick, club-shaped, crimson, head-like or spike-like spadix with many rather inconspicuous, unisexual, little flowers. The plant was originally called "*Fungus melitensis*" because of its fungus-like growth and general appearance and was supposed to grow only on the island of Malta. Actually, it is native to the whole of the Levant, southern Europe, and northern Africa (158). On Malta in ancient days it was so highly prized for its supposed medicinal virtues in the treatment of dysentery (158) that a military sentry was placed on guard at every spot where it was known to occur (299). It is frequently eaten in times of food scarcity, especially on the Canary Islands, so it is quite probable that this is actually the plant referred to by JOB (299).

91. Cyperus papyrus L.
(FIGURE 59; FIGURE 63)

EXODUS 2: 3 & 5—And when she could no longer hide him, she took for him an ark of *bulrushes*, and daubed it with slime and with pitch, and put the child therein; and she laid it in the *flags* by the river's brink . . . And the daughter of PHARAOH came down . . . and when she saw the ark among the *flags*, she sent her maid to fetch it.

JOB 8: 11—Can the *rush* grow up without mire? can the flag grow without water?

ISAIAH 18: 2—That sendeth ambassadors by the sea, even in vessels of *bulrushes* upon the waters . . .

ISAIAH 19: 6-7—And they shall turn the rivers far away; and the brooks of defence shall be emptied and dried up: the *reeds* and flags shall wither. The *paper reeds* by the brooks, by the mouth of the brooks, and every thing sown by the brooks, shall wither, be driven away, and be no more.

ISAIAH 35: 7—And the parched ground shall become a pool, and the thirsty land springs of water: in the habitation of dragons, where each lay, shall be grass with *reeds* and rushes.

ISAIAH 58: 5—Is it such a fast that I have chosen? a day for a man to afflict his soul? is it to bow down his head as a *bulrush*, and to spread sackcloth and ashes under him? . . .

The Egyptian bulrush or papyrus, *Cyperus papyrus*[55], referred to in the passages cited above, has smooth 3-sided stems attaining a height of 8 to 10 or even 16 feet and a thickness of 2 or 3 inches at the base, tapering upwards gradually and terminating in a sheath from which issues a large tuft of numerous grass-like panicles, each bearing many small florets like those of a grass (299). The whole inflorescence, when perfect, assumes the form of a pendulous umbel, comparable to that of a common loose house-mop standing on its handle. The Hebrew words which are thought to refer to this plant are "gômê" or "gomeh" (184, 299) or "gomé" (267) (Greek, $θίβη$, meaning "ark or wicker basket", $πάπυρος$, meaning "papyrus") and "aroth" or "arot" (Greek, $τὸ ἄχι τό χλωρόν$), rendered "meadows" in the Revised Version. The papyrus formerly grew in great abundance along the banks of the Nile, forming a dense jungle (some writers use the term "forest"!), but is now practically extinct in lower Egypt, although still found along the White Nile and in Nubia (184) as subsp. *antiquorum* (Willd.) Chiov. (203b).

Even though it has been exterminated from the Scriptural parts of Egypt, it is still found—as var. *palaestinae* Chiov. (203b)—in Palestine, and TRISTRAM says that he found it "growing luxuriantly in a swamp at the north end of the plain of Gennesaret, and it covers many acres of the inaccessible marshes of the Huleh, the ancient Merom . . . The whole marsh is marked on the maps as 'impassable', and most truly it is so. I have never anywhere else met with a swamp so vast and so utterly impenetrable. First there is an ordinary bog, which takes one up to the knees in water; then, after a mile, a belt of deeper water, where the yellow water lily flourishes. Then a belt of tall reeds, the open water covered with white water lily; and beyond, again, an impenetrable wilderness of Papyrus, extending right across to the east side. In fact, the whole is simply a floating bog of several miles square—a very thin crust of vegetation over an unknown depth of water, and if the weight of the explorer breaks through this, suffocation is imminent" (299). Dr. THOMSON describes this area in a similar manner and calls it a "ten-mile marsh" of "utterly impassable slough" (299). It seems, however, from more recent writers, that the papyrus is found in only a few such spots in Palestine and Syria today. In fact, Miss HORNER says that "it is now completely extinct in Egypt in the wild state and exists only locally in Africa and in a few spots in Palestine" (178). POST records it from Râs-ul-'Ayn, Khân-Minyah, Hulah, Hadera, and ul-Ghawr (267). It is said to grow abundantly about the lake of Tiberias (306). OPPENHEIMER, writing as late as 1938 (255a), describes a dense jungle of it in

---
[55]Also called *Papyrus antiquorum* Willd.

the Huleh swamps along the Jordan river. The papyrus plant is called "babeer" by the Arabs (184).

Besides being used for making small vessels for floating on the water, as described in Exodus 2: 3, and for mats and various other domestic purposes, it is famous as the source of the material from which the paper of the ancients was made (268, 293, 299). Most of the invaluable records found in the Egyptian tombs and those in the excavated cities of Herculaneum and Pompeii (which were suddenly buried in ashes thrown out by Mount Vesuvius in 79 A.D.)—were written on paper made from these papyrus stems, attesting to the tremendous importance of this plant in early times—and to us today in supplying records of those times (299).

The method of manufacturing paper from papyrus seems to have been very simple. The stems of the plant were first peeled and the pith then cut longitudinally into thin slices, which were laid side by side with their edges just touching one another. These were then sprinkled with gummy water, or, as some say, with the muddy waters of the Nile. A heavy pressure was then applied, and thus the whole became united into one piece. The sheet was then dried and cut up into pieces of the required size. In the better grades of papyrus paper several layers of stem-slices were laid crosswise upon each other (158).

The lower part of the young papyrus plant was eaten by the natives. Shoes were also made of it. It was apparently once a very common plant in Egypt, Abyssinia, and Syria. When Egypt was captured by the Greeks, its use for paper was adopted by the Greeks. The Romans also employed it. In that rather late period, even though linen parchment and other materials were in use for writing purposes, papyrus was yet greatly preferred. It was, in fact, employed in Italy until the 11th and 12th centuries, when cotton paper first succeeded in superceding it.

The pale fawn-colored tassel-like inflorescences at the summit of the stems were used to adorn temples and to crown the statues of gods. They were also worn as a crown by famous and illustrious men and national heroes. The woody rootstocks were chewed—as they still are today in Abyssinia—because they contain a sweet juice resembling licorice. Because of the scarcity of timber trees in Egypt, papyrus stems were used as fuel.

Some commentators believe that in addition to the papyrus some of the passages cited at the head of this chapter may refer also to *Scirpus lacustris* (which see) and various species of *Juncus*, which grow in wet places in the Holy Land and Egypt. STEARN has aptly pointed out that Victorian illustrations of MOSES among the "bulrushes" show him not among papyrus or even the plant now usually referred to as bulrush in Europe (*Scirpus lacustris*), but rather among reed-mace or cattail (*Typha*). This mistake of early painters, he notes, has led to the English vernacular name of "bulrush" being transferred in common usage from *Scirpus* to *Typha*.

The MOFFATT translation of Exodus 2: 3 is particularly lucid: "When she could hide him no longer, she took a creel made of *papyrus reeds*, daubed it over with bitumen and pitch, and put the child in it, laying it among the *reeds* at the side of the Nile". The GOODSPEED translation also uses "papyrus reed" here, but the Douay version says "bulrushes". Both MOFFATT and the GOODSPEED versions use "papyrus" in place of the "rush" of the King James and Douay versions of Job 8: 11. Both, however, continue to use "bulrush" in Isaiah 58: 5 and "reeds" in Isaiah 35: 7 and the first part of 19: 6-7. In the second part of the latter passage the "paper reeds" of the King James version is rendered "meadow grass" by MOFFATT and "sedge-grass" by the GOODSPEED version. Isaiah 18: 2 is rendered: "That sends ambassadors by sea, In *papyrus vessels* on the face of the waters" by the GOODSPEED version, while MOFFATT says: "that sends its envoys overseas in *light skiffs* down the stream." The Douay version completely misses the symbolism of Isaiah 58: 5 by saying: "to wind his head about like a *circle*", instead of "To bow down his head like a *bulrush*" as the GOODSPEED and MOFFATT translations so beautifully render it. The Douay version uses "reed" in Isaiah 35: 7 and the first part of 19: 6-7 and "bulrush" for the "rushes" of the former verse and "bulrush" for the "flags" of the latter. The JASTROW version follows the King James in regard to the "bulrushes", "flags" and "rushes" of Exodus and Job. In Isaiah 18: 2, however, it uses "papyrus" for "bulrushes". In Isaiah 19: 6-7 it agrees on "reeds", but substitutes "mosses" for "paper reeds"! In the other two Isaiah passages it agrees with the Authorized Version.

Some commentators consider the "aroth" or "arot", translated "paper reeds" in the Authorized Version, to be "green herbage" in general (299). The Revised Version renders it "meadows" (27). LEESSER gives a similar translation: "well rooted plants" (9). According to SMITH, "there is not the slightest authority" for the "paper reeds" rendition of the King James version (306). KIMCHI says that "aroth" is the word referring to pot-herbs and green plants in general, and SMITH maintains that "it probably denotes the open grassy land on the banks of the Nile" (306). According to BRUCE, modern Abyssinians still make small boats out of papyrus.

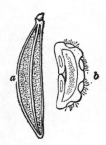 

FIGURES 17-18. — Fig. 17 (*left*) shows the seed of *Cuminum cyminum*, the cummin, in longitudinal view and in cross-section. — Fig. 18 (*right*) similarly depicts the seed of *Coriandrum sativum*, the coriander (see also fig. 57). Both these seeds are highly aromatic and pungent and were used in Bible times much as we today use the seeds of poppy, caraway, and sesame. They were sprinkled on bread, cakes, and pastry, were eaten with fish and meats, especially in stews, as a stimulant to the appetite, and were ground up with grain to make a more palatable flour. (Lindley, Medical and Oeconomical Botany, 1856).

92. Diospyros ebenaster Retz. 93. Diospyros ebenum König
94. Diospyros melanoxylon Roxb.

EZEKIEL 27: 15—... many isles were the merchandise of thine hand: they brought thee for a present horns of ivory and *ebony*.

The Hebrew word "hodnim" or "havnim" is the one rendered "ebony" in the above passage, probably correctly so, although it is completely omitted from the Septuagint. The ebony here referred to was without doubt the wood of *Diospyros ebenaster* and *D. melanoxylon*, two species of "date-plum" or "date-tree" (184; not to be confused with the date palm, *Phoenix dactylifera*) from India (299), and perhaps also of *D. ebenum* of Ceylon (184, 299). It was conveyed by Phoenician ships across the Arabian Sea and up the Red Sea or across the Gulf of Oman and the Persian Gulf to the market at Tyre (184), whence it was transported overland by camel caravan. The outer wood of these date-plum trees is white and soft, but when old the interior wood becomes hard, black, heavy, and durable, and still constitutes most of the costly ebony of commerce. Ebony takes on a fine polish and is highly valued for cabinet work, turnery, the manufacture of fancy ornamental articles (299) and instruments, and as a veneer for other woods (158). The trees producing this wood are large, but slow-growing, and have simple entire-margined leaves and small bell-shaped flowers. The usable heartwood attains a maximum diameter of two feet in well-developed trees (184).

The prophet EZEKIEL mentioned (about 588 B.C.) ivory and ebony together because they were probably used in conjunction with each other by the Hebrews (268). Ebony was, among the ancients, and still is today, frequently inlaid with ivory, with which it contrasts so strikingly in color.

JONES suggests that the Tyrian merchants also handled ebony from Ethiopia (306). He admits that "it is not known what tree yielded the Ethiopian ebony" It seems far more probable to us that only Indian and Ceylonese ebony were involved in the Scriptural reference, and that the 200 logs of ebony presented to the kings of Persia every year by the Ethiopians were originally from India or Ceylon (95). All the modern English versions of the Bible agree on the use of the term "ebony" in the Ezekiel passage.

According to classical mythology, the royal throne of PLUTO, lord of the underworld, was made of ebony. The Pythian APOLLO and the statues of many Egyptian gods and goddesses, especially those of Night, Darkness, and Sorrow (95), were carved of ebony (298).

95. Eberthella typhi (Schröt.) Buchanan
96. Rickettsia prowazeki da Rocha-Lima
97. Streptococcus erysipelatis Fahleisen

LEVITICUS 26: 16—I also will do this unto you; I will even appoint over you terror, consumption, and the *burning ague*, that shall consume the eyes, and cause sorrow of heart . . .
DEUTERONOMY 28: 22—The Lord shall smite thee with a consumption, and with a *fever*, and with an inflammation, and with an *extreme burning* . . .
MATTHEW 8: 14-15—And when JESUS was come into PETER'S house, he saw his wife's mother laid, and sick of a *fever*. And he touched her hand, and the *fever* left her . . .
JOHN 4: 52—Yesterday at the seventh hour the *fever* left him.

If it is true that the commonest fever of Biblical days was typhoid fever, as many assert, then *Eberthella typhi* may be included as one of the organisms to be treated in this work, for it is the bacterium which causes typhoid fever, and bacteria are usually included in the plant kingdom. The Hebrew words applying to fevers in the Bible are "kâdâchât" or "kaddachath" of the Old Testament, "dalleketh" or "daleket" and "charchur" of the New Testament, the last-mentioned perhaps referring to erysipelas, caused by another bacterium, *Streptococcus erysipelatis* (306).

The "fever" of Acts 28: 8, however, is generally regarded as that regularly accompanying the "bloody flux" or what is now called dysentery. While there is a bacillary type of dysentery, caused by a bacillus (plant) in the intestinal canal, it is more probable that the serious and often fatal (306) type of dysentery found in the Holy Land was of the amebic kind, caused by the presence of amebae (animals) in the intestines, and thus would not properly enter into the scope of a work on the plants of the Bible. Intermittent fevers of the malarial sort, also caused by animal organisms, are common in Arabia along with dysentery and inflammatory fevers. Putrid fever, now known as typhus, is common in the region and probably also was so in Biblical days. It is caused by the organism *Rickettsia prowazeki*. MOFFATT, GOODSPEED, and JASTROW use the direct term "fever" in place of the "burning ague" of Leviticus 26: 16, but the Douay version there says "poverty, and *burning heat*." Other Biblical diseases of plant origin are described herein under *Epidermophyton*, *Mycobacterium*, *Pasteurella*, *Shigella*, and *Staphylococcus*.

98. Elaeagnus angustifolia L.

I KINGS 6: 23 & 31-33—And within the oracle he made two cherubims of *olive tree*, each ten cubits high . . . And for the entering of the oracle he made doors of *olive tree*: the lintel and the side posts were a fifth part of the wall. The two doors also were of *olive tree*; and he carved upon them carvings of cherubims and palm trees and open flowers, and overlaid them with gold, and spread gold upon the cherubims, and upon the palm trees. So also made he for the door of the temple posts of *olive tree*, a fourth part of the wall.
I CHRONICLES 27: 28—And over the *olive trees* and the sycomore trees that were in the low plains was BAAL-HANAN the Gederite . . .
NEHEMIAH 8: 15—And that they should publish and proclaim in all their cities, and in Jerusalem, saying, Go forth unto the mount, and fetch olive branches, and *pine branches*, and myrtle branches, and palm branches, and branches of thick trees, to make booths, as it is written.
ISAIAH 41: 19—I will plant in the wilderness the cedar, the shittah tree, and the myrtle, and the *oil tree*.
MICAH 6: 7—Will the Lord be pleased with thousands of rams, or with ten thousands of rivers of *oil*?

The Hebrew word involved here is "êtz shamen", "ets shemen", or "âtzai-shemen". It has been variously translated, usually as "oil tree" or "olive tree" in the Authorized Version, but in the Nehemiah reference is rendered "pine

---

FIGURE 19. — *Cupressus sempervirens* var. *stricta*, the columnar cypress, planted since ancient times in cemeteries throughout Mediterranean and Near Eastern lands. The Biblical "cypress", source of the "gopher wood", of which NOAH'S Ark was built, is very similar, but has wide-spreading horizontal branches. (Wood engraving of a view in Jerusalem in the 1870's from C. W. Wilson's Picturesque Palestine, 1883).

branches", apparently in deference to the Septuagint, which, for some strange reason, has "cypress" in this passage. The Douay version uses "beautiful wood", the Revised, JASTROW, and GOODSPEED versions "wild olive", LEESSER and MOFFATT "oleaster" for the same passage! Many authorities believe that the narrow-leaved oleaster, *Elaeagnus angustifolia*, is the plant referred to not only in the Nehemiah verse, but in all the passages cited at the head of this chapter (299). The MOFFATT, GOODSPEED, and Douay versions, however, all seem to regard the olive, *Olea europaea* (which see), as the plant referred to in the other passages, except that the Douay version translates the second phrase of Micah 6: 7 entirely differently: "or with many thousands of fat *he goats*". Obviously the olive cannot be meant in the Nehemiah reference, for it is also mentioned specifically in the same clause. The Hebrew word for the olive is quite uniform and distinct—"zait" or "za'tim". Thus, it seems quite probable that SMITH, GROSER, JONES, and other commentators are correct when they render as "oleaster" all of the "êtz shamen" references cited above.

The narrow-leaved oleaster is a small, stiff-branched tree or graceful shrub, 15 to 20 feet tall, common in all portions of Palestine except, perhaps, the Jordan valley. It is said by some authorities to be especially frequent about Mounts Tabor and Hebron and in Samaria, but Dr. EVENARI reports it as rare now. The leaves are small, narrowly elongated, bluish above, and silvery beneath. The flowers are small and tubular, silvery on the outer surface, yellow within, and extremely fragrant. The fruit is either small and tasteless, or, in some varieties, as large as an olive and quite edible, although bitter. The wood is hard and fine-grained, and therefore would be well-suited to the carving of images and figures. Although still sometimes called "wild olive", it has no botanical relationship with the true olive. It yields a rather inferior kind of oil, which is used as a medication, but not for food, and which is thought by some to constitute the "oil" referred to in Micah 6: 7. The fruits of the oleaster, known as "Trebizond dates", are dried and pounded and made into a kind of bread by the Arabs (299).

Some writers have tried to identify the "oil tree" of Isaiah with *Balanites aegyptiaca* (which see), the Jericho balm—"zukkum" or "zackum" of the Arabs—which is also an oil-producing tree, but TRISTRAM, whose conclusions are usually very dependable, insists that it was the oleaster, which is far more common in Palestine and would be better known. In fact, TRISTRAM makes the categoric statement, contradicted by other writers, that *Balanites* "does not exist except in the tropical region of the Jordan Valley" (299). Zackum oil, too, is held in high repute by the Arabs for its medicinal properties (306).

Because of the misleading translation "pine branches" in Nehemiah, some commentators have thought that the "oil tree" was actually a pine and that the "oil" was the tar and turpentine obtained from pine trees (299). There is no evidence to support these assumptions. Marginal Bibles give the alternate translation "oily trees" and "trees of oil" for I Kings 6: 23, while the Revised Version says "olive wood". In Isaiah 41: 19 the Revised Version has a marginal alternate translation of "oleaster".

99. **Epidermophyton rubrum** Castellani
100. **Favotrichophyton violaceum** (Sabouraud) Dodge
101. **Trichophyton rosaceum** Sabouraud

LEVITICUS 13: 30-37—Then the priest shall see the plague: and, behold, if it be in sight deeper than the skin; and there be in it a yellow thin hair; then the priest shall pronounce him unclean: it is a dry *scall*, even a leprosy upon the head or beard. And if the priest look on the plague of the *scall*, and, behold, it be not in sight deeper than the skin, and that there is no black hair in it; then the priest shall shut up him that hath the plague of the *scall* seven days: And in the seventh day the priest shall look on the plague: and, behold, if the *scall* spread not, and there be in it no yellow hair, and the *scall* be not in sight deeper than the skin; He shall be shaven, but the *scall* shall he not shave; and the priest shall shut up him that hath the *scall* seven days more: And in the seventh day the priest shall look on the *scall*; and, behold, if the *scall* be not spread in the skin, nor be in sight deeper than the skin; then the priest shall pronounce him clean: and he shall wash his clothes, and be clean.

But if the *scall* spread much in the skin after his cleansing; Then the priest shall look on him: and, behold, if the *scall* be spread in the skin, the priest shall not seek for a yellow hair; he is unclean. But if the *scall* be in his sight as a stay, and that there is black hair grown up therein; the *scall* is healed, he is clean: and the priest shall pronounce him clean.

LEVITICUS 14: 54—This is the law for all manner of plague of leprosy, and *scall*.

According to modern authorities the "scall" (Hebrew, "netek") of the Old Testament is what we now call ringworm, and, indeed, the MOFFATT and GOODSPEED versions so translate the word. JASTROW still uses the term "scall" and describes it as "a leprosy of the head and beard." Ringworm is a contagious infection of the skin of man and domestic animals due to several fungus parasites, chiefly the three listed at the head of this chapter. These parasites form ring-shaped discolored patches covered with vesicles or powdery scales. These patches may occur anywhere on the body, face, or scalp, but especially in the hairy parts. Medical men differentiate several varieties under the names "tinea circinata", "tinea tonsurans", etc. (158).

FIGURE 20.—*Lawsonia inermis,* the henna-plant. Referred to in the Song of Solomon as "camphire", this was an important cosmetic plant of the ancients. Its leaves produced a dye for painting finger- and toe-nails, beards, mustaches, etc. (After Blanco, Flora de Filipinas, 1877).

102. Faba vulgaris Moench

II SAMUEL 17: 27-28— . . . and BARZILAI the Gileadite of Rogelim, Brought beds, and basons, and earthen vessels, and wheat, and barley, and flour, and parched corn, and *beans*, and lentiles, and parched pulse.

EZEKIEL 4: 9—Take thou also unto thee wheat, and barley, and *beans*, and lentiles, and millet, and fitches, and put them in one vessel, and make thee bread thereof, according to the number of the days that thou shalt lie upon thy side, three hundred and ninety days shalt thou eat thereof.

The Hebrew word "pôl" (Greek, κύαμος) is rendered "beans" in the above verses by all translators and unquestionably refers to the broad bean, *Faba vulgaris*[56]. This species, an annual plant, is thought to be originally indigenous to northern Persia, but was extensively cultivated in western Asia in very early times as a food plant, so that the actual land of its origin is now a question that is hopelessly obscured. It was early cultivated in Egypt, for beans of this sort have been found in the mummy coffins in Egyptian tombs. It was also cultivated by the Greeks and Romans (299). The Romans introduced it into England (126), where it is now chiefly used as a fodder for horses[56a]. However, the variety known as the Windsor bean (f. *megalosperma* (Alef.) Beck) is occasionally cultivated in gardens in Europe and America as a vegetable. Broad beans are very extensively cultivated in Palestine and their meal is made into bread now even as it was in Biblical days. They are also boiled and eaten there as a vegetable by the poorer classes.

The broad bean is a hardy, erect, simple-stemmed plant, which grows to about 3 feet tall and is very smooth throughout. Its compound leaves consist of only 2 to 6 leaflets and there are no tendrils. The pea-like flowers are borne singly or in small clusters in the leaf-axils or on very short peduncles and are all white or pencilled with lilac streaks and marked with a black or purplish spot (267, 268). The pods are large and thick and the beans themselves large and compressed (in f. *megalosperma*) or small and nearly round (in f. *equina* L.), brown or black in color. BURCKHARDT states that the shores of the Nile are often beautifully fragrant for long distances with the sweet perfume of the bean fields adjacent to them. The broad bean is still widely cultivated in Syria, mowed by means of the scythe at the time of the wheat harvest, and fed to cattle for fodder. According to SMITH, beans are in blossom in Palestine in January and continue to flower until March. In Egypt they are sown in November and reaped in the middle of February (306).

Strange to say, the bean was held in great disrepute among the higher classes in classical times. According to beliefs then current it caused nightmares and insanity. To dream of beans presaged trouble; "even ghosts fled shuddering from the smell of beans" (298). The earth goddess, CERES, refused to include beans among her gifts to man. The seers, prophets, and oracles of those days refused to eat them, believing that to do so would becloud their vision and dim their foresight. Even HIPPOCRATES (460-359 or 377 B.C.), the

---

[56] Also referred to as *Vicia faba* L. and *Faba faba* (L.) Pollard.

[56a] W. T. STEARN states that the wild origin of this plant is uncertain "but is apparently Mediterranean." It was cultivated in Neolithic and later prehistoric time (77a). He points out that "It is most improbable that a plant of such economic importance would have to wait until Roman times for introduction into Britain. The early beans belonged to a small-seeded type, comparable to var. *minor* Beck, with seeds 0.65—1.25 cm. long. Such beans have come to light from Bronze Age lake dwellings in Switzerland. Similar beans have been found at Glastonbury, Somerset, England, on the site of lake dwellings belonging to about the time of Christ and then inhabited by a Celtic people. I should certainly not credit the Romans with their introduction. Very small beans have been excavated in Palestine at Beth-Shean and attributed to about 3000—2000 B.C." (116a, 273a).

Greek physician, usually regarded as the Father of Medicine, believed that to eat beans was injurious to one's vision. CICERO avoided them because they "corrupted the blood and inflamed the passions". Roman priests regarded them as unholy and unclean. The Greek philosopher PYTHAGORAS (582-after 507 B.C.) believed that some human souls after death entered bean plants. Tradition tells us that, pursued by enemies, the philosopher came to a vast bean field which he would have to cross in order to escape his pursuers, but which he refused to cross because he believed the bean plants to be animated by human souls and therefore not to be tramped upon. So he was overtaken and slain by his enemies (298).

103. Ferula galbaniflua Boiss. & Buhse
(FIGURE 61)

EXODUS 30: 34—Take unto thee sweet spices, stacte, and onycha, and *galbanum*; these sweet spices with pure frankincense . . .
ECCLESIASTICUS 24: 15— . . . a pleasant odour like the best myrrh, as *galbanum*, and onyx, and sweet storax . . .

The Hebrew word here rendered "galbanum" is "chelbenah" (Greek, χαλβάνη) and it is very probable that it refers to the plant *Ferula galbaniflua*[57]. Galbanum is a fetid yellowish or brownish gum resin containing the chemical substance umbelliferone (158), obtained from several species of plants allied to fennel, in the carrot family, natives of Syria and Persia. It is not certain just exactly what species today yield the galbanum of commerce, so it is naturally even less certain what one or ones were used by the Israelites (299). POST records nine species and varieties of *Ferula* from the Holy Land area (265, 266), but, curiously enough, does not mention the one suggested by SMITH. Dr. EVENARI claims that this species is not Palestinian, and, further, that none of the Palestine species produce galbanum. Like others of the species of the ancient Jews, this product was imported. The plants producing galbanum are strong-rooted perennials, their flowering stems reaching several feet in height, with alternate, partially sheathing, finely divided leaves, terminated by umbels of yellow or greenish-white flowers. The gum is the natural exudation of the stem or is obtained by making a transverse incision in the young stem a few inches above the ground. The milky juice which exudes from the wound soon hardens and forms one of the kinds of commercial galbanum. Its odor is strongly balsamic and pungent, disagreeable when burned, although not as bad as that of asafetida, the product of three allied Persian and East Indian species of the same genus (158). Galbanum is at present used medicinally as an antispasmodic and in the manufacture of varnish (158).

CARRUTHERS suggests *Galbanum officinale* D. Don as the source of the ingredient of the holy incense (184), stating that this plant grows in Syria, and JONES adds *Opoidia galbanifera* Lindl. as another source (184). These species are not listed by POST (266) and there is considerable question as to their taxonomic validity. A related plant, *Bubon galbanum* L. [58], native to the Cape of Good Hope, also produces a kind of galbanum, but, although seriously suggested by some authors, can have nothing to do with the galbanum of the Bible (299). SPRENGEL was of the opinion that the plant referred to in the Scriptures was *Ferula ferulago* L., a species native to northern Africa, Crete, and Asia Minor. Dr. HOWES of Kew, widely recognized as an authority on gum-producing plants, tells us that "Palestine or the Holy Land has never been regarded as a source of galbanum. Species considered as alternative sources of 'gum galbanum' are *Ferula schair* Borcz (Turkestan) and *F. rubricaulis* Boiss. (Persia). Other species mentioned in the older literature, are given as *Ferula erubescens* Boiss., *F. gummosa* Boiss. and *F. persica* Sims., but these are generally regarded as synonyms of *F. galbaniflua* Boiss. & Buhse."

---
[57] Also referred to as *Ferula persica* Sims.
[58] Also called *Peucedanum galbanum* (L.) Benth. & Hook. f.

104. Ficus carica L.
104a. F. carica var. silvestris Nees

GENESIS 3: 7—And the eyes of them both were opened, and they knew that they were naked; and they sewed *fig leaves* together, and made themselves aprons.
NUMBERS 13: 23— . . . and they brought of the pomegranates, and of the *figs*.
DEUTERONOMY 8: 8—A land of wheat, and barley, and vines, and *fig trees* . . .
JUDGES 9: 10-11—And the trees said unto the *fig-tree*, Come thou, and reign over us. But the *fig tree* said unto them, Should I forsake my sweetness, and my good fruit, and go to be promoted over the trees?
I SAMUEL 25: 18— . . . and two hundred cakes of *figs* . . .
I SAMUEL 30: 12—And they gave him a piece of a cake of *figs* . . .
I KINGS 4: 25—And Judah and Israel dwelt safely, every man under his vine and under his *fig tree*.
II KINGS 18: 31— . . . and then eat ye every man of his own vine, and every one of his *fig tree* . . .
II KINGS 20:7—And ISAIAH said, Take a lump of *figs*. And they took and laid it on the boil and he recovered.
I CHRONICLES 12: 40—Moreover they that were nigh . . . brought . . . cakes of *figs* . . .
PSALMS 105: 33—He smote their vines also and their *fig trees* . . .
PROVERBS 27: 18—Whoso keepeth the *fig tree* shall eat the fruit thereof . . .
SONG 2: 13—The *fig tree* putteth forth her green *figs* . . .
ISAIAH 28: 4—And the glorious beauty which is on the head of the fat valley, shall be a fading flower, and as the *hasty fruit* before the summer, which when he that looketh upon it seeth, while it is yet in his hand he eateth it up.
ISAIAH 36: 16— . . . and eat ye every one of his vine, and every one of his *fig tree* . . .
ISAIAH 38: 21— . . . Let them take a lump of *figs*, and lay it for a plaister upon the boil, and he shall recover.
JEREMIAH 5: 17— . . . they shall eat up thy vines and thy *fig trees* . . .
JEREMIAH 8: 13— . . . there shall be no grapes on the vine, nor *figs* on the *fig tree* . . .
JEREMIAH 24: 1-8— . . . behold, two baskets of *figs* were set before the temple of the Lord . . . One basket had very good *figs*, even like the *figs* that are first ripe: and the other basket had very naughty *figs*, which could not be eaten, they were so bad . . . *Figs*: the good *figs*, very good; and the evil, very evil, that cannot be eaten, they are so evil . . . Like these good *figs*, so will I acknowledge them that are carried away captive of Judah . . . And as the evil *figs*, which cannot be eaten, they are so evil.
HOSEA 2: 12—And I will destroy her vines and her *fig trees* . . .
HOSEA 9: 10— . . . I saw your fathers as the *firstripe* in the *fig tree* at her first time . . .
JOEL 1: 7 & 12—He hath laid my vine waste, and barked my *fig tree*: he hath made it clean bare, and cast it away; the branches thereof are made white . . . The vine is dried up, and the *fig tree* languisheth.
JOEL 2: 22— . . . the *fig tree* and the vine do yield their strength.
AMOS 4: 9— . . . when your gardens and your vineyards and your *fig trees* and your olive trees increased, the palmerworm devoured them.
AMOS 8: 1-2— . . . behold a basket of *summer fruit*. And he said, AMOS, what seest thou? And I said, A basket of *summer fruit* . . .
MICAH 4: 4—But they shall sit every man under his vine and under his *fig tree*: and none shall make them afraid . . .
NAHUM 3: 12—All thy strong holds shall be like *fig trees* with the *firstripe figs*: if they be shaken, they shall even fall into the mouth of the eater.
HABAKKUK 3: 17—Although the *fig tree* shall not blossom . . .
HAGGAI 2: 19—Is the seed yet in the barn? Yea, as yet the vine, and the *fig tree*, and the pomegranate, and the olive tree, hath not brought forth . . .
ZECHARIAH 3: 10—In that day, saith the Lord of hosts, shall ye call every man his neighbour under the vine and under the *fig tree*.
MATTHEW 7: 16— . . . Do men gather grapes of thorns, or *figs* of thistles?
MATTHEW 7: 17-20—Even his not every good *tree* bringeth forth good *fruit;* but a corrupt *tree* bringeth forth evil fruit. A good *tree* cannot bring forth evil *fruit*, neither can a corrupt *tree* bring forth good fruit. Every *tree* that bringeth not forth good *fruit* is hewn down, and cast into the fire. Wherefore by their *fruits* ye shall know them.
MATTHEW 21: 1 & 19-21— . . . they drew nigh unto Jerusalem, and were come to *Bethphage* . . . And when he saw a *fig tree* in the way, he came to it, and found nothing thereon, but leaves only, and said unto it, Let no fruit grow on thee henceforward for ever. And presently the *fig tree* withered away. And when the disciples saw it, they marvelled, saying, How soon is the *fig tree* withered away! JESUS answered and said unto them, Verily I say unto you, If ye have faith, and doubt not, ye shall not only do this which is done to the *fig tree* . . .
MATTHEW 24: 32—Now learn a parable of the *fig tree;* When his branch is yet tender, and putteth forth leaves, ye know that summer is nigh.
MARK 11: 13 & 20—And seeing a *fig tree* afar off having leaves, he came, if haply he might find anything thereon: And when he came to it, he found nothing but leaves; for the time of *figs* was not yet . . . And in the morning, as they passed by, they saw the *fig tree* dried up from the roots.
LUKE 6: 44—For every tree is known by his own fruit. For of thorns men do not gather *figs*, nor of a bramble bush gather they grapes.

LUKE 13: 6-9—... A certain man had a *fig tree* planted in his vineyard; and he came and sought fruit thereon, and found none. Then said he unto the dresser of his vineyard, Behold, these three years I come seeking fruit on this *fig tree*, and find none: cut it down; why cumbereth it the ground? And he answering said unto him, Lord, let it alone this year also, till I shall dig about it, and dung it: And if it bear fruit, well: and if not, then after that thou shalt cut it down.
LUKE 19: 29—And it came to pass, when he was come nigh to *Bethphage* ...
LUKE 21: 29-30—... Behold the *fig tree*, and all the trees; When they now shoot forth, ye see and know of your own selves that summer is now nigh at hand.
JOHN 1: 48 & 50—Before that PHILIP called thee, when thou wast under the *fig tree*, I saw thee ... Because I said unto thee, I saw thee under the *fig tree*, believest thou? ...
JAMES 3: 12—Can the *fig tree*, my brethren, bear olive berries? either a vine, *figs*? ...
REVELATION 6: 13—And the stars of heaven fell unto the earth, even as a *fig tree* casteth her untimely *figs*, when she is shaken of a mighty wind.

The common fig, *Ficus carica*, is unquestionably one of the most important plants of the Bible, in which it is mentioned no less than 57 times. It is the very first plant to be mentioned by name in the Scriptures[58a], although SMITH does not think (299) that fig leaves are really meant in the story of the aprons made in the garden of Eden (Genesis 3: 7). He bases his contention on the fact that the fig is not mentioned again in the Scriptures over a period of 2500 years [58b]. When we stop to consider, however, that the garden of Eden story is only a legend or tradition, probably not written down in its present form for hundreds or even thousands of years after the events in it were supposed to have taken place, it does not seem so improbable that the fig was indeed intended, for the fig was by that time probably commonly cultivated by the Hebrews[58c]. The Geneva Bible (1560) renders this verse: "They sewed *fig leaves* together and made themselves breeches," and so has become known also as the Breeches Bible (158).

The Hebrew word used in this reference, as well as in Deuteronomy 8: 8 and elsewhere, is "tĕênâh" or "t'aynah" (Greek, σῦκον or συκῆ, meaning, "a fortified place at Syracuse in Sicily," so called, no doubt, because of the fig trees growing there). This is the word used in references to the fig tree as such. The four other words used in the other passages allude, not to the tree as such, but to different stages or conditions of the fruit: (1) "tĕênîm" or "t'anim" (plural form of "tĕênâh"), used in Jeremiah 8: 13 and elsewhere, indicates figs as fruit, (2) "pag" or "pageha" (Greek, ὄλυνθος), used in Song 2: 13 and elsewhere, is the green or unripened fruit, which remains on the tree through the winter, (3) "bikkûrah" or "bi'kurah" (Greek, οκοπός, literally "one who watches"), used in Hosea 9: 10 and elsewhere, is the "firstripe" or "early fig", and (4) "debelah" or "d'velet" and "d'velim" (Greek, παλάθη), used in I Samuel 25: 18, II Kings 20: 7, and elsewhere, is a cake of dried figs, that is, the main produce of the tree kept for winter use, often mentioned in the Old Testament as a staple article of food (184). The Arabic name for the fig is "tîn". The "Bethphage" mentioned in the New Testament signifies literally a "house of green figs", probably a figurative name for a sunless ravine near

---

[58a]Dr. CONDIT in his recent book on the fig is inclined to believe that the fig of this passage may be some species of *Ficus* other than *F. carica* because he thinks that the "thin, rough leaves do not seem suitable for sewing together into aprons," and because "the word 'fig' may be and commonly is used for any one of several distinct species of the genus *Ficus* [possibly *F. bengalensis* or *F. indica*]". Since the former is not mentioned anywhere else in the Scriptures, and since the latter is native only to such distant places as Burma and Malaya, the authors do not think it likely that either could have been involved in this story (93a).

[58b]MILTON's "Paradise Lost" describes this fig as the banyan, *F. benghalensis* L., in these words:

"*The fig tree, not that kind for fruit renowned,
But such as at this day, to Indians known
In Malabar of Deccan spreads her arms,
Branching so broad and long that in the ground
The bended twigs take root and daughters grow
about the mother tree.*" (93a)

[58c]"W. R. PATON suggested that the story of ADAM and EVE and their aprons is reminiscent of an ancient custom of fertilizing fig trees by a pair of human scapegoats who, like the victims of the Thargelia, associated themselves with the tree by wearing its foliage or fruit" (93a).

Jerusalem (184). All the English versions of the Bible agree substantially with the Authorized Version in regard to the references to the fig in the Bible. The JASTROW version renders the "hasty fruit" of Isaiah 28: 4 "first-ripe figs". The GOODSPEED, JASTROW, and MOFFATT versions of the Genesis story say that the fig leaves were used to make "girdles", whereas the Douay and King James versions say "aprons" and the Geneva Bible says "breeches".

The fig is generally regarded as a native of southwestern Asia (158) or Syria (184), but in early times, as now, it was extensively cultivated in Egypt, Palestine, and Syria, and certainly formed one of the principal articles of food of the people[58d]. In Samuel we read that part of the present sent to King DAVID by ABIGAIL in 1060 B.C. consisted of two hundred cakes of figs. The fig tree is common throughout Palestine now, both wild and cultivated. There are today extensive fig orchards near Jerusalem and one or more fig trees in every private garden (299). This explains several Biblical passages such as the ones about every man sitting "under his own fig tree" and "eating of his own fig tree". The tree is very variable in habit of growth. Often it is found in the form of a long, straggling, branching shrub, even sprawling vine-like over rocks in stony places, as may be seen about Jerusalem and other parts of Palestine. In favorable situations, however, it assumes the character of a tree, usually not more than 20 or 30 feet tall. When thus standing singly it often forms a conspicuous object in the landscape (cfr. Mark 11: 13), its stem measuring 2 to 3 feet in diameter. In Palestine such trees are often seen overshadowing wells (299). It is frequently planted in the corners of vineyards even today, as it apparently was in Biblical times according to the story in the 13th chapter of Luke.

The fig tree has a very peculiar type of fruit, called a syconium (158, 293), which is actually a very much enlarged and fleshy receptacle, varying from round (in var. *globosa* Boiss.) to nearly conical, depending on the variety (146). This fruit is attached by the narrow end, like a pear, to the branches, the broad end having a small opening like a pore. The flowers are numerous, small, and diclinous, attached around the interior surface of the cavity of the receptacle. The gritty particles that one feels between one's teeth on eating a fig are the true fruits and seeds (299).

While tradition usually holds that it was the "apple" (that is, apricot) which was the "tree of knowledge" in the garden of Eden, other legends say that this mythical tree was the fig (125). The latter claim is doubtless based on the statement in Genesis that ADAM and EVE took fig leaves with which to clothe themselves immediately after eating of the forbidden fruit (219). The fruit of the fig is said to be poisonous for a short time before ripening, but when ripe the poisonous principle disappears and is replaced by sugar. The tree yields copious shade when it grows in the arborescent form and is still extensively planted in the courtyards of Oriental houses (268), where its shade is said to be fresher and cooler than that of a tent. Normally, two crops of figs are produced per year, the first or winter figs ripening in June, the second or summer figs on the new wood in August and September (267). Often when the summer figs are just starting some ripe winter figs may still be found lingering on the branches, half-hidden by the foliage. Figs are used medicinally in the East and are commonly employed as a remedy for boils and other cutaneous eruptions in the same way as the prophet ISAIAH used them to heal the malady of King HEZEKIAH in II Kings 20: 7. It is the summer crop of figs which is preserved for use in the winter, and these when dried, are made into cakes or are kept hung on strings. Baskets, dishes, and umbrellas are made by sewing fig leaves together; and a fig leaf is still the traditional apron used by sculptors on their statues of the human form. Besides being so extensively cultivated, it is one of the few plants found wild in all parts of the Holy Land. It puts out its earliest fruit-buds before its leaves, the former in February, the latter in April and May. When the leaves are out the fruit ought to be ripe,

---

[58d] CONDIT's studies show that "the fig tree was probably first cultivated in the fertile part of southern Arabia, where wild specimens, such as those reported in 1923 by C. M. DOUGHTY, are still found" (93a).

which fact helps explain the story in Matthew 21: 19.

While tradition usually states that it was a Judas-tree, *Cercis siliquastrum* (which see), on which JUDAS hanged himself, other legends state that it was a fig tree (125).

The fig has recently been shown to depend on one specific kind of insect to secure fertilization, and one cannot introduce the fig into a new region and hope to secure fruit without also introducing the insect on which it depends. When figs are grown outside their native haunts, it usually becomes necessary to resort to a process known as caprification in order to secure fruit (93a). This is accomplished by suspending branches of the wild caprifig, *F. carica* var. *silvestris* Nees, with fruits containing the fig wasp (*Blastophaga grossorum*), on the branches of the cultivated trees. The insects emerge from the inedible caprifigs and enter the edible ones, effecting cross-pollination. The resultant ripening of the seeds prevents the fig fruits from dropping prematurely and improves their flavor. Some few varieties of figs mature without caprification, but caprification is practiced in Smyrna, Spain, Portugal, and California with imported caprifigs and fig wasps.

The importance and value of the fig to the Israelites is illustrated very graphically in the fact that whenever the prophets of old berated the people for their wickedness, they always threatened that the vine and fig crops would be destroyed, and when they held out the promise of great rewards they said that the vine and fig crops would be restored. "To sit under one's vine and one's own fig tree" became a proverbial expression among the Jews to denote peace and prosperity (93a).

PLINY states that "figs are the best food that can be taken by those who are brought low by long sickness, and are on the recovery". Mount Olivet was famous for its fig trees in ancient times, and they are still found there (306). The fig is mentioned abundantly in tradition and legendry. An old tradition states that when MARY sought shelter for the baby JESUS from the pursuing soldiers of HEROD bent on killing him, a fig tree opened its trunk so that they could enter and be hidden until the soldiers had passed (298). In classical mythology LYCEUS, a Titan, was transformed into a fig tree by RHEA, the "Mother of the Gods". Another story states that the fig was created by BACCHUS, the god of wine. A fig tree is said to have been growing on the site of what was to become Rome when the cradle of ROMULUS and REMUS floated by and got stranded in its branches. For this reason the fig was venerated by the Romans even as late as the time of the Empire (B.C. 29 to A.D. 395). During the Bacchanalian feasts Roman women wore collars of figs as symbols of fecundity and the men carried statues of PRIAPUS carved from fig wood (29a, 298).

105. Ficus sycomorus L.

(FIGURE 62)

I KINGS 10: 27—And the king made silver to be in Jerusalem as stones, and cedars made he to be as the *sycomore trees* that are in the vales, for abundance.
I CHRONICLES 27: 28—And over the olive trees and the *sycomore trees* that were in the low plains was BAAL-HANAN the Gederite ...
II CHRONICLES 1: 15—And the king made silver and gold at Jerusalem as plenteous as stones, and cedar trees made he as the *sycomore trees* that are in the vale for abundance.
II CHRONICLES 9: 27—And the king made silver in Jerusalem as stones, and cedar trees made he as the *sycomore trees* that are in the low plains in abundance.
PSALMS 78: 47—He destroyed their vines with hail, and their *sycomore trees* with frost.
ISAIAH 9: 10— ... the *sycomores* are cut down, but we will change them into cedars.
AMOS 7: 14— ... I was an herdsman, and a gatherer of *sycomore fruit*.
LUKE 19: 4—And he ran before, and climbed up into a *sycomore tree* to see him: for he was to pass that way.

The Hebrew word translated "sycomore" in the above references is "shikmim" or "shikmoth" (Greek, συκομορέα) and there is no doubt at all as to the identity of the plant in question. It is the well-known sycomore-fig,

Ficus sycomorus[59], also sometimes called "mulberry-fig" or "fig-mulberry" (306). It should not be confused with the common "sycamore" of eastern North America, *Platanus occidentalis* L., or the Old World *P. orientalis* L. and related species, which ought, to avoid confusion, be referred to by their more correct name of "planetrees". The sycomore-fig of the Bible is a strong-growing, robust, wide-spreading tree of great importance and extensive use (306), growing 30 or 40 feet tall (299), sometimes attaining a trunk circumference of 20 or more feet, with a crown to 120 feet in diameter. The main stem or trunk is short, dividing into usually twisted and gnarled main branches near the ground. The lateral branches fork outward in every direction (178). It is thus a tree which is very easily climbed and is frequently planted along roadsides, which accounts for the fact that ZACCHAEUS chose it to secure a vantage point from which he might see JESUS. It is an evergreen tree with unlobed, somewhat heart-shaped, fragrant leaves, smooth above, slightly hairy beneath, and smaller than those of the common fig, *F. carica* (which see). The fruit is produced abundantly in clusters on all parts of the tree, both on young and old branches, and even on the old limbs and trunk itself. It is very similar to that of the common fig, only smaller and much inferior in quality. Nevertheless, being very sweet to the taste, it is used extensively for food in Palestine and Egypt, especially among the poorer people. In Cairo it is offered for sale by street-hawkers (299).

The wood of the sycomore-fig is very soft and porous, but in spite of this fact is very durable. Mummy-cases made of it over 3000 years ago have been found in good condition in Egyptian tombs. The sycomore-fig is very abundant in Egypt and is the largest exogenous tree found in that country. Its wood is there employed for making furniture, doors, and boxes, as well as sarcophagi (184). Roving nomads pitch their tents in its shade and its fruit still forms an important part of the meager diet of these wandering people (298). It is about as sacred to the primitive tribes of that region as the oaks were to the ancient Druids (298). It is frequently planted for shade and attains to a great age. According to SMITH, one tree of this species at Matharee, near Cairo, dates back, if one is to believe popular legend, to the time of the flight of JOSEPH and MARY to Egypt with the infant JESUS in B.C. 4, tradition saying that they rested under its welcome shade (299).

The sycomore-fig of Palestine and Egypt differs from the common fig in several respects, notably in that it is evergreen and has smaller unlobed leaves. Like the common fig, it also produces fruit several times during the year (219). It grows abundantly in the valleys and lowlands, and because of its low wide-spreading branches affords a delightful shade along roadsides and caravan routes in those hot and parched lands. In Palestine it was intimately connected with the rites and mysteries of ancient Nature-worship, against which the Hebrew prophets inveighed so often. It will not survive the inclement seasons of the mountainous regions (*cfr.* Psalms 78: 47), but abounds only in the lowlands today even as in Biblical times. It was one of the most valuable fruit trees of Jericho and Canaan (268), but was never very common in Palestine (184).

The Amos reference is incorrectly translated in the Authorized and Douay versions—in the former the prophet is described as "a *gatherer* of sycomore fruit", in the latter as "a herdsman *plucking* wild figs". The GOODSPEED, JASTROW, and MOFFATT versions are much more accurate botanically. MOFFATT says: "I am only a shepherd, and I *tend* sycomores", JASTROW says "I was a herdsman, and a *dresser* of sycomore-trees", while GOODSPEED makes it: "I am a shepherd and a *dresser* of sycomores." It is customary for the cultivators of the sycomore-fig three or four days before gathering, when the "fruit" is about an inch long, to pare or scrape off a part at the center point or to make a puncture there with the fingernail or a sharp-pointed instrument (306). Unless this cutting or piercing operation is performed on

---

[59] Also known as *Sycomorus antiquorum* Gasp.; referred to by some writers as "*Ficus sycomora*" (261) and "*Ficus sycamorus*" (184).

every fig, the "fruit" will secrete a quantity of watery juice and will not ripen. AMOS was apparently employed in making these incisions in the sycomore-figs—he was, thus, a "tender" or "dresser" of the sycomore, not a "gatherer" or "plucker".

So great was the value of these trees to the ancient Jews that King DAVID appointed a special overseer for them, as he did for the olives (*cfr.* I Chronicles 27: 28), and it is mentioned as one of Egypt's most serious calamities when her sycomores were destroyed by frost (Psalms 78: 47).

The "sycomore" of the Bible should not be confused with the English "sycamore", which is a maple, *Acer pseudoplatanus* L., and with which it was identified by early English writers and clergymen and with which some non-botanists still confuse it there! The spelling "sycamore" used by the GOODSPEED version and suggested by SMITH (184) is incorrect and misleading when applied to the Biblical plant. It is of interest to note that the Septuagint confused the "sycomore" with the "sycamine" (306), the latter being an entirely different tree, *Morus nigra* (which see).

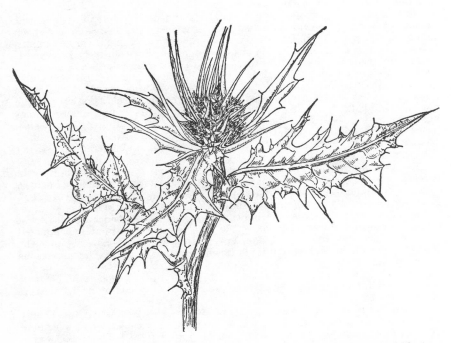

FIGURE 21. — *Gundelia tournefortii*, the gundelia. This plant is thought by some authorities to be the plant referred to as a "wheel" or "rolling thing" in the Old Testament. With the aspect of an eryngo, it rolls over the ground when dry and gathers in huge heaps in hollows. (Crowfoot and Baldensperger, From Cedar to Hyssop, 1932).

106. Gossypium herbaceum L.

ESTHER 1: 5-6—And when these days were expired, the king made a feast unto all the people that were present in Shushan the palace, both unto great and small, seven days, in the court of the garden of the king's palace . . . where were white, *green*, and blue hangings, fastened with cords of fine linen and purple to silver rings and pillars of marble . . .

The Hebrew word for "cotton" is "karpas" (266) (Greek, κάρπασος). This word appears in the original of the passage cited above, but was mis-translated as "green" in the King James, LEESSER, and Douay versions. It is more correctly rendered "cotton", as has been pointed out by CELSIUS and by SMITH (299), and the Revised Version suggests this translation in a marginal note. The GOODSPEED translation is: ". . . There was white stuff of *cotton*, and there were blue hangings . . ." MOFFATT renders the passage "When this was over he gave a banquet to all the men within the citadel of Susa, to high and low alike, for seven days, in the park belonging to the royal palace; there were hangings of white and violet *cotton*, corded with white and purple linen, caught up on silver rings and marble columns." The JASTROW version says: "there were hangings of white, fine *cotton*, and blue, bordered with cords of fine linen and purple."

Cotton, especially the species *Gossypium herbaceum*, was cultivated since time immemorial in the Far East, records of it in India extending as far back as 800 B.C. (306). It was not introduced to the western civilization of Europe until about 330 B.C., when ALEXANDER THE GREAT brought it back with him from India. We have no direct evidence to show when it was introduced into Palestine, but PAUSANIAS, the Greek traveler and topographer who flourished about the second century (158) indicates that it was already cultivated then in the Holy Land and in Egypt, because he comments on the cotton of Judea being yellower in color than that of Egypt (SMITH believes (299) that this statement was made in about 480 B.C., apparently confusing the topographer with PAUSANIUS, king of Sparta). It is probable that the Jews became acquainted with cotton during the period of their captivity in Persia under King AHASUERUS (598-536 B.C.), whose kingdom, we read, extended from India to Ethiopia (299). This wide extent of AHASUERUS' kingdom makes it very plausible that the highly colored cotton cloth of India should have been a part of the ornamentation of his palace, wherein the event described in the book of Esther is supposed to have taken place in about 521 B.C. (27). Cotton is now Egypt's chief commercial crop and one of the staple articles of commerce of Palestine (299). HERODOTUS in about 450 B.C. wrote that the inhabitants of India "had a plant that bore, instead of fruit, a wool like that of sheep, but finer and better, of which they made clothes", and THEOPHRASTUS, in 300 B.C., tells of cotton growing in Ethiopia (299). VARRO (116-27 B.C.) wrote of "tree-wool" on the authority of CTESIAS, a contemporary of HERODOTUS. PLINY in 115 A.D. described cotton as growing in Egypt, a statement confirmed by JULIUS POLLUX about a century later (306). DALMAN states that cotton was probably first cultivated in Alexandria (98). COLUMBUS found cotton being used in 1492 by the Amerinds in Cuba, and CORTEZ and PIZARRO soon thereafter found it grown and utilized in Mexico (299).

There are numerous species of cotton, the original Arabian and Indian type being the so-called Levant cotton, *G. herbaceum* (266). This is grown as an annual, but in its native haunts is a perennial, growing to become an erect freely branching shrub to as much as 6 feet tall. The leaves are alternate, heart-shaped, shallowly 3-7-lobed, and soft in texture. The flowers are large, very showy, mallow-like, yellow with a purple center, fading to pink or red, subtended by a cup-shaped involucre composed of several large fringed bracts. The fruit or "boll" is a 3-5-celled valvate capsule, "about the size and shape of a fig" (299). When it is fully ripe it splits through the middle of each valve, and a mass of fine white filaments (the cotton of commerce) extrudes. The seeds are attached to these cotton filaments or fibers ("staple"). *G. herbaceum*, although originally Asiatic, is extensively grown in the southern United States (266). The tree cotton, *G. arboreum* L., is a species native to northern Africa, now much cultivated in upper Egypt. The sea-island cotton, *G. barbadense* L., native to the West Indies, is now extensively cultivated in the southern United States and in Egypt. The southern United States now raises about 85% of the world's cotton, the remainder being grown chiefly by Egypt, India, China, and Brazil. Because of its cheapness and considerable durability it is now generally regarded as the most important of all the textiles (158). Cotton is extensively employed in medicine, surgery, and the arts, and, in the form of guncotton, in warfare. Chemically it is almost pure cellulose (158).

The Vulgate rendered the "carpas" of the Esther reference as "carbasini coloris", implying that a color, not a material, was intended. This is doubtless the basis of the Authorized and Douay versions' translation of "green" (306). Some commentators are of the opinion that cotton was also included in the "shêsh" of the earlier and the "bûts" or "bootz" of the later books of the Old Testament, rendered "white linen" and "fine linen" in most translations (306). The dress of Egyptian priests, as of those of the Hebrews, was always made of linen, although PLINY asserts that the former preferred cotton. Cotton garments for temple attendants are said to be mentioned on the Rosetta stone (306) and some commentators claim that the Hebrew ephods, originally commanded to be made only of linen (Exodus 28), later were made with cotton. In view of the scrupulosity with which the Hebrew priests carried out the letter of the law, this seems highly improbable. Egyptian mummy-cloths, formerly regarded as having been made of cotton, are now known to have been linen. Cotton, however, now forms almost the entire apparel of Palestinian women (27).

107. Hedera helix L.

II MACCABEES 6: 7— . . . and when the feast of BACCHUS was kept, they were compelled to go in procession to BACCHUS carrying *ivy*.

The Greek word here translated "ivy" is "κισσός", and there is no doubt that the common English ivy, *Hedera helix*, is the plant to which reference is made. The Hebrew is similar—"k'soos". The GOODSPEED version of this verse is: " . . . and when the festival of DIONYSUS was celebrated, they were compelled to wear wreaths of *ivy* and march in procession in his honor". The carrying or wearing of the ivy was in itself probably not distasteful to the Hebrews (101), but what they objected to was the fact that they were forced to enter the temple of a heathen god and do homage to him. The event here described is thought to have taken place at about 168 B.C.

The ivy and the holly, *Ilex aquifolium* L., have always been associated with religious rites and festivals. Ivy was one of the plants dedicated to BACCHUS by the ancient Greeks (126). Holly was used by the Romans during their Saturnalian festivals (673-640 B.C.). Sprigs of it, sent to acquaintances and associates, were a symbol of friendship. The use of holly and ivy was taken over by the Christian church, just as so many other features and elements of heathen worship were taken over from time to time on the time-tested principle that it is easier to absorb popular customs than to eradicate them by condemnation (125). English churches began to use holly and ivy at Christmas time in the reign of HENRY VI (1422-1461, 1470-1471) (299). With the passage of the centuries, holly and ivy have, thus, lost completely their original heathen connotations and are today welcome both in the church and the home, wherever they are available, as evergreen symbols of everlasting life.

The English ivy is a high-climbing vine consisting of many varieties and forms. In its typical form it climbs over rocks and walls by means of numerous, short, adhesive, aerial rootlets from along its stems and branches. Its evergreen leaves are thick-textured, angled or 5-lobed, or the uppermost ovate and unlobed, bright glossy green on both surfaces. The flowers are small, yellowish, borne in panicled umbels. The fruit is a small, top-shaped or globular, black berry, a favorite of birds. The species is found on walls and cliffs sparingly throughout the Holy Land, and is regarded by some as being indigenous there (266). Dr. EVENARI writes us that "it may be a relic from pluvial times."

A garland or bush of ivy, or the representation of one or the other, was formerly placed outside the door of every tavern or roadhouse as an indication that wine was sold therein (158). This custom is obviously a direct outgrowth of the Roman and Greek association of the plant with BACCHUS, the god of wine and conviviality (126).

108. Hordeum distichon L.   109. Hordeum hexastichon L.
110. Hordeum vulgare L.

EXODUS 9: 31—And the flax and the *barley* was smitten: for the *barley* was in the ear, and the flax was bolled.
LEVITICUS 27: 16— . . . an homer of *barley* seed shall be valued at fifty shekels of silver.
NUMBERS 5: 15—Then shall the man bring his wife unto the priest, and he shall bring her offering for her, the tenth part of an ephah of *barley* meal . . .
DEUTERONOMY 8: 8—A land of wheat, and *barley*, and vines, and fig trees, and pomegranates . .
JUDGES 7: 13— . . . Behold, I dreamed a dream, and, lo, a cake of *barley* bread tumbled into the host of Midian, and came unto a tent, and smote it that it fell . . .

RUTH 1: 22—... and they came to Bethlehem in the beginning of *barley* harvest.
RUTH 2: 17 & 23—So she gleaned in the field until even, and beat out what she had gleaned: and it was about an ephah of *barley* ... So she kept fast by the maidens of BOAZ to glean unto the end of *barley* harvest and of wheat harvest, and dwelt with her mother in law.
RUTH 3: 2 & 15-17—... Behold, he winnoweth *barley* to night in the threshingfloor ... And when she held it, he measured six measures of *barley*, and laid it on her ... These six measures of *barley* gave he me ...
II SAMUEL 14: 30—... See, JOAB'S field is near mine, and he hath *barley* there ...
II SAMUEL 21: 9—... in the days of harvest, in the first days, in the beginning of *barley* harvest.
I KINGS 4: 28—*Barley* also and straw for the horses and dromedaries brought they unto the place where the officers were ...
II KINGS 4: 42—And there came a man ... and brought the man of God bread of the firstfruits, twenty loaves of *barley*, and full ears of corn in the husk thereof ...
JOB 31: 40—Let thistles grow instead of wheat, and cockle instead of *barley* ...
ECCLESIASTES 11: 1—Cast thy *bread* upon the waters: for thou shalt find it after many days.
EZEKIEL 4: 9 & 12—Take thou also unto thee wheat, and *barley*, and beans, and lentiles, and millet, and fitches, and put them in one vessel, and make thee bread thereof ... And thou shalt eat it as *barley* cakes ...
EZEKIEL 13: 19—And will ye pollute me among my people for handfuls of *barley* and for pieces of bread ... ?
HOSEA 3: 2—So I bought her to me for fifteen pieces of silver, and for an homer of *barley*, and an half homer of *barley*.
JOHN 6: 9 & 13—There is a lad here, which hath five *barley* loaves, and two small fishes ... Therefore they gathered them together, and filled twelve baskets with the fragments of the five *barley* loaves ...
REVELATION 6: 6—... A measure of wheat for a penny, and three measures of *barley* for a penny ...

The common barley, *Hordeum distichon*, winter barley, *H. hexastichon*, and spring barley, *H. vulgare*[60], have been cultivated in temperate regions of the world since time immemorial (299) and are still today one of the principal grain foods or cereal crops of man. Barley and wheat were the two staple cereal crops of Egypt and Palestine. Barley, being the less expensive, was most used for feeding cattle, although it was also used by itself or mixed with wheat or other seeds as food for man (*cfr.* Ezekiel 4: 8-12) (98, 179). It is mentioned in the Bible some 32 times, either as a plant growing in the fields or in reference to products made from it, such as barley meal, barley bread, barley cakes, and barley loaves. This is a fair indication of how extensively it was grown and used in ancient Palestine and other parts of Bible lands, where today it still forms the chief part of the food of the poor. It is said to have been practically the sole food for horses, and was commonly used also for feeding to asses, draught oxen, and cattle. Oats were unknown (*cfr.* I Kings 4: 28) (184). Barley bread then, as now, was the common food of the poorer people of Palestine.

No grain which has been brought under cultivation by man equals barley in the extent of climatic variation under which it will grow successfully. It will survive heat and drought better than any other cereal, and ripens so rapidly that the short summers of quite northern latitudes are sufficient for it. These same northern short summers are not of sufficient length to ripen wheat; therefore barley becomes a far more important crop than wheat in northern lands. In the irano-turanian region of Palestine barley is the typical crop. This country is steppe-like and has only a small annual rainfall.

The exact land of barley's origin is not known. It has been described as "the most universal cereal in the world" (184). The ancient Egyptians believed that their goddess of fecundity, ISIS, taught man its cultivation. The Hebrew words in the Scriptures rendered "barley" are "sĕôrâh" or "s'ôrâh" and "s'o'rim" (Greek, κριθη or κριθαί; the latter is the plural form). Barley harvest takes place in March or April in the Holy Land, depending on the locality (184), and in hilly districts even as late as May (306). Barley ripens about a month earlier than wheat in Egypt today (306), even as it did in Biblical days when a hailstorm destroyed PHARAOH'S barley but not his wheat (*cfr.* Exodus 9: 22-35) in about 1941 B.C. (268). The barley was already in the ear when the wheat was just sprouting from the ground, according to this story. The latter was thus spared the fatal injury which the storm inflicted

---

[60]Also called *Hordeum sativum* Pers.

on the more advanced spring crops. In Palestine, too, barley is always harvested before the wheat, in some places a week, in other places as much as three weeks earlier. It is sown at any time between November and March. Its green ears boiled and served with milk, form a dish frequently eaten in Egypt, and an intoxicating liquor is now made from it, much relished by the poorer classes.

Barley was so well known in ancient times, according to Mrs. BARNEVELD, that it even supplied the Hebrews with a unit of linear measurement (47)—two barley-corns making a "finger-breadth", 16 a "hand-breadth", 24 a "span", and 48 a "cubit" or 16 inches. Three barley-corns, laid end to end, made an inch. This system of measurement is still employed in measuring the length and breadth of shoes (158). Ten kinds of wild barley grow in the Holy Land (267), one of which, *H. spontaneum* C. Koch[61], may have been the ancestor of the cultivated species and varieties. According to POST, *H. hexastichon* was the species most commonly cultivated by the ancients, with *H. vulgare* also common since prehistoric times (267). The former is now much sown in the Jordan valley and Moab, the latter in Lebanon, and *H. distichon* in the Hauran and Syria.

The "barley" of I Chronicles 11: 13 is a mis-translation for "lentils", as will be pointed out under *Lens esculenta* (which see). The "bread" of Ecclesiastes 11: 1 has been considered by some writers to refer to barley bread, which is most probably so. The only version to regard it other than as bread is that of MOFFATT, who regards it as symbolic and gives to the entire verse a new and much broader meaning: "Trust your *goods* far and wide at sea, till you get good returns after a while." The GOODSPEED and MOFFATT versions, which give Biblical measures and valuations in their modern equivalents, shed interesting light on the comparative value of barley in those days. Revelation 6: 6 is rendered by GOODSPEED: "Wheat at a dollar a quart, and *barley* at three quarts for a dollar" (the WEYMOUTH version says: "A whole day's wage for a loaf of bread, a whole day's wage for three *barley cakes*"). Leviticus 27: 16 is rendered by MOFFATT: "land sown with ten bushels of *barley* shall be valued at seven pounds of silver" or, by present standards, about $76. The JASTROW version uses the word "barley" in all the passages where it occurs in the Authorized Version.

Since barley was the commonest food of the poor and was not too greatly esteemed even by them (306), it was made use of in Biblical days in parables as a symbol of poverty and of cheapness or worthlessness (*cfr.* Hosea 3: 2) (184). This fact also explains the use of a small amount of barley meal, instead of wheat meal, in the jealousy offerings described in Numbers 5: 15, indicating plainly to the Hebrews, whose minds are geared to see symbolism in everything, the low regard in which the implicated parties were held (306). It also renders more forceful the exclamation of shocked insult in Ezekiel 13: 19. Present-day Bedouins refer to their enemies as "cakes of barley bread" to indicate their utter scorn of them. Understanding this fact, makes it easier to interpret the dream of GIDEON in Judges 7: 13-15, GIDEON having been a very poor and humble man of the type that would most certainly have been scornfully referred to as a "cake of barley bread" by the haughty Midianites (306).

Barley is extensively grown now in the western world for the production of malt, from which by fermentation and distillation ale and beer are produced. This intoxicating property of the grain is not mentioned or alluded to anywhere in the Bible, so it seems logical to conclude that it was unknown to the ancients (299). So-called green malt is produced by allowing barley grain to soften in water and begin to sprout. The diastase enzyme is developed during the process of sprouting, and this is capable of transforming the starch of the malt into sugar during fermentation (158).

---

[61]Also known as *Hordeum vulgare* var. *spontaneum* (C. Koch) Koern.

111. Hyacinthus orientalis L. 112. Lilium candidum L.
(PAGE 17; FIGURE 64; FIGURE 65; FIGURE 89)

SONG 2: 1-2 & 16—I am the rose of Sharon, and the *lily* of the valleys. As the *lily* among thorns, so is my love among the daughters . . . he feedeth among the *lilies*.
SONG 4: 5— . . . like two young roes that are twins, which feed among the *lilies*.
SONG 6: 2-4—My beloved is gone down into his garden, to the beds of spices, to feed in the gardens, and to gather *lilies*. I am my beloved's, and my beloved is mine: he feedeth among the *lilies*. Thou art beautiful, O my love, as Tirzah, comely as Jerusalem . . .

There has been much discussion about the passages cited above in reference to the Hebrew word "shoshanah". Some commentators have supposed that the *"lily* of the valleys" of SOLOMON was *Lilium chalcedonicum* (which see), but this is not possible because it is not a plant of the lowlands and is very rare. SPRENGEL considered it to be *Narcissus jonquilla* L., which, however, is not recorded from the Holy Land by POST except in cultivation (267). Others have identified it with *Sternbergia lutea* (Herb.) Ker[62]. This species is not known from Palestine (267), but related species of these autumn-flowering bulbs, like *S. pulchella* Boiss. & Bl., *S. fischeriana* (Herb.) Roem., and *S. aurantiaca* Dinsm., with large yellow flowers, are native there and it is probable that it was one of these to which these authors referred. The plant known to us at present as "lily-of-the-valley", *Convallaria majalis* L., definitely does not occur in Palestine and has nothing whatever to do with the problem.

Numerous authors have held that the "lily of the valleys" was identical with the "lily of the fields", *Anemone coronaria*, and some have even tried to identify it as a violet, a jasmine, or a buttercup! Modern experts do not appear to hold to any of these theories. The GOODSPEED version seems to be botanically the most plausible in its translation of these Song of Solomon passages: 2: 1-2, "I am a saffron of the plain, a *hyacinth* of the valleys. Like a *hyacinth* among thistles, so is my loved one among the maidens"; 4: 5, " . . . that pasture among the *hyacinths*"; and 6: 2-4, "My beloved has gone down to his garden, to the beds of spices, To pasture his flock in the gardens, and gather *hyacinths*. I belong to my beloved, and my beloved to me, who pastures his flock among the *hyacinths* . . . " MOFFATT, who continues to use the much misunderstood word "lilies" in most of these passages, renders the "comely as Jerusalem" in Song 6: 4 of the Authorized, Douay, and GOODSPEED versions as "lovely as a *lily* of the valley", apparently again referring to the hyacinth. The JASTROW version follows the King James closely in all these texts.

The garden hyacinth, *Hyacinthus orientalis*, is indigenous and very common in fields and rocky places in Palestine, Lebanon, and northward. Its flowers in the wild form are always deep blue and very fragrant (267). In the springtime some hillsides in Galilee are literally covered blue with the fragrant, exquisite blooms.

Interesting legends are associated with the hyacinth. In classical mythology HYACINTHUS was a very handsome boy who was well-loved by both APOLLO, the god of manly youth and beauty, poetry, music, and the wisdom of oracles, and by ZEPHYRUS, god of the winds. When the boy seemed to prefer APOLLO, the wind god became very jealous. In a game of quoits between APOLLO and HYACINTHUS, the wind god deflected one of the missiles so that it struck and killed the boy. APOLLO at once transformed the youth into a blue hyacinth plant whose beauty and fragrance should be an eternal memorial to the fallen boy. Over it the grieving god sighed "Ai, Ai" (the sound of grief most universal among all eastern peoples), and the letters "AI" in Greek character are said to be inscribed even today on the tepals of this flower (125, 298). OVID refers to this legend when he speaks of "Languid Hyacinth, who wears His bitter sorrows painted on his bosom" and MOSCHUS when he says "Now tell your story, Hyacinth, and show Ai! Ai! the more, amidst your sanguine woe!" (95, 125).

The sound of the wail of grief, "Ai", however, is very similar to that of the Greek word "AEi", meaning "eternal", and it is supposed that from this the hyacinth came to signify remembrance to the Greeks (125). The representa-

---
[62]Also called *Oporanthus luteus* Herb.

tion of a hyacinth was often sculptured on tombs even in rather recent times (298).

The native names for the Madonna lily, on the other hand, in all the areas neighboring the Holy Land are remarkably similar, not only to each other but also to the word used in the Bible and rendered "lily" in the King James version. Yet, as we have seen, commentators have suggested several different lilies, lily-like plants, and even just attractive flowers in general for the Biblical word. The white Madonna lily has been the supposed Biblical lily among the vast majority of ancient and medieval writers and even a number of modern authors. TRISTRAM lists this lily, *Lilium candidum*, as rare in Lebanon (326). BOISSIER records it from Palestine on the basis of KUNTH's statement as well as from Lebanon (61). PLINY mentioned it as occurring in Palestine, but may actually have meant only nearby Phoenicia, Lebanon, and Syria. CELSIUS, and later LINNAEUS, accepted PLINY as the authority for its occurrence in Palestine and it has been generally regarded as a Palestinian plant in the popular mind ever since. Yet HASSELQUIST found no true lily in his Palestinian explorations and none of the 18th and 19th century botanists and travelers in the Holy Land found the Madonna lily there. And so it seemed to numerous critical workers that all the references in literature and art to the occurrence of the white Madonna lily in Palestine were based on some error in identification or of geographic location. Categoric statements were made in recent years that this lily definitely is not a native Palestinian plant (233). However, on June 24, 1925, the entire picture was changed, for on that day NAFTOLSKY, on a field trip with students of the Hebrew University in Jerusalem, found a genuinely wild plant of *Lilium candidum* in a deep moist limesink at the northern edge of Palestine. The following year, ten more plants, a few with inflorescences, were found in this same locality, and there seems no doubt now of the correctness of the identification and of the native character of the colony. These lilies and similar shade-loving plants are rare here and entirely absent in the rest of Palestine, but are abundant in nearby Lebanon (341). Although the Lebanon records were for a long time also regarded as erroneous, or to refer to escaped cultivated plants, it is now believed that the plant is truly native there (341). It seems probable that the Madonna lily, even though it was not found during some 200 years of searching in Palestine, is not a recent introduction. Localities where it has been found are virtually inaccessible, isolated, and quite distant from one another. They occur in a geological formation similar to that of the Lebanon stations. These places are probably the only remnant outposts that the inhospitably dry country has left to them. WARBURG suggests that back in Pliocene times *Lilium candidum*, along with such things as the cedar-of-Lebanon and *Pinus peuce*, migrated from the Himalayan region to Palestine. This was before the great intervening steppes and deserts were established and checked all further migration. Evidence for this is the fact that the Madonna lily is much more closely related to the lilies of eastern Asia than to those of the Mediterranean area, although botanists who felt that the weight of evidence was against its being native to Palestine were forced to assume a Mediterranean origin for the species. In these Pliocene and subsequent ancient times undoubtedly great forests covered the most of what is now Palestine and Syria. Remnants of these forests have been found even in historic times. In such extensive wooded and shady areas *L. candidum* must have flourished even into and perhaps through early historic times. Then through the well-known ignorance and rapaciousness of man in his dealings with natural forests, these wooded areas were continually decimated (341). The lilies, WARBURG thinks, must have been plucked and even uprooted in vast quantities before and during the many religious festivities of the Jews and Christians up to the time of the Arab conquest in 636 A.D. The hot dry summers of Palestine today do not suit the growth requirements and the late-summer blooming habit of this lily, so it has been long since exterminated everywhere save in the cool wet limesinks where NAFTOLSKY discovered it (341).

In view of this far-reaching discovery it now seems probable that the

Madonna lily may be the "lily" referred to in Song 6: 2, at least. Löw believes that only a cultivated lily of outstanding beauty would be found in such a specialized garden as was possessed by the wealthy and powerful SOLOMON. The hyacinth, tulip, and narcissus, according to this authority, all bloom before the time of grape harvest and were too common and well known to be cultivated. HEHN agrees that only a lily could be referred to in these Song of Solomon passages because the context seems to indicate plainly that the time of grape harvest was the time of year alluded to here (170). KAUTZSCH was of the opinion that the "lily" of these passages was probably not white, thus eliminating *L. candidum*, but was probably one of the Palestinian irises (185). Similarly, CHRIST is inclined to favor some Palestinian iris (90). HEHN concludes that it was either *Lilium chalcedonicum*, *L. bulbiferum*, or more probably a species of *Fritillaria*. He seems to base this conclusion on PLINY'S statement: "Est et rubens lilium quod Graeci κρίνον vocant". However, *L. bulbiferum* is not a native of Palestine and *L. chalcedonicum* is very rare, if, indeed, it occurs there at all now (however, there is no reason to suppose that it might not have been in the ancient forests like the Madonna lily). WARBURG says that the only common fritillary, *Fritillaria libanotica* (Boiss.) Baker, has very small flowers and is not worthy of cultivation (341). KAUTZSCH states that even today at the time of wheat harvest the threshingfloors are ornamented with pretty wild flowers (185), so it seems very logical that showy lilies be spoken of in the poet's allusions to the harvest season. HEHN emphasizes that the *Colchicum* blooms in late autumn, and the iris, crocus, gladiolus, and narcissus all are early spring-bloomers, and thus all these are unsuited to the context (170).

Löw is convinced from etymologic studies and Jewish cultural evidences that the "susanna" of the Bible is unquestionably the Madonna lily and not the iris as first identified by early Bible translators and as maintained by commentators like KAUTZSCH and CHRIST (185). ASCHERSON, on the other hand, thinks that there is little likelihood that *Lilium candidum* is the "lily of the field" of Song 2: 1 (33). He is doubtful whether the author of the Song of Solomon had any definite lily, amaryllid, or iris species in mind. "Schüschan" and "schôschanna" are common names that apply to all large, lily-like, handsome flowers in that area today, he says (31), and are not restricted to the white Madonna lily, as some writers imply.

WARBURG argues against the iris, at least in the references to SOLOMON'S garden, because the large-flowered native irises, he says, are too difficult to transplant to make subjects for cultivation. The iris now so widely cultivated, *I. germanica*, was certainly not known there then (326). Evidences from vases and other art objects of great antiquity indicate that the Madonna lily was well distributed as a cultivated plant already at about the year 300 A.D. (170, 341).

Whatever may be the final decision regarding the "lily" of SOLOMON'S garden at the time of grape or wheat harvest, it seems to us that GOODSPEED is probably correct in regarding as the hyacinth the "lilies" among which browsing animals grazed, for these are plants of the grassy fields, not, like the true lilies, of the dark woods.

113. Iris pseudacorus L.

HOSEA 14: 5—I will be as the dew unto Israel: he shall grow as the *lily*, and cast forth his roots as Lebanon.
ECCLESIASTICUS 39: 14— . . . And flourish like a *lily* . . .
ECCLESIASTICUS 50: 8—And as the flower of roses in the spring of the year, as *lilies* by the rivers of waters, and as the branches of the frankincense-tree in the time of summer.

There has been much discussion about the "lily" (Hebrew, "shushan") of the above references, as, indeed, there has been about this word wherever it occurs in the Scriptures. Of considerable importance are the modern translations of the Hosea passage. MOFFATT renders the second half of this verse: "he shall blossom like a *lily*, and strike roots down like a poplar". The GOODSPEED version agrees in this new translation by saying: "so that he will blossom like the *lily*, And his roots will spread like the poplar"; but the JASTROW translation still follows the Authorized Version. It seems probable that of the many "lily" references cited by various commentators to species of *Iris*, the three given above are the only ones which actually belong there. Unlike the other "lily" references in the Bible, these three allude to some showy flower which was common and well-known, grew in or along the sides of streams, and had extensive poplar-like root-systems. This description does not apply to any of the many other plants suggested as "lilies" of the Bible by various authors: *Lilium chalcedonicum, L. candidum, Narcissus jonquilla, N. tazetta, Cyclamen persicum, Fritillaria imperialis, Hyacinthus orientalis, Sternbergia lutea, Tulipa montana, Convallaria majalis, Anemone coronaria, Ranunculus asiaticus, Adonis palestina,* or *Anthemis palaestina.*

Of the fifty or more kinds of *Iris* recorded from the Holy Land (267) the yellow flag, *I. pseudacorus,* seems likely. The POST herbarium has a specimen from Hûlah in Palestine. Other specimens are reported from Anti-lebanon and Damascus (267). Most of the other species of this genus inhabit hillsides, mountainous country, fields and grainfields, and even desert places, but the yellow flag grows in shallow water at the margins of ponds and streams, often making extensive masses. Its sword-shaped leaves and slightly flattened stems are 3 feet or more tall. The flowers are a deep-yellow, the 3 outer divisions reflexed, with tawny spots and radiating purple veins, the 3 inner divisions shorter and erect. Dr. EVENARI does not believe that *I. pseudacorus* is a Palestinian plant. He suggests, instead, that any of a "few species of wonderful black-brown *Iris* as *I. bismarkiana* Regel" may be meant.

"Sûsân" is at present the Arabic name applied to species of *Iris* in the Holy Land (267). For this reason it is very possible that the "Shushan" of Nehemiah 1: 1, Esther 1: 2, and Psalms 60: title, and the "Susa" of Esther 11: 3 & 16: 18 (GOODSPEED version), and the many other places in the Bible where this and similar words appear, may apply to *Iris*, as many writers have supposed. However, as has been pointed out under *Anemone coronaria* (which see), the word is applied not only to *Iris*, but to any brilliantly colored flower which vaguely resembles a lily (31). Thus, it seems to us entirely unwarranted to refer categorically all these passages to *Iris*, when the context in most of them points much more strongly to other plants.

KAUTZSCH writes that the "lily" of Song 5: 13 and 7: 3 was probably one of the many Palestinian irises (185). CHRIST holds that the evidence does not warrant one to say whether the Old Testament "lily" was a white lily or just a lily-like plant such as the Palestinian irises that grow there in tremendous quantities (90). *Iris palaestina* (Bak.) Boiss., recently suggested for the passages cited at the head of this chapter (233), is not very satisfactory in this context because it inhabits dry hillsides rather than the wet margins of streams.

The iris was recommended by herbalists for all manner of human ailments, including bruises, coughs, dropsy, fits, snake-bite, "spleens", and violent anger or temper tantrums. A bruise could be cured, so they asserted, if one laid some iris tepals on the black-and-blue spot for a few days. The root was chewed by infants during the period of teething and worn as a necklace both by the infants and their elders. The cities of Paris and Leghorn at one time exported 20 million iris root "beads" per year for this purpose. It was thought that scrofula and blood infections could be cured by making an incision and inserting a bead of iris root (298). Orris root, obtained chiefly from *I. florentina* L., was thrown into fires to produce a pleasant odor, and, chewed, to neutralize the smell of liquor, garlic, or tobacco on the breath. It was also crushed and used as a substitute for dried violet flowers in sachet powders (298).

The fleur-de-lis, *I. germanica* L., was adopted by King CLOVIS (465?-511) of France on his battleshield before his great battle with the Huns. After his victory he embraced Christianity, and the fleur-de-lis, conventionalized, became a part of the royal arms and standards of the nation (95, 298). The Empress THEODORA, wife of the Byzantine JUSTINIAN, in 527 A.D. included a fleur-de-lis in her crowns (158). Its three-partite form became the symbol of Christianity in the Crusade of 1137 under LOUIS VII, and it became known as the "fleur de Louis" (flower of Louis), which by contraction became "fleur de luce" (flower of light) and then "fleur-de-lys" and now "fleur-de-lis" (95,126). The coat-of-arms of ancient France in 1179 was a blue field sprinkled with many fleurs-de-lis. Their number was reduced to three in 1364. England included them in her coat-of-arms from 1340 to 1801 (158).

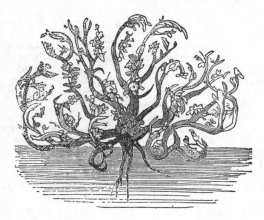

FIGURE 22. — *Anastatica hierochuntica*, probably the "rolling thing" of ISAIAH. (La Belgique Horticole, 1860). See also fig. 36.

114. Juglans regia L.

SONG 6: 11—I went down into the garden of *nuts* to see the fruits of the valley ...

It is now generally agreed that the Hebrew word "egôz", meaning "walnut" (Greek, καρύα, meaning "the walnut tree"), translated "nuts" in this passage, refers to the Persian or common walnut, *Juglans regia*. This species is usually regarded as indigenous to northern Persia, but is actually found wild in many parts of northern India, eastward through the Himalayas to China, and westward through Persia (299). In very olden times it seems to have been introduced into, and even naturalized in, Lebanon and some parts of Gilead. In fact, some think that it may even have been indigenous to the more mountainous parts of these lands (27, 299), although POST denies this (267). Even in the time of SOLOMON it was widely cultivated for its fruit in the Orient. JOSEPHUS (37?-95? A.D.) states that in his day extremely old trees of this species were abundant in Palestine, especially around the Lake of Gennesaret (299, 306). POST states that it is now cultivated everywhere in the higher parts of the region, usually in woods, near watercourses, and by the village fountains and wells, for its delicious nuts and valuable timber (265). He calls it "one of the noblest of trees" (267).

MOFFATT renders the verse at the head of this chapter: "Down I went to the *walnut-bower*, to see the green plants of the dale." SOLOMON's "nut garden" (GOODSPEED version) or walnut-bower is supposed to have formed part of his extensive gardens at Etham, six miles from Jerusalem (299). Although no trace remains of SOLOMON's actual gardens, the locality is still a beautiful spot. Dr. BONER says of it: "It is one of the sweetest valleys into which the eye can look, a well-watered orchard covered with every goodly fruit-tree that Syria nourishes" (299).

The Persian walnut is a tree which may attain a height of 30 feet or more, with a hemispheric crown, producing a dense shade (265, 293). Its leaves are pinnately compound and glabrous, consisting of 2-6 pairs of oblong, entire, resinous, very fragrant leaflets (267). The nuts are ovate or globular, with four partitions. An oil is extracted from them which is only slightly inferior to olive oil and is used extensively for soap manufacture in Europe. Although not introduced into England until about 350 years ago, the walnut is now popular there and is often known to us as the "English" walnut—another example of a popular misnomer like the "Irish" potato, "Bermuda" onion, and "French" marigold. Its fine shade, fragrant leaves, and delicious fruit doubtless made it a prime favorite in SOLOMON's garden. It is widely cultivated in middle and southern Europe today (299), as well as in California and the southern United States. Its fruit is one of the most important of all present-day commercial nuts (158).

Some commentators believe that the "algum trees" of II Chronicles 2: 8 may also have been walnuts, but it is now more generally believed that they were either *Cupressus sempervirens* var. *horizontalis* or *Sabina excelsa*, which see. The "nuts" of Genesis 43: 11, represented by the Hebrew word "botnîm" are now regarded as having been pistachio nuts, and Hebrew dictionaries now so translate this word (see under *Pistachia vera*).

Many superstitions and legends are associated with the walnut. The Greeks held it sacred to DIANA. Both they and the Romans regarded it as a symbol of fecundity, and scattered its nuts about at weddings. In the Middle Ages it was believed that evil spirits and even the devil himself lurked in its branches, and even today there is a widespread belief that it exudes a poisonous essence which will kill all other vegetation beneath and adjacent to it (298). Because of its connection with evil spirits, its nuts were thought to be useful in warding off lightning, fevers, and epileptic fits. A walnut placed beneath a chair on which a witch is seated will effectively root her there (298). In some lands the country folk believe that walnuts bear better if they are abused and beaten to drive out the evil spirits. In Russia there is a proverb stating that "A dog, and a wife, and a walnut tree: the more you beat them, the better they be." In Lithuania there is a curious legend about the Deluge. According to this version, God was eating walnuts when the waters began to flood the earth, casting the shells to the earth beneath him. The righteous men and women climbed into these shells, which each expanded to a size sufficient to carry them, and thus made individual "arks" which saved their lives. SAINT AGATHA is said to cross the Mediterranean annually from Catania to Gallipoli in a walnut shell (298).

115. **Juncus effusus** L.   116. **Juncus maritimus** Lam.
117. **Scirpus holoschoenus** var. **linnaei** (Reichenb.) Asch. & Graebn.
118. **Scirpus lacustris** L.   119. **Scirpus maritimus** L.

GENESIS 41: 2—And, behold, there came up out of the river seven well favoured kine and fat fleshed; and they fed in a *meadow*.
JOB 8: 11—Can the rush grow up without mire? can the *flag* grow without water?
ISAIAH 9: 14—Therefore the Lord will cut off from Israel head and tail, branch and *rush*, in one day.
ISAIAH 19: 6 & 15—And they shall turn the rivers far away; and the brooks of defense shall be emptied and dried up: the reeds and the *flags* shall wither . . . Neither shall there be any work for Egypt, which the head or tail, branch or *rush*, may do.

In the Job passage quoted above the "rush" probably refers to *Cyperus papyrus* and *Arundo donax*, which see, but the "flag", according to some commentators, may refer to true rushes and bulrushes (27). The word in the original text of this verse is "achú" (Greek, ἄχι βούτομον). The same word occurs in Genesis 41: 2, where it is translated "meadow" by the King James version, but where the Revised Version, JASTROW, and MOFFATT substitute "reed-grass", while the GOODSPEED version says "sedge". The word is Egyptian, not Hebrew, and the plant referred to is described by the Oxford Bible as "a luxuriant and nutritious grass growing by rivers" (27). In the Isaiah passages the word rendered "rush" by the Authorized Version is "agmon" or "ahgmon" (Greek, κάλαμος). The context seems to call for a low-growing (symbolically, humble) plant as contrasted to a tall-growing (proud) one. MOFFATT, JASTROW, and the GOODSPEED versions agree that the tall one was the date palm. For the low one MOFFATT suggests "rush" and the Goodspeed version "reed". MOFFATT'S translation of these two passages renders them far more understandable: "so he lopped off head and tail, palm-branch and *rush*, in a single day" and "high or low, palm or *rush*, none can do anything for Egypt." It seems more plausible to us that the low-growing plant was a species of rush or bulrush, than that the tall reed, *Arundo donax*—which grows to 18 or more feet in height—was intended, as the GOODSPEED version implies.

Numerous species of rush and bulrush grow in the Holy Land area. Of bulrush or clubrush, *Scirpus*, there are 15 kinds, and of rush, *Juncus*, there are 21 (267). It is hard to say which of these would be the most likely for our passages, but it seems reasonable to suppose that rare species and species known only from the mountains of Lebanon and Syria or from the seacoast marshes would not be as likely to be involved as would commonly distributed ones from the valleys and plains of Palestine. The common soft or bog rush, *Juncus effusus*, is said by POST (265, 267) to be found in "wet places, common throughout, even in Sinai and other deserts". The sea hard rush, *J. maritimus*[63], is found in "damp places" from Latakia in the northern region, through

---

[63]Also known as *Juncus rigidus* Desf.

Palestine, to Sinai. *J. glaucus* Ehrh.[64] and *J. lamprocarpus* Ehrh.[65], sometimes suggested, are found only in mountain regions of Lebanon, Coelesyria, and the Antilebanons, not "common from coast to subalpine regions" or "common throughout" as previously stated (233), and so are probably not involved here. The cluster-headed clubrush, *Scirpus holoschoenus* var. *linnaei*[66], is "common" in damp places from Lebanon and Coelesyria, through Palestine, to Sinai. The lake clubrush or tall bulrush, *S. lacustris* [67] is found in swamps and ditches from Tripoli to the Dead Sea. The sea or saltmarsh clubrush, *S. maritimus*[68], despite its name, is "common throughout" in "ditches and swamps" from Latakia, Beirut, and Tripoli, Antilebanon, Coelesyria, and Damascus, through Palestine to the lower Jordan valley and beyond. All these species may well be taken as representing the "rush" and "flag" of Isaiah and Job, but that any one of them is the "nutritious" river grass of Egypt is doubtful.

120. Juniperus oxycedrus L.

JEREMIAH 17: 6—For he shall be like the *heath* in the desert . . .
JEREMIAH 48: 6—Flee, save your lives, and be like the *heath* in the wilderness.

The Hebrew words here translated "heath" in the King James version (306) are "'ar-âr" (in Jeremiah 17: 6) and "'arô-êr" or "aro'air" (in 48: 6) (Greek, $\dot{\alpha}\gamma\rho\iota o\mu\nu\rho\iota\kappa\eta$), but probably do not refer to true heaths, most of which do not occur in the Holy Land, although *Erica vagans* L. is said to occur sparingly on the coastal plains of Syria and *E. orientalis* R. A. Dyer in the Lebanon mountains (299). POST does not list either of these heaths from the Holy Land, so, if they occur in those areas it is probably only in cultivation. The latter, in fact, is a South African species. But POST does record *E. verticillata* Forsk. from sandstone and chalky rocks in the mountains of the northern regions, Syria, and Lebanon (267). He also records the related *Pentapera sicula* var. *libanotica* Barb. from Lebanon, growing in fissures of rocks. TRISTRAM asserts categorically that "there is no heath south of the Lebanons" (27, 184).

Most authorities believe that the plant referred to by JEREMIAH was not a true heath, but a savin or juniper. As CELSIUS first pointed out (86, 306), the Arabic word applied to the Phoenician juniper is "ar'ar" (267)—exactly the same as the Hebrew word rendered "heath" by the King James Bible. The Phoenician juniper, *Sabina phoenicia* (which see), however, is listed by POST only from hills and rocky places in Arabia Petraea. But the name was probably applied to other junipers of similar habit. JOHN SMITH was of the opinion that JEREMIAH's plant was the savin juniper, *Sabina vulgaris*[69], which he and other writers (27,184,306) say is common throughout the deserts, plains, and rocky places of Syria and is an abundant plant in Palestine, growing even up to the very summit of Mount Hor (101, 268, 299). POST, however, disputes these writers and maintains that the savin has been recorded from the region "probably through error" in identification (267). According to POST the Arabic name "ar'ar" is now also applied to the brown-berried cedar, *Juniperus oxycedrus*, and it is probably this plant to which SMITH and the other writers actually referred when they spoke of the savin.

The brown-berried or sharp cedar usually is a hemispheric or even prostrate shrub, occasionally in especially favorable habitats attaining the stature of a tree 15 to 20 feet tall (267). The needles are all of one type, linear-lanceolate in shape, ternately whorled, and wide-spreading, keeled on the lower surface. The staminate flowers are borne in solitary, nearly sessile, ovate catkins. The fruits (known technically as galbules) are globular, about half an inch in diameter, and (in spite of the common name) red in color, eventually becoming

---

[64]Also known as *Juncus glaucus* Sibth. and *J. inflexus* L.
[65]Also known as *Juncus compressus* Roth.
[66]Also known as *Scirpus linnaei* Reichenb.
[67]Also known as *Schoenoplectus lacustris* (L.) Palla.
[68]Also known as *Scirpus corymbosus* Forsk. and *Bulboschoenus maritimus* (L.) Palla.
[69]Also called *Juniperus sabina* L.

quite glossy (267). The species is common in mountainous regions in Syria, Lebanon, the Antilebanons, Galilee, Gilead, and Bashan, often inhabiting the most barren and rocky parts of the desert (184), lonely crags, and the inaccessible fissures of rocky cliffs. Thus, being so often found on high desolate mountains and isolated rocks, inaccessible to all save the bounding gazelle, this plant was quite an appropriate one for JEREMIAH to use as a symbol when he said "Flee, save your lives, and be like the *heath* in the wilderness."

The common juniper, *Juniperus communis* L., and various species of tamarisk, *Tamarix*, are also said by certain commentators to be common shrubs in the desert places (that is, the "wilderness") of the Holy Land and therefore perhaps involved in these passages (299). The JASTROW version uses the "tamarisk" translation in both of these passages. The Revised Version suggests "tamarisk" in a marginal note (27). However, of the various species of tamarisk suggested, only *T. mannifera* is typically a desert plant; the others are inhabitants of coastal regions, the banks of streams, and wet maritime sands. It is our opinion that the tamarisks may all safely be eliminated as possibilities for JEREMIAH's "heath". POST also asserts that *Juniperus communis* does not occur in the region (267).

It is interesting to note in this connection that the Douay version uses "tamaric" in Jeremiah 17: 6 and "heath" in 48: 6, while the GOODSPEED version translates the Hebrew words of both Jeremiah passages as a noncommittal "scrub". LEESSER translates "arar" as "lonely tree" and "aro'air" as "solitary tree". MOFFATT does the same in Jeremiah 17: 6, but in 48: 6 says: "Run for your lives, get away like a *wild ass* to the wold." This rendition is not nearly as expressive or meaningful nor, we believe, as accurate, as the literal translation which implies that the Moabites should flee to the mountains and hide themselves in the remote and inaccessible crevices of the rocks like the juniper if they wanted to escape the wrath of God.

121. Laurus nobilis L.
(FIGURE 47)

PSALMS 37: 35—I have seen the wicked in great power, and spreading himself like a *green bay tree*.

Although the Douay, MOFFATT, and GOODSPEED versions all regard the tree here referred to as the cedar-of-Lebanon, *Cedrus libani* (which see), based apparently on the fact that the Septuagint used the expression κέδρος τοῦ λιβάνον, literally "the cedar of Lebanon" here, almost all the botanical authorities regard it as the sweet-bay, bay-laurel, laurel, or bay-tree, *Laurus nobilis*, and this seems to be the more reasonable view. The Hebrew word used for the cedar is "erez", while the word employed in the Psalms reference here cited is "ezrâch". This word "ezrâch" occurs in 14 other places in the Bible, but in all the other places it is used to signify "native" as opposed to "a stranger" or "a foreigner" (299). If it was any particular kind of native *plant* that is referred to here, as most translators agree that it was, then it must have been a green shrub or tree of considerable size growing by streamsides, such as the sweet-bay (184). Of course, the oleander, *Nerium oleander* (which see), comes to mind at once as the most gloriously beautiful of all the woody plants growing in profusion along the sides of streams and lakes in the Holy Land, but the oleander is not considered for this passage because it is invariably referred to as a "rose" in the Old Testament and it does not meet the other requirements of the text and context. The Revised and LEESSER Versions render the word as "a *green tree* in its native soil" and this is the translation adopted by some commentators (27). The JASTROW version is "*leafy tree* in its native soil."

The sweet-bay is a native, according to POST (265), inhabiting thickets and woods from the coast to the middle mountain zones (267). It is an evergreen tree, attaining a height of 40 to 60 feet. Since it is of compact growth, the term "spreading" does not apply well to it. However, both the MOFFATT and the GOODSPEED translations have replaced the adjective "spreading" of the King James version with the word "towering", which is more appropriate (although they apply it to the cedar-of-Lebanon). The sweet-bay grows throughout southern Europe and the Mediterranean region (101, 178, 268, 293). Its leaves are oblong to lanceolate, thick in texture, glossy on the upper surface, somewhat wavy-margined. The flowers are small, greenish-white, borne in lateral or terminal umbels or in dense many-flowered cymes. The fruits are one-seeded black berry-like drupes, about the size of a small grape.

It was probably on account of its never-changing greenness, as well as for the pleasant spicy fragrance of its leaves, that DAVID selected the sweet-bay as the symbol of prosperity. The same reasons caused the Greeks and Romans to prize it and to adorn with it the brows of their priests, poets, heroes, and the victors in the Pythian and Olympian games. It was also a mark of distinction for certain high offices and political functions (158). It was the symbol of triumph when borne in the hand, and generals in those days sent to their emperor reports of their successful campaigns wrapped in laurel leaves (298). A crown of bay leaves was the reward of genius. As DAVID looked at the fragrant and evergreen bay tree, unaffected by winter's cold and storms, it seems natural that he should have been reminded of people who live in continuous prosperity and even wealth while those about them suffer from the bitter winds of adversity.

Even though it is said to be abundant on Mount Carmel and around Hebron (299), it is in general not common in the Holy Land—just as very wealthy people probably also were not common. ROYLE comments that "The reason why the laurel is not more frequently mentioned in Scripture is probably because it was never very common in Palestine, as otherwise, from its pleasing appearance, grateful shade, and the agreeable odor of its leaves, it could hardly fail to attract attention." Its leaves are still used as a condiment, and its fruit, leaves, roots, and bark have long been employed in medicine. The leaves yield a fragrant oil called "oil of bay", which is not to be confused with the commercial oil used in making bay-rum (158).

As might be expected, the sweet-bay or laurel enters abundantly into classical mythology. It is supposed to have given oracles and soothsayers the power of prophecy; it kept away misfortune, and protected a house from lightning. The Roman emperor TIBERIUS (42 B.C.-37 A.D.) is said to have had such faith in the efficacy of the laurel that during electrical storms he would place a wreath of laurel leaves on his head and hide under his bed (298). Standing under a bay tree was protection against witches and wizards. Laurel berries were supposed to keep off various diseases. The emperor NERO (37-68 A.D.) fled to Laurentium during a pestilence "that he might save his precious health by breathing air that the laurels had purified" (298). The nymph, DAPHNE, pursued by APOLLO, was transformed by the gods into a laurel tree; thereupon APOLLO commanded that the laurel tree should bear its leaves both in summer and winter and that these be used to "crown all who excelled in courage, service, or the creation of beauty" (126, 298). Placing laurel leaves beneath the pillow of a person striving to become a poet would make him one. Our present word "baccalaureate" means "laurel berries" and alludes to the wearing of bay wreaths by poets and scholars addressing the university or receiving academic honors (126, 158, 298).

122. Lawsonia inermis L.
(FIGURE 20)

SONG 1: 14—My beloved is unto me as a cluster of *camphire* in the vineyards of Engedi.
SONG 4: 13—Thy plants are an orchard of pomegranates, with pleasant fruits: *camphire*, with spikenard.

The Hebrew word "copher" or "kopher" (Greek, κύπρος, dictionary translation "a tree growing in Cyprus") is translated in the Authorized Version as "camphire". The Douay version and marginal Bibles give the alternative translation of "cypress", but the Hebrew words regularly employed for that tree are quite different (see under *Cupressus*). It is also now generally agreed that the true camphire or camphor, *Camphora officinarum* Nees[70], has nothing whatever to do with the plant referred to in the Song of Solomon, in spite of the fact that SHAW and others considered it thus (293). The true camphor is a Chinese plant and it is extremely doubtful whether it was known in western Asia in SOLOMON's time (299). Instead, it is now generally believed that the plant referred to by SOLOMON was the henna-plant, *Lawsonia inermis*[71] (98), made famous by Arabian naturalists and writers (178, 179, 268), also called "alhenna" and "Egyptian privet". The Arabic name for this plant is "henna" (98). The Revised, MOFFATT, JASTROW, and GOODSPEED versions all use "henna" instead of "camphire". MOFFATT renders the cited verses as: "My darling is my bunch of *henna-blossom* from the garden of Engedi" and "Your charms are a pomegranate paradise—with *henna* and roses and spikenard." The JASTROW version says "My beloved is unto me as a cluster of *henna*". The Chaldee version of Song 1: 14 says "cluster of *grapes*" instead of "cluster of *camphire*," apparently because of the word "vineyard" which follows. Several old versions merely retain the Hebrew word "copher" untranslated (306), as does the more recent LEESSER.

The henna-plant is said to be originally a native of northern India (158), but is a well-known shrub throughout the Orient, having escaped from cultivation and now growing wild in Nubia, Egypt, Arabia, Syria, Lebanon, and Palestine. It grows from 4 to 12 feet tall, with opposite, terete or somewhat 4-sided, and eventually spinescent branches, dark bark, pale-green, elliptic-lanceolate, privet-like leaves, and clusters of small, white or yellow, power-

---
[70]Also known as *Laurus camphora* L. and *Cinnamomum camphora* (L.) Nees & Eberm.
[71]Also known as *Lawsonia alba* Lam. and *L. spinosa* L.

fully fragrant flowers at the tips of the branches (299). Eastern belles take large bouquets of these flowers to their bath, and consider a nosegay of henna one of the most elegant gifts that can be received from a friend.

Henna leaves are dried, crushed into a powder, mixed with water, and made into a paste. This paste has been used since time immemorial as a cosmetic, as is attested by numerous Egyptian mummies. By means of this substance a bright-yellow, orange, or red color was imparted to finger-nails, toe-nails, the tips of the fingers, the palms of the hands, and the soles of the feet of young girls. It was also used by the men for coloring their beards and the manes and tails of their horses. The dye was renewed once in every two or three weeks. This use of henna as a cosmetic was practiced in Egypt during the captivity of the children of Israel in that land, so they undoubtedly became familiar with it there (299). Apparently the custom of dyeing the nails and hair and skin was frowned upon by the Jewish leaders as something pagan. This is indicated by the fact that Levitical law required that captive women who were so decorated must have their hair shaved off, their nails pared, their palms and soles thoroughly scrubbed before they could be married to an Israelite. For instance, in Deuteronomy 21: 11-12, MOFFATT translation, we read: "If you see among the prisoners a beautiful woman whom you desire and long to marry, take her home, let her shave her head and pare her nails and throw off her prisoner's robe." Today henna paste in the Orient is made by mixing the powdered henna leaves with catechu, an extract of the wood of the eastern Indian *Acacia catechu* Willd. and *A. suma* Kurz and the Malayan *Ourouparia gambier* (Roxb.) Baill. (158). While this may have been done in India in Biblical days, it is doubtful if catechu was available to the Jews and neighboring peoples.

The scent of henna flowers is much like that of roses, and bouquets of the flowers are sold in the streets of Cairo today (299). In India the flowers are used as offerings in the temple of Buddha. Mohammedans also employ them in their religious ceremonies (158). Older authors claim that in the Holy Land henna is found "at Engedi only" (27, 184), but POST reports it from Beirut, Tripoli, Engedi, Jericho, Jaffa, Gaza, etc. (266).

HASSELQUIST, speaking of this plant in 1766, says: "the leaves are pulverized and made into a paste with water; the Egyptians bind this paste on the nails of their hands and feet, and keep it on all night: this gives them a deep yellow, which is greatly admired by Eastern nations. The color lasts for three or four weeks before there is occasion to renew it. The custom is so ancient in Egypt that I have seen the nails of the mummies dyed in this manner" (306). SONNINI DE MANONCOURT (1751-1812), the French naturalist, reports that Egyptian women in his day were fond of decorating themselves with bouquets or sprigs of henna flowers, carrying them in their hands and on their bosoms (306).

123. Lecanora affinis Eversm.   124. Lecanora esculenta (Pall.) Eversm.   125. Lecanora fruticulosa Eversm.   126. Nostoc spp.
(FIGURE 27; FIGURE 93)

EXODUS 16: 4 & 13-35—Then said the Lord unto MOSES, Behold, I will rain *bread* from heaven for you; and the people shall go out and gather a certain rate every day ... In the morning the dew lay round about the host. And when the dew that lay was gone up, behold, upon the face of the wilderness there lay a small round thing, as small as the hoar frost on the ground. And when the children of Israel saw it, they said one to another, It is *manna*: for they wist not what it was. And MOSES said unto them, This is the *bread* which the Lord hath given you to eat ... Gather of it every man according to his eating, an omer for every man, according to the number of your persons; take ye every man for them which are in his tents. And the children of Israel did so, and gathered, some more, some less. And when they did mete it with an omer, he that gathered much had nothing over, and he that gathered little had no lack; they gathered every man according to his eating. And MOSES said, Let no man leave of it till the morning. Notwithstanding they hearkened not unto MOSES; but some of them left of it until the morning, and it bred worms, and stank: and MOSES was wroth with them. And they gathered it every morning, every man according to his eating: and when the sun waxed hot it melted. And it came to pass that on the sixth day they gathered twice as much *bread*, two omers for one man: and all the rulers of the congregation came and told MOSES. And he said unto them, This is that which the Lord hath said, To morrow is the rest of the holy sabbath unto the Lord: bake that which ye will bake today.

and seethe that ye will seethe; and that which remaineth over lay up for you to be kept until the morning. And they laid it up till morning, as MOSES bade: and it did not stink, neither was there any worm therein. And MOSES said, Eat that to day; for to day is a sabbath unto the Lord: to day ye shall not find it in the field. Six days ye shall gather it; but on the seventh day, which is the sabbath, in it there shall be none. And it came to pass, that there went out some of the people on the seventh day for to gather, and they found none . . . he giveth you on the sixth day the *bread* of two days . . . And the house of Israel called the name thereof *Manna*: and it was like coriander seed, white; and the taste of it was like wafers made with honey. And MOSES said, This is the thing which the Lord commandeth, Fill an omer of it to be kept for your generations; that they may see the *bread* wherewith I have fed you in the wilderness, when I brought you forth from the land of Egypt. And MOSES said unto AARON, Take a pot, and put an omer full of *manna* therein, and lay it up before the Lord, to be kept for your generations. As the Lord commanded MOSES, so AARON laid it up before the Testimony, to be kept. And the children of Israel did eat *manna* forty years, until they came to a land inhabited; they did eat *manna* until they came unto the borders of the land of Canaan.

NUMBERS 11: 6-9—But now our soul is dried away; there is nothing at all, beside this *manna*, before our eyes. And the *manna* was as coriander seed, and the color thereof was as the color of bdellium. And the people went about and gathered it, and ground it in mills, or beat it in a mortar, and baked it in pans, and made cakes of it; and the taste of it was as the taste of fresh oil. And when the dew fell upon the camp in the night, the *manna* fell upon it.

DEUTERONOMY 8: 3 & 16—And he humbled thee, and suffered thee to hunger, and fed thee with *manna*, which thou knewest not, neither did thy fathers know . . . Who fed thee in the wilderness with *manna*, which thy fathers knew not . . .

JOSHUA 5: 12—And *manna* ceased on the morrow after they had eaten of the old corn of the land; neither had the children of Israel *manna* any more; but they did eat of the fruit of the land of Canaan that year.

NEHEMIAH 9: 20—Thou gavest also thy good spirit to instruct them, and withheldest not thy *manna* from their mouth.

PSALMS 78: 24-25—And had rained down *manna* upon them to eat, and had given them of the *corn of heaven*. Man did eat *angels' food*: he sent them meat to the full.

WISDOM OF SOLOMON 16: 20-21—Instead whereof thou feddest thine own people with *angels' food*, and didst send them from heaven *bread* prepared without their labour, able to content every man's delight, and agreeing to every taste. For thy sustenance declared thy sweetness unto thy children, and serving the appetite of the eater, tempered itself to every man's liking.

JOHN 6: 31—Our fathers did eat *manna* in the desert; as it is written, He gave them *bread from heaven* to eat.

HEBREWS 9: 4— . . . wherein was the golden pot that had *manna* . . .

REVELATION 2: 17— . . . To him that overcometh will I give to eat of the hidden *manna* . . .

There are three distinct types of "manna" mentioned in the Bible. One was a type which was secured by purchase and trade as recounted in Baruch 1: 10. This type consisted of the gummy exudations of certain desert trees, particularly *Tamarix mannifera*, *Alhagi maurorum*, and *Fraxinus ornus* (see under *Tamarix mannifera*).

The second type of "manna" was that which grew up during the night, when the ground was moist, but which "withered away" and "stank" when the heat of the sun fell upon it (Exodus 16: 13-21). This type is thought to have been produced by species of the algal genus *Nostoc*. These tiny algae have been known to grow quite regularly with unbelievable rapidity during the night, when there is an abundant fall of dew and the surface of the ground is quite moist. Being very soft and gelatinous, these algal growths which spring up during the night disappear the next morning when the sun evaporates the dew and dries up the surface of the ground, only to reappear the next night that there is abundant dew. This is the type of "manna" referred to in the phrase "when the sun waxed hot it melted". This statement can apply only to the algal type (53).

The third type of "manna" is the type that "fell from heaven" (Numbers 11: 9, GOODSPEED version: "Whenever the dew fell on the camp at night, the *manna* used to fall with it"). Considerable discussion and heated argument are to be found in literature on this subject. Modern opinion among botanists is that the "manna" which fell so regularly from the skies was, in major part, at least, composed of several lowly lichens, *Lecanora affinis*[72], *L. esculenta*[73],

---

[72]Also called *Parmelia esculenta* var. *affinis* Bischoff and *Aspicilia alpinodesertorum* f. *affinis* (Eversm.) Elenk.

[73]Also called *Lichen esculentus* Pall., *Lecanora desertorum* Krmphbr., and *L. desertorum* var. *esculenta* (Pall.) Krmphbr.

and *L. fruticulosa* [74]. These species of *Lecanora* occupy vast tracts of barren plains and mountains in many parts of western Asia and northern Africa, much as the reindeer-moss, *Cladonia rangiferina* (L.) Web., occupies vast stretches of the arctic tundra. Unlike the reindeer-moss, the *Lecanora* lichens after long periods of drought, curl up and break loose from the ground. Being extremely light, they are carried up by the winds and often transported great distances in the air, ultimately falling to the ground again and sometimes forming layers several inches deep (89). Sheep eat these lichens with considerable satisfaction and the local tribesmen still make a kind of bread from them. Often, depending on the strength and direction of the air currents, these lichens fall to earth at long distances from where they originally grew, and in places where they are unknown to the natives in any state except that in which they find them after they have dropped so mysteriously from the skies. The superstitious tribesmen, thus, very understandably still regard them as having truly "fallen from heaven". In or about the year 1854 a shower of these lichens fell in Persia during a great famine, much to the joy and thanksgiving of the starving inhabitants! A similar lichen grows on the Sahara Desert and is gathered regularly by the nomad tribesmen for food in times of scarcity. It can be cooked in various ways and made into bread (53, 54, 55).

It is very probable, however, that plant mannas of the gummy type mentioned above, and even other substances were included in the term "manna" besides the lichen and algal mannas here described, and that the word came to have a symbolic significance for any food "miraculously" produced in time of scarcity or on the barren deserts. It is estimated that between 2 and 2½ million Israelites were living in their tents in the "wilderness" of Sinai at the time that these events took place (1491-1451 B.C.). The allowance to each man, according to MOFFATT's translation, was about 7 pints a day. Each pint of substances like this weighs about a pound. This would mean that about 9250 tons of the material would have to be collected daily—and it is said that this continued (although perhaps not exclusively) for forty years! Other authorities figure the amount collected per day as only 2¼ pints per person, but this would still mean about 2000 tons per day.

The etymology and real meaning of the word "manna" are indicated plainly in the Septuagint and Vulgate, and by old writers like JOSEPHUS. The Hebrew word by which this substance was designated in the Old Testament is "mân". This is actually the neuter interrogative pronoun, meaning "What?" The name that we use is a derivation of "mân hu", meaning "What is this?", which the Jews presumably asked when they first saw the substance on the ground (306). In one passage where the Authorized Version says "It is manna" the JASTROW version says "What is it?" In Psalms this mysterious food is spoken of as "angels' food" and "corn of heaven" by the King James Version, while JASTROW calls it "bread of the mighty".

The theory has been advanced by some Bible students that the manna which appeared so suddenly during the night while the children of Israel were encamped in the desert of Sinai actually consisted of quails or quail-dung, or both. In Exodus 16: 12-13 quails are mentioned as having appeared at the same time as the "manna". The quail (*Coturnix vulgaris*) is a well-known migratory bird found all over Europe, Asia, and Africa (2, 3). The bird is about the size of a small partridge. Quails breed in large numbers in Palestine, and ordinarily a few pairs remain there throughout the winter. The vast majority, however, makes its appearance quite suddenly in March. The birds migrate to the warmer parts of Africa in the winter. Aided by favorable winds, they return from Africa quite suddenly—often myriads appearing during a single night. The Hebrew word for this bird, however, is "shĕlâv", and it does not seem at all likely that the Israelites would have referred to the bird as "manna"—a word which by its very nature implies something with which they had never had any previous experience. Nor does it seem likely that the dung of the

---

[74] Also known as *Sphaerophorus gelatinosus* Trev. and *Aspicilia alpinodesertorum* f. *fruticulosa* (Eversm.) Elenk.

quails would have elicited such a term from them. Surely someone among the vast horde of Israelites would have observed the origin of the dung! Yet it is not at all unlikely that both the flesh and the dung of these birds were used as food along with the manna, thus providing the nitrogenous and other sustaining elements necessary for a more balanced diet. That migrating birds make use of wind and air currents in their long flights is a well-established fact. The fact that the manna appeared along with the quails lends support to the theory that the manna consisted in part, at least, of lichens also transported by high winds from the African mainland.

127. Lens esculenta Moench
(FIGURE 90)

GENESIS 25: 29-34—And JACOB sod pottage: and ESAU came from the field, and he was faint: And ESAU said to JACOB, Feed me, I pray thee, with that same red pottage; for I am faint: therefore was his name called EDOM. And JACOB said, Sell me this day thy birthright. And ESAU said, Behold, I am at the point to die: and what profit shall this birthright do to me? And JACOB said, Swear to me this day; and he sware unto him: and he sold his birthright unto JACOB. Then JACOB gave ESAU bread and pottage of *lentiles* . . .

II SAMUEL 17: 27-29— . . . and BARZELLAI the Gileadite of Rogelim, Brought beds, and basons, and earthen vessels, and wheat, and barley, and flour, and parched corn, and beans, and *lentiles*, and parched pulse.

II SAMUEL 23: 11— . . . And the Philistines were gathered together into a troop, where was a piece of ground full of *lentiles* . . .

I CHRONICLES 11: 13— . . . there the Philistines were gathered together to battle, where was a parcel of ground full of *barley* . . .

EZEKIEL 4: 9—Take thou also unto thee wheat, and barley, and beans, and *lentiles*, and millet, and fitches, and put them in one vessel, and make thee bread thereof . . .

The Hebrew word translated "lentiles" in these passages is "adasha" or "ădâshîm" (Greek, φακός), and there is no doubt whatever about the identity of the plant. The lentil plant, *Lens esculenta*[75], (98) is a small, erect, annual, vetch-like plant with slender stems and pinnately compound tendril-bearing leaves, producing small, white, violet-striped flowers and flat pea-like pods in which the lentils are borne. The lentils themselves are about the size of small peas, but are flattened and lens-shaped, convex on both surfaces (299). There are three or four kinds of lentils, all of which are highly regarded in those countries where they are grown—southern Europe, Asia, and northern Africa. There is a "red" variety which was doubtless the one of which JACOB's "red pottage", "omelet" (according to MOFFATT), or "stew" (according to GOODSPEED) was made. PRATT points out (268) that this lentil pottage is not actually red, but is called "red" because Orientals call anything "red" which is, strictly speaking, a yellow-brown—as, for instance, "red" cattle. Travelers in the Orient from the West describe this pottage as of a reddish-chocolate color when boiled.

Lentils continue to be widely cultivated in Egypt and Palestine for their nutritious seeds, and even with us lentil soup is often found on the menu. A fairly good bread is made from lentils and barley. This kind of bread is quite common in some parts of Egypt, especially in the country toward the cataracts of the Nile, where it is the only kind of bread known. It is widely eaten by the poor. The Arabic name of the lentil is "adas" (184). The "red" variety is the one most frequently cultivated. Lentils are cut and threshed like wheat, then stewed like string beans, and made into pottage (184).

Some commentators assert that the use of the word "barley" in I Chronicles 11: 13 is a mistake in the original, and that the word there should be "lentils" as in II Samuel 23: 11, which refers to the same place and event. The MOFFATT, GOODSPEED, JASTROW, LEESSER, and Douay versions, however, all continue to use the word "barley", which is "s'orim" in Hebrew. Mrs. BARNEVELD states that on the borders of Egypt stands a town called Phakussa, literally "the lentil town" (47).

Lentil meal is considered very nutritious, and the plants are still extensively cultivated, especially in poorer soil, in France and other parts of Europe, large

---

[75]Also called *Ervum lens* L. (Mrs. BARNEVELD misspells it *"Ervun lens"* (47)) and *Lens culinaris* Medic.

quantities being imported by England (299). The patented products "Ervalenta" and "Revalenta" consist of lentil meal, the names being merely anagrams of *Ervum lens* with the addition of the suffix "-ta" (299). Decorticated lentil seeds are commonly sold in the bazaars of India. The lentil is thought to be a native of western Asia, but is now common in fields throughout the Holy Land (266).

128. Lilium chalcedonicum L.

SONG 5: 13—His cheeks are as a bed of spices, as sweet flowers: his lips like *lilies*, dropping sweet smelling myrrh.

A vast amount of discussion has centered about the identification of the plant or plants referred to in the Bible as "lily" or "lilies" (Hebrew, "shoshanim"). A great many references to Scriptural passages are often cited to *Lilium chalcedonicum*, but the one in Song 5: 13 seems to us to be the only one strictly applicable. MOFFATT renders the verse: "his cheeks are beds of balsam-flower, banks of perfume, his lips are *lilies* red, breathing liquid myrrh." Some authors have thought that *Anemone coronaria* is intended here, the typical form of which has scarlet flowers (27). However, the anemony is found "everywhere in Syria and Palestine below sub-alpine regions" according to POST (265). The context of the passage from SOLOMON's Song calls for some rare plant of exceptional beauty, not a plant as common as the anemony, even though the latter is of striking beauty. The scarlet or Martagon lily, *Lilium chalcedonicum*, of the Levant, on the other hand, is very rare in the Holy Land. In fact, POST does not even record it (265, 267), but J. SMITH, the famous botanist of the Royal Botanic Gardens at Kew, states definitely that it is native there (27, 299). TRISTRAM mentions it from the catalogue of Bible Land plants at Kew, even though he did not see it himself during his explorations (326). POST lists the white Madonna lily, *L. candidum*, as doubtfully native, but a white flower obviously does not fit the context of this passage, nor does the blue-flowered hyacinth, *Hyacinthus orientalis*, adopted by the GOODSPEED version! The JASTROW translation uses the term "lily" just as it is used in the Authorized Version. HEHN mentions *L. chalcedonicum* as a possible choice for the Song of Solomon plant, but he gives more credence to *Fritillaria libanotica* (170). LAGARDE and WESSELY believe that *L. chalcedonicum* is the "lily" of the New Testament (100), but almost all other writers omit any true lily from consideration. W. T. STEARN states that *L. chalcedonicum* "is a Greek species, not a Palestinian one. The lip-red flowers of the Song of SOLOMON were probably red anemones."

The various other Biblical references to "lilies" will be found discussed herein under *Anemone coronaria*, *Hyacinthus orientalis*, *Iris pseudacorus*, and *Nymphaea alba*, which see.

129. Linum usitatissimum L.
(PAGE 135)

GENESIS 41: 42—And PHARAOH took off his ring from his hand, and put it upon JOSEPH's hand, and arrayed him in vestures of fine *linen*, and put a gold chain about his neck.
EXODUS 9: 31—And the *flax* and the barley was smitten: for the barley was in the ear, and the *flax* was bolled.
EXODUS 25: 4—And blue, and purple, and scarlet, and fine *linen*, and goats' hair.
EXODUS 26: 1, 31, & 36—Moreover thou shalt make the tabernacle with ten curtains of fine twined *linen*, and blue, and purple, and scarlet . . . And thou shalt make a vail of . . . fine twined *linen* . . . And thou shalt make an hanging for the door of the tent, of . . . fine twined *linen*, wrought with needle-work.
EXODUS 27: 16 & 18—And for the gate of the court shall be an hanging of . . . fine twined *linen* . . . and the height five cubits of fine twined *linen* . . .
EXODUS 28: 5-6, 8, 15, 39, & 42—And they shall take gold, and blue, and purple, and scarlet, and fine *linen*. And they shall make the ephod of gold, of blue, and of purple, of scarlet, and fine twined *linen*, with cunning work . . . And the curious girdle of the ephod . . . shall be of . . . fine twined *linen* . . . And thou shalt make the breastplate of judgment . . . of fine twined *linen* . . . And thou shalt embroider the coat of fine twined *linen*, and thou shalt make the mitre of fine *linen* . . . And thou shalt make them *linen* breeches to cover their nakedness; from the loins even unto the thighs they shall reach.

EXODUS 35: 6, 23, 25, & 35—And blue, and purple, and scarlet, and fine *linen* ... And every man, with whom was found ... fine *linen* ... brought them ... And all the women that were wise hearted did spin with their hands, and brought that which they had spun, both of blue, and of purple, and of scarlet, and of fine *linen* ... Them hath he filled with wisdom of heart, to work ... in fine *linen* ...

EXODUS 36: 8, 35, & 37—And every wise hearted man among them that wrought the work of the tabernacle made ten curtains of fine twined *linen* ... And he made a vail of ... fine twined *linen* ... And he made an hanging for the tabernacle door of ... fine twined *linen* ...

EXODUS 38: 9, 18, & 23— ... the hangings of the court were of fine twined *linen* ... And the hanging for the gate of the court was ... fine twined *linen* ... And with him was AHOLIAB ... an embroiderer in ... fine *linen*.

EXODUS 39: 2, 3, 5, 8, 24, & 27-29—And he made the ephod of ... fine twined *linen* ... And they did beat the gold into thin plates, and cut it into wires, to work it in ... the fine *linen*, with cunning work ... And the curious girdle of his ephod ... was of ... fine twined *linen* ... And he made the breastplate of cunning work ... and fine twined *linen* ... And they made upon the hems of the robe pomegranates of ... twined *linen* ... And they made coats of fine *linen* of woven work for AARON, and for his sons: And a mitre of fine *linen*, and goodly bonnets of fine *linen*, and *linen* breeches of fine twined *linen*, And a girdle of fine twined *linen*.

LEVITICUS 6: 10—And the priest shall put on his *linen* garment, and his *linen* breeches.

LEVITICUS 13: 47-48, 52, & 59—The garment also that the plague of leprosy is in, whether it be a woollen garment, or a *linen* garment; Whether it be in the warp, or woof; of *linen*, or of woollen ... He shall therefore burn that garment, whether warp or woof, in woollen or in *linen* ... This is the law of the plague of leprosy in a garment of woollen or *linen*.

LEVITICUS 16: 4, 23, & 32—He shall put on the holy *linen* coat, and he shall have the *linen* breeches upon his flesh, and shall be girded with a *linen* girdle, and with a *linen* mitre shall he be attired ... And AARON shall come into the tabernacle of the congregation, and shall put off the *linen* garments ... And the priest ... shall put on the *linen* clothes, even the holy garments.

LEVITICUS 19: 19— ... thou shalt not sow thy field with mingled seed: neither shall a garment mingled of *linen* and woollen come upon thee.

DEUTERONOMY 22: 11—Thou shalt not wear a garment of divers sorts, as of woollen and *linen* together.

JOSHUA 2: 6—But she had brought them up to the roof of the house, and hid them with the stalks of *flax*, which she had lain in order upon the roof.

JUDGES 14: 12-13— ... then I will give you thirty *sheets* and thirty change of garments ... then shall ye give me thirty *sheets* and thirty change of garments ...

JUDGES 15: 14— ... and the cords that were upon his arms became as *flax* that was burned with fire ...

I SAMUEL 2: 18—But SAMUEL ministered before the Lord, being a child, girded with a *linen* ephod.

I SAMUEL 22: 18— ... and slew on that day fourscore and five persons that did wear a *linen* ephod.

II SAMUEL 6: 14— ... And DAVID danced before the Lord with all his might; and DAVID was girded with a *linen* ephod.

I KINGS 10: 28—And SOLOMON had horses brought out of Egypt, and *linen* yarn: the king's merchants received the *linen* yarn at a price.

I CHRONICLES 15: 27—And DAVID was clothed with a robe of fine *linen* ... DAVID also had upon him an ephod of *linen*.

II CHRONICLES 1: 16—And SOLOMON had horses brought out of Egypt, and *linen* yarn: the king's merchants received the *linen* yarn at a price.

II CHRONICLES 3: 14—And he made the vail of blue, and purple, and crimson, and fine *linen* ...

II CHRONICLES 5: 12— ... being arrayed in white *linen* ...

ESTHER 1: 6—Where were white, green, and blue, hangings, fastened with cords of fine *linen* ...

ESTHER 8: 15— And MORDECAI went out from the presence of the king in royal apparel of blue and white, and with a great crown of gold, and with a garment of fine *linen*.

PROVERBS 7: 16—I have decked my bed with coverings of tapestry, with carved works, with fine *linen* of Egypt.

PROVERBS 31: 13, 22, & 24—She seeketh wool, and *flax*, and worketh willingly with her hands ... her clothing is *silk* and purple ... She maketh fine *linen* and selleth it ...

ISAIAH 3: 23—The glasses, and the fine *linen*, and the hoods, and the vails.

ISAIAH 19: 9—Moreover they that work in fine *flax*, and they that weave networks, shall be confounded.

ISAIAH 42: 3—A bruised reed he shall not break, and the smoking *flax* he shall not quench.

JEREMIAH 13: 1— ... Go and get thee a *linen* girdle ...

EZEKIEL 9: 2-3— ... and one man among them was clothed with *linen* ... And he called to the man clothed with *linen* ...

EZEKIEL 10: 2 & 6-7—And he spake unto the man clothed with *linen* ... when he had commanded the man clothed with *linen* ... and put it into the hands of him that was clothed with *linen* ...

EZEKIEL 16: 10 & 13— ... I girded thee about with fine *linen* ... and thy raiment was of fine *linen*.

EZEKIEL 27: 7 & 16—Fine *linen* with broidered work from Egypt was that which thou spreadest forth to be thy sail ... Syria was thy merchant by reason of the multitude of the wares of thy making: they occupied in thy fairs with emeralds, purple, and broidered work, and

fine *linen*, and coral, and agate.
EZEKIEL 40: 3— . . . there was a man whose appearance was like the appearance of brass, with a line of *flax* in his hand . . .
EZEKIEL 44: 17-18—And it shall come to pass, that when they enter in at the gates of the inner court, they shall be clothed with *linen* garments; and no wool shall come upon them, whiles they minister in the gates of the inner court, and within. They shall have *linen* bonnets upon their heads, and shall have *linen* breeches upon their loins; they shall not gird themselves with anything that causeth sweat.
DANIEL 10: 5— Then I lifted up mine eyes, and looked, and behold a certain man clothed in *linen* . . .
HOSEA 2: 5 & 9— . . . that give me . . . my *flax* . . . Therefore will I return . . . and will recover my wool and my *flax* given to cover her nakedness.
ECCLESIASTICUS 40: 4—From him that weareth purple and a crown, unto him that is clothed with a *linen* frock.
MATTHEW 27: 59—And when JOSEPH had taken the body, he wrapped it in a clean *linen* cloth.
MARK 14: 51-52—And there followed him a certain young man, having a *linen* cloth cast about his naked body; and the young men laid hold on him: And he left the *linen* cloth, and fled from them naked.
MARK 15: 46—And he brought fine *linen*, and took him down, and wrapped him in the *linen*, and laid him in a sepulchre . . .
LUKE 16: 19—There was a certain rich man, which was clothed in purple and fine *linen*, and fared sumptuously every day.
LUKE 23: 53—And he took it down, and wrapped it in *linen* . . .
JOHN 19: 40—Then took they the body of JESUS, and wound it in *linen* clothes with the spices, as the manner of the Jews is to bury.
REVELATION 15: 6—And seven angels came out of the temple, having the seven plagues, clothed in pure and white *linen* . . .
REVELATION 18: 12 & 16—The merchandise of gold, and silver, and precious stones, and of pearls, and of fine *linen*, and purple, and silk, and scarlet . . . Alas, alas that great city, that was clothed in fine *linen* . . .
REVELATION 19: 8 & 14—And to her was granted that she should be arrayed in fine *linen*, clean and white: for the fine *linen* is the righteousness of saints . . . And the armies which were in heaven followed him upon white horses, clothed in fine *linen*, white and clean.

Flax is the oldest known of textile fibers. Cotton is mentioned only once in the Bible in connection with an event which took place during the Babylonian captivity, and it is never mentioned along with wool and linen when materials for making clothing are mentioned. Since there is no mention of hemp or any other fiber plant having been cultivated in Egypt or Palestine in Biblical days, it must be inferred that linen in various qualities was the material out of which clothes which were not woolen were made. It was also in general use for all domestic purposes, such as towels (John 13: 4-5), napkins (John 11: 44), girdles and under-garments (Isaiah 3: 23; Mark 14: 51), nets (Isaiah 19: 9), and measuring-lines (Ezekiel 40: 3), as well as for sails, pennons, and flags (179, 299). It was an important crop in Egypt, and was known and used in Canaan before the arrival of the Israelites (Joshua 2: 6). It was woven into garments by the ancient Egyptians and was wrapped around their bodies after they were embalmed. The most ancient of known mummies are wrapped in linen shrouds (27, 184). The Greeks and Hebrews also employed it for the winding-sheets of the dead (Matthew 27: 59; Mark 15: 46; Luke 23: 53), as is attested also by HOMER and EURIPIDES.

The custom of blanching the flax fibers by exposure to the sun and air, often on the flat roofs of the houses, is practiced to this day in many parts of Europe as well as in the Holy Land. The priests serving in the temple had to wear nothing but linen clothes, and a mixed cloth of wool and flax together was strictly forbidden to the Jews (Leviticus 19: 19; Deuteronomy 22: 11).

There were three distinct kinds of linen used in Biblical times, and apparently there were specified uses for each kind. The ordinary linen of coarsest texture is mentioned in Leviticus 6: 10, Ezekiel 9: 2, Daniel 10: 5, and Revelation 15: 6. The second type was linen of superior fabric, mentioned in Exodus 26: 1 and 39: 27. A third type was linen of finest texture and costliest quality, mentioned in Esther 8: 15, I Chronicles 15: 27, and Revelation 19: 8. No less than seven Hebrew words are involved. The words "pishtah" (27, 184) or "pista" (98) and "pishtim", and "gristah" (299) (Greek, λίνον, meaning "linen or anything made from flax," according to Greek dictionaries) refer to the flax plant as such, *Linum usitatissimum*[76], and is used in Exodus 9: 31 and

---

[76]Also referred to as *Linum sativum* Hasselq.

Isaiah 19:9. In Genesis 41:42 the Hebrew word used is "shêsh" (Greek, βύσσος), rendered "fine linen" in the Authorized, Revised, and MOFFATT versions ("linen" by GOODSPEED and LEESSER; "cotton" in a marginal note in the Revised; "silk" by the Douay and in a marginal note in the Authorized versions). In Exodus 25:4 and 35:6 and Ezekiel 27:7 the same word is used, but rendered "fine linen" by all versions, except that the latter is omitted by MOFFATT and the Authorized gives "silk" as a marginal suggestion for Exodus 25:4. The word "shêsh" is probably originally an Egyptian word and may actually refer to the linen yarn only. JOSEPH when promoted to the dignity of being a ruler of Egypt was arrayed in robes made of this type of linen (Genesis 41:42). Among the offerings for the tabernacle of the things that the Jews had brought with them out of Egypt was also linen of this sort (Exodus 25:4 and 35:6). In Ezekiel 27:7 this type of linen is enumerated among the products of Egypt which the Syrians imported and used for the sails of their ships. In Exodus 28:42 and Leviticus 6:10 and 16:24 the word "bad" or "bŏd" (Greek, λίνον) is used. It seems to be synonymous with the Egyptian "shêsh" or may refer to the cloth made from "shêsh" yarn (184). In no case is "bad" used for other than the linen employed in making garments worn in religious ceremonies (306).

"Butz", "bootz", or "buts" (Greek, βύσσος), used in Esther 8:15, I Chronicles 15:27, II Chronicles 3:14 and 5:12, and many other passages is rendered "fine linen" by the Authorized, Douay, MOFFATT, and GOODSPEED versions except that the Esther reference is translated "silk" in the Douay and the I Chronicles reference is omitted by MOFFATT. This was the material of which the robes of kings, rich men, and the temple choir were made, as was also the temple veil. It is the original of the βύσσος of the New Testament, the dress of DIVES (Luke 16:19), and of the bride of the Lamb (Revelation 19:8). The word is considered to be Assyrian and was applied to linen imported from the Orient, while "shêsh" was applied to that imported from Egypt. In Judges 14:12 and Isaiah 3:23 another word appears, "sadin" or "s'deenim" (Greek, σινδών,) the material of which sheets and ordinary clothes were made. In the Authorized Version the Judges passage is rendered "sheets", in the JASTROW and Revised "linen garments", in MOFFATT "fine linen shawls", in the Douay and LEESSER "shirts", and in the GOODSPEED "linen robes". In the Isaiah passage the Authorized Version says "fine linen", GOODSPEED "linen vests", MOFFATT "linen turbans", and the Douay "lawns". The word "ētun" is used only once, in Proverbs 7:16, rendered "fine linen of Egypt" in the King James and "yarn of Egypt" in the Revised and JASTROW versions, omitted completely by the Septuagint, but a very similar Greek word, ὀθόνη, is the "great sheet" in PETER's vision and the gravecloth of JESUS (John 19:40). In I Kings 10:28 and II Chronicles 1:16 still another word is used, "mikvay" or "mikveh". This is the most controversial of all. The Authorized Version, following JUNIUS and TREMELLIUS (306), translates it "linen yarn" and regards it as an import from Egypt in SOLOMON's day; the Septuagint and other old versions retain it as a proper name; the Revised Version makes it "drove"; GESENIUS translates it "troop" and BOCHART "tax" (184); the Douay version regards it as a place name "Coa"; the JASTROW translation considers it the place "Kiveh" or "Keve", respectively; and GOODSPEED makes it "Kuë". MOFFATT renders the verse: "SOLOMON's horses were imported from Muzri and from *Kuê*; the royal dealers used to bring a troop of horses from *Kuê*, paying cash for them." The weight of opinion, therefore, seems to be against the last-mentioned word—"mikveh"—having any connection with flax. In Isaiah 42:3 Jastrow changes the "smoking flax" of the King James Version to "dimly burning wick."

Some writers cite the approximately 20 Old Testament passages referring to cord, rope, and string, based on various Hebrew words. However, this is probably an error, for the best modern opinion is that cord, rope, and string in Old Testament days were made from the hides of camels and other animals, as is done today by the Bedouin Arabs, or from twisted reeds, rushes, osiers, and willow twigs.

Flax was one of the earliest plants cultivated in Egypt and Palestine. The Talmud and Rabbinical tracts abound with comments concerning the sowing and gathering of this plant, as well as on the methods of bleaching and dressing it afterwards and manufacturing with it (268). Old Egyptian paintings and inscriptions represent it, and even today springtime in Egypt finds the fields blue with the handsome flowers of the flax. The vaunted "fine linen" of Egypt, however—the finest available to the Israelites and greatly preferred by all the ancients over the coarser types of Syria—was still a very coarse cloth by our standards today.

The common flax plant grows from 1 to 4 feet tall, has a slender, simple, erect, wiry stem, with numerous small, pale-green, linear or lanceolate, acute-tipped leaves, and large, deep- or pale-blue (rarely white) flowers, about ½ inch across, at the top. The fruits are 5-celled capsules, about the size of a pea. The species is thought to be originally from Mesopotamia, now not as extensively cultivated in the Holy Land as it is in Egypt and Abyssinia (266). The failure of the flax crop is listed as one of God's punishments (Hosea 2: 9). The manufacture of linen from flax fibers was a domestic industry of Jewish women (Proverbs 31: 13 & 19) (184). It was made by them into the robes and aprons worn by the priests and temple attendants. It was used for wicks in lamps (Isaiah 42: 3). For centuries it was the only and universal textile fabric (184) and in Biblical days it was one of the most important crops of Palestine (306). At present its use has been largely superseded by cotton (27).

Flax capsules are often called "bolls", and the expression "the flax was bolled" in Exodus 9: 31 means that it had arrived at a state of maturity (299). When the bolls are ripe the flax plants are harvested and tied into bundles or sheaves. These bundles are then immersed in water for several weeks, which causes them to "ret", that is, causes the fibers to separate from the non-fibrous portions of the stems. The bundles are then opened and the retted stems spread out to dry (Joshua 2: 6), after which they are combed or "hackled" to remove the fibers. Representations of the cultivation and dressing of flax are found in the ancient Egyptian tombs and papyri, yet it seems probable that the cultivation of flax for its fiber did not start in Egypt, but originated in Mesopotamia and India and then spread over the entire ancient world at a very early period of antiquity (306). Its fiber has been found in the ancient lake-dwellings of Switzerland (299). Linseed oil is expressed from its seeds, and the remaining parts of the seed after the oil is extracted are compressed and made into a cattle feed. Many kinds of twine and rope are now made of linen fiber.

In Teutonic mythology the earth goddess, HILDA, taught mankind the art of growing and weaving flax. She visits every home twice a year to see if the men have planted enough in their fields, and, later, to see if the women have enough of the fiber in their homes and are properly industrious with it (125). If she is not satisfied with what she finds it is taken as a sign of thriftlessness and laziness on the part of the family and the next year's crop will be blighted in punishment (298). "Because HILDA is the goddess of plenty, flax, in the regard of some of the northern people, has become the type of life. When a German baby does not thrive they place him naked on the grass and scatter flaxseed over him, in the belief that such of the seed as, falling on the earth, takes root and flourishes, will join his fortunes to the plentiful life that is everywhere about him; so he must begin to grow when the little plants appear" (298).

130. Lolium temulentum L.
(FIGURE 41; FIGURE 72)

MATTHEW 13: 24-30—... The kingdom of heaven is likened unto a man which sowed good seed in his field: But while men slept, his enemy came and sowed *tares* among the wheat, and went his way. But when the blade was sprung up, and brought forth fruit, then appeared the *tares* also. So the servants of the householder came and said unto him, Sir, didst not thou sow good seed in the field? from whence then hath it *tares*? He said unto them, An enemy hath done this. The servants said unto him, Wilt thou then that we go and gather them up? But he said, Nay; lest while ye gather up the *tares*, ye root up also the wheat with them. Let both grow together until the harvest: and in the time of harvest I will say to the reapers, Gather ye together first the *tares*, and bind them in bundles to burn them: but gather the wheat into my barn.

The Greek word translated "tares" in the Authorized, Revised, and LAMSA versions is ξιξάνια. The dictionary translation for this word is "darnel", and it is similar to the word ξιξάνιον, said to be "a weed that grows in wheat". It is generally agreed now that the plant referred to is not the plant known as "tares" in Europe today (*Vicia sativa* L.), but is the annual or bearded darnelgrass, *Lolium temulentum* (299), a strong-growing grass very closely resembling wheat or rye in appearance and from which it is extremely difficult to distinguish in its early stages, although its seeds are much smaller (184, 306). If it is not eradicated early, but is left until the time of harvest, it is cut down with the wheat and is then very difficult to separate from it. More or less of the darnel is then, naturally, ground up with the wheat (98, 299). Its seeds have very deleterious effects, and even poisonous properties have been ascribed to them, either due to some chemicals naturally in them or to those of some fungus growing within the seed (180, 205, 253). STEIN, in an article entitled "Poisoning of human beings by weeds contained in cereals", states that at present it is generally held that darnel is poisonous only when infested with these fungi (310). SMITH states that death may result from eating bread with too much darnel seed in its flour. THEOPHRASTUS and other early Greek writers were well acquainted with the poisonous effects of darnel. GERARDE says that "the new bread wherein darnel is, eaten hot causeth drunkenness" (299). This is doubtless why in some books the plant is referred to as "drunken darnel". It is also said to cause blindness (299). Although there are many thousands of kinds of grasses in the world, there are only two or three known to possess poisonous properties. Two of these are species of *Stipa* and the third is the darnel. Obviously darnel has nothing whatever to do with the wild-rice, *Zizania aquatica* L., to which LINNAEUS applied the same Greek word "zizania" and which is a valuable American food plant (299). About all that can be said to the credit of darnel is that it can serve as a chicken and pigeon feed (98), and even this is not recommended.

LUTHER translates the word merely as "weeds", and in this he is followed by GOODSPEED, O'HARA, MOFFATT, and WEIGLE. The Revised Version suggests "darnel" in a marginal note, and the Douay version "cockle"; WEYMOUTH uses "darnel"; and the Basic English version says "evil seeds" and "evil plants". The word "zizania" is apparently of Oriental, not Greek, origin (306), and is not used by any of the ancient or classic Greek authors, but still darnel-grass is called "zizanion" by the Spaniards and "siwan", "zawân", or "zuwân" by the Arabs (267, 268). It is very frequent in grainfields in Palestine, Lebanon, and Syria, as well as in all other countries bordering on the Mediterranean (184) and in Great Britain (27). The poorer people do not clean it out of their grainfields, lest in so doing they accidentally pull up and thus lose a single grain plant. So still today the farmer's directions would be the same: not to tear up the darnel lest the grain also be torn up, but wait until harvest time and then separate them. DALMAN and EVENARI agree with STEIN that the poisonous effects of this plant are due to a fungus growing beneath the seed-coats and not to the darnel plant itself (98). Since this fungus seems to be so generally associated with the darnel, the net result is virtually the same. It is of interest to note, in passing, that the farmers of Palestine believe that the darnel is merely a degenerate or even "bewitched" wheat, and that in wet seasons the wheat actually turns into darnel (98, 306).

131. Lycium europaeum L.

JUDGES 9: 14-15—Then said all the trees unto the *bramble*, Come thou, and reign over us. And the *bramble* said unto the trees, If in truth ye anoint me king over you, then come and put your trust in my shadow: and if not, let fire come out of the *bramble*, and devour the cedars of Lebanon.

About twenty-two different Hebrew and Greek words are used in the Bible in referring to spiny or prickly shrubs or weeds, and these words are all indifferently and very inconsistently translated "bramble", "brier", "thorn", and "thistle" by the Authorized Version. In many cases we have little or nothing to guide us in identifying the actual plants meant. In other cases present Arabic or Hebrew names for certain plants, or the context of the passages, help us in identifying them.

Although the story recorded in Judges 9: 8-15 is obviously an allegorical one, wherein the various trees and shrubs and herbs symbolize nations, the Hebrew word "atâd" (Greek, ῥάμνος) used in Judges 9: 14 is now regarded (184) as definitely referring to the European boxthorn or desert-thorn, *Lycium europaeum*[77] (81, 98). This is a thorny shrub, 6 to 12 feet tall, with clustered, oblanceolate, oblique, acutish or obtuse leaves, and small violet flowers, eventually producing small globular red berries (267). It is native to and common throughout the Holy Land area, especially in the region from Lebanon to the Dead Sea (184). It has been found at Aleppo, Latakia, Tripoli, Jerusalem, the upper Jordan valley, Jaffa, Gaza, Jericho, and south as far as Sinai (267). It is frequently used for hedges.

---
[77]Also called *Lycium mediterraneum* Dun. and *L. spinosum* Hasselq.

FIGURE 23. — *Morus nigra*, the red mulberry. An old tree at the Lower Pool of Siloam, said to mark the spot where ISAIAH was sawn asunder. This tree is referred to as the "sycamine" by LUKE. Its leaves were used in the cultivation of silkworms, but do not serve as well as those of the Chinese white mulberry, a species not known in Palestine in Bible time. Silk culture, therefore, did not flourish in Bible lands until the introduction of the white mulberry. (Wood engraving from C. W. Wilson's Picturesque Palestine, 1883).

132. Mandragora officinarum L.
(FIGURE 66; FIGURE 67)

GENESIS 30: 14-16—And REUBEN went in the days of wheat harvest, and found *mandrakes* in the field, and brought them unto his mother LEAH. Then RACHEL said to LEAH, Give me, I pray thee, of thy son's *mandrakes*. And she said unto her, Is it a small matter that thou hast taken my husband? and wouldest thou take away my son's *mandrakes* also? And RACHEL said, Therefore he shall lie with thee tonight for thy son's *mandrakes*. And JACOB came out of the field in the evening, and LEAH went out to meet him, and said, Thou must come in unto me; for surely I have hired thee with my son's *mandrakes* . . .

SONG 7: 13— The *mandrakes* give a smell, and at our gates are all manner of pleasant fruits, new and old, which I have laid up for thee, O my beloved.

The Hebrew word translated "mandrakes" in the Authorized, Douay, JASTROW, and GOODSPEED versions is "dudâim" (Greek, μανδραγόρας or "mandrake"), and most commentators are agreed that the plant referred to is the love-apple or mandrake, *Mandragora officinarum*[78]. The Septuagint, Vulgate, Syriac, and Arabic versions, the Targums, and most of the learned Rabbinical tracts all agree with this determination (306). It probably has nothing whatever to do with *Cucumis dudaim* L., which, unfortunately, was given the same Hebrew word as its specific name by LINNAEUS and which SPRENGEL and others have supposed to be the Biblical "mandrakes" (299).

The love-apple is a stemless herbaceous perennial, related to the nightshade, potato, and tomato. It has a large beet-like tap-root, which is often forked, from the top of which arise many lanceolate, oblong, or ovate, wrinkled, dark-green leaves about a foot long and to 4 inches wide, lying flat on the ground in the form of a rosette, much like those of an English primrose. From the center of this rosette of leaves arise the flower-stalks, each bearing a single, purple, bluish, or greenish-white flower similar in shape and size to that of the potato, followed in due time by a subglobose yellowish berry about the size of a large plum (184). When perfectly developed the fruits lie in the center of the rosette of leaves like yellow bird eggs in a shallow nest, and have a fleshy pulp possessing a peculiar but not unpleasant smell and sweetish taste. The plant is slightly poisonous, but not nearly as much so as some of its relatives, being principally an emetic, purgative, and narcotic. It was much employed in medicine in olden times (126, 179, 299). Its efficacy, however, lay more in the superstitious notions regarding it than in its actual properties. These superstitions arose from the fact that its thick tap-roots have some resemblance in shape to the lower portions of the human body (124a, 318a). Therefore to it were ascribed certain amorous or aphrodisiac properties, as the Genesis story of LEAH and RACHEL attests. The Arabs call it "devil's apples" because of its supposed power to excite voluptuousness. It is also thought to stimulate fruitfulness (184) and has long been famous for its use in love-potions and incantations (27, 126).

JOSEPHUS states that during and before his time the mandrake was held in great superstitious awe by the Jews and the Greeks (126) and that "he who would take up a plant thereof, must tie a dog thereunto, to pull it up, otherwise if a man should do it, he should surely die in a short space after" (95, 299). It is further stated that the dog will die shortly after pulling it up! The ancient Romans considered it so potent and valuable in medicine as a narcotic and restorative that the collecting of the root was made a special ceremony. According to PLINY, the collector stood with his back to the wind, drew three concentric circles around the plant with the point of a sword, poured a libation on the ground, and, turning to the west, began to dig it up with his sword (95, 299).

---

[78] Also called *Atropa mandragora* L. and *Mandragora officinalis* Mill.

GERARDE in his herbal of 1597 says: "There hath been many ridiculous tales brought up of this plant, whether of old wives, or some runagate surgeons, or physique-mongers, I know not. They add, that it is never or very seldome to be found growing naturally, but under a gallows, where the matter that has fallen from a dead body hath given it the shape of a man, and the matter of a woman the substance of a female plant, with many other such doltish dreams" (299). SHAKESPEARE was acquainted with the legend that the roots give out a scream of agony when pulled out of the earth, for he says: "And shrieks like mandrakes torn out of the earth, that living mortals hearing them run mad" (95, 299). He also knew the story that this shriek meant death to the person or animal so foolhardy as to attack the plant, for he says: "Could curses kill, as doth the mandrake's groan" (95). MOORE, in his "Light of the Harem", says: "The phantom shapes — Oh! touch them not — Which appal the murderer's sight, Lurk in the fleshly mandrake's stem That shrieks when touched at night". BEN JOHNSON, in his "Masque of Queens" says: "I last night lay all alone On the ground, to hear the mandrake groan" (95).

The love-apple is a common plant in deserted fields throughout the Biblical area (184, 265), being a native of the Mediterranean region, southern Europe, and the Levant. HASSELQUIST observed it near Nazareth, POST records it from Sidon, Megiddo, Mount Carmel, Gaza, Moab, Judea, Gilead, and the Jordan valley (267), and THOMSON saw it most abundantly on the lower ranges of Lebanon and Hermon (306). He says: "The apples when ripe are of a pale yellow colour, soft, and of an insipid sickish taste. They are said to produce dizziness, but I have seen people eat them without experiencing any such effect. The Arabs, however, believe them to be exhilarating and stimulating even to insanity, hence the name 'apples of jan' (evil spirits)" (299). Its leaves are put forth in very early spring and the fruit can be gathered by the time of the May wheat harvest (95).

Because some authorities have stated that neither the flowers nor fruit of the mandrake "give a smell", either pleasant or unpleasant (299), some commentators have believed that the citron, *Citrus medica* L., or the common edible field mushroom, *Agaricus campestris* L., or even merely "flowers" were intended in the Song of Solomon verse quoted at the head of this chapter. Jasmine has been suggested for the "mandrakes" by Dr. CHARLES J. BRIM's work (73), based on RASHI's commentary. The GOODSPEED version renders the "mandrakes" of both the Genesis and the Song of Solomon passages as "love-apples" or "love's apples", and so does the Revised Version in a marginal note. COTES states that the fruit does have an odor, that it "smells sweet and fresh like an apple" (95), and SMITH informs us that the entire plant is "very fetid" (306). Since strong-smelling plants and substances are generally known to be far more attractive to Oriental peoples than to Occidental, there seems to be no discrepancy in regarding the mandrake as the plant of Song 7:13 (306).

The Scriptural mandrake must not be confused with *Lycopersicum esculentum* Mill., the tomato, which is also known as "love-apple", nor with our American mandrake, *Podophyllum peltatum* L., both New World species. The thick roots of white bryony, *Bryonia alba* L., were sometimes substituted for mandrake roots by quacks in medieval England (95, 299) and an old English law severely punished with fines and imprisonment any persons who thus "deceived" the public!

As is to be expected, the mandrake figures abundantly in folklore. In the "lost books" of SOLOMON, said to treat of magic and therefore destroyed by King HEZEKIAH lest their contents do harm (95), there is said to be described a root called "baharas" or "baara" which was supposed to possess most amazing properties. This root is thought to be the mandrake. JOSEPHUS would have us believe that mandrake leaves shine in the dark, but if anyone tries to pluck them they rise in the air and fly away like will-o'-the-wisps. This is still a current belief and the Arabs call the plant "devil's candles" (95). The German name for the plant is "Zauberwurzel" or "sorcerer's root", and German girls wear it as a charm much as we do a rabbit-foot. Similar charms were sold in medieval England, especially at the time of HENRY VIII (1509-

1547), and Lord BACON writes in his work on natural history: "Some plants there are—but rare—that have a mossie or downie root, and likewise that have a number of threads, like beards; as the Mandrakes, whereof witches and impostours do make ugly images, giving it the form of a face at the top of the root, and leave those strings to make a broad beard downe to the foot" (95, 126). The word "alruna" has applied both to witches and to mandrakes since the time of the Goths, thus attesting again to the close connection supposed to exist between evil spirits and the mandrake plant (126). The little images or charms cut from the roots in medieval England were called "alrunen", and medieval Germans dressed these every day lest they be offended and do harm to their owners (95). Among the French peasants the mandrake was supposed to be the abode of a little elf, called "main-de-gloire" or "maglore", who had to be propitiated with daily offerings of food.

A very old writer implies that mandrake would act as an anaesthetic which could be used for those nightmarish operations during which in olden days the poor sufferer had no means of relief from his humanly unbearable agony until sheer pain either produced the blessed respite of unconsciousness or death: "When the sicke manne is in anguishe, this potion boiled from the mandrake his root will stille the paine—nor will hee suffer aught should the Leeche probe his wounde, or e'en sawe through his legge or arme, syne his nervies are ded while this potion lasteth" (95). It was also believed to cure cramps, nightmares, sterility, and toothache, and protected the owner against bad weather and robbers (298). More than twenty books have been written on the "medicinal, spiritual, and diabolical" properties of the mandrake (298). The possession of mandrake roots was cause for suspicion of witchcraft in some places and as late as 1630 three women were put to death in Hamburg, Germany, on no other charge than that of having mandrake roots in their homes (298).

133. Mentha longifolia (L.) Huds.

MATTHEW 23: 23—Woe unto you, scribes and Pharisees, hypocrites! for ye pay tithe of *mint* and anise and cummin, and have omitted the weightier matters of the law . . .
LUKE 11: 42—But woe unto you, Pharisees! for ye tithe *mint* and rue and all manner of herbs, and pass over judgment and the love of God . . .

Quite a few mints are common in Palestine (126), but the horse mint, *Mentha longifolia*[79], is most probably the one here referred to (299). It is found in ditches and on the banks of streams, common even to the alpine regions (265, 299). The word here translated "mint" is the Greek ἡδύοσμον, meaning, according to the dictionary, "a sweet smelling herb or mint"— "becaim" in Hebrew[80], according to SMITH (299). It is rendered "mint" in all the versions except the Basic English, which substitutes the inclusive phrases "all sorts of sweet-smelling plants" and "every sort of plant". PRATT believes that *M. arvensis* L. was also cultivated by the Jews in New Testament times (268), and other authorities maintain that the mint was *M. sativa*[81] L. (27, 98, 184). However, neither of these two mints is now known from the Holy Land, either wild or cultivated, so it does not seem probable to us that they were there in Biblical days (267). The horse mint, on the other hand, is found today throughout the region and is still commonly planted at Aleppo (306). It is a much larger plant than the ordinary garden mints (178), attaining a height of three feet or more. Its small leaves are nearly sessile, varying from ovate or lanceolate to oblong, with toothed and sometimes wavy margins. The lilac flowers are clustered in conic or cylindric terminal spikes (267).

The warm flavor of mints, due to the presence of characteristic essential oils (158), is well-known to all of us, but the ancient Hebrews, Greeks, and Romans employed it for flavoring, as a carminative in medicine, and as a condiment in cookery far more than we do (306). ROSENMÜLLER remarks that

---
[79]Also known as *Mentha spicata* var. *longifolia* L. and *M. sylvestris* L.
[80]The Hebrew word for "mint" is "dandanah".
[81]Now usually called *Mentha gentilis* L.

in a cookery book of the Roman epicure, MARCUS GABIUS APICIUS, who flourished from 14 to 37 A.D., green as well as preserved mint is alluded to on almost every page. DIOSCORIDES lists it as having been used as a stomachic in his day. The Hebrew synagogues used to have fragrant mint stems and leaves scattered over the floors, yielding their fragrance as they were stepped upon (293).

Many authorities include the "bitter herbs" of Exodus 12: 8 and Numbers 9: 11 in references to mint, but, as has been previously shown (see under *Cichorium endivia*), it seems more likely to us that these were the leaves of endive, chicory, lettuce, watercress, sorrel, and dandelion, eaten as a salad. Mint is, however, today one of the "bitter herbs" of the Paschal feast (27, 184).

In classical mythology the nymph, MINTHO, was transformed into a mint plant by PROSERPINE, wife of PLUTO, because she had attracted the attention and affections of the god of the underworld (125). As an humble mint plant the nymph lost some of her beauty, yet continues to attract men by her freshness and fragrance (298).

134. Morus nigra L.
(FIGURE 23)

EZEKIEL 16: 10 & 13— ... and I covered thee with *silk* ... and thy raiment was of fine linen, and *silk*, and broidered work ...
I MACCABEES 6: 34—And to the end they might provoke the elephants to fight, they showed them the blood of grapes and *mulberries*.
LUKE 17: 6— ... If ye had faith as a grain of mustard seed, ye might say unto this *sycamine tree*, Be thou plucked up by the root, and be thou planted in the sea; and it should obey you.
REVELATION 18: 12— ... and purple, and *silk*, and scarlet ...

Early commentators regarded the "sycomore" and the "sycamine" tree as one and the same (27, 299), but it is now agreed that only the "sycomore" of the Bible is *Ficus sycomorus* (which see), while the "sycamine", (Hebrew, "shikmah") of Luke 17: 6 is actually the black mulberry *Morus nigra* (Hebrew, "toot"). Although originally a native of northern Persia, the black mulberry is now cultivated everywhere in the Holy Land for its delicious fruit, and POST actually gives "sycamine" as one of its common names (265, 267). The Greek word used in the Luke reference is συκάμινος, for which the dictionary translation is "mulberry", and the tree is still called "sycaminos" or "sycamenea" in Greece today (27, 184).

The Chinese and Indian species, *M. alba* L., is now very widely cultivated in Syria and Palestine for growing silkworms (27, 184, 306), but botanists agree that it is a recent introduction, supplanting the inferior *M. nigra* which was formerly used for that purpose (265, 267, 268, 293, 299). Silk does not seem to have been known to the Jews until late in their history, perhaps about 600 B.C. The word "silk" employed in the Authorized Version's rendition of Proverbs 31: 22 (1015 B.C.) is now regarded by all translators as an error for "linen". The original Hebrew uses the word "shêsh". Similarly, the Douay version's use of "silk" in Genesis 41: 42 (1715 B.C.) and the marginal suggestion of the Authorized Version for the same verse and for Exodus 25: 4 (1491 B.C.) are thought to be errors, because silk was not known to the Jews at those early dates. The material referred to in these passages was undoubtedly linen. The first undisputed mention of silk in the Bible is in the time of EZEKIEL, in about 594 B.C., about 420 years after SOLOMON (although some commentators still feel that SOLOMON must have been acquainted with silk). The Hebrew word here used is "meshi". Apparently EZEKIEL had become acquainted with silk during the period of his captivity in Babylon (299). THEOPHRASTUS and other ancient Greek writers speak of silk, which appears to have been introduced into Greece from Persia at about 325 B.C. (299). In the book of Revelation we read that silk was one of the valuable commodities of Babylon (299).

There is no evidence to show just when mulberries first began to be cultivated in Palestine and Lebanon for rearing silkworms, although silk is now a staple product of these areas, especially on the slopes of Mount Hermon and Mount Lebanon and other parts of the northern sections (299). *Morus alba* was cultivated for rearing silkworms in China as far back as 4000 B.C., but has only comparatively recently been introduced into the Holy Land for that purpose. The trees which JESUS saw were undoubtedly *Morus nigra*. JEWETT has remarked concerning the country of the Druses in Lebanon: "the country here is as remarkable for the innumerable multitudes of its mulberry trees, as Egypt is for its palm trees." During the greater part of the year these trees clothe the landscape in every direction with a delightful verdure. On the rock of Ophel at the entrance to the valley of Hinnom, not far from Jerusalem, stands an old mulberry tree which is said to mark the spot where King MANASSEH had the prophet ISAIAH sawed in half.

The "mulberry trees" of II Samuel 5: 23-24 and I Chronicles 14: 14-15 are now regarded as having been aspens (see under *Populus euphratica*). It is interesting to note that only the Authorized and the WEIGLE versions use the word "sycamine" in Luke 17: 6; the Douay, MOFFATT, GOODSPEED, O'HARA, and LAMSA versions all say "mulberry"; WEYMOUTH says "black-mulberry"; and the Basic English, as usual, sidesteps with merely "this tree".

The black mulberry is a low-growing, thick-crowned, stiff-branched tree growing from 24 to 35 feet tall, but only seldom more than 30 feet tall, forming a stout trunk, with deciduous, glabrous, cordate-ovate, lobed or unlobed leaves, small, subsessile, greenish heads or spikelets of flowers, and black edible fruit (267). It is said to afford a dense shade in the summer and to live to a great age (299).

Silkworms were first introduced into western Europe in the sixth century A.D., but the silk industry made very slow progress there. In 1146 it was a well-developed industry in Sicily, whence it spread to Italy, Spain, and southern France in about 1510. The black mulberry was introduced into England in 1548 and the culture of silkworms was greatly encouraged by JAMES I (1603-1625), but, because of the adverse climate, the industry has never succeeded there (299).

The juice of mulberries, being red (somewhat like blood) (125), was used to incite the elephants of Antioch to battle (I Maccabees 6: 34). In Burma the mulberry is worshipped, while in Europe there is a superstition that the devil uses black mulberries to blacken his boots (298). In some parts of China a thick preserve is made from mulberries on the fifteenth day of their first month because of a legend that a kind fairy once was so delighted with the gift of some of this preserve that she promised to make the Chinese mulberry trees yield a hundred times as much silk as before. In classic mythology the ill-starred lovers, PYRAMUS and THISBE, died in each other's arms in a strikingly Romeo-and-Juliet fashion, under a white mulberry tree, and their blood, splashing on the white fruit of this tree, stained it red, and thus the red mulberry was created (298).

135. Mucor mucedo L.

JOSHUA 9: 12—This our bread we took hot for our provision out of our houses on the day we came forth to go unto you; but now, behold, it is dry, and it is *mouldy*.

The Hebrew word rendered "mouldy" in the verse quoted above is "nikoodim". There are numerous references in the Bible to various fungi. Some of these can be identified with a fair degree of certainty. Among these are the several disease-producing bacteria and related schizophytes discussed under *Eberthella, Epidermophyton, Shigella, Pasteurella, Staphylococcus, Mycobacterium*, etc. The yeasts used in the manufacturing of bread and of wine are also identifiable (see under *Saccharomyces*). Another of the fungi concerning whose identification there is little doubt is the one mentioned in this passage from the book of Joshua. The common gray bread-mold, *Mucor mucedo*, is the commonest mold attacking old stale bread practically all over the world. There is little doubt that this was the mold which spoiled the bread of the Hivites.

It is worthy of note that the more modern translations of Joshua 9: 12 give an entirely different concept of the meaning. GOODSPEED says that the bread was "dry and *crumbled*", MOFFATT "dry and *crumbling*", JASTROW "it is dry, and is become *crumbs*", while the Douay version is "now they are become dry, and broken in *pieces*, by being exceeding old."

In Haggai 2: 17 there is a reference to "mildew" (the Hebrew word is "yirakon") and in Amos 4: 9 another similar reference. In both these verses a "blasting" is also spoken of in conjunction with the mildew. It seems most probable that both terms referred to some fungous diseases of cultivated crops, undoubtedly similar to our present-day blights and mildews. Which ones, however, cannot yet be stated with any degree of accuracy. It is also practically certain that some of the "leprosy" spoken of in the 13th and 14th chapters of Leviticus, especially that described as attacking walls and stones, was mildew of some sort.

136. **Mycobacterium leprae** (A. Hansen) Lehm. & Neum.

LEVITICUS 13 & 14—... When a man shall have in the skin of his flesh a rising, a scab, or bright spot, and it be in the skin of his flesh like the plague of *leprosy*; then he shall be brought unto AARON ... and when the hair in the plague is turned white, and the plague in sight be deeper than the skin of his flesh, it is a plague of *leprosy* ... And if the priest see that ... the scab spreadeth in the skin, then the priest shall pronounce him unclean: it is a *leprosy*. When the plague of *leprosy* is in a man, then he shall be brought unto the priest ... And ... if the rising be white in the skin, and it have turned the hair white, and there be quick raw flesh in the rising; It is an old *leprosy* ... And if a *leprosy* break out abroad in the skin, and the *leprosy* cover all the skin ... and ... if the *leprosy* have covered all his flesh, he shall pronounce him clean ... But when raw flesh appeareth in him, he shall be unclean ... for the raw flesh is unclean: it is a *leprosy* ... The flesh also, in which ... was a boil ... And if ... it be in sight lower than the skin, and the hair thereof be turned white ... it is a plague of *leprosy* broken out of the boil ... or if there be any flesh in the skin whereof there is a hot burning ... and, behold, if the hair in the bright spot be turned white, and it be in sight deeper than the skin; it is a *leprosy* broken out of the burning ... it is the plague of *leprosy* ... it is the plague of *leprosy* ... even a *leprosy* upon the head or beard ... And if there be in the bald head, or bald forehead, a white reddish sore; it is a *leprosy* ... as the *leprosy* appeareth in the skin of the flesh; He is a *leprous* man, he is unclean ... And the *leper* in whom the plague is, his clothes shall be rent, and his head bare, and he shall put a covering on his upper lip, and he shall cry, Unclean, unclean ... The garment also that the plague of *leprosy* is in ... it is a plague of *leprosy* ... And ... if the plague be spread in the garment ... the plague is a fretting *leprosy*; it is unclean ... for it is a fretting *leprosy*; it shall be burnt in the fire ... This is the law of the plague of *leprosy* ... This shall be the law of the *leper* ... and ... if the plague of *leprosy* be healed in the *leper* ... he shall sprinkle upon him that is to be cleansed from *leprosy* seven times, and shall pronounce him clean ... This is the law of him in whom is the plague of *leprosy* ... And ... When ... I put the plague of *leprosy* in a house ... This is the law for all manner of plague of *leprosy*, and scall, And for the *leprosy* of a garment, and of a house, And for a rising, and for a scab, and for a bright spot: To teach when it is unclean, and when it is clean: this is the law of *leprosy*.

LEVITICUS 22: 4—What man soever of the seed of AARON is a *leper* ...
NUMBERS 5: 2—Command the children of Israel, that they put out of the camp every *leper* ...
NUMBERS 12: 10—... and, behold, MIRIAM became *leprous*, white as snow: and AARON looked upon MIRIAM, and, behold, she was *leprous*.
DEUTERONOMY 24: 8—Take heed in the plague of *leprosy* ...
II KINGS 5: 1, 3, 6, 11, & 27—Now NAAMAN ... was a *leper* ... Would God my lord were with the prophet that is in Samaria! for he would recover him of his *leprosy* ... that thou mayest recover him of his *leprosy* ... and strike his hand over the place, and recover the *leper* ... The *leprosy* therefore of NAAMAN shall cleave unto thee, and unto thy seed forever. And he went out from his presence a *leper* as white as snow.
II KINGS 7: 3 & 8—And there were four *leprous* men ... And when these *lepers* came to the uttermost part of the camp ...
II CHRONICLES 26: 19-21 & 23—... and while he was wroth with the priests, the *leprosy* even rose up in his forehead ... and, behold, he was *leprous* in his forehead ... And UZZIAH the king was a *leper* unto the day of his death, and dwelt in a several house, being a *leper* ... He is a *leper*.
MATTHEW 8: 2—And, behold, there came a *leper* ...
MARK 1: 40—And there came a *leper* to him ...
MARK 14: 3—And being in Bethany in the house of SIMON the *leper* ...
LUKE 5: 12-13—And it came to pass, when he was in a certain city, behold a man full of *leprosy*: who seeing JESUS fell on his face, and besought him, saying, Lord, if thou wilt, thou canst make me clean. And he put forth his hand, and touched him, saying, I will: be thou clean. And immediately the *leprosy* departed from him.
LUKE 17: 12—And as he entered into a certain village, there met him ten men that were *lepers*, which stood afar off.

The very much abbreviated references given above indicate how prevalent a disease leprosy was in Biblical days. The entire 13th and 14th chapters of Leviticus are devoted to detailed directions for the recognition and treatment of various forms of the disease. It is probable that certain other diseases are included in the term "leprosy" of the Old Testament (Hebrew "tzaraat", meaning "leprosy", and "m'tzorah", meaning "leprous" or "leper"). For instance, the "scall" of Leviticus 13: 30-37 and 14: 54 is now recognized as being ringworm (see under *Epidermophyton rubrum*). Psoriasis was probably also included, as well as other skin afflictions (158). The "leprosy" of garments and houses, of course, must have been something else again. Some have suggested that the "leprosy" of stone walls may have been the "nitrous efflorescence on the surface of the stone, produced by saltpetre, or rather an acid containing it, and issuing in red spots", also the exfoliation of stone from other causes (306). The growth of molds, mildews, and even algae and lichens on the walls of houses, and of molds and mildews on garments, would probably also have been treated as a "leprosy" by the priests, but there is not sufficient evidence to warrant identification of these plant organisms. The same applies to the "blight" and "mildew" of Haggai 2: 17 and Amos 4: 9.

True leprosy (158) is a chronic highly infectious disease caused by the bacillus, *Mycobacterium leprae*. It is characterized by the formation of tubercular nodules, ulcerations, and disturbances of the sensory organs, and is usually fatal. It is one of mankind's most horrible diseases. Although old authors, like HIPPOCRATES (460-359 or 377 B.C.), distinguish three types, "lepra alphoides", "lepra vulgaris", and "lepra nigricans", of which the first was easily healed, modern authorities classify the many forms of the disease into two main varieties: (1) tubercular or nodular, and (2) anaesthetic leprosy. The former is characterized by the appearance of small red areas on the skin, which later become pigmented and develop into tubercles. The mucous membranes of the throat, mouth, and larynx are also affected, the hair and nails may fall out, the hands and feet become distorted, and there is gradual destruction of the joints and bones. When the disease is fully developed the face takes on a characteristic leonine appearance from the thickening of the skin ("elephantiasis graecorum"). The anaesthetic variety, on the other hand, the most common tropical form, is characterized by atrophy of the peripheral sensory nerves, resulting first in greatly excited sensitivity, followed soon by complete loss of sensation. Some Indian fakirs who lie on spikes or walk on broken glass without feeling any pain have this type of the disease. An incurable form among the Jews was characterized by the presence of smooth, shining, depressed, white patches or scales, the hair in which participated in the whiteness, while the skin and adjacent flesh became insensible (158).

137. Mycobacterium tuberculosis var. hominis (Koch) Lehm. & Neum.

LEVITICUS 26: 16— . . . I will even appoint over you terror, *consumption*, and the burning ague . . .
DEUTERONOMY 28: 22—The Lord shall smite thee with a *consumption* . . .

That the "consumption" of the Bible was the same disease as we now know under this name, namely, tuberculosis, caused by the bacillus, *Mycobacterium tuberculosis* var. *hominis*, is not at all certain, but seems probable. Many other diseases are mentioned or referred to in the Bible, but all seem to be either of virus or animal-parasite origin, or of vitamin deficiency, or are diseases, like cancer, palsy, epilepsy, and insanity, whose exact causes have not yet been ascertained. Others are referred to in such general terms as to be unidentifiable.

138. Myrtus communis L.
(PAGE 145; FIGURE 68)

NEHEMIAH 8: 15—And that they should publish and proclaim in all their cities, and in Jerusalem' saying, Go forth unto the mount, and fetch olive branches, and pine branches, and *myrtle* branches, and palm branches, and branches of thick trees, to make booths, as it is written.
ESTHER 2: 7—And he brought up *Hadassah*, that is, ESTHER, his uncle's daughter . . .

ISAIAH 41: 19—I will plant in the wilderness the cedar, the shittah tree, and the *myrtle*, and the oil tree . . .

ISAIAH 55: 13—Instead of the thorn shall come up the fir tree, and instead of the brier shall come up the *myrtle* tree.

ZECHARIAH 1: 8 & 10-11—I saw by night, and behold a man riding upon a red horse, and he stood among the *myrtle* trees that were in the bottom; and behind him were there red horses, speckled, and white . . . And the man that stood among the *myrtle* trees answered and said, These are they whom the Lord hath sent to walk to and fro through the earth. And they answered the angel of the Lord that stood among the *myrtle* trees, and said, We have walked to and fro through the earth, and, behold, all the earth sitteth still, and is at rest.

The myrtle, *Myrtus communis*, is believed to be native to western Asia (101, 268, 293, 299). It is common in Palestine and Lebanon, especially about Bethlehem, Hebron, and on the slopes of Mount Carmel and Mount Tabor (184), and has become naturalized in most of the countries bordering on the Mediterranean. HASSELQUIST observed it on some of the hillsides near Jerusalem, and HOOKER says that it is not uncommon in Samaria and Galilee (306). The Hebrew word translated "myrtle" in all versions of the Bible is "hadas" (Greek, μυρσίνη). In good enviroments this plant grows up into a small evergreen tree, 20 or 30 feet tall (27, 184, 299), but it is more often a straggling bush 1½ to 4½ feet tall (267). Its wood is hard and mottled, often knotty, and is highly valued in turnery. Its dark, leathery, ovate-lanceolate, glossy, sweet-smelling leaves and fragrant white or rosy flowers and black or blackish-blue berries are used in perfumery and for making sachet powders. A fragrant oil is also produced from the myrtle. It is still used today by the Jews at the Feast of the Tabernacles, when it is procurable, and sprigs with three leaves in a whorl (which are not common) are especially esteemed for this occasion (299, 306).

Some commentators have believed that the "branches of thick trees" in the Nehemiah passage quoted above also refer to the myrtle, but this is not at all probable since myrtles are also specifically mentioned in the same sentence. MOFFATT translates the phrase as "evergreens" and this is probably more correct.

In the Scriptures the myrtle is referred to chiefly as a symbol of divine generosity. DONEY states that when "ADAM was expelled from Paradise he was allowed to take with him wheat, chief of foods; the date, chief of fruits; and the myrtle, chief of scented flowers" (101, 298). Because of its evergreen character the Greeks considered it as a symbol of love and immortality, and used it for crowning their priests, heroes, and outstanding men (95). The Romans wore it entwined with laurel after bloodless victories on the battlefield (95, 125). Because in their common human conceit they regarded the work of their great men as immortal, the Greeks and Romans placed wreaths of myrtle on the brows of their successful poets and playwrights (298). Throughout the Bible it is emblematic of peace and joy. It was a favorite plant throughout the Orient from very early times and was highly prized by the Hebrews from those days when they gathered its boughs, with many others, and brought them to shade their outdoor dwellings at the original Feast of the Tabernacles in 445 B.C., when all the children of Israel sat beneath their thick-leaved trees, or in booths erected on the flat roofs of their houses, and called on the widow, the orphan, and the stranger to put away their sorrows and rejoice in the Lord. Branches of myrtle trees were included among those which NEHEMIAH ordered to be gathered for this occasion, and still today the Jews gather myrtle from the valleys among the hills of Lebanon and by the rivers of Galilee. To the ancient Jews it was symbolic not only of peace, but also of justice. HADASSAH, the original name of ESTHER, is very similar to the Hebrew word for myrtle and the Targums say: "They call her HADASSAH, because she was just, and those that are just are compared to myrtles". In the bazaars of Jerusalem and Damascus, the flowers, leaves, and fruit of the myrtle are sold for making perfume (178). The bark and roots are used for tanning the finest Turkish and Russian leather, to which they impart a distinctive delicate scent.

The "myrtle" of the Bible should not be confused with the various American and European plants now known by that name, especially the periwinkle,

Vinca minor L., whose leaves somewhat resemble it. In old French the name was applied to species of bilberry and whortleberry (myrtleberry).

Because it was a symbol of sensual love and passion, originally sacred to VENUS (125, 158), it was looked upon with disfavor by the extremely pious in many lands and ages, but in England the myrtle is a symbol of peace, home, and restfulness, and in Germany brides wear a myrtle wreath (125). In England a sprig of myrtle is also often added to the bride's orange blossoms as a symbol of peace and immortality, and in some sections of the country it is regarded as a bringer of good luck. In Bohemia it is placed on caskets in funerals as a symbol of the hope for immortality (298). In the Oriental fables of AZZ EDDIN the rose bows her head before the myrtle, yielding to the latter precedence as the prince of scented flowers (179). In classic mythology the myrtle tree is said to have had its origin when MYRTILUS, the son of MERCURY, following his traitorous betrayal of his master, was transformed into a myrtle tree after the sea refused to accept his body and had cast it ashore (298). Another legend states that the girl MYRENE was changed into a myrtle by VENUS in anger over her desertion for a human love. Still another legend says that MYRSINE, a fleet-footed girl who beat MINERVA in a foot race, was transformed into a myrtle by that outraged goddess (125, 298). When VENUS found that her son, CUPID, had fallen in love with PSYCHE, she beat that weeping nymph with a myrtle rod. When pursued by satyrs, it was in a myrtle grove that VENUS escaped (298). One of the wreaths of myrtle carried in the procession of EUROPA at Corinth is said to have measured ten feet in diameter (298).

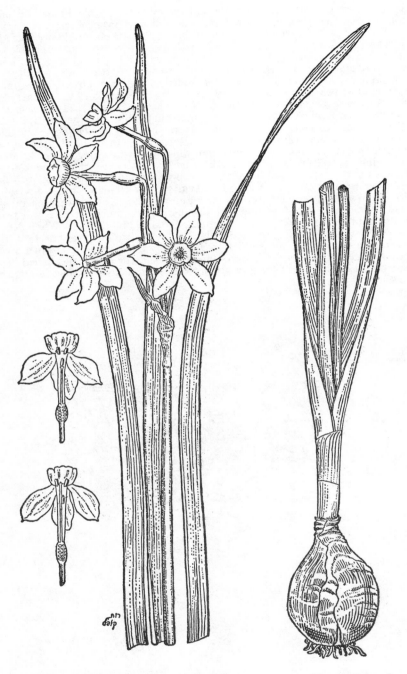

FIGURE 24. — *Narcissus tazetta*, the polyanthus narcissus. Several Hebrew and Aramaic words were translated as "rose" by the early English translators who were not botanically trained. The "rose", thus, of ISAIAH was this narcissus, an abundant Palestinian plant. (Feinbrun and Zohary, Iconographia Florae Terrae Israëlis, 1949).

139. Narcissus tazetta L.
(FIGURE 24)

ISAIAH 35: 1—The wilderness and the solitary place shall be glad for them; and the desert shall rejoice, and blossom as the *rose*.

There has been a tremendous amount of argument concerning the identity of the "roses" of the Bible. It seems almost certain that a number of different plants were loosely referred to under this name by the writers of the original texts and by the translators. Many authorities feel that the "rose of Sharon" of Song 2: 1 was *Narcissus tazetta* (101, 267). Others identify it as *Cistus, Tulipa montana, Anemone coronaria, Colchicum autumnale, Anemone fulgens,* or *Corchorus olitorius*. It is herein discussed by us under *Tulipa montana,* which seems to be the most plausible identification. SMITH comments that "there is nothing in the text to lead us to suppose that the 'rose of Sharon' was sweet-smelling, and therefore any abundant and pretty-flowering plant growing in Sharon may be . . . the one meant", as long as it is bulbous-rooted (299). DONEY thinks that the Isaiah reference cited above may apply to a true species of *Rosa*, but he is virtually alone in that belief (101).

The use of the word "rose" in the Authorized and JASTROW versions of Isaiah 35: 1 may have originated from the fact that the Hebrew word for "bulb"—"chablat zeleth", "chavatzelet", or "habazzeléth"—appears in the original text (181). True roses do not grow from bulbs, but do produce showy fruits, called hips, which may have resembled bulbs to the non-botanical translators of the Authorized version. It is, however, certainly much more probable that a bulbiferous plant was meant in this passage. A writer in "The Rose Annual for 1946" says concerning the "rose" of this passage and of Song 2: 1: "Though the Hebrew word is the same in both passages, the Septuagint translators—whose rendering of the Bible, it must be remembered, extends over a very large number of years—turn it by two different words; in Isaiah by *krinon,* the 'lily', in Canticles by *anthos,* which simply means 'blossom'. The Vulgate, or Latin translation, as usual follows the Septuagint; but the Targum on Canticles, though not on Isaiah, renders [it] by *narkos,* the narcissus, and it is possible that this is what the Septuagint and Vulgate meant by 'blossom'" (181). According to A. I. MACKAY the Hebrew word "habazzeleth" is apparently a compound form of two words meaning "acrid bulb". The German botanist, GESENIUS, favored the autumn-crocus for the "rose" of Song 2: 1 and Isaiah 35: 1, and the Revised Version suggests it in a marginal note for both passages. *Colchicum autumnale,* our autumn-crocus, is not known from Palestine, although POST records 3 other species from that area. The LEESSER and Douay versions translate it "lily", GOODSPEED "crocus", and MOFFATT "narcissus". The tulip has also been suggested (233). The "rose" of Ecclesiasticus was a spring-flowering bulb, thus eliminating true roses. It might have been either crocus, tulip, or narcissus. We are inclined to follow MOFFATT and to regard the narcissus as the "rose" of both these passages.

The polyanthus narcissus, *Narcissus tazetta,* in several varieties, grows abundantly on the plain of Sharon, which is a rich tract of land lying between the mountains of the central part of Palestine and the Mediterranean, supposed to be the region between Caesarea and Joppa in particular. It also grows on the slopes of the adjacent hills and at Jerusalem, Jericho, Mount Ebal, Esdraelon, Amanus, Aleppo, Beirut, and Sidon. Being sweet-smelling, it is a great favorite of the people. During its flowering season bouquets of narcissus are to be found in almost every house (299), especially in Damascus. Its white or cream-colored flowers are borne in clusters of 3 to 10 at the apex of the naked scapes. The 2 to 6 strap-shaped leaves are flattish, obtuse at the apex, appear at the same time as the flowers, and are shorter than the flowering scapes.

According to Greek mythology NARKISSOS was a particularly handsome young man who was loved by ECHO, a nymph, the daughter of Air and Earth, but he did not return her love. In her grief and despair the hapless maid pined and faded away until nothing was left of her but her voice, which one can still hear, sadly answering when one calls her name (126, 158). Meanwhile NARKISSOS fell in love with his own reflection in a pool of water, tumbled in, and was drowned, and where his body floated to the shore a white narcissus flower grew up (126, 298). The same flower was used by PLUTO, god of the underworld, to lure PROSERPINE to share his kingdom. The narcissus was believed to give off an evil emanation, producing dullness of the intellect, insanity, and even death. Narcissi wreathed the brows of the goddesses on Mount Olympus (126), of the avenging Furies or Eumenides (158, 298), of the three Roman Fates, and of DIS or PLUTO himself. "When the dead went into the presence of the gods of the underworld, they carried crowns of narcissus that those who mourned had placed in their white hands when the last good-byes were said" (298).

140. Nardostachys jatamansi (Wall.) P. DC.
(FIGURE 34)

SONG 1: 12—While the king sitteth at his table, my *spikenard* sendeth forth the smell thereof·
SONG 4: 13-14—Thy plants are an orchard of pomegranates, with pleasant fruits: camphire, with *spikenard*, *spikenard* and saffron . . .
MARK 14: 3—And being in Bethany in the house of SIMON the leper, as he sat at meat, there came a woman having an alabaster box of ointment of *spikenard* very precious; and she brake the box, and poured it on his head.
JOHN 12: 3—Then took MARY a pound of ointment of *spikenard*, very costly, and anointed the feet of JESUS, and wiped his feet with her hair: and the house was filled with the odour of the ointment.

It is quite generally agreed that the "spikenard" of these references in the Authorized and Douay versions was from the plant known as *Nardostachys jatamansi*[82]. The word used in the texts is, in the Hebrew "nêrd", "naird", or "nard", and in the Greek, νάρδος. The nard plant is a perennial herb of the valerian family, related to the well-known *Valeriana officinalis* L., but with even stronger and more pleasantly scented roots (101, 268, 299). It is a native of high altitudes in Nepal, Bhutan, and other parts of the Himalaya Mountains, and its range extends from there into western Asia. The roots and spike-like woolly young stems, before the leaves unfold, are dried and used for making the perfume (299). In India it is still used as a perfume for the hair, and there is every reason to believe that the precious ointment spoken of in the Scriptures as "spikenard" (126) came originally from far-off India. In ancient times it was a favorite perfume of the belles of Rome, but their tastes in perfume must have been quite different from those of the ladies of today. Nard is now no longer used by the ladies of our western civilization, and its scent would, indeed, be considered quite disagreeable by many (299). It has nothing whatever to do with the English matgrass, *Nardus stricta* L.

Some commentators state that "nardas" were plants with fragrant roots in general, and spike-nard was a kind of nard of spike-like form imported from India and highly prized by the Hebrews as a perfume and stimulant. This description fits *Nardostachys* perfectly. Because of the long distance from which it had to be imported, it was understandably expensive. The best spikenard ointment was commonly imported in sealed boxes of alabaster and thus stored, opened only on very special occasions. When the master of a house received distinguished guests he not only crowned them with flowers, but he broke the seal on an alabaster box and anointed them with spikenard. The plant grows naturally in the cold, dry, alpine pastures of the Himalayas and is brought from there to the plains of India, where it forms a considerable

---
[82]Also known as *Valeriana jatamansi* Wall.

article of commerce. The Arabs compare the plant to the tail of an ermine, because of the shaggy appearance imparted by the woody fibers of the plant's stems and branches, which remain as a protective coat for the plant in the cold and comparatively dry climate where it grows. The Hebrews and Romans used the fragrant ointment of this plant in the burial of their dead. Arabian authors refer to the plant under the name of "sunbul", the Hindus call it "jatamansee", and DIOSCORIDES refers to it as γαγγιτις or "Ganges nard" (306).

Some writers, like Sir GILBERT BLANE, have thought that the "calamus", "sweet calamus", and "sweet cane from a far country" of Exodus 30: 23, Ezekiel 27: 19, Song 4: 14, and Jeremiah 6: 20 were *Nardostachys jatamansi* also, but this is very improbable since both "calamus" and "spikenard" are mentioned in the same sentence in Song 4: 14. The "calamus", "sweet calamus", and "sweet cane" are discussed by us under *Andropogon aromaticus* (which see). MOFFATT, O'HARA, LAMSA, and WEIGLE all use the word "nard" in Mark 14: 3 and John 12: 3, but the WEYMOUTH and Douay versions say "spikenard" in both places, and the Basic English uses the phrase "perfumed oil". MOFFATT, in Song 4: 13-14, translates the first "spikenard" of the King James version as "roses" and the second as "spikenard": "Your charms are a pomegranate paradise—with henna and roses, and spikenard and saffron." The GOODSPEED version continues to use "nard" twice there, and also in the Mark passage, but substitutes "choice perfume" for it in John 12: 3. Regarding Song 1: 12 there is a difference of opinion—GOODSPEED says "While the king was on his couch, his *nard* gave forth its fragrance", but MOFFATT says "When my king is on his diwan, my *charms* breathe out their fragrance." The Revised Version gives a marginal translation of "pistic nard" for Mark 14: 3.

The "spikenard" of the Bible should not be confused with the American spikenard, *Aralia racemosa* L., an entirely different plant, which was, of course, unknown in Biblical times.

141. Neisseria gonorrhoeae Trevis.
142. Treponema pallidum (Schaud. & Hoffm.) Schaud.

EXODUS 20: 5— . . . visiting the iniquity of the fathers upon the children unto the third and fourth generation . . . .
LEVITICUS 13: 2—When a man shall have in the skin of his flesh a *rising*, a *scab*, or *bright spot*, and it be in the skin of his flesh like the plague of leprosy . . .
LEVITICUS 15: 2-7— . . . When any man hath a *running issue* out of his flesh, because of his *issue* he is unclean. And this shall be his uncleanness in his *issue*: whether his flesh run with his *issue*, or his flesh be stopped from his *issue*, it is his uncleanness. Every bed, whereon he lieth that hath the *issue*, is unclean: and every thing, whereon he sitteth, shall be unclean. And whosoever toucheth his bed . . . shall be unclean . . . And he that sitteth on any thing whereon he sat that hath the *issue* shall . . . be unclean . . . And he that toucheth the flesh of him that hath the *issue* shall . . . be unclean . . .
LEVITICUS 22: 4-5—What man soever . . . hath a *running issue* . . . And whoso toucheth any thing that is unclean . . . or a man of whom he may take *uncleanness* . . .
NUMBERS 5: 2-3—Command the children of Israel, that they put out of the camp . . . every one that hath an *issue* . . . Both male and female shall ye put out, without the camp shall ye put them; that they defile not their camps . . .
NUMBERS 25: 8-9 . . . So the *plague* was stayed from the children of Israel. And those that died in the *plague* were twenty and four thousand.
DEUTERONOMY 4: 3—Your eyes have seen what the Lord did because of BAAL-PEOR: for all the men that followed BAAL-PEOR, the Lord thy God hath destroyed them from among you.
II SAMUEL 3: 29— . . . and let there not fail from the house of JOAB one that hath an *issue* . . .
PSALMS 38: 3-7 & 11—There is no soundness in my flesh . . . My wounds stink and are corrupt because of my foolishness . . . For my loins are filled with a loathsome *disease*: and there is no soundness in my flesh . . . My lovers and my friends stand aloof from my *sore* . . .
PSALMS 106: 28-30—They joined themselves also unto BAAL-PEOR, and ate the sacrifices of the dead . . . and the *plague* brake in upon them. Then stood up PHINEHAS, and executed judgment: and so the *plague* was stayed.
JOHN 9: 1-3—And as JESUS passed by, he saw a man which was blind from his birth. And his disciples asked him, saying, Master, who did sin, this man, or his parents, that he was born blind? JESUS answered, Neither hath this man sinned, nor his parents . . .

In spite of many popular beliefs to the contrary, venereal diseases have scourged mankind from the earliest days of antiquity, and in spite of every nation's attempts to blame them on other nations or tribes, they are and always have been practically universal. The two most important organisms, formerly regarded as belonging to the animal kingdom, but now regarded as members of the plant kingdom (52), are *Treponema pallidum*[83], which causes syphilis, and *Neisseria gonorrhoeae*[84], which causes gonorrhea in man exclusively.

Syphilis is a word introduced into technical usage from Syphilus, a character in a poem entitled "Syphilus sive Morbi Gallici", written by the Italian physician and poet, H. FRACASTOR in 1530, which he begins with a paraphrase from VIRGIL: "I will sing of that terrible disease . . ." The first part of his work gives a quite modern description of the symptoms, and the second part a statement of the treatment, wherein the mercury cure is praised (277). However, our records of the disease extend much farther back into antiquity. The curse in Exodus 20: 5 at the time of the giving of the ten commandments to MOSES (about 1491 B.C.) seems to be a plain reference to venereal diseases and their disastrous effects "unto the third and fourth generation". The long chapters 13 and 15 in the book of Leviticus (about 1490 B.C.) are believed to refer in part, at least, to venereal diseases. Of chapter 13 BECKET has said that "under the widely comprehensive notion of leprosy were included other forms of the skin diseases, owing their existence to some previous affection of the genital organs" (277). The first 18 verses of the 15th chapter of Leviticus deal with "issues" in man, some apparently so highly contagious that the most elaborate precautions were taken to prevent their spread. These "issues" (Hebrew, "zov") can be interpreted only as having been venereal ones, probably gonorrhea (306). Commentators also seem to be fairly well agreed that the descriptions in Leviticus 22: 4-5 (1490 B.C.), Numbers 5: 2-3 (1490 B.C.), and II Samuel 3: 29 (1048 B.C.) are to be interpreted as referring to gonorrhea (306), which, while not as serious as syphilis, is far more highly contagious. The man born blind, healed by JESUS, as recorded in John 9: 1-3, seems to be an obvious case of gonorrheal opthalmia (20, 140).

Although the references in Numbers 25: 8-9, Deuteronomy 4: 3, and Psalms 106: 28-30 are often cited as applying to bubonic plague, it seems rather certain to us that the "plague" (Hebrew, "magaypha") here referred to was a venereal one, apparently one of extreme virulence and malignancy such as have been frequently recorded in medical history. The worship of BAAL-PEOR was one connected with extremely licentious rites. In fact, BAAL-PEOR was identified by the Rabbis and early church fathers with PRIAPUS, the Greek god of male generative power, son of DIONYSUS and APHRODITE, regarded as the protector of vineyards, gardens, herds, bees, and fish. Lampsacus, on the Hellespont, was the center from which his worship spread in classical days (158). In Biblical times this pagan divinity was referred to as BAAL, and was the supreme male divinity of the Phoenician and Canaanite nations, as ASHTARTE was their chief female divinity (306). Since both names are frequently used in the plural, it is supposed that there were many local modifications of these divinities, corresponding to our modern sects, some accompanied by more licentious orgies than others. The worship of these gods of reproduction is one of very great antiquity. It was apparently well established among the Moabites and Midianites at the time of MOSES, about 1452 B.C. (Numbers 22: 41). In spite of the fearful punishment visited on the Israelites in that year for worshipping BAAL, they continued to do so (Judges 2: 10-13) with the exception of a period when GIDEON was judge (Judges 6: 25 & 8: 33) up to the time of the prophet SAMUEL (Judges 10: 10, I Samuel 7: 4) in about 1120 B.C. In the time of the kings this worship was widespread and became the official religion of the court and the people of the ten tribes (I Kings 16: 31-33 & 18: 19-22), and, though occasionally suppressed, was never completely abolished among them (II Kings 17: 16). It was also the official court

---

[83] Also called *Spirochaete pallidum* Schaud. & Hoffm., *Spirochaeta pallida* Hoffm. & Prowaz., *S. pallida* Schaud., *Spironema pallidum* (Schaud. & Hoffm.) Vuill., *Microspironema pallidum* (Schaud. & Hoffm.) Stiles & Pfender, and *Trypanosoma luis* Krzys.

[84] Also called *Merismopedia gonorrhoeae* (Trevis.) Zopf, *Diplococcus gonorrhoeae* (Trevis.) Lehm. & Neum., *Micrococcus gonorrheae* Flügge, *M. gonococcus* Schröt., and *M. gonorrhoeae* (Trevis.) Lehm. & Neum.

religion for a time in the kingdom of Judah (II Kings 8: 27, 11: 18, 16: 3, 21: 3, and II Chronicles 28: 2).

That the "plague" which decimated the men of Israel on their first contact with BAAL-PEOR was a disease which they obviously received on intercourse with Moabite women and which struck them down very rapidly is apparent from the text. That it was bubonic plague is not at all plausible, for Israelite men would obviously not cohabit with women afflicted with bubonic plague, nor would the latter be in a mood for such action. That it was a particularly virulent form of gonorrhea, not at present known, which killed the Hebrew men so rapidly because they had not built up any immunity to it by previous contact, seems far more likely. That no further epidemic of this sort is recorded for the 300 years that followed, during which the Hebrews continued to worship BAAL to greater or lesser extent, seems to indicate either the building up of a natural immunity or the disappearance of that particularly virulent strain of the disease organism, a situation not without parallel in medical history. Other forms of BAAL worship, also, were not accompanied by as much licentiousness as was that of BAAL-PEOR (306). His worship in some of the forms was accompanied by much pomp and ceremony. Images were set up, and his altars on hilltops and the roofs of houses were numerous. Even temples were dedicated to him. Priests of BAAL were everywhere, of various social classes, and arrayed in characteristic robes. His worship included the burning of incense, offering of burnt sacrifices, and occasionally the sacrifice of human offerings (Jeremiah 19: 5). The officiating priests danced around the altars with frantic shouting and often mutilated themselves with knives (306). It is worthy of note that JASTROW substitutes "plague" for the word "sore" (Hebrew, "nehga") used in Psalms 38: 11 by the King James Version[85].

143. Nerium oleander L.
(FIGURE 69)

ECCLESIASTICUS 24: 14—I was exalted like a palm-tree in Engaddi, and as a *rose plant* in Jericho, as a fair olive-tree in a pleasant field, and grew up as a plane-tree by the water.
ECCLESIASTICUS 39: 13—Hearken unto me, ye holy children, and bud forth as a *rose* growing by the brook of the field.

A great amount of discussion and argument has centered around the identification of the "rose" of the Bible. Various authors, discussing the two Apocryphal passages cited above, have suggested *Rosa phoenicia* Boiss., *Rosa canina* L., *Hibiscus syriacus* L., *Cistus* spp., etc., but, in general, have not suggested any of the bulbous-rooted species discussed under *Narcissus tazetta* and *Tulipa montana* (which see). The consensus now seems to be that the "rose" of Ecclesiasticus 24: 14 and 39: 13 was the oleander, *Nerium oleander* (101, 268).

The oleander is regarded by some authorities (158) as originally native to the East Indies, but it has been cultivated throughout the warm regions of the world for many centuries. It has even become naturalized in many regions. POST regards it as a native of the Holy Land (267), and there seems to be no valid reason for supposing that it was not present in the Holy Land at 200 B.C. when the above passages are supposed to have been written (98). The oleander certainly flourishes now and forms dense thickets in some parts of the Jordan valley. So abundant is it now in Palestine that it is mentioned by every traveler who has any eye at all for flowers, as one of the outstanding features of the scenery. Watercourses and wadis in the Holy Land are often lined with groves and thickets of oleander. It is usually a shrub, 3 to 12 feet tall. The leaves are leathery, evergreen, opposite or ternate on the stems, oblong-lanceolate, and entire-margined. The flowers are borne in dense corymbose-cymose clusters at the tops of the stems, and are pink or white in color, often more or less "doubled" and very rose-like in appearance (267). Every part of the plant is dangerously poisonous (158). It is found in great abundance about the Lake of Galilee, the Lake of Tiberias, in Samaria, and along the

---

[85]"Sayt" is the Hebrew word for "swelling or rising", "sa'pachat" for "scab", and "baheret" for "bright spot".

brooks running into the Dead Sea and river Jordan from the east (267). From the frequency with which it is found by watercourses it seems the most likely plant to have been the "rose growing by the brooks." Also because of this wet habitat it has been thought by some authors to be one of the "willows of the brook". Its leaves are, indeed, willow-like in shape, yet hardly anyone would confuse it with a willow (see under *Populus euphratica* and *Salix*). Because of the vigor with which it grows and the abundance of its rich evergreen foliage, it has been thought by some to represent the "green bay tree" of Psalms 37: 35 (126), but this is not considered very probable by us either (see under *Laurus nobilis*). The GOODSPEED version considers the "gopher wood" of Genesis 6: 14, of which NOAH'S ark was built, as oleander, but considering the small diameter of the stems of oleander this seems to us to be a very far-fetched theory. The cypress, as MOFFATT has suggested, seems far more likely (see under *Cupressus*).

The oleander is the floral emblem of SAINT JOSEPH, but because of its poisonous nature it is known as "horse-killer" in India. In Greece, Italy, and India it is a funeral plant, but it is also used to decorate Hindu temples (298). Dr. EVENARI writes us that the Palestinians' "pharmaceutical industries prepare from the oleander a very active cardiac glucoside used abundantly in medicine."

144. Nigella sativa L.
(PAGE 249)

ISAIAH 28: 25 & 27—When he hath made plain the face thereof, doth he not cast abroad the *fitches* and scatter the cummin? . . . For the *fitches* are not threshed with a threshing instrument, neither is a cart wheel turned about upon the cummin; but the *fitches* are beaten out with a staff, and the cummin with a rod.

Singular confusion has surrounded the identity of the plant of these references. The Hebrew word used in the original texts is "ketyaeh" (299) or "ketzach" (184) (Greek, μελάνθιον, which recent Greek dictionaries translate as "an herb whose seeds were used as spice"). It was unfortunately translated "fitches" in the Authorized Version, but has nothing whatever to do with true fitches or vetches, which are species of the genus *Vicia*. MOFFATT and LEESSER, wishing to correct this obvious error, rendered it "fennel", but here again, our plant bears no relation whatever to the true fennel, *Foeniculum vulgare* Mill., which is not a Bible plant. To render the situation still more of a comedy of errors, the Authorized Version translates an entirely different Hebrew word, "cussemoth", in Ezekiel 4: 9, as "fitches" and this has now been proved to refer actually to spelt, *Triticum aestivum* var. *spelta* (which see)! The GOODSPEED version adds to the confusion by identifying the plant of Isaiah as "dill"; the Douay version says "gith".

The "fitches" of Isaiah are now generally agreed (158) to have been *Nigella sativa*, the so-called "nutmeg-flower" (no relation whatever to the nutmeg!), an annual plant of the buttercup family related to our cultivated love-in-the-mist, *Nigella damascena* L. (179, 265, 268). It grows a foot or more tall and has finely cut fennel-like leaves, white or light-blue buttercup-like flowers and 5-celled capsular fruit pods containing numerous black pungent seeds (299). The nutmeg-flower grows wild in southern Europe, Syria, Egypt, northern Africa, and other Mediterranean lands, where it is also extensively cultivated for its strongly pungent aromatic seeds (299). POST states that at present it exists in the Holy Land only in cultivation, under the Arabic name of "kazha" and the Hebrew name "ketsah", although a hairy variety, var. *brachyloba* Boiss., occurs wild there (266). The black seeds are sprinkled over some kinds of bread and cakes in the Orient, and are used for flavoring curries and other dishes in Palestine and Egypt. Egypt's ladies eat them to produce stoutness, which is considered an attribute of beauty in those lands (299).

It is thought that the early English translators of the Bible rendered the Hebrew word as "fitches" in the mistaken belief that vetches (*Vicia*) actually were the plants to which reference was made. Later commentators substituted "peas" and even "poppies". Marginal Bibles and JASTROW suggest "black cummin". PLINY states that in his day (23-79 A.D.) nutmeg-flower seeds were used for seasoning bread. Their pungency is equal to that of pepper, and they were used as a seasoning for foods long before pepper was known. In modern Egypt one finds a sort of loaf or cake, finer textured than common bread, which is covered with these seeds, called "black seed" or "blessed seed" by the Arabs, much as we use caraway seeds. They impart to the loaves an aromatic and not unpleasant taste, and are said to render the bread more wholesome and stimulating to the appetite.

It is interesting to note that cummin and nutmeg-flowers are still gathered in Palestine in the same way as the prophet ISAIAH described in 725 B.C. Were a wheel passed over these plants they would be crushed and their valuable carminative oil wasted. To avoid this, the seeds are beaten out with a staff or a flail. The seeds of cummin (see under *Cuminum*) are much more easily detached from the stems than those of the nutmeg-flower, for the latter are in large thick capsules. Therefore the cummin seeds can be harvested by being beaten with a short rod, but for the "fitches" a longer and stronger "staff" or flail is needed (299).

The very similar yellow fennel-flower, *Nigella orientalis* L., has been suggested as one of the "fitches" of Isaiah (184), and perhaps MOFFATT and LEESSER were referring to this plant when they used the word "fennel". It is a native plant of the area, with yellow flowers, but does not meet the requirements of our textual references, although TRISTRAM states that its decidedly inferior quality seeds are sometimes used today for adulterating pepper (27, 178).

145. Notobasis syriaca (L.) Cass.   146. Scolymus maculatus L.

JOB 31: 40—Let *thistles* grow instead of wheat, and cockle instead of barley . . .
ISAIAH 34: 13—And thorns shall come up in her palaces, nettles and *brambles* in the fortresses thereof . . .

The word usually employed for thistles in the Bible is the Hebrew "dardar", but in the Job reference cited above the word used is "choach". In the Greek Septuagint this has been rendered τρίβολος just like "dardar" is, and many commentators have supposed for this reason that the same plants were intended. Recent authorities, however, have concluded (184) that "choach" refers more specifically to the Syrian thistle, *Notobasis syriaca*[86], most common in JOB's country. The MOFFATT and GOODSPEED versions substitute the word "thorns" in this verse, although they continue to say "thistles" in all the "dardar" passages (see under *Centaurea calcitrapa*).

CARRUTHERS (184) is of the opinion that "choach" is a generic term applied to any spiny herb or shrub, including thistles and knapweeds, while "dardar" applies only to *Centaurea calcitrapa*. The "bramble" of Judges 9: 14-15 is discussed by us under *Lycium europaeum* (which see).

The Syrian thistle is a spiny herb, growing up to 3 feet tall, with erect stems, branching above, spiny-margined leaves that are smooth above and pubescent beneath, and terminal, ovate, purple heads of flowers subtended by many-parted floral leaves which are modified into sharp spines. It is very common in fields and along roadsides throughout the Holy Land region.

The Isaiah reference cited at the head of this chapter is thought to include not only the Syrian thistle, but also the spotted golden thistle, *Scolymus maculatus*. The latter is a spiny herb to 3 feet tall, with spinose white-margined leaves, and terminal heads of yellow flowers. It is also common in fields and waste places throughout the Holy Land. *Carthamus oxyacantha* Bieb. has also been suggested (184), but according to POST this species is known from the region through only a single report (267). It does not seem likely, therefore, that it is involved here or in any other Biblical reference. The word "brambles" used by the Authorized Version in this verse is replaced by "thistles" in MOFFATT, GOODSPEED, Douay, and JASTROW.

---

[86] Also known as *Cirsium syriacus* (L.) Gaertn., *Cnicus syriacus* (L.) Roth, and *Carduus syriacus* L.

147. **Nymphaea alba** L. 148. **Nymphaea caerulea** Sav.
149. **Nymphaea lotus** L.
(PAGE 155)

I KINGS 7: 19, 22, & 26—And the chapiters that were upon the top of the pillars were of *lily* work in the porch, four cubits ... And upon the top of the pillars was *lily* work ... and the brim thereof was wrought like the brim of a cup, with flowers of *lilies* ...

II CHRONICLES 4: 5— And the thickness of it was an handbreadth, and the brim of it like the work of the brim of a cup, with flowers of *lilies* ...

It has been thought by numerous authorities (268, 299) that the "lily work" (Hebrew, "shushan") of the I Kings and II Chronicles passages cited above was patterned after the flowers of the waterlily. This carved ornamentation on SOLOMON's temple was probably in the form of rosettes, and the many-petaled and rose-like flowers of the Egyptian waterlilies might very well have served as the pattern. There are numerous sculptured representations of these plants found in the ancient Egyptian tombs, showing very definitely that waterlily flowers were used for such purposes in early times, at least by the Egyptians. That the Phoenician architects who worked on SOLOMON's temple should make the capitals of the columns to resemble the lotus-headed capitals of Egyptian architecture is very plausible (306). Yet MOFFATT's new translation of the "brim curling ... like the petals of a lily" does not apply well to the waterlily, but applies better to a true lily or even a tulip. The GOODSPEED version, also, introduces a new idea into the concept of the form of the ornamentation, saying: "And its brim was in workmanship like the brim of a cup, similar to the flower of a *lily*."

PRATT is of the opinion that all the Old Testament references to "lily" apply to *Nymphaea lotus*. This does not seem very probable to us, but it might help explain the fact that five times in the Song of Solomon reference is made to "feeding among the lilies" (see under *Hyacinthus*), for the seeds, roots, and stalks of *Nymphaea* are common articles of food in Egypt. Few flowers can equal the Egyptian lotus or waterlily, *Nymphaea lotus*, in beauty. It is the "white lotus" of Egypt, the "bride of the Nile" (298). In form like a big white rose, it used to float in profusion on the waters of the Nile, but is today very rare on that ancient river. Its leaves are 12 to 20 inches in diameter and its white flowers (with the outermost petals sometimes pinkish) 5 to 10 inches across. It is still much admired by the Egyptians, whose belles wear it as a headdress (126). The seeds are ground into flour for bread, or are eaten roasted.

Nymphaea alba is also given by commentators as probably another of the waterlilies with which the children of Israel were acquainted. It is the common European white waterlily, but occurs also in the Holy Land (266, 326), and northern Africa. It is, however, not as characteristic of Egypt as the white lotus, *N. lotus*. The other classic waterlily of Egypt is *N. caerulea*, the "blue lotus". Its leaves are 12 to 16 inches across and its light-blue flowers 3 to 6 inches in diameter.

The lotus has always figured large in Indian and other Oriental poetry and religion (125). The Hindus believe that the original Great Spirit, OM, moved over the surface of the vast sea which existed before the earth was created, and "quickened into life a wondrous golden lotus, resplendent as the sun, which floated upon the lonely waters. From OM proceeded the great deities— BRAHMA, the creator; VISHNU, the preserver; and SIVA, the destroyer. Each of these gods is represented as seated upon a golden lotus. BUDDHA, an emanation from VISHNU, was supposed to have first appeared in the world floating upon an enormous lotus, which spread itself over the ocean" (95). Seated upon his lotus flower, BRAHMA created the world (298). LAKSHMI, the Hindu goddess of love, reclines on a lotus, and KAMADIVA, the Hindu counterpart of CUPID, floats down the river Ganges on a lotus blossom. FUDO, the Japanese counterpart of MERCURY, flies through the skies on lotus sandals (298).

A Greek legend states that a lovely nymph, deserted by HERCULES, flung herself into the Nile and her body was transformed into a lovely white lotus (298). It was a sacred flower to the Egyptians 4000 years ago. They wound its long stem about a honored guest's head with the bud hanging down over his forehead. According to Goodyear's famous "Grammar of the Lotus", it was an important model in art 3000 years before Christ, the inspiration of the Ionic capitals (126), the Greek fret or meander, and the doubled fret or swastika, "earliest of symbols and ornaments, to be found on temple fronts of the old world and the new, where it represents light and dark, death and life, male and female, good and evil" (298). The lotus is the national flower of Siam. The Japanese decorate their temples with it, and wrap up in its leaves their food offerings to the departed (298). In Homer's Odyssey a mythical nation is described where the people subsist entirely on the lotus and live in a state of dreamy indolence which this diet is supposed to induce; in later stories these became a people in northern Africa called the "Homeric lotus-eaters" or lotophagi (158). Osiris, the great Egyptian god of the underworld and judge of the dead, is represented as wearing a crown of sacred lotus flowers. Horus, the god of silence, sits, like Buddha, on a lotus, finger to his lip, commanding silence (298).

It is interesting to note that while the lotus of the ancient Egyptians was a true waterlily, the name is now applied to a related plant whose flowers and leaves stand upright above the water, *Nelumbium speciosum* Willd., with white, pinkish, or red flowers, and edible nut-like fruits borne in a saltshaker-like fruiting-case (158), and this is widely regarded, though erroneously, as being the sacred lotus of the Egyptians (95). Actually it is a native of eastern Asia and the East Indies (158, 266).

FIGURE 25. — *Olea europaea*, the olive. A view in the Garden of Gethsemane, with two Franciscan monks and an Arabian gardener. Biblical "gardens" were usually olive orchards where one could retire during the heat of the day and find blessed shade. The olive was valuable also as a source of fruit, oil, and lumber. Most of the ceremonial "anointing" was performed with olive oil. Since the time of NOAH (see p. 15 and fig. 26), the olive has been a symbol of peace. (Wood engraving from C. W. Wilson's Picturesque Palestine, 1883).

150. Olea europaea L.
(PAGE 15; FIGURE 25; FIGURE 26)

GENESIS 8: 11—And the dove came in to him in the morning; and, lo, in her mouth was an *olive leaf* pluckt off . . .
GENESIS 28: 18— . . . and poured *oil* upon the top of it.
EXODUS 23: 11— . . . In like manner thou shalt deal with thy vineyard, and with thy *oliveyard*.
EXODUS 27: 20—And thou shalt command the children of Israel, that they bring thee pure *oil olive* beaten for the light, to cause the lamp to burn always.
LEVITICUS 2: 1-7 & 15— . . . and he shall pour *oil* upon it . . . and of the *oil* thereof . . . unleavened cakes of fine flour mingled with *oil*, or unleavened wafers *anointed* with *oil* . . . it shall be of fine flour unleavened, mingled with *oil* . . . and pour *oil* thereon . . . it shall be made of fine flour with *oil* . . . and thou shalt put *oil* upon it . . .
LEVITICUS 5: 11— . . . he shall put no *oil* upon it . . .
LEVITICUS 6: 21—In a pan it shall be made with *oil* . . .
DEUTERONOMY 6: 11— . . . vineyards and *olive trees*, which thou plantedst not . . .
DEUTERONOMY 7: 13— . . . thy corn, and thy wine, and thine *oil* . . .
DEUTERONOMY 8: 8— . . . A land of *oil olive*, and honey.
DEUTERONOMY 24: 20—When thou beatest thine *olive tree*, thou shalt not go over the boughs again: it shall be for the stranger, for the fatherless, and for the widow.
DEUTERONOMY 28: 40—Thou shalt have *olive trees* throughout all thy coasts, but thou shalt not *anoint* thyself with the *oil*; for thine *olive* shall cast his fruit.
DEUTERONOMY 32: 13— . . . and *oil* out of the flinty rock.
DEUTERONOMY 33: 24— . . . and let him dip his foot in *oil*.
JOSHUA 24: 13— . . . of the vineyards and *oliveyards* which ye planted not do ye eat.
JUDGES 9: 8-9—The trees went forth on a time to *anoint* a king over them; and they said unto the *olive tree*, Reign thou over us. But the *olive tree* said unto them, Should I leave my fatness, wherewith by me they honour God and man . . .
JUDGES 15: 5— . . . and burnt up both the shocks, and also the standing corn, with the vineyards and *olives*.
I SAMUEL 8: 14—And he will take your fields, and your vineyards, and your *oliveyards* . . .
I SAMUEL 10: 1—Then SAMUEL took a vial of *oil* . . .
I SAMUEL 12: 3 & 5— . . . and before his *anointed* . . . and his *anointed* is witness this day . . .
II SAMUEL 15: 30—And DAVID went up by the ascent of mount *Olivet* . . .
I KINGS 5: 11—And SOLOMON gave HIRAM twenty thousand measures of wheat for food to his household, and twenty measures of pure *oil* . . .
I CHRONICLES 27: 28—And over the *olive trees* and the sycomore trees that were in the low plains was BAAL-HANAN the Gederite: and over the cellars of *oil* was JOASH.
II CHRONICLES 2: 10— . . . and twenty thousand baths of *oil*.
NEHEMIAH 5: 11—Restore, I pray you, to them . . . their *oliveyards* . . . and the *oil* . . .
NEHEMIAH 9: 25— . . . and *oliveyards* . . .
JOB 15: 33— . . . and shall cast off his flower as the *olive*.
JOB 29: 6— . . . and the rock poured me out rivers of *oil*.
PSALMS 23: 5— . . . thou *anointest* my head with *oil* . . .
PSALMS 52: 8—But I am like a green *olive tree* in the house of God . . .
PSALMS 128: 3— . . . thy children like *olive plants* round about thy table.
PROVERBS 27: 9—*Ointment* and perfume rejoice the heart . . .
ECCLESIASTES 10: 1—Dead flies cause the *ointment* of the apothecary to send forth a stinking savour . . .
ISAIAH 1: 6— . . . neither mollified with *ointment*.
ISAIAH 17: 6—Yet gleaning grapes shall be left in it, as the shaking of an *olive tree*, two or three berries in the top of the uppermost bough, four or five in the outermost fruitful branches thereof . . .
ISAIAH 24: 13— . . . there shall be as the shaking of an *olive tree* . . .
EZEKIEL 27: 17 — . . . they traded in thy market wheat of Minnith, and Pannag, and honey, and *oil*, and balm.
HOSEA 2: 5 & 8— . . . I will go after my lovers, that give me . . . mine *oil* . . . For she did not know that I gave her corn, and wine, and *oil* . . .
HOSEA 12: 1— . . . and *oil* is carried into Egypt.
HOSEA 14: 6— His branches shall spread, and his beauty shall be as the *olive tree* . . .
JOEL 1: 10— . . . the *oil* languisheth.
JOEL 2: 24— . . . and the fats shall overflow with wine and *oil*.
AMOS 4: 9— . . . when your gardens and your vineyards and your fig trees and your *olive trees* increased, the palmerworm devoured them . . .
MICAH 6: 15— . . . thou shalt tread the *olives*, but thou shalt not *anoint* thee with *oil* . . .
HABAKKUK 3: 17— . . . the labour of the *olive* shall fail . . .

Moldenke   — 158 —   Bible Plants

ZECHARIAH 4: 3 & 11-14—And two *olive trees* by it... What are these two *olive trees*... What be these two *olive branches* which... empty the golden *oil* out of themselves?... These are the two *anointed* ones...
ZECHARIAH 14: 4—And his feet shall stand in that day upon the mount of *Olives*, which is before Jerusalem on the east, and the mount of *Olives* shall cleave in the midst thereof...
II ESDRAS 16: 29—As in an orchard of *olives* upon every tree there are left three or four *olives*.
MATTHEW 6: 17—But thou, when thou fastest, *anoint* thine head, and wash thy face.
MATTHEW 25: 3-4 & 8—They that were foolish took their lamps, and took no *oil* with them: But the wise took *oil* in their vessels with their lamps... And the foolish said unto the wise, Give us of your *oil*; for our lamps are gone out.
MARK 6: 13—... and *anointed* with *oil* many that were sick...
LUKE 10: 34—And went to him, and bound up his wounds, pouring in *oil* and wine...
LUKE 19: 29—... at the mount called the mount of *Olives*...
LUKE 21: 37—... and at night he went out, and abode in the mount that is called the mount of *Olives*.
ACTS 1: 12—Then returned they unto Jerusalem from the mount called *Olivet*...
ROMANS 11: 17 & 24—And if some of the branches be broken off, and thou, being a wild *olive tree*, wert graffed in among them, and with them partakest of the root and fatness of the *olive tree*... For if thou wert cut out of the *olive tree* which is wild by nature, and wert graffed contrary to nature into a good *olive tree*: how much more shall these, which be the natural branches, be graffed into their own *olive tree*?
JAMES 3: 12—Can the fig tree, my brethren, bear *olive* berries?
JAMES 5: 14—Is any sick among you? let him call for the elders of the church; and let them pray over him *anointing* him with *oil* in the name of the Lord.
REVELATION 11: 4—These are the two *olive trees*...
REVELATION 18: 13—... and *ointments*... and oil, and fine flour, and wheat...

The olive was unquestionably one of the most valuable trees of the ancient Hebrews. Its importance to them may be judged from the passages cited above, which are only a partial listing of the many places in the Bible where the words "olive" (Hebrew, "zayit"), "olive tree" (Greek, ἐλαία), "oliveyard" (Hebrew, "zaytim"), "oil" (Hebrew, "shemen"), or "ointment" (Hebrew, "shemen") are mentioned. Also, the many passages where the words "anoint" (Hebrew, "m'shoach"), "anointing", and "anointed" are mentioned should be included here, for most of this anointing was done with olive oil. The olive, *Olea europaea*, abounds in Palestine, and in many places is the only tree of any size to be seen. TRISTRAM says that "the most extensive olive yards are on the borders of the Phoenician plain. But they are scarcely less important in the county of Ephraim, and all the valleys from the plain of Esdraelon to Benjamin, the patrimony of Manassah and Ephraim, are clad with olives to this day. The vale of Shechem is one noble olive grove. The plain of Moreh is studded with them. They form the riches of Bethlehem and cover the lower slopes of the valleys round Hebron. The plains of Gilead, and all the lower slopes, as well as the more fertile portions of Bashan, form a long series of olive groves, neglected indeed, but still ready to yield their fatness in return for the most trifling culture; and they are the wealth of the regions of Philistia and Sharon" (299). A strikingly beautiful description of the olive tree is given by MACMILLAN in his book entitled "The poetry of plants" (219).

As Miss HORNER and many others have so truly pointed out, "no tree is more closely associated with the history of man and the development of civilization than the olive" (178). It is a characteristic tree of the Holy Land, existing there in four forms (267). The branches of the wild plant are rather stiff and spinescent. The typical cultivated tree is much branched and evergreen, 20 or more feet tall, with a gnarled trunk and smooth ash-colored bark (306). Its leaves are leathery, oblong, lanceolate, or nearly round, entire-margined, and white-scurfy on the under surface. The small whitish or yellow flowers are borne in axillary racemes (267). The fruits are large, ovate or ellipsoid, black or violet-blue drupes, the outer fleshy parts of which yield the valuable olive oil of commerce. The ripe fruit is eaten raw, as also is the green or unripe fruit, the latter usually as a pickle or relish (158). The hard wood of the trunk and limbs is of a rich yellow or amber color and fine grain, often handsomely variegated (158), and is still today employed for the finest cabinet work and turnery. The tree grows slowly, but attains to a great age (306). Some trees on the Mount of Olives and at Gethsemane are said to have been there since the time of CHRIST (299); but this is not very probable since the Roman emperor, TITUS VESPATIAN in 70 A.D. is reported by contemporary historians as having cut down all the olive trees there (299, 306). It is very difficult, however, to kill an olive tree by cutting it down, because new sprouts are sent up from the root and all around the margins of the old stump, often forming a grove of 2 to 5 trunks, all from a single root, where originally was only one tree (268, 293, 299).

Some authorities are of the opinion that the "olive leaf" of Genesis 8: 11, in the story of NOAH's ark, may have been from *Tamarix mannifera* (which see), but most commentators dismiss this belief as entirely unjustified and claim that the translators are correct in rendering it an olive leaf. The leaf would have had to come from a tree which was so well known that it could be identified from just a single leaf. The olive, being of such tremendous economic significance to the people of that time and region, would be such a tree, while the tamarisk would not. That the olive was not yet known at the time of NOAH seems difficult to believe, although that is the claim made by those who maintain the tamarisk for this passage.

There has also been argument about the "wild olives" ("olive branches" and "olive tree" in the Authorized Version) of Nehemiah 8:15 and I Kings 6:23 and 31-33. It is maintained by many commentators that the tree of these three references could not have been the true olive, but was the narrow-leaved oleaster, *Elaeagnus angustifolia* (which see). Since the cherubims of SOLOMON's temple, as described in the I Kings reference, were 18 feet high and the spread of each wing was 9 feet, it seems certain that they must have been made of numerous pieces of wood joined together (299). If this was oleaster wood, a tremendous number of pieces would have to be joined together, for the oleaster is not a very large tree and has a very slender trunk. However, if it were true olive wood, several pieces would still have been required, for, while the olive attains greater size and dimensions than the oleaster, it is still not a timber tree. Its trunk is usually short and often much contorted. The Hebrew word for olive is "zaith", "zayit", or "zaytim", and the Greek is 'ελαία.

It is worthy of note that the fact that the dove in the story of NOAH's ark is said to have brought back an olive leaf as an indication that God's wrath (in the form of the flood) was abating, has caused both the dove and the olive to be employed ever since as symbols of peace and friendship (126).

The tree is thought to be a native of western Asia, and not of southern Europe as its specific name might lead one to suppose, but was early introduced into Mediterranean Europe (299). To the Oriental nations it is a symbol of prosperity and divine blessing, beauty, luxuriance, and strength (306). It was so abundantly cultivated that we find the expression "oliveyards" in the Bible often and quite naturally coupled with vineyards and grainfields in descriptions of Palestine. Almost every village has its olive grove or orchard. As an emblem of sovereignty, olive oil was used at coronations. The oil was also employed in sacrificial offerings, as fuel for lamps, as a tonic for the hair and skin, and medicinally in surgical operations (306). It formed the base of the perfumed ointments sold in classic Rome and Athens (298). The fruit was normally gathered by shaking or beating the tree, but a few fruits were always left on the boughs for the poor, the stranger, the orphan, and the widow to gather. The "palmerworm" of Amos 4: 9 was the locust, which constitutes a dread enemy of the olive, as indeed, of all green vegetation in those lands.

The olive requires grafting. Ungrafted suckers produce a small worthless fruit (184). This explains the powerful allegory of PAUL in Romans 11: 17-24. The Greek word there employed for "of a wild olive" is ἀγριέλαιος.

A word should perhaps be said here concerning the "gardens" of the Bible. Gardens in Biblical days—and to a large extent even today in Oriental lands—were not flower gardens in our sense. They were merely orchards enclosed by a hedge of thorn or a wall of stone (27c, 218a, 306). Usually the trees in this orchard were olive trees, with perhaps a fig at each corner. Richer persons and kings also had almonds, walnuts, pistachios, and various spices growing in their gardens. Usually a tower was built in each such orchard "garden" where a guard kept watch and drove away wild beasts and pilferers. These Oriental "gardens" have been described by travelers as "a confused miscellany of trees jumbled together, without either posts, walks, arbors, or anything of art or design, so that they seemed like thickets rather than gardens" (306). Gethsemane was such an olive orchard "garden" at the foot of the Mount of Olives, where the oil-presses were located. The prime attractions of the garden in lands like those of Biblical history are not beauty or artistry, but the strictly utilitarian fruit which they produce and the shade which they provide, in which a person may obtain blessed relief from the scorching sun.

In Greek and Roman mythology the olive was the symbol of ATHENE or MINERVA, goddess of medicine and health. Her gift of the olive to mankind was judged greater than POSEIDON'S gift of the horse, and for this reason Athens was named in her honor (298). In modern Italy an olive branch hung over a door is supposed to keep out devils, witches, and evil spirits (126). Many legends concerning the olive occurred among medieval Christians and are still believed in many sections, among which is the following, which exists in many variations. A seed each of the olive, cypress, and cedar is said to have been given to SETH for his dying father, ADAM, by the angel guarding the Garden of Eden. Planted in ADAM'S mouth, they eventually grew up into a single tree of three trunks, one of olive, one of cypress, and one of cedar wood. Beneath this tree DAVID wept in contemplation of his sins. The tree was felled by SOLOMON, but the timber could not be hewn and was therefore cast into a marsh, where it floated and formed a bridge for the Queen of Sheba. Finally, the wood was fashioned into the cross on which JESUS was crucified (298).

151. Origanum maru L.
152. Origanum maru var. aegyptiacum (L.) Dinsm.

EXODUS 12: 22—And ye shall take a bunch of *hyssop*, and dip it in the blood that is in the bason, and strike the lintel and the two side posts with the blood that is in the bason . . .
LEVITICUS 14: 4, 6, & 52—Then shall the priest command to take for him that is to be cleansed two birds alive and clean, and cedar wood, and scarlet, and *hyssop* . . . As for the living bird, he shall take it, and the cedar wood, and the scarlet, and the *hyssop*, and shall dip them and the living bird in the blood of the bird that was killed . . . And he shall cleanse the house with the blood of the bird, and with the running water, and with the living bird, and with the cedar wood, and with the *hyssop*, and with the scarlet.
NUMBERS 19: 6 & 18—And the priest shall take cedar wood, and *hyssop* . . . And a clean person shall take *hyssop*, and dip it in the water, and sprinkle it upon the tent.
I KINGS 4: 33— And he spake of trees, from the cedar tree that is in Lebanon even unto the *hyssop* that springeth out of the wall . . .
PSALMS 51: 7—Purge me with *hyssop*, and I shall be clean: wash me, and I shall be whiter than snow.
HEBREWS 9: 19— . . . he took the blood of calves and of goats, with water, and scarlet wool, and *hyssop*, and sprinkled both the book, and all the people.

The word "hyssop"—Hebrew "êzôb" or "ezov" (Greek, ὕσσωπος)—is unquestionably the most puzzling and controversial of all the words in the Bible applying, or thought to apply, to plants and plant products (306). A tremendous amount has been written on the subject and many diverse views have been advanced and hotly argued. CELSIUS (86) devotes 42 pages to this subject and discusses 18 different kinds of plants which had been suggested already in his day (ca. 1747) as the "hyssop", without arriving at any satisfactory conclusion. CALLCOTT (78) and many non-botanically trained readers of the Bible have innocently supposed that it was the plant which LINNAEUS named *Hyssopus officinalis* L., the well-known garden herb now called "hyssop". CELSIUS also reluctantly came to this conclusion (306), and even Mrs. BARNEVELD in her booklet "Plants and Flowers found in the Bible" (1935) adopts this plant as the Scriptural "hyssop" (47). However, the hyssop of our gardens is not native to either the Holy Land or Egypt, being indigenous only to southern Europe. It does not in any manner fit the requirements of the Biblical plant (299).

ROYLE shared the opinion of many writers (282) that the "hyssop" was the common caper, *Capparis sicula* Duham., a spiny shrub, 4 or 5 feet tall, or decumbent, common in desert places, the more rocky parts of Palestine, and on old walls and foundations, especially in and about Jerusalem. Its present Arabic name "asaf" or "ezzof" is strikingly similar to the Hebrew word "êzôv" used in our texts. Yet it seems hardly likely to us that it was this plant to which reference is made in these passages. TRISTRAM favored the related thorny caper, *C. spinosa* L. (184), but suggests also the whorled savory, *Satureja thymbra* L.[87].

KITTO (202) considered that the pokeweed, *Phytolacca decandra* L., filled all the textual requirements, but he overlooked the all-important fact that this is an American species only recently introduced into Palestine and, of course, unknown in Biblical days. Probably the plant that he actually had in mind was the related *P. dodecandra*, an Abyssinian species which is well-known from Egypt, but not recorded from Palestine, although *P. pruinosa* Fenzl is native there. However, the opinion that the "hyssop" was a pokeweed is not held by any modern authority.

Other writers, like SMITH (299), have been of the opinion that the sorghum, *Sorghum vulgare* (which see), is the most likely plant. They claim that only this plant of the many suggested would have been abundant enough to provide the approximately 133,000 Israelites with a bunch of "hyssop" on immediate notice as required by MOSES on the morning of the Passover. They believe that it was the culms and inflorescence-panicles of this sorghum grass which furnished the brushes for sprinkling the Passover blood, since it is well known that they were commonly used by the Israelites as brushes, brooms, and even scrub-brushes in ordinary domestic use. Others, however, feel that a plant used so commonly in ordinary and menial domestic work would not be employed in so sacred a rite. It is worthy of note, in this connection, that even today in some Roman Catholic churches the brush used for sprinkling holy water is called a "hyssop" (158). In Matthew 27:48 and Mark 15:36 we read that one of the soldiers put a sponge of vinegar on a "reed" to offer to the crucified JESUS. John 19:29, in describing the same incident, says that it was a "hyssop" to which the sponge was attached. SMITH believes that the sorghum would reconcile these statements, for the stem of this grass is truly reed-like and 5 or more feet tall.

NEWTON (211) has suggested the wall rue, *Asplenium ruta-muraria* L., for the "hyssop that springeth out of a wall", while REDGROVE (272) has suggested the maidenhair spleenwort, *A. trichomanes* L., since he says that it is much more abundant in the Holy Land. These writers share the opinion of MOFFATT that the passage in I Kings 4:33 in which SOLOMON's botanical knowledge is described refers to plants of both woody and herbaceous nature, not merely to "trees" as rendered by the King James, Douay, JASTROW, and GOODSPEED versions. MOFFATT translates the verse: "He could talk about any plant, from a cedar in Lebanon to a *hyssop* in the wall." To these authors the two ferns mentioned above serve well in this context, since they are common on old walls and since they furnish the proper contrast in size and importance to the lofty cedar-of-Lebanon. Those who believe, with the King James, Douay, JASTROW and GOODSPEED versions, that it was only trees to which reference is here made, favor the caper. It and the fig would be the only "trees" commonly found clambering over walls.

It seems most probable that several different plants are referred to under the Biblical term "hyssop", and that it is futile to attempt to find a single one to fit all the passages. The "hyssop" of the crucifixion passages is herein regarded as sorghum. For the "hyssop" of the Old Testament the Syrian marjoram, *Origanum maru*[88], and Egyptian marjoram, *O. maru* var. *aegyptiacum*[89], are most likely, and it is interesting to note that BOCHART, SMITH (306), and, most recently, MOFFATT agree with this determination, at least for the Exodus, Leviticus, Numbers, and Psalms references, although MOFFATT seems

---

[87] Referred to by him as "*Satureia thymbia*".
[88] Also known as *Origanum syriacum* L. and *O. syriacum* Sieb.
[89] Also known as *Origanum aegyptiacum* L., *O. nervosum* Vogel, and *O. maru* var. *sinaicum* Boiss.

still to be in doubt about the "hyssop that springeth out of the wall" where he still uses the word "hyssop" instead of "marjoram". The marjoram is very common among rocks and on terrace walls (267) and thus fits the context of this passage perfectly. MOFFATT also fails to use the word "marjoram" in Hebrews 9: 19, although it is perfectly obvious that the reference there is to the same plant that MOSES used. WEYMOUTH, LAMSA, WEIGLE, and even the Basic English translations also all continue to use the word "hyssop" there.

The marjorams are mints, growing, under favorable conditions, 1½ to 3 feet tall, but are much more dwarfed when growing in rock crevices and walls. They are slightly shrubby at the base, with erect, stiff, hairy branches, ovate to ovate-oblong, nearly sessile, untoothed, thick-textured, hairy leaves, and terminal panicles of spicate white flowers (267). An aromatic substance is obtained from the crushed and dried leaves (98). "Hyssop" is given as the accepted common name of *O. maru* by POST, but an unrelated plant, *Bacopa monnieria* (L.) Hayata & Matsum., is known as "water hyssop" (267).

It is worth noting that while the hairy stems of the marjoram, if assembled in a bunch, with their leaves and flowers, would hold water very well and would make an excellent sprinkler, the smooth, crooked, woody, prickly stems of the caper, with their smooth and widely scattered leaves, would not serve at all as a sprinkler (27). The "hyssop", according to the Scriptures, was used to sprinkle the doorposts of the Israelites in Egypt with the blood of the paschal lamb so that the angel of death would pass by that house (Exodus 12: 22). It was employed in the purification of lepers and "leprous" houses (Leviticus 14: 4-52) and in the sacrifice of the red heifer as a propitiation for sin (Numbers 19: 6-18). In allusion to this purificatory use, or perhaps because of some real or fancied detergent property of the plant, the psalmist says "Purge me with hyssop" (Psalms 51: 7).

It seems probable that the original identification of the Hebrew word "êzôb" or "ezov" with the Greek "hyssopus" was due to the similarity in sound of the two words and therefore we can blame the writers of the Septuagint for most of the confusion in which this has resulted (306). The ὕσσωπος of DIOSCORIDES seems to have been *Satureja graeca* L. and *S. juliana* L., neither of which could have been the Biblical plant. KUHN believed that the Hebrews used *Origanum maru* var. *aegyptiacum* as "hyssop" in Egypt and *O. maru* in Palestine, and that the plant of DIOSCORIDES was *O smyrnaeum* L. (306).

153. Ornithogalum umbellatum L

II KINGS 6: 25—And there was a great famine in Samaria: and, behold, they besieged it, until an ass's head was sold for four score pieces of silver, and the fourth part of a cab of *dove's dung* for five pieces of silver.

The Hebrew word here translated "dove's dung" is "chiryonim" (Greek κόπρος περιστερῶν). Some commentators—and apparently also MOFFATT and the translators of the GOODSPEED and Douay versions—have thought that doves' (or pigeons', according to Douay) dung was actually eaten, used, perhaps, as a substitute for salt as JOSEPHUS suggests (299). It seems more probable, however, that "dove's dung" was the vernacular or popular name for a wild plant of the area. It has been suggested, with good reason, that this plant may have been the one which we now call the umbelled star-of-Bethlehem, *Ornithogalum umbellatum*. This and numerous related species and varieties are very abundant in Palestine. Fields, hillsides, and stony places in spring are rendered white with the flowers of these plants. This whitening of the rocks and hillsides is suggestive of the whitening of the sides of buildings and the ground by the excrement of doves and pigeons, and it is even possible that the common name "bird-milk" and the scientific name *Ornithogalum*, proposed by LINNAEUS, allude to this same similarity, for it is known that LINNAEUS believed that this was the plant to which reference was made in the Bible (306). "Dove's dung" is now given as a common name for *O. umbellatum* (118), and it is definitely known that its bulbs were used as food in Syria. DIOSCORIDES reports that even in his day the bulbs were commonly gathered, dried, ground up into meal, and then mixed with flour to make bread. Modern Italians in time of scarcity eat the bulbs. These apparently authentic reports are remarkable since chemical analysis shows that the entire plant is intensely poisonous. Grazing animals avoid it, or, if they do eat of it, are poisoned. FERNALD & KINSEY (118) state that the bulbs are edible only after being thoroughly roasted or boiled.

The umbelled star-of-Bethlehem is a bulbous plant 6 to 12 inches tall, with numerous, narrow, grass-like leaves which have a central white band along the midrib. The numerous white flowers are star-shaped, borne in loose umbel-like corymbs at the apex of a naked scape. The tepals each have a median green band on their under surface.

BOCHART is of the opinion that the "dove's dung" was a species of chickpea, perhaps *Cicer arietinum* L. or *C. pinnatifidum* Jaub. & Spach, which he says that the Arabs call "usnan" and also "dove's or sparrow's dung" (306). CELSIUS, however, disputes this identification and shows that it is based on a complete error (306).

FIGURE 26. — In the early days of the Church artists delighted in portraying the dramatic scenes of the Old Testament. As part of a fresco in the catacombs of Rome, where early Christians used to meet, we find this painting of NOAH standing in a roofless box-like Ark and receiving the sprig of olive leaves from the dove. This incident, indicating that the storms had ended, the flood abating, and God's wrath appeased, has caused both the olive and the dove to become symbols of peace. The juxtaposition of this scene, of most austere simplicity, between panels depicting peacocks with their tails expanded, symbolizing the pride and vanity of irreligious man, is most interesting. (A. Coutance, L'Olivier, 1877; see also p. 15 and fig. 25).

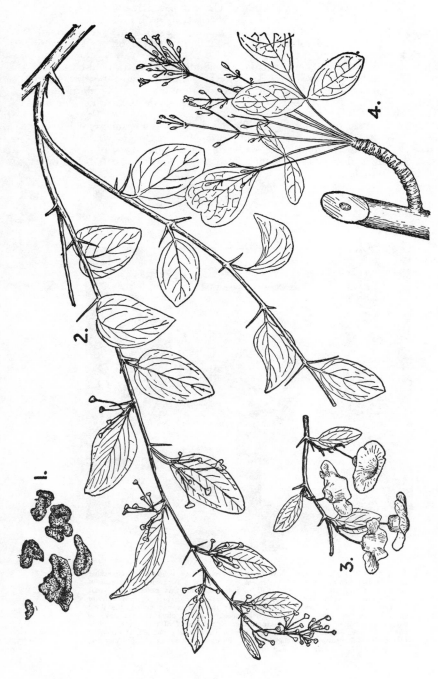

FIGURE 27. — (1) *Lecanora esculenta*, the manna lichen, probably chief constituent of the "manna" that fell from the skies in Sinai (see also fig. 93). (2 & 3) *Paliurus spina-christi*, now regarded as the plant whose flexible, viciously spiny branches were used by the Roman soldiers to make the crown of thorns pressed on JESUS' brow. (4) *Commiphora kataf*, one of the sources of Biblical "myrrh", is native to Arabia, Abyssinia, and the Somali Coast. (John L. Blum, in Science Counselor, 1941).

154. Paliurus spina-christi Mill.
(FIGURE 27)

MATTHEW 27: 29—And when they had platted a crown of *thorns*, they put it upon his head ...
JOHN 19: 2—And the soldiers platted a crown of *thorns*, and put it on his head ...

There has been considerable discussion and diversity of opinion regarding the "thorns" of the Bible. Since the term is so obviously a generic one, applied to spiny plants in general, it seems certain that numerous and quite different plants were thereby referred to. At least five distinct Hebrew words have been so translated in the Authorized Version— "chedek", "kôtz" or "kotzim", "kimmeshonim", "kimsonim", or "kimmesonim", "choach", and "na'atzootz" or "na'atzootzim". The "thorns" of Isaiah 7: 19 and 55: 13 and of Matthew 7: 16 are by us regarded as *Zizyphus spina-christi* (which see); those of Numbers 33: 55, Judges 8: 7, and the "brambles" and "briers" of other passages are herein regarded as *Rubus sanctus* (which see); those of Hosea 9: 6 and Isaiah 34: 13 are *Xanthium spinosum* (which see); those of Matthew 13: 7 and Hebrews 6: 8 are *Centaurea calcitrapa* (which see); and those of Genesis 3: 18, Proverbs 15: 19, Isaiah 33: 12, and Psalms 58: 9 are herein discussed under *Rhamnus palaestina* (which see).

Many commentators believe that the "thorns" of which the "crown" (or "wreath" according to WEYMOUTH's and GOODSPEED's better rendition) was made, were from *Zizyphus spina-christi* (125) and this belief has, in fact, led to its specific name. This spiny plant often grows up to become a tree to 40 feet tall and it is possible that it was the plant employed by the soldiers, although the weight of evidence seems to be more in favor of *Paliurus spina-christi*[90] which does not grow as tall and would have been easier to gather (125, 267, 299). Some have suggested that branches of both may have been entwined, but this does not seem probable because of the extra labor which such a course would have involved. POST records *Zizyphus* from Jerusalem, but not *Paliurus*, although he does record the latter from many parts in and about the Holy Land, including Alexandretta, Antioch, Beirut, Tripoli, Nazareth, Mount Tabor, Ramah in Galilee and Jaffa (99b)[90a]. He says that it does not now occur in Judea (267). Dr. EVENARI agrees with POST that *Paliurus* does not grow in Judea, and is rare in Palestine. *Zizyphus* grows around the old city of Jerusalem and especially near the place which is thought to have been Golgotha.

One thing, however, is certain—the plant now called "crown-of-thorns", *Euphorbia milii* Desmoul., and popularly cultivated under this name in Great Britain and America, has nothing whatever to do with the plant of the Bible. It is a native of Madagascar and was totally unknown in JESUS' day.

The Christ-thorn, *Paliurus spina-christi*, is a straggling shrub 3 to 9 feet tall, with zigzag alternately 2-ranked branches, and ovate or round, 3-nerved, leathery leaves. The flowers are inconspicuous and are followed by red, capsular, winged fruits. The flexible branches are armed at the base of each leaf with a pair of unequal, very stiff, and sharp stipular spines—one of each pair being straight and about half an inch long, the other curved and considerably shorter. The unusually pliable texture of the young branches renders it particularly easy to plait into a crown-like wreath.

[90] Also known as *Paliurus aculeatus* Lam., *P. australis* Gaertn., *Rhamnus paliurus* L., and *Zizyphus paliurus* (L.) Willd.
[90a] Rev. HENSLOW writes in answer to DINSMORE's denial of *Paliurus* in Palestine: "with regard to *Paliurus*, if it does not *now* grow around Jerusalem, it must have been exterminated since 1889, for Dr. TRISTRAM says (The Natural History of the Bible, p. 428) :- 'The Arabs of the Jordan Valley at the present day ... confine the name Samur to the *P. aculeatus*, or Christ's Thorn ... It is common about Jerusalem.' " (170a).

WEYMOUTH'S translation of Matthew 27: 28-29 is particularly lucid: "Stripping off His garments, they put on Him a scarlet cloak. They twisted a wreath of *thorny twigs* and put it on His head, and as a sceptre they put a cane in His right hand, and kneeling to Him they shouted in mockery, 'Hail, King of the Jews!' "

155. Panicum miliaceum L.

EZEKIEL 4: 9—Take thou also unto thee wheat, and barley, and beans, and lentiles, and *millet*, and fitches, and put them in one vessel, and make thee bread thereof . . .
EZEKIEL 27: 17—Judah, and the land of Israel, they were thy merchants: they traded in thy market wheat of Minnith, and *Pannag,* and honey, and oil, and balm.

It is thought by some authorities that the "pannag" of the second reference may refer to the European millet, *Panicum miliaceum,* or else to many kinds of small wares carried in baskets or panniers on the backs of pack-animals in Palestine and known as "pannag" today (299). The MOFFATT and GOODSPEED versions, however, differ from these and all other previous authors in regarding it as "wax". The Douay version calls it "rosin", while JASTROW says "balsam."

The other reference seems much more certain. The seeds of millet are the smallest of all the grass seeds cultivated as food by man (268), but each panicle produces a tremendous number of seeds. In fact, each plant produces so many grains that the plant's specific name *"miliaceum"* is said to have been applied by LINNAEUS in allusion to its "thousand seeds". Large fields of this grain are found in Egypt and Palestine. It is said by POST to be a native of India (267) and by BAILEY as probably originally indigenous to the East Indies (312), but it certainly has been cultivated by man since time immemorial. It is definitely the "millet" of history, grown both for the grain and for forage (27). It is an annual grass, seldom growing more than about 2 feet tall (267). The linear leaves are flat and harshly hairy. The flowers and seeds are borne in compound, much-branched, spreading, and at length nodding panicles. It is widely cultivated in Europe and Asia today for its grain, which is used as food for man and as a bird food. In the United States it is commonly cut for hay (158).

Some writers think that the sorghum, *Sorghum vulgare* (which see)— called "millet" by SMITH (299)—may be included in one or the other of these references, or even the Italian or German millet, *Setaria italica* (L.) P. Beauv.[91] (98), but this does not seem probable.

The Hebrew word used for "millet" in Ezekiel 4: 9 is "dôchan" (Greek, κέγχρος). TRISTRAM reports that this name—"dukhn" according to POST— is still used by the Arabs for the two kinds of millet mentioned above, both of which are "largely grown in the East" (184). The small seeds of both are used on cakes, but are also eaten uncooked by the very poor of the land (184). PAXTON writes that it is almost the only food eaten by the common people of Arabia Felix. "Disagreeable" is about the most charitable description that could be given of the bread made from it—and here we have the reason why millet was given to the prophet EZEKIEL as part of his hard fare (262).

"Pannag" is the Hebrew word actually used in Ezekiel 27: 17, left untranslated by the King James and the Revised versions. The Syriac version renders it "millet", while the Septuagint treats it as the proper name of a place. Some Greek translations use "cassia" (κασία) for it. A marginal note in the Revised Version follows the Targum and suggests that it is "perhaps a kind of confection" (27). The Vulgate and the LEESSER Version regard it as "balsam".

SMITH has suggested that the word "pannag" may be related to the Greek "panaxeia", meaning a universal medicine or panacea, considered by the Greek physicians as a cure for a great many human diseases (299). It is not definitely known from what plant the ancients obtained their "panaxeia", but some writers have supposed that it was the famous ginseng of China, *Panax ginseng* C. A. Mey.[92] The Chinese believe that the roots of this plant resemble in

---
[91]Also known as *Panicum italicum* L. and *Chaetochloa italica* (L.) Nash.
[92]Also referred to as *Panax schinseng* Nees.

form the outlines of the human body and therefore will cure almost any disease of the human body, and, in addition, will restore youth and vigor to old people (299). It is, however, extremely doubtful whether the trade of Greece and Palestine extended as far east as China in EZEKIEL's day (595 B.C.) (299).

ROYLE maintains that the true "dukhun" of the Arabs is *Panicum miliaceum* which is "universally cultivated in the East", as well as in Europe and tropical countries (306). SMITH says that in all probability the grains of *Panicum miliaceum*, *Setaria italica*, and *Sorghum vulgare* may all be comprehended in the Hebrew word "dôchan" (306), as they were all used by the ancient Egyptians and Hebrews.

156. Pasteurella pestis (Lehm. & Neum.) Bergey

NUMBERS 12:10—And the cloud departed from off the tabernacle; and, behold, MIRIAM became *leprous*, white as snow: and AARON looked upon MIRIAM, and, behold, she was *leprous*.
NUMBERS 13: 32—And they brought up an *evil report* of the land which they had searched unto the children of Israel, saying, The land, through which we have gone to search it, is a land that *eateth up* the inhabitants thereof . . .
NUMBERS 14: 37—Even those men that did bring up the *evil report* upon the land, died by the *plague* before the Lord.
NUMBERS 16: 46-50— . . . the *plague* is begun. And AARON took as MOSES commanded, and ran into the midst of the congregation; and, behold, the *plague* was begun among the people . . . And he stood between the dead and the living; and the *plague* was stayed. Now they that died in the *plague* were fourteen thousand and seven hundred . . . and the *plague* was stayed.
DEUTERONOMY 28: 27—The Lord will smite thee with the botch of Egypt, and with the *emerods* . . .
I SAMUEL 5: 6, 9, 11, & 12—But the hand of the Lord was heavy upon them of Ashod, and he destroyed them, and smote them with *emerods*, even Ashod and the coasts thereof . . . The hand of the Lord was against the city with a very great destruction: and he smote the men of the city, both small and great, and they had *emerods* in their secret parts . . . for there was a deadly destruction throughout all the city; the hand of God was very heavy there. And the men that died not were smitten with the *emerods*: and the cry of the city went up to heaven.
I SAMUEL 6: 4, 11, 17, & 19— . . . Five golden *emerods*, and five golden mice . . . for one *plague* was on you all, and on your lords. Wherefore you shall make images of your *emerods*, and images of your mice that mar the land . . . And they laid the ark of the Lord upon the cart, and the coffer with the mice of gold and the images of their *emerods* . . . And these are the golden *emerods* which the Philistines returned for a trespass offering unto the Lord . . . And he smote the men of Beth-shemesh . . . even he smote of the people fifty thousand and threescore and ten men: and the people lamented, because the Lord had smitten many of the people with a great slaughter.
JEREMIAH 21: 9—He that abideth in this city shall die . . . by the *pestilence* . . .
I CORINTHIANS 10: 10—Neither murmur ye, as some of them also murmured, and were destroyed of the *destroyer*.
HEBREWS 3: 17—But with whom was he grieved forty years? was it not with them that had sinned, whose *carcases fell in the wilderness*?
JUDE 5— . . . the Lord, having saved the people out of the land of Egypt, afterward *destroyed* them that believed not.

There is no unanimity among medical writers in regard to the many diseases referred to in the Bible. In most cases the recorded symptoms are so vague or so general that they could well be indicative of several or many specific human ailments. This has been pointed out in our discussion of the "plague" of leprosy (see under *Mycobacterium leprae*). Other types of "plague" are described in the Bible. One of these was apparently venereal (see under *Neisseria*); another is generally regarded as having been bubonic plague. For instance in the 5th and 6th chapters of I Samuel we read of a plague with which "mice" were connected. These "mice" are described as marring the land. The Hebrew word "achbar" refers both to mice and rats. It is probable that rats, rather than mice, are here referred to (250). We read that the ark of the tabernacle was captured by the Philistines and that the inhabitants of each city to which it was brought were attacked by a plague which killed thousands. In the King James and Douay versions this "plague" (Hebrew, "magepha") is called "emerods", but the Revised Version of 1885 substitutes "tumors" and adds a marginal note reading: "or plague—boils; as read by the Jews emerods" (250). The symptoms of this plague are given in part in I Samuel 5:9: "and he smote the men of the city, both small and great, and they had *emerods* in their secret parts." The GOODSPEED version says "... so that *plague-boils* broke out upon them." The word emerod (Hebrew, "bapolim") is an older one for what we now call hemorrhoids, but the extremely high fatalities recorded in the Bible, together with the allusion to rats and boils in the groin, point to some plague far more serious than hemorrhoids. True hemorrhoids, however, seem to be referred to in Psalms 78: 66.

Regarding the "plague" of SAMUEL, ALLEN says: "The disease was very fatal and accompanied by tenesmus and evacuations of blood. It might have been a form of dysentery, but that has not swellings. Bubo, however, does have tenesmus and bloody stools, and as the swellings are in the secret parts, in all probability these emerods were bubo. There was no sanitary precaution in the Levitical Code to prevent spreading of this disease because of the belief that it was due to divine judgment supernaturally inflicted. Had the Israelites kept themselves free from intercourse with other nations they would probably have been free" (20).

NEUSTATTER says: "In view of the fierce, contagious sickness with quick and high mortality as described in those chapters (I Samuel, 5 and 6) one city after another to which the ark was taken showed an outbreak of the disease; death and excruciating pain befell the people, and the cry of the cities went up to heaven. Tumors do not produce such symptoms" (250).

The earliest known advocate of this bubonic plague identification was THENIUS, who is generally regarded as the originator of Biblical criticism. It was further upheld by the famous Swiss scientist, J. J. SCHEUCHZER (1672-1733). It is regarded as an accepted fact by PIEL, writing in the November 18, 1946, issue of "Life" magazine (264).

Various writers, including ALLEN (20), have also identified—apparently correctly—as bubonic plague the pestilential punishment inflicted upon the Israelites during and after the rebellion of KORAH (Numbers 16, 26: 9-11, 27: 3), as well as the "plague" which killed the spies who returned to MOSES after exploring the land of Canaan. However, in the case of the "plague" which came as a punishment to the men of Israel after having worshipped in the temples of BAAL-PEOR, the context points strongly to a more strictly venereal affliction (see under *Neisseria*).

The causal organism of bubonic plague in man—also affecting rats, mice, guinea-pigs, ground-squirrels, cats, monkeys, and rabbits—is *Pasteurella pestis*[93]. It is carried from rat to rat or from rat to man by fleas. The disease is an acute, systemic, malignant, highly contagious one which is very prevalent in central Asia and has periodically visited Asia Minor, northern Africa, and the large cities of Europe with frightful mortality. It is characterized by an inflammatory swelling and ulceration of the lymphatic glands of the neck, armpits, and especially in the groin (158). The swollen parts become extremely sensitive to the touch, and the afflicted person suffers from headache, vertigo, high fever, vomiting, excruciating pains, and great prostration (373). The skin becomes mottled with purple spots. Mortality is even today from 60 to 90 percent, and death usually ensues within 48 hours, but in some forms within 5 hours (373). In another form, known as pneumonic plague, it is a galloping respiratory infection resembling acute pneumonia and is 100% fatal (264).

Bubonic plague has ravaged Europe and Asia Minor in 195 recorded epidemics from the time of the prophet SAMUEL (about 1141 B.C.) to the 18th century. In the Black Death decades of the 14th century it took the lives of 25 million Europeans, one-fourth of the continent's total population (264). It is often called the Levantine, Oriental, black, and "the poor's" plague—the last-mentioned appellation because of the fact that it attacks so regularly the poor half-starved masses that congregate in the slums of cities. It existed in northern Africa in the 3rd and 4th centuries B.C. according to RUFUS OF EPHESUS (373). In 542 A.D. it appeared in Egypt and in a year spread to

---

[93]Also known as *Bacterium pestis* Lehm. & Neum., *Bacillus pestis* (Lehm. & Neum.) Migula, *B. pestis bubonicae* Kruse, and *Eucystia pestis* (Lehm. & Neum.) Enderl.

Constantinople, where it is said to have caused the death of 10,000 persons in one day. The LEESSER version translates the "emerods" of the Authorized Version as the "inflammatory disease of Egypt". In the 20-year Black Death epidemic in the 14th century 40 million persons died. Out of 2 million inhabitants of Norway only 300,000 survived. During the great plague of London in 1665, so vividly portrayed in a recent popular historic novel, 63,596 persons died out of a total population of 460,000.

The plague is not common in the tropics, since it cannot develop in hot dry air. Warm moist air is very favorable to it, and it thrives on filth and famine. The usual incubation period is from 3 to 6 days, and the early symptoms are similar to those of typhus. The final seizure is abrupt, accompanied by chills, great depression, blunted intellect, pains in the bones, and high fever, followed by the appearance of painful bubos in the groin, neck, and face (373). The germs can be carried in rags, general merchandise, and clothing. In the Orient it is firmly believed that they occur also in the soil.

157. Phoenix dactylifera L.
(FIGURE 9; FIGURE 28; FIGURE 70; FIGURE 92)

GENESIS 14: 7—... that dwelt in Hazezon-*tamar*.
GENESIS 38: 6—And JUDAH took a wife for ER his firstborn, whose name was *Tamar*.
EXODUS 15: 27—And they came to Elim, where were ... threescore and ten *palm trees* ...
LEVITICUS 23: 40—And ye shall take you on the first day the boughs of goodly trees, branches of *palm trees* ...
NUMBERS 33: 9 — ... in Elim were ... threescore and ten *palm trees* ...
DEUTERONOMY 2: 8— ... through the way of the plain from *Elath* ...
DEUTERONOMY 34: 3— ... and the plain of the valley of Jericho, the city of *palm trees* ...
JUDGES 1: 16—And the children of the Kenite, MOSES' father in law, went up out of the city of *palm trees* ...
JUDGES 3: 13— ... and possessed the city of *palm trees*.
JUDGES 4: 5—And she dwelt under the *palm tree* of DEBORAH ...
JUDGES 20: 33—And all the men of Israel rose up out of their place, and put themselves in array at Baal-*tamar* ...
II SAMUEL 13: 1— ... ABSALOM ... had a fair sister, whose name was *Tamar* ...
II SAMUEL 14: 27—And unto ABSALOM were born three sons, and one daughter, whose name was *Tamar*: she was a woman of fair countenance.
I KINGS 6: 29 & 32—And he carved all the walls of the house round about with carved figures of cherubims and *palm trees* ... and he carved upon them carvings of cherubims and *palm trees* ...
I KINGS 9: 26—And king SOLOMON made a navy of ships in Ezion-geber, which is beside *Eloth* ...
II KINGS 14: 22—He built *Elath*, and restored it to Judah ...
II KINGS 16: 6—At that time REZIN king of Syria recovered *Elath* to Syria, and drave the Jews from *Elath* ...
I CHRONICLES 3: 9— ... and *Tamar* their sister.
II CHRONICLES 8: 4 & 17—And he built *Tadmor* in the wilderness ... Then went SOLOMON to Ezion-geber, and to *Eloth*, at the sea side ...
II CHRONICLES 20: 2— ... and, behold, they be in Hazazon-*tamar*, which is En-gedi.
II CHRONICLES 26: 2—He built *Eloth*, and restored it to Judah ...
II CHRONICLES 28: 15— ... and brought them to Jericho, the city of *palm trees* ...
II CHRONICLES 31: 5— ... the children of Israel brought in abundance the firstfruits of corn, wine, and oil, and *honey* ...
NEHEMIAH 8: 15— ... Go forth unto the mount, and fetch ... *palm branches* ...
PSALMS 1: 3—And he shall be like a *tree* planted by the rivers of water, that bringeth forth his fruit in his season; his leaf also shall not wither ...
PSALMS 92: 12-14—The righteous shall flourish like the *palm tree* ... Those that be planted in the house of the Lord shall flourish ... They shall bring forth *fruit* in old age; they shall be fat and flourishing.
SONG 7: 7-8—This thy stature is like unto a *palm tree* ... I said, I will go up to the *palm tree*, I will take hold of the *boughs* thereof ...
JEREMIAH 10: 5—They are upright as the *palm tree* ...
EZEKIEL 40: 31— ... and *palm trees* were upon the posts thereof ...
EZEKIEL 47: 19—And the south side southward, from *Tamar* even to the waters of strife ...
EZEKIEL 48: 28— ... the border shall be even from *Tamar* unto the waters of strife ...
I MACCABEES 13: 51—And they entered into it ... with thanksgiving, and branches of *palm-trees* ...
II MACCABEES 10:7—Therefore they bare branches, and fair boughs, and *palms* also ...
II MACCABEES 14: 4— ... presenting unto him a crown of gold, and a *palm* ...
MARK 11: 1—And when they came nigh to Jerusalem, unto Bethphage and *Bethany* ...
LUKE 19: 29—And it came to pass, when he was come nigh to Bethphage and *Bethany* ...
JOHN 11: 18—Now *Bethany* was nigh unto Jerusalem ...
JOHN 12: 13—Took branches of *palm trees*, and went forth to meet him ...

Acts 11: 19—Now they ... travelled as far as *Phenice* ...
Acts 15: 3— ... they passed through *Phenice* and Samaria ...
Acts 27: 12— ... if by any means they might attain to *Phenice* ...
Revelation 7: 9—After this I beheld ... a great multitude ... with ... *palms* in their hands.

The "palm tree" (Hebrew, "elot") of the Bible is unquestionably the date palm, *Phoenix dactylifera*. It was at one time as characteristic of Palestine as of Egypt. Being so peculiar in the erect mode of its growth, with its branchless tapering stem to 80 feet or more in height and its great terminal cluster of feathery leaves, each 6 to 9 or more feet long, it stood out conspicuously from all the other trees of the land (179). In some districts it was found in dense groves; in other places isolated trees served as well-known landmarks. It was natural that it should be used as a form of ornamentation in Oriental architecture (306). Its stems and leaves were favorite subjects for architectural embellishment, usually in relief, from the time of the building of Solomon's temple onwards (I Kings 6: 29 & 32, Ezekiel 40: 31). The capitals of Egyptian temple and palace columns were conventionalized from the terminal crown of palm leaves, and this form of capital persists to the present day (179). The immense branch-like leaves (referred to as "branches" in the Bible) were symbols of triumph and were used on occasions of great rejoicing (John 12: 13, Revelation 7: 9). The large leaves are still used to cover the roofs and sides of houses and to give solidity to reed fences (144, 219, 265, 268, 293). Mats, baskets, and even dishes are made of them. Small leaves are used as dusters and the wood of its trunk for timber. Rope is made from the web-like integument in the crown. Its fruit, borne in immense drooping clusters, weighing 30 to 50 pounds (299), from among the leaves, is the chief article of food for innumerable tribes of Arabia and northern Africa.

The kernel of the date is ground up or soaked in water for several days and used as food for camels, cows, and sheep (299). It is said to be more nutritious than barley. Date seeds are also employed for stringing as beads. A liquor is secured by piercing the spathe surrounding the flowers, whereupon a syrupy liquid exudes. On the testimony of Herodotus, Strabo, and Pliny (306) this was made into an intoxicating liquor by the ancient Babylonians, and it is supposed that this is the liquor referred to in some verses where "strong drink" is mentioned in the Bible as distinct from wine (Leviticus 10: 9; Numbers 6:3; Deuteronomy 14: 26 & 29: 6; Judges 13: 4; Proverbs 20: 1 & 31: 4 & 6; Isaiah 5: 11, 24: 9, and 28: 7; Micah 2: 11; and Luke 1: 15) (299). It is also what the Jews usually referred to when they spoke of "honey" (Hebrew, "d'vash") (262) (Genesis 43: 11; I Samuel 14: 25; Psalms 19: 10; Proverbs 24: 13, 25: 16, and 27: 7; Isaiah 7: 15; Song 4: 11; Revelation 10: 9). It has been reported that bees are mentioned in the Bible but four times, while "honey" is mentioned forty-nine times. This tends to substantiate the claim that the word "honey" was not always used in the sense that we use it today. Herodotus speaks of the palm tree as producing bread, wine, and honey. Herodotus, along with Strabo and Pliny, mention the fact that the ancient Orientals made both wine and honey from date palms (119). The Arabs have a saying that the palm tree has as many uses as there are days in the year, and this seems to be almost literally true.

The Hebrew word for the date palm is "tàmâr" or "temarim" (Greek, φοῖνιξ). It became the Jews' symbol of grace and elegance and was often bestowed by them to women, as, for instance, to the sister of Absalom, in allusion to their graceful upright carriage. Even today in Palestine "Tamar" and "Tamarah" are often used as girls' names. Although it was abundant throughout the entire Levant, the palm tree has always been most intimately associated with Palestine (306). The name "Phoenicia" (Hebrew, "Tzidon") by which part of the Levant, particularly the portion including Tyre and Sidon, was known to the Greeks and Romans, means "land of palms". Some of the ancient coins of Tyre and Sidon bear the figure of the palm, as does also a Jewish coin issued at the time of Judas Maccabaeus, 175 B.C. (299). Some authors imply that "Phoenicia" was applied to the whole of Palestine. It is the "Phenice" of the King James and Douay versions of Acts 11: 19 and 15: 3, rendered "Phoenicia" by the Moffatt, Lamsa, O'Hara, Weymouth,

Weigle, and Basic English versions, but that of Acts 27: 12 was a harbor on the island of Crete now called Phoenix, according to Weigle, Goodspeed, Moffatt, Weymouth, and the Basic English versions ("Phoenis" in O'Hara and "Phenice" in Lamsa). It is thought that at one time the whole Jordan valley from the shores of Gennesaret to the end of the Dead Sea was covered with date palms. Josephus states that in his day (about 37-95 A.D.) there was still a grove of palms seven miles long near Jericho, that palms were abundant around the lake of Galilee, in the lower Jordan valley, on the Mount of Olives, and about Jerusalem (299). They grew luxuriantly in the vale of Shechem, on the maritime plains, and about Beirut (184). Palms require a good deal of careful cultivation, and it is probably owing to the lack of this cultivation that they perished (119). They must grow thirty years to reach full maturity, which then lasts until the century mark is passed. They then go into a gradual decline until the end of their second century, when they die (119).

The palm became symbolic of Palestine, and to commemorate the conquest of the Jews and the destruction of Jerusalem by Titus in 70 A.D., the Roman emperor Vespasian issued a coin showing a weeping woman sitting beneath a palm tree (299). Now, however, the palm is not nearly as abundant in the Holy Land as it was then. "The absence of the date palm from the Jordan River valley is one mark of the complete desolation, if not desecration, of the Holy Land" (119). Various authors report that the remains of the formerly abundant palms of that region are seen in the petrified tree trunks piled up at the northern end of the Dead Sea[94] and in the half-fossilized palm fronds seen in recently formed limestone at Ain Jidy, the ancient Engedi. Borah says that the palm is actually "gone" now from the Jordan valley and that where once it grew so abundantly in thick forests, drifting sand has covered every trace, although restoration and reforestation are now being attempted (69). So deep is the sand in these places that the site of ancient Jericho, the "city of palms" (II Chronicles 28: 15), could scarcely be located a few years ago; although now Dr. Evenari informs us that it has been found and excavated, and the skilled agriculturists in the newer Jewish settlements have been quite successful so far in their attempts to establish new and large plantations.

The date palm has a very extensive geographic range, from India through western Asia and the Levant to Arabia Petraea, Sinai, and Egypt, and all of northern Africa west to the Atlantic. It is the chief food plant of the desert, furnishing food, oil, and shelter to millions of people (299).

The use of palm leaves in processions seems to have originated at the time of the restoration of the temple by Judas Maccabaeus and has been continued to this day by Jews at the Feast of the Tabernacles and by Christians on Palm Sunday and Easter (125, 298, 299). The palm leaf ("lulab" or "lulav"), bound with myrtle on the right and citron on the left hand, was the triple badge of desert life, carried by the Jews and shaken at the Feast of the Tabernacles, after which it was carefully laid up at home (184).

The date palm is dioecious, a fact which was apparently known to the ancients, for we see representations in ancient Egyptian and Babylonian inscriptions of bunches of the male flowers being removed from the staminate trees and hung on the pistillate trees. Löw calls attention to a story in the Talmud of a female palm tree in Jericho weeping for its male companion until a branch of the male was brought over (218). Most authorities now believe, however, that the ancients did not realize the sexual difference in the flowers, but regarded the placing of the staminate flowers on the pistillate tree as a form of fertilizing, akin to placing manure in the soil. This old concept is, in fact, the source of our word "fertilization" for sexual union (299). In inter-tribal warfare the greatest calamity that could be inflicted upon the conquered was the destruction of their male date palms (299).

Numerous places are identified in the Bible by the abundance of the date palms there, for instance, Jericho (Hebrew, "Y'richo"), Hazazon-tamar

---

[94] Dr. Evenari states that he has not been able to locate this petrified wood at this locality in recent years.

(Hebrew, "chatz'tzon tamar"; meaning "the felling of the palm trees"), Baal-tamar (Hebrew, "baal-tamar"; meaning "male palm tree"), and SOLOMON'S Tadmor (afterwards known as Palmyra) (184). MOFFATT, in fact, has carried this common linguistic practice into his modern translation of the Bible when he says "Palmtown" for "the city of palm trees" in Judges 1: 16 and 3: 13. Bethany (Hebrew, "bet teainah") means "the house of dates", and doubtless received its name from their abundance at that locality (27). SMITH mentions that the place name Elim (Hebrew, "Eh'limah"), one of the stopping-places of the Jews between Egypt and Sinai, was so named because of its 70 palm trees (Exodus 15: 27, Numbers 33: 9), and that Elath (Hebrew, "eh'lit") or Eloth (Hebrew, "e'lot") of Deuteronomy 2: 8, I Kings 9: 26, II Kings 14: 22 and 16: 6, and II Chronicles 8: 17 and 26: 2 is another plural form of the same Hebrew word "elim", and may likewise mean "palm-trees" (306).

In classic days the date palm symbolized worldly riches, procreation, victory, and light. It was dedicated to APOLLO, god of manly beauty, youth, poetry, music, and wisdom (126, 158). Mohammedans (126) believe that the palm was created by MOHAMMED (570?-632 A.D.). To the ancient and medieval Christian martyrs, angels were said to bring palm "branches" to convey their souls from the tortures of rack, cross, and flame to heaven, and so the palm became a symbol of martyrdom (298). On All Souls Day (November 2) palm leaves are cast into a fire and their rising into the sky as smoke is taken as proof of victory by the souls which are that day released from purgatory (298). The traditions of some countries state that the palm, not the fig or "apple", was the "tree of knowledge" in the garden of Eden—a belief perpetuated in the coat-of-arms of the state of South Carolina where there is a palm entwined by a serpent (126). ST. CHRISTOPHER is said to have used a palm staff when he carried the small and weak, including the infant JESUS, across the raging river. JESUS told him to thrust his staff into the ground and it sprouted into a date palm, a miracle which resulted in the conversion of CHRISTOPHER. SAINT CLARA renounced the world on Palm Sunday and received a palm from ST. FRANCIS OF ASSISI as a symbol of sanctity (298). Other legends state that during the flight to Egypt, MARY commanded the palm to bend down its head to shade the infant JESUS, and at another time JESUS commanded a palm to bend down so that he could pick the dates for his mother after JOSEPH had refused to do so. It did this "so willingly that he blessed it and chose it as a 'symbol of salvation for the dying', promising that when He entered Jerusalem in triumph it should be with a palm in His hand" (298). Angels are said to have carried a palm from heaven to MARY, mother of JESUS, after the crucifixion. MARY gave it to ST. JOHN to be carried by him before her bier to the grave three days hence. In medieval days it was thought to prevent sunstroke, avert lightning, cure fevers, and drive away mice and fleas (298).

158. **Phragmites communis** Trin.

III MACCABEES 4: 20— . . . when they told him and proved that even the paper manufactory and the *pens* which they used for writing had already given out.
III JOHN 13—I had many things to write, but I will not with ink and *pen* write unto thee.

The "reeds" of the Bible have presented considerable problems to botanical students. In most cases it is believed that the plant referred to by this term was *Arundo donax* (which see). In other cases *Sorghum vulgare* (which see) is involved. In the famous dream of PHARAOH, however, as recorded in Genesis 41: 2, some commentators feel that the common reed, *Phragmites communis*[95], is alluded to. MOFFATT and JASTROW, indeed, render as "reed-grass" the "meadow" of the Authorized Version's translation of this verse.

---
[95]Also known as *Arundo phragmites* L., *A. vulgaris* Lam., *Trichoon phragmites* (L.) Rendle, *Phragmites phragmites* (L.) Karst., and *P. scriptorum* J. E. Sm. It has also been incorrectly referred to as *Arundo maximus* Forsk. and *Phragmites maximus* (Forsk.) Chiov.

The GOODSPEED version says "sedge", while the Douay version is "they fed in *marshy places*". By us this passage is discussed under *Butomus umbellatus* (which see).

Some writers have identified the New Testament "reed" of Matthew 27: 48 and Mark 15: 36 with *Phragmites*, but this identification does not seem likely (see under *Sorghum vulgare*). DINSMORE feels that *Phragmites* may have been the "reed shaken by the wind" of Matthew 11: 7 (267), but we are following POST in referring this passage to *Arundo donax*. However much doubt there may be concerning the accuracy of these identifications mentioned above, there is no doubt whatever that the references in III John 13 and III Maccabees 4: 20 are to the common reed. Pens for writing on parchment or skins were quite uniformly made from the stems of this reed. The oldest known documents with writing of a Semitic race are probably the bricks of Nineveh and Babylon (306). The oldest reference to writing in the Bible is in Numbers 17: 3 (about 1471 B.C.), from which we learn that the writing was done on wood. In II Esdras 14: 24 writing tablets of boxwood are mentioned. In Job 19: 24 is described a method of carving words in rock and then filling them with molten lead. In II John 12 and III Maccabees 4: 20 paper of papyrus is mentioned. For ordinary purposes wood tablets covered with wax served (Luke 1: 63). To write on these a pointed stylus was employed, often made of iron. For harder materials a graver was used. Only for writing on parchment and skin were reed pens serviceable. The ink used in reed pens was lamp-black "dissolved in gall-juice" (306). It was carried in an inkwell suspended from the girdle (Ezekiel 9: 2-3), as is done even at the present time in the Orient. To professional scribes we find references in Psalms 45: 1, Ezra 7: 6, and II Esdras 14: 24.

HERODOTUS states that the Ionians learned the art of writing from the Phoenicians and that their books were called "skins" because they used dressed sheep and goat skins when short of papyrus paper. In JOSEPHUS' day parchment was used for the manuscript of the Pentateuch. The "membranae" of II Timothy 4: 13 were parchment skins. The Talmud provided that the Law be written only on the skins of "clean" mammals or birds. These skins, when written upon, were rolled up on one or two sticks and fastened with a thread, the ends of which were sealed (Psalms 40: 7-8, Isaiah 29: 11 & 34: 4, Jeremiah 36: 14, Ezekiel 2: 9-10, Daniel 12: 4, Zechariah 5: 1, Revelation 5: 1). The rolls were usually written on only one side, rarely on both sides (Ezekiel 2: 9-10, Revelation 5: 1).

The common reed, from which reed pens were made—and still are in Oriental lands—is a tall grass, 3 to 15 feet in height, with its thick leafy stems woody at the base, growing from long horizontal rootstocks. The 2-ranked leaves are lanceolate, broad, and roughened at the margins, 6 to 12 inches long and ½ to 2 inches wide. The flowers are borne in dense fluffy panicles, 6 to 12 or more inches in length, at the apex of the stems. The species is common in marshes and swamps throughout the Holy Land, Arabia Petraea, and Sinai, as well as much of Europe, Asia, and America (267).

159. **Pinus brutia** Tenore

LEVITICUS 23: 40—And ye shall take you on the first day the boughs of goodly trees, branches of palm trees, and the boughs of *thick trees*, and willows of the brook . . .
NEHEMIAH 8: 15— . . . Go forth unto the mount, and fetch olive branches, and pine branches, and myrtle branches, and palm branches, and branches of *thick trees*, to make booths . . .
PSALMS 74: 5-6—A man was famous according as he had lifted up axes upon the *thick trees*. But now they break down the carved work thereof at once with axes and hammers.
ISAIAH 41: 19— . . . I will set in the desert the fir tree, and the *pine*, and the box tree together.
ISAIAH 60: 13—The glory of Lebanon shall come unto thee, the fir tree, the *pine tree,* and the box tree together, to beautify the place of my sanctuary . . .

The words "fir", "pine", "cypress", "juniper", and even in a few cases "cedar", are used so loosely in the Authorized Version that it is next to impossible to determine exactly what trees are referred to in each specific passage. There has always been, and there still is today, a great amount of difference of opinion on the identity of the evergreens of the Bible. The newer translations have done little to clear up this confusion. MOFFATT, for instance, translates the "fir" of the Authorized Version as "cypress" in some places, as "pine" in others, and leaves it "fir" in a few passages—on what basis is not clear. The same applies to the Douay and GOODSPEED versions (see under *Apinus pinea* and *Pinus halepensis*).

In Nehemiah 8:15 we find a reference to "pine branches" and to "branches of thick trees" for use in building the booths for the Feast of the Tabernacles. The "pine branches" of this verse are now generally regarded as having been branches of the narrow-leaved oleaster, *Elaeagnus angustifolia* (which see)—"wild olive" in GOODSPEED, "oleaster" in MOFFATT, "branches of beautiful wood" in Douay. The "branches of thick trees" in the same verse, however, are now regarded by many as having been pine branches—"evergreens" in MOFFATT. In Isaiah 41:19 the Authorized Version speaks of "fir ... pine ... and box"—the Douay version renders this "fir ... elm ... and box", MOFFATT says "fir ... plane ... and cypress," and the GOODSPEED and JASTROW rendition is "cypress ... plane ... and larch." In Isaiah 60:13 the Authorized Version again speaks of "fir ... pine ... and box". The Douay version changes this to "fir ... box ... and pine", MOFFATT to "pines ... planes ... and cypresses", GOODSPEED to "cypress ... pine ... and larch", and JASTROW to "cypress ... plane-tree ... and larch." This gives a general idea of the confusion surrounding the conifers of the Bible! It seems apparent, though, that pines are involved somewhere in these passages. These pines are thought now to be the Brutian pine, *Pinus brutia*[96], although other evergreens have been suggested, for instance, *Pinus halepensis* (by Evenari), *Picea orientalis* Carr., *Sabina thurifera* (L.) Antoine[97], and other species not recorded at all by POST from the Holy Land area (265, 267). Other commentators have suggested *Abies cilicica* (Ant. & Ky.) Carr., *Juniperus drupacea* Labill., and *J. macrocarpa* Sibth. & Sm. These three latter trees, however, are recorded by POST (267) only from Lebanon and other northern areas of the Levant which were not occupied by the Israelites at the time of ISAIAH and NEHEMIAH, and, of course, not at the time of MOSES (306), so that it is not at all probable that the branches of these trees would have been available for the construction of the booths. *Pinus brutia*, on the other hand, while it is also a mountain-inhabiting species of the northern regions and Lebanon, occurs likewise in Palestine[98] and therefore would have been available after the children of Israel were led into Palestine by JOSHUA. It is noteworthy in this connection that the instructions to the Jews as recorded by NEHEMIAH began with the directive "Go to the mount ... ", or, as MOFFATT says, "Go to the hill-country ... " In the instructions given by MOSES in the book of Leviticus, spoken before the Israelites entered Palestine, the reference must have been to species of coniferous evergreens with which MOSES was acquainted from the sojourn in the Sinai peninsula. The only two recorded by POST from that area are the evergreen cypress, *Cupressus sempervirens* var. *horizontalis*, and the Phoenician juniper, *Sabina phoenicia*, which see.

The Hebrew word used in the Isaiah passages for "pine" is "tidhar" (Greek βραθυδαάρ, πεύκη). The linguistic root of this word signifies "to revolve" and it is possible that it refers to the fact that the branches of pines and firs are borne in whorls (299). The word used for "pine branches" in Nehemiah is "etz shamen" or "etz shemen", the very same as is used for "oil tree" in Isaiah 41:19, hence the modern relegation of this reference to the oleaster. The words used for "thick trees", on the other hand, are "etz a'vot" or " 'etz 'aboth" (Greek, κλάδοι, meaning "a young shoot", and δασεῖς, meaning "thickly covered"). The Leviticus reference to "thick trees" in the Authorized, JASTROW, and Douay versions is rendered "leafy trees" by both the MOFFATT and

---

[96] Also called *Pinus maritima* Lamb., *P. pyrenaica* Lepeyr., *P. carica* D. Don, and *P. halepensis* var. *brutia* (Tenore) Elwes & Henry.
[97] Also known as *Juniperus thurifera* L.
[98] Evenari states that *Pinus brutia* is not found in Palestine, but it is listed from Palestine by POST, and from Moab and Gilead by DINSMORE (100).

Goodspeed versions and "three-leaved myrtle" by Leesser. Moffatt regards the "fir tree" of Zechariah 11: 2 as "pine-tree", but the Goodspeed version renders it "cypress".

The Brutian pine attains a height of 10 to 35 feet. Its growth is rather diffuse; the branches are somewhat whorled. The leaves are in clusters of two, rather thick, longer, darker, and more rigid than those of the Aleppo pine. The staminate catkins are in capitate clusters, and the ascending oblong-conic cones are whorled in groups of 3 to 6 (267).

John Smith gives an interesting commentary on the "pines" of these much-discussed passages (299): "After Nehemiah the word pine does not again occur in the Bible, but about five hundred years later it is mentioned by Josephus, who says Solomon had pine wood brought in ships from Ophir, which 'was made use of partly for pillars and supports to the king's temple and palace, partly for musical instruments, as harps, cymbals, psalteries, and the like, for the Levites to glorify God upon. It is to be noted, that for size and beauty, Solomon had never seen any of this sort of wood comparable to it before. This was none of the wood that passes commonly upon the world for pine in the way of trade. This was somewhat of the grain of a fig tree, only a little whiter, and more glossy.' On considering that Josephus wrote eleven hundred years after Solomon had trees from Ophir" [see under *Pterocarpus santalinus*] "and as we have no account of any being imported after that, Josephus's description must be received with some degree of reservation, but what he says is sufficient to show that in his time a kind of wood was known in the 'way of trade' by the name of Pine. In a recently published edition of Josephus, instead of the word Ophir it is said that 'Solomon had pine trees from Aurea Chersonesus'; if Chersonesus means Cherson, in the Crimea, then we are induced to believe that king Hiram sent ships into the Black Sea to bring timber trees for Solomon, but if such is the case it is not alluded to in the Bible" (299).

Tristram believed that the word "tidhar" may have referred to the common elm, *Ulmus campestris* L.—and in this he is followed, in one passage, at least, by the Douay version—but there is no evidence to support this contention (184, 299). The Authorized Version uses "elms" for the word "elah" in Hosea 4: 13, but this same word is elsewhere translated "oak", "plane-tree", "terebinth", and "teil-tree" by the same version. The common elm is known from Aleppo, Aintab, and Lebanon (184), and Post records it questionably from Palestine (267). It is not believed to be indigenous, certainly, to southern Palestine (184), and the translation is now regarded as wholly erroneous.

The plane-tree, *Platanus orientalis*, suggested by a marginal note in the Revised Version, as well as by Jones (184), Moffatt (12), and others (27), is a common tree of valleys and lowlands, along the sides of streams and lakes. Its abundance would make it a likely tree to be used by the Jews, but it is not a montane species and thus would not fill the Nehemiah context, although this would not eliminate it from the Isaiah texts. Still, even in the Isaiah references, it does not seem as likely as the pine because it would hardly be spoken of as a part of "the glory of Lebanon", as coniferous evergreens would.

160. **Pinus halepensis** Mill.
(Figure 73)

II Samuel 6: 5—And David and all the house of Israel played before the Lord on all manner of instruments made of *fir wood*, even on harps, and on psalteries, and on timbrels, and on cornets, and on cymbals.
I Kings 5: 8 & 10— . . . I will do all thy desire concerning timber of cedar, and concerning timber of *fir* . . . So Hiram gave Solomon cedar trees and *fir trees* according to all his desire.
I Kings 6: 34—And the two doors were of *fir wood* . . .
II Kings 19: 23— . . . With the multitude of my chariots I am come up to the height of the mountains, to the sides of Lebanon, and will cut down the tall cedar trees thereof, and the choice *fir trees* thereof . . .
II Chronicles 2: 8—Send me also cedar trees, *fir trees*, and algum trees, out of Lebanon.
Psalms 104: 17— . . . as for the stork, the *fir trees* are her house.

SONG 1: 17—The beams of our house are cedar, and our rafters of *fir*.
ISAIAH 14: 8—Yea, the *fir trees* rejoice at thee . . .
ISAIAH 37: 24—. . . I will cut down the tall cedars thereof, and the choice *fir trees* thereof . . .
ISAIAH 41: 19—. . . I will set in the desert the *fir tree*, and the pine, and the box tree together.
ISAIAH 44: 14—. . . he planteth an *ash* and the rain doth nourish it.
ISAIAH 55: 13—Instead of the thorn shall come up the *fir tree* . . .
ISAIAH 60: 13—The glory of Lebanon shall come unto thee, the *fir tree*, the pine tree, and the box together, to beautify the place of my sanctuary . . .
EZEKIEL 27: 5—They have made all thy ship boards of *fir trees* of Senir . . .
EZEKIEL 31: 8—. . . the *fir trees* were not like his boughs . . . .
NAHUM 2: 3—. . . and the *fir trees* shall be terribly shaken.
ZECHARIAH 11: 1-2—Open thy doors, O Lebanon, that the fire may devour thy cedars. Howl, *fir tree*; for the cedar is fallen . . .

There has been, and still is, considerable divergence of opinion on the identity of the various evergreens of the Bible. The Hebrew words usually translated "fir" or "fir tree" by the King James version are "bĕrôth" or "b'rotim" and "b'rosh", "bĕrôsh", or "b'roshim" (Greek, κυπάρισσος, for which dictionaries say "cypress"). Most commentators agree (184) that these words refer to the Aleppo pine, *Pinus halepensis*[99]. This is a tree 9 to 60 feet tall, with diffuse, somewhat whorled, ascending branches and yellowish or brownish branchlets. The bark is gray and smooth, becoming fissured in age. The leaves are in fascicles of two or sometimes three, light-green, and 2½ to 6 inches long. It is a native of the Mediterranean area, abundant on dry hills in Palestine, Lebanon, and northward (267). Its wood is said to be "scarcely inferior to the cedar" of Lebanon (184).

MOFFATT feels that the "fir" of II Kings 19: 23, I Kings 6: 34, Isaiah 14: 8 & 37: 24, and Ezekiel 27: 5 was the cypress, *Cupressus sempervirens* var. *horizontalis* (which see), while that of Isaiah 60: 13 and Zechariah 11: 1-2 was a "pine" (see under *Pinus brutia*). He retains "fir" only for Isaiah 41: 19 and 55: 13, II Samuel 6: 5, Song 1: 17, II Chronicles 2: 8, and I Kings 5: 8 & 10. In the II Chronicles passage he has to retain the word "fir" because he renders the next word—"algum trees" in the Authorized Version—"cypress logs". In II Samuel 6: 5 he substitutes "lutes" for the Authorized Version's "all manner of instruments made of *fir wood*, even on harps." His translation of the verse is especially interesting: "DAVID and all the house of Israel were dancing lustily before the Eternal and singing with lutes, with lyres, with drums, with rattles, and with cymbals." The GOODSPEED version says: "were reveling before the Lord with all their might with songs and harps and lyres and with tambourines and castanets and cymbals." The GOODSPEED version uses "cypress" in all the passages cited at the head of the chapter except the one just quoted and in II Chronicles 2: 8 where "fir" is retained. JASTROW says "cypress" in all save in Isaiah 44: 14, where he says "bay-tree".

Some writers believe that the Hebrew word "ôren" (Greek, πίνυς; dictionary translation "the pine tree"), rendered "ash" by the Authorized Version of Isaiah 44: 14 actually also refers to the Aleppo pine. Others have identified it with *Fraxinus ornus* L., "*F. parviflora* Tenore" (probably an error for *F. parvifolia* Lam.), and *F. syriaca* Boiss., but most authorities now agree that a coniferous evergreen was alluded to in this passage, thus eliminating all these true ashes. MOFFATT apparently disregards the reference entirely. The GOODSPEED version plays safe with "some other tree of the forest", the Septuagint, Vulgate, and Douay versions render the word "pine tree", JASTROW says "bay-tree", and the Revised Version says "fir tree" (27). This is the only passage where the word "ôren" occurs in the Bible, and it is mentioned as a tree from whose wood idols were made (184). It seems strange that the word "bĕrôsh" should be used in the many other Isaiah references to the Aleppo pine and "ôren" only once. The identification of the latter seems, therefore, to be still a matter of considerable doubt. TRISTRAM has suggested that it may be the tree known as "aran" in Arabic, native to Arabia Petraea, which he does not further identify save to say that it resembles "our mountain ash" (184). POST records no species of *Sorbus* from Arabia Petraea, but "aran" is recorded as the Arabic name for the flowering ash, *Fraxinus ornus* (which see) (27).

---

[99]Also known as *Pinus maritima* Desf., *P. maritima* Mill., and *P. hierosolymitana* Duham.

The "fir" is often mentioned in connection with the cedar-of-Lebanon, as a "choice" and "goodly" tree. One may assume that it grew in Lebanon with the cedar. Its timber was used for flooring, ceiling, and doors in building SOLOMON's temple, for the rafters of ships' decks, and for musical instruments, especially harps and lutes. In the Septuagint the word is replaced by "pine", "cypress", and "juniper". Some writers state that it may have included all these in its connotation, similar to our term "conifer", and suggest in addition the stone pine, *Apinus pinea* (27) and Brutian pine, *Pinus brutia* (184), which see. WILLIAM SMITH has suggested (306) the European larch, *Larix decidua* Mill.[100], both for the "ash" and the "fir", but this is out of the question as no species of larch is native to that section of the world (as has been brought out under *Buxus longifolia*, which see). CELSIUS was of the opinion that "bĕrôsh" referred to the cedar-of-Lebanon (306), but this seems most unlikely since it is several times mentioned along with and in contradistinction to the cedar. The Scotch pine or Scotch fir, *Pinus sylvestris* L., also suggested by SMITH (306) is likewise impossible because it is not native to that region.

Hosea 14: 8 is often cited as another reference to *Pinus halepensis*. The Authorized, Douay, and GOODSPEED versions use "green fir tree" for this passage and by MOFFATT and JASTROW render it "cypress". By us it is considered to refer to *Apinus pinea* (which see). The use of the expression "fir trees" in Nahum 2: 3 by the Authorized Version is now regarded as totally erroneous. The reference is actually to horses or their drivers—MOFFATT says: "and their *horses* prance at the muster", GOODSPEED: "and the *chargers* will prance", Douay: "and the *drivers* are stupefied."

161. Pistacia lentiscus L.
(FIGURE 74)

GENESIS 43: 11—... If it must be so now, do this; take of the best fruits in the land in your vessels, and carry down the man a present, a little *balm*, and a little honey, spices, and myrrh, nuts and almonds.
SUSANNAH 54—Now then, if thou hast seen her, tell me, Under what tree sawest thou them companying together? Who answered, Under a *mastic-tree*.

The "balm" of Genesis 43: 11 must have been a native product of the land of JACOB, i.e., Canaan, unknown in Egypt at the time, and therefore could not have been *Commiphora opobalsamum* as has been suggested by so many writers, since that species is native to the mountainous portions of southern Arabia and trade in spices had not yet been established in JACOB's day. The same reason excludes the true myrrh, *Commiphora myrrha*, from this passage. It is now thought that the "balm" of JACOB was the product of the lentisk or mastic tree, *Pistacia lentiscus*. This is a bushy or shrubby tree 3 to 10 feet tall, with evergreen abruptly pinnate leaves consisting of a winged petiole and 3 to 5 pairs of leathery, oblong-lanceolate to obovate, obtuse, mucronulate leaflets. The axillary panicles of flowers and fruit are small and stiff. The "balm" (Hebrew, "tzrai") is a fragrant, terebinthine, gummy exudation of the sap secured by making incisions in the stem and branches, usually in the month of August (306). It is known in commerce as "mastic", "mastick", or "mastich", and has been an article of trade since the earliest times (27, 184). The best grades are in the form of yellowish-white translucent tears or drops and are employed in medicine as an astringent and aromatic (158). The poorer grades are used extensively as a varnish. It was also used "to strengthen the teeth and gums" by the ancients, who prized it highly for this attribute (306). The children in eastern lands often spend their coins for this material, which they use like chewing-gum. The Greeks make a liquor flavored with this material from grape skins and call it "mastiche". The Syrians prepare a similar drink this way, too. The lentisk tree is native to southern Europe, the Levant, and, in fact, all countries bordering on the Mediterranean (184). It is indigenous to and common in Lebanon and throughout Palestine (27, 184, 266).

---

[100]Referred to by him as *"Laryx Europoea"*; also called *Pinus larix* L., *Larix europaea* P. DC., and *L. larix* (L.) Karst.

Some commentators (184, 299) feel that the lentisk may also be involved in the "balm" of Ezekiel 27: 17, Genesis 37: 25, Jeremiah 8: 22, 46: 11, and 51: 8, but that of Ezekiel is herein regarded as *Commiphora opobalsamum* and that of the other references as *Balanites aegyptiaca*, which see.

The mastic of Susannah is sometimes said to have been originally regarded as the "holm tree", but this reference is doubtless to verse 58 instead of 54. The holm or holly, *Ilex aquifolium* L., occurs in our area only in the mountains of Syria, and is not now believed to have been the tree referred to in Susannah 58. Instead, the holly oak, *Quercus ilex* (which see), or a related species, is herein regarded as the "holm" of this Apocryphal reference. The Greek word used for the mastic tree in Susannah 54 is σχῖνος (184), and marginal Bibles give the alternative translation of "lentisk", which is equally correct, although not recorded in Webster's dictionary (156). There is no doubt that the Greek word is correctly rendered, as is quite evident from the descriptions of the tree by THEOPHRASTUS, PLINY, DIOSCORIDES, and other early writers (306). The best mastic has always come from the island of Chios (the "Scio" of Tournefort) (158, 306).

162. **Pistacia terebinthus** var. **palaestina** (Boiss.) Post
(FIGURE 75)

GENESIS 18: 8—And he took butter, and milk, and the calf which he had dressed, and set it before them; and he stood by them under the *tree*, and they did eat.
GENESIS 36: 41— ... duke *Elah* ...
I SAMUEL 17: 2 & 19—And SAUL and the men of Israel were gathered together, and pitched by the valley of *Elah* ... Now SAUL, and they, and all the men of Israel, were in the valley of *Elah*, fighting with the Philistines.
I KINGS 16: 8—In the twenty and sixth year of ASA king of Judah began *Elah* the son of BAASHA to reign over Israel ...
ISAIAH 6: 13—But yet in it shall be a tenth, and it shall return, and shall be eaten: as a *teil tree*, and as an oak, whose substance is in them, when they cast their leaves ...
HOSEA 4: 13—They sacrifice upon the tops of the mountains, and burn incense upon the hills, under oaks and poplars and *elms*, because the shadow thereof is good ...
ECCLESIASTICUS 24: 16—As the *turpentine-tree* I stretched out my branches, and my branches are the branches of honour and grace.

The Palestine terebinth,[100a] *Pistacia terebinthus* var. *palaestina*, is a large deciduous tree with straggling boughs and much of the appearance of an oak when it is in the winter leafless condition (184, 299). It grows from 12 to 25 feet tall. Its pinnately compound leaves, composed of 5 to 7 pairs of leaflets, rarely with a small odd one at the apex, are similar to those of an ash, but are smaller and reddish-green in color. The petioles are hairy and the leaflets ovate or oblong in shape, acuminate at the apex, somewhat hairy along the margins. The inconspicuous apetalous flowers and obovate red fruits are borne in branched axillary panicles (266). Every part of the tree contains a fragrant resinous juice. It is common on the lower slopes of hills throughout Syria, Lebanon, Palestine, and Arabia Petraea (326), generally growing solitary, seldom in thickets or forests, and found mostly in localities too warm or dry for the oak, which it in general replaces (184). Being a tree of considerable size and longevity, it was venerated like the oak in ancient times. Incisions in the stem and branches yield the so-called Chio, Chian, or Cyprus turpentine of commerce, and this is doubtless the reason for its being called the "turpentine-tree" in Ecclesiasticus and by the Douay version. Since it is native to Gilead, it is quite probable that its resinous juice formed part of the "spicery" which the Ishmaelites carried into Egypt from Gilead (Genesis 37: 25).

In the time of JOSEPHUS (about 37-95 A.D.) there was a giant terebinth tree near Hebron, which legend stated had been there "since the creation of the world". Under this tree the captive Jews were sold by TITUS VESPATIAN in 69 A.D., and it is supposed to be the tree under which ABRAHAM entertained the three angels (Genesis 18: 8). The tree died about 330 A.D. and has since been replaced by an oak. The Hebrew word in the text is "etz", the literal meaning of which is simply "tree".

Some commentators, as well as the translators of the Revised Version (in

---

[100a]Commonly called "butm" by the natives (124d).

marginal notes), think that the "oak" of Judges 6: 11 and Amos 2: 9, as well as the "oak tree" on which ABSALOM was hanged (II Samuel 18: 9), were the terebinth (184), but this is not very probable. The terebinth never forms such dense forests as are described in the ABSALOM story (124a). In the famous "Scripture Herbal" (78) the Isaiah 6: 13 reference to "teil tree" is said to refer to *Tilia europaea* L., but that is impossible for that tree does not occur in the area (266, 299). This misidentification arose from the fact that "teil" is an obsolete name for the "lime" or linden tree (27, 158) and illustrates again the innocent naivety with which non-botanically trained writers and preachers identified Bible plants with those of their own homeland!

SMITH's elaborate argument for *Ulmus campestris* L. as the "teil tree" of Isaiah and the "elms" of Hosea (299) is far-fetched and highly improbable. The common elm is known definitely only from the mountainous regions of Syria and Lebanon, with one questionable record from Palestine (267), whereas the trees referred to by ISAIAH and HOSEA were obviously common in Palestine. SHAW made the same error in regard to the Hosea passage (293). There is no doubt in our minds that the terebinth was intended in both of these Scriptural references, and the word is so rendered by the GOODSPEED, MOFFATT JASTROW, and Revised versions. The Douay version says "turpentine tree" in the Isaiah, Hosea, and Ecclesiasticus passages, obviously referring to the same species. CELSIUS ventured no definite opinion on the "elm" of Hosea, but TRISTRAM is convinced that it was the terebinth (265). The Hebrew word used for the terebinth in these references is "elâh" [Greek, τερέβινθος (27, 184) or "ptelea" (299)]. The Septuagint generally uses "terebinth" for all the "elâh" references (184).

According to PRATT (268) there is no doubt that the valley of Elah, where DAVID went when SAUL and his army were encamped there, and where, with the smooth pebbles of the brook, he killed the giant GOLIATH (I Samuel 17: 2-49), received its name from the terebinth trees growing there. Dr. ROBINSON considers this valley to lie along the ancient road between Jerusalem and Gaza, and remarks that the largest terebinths which he ever encountered in the Holy Land stand there. "Elah" was also used as a personal name, as we see from Genesis 36: 41 and I Kings 16: 8, probably in allusion to the fragrance and value of the tree (27). A legend states that JUDAS hanged himself on a terebinth tree (184) (see under *Cercis*). CELSIUS has endeavored to show that the Hebrew words "êl", "êlîm", "êlôn", "êlâh", and "allâh", wherever, they occur in the Scriptures, all refer to the terebinth (306). ROSENMÜLLER on the other hand, refers "êl" and "êlâh" to the terebinth, but "âllâh", "allôn", and "êlôn" to the oak (see under *Quercus*) (306).

163. Pistacia vera L.

GENESIS 43: 11— . . . Carry down the man a present, a little balm, and a little honey, spices, and myrrh, *nuts*, and almonds.
JOSHUA 13: 26—And from Heshbon unto Ramath-mizpah, and *Betonim* . . .

The Hebrew word employed in the Genesis passage cited above is "botnîm", and was translated "nuts" in the Authorized and JASTROW versions. The Greek versions say τερέβινθος, obviously an error, for "terebinthos" applies to an entirely different tree (see the preceding chapter). Most commentators now agree that these "nuts" of JACOB were pistachio nuts, *Pistachia vera*[101] (299, 306), called "batam" by the Arabs (184). MOFFATT, as well as the GOODSPEED, LEESSER, and Revised versions (the latter in a marginal note), all say "pistachio nuts", but the Douay version, curiously, says "turpentine".

The pistachio tree attains a height of 10 to 30 feet, with a spreading top. The deciduous leaves are at first velvety, later smooth except for the margins, usually odd-pinnate with one or two pairs of broadly ovate, leathery, obtuse or mucronulate leaflets, sometimes 1-foliolate (266). It is native to western Asia and Asia Minor and, contrary to the implication of Webster's dictionary

---

[101] Also known as *Pistacia trifolia* L.

(158), was introduced into Mediterranean Europe at approximately the beginning of the Christian era (299). It is found wild in many rocky parts of Lebanon and Palestine (266). It is also widely cultivated in southern Europe and Syria, especially about Damascus, for its edible nuts (266, 299, 306). WILLIAM SMITH says that "Syria and Palestine have been long famous for pistachio-trees", and the town of Batna in the Aleppo region is thought to have had its name derived from the abundance of pistachios there (306). Modern travelers, however, do not report the tree as very abundant now. TRISTRAM says that it is now "somewhat rare" in Palestine, "but the fruit is very abundant" (184). HOOKER found only two or three trees in all Palestine (306). The nut has a light-colored woody shell, and the kernel, which is light yellowish-green ("pistachio green") even when ripe, has a sweet and delicate flavor much relished in all countries where the tree grows. The kernels are eaten raw or are fried with pepper and salt (268), and form a popular dessert in the Orient and Europe. It is also used as a flavoring substance in cookery and confectionery (156), the pistachios from Aleppo being especially celebrated for their fine quality (299, 306).

It is believed that the proper place name "Betonim" in the book of Joshua was applied in allusion to the abundance of pistachio trees in that locality.

164. Platanus orientalis L.

GENESIS 30: 37—And JACOB took him rods of green poplar, and of the hazel and *chestnut tree*; and pilled white strakes in them, and made the white appear which was in the rods.
EZEKIEL 31: 8—The cedars in the garden of God could not hide him: the fir trees were not like his boughs, and the *chestnut trees* were not like his branches; nor any tree in the garden of God was like unto him in his beauty.
ECCLESIASTICUS 24: 14—I was exalted like a palm-tree in Engaddi, and as a rose plant in Jericho, as a fair olive-tree in a pleasant field, and grew up as a *plane-tree* by the water.

Although the common chestnut, *Castanea sativa* Mill.[102], is native to the Caucasus region, Asia Minor, and adjacent portions of western Asia (299) and although it has been reported by TRISTRAM from the Lebanon and Antilebanon mountains (267), it is not known to be indigenous to Palestine. The words "chestnut trees" in the Authorized Version and rabbinical writings (306) are therefore now usually regarded as an erroneous translation of the Hebrew word "'armôn" (Greek, $\pi\lambda\acute{a}\tau\alpha\nu o\varsigma$). Modern authorities believe that the tree referred to was the Oriental planetree, *Platanus orientalis*, and the Septuagint, MOFFATT, GOODSPEED, Douay, JASTROW, and Revised versions all use the words "plane", "plane-tree", or "platanus" in these verses. In fact, MOFFATT and the GOODSPEED version both regard the "pine" of Isaiah 41: 19 and the "cypress" of Isaiah 44: 14[103] as the planetree. MOFFATT also regards the "pine tree" of Isaiah 60: 13 as the planetree, but the GOODSPEED version continues to regard it as a pine.

The Oriental planetree is a massive tree to 60 or more feet tall, with a trunk often of vast circumference—sometimes to 40 feet (299)—whose outer bark peels off in sheets or scales, thus exposing a smooth whitish or yellowish inner bark. The large, alternate, palmately veined leaves are cuneate, truncate, or subcordate at the base, and more or less deeply 3-5-cleft into lanceolate lobed or dentate segments, at first woolly beneath, later smooth. The small greenish flowers are borne in 3 to 7 spherical heads on elongated pendulous terminal peduncles (267). It is common throughout Lebanon, Syria, and Palestine, ascending even to the subalpine regions (268, 293). It is, however, chiefly a tree of the plains and lowlands, growing along the edges of streams and lakes and in marshy places—as do our American species. It is said to be especially common along the shores of the upper Jordan (299). In the Bible it is usually mentioned along with willows and poplars, which grow normally only in moist low ground (184). It is very probable that planetrees were more abundant in the Holy Land in Biblical days than they are now (306).

---
[102]Also known as *Castanea vulgaris* Lam., *C. vesca* Gaertn., *C. castanea* (L.) Karst., and *Fagus castanea* L.
[103]LEESSER retains the term "cypress" in his translation of the Hebrew word "tirzah".

One of the evidences that the planetree is actually the tree referred to in the passages cited at the head of this chapter, is the fact that the Hebrew word "'armôn" is derived from a linguistic root which literally means "nakedness" (27). It is a well known fact that the outer bark of planetrees quite regularly falls off in layers or small strips, sections, or scales, leaving the trunk and larger branches quite smooth and white or yellowish and thus "naked". It is this habit of periodically shedding its outer bark—dusty, dirty, and soot-covered—which enables planetrees to thrive so well in smoky cities, where most other trees cannot survive (178, 299). The Hebrew name "'armôn" and the Arabic "dilb" are recorded as common names for this tree by POST (267). When it grows singly in the open its branches are very wide-spreading and it is a highly ornamental tree for gardens and parks (299).

The Oriental planetree is the celebrated "chenar" whose praises were sung so eloquently by the Persian poets, and is the tree which the Greeks and Romans valued so highly for its shade. Of this tree the ancients planted groves around their dwellings, since no other afforded them so fine a shade. The great beauty of the planetree in Assyria has been noted and commented upon by travelers even in modern days, and rich groves of its massive crowns and dense foliage, as well as individual trees of so great a size as to be remarkable even there, render the description of the prophet EZEKIEL as appropriate now in that land as it was when first applied.

Although the planetree grows in the mountains of Lebanon, it cannot be called a characteristic plant of that region, as can the cedar-of-Lebanon, the cypress, and the pine. It is therefore our opinion that MOFFATT and the GOODSPEED versions are in error when they consider the "pine" and "cypress" of Isaiah as planetrees. These trees are spoken of by ISAIAH as part of "the glory of Lebanon", along with the cedar and other coniferous evergreens, and it is our opinion that they were also conifers (see under *Pinus brutia* and *P. halepensis*).

Some commentators make the statement (27) that the Oriental planetree is the commonly cultivated tree of our city streets, especially the streets of London. This, however, is not correct. The cultivated or so-called London planetree is *P. acerifolia* Willd., a hybrid between the Oriental planetree and our eastern American species, *P. occidentalis* L. JOHN SMITH'S statement that *P. acerifolia* is a native of Syria is erroneous (299). The planetree of the Bible has nothing to do with the wild "plane tree" of Great Britain, which is a maple, *Acer pseudoplatanus* L. (299).

165. Populus alba L.

GENESIS 30: 37—And JACOB took him rods of green *poplar* . . .
ISAIAH 65: 3—A people that provoketh me to anger continually to my face, that sacrificeth in gardens, and burneth incense upon *altars of brick*.
HOSEA 4: 13—They sacrifice upon the tops of the mountains, and burn incense upon the hills, under oaks and *poplars* and elms, because the shadow thereof is good . . .
HOSEA 14: 5— . . . he shall grow as the lily, and cast forth his roots as *Lebanon*.

The poplars of Palestine, where there are said to be 3 or 4 native species (267, 299), grow commonly with willows and planetrees along streams and the margins of bodies of water. Dr. EVENARI, however, considers only *Populus euphratica* as truly indigenous there now. Be that as it may, Palestinian poplars are mostly fast-growing, especially when young, producing long straight shoots which might well have furnished the rods used by JACOB (299). Some commentators have thought that JACOB's rods were shoots of the storax tree, *Styrax officinalis* (which see), because the Hebrew words here employed are "libneh" or "livneh" and "livenim" (Greek, λεύκη; dictionary translation "the white poplar"), meaning "white", and the leaves of the storax are pale or whitish and its flowers are white. Some Greek versions use the word στύραξ (184), and the Revised Version, in a marginal note, suggests "storax tree" in Genesis 30: 37 (27). The leaves of the white or silver poplar, *Populus alba*, however, are also very white beneath, and its young shoots would suit JACOB's purpose admirably. The storax tree, on the other hand, is of scrubby growth, with short branches, and not at all suited to his purpose (299), and certainly could not be ranked with the oak and terebinth as a shade-producing tree under which altars might be built (27). POST states that the white poplar is common in wet places in Syria, Lebanon, Palestine, and Sinai (265), although in a later publication (267) he states that in Palestine it is now only cultivated. It is so common about Damascus that, along with the black poplar, *P. nigra* L., it forms quite a forest (268). It attains a height of 30 to 60 or more feet, with spreading branches. The deciduous leaves are shiny-green above, snowy white-woolly beneath, varying from rounded heart-shaped to ovate or ovate-oblong, sinuate-dentate, angulate, or lobed along the margins. The flowers are produced in catkins, appearing before the leaves. The young buds are covered with a resinous varnish and give off a sweet balsamic odor in the spring. On being bruised, a fragrant resin is produced. This may well have been the "incense" burned by EPHRAIM in the groves of poplars, although the "incense" of the Bible is usually regarded as having been frankincense (see under *Boswellia thurifera*). CELSIUS and most modern authorities have no doubt that the white poplar is the "poplar" of the Old Testament (306).

Because of the dense shade which it produces, the white poplar was always extensively cultivated in the Holy Land, and still is today (along with about 5 other kinds of poplar) (267). It is stated by numerous commentators that the altars of the various pagan religions which existed from time to time in that part of the world were usually erected on the top of a hill and in the shade of a white poplar or in a poplar grove. MOFFATT translates Isaiah 65: 3: "a people who provoke me to my face continually, by sacrificing in their groves, and burning incense under the *white poplars*." Hosea states that these pagan altars were built in the shadow of oaks, poplars, or terebinths (Hosea 4: 13).

Modern opinion holds that the Authorized Version's translation of Hosea 14: 5—"cast forth his roots as Lebanon"—is erroneous, the Hebrew word being "livneh", not "l'vanon". MOFFATT says, instead: "and strike roots down like a *poplar*", while the GOODSPEED version is: "And his roots will spread like the *poplar*." It is, of course, well known that poplar roots grow rapidly toward any water in the vicinity. Many cities have ordinances against the planting of poplars along their streets because of the fact that their roots so quickly seek out and clog up underground water facilities and sewers.

The story of JACOB and the rods is most interesting in that it shows so clearly the existence of the belief in the effect of prenatal influences on pregnant females as far back as 1745 B.C. The description of this naive belief as given in the GOODSPEED version of Genesis 30: 37-39 is worthy of repetition here: "Then JACOB procured some fresh boughs of *poplar*, almond, and plane, and peeled white stripes in them, thus laying bare the white on the boughs. He then placed the boughs which he had peeled in front of the sheep in the troughs, that is, the watering-troughs, where the sheep came to drink. Since they bred when they came to drink, the sheep bred among the boughs, and so had lambs that were striped, speckled, and spotted."

In classic Greece (298) the groves of Academus were of poplar trees. HERCULES, after being bitten by a venomous snake, found an antidote in poplar leaves, and from that time on the poplar was considered sacred to him. Another legend in Greek mythology states that after HERCULES had found the oxen of Geryon and had slain CACUS, the giant, he crowned himself with a wreath of poplar leaves. His next labor took him to Hades, where the upper surface of the poplar leaves was blackened by the smoke and fire, while the under surface remained fresh from the perspiration of his brow. Since then, the legend states, poplar leaves have been dark above and silver beneath (298). Another Greek myth states that the Heliades, mourning their slain brother PHAETON, were transformed into poplar trees and their tears into amber. Still another beautiful Greek story tells of GANYMEDE, cupbearer of the gods, after falsely accusing many other trees, finally came to the poplar, whose branches then hung low to the ground, and demanded the silver spoons stolen from ZEUS. The poplar lifted its branches to show that nothing was concealed thereunder, whereupon the stolen silver fell to the ground and was seized by GANYMEDE. The poplar trembled in apprehension and its leaves became blanched from its terror. In punishment ZEUS sentenced it to forever after hold its branches upright into the sky and to bear silvery trembling leaves (298).

166. Populus euphratica Oliv.

LEVITICUS 23: 40—And ye shall take you on the first day the boughs of goodly trees, branches of palm trees, and the boughs of thick trees, and *willows* of the brook . . . [104]
LEVITICUS 26: 36—And upon them that are left alive of you I will send a faintness into their hearts in the lands of their enemies; and the sound of a *shaken leaf* [105] shall chase them; and they shall flee, as fleeing from a sword; and they shall fall when none pursueth.
II SAMUEL 5: 23-24— . . . but fetch a compass behind them, and come upon them over against the *mulberry trees*. And let it be, when thou hearest the sound of a going in the tops of the *mulberry trees*, that then thou shalt bestir thyself . . .
I CHRONICLES 14: 14-15— . . . turn away from them, and come upon them over against the *mulberry trees*. And it shall be when thou shalt hear the sound of a going in the tops of the *mulberry trees*, that then thou shalt go out to battle . . .
PSALMS 84: 6—Who passing through the valley of *Baca* make it a well.
PSALMS 137: 2—We hanged our harps upon the *willows* in the midst thereof.
ISAIAH 7: 2— . . . And his heart was moved, and the heart of his people, as the *trees of the wood* are moved with the wind.

The Hebrew word translated "mulberry trees" in the Authorized and JASTROW versions is "becâîm" (Greek, ἄπισι, συκάμιγος, or συκαμίνος; dictionary translation "mulberry tree"). Some versions of the Bible retain the Hebrew word untranslated (306). Older commentators and most of the rabbinical writers thought that a true mulberry, *Morus nigra* (which see)—the tree called "sycamine" in the New Testament—was the "mulberry" of the Old Testament. This, however, is not at all probable and is not believed today (306), because mulberry leaves are soft-textured and are borne on a firm round petiole, and therefore do not make a rustling sound when stirred by the breeze. Modern commentators agree that it was a species of aspen that was referred to in these passages, and of the several possible species the Euphrates poplar, aspen, or balsam-tree, *Populus euphratica*, is the most likely. This is a tree 30 to 45 or more feet tall, with spreading branches and puberulent twigs. The leaves are glabrous and glaucous, stiff-textured, borne on weak flattened petioles. The leaf-blades on young plants and the sterile lower branches of older plants are linear to lanceolate and entire-margined, those of fertile branches are elliptic, oblong, ovate, rhomboid, or deltoid in shape, cuneate to truncate at the base, and often more or less irregularly dentate (267). The Euphrates aspen is found along rivers and on streambanks throughout the area from Syria through Palestine to Arabia Petraea, especially in the Jordan valley (95). POST is of the opinion that this species was the "mulberry" of II Samuel and I Chronicles, and in this view we concur (27, 267). ROYLE, TRISTRAM, JONES, SMITH, and SHAW all favor *P. tremula* L. (178, 184, 293), but, according to POST, this tree occurs in the region only in cultivation.

The Leviticus reference to "willows" (Hebrew, "aravim") is still regarded by MOFFATT, JASTROW, and the Douay and GOODSPEED versions as referring to true willows or "water-willows", but most other modern commentators refer them to *Populus euphratica*. The famous "willows" on which the Jews hung their harps as they wept at the thought of suffering Zion, are now believed by most authorities to have been Euphrates aspens (299). Both MOFFATT and the GOODSPEED version say "poplars" in this passage, but the Douay and JASTROW versions still use the word "willows".

Joshua 15: 8 is sometimes also cited as another reference to "mulberry", that is, aspen, but on what basis is not evident to us. ROSENMÜLLER and the Douay version follow the Septuagint in identifying the "mulberry trees" of II Samuel and I Chronicles as "pear trees". The common pear, *Pyrus communis* L., native to Russia, is now cultivated in the Holy Land and even occurs

---
[104] The Hebrew for "willow(s) of the brook" is "arvay nachal".
[105] The Hebrew for "shaken leaf" is "nidaf aleh".

there subspontaneously, but could not have been known in Biblical days. The Syrian pear, *P. syriaca* Boiss., occurs on rocky hillsides throughout the area, but does not fill the requirements of our texts. Its petioles are not weak and flattened, so it does not produce any decided rustling sound; it does not make up forests in which an army could hide; nor does it grow along streams and lakes like a willow.

There is some argument concerning the passage in Psalms 84: 6. PRATT comments that "Baca" is practically the same word as that used for the aspen, and therefore the "valley of Baca"[106] probably refers to an abundance of aspens there. Dr. ROYLE refers the Hebrew "bâcâ" to the so-called "gnat-tree", known to the Arabs as "shajrat-al-bak" and identified by him only as a species of poplar (306). CELSIUS and PARKHURST believed that the Hebrew "bâcâ" or "bâchâ" is identical with the Arabic word "baca" applied, according to Sprengel (306), to *Commiphora opobalsamum* (which see), a shrub yielding an odoriferous gum that drops from its branches like tears, thus perhaps accounting for the literal translation of the word, which is "weeping" (306). This interpretation is rejected by us because *Commiphora opobalsamum* is native only to Arabia, particularly the southwestern portion known as Yemen, and does not occur at all in Palestine (306). WILLIAM SMITH states categorically that "the explanation given by ROYLE . . . is untenable; for the Hebrew 'bâcâ' and the Arabic 'baka' are clearly distinct both in form and signification, as is evident from the difference in the second radical letter in each word" (306). MOFFATT invents for "the valley of Baca" the new proper name "Weary-glen". Yet it must be recalled, in this connection, that the leaves of aspens, being borne on flat instead of normal round petioles, always hang down or droop, and are continually swaying back and forth, like wailing women, so that here, too, the application of a word meaning "weeping" is not at all inapplicable. Speaking of the aspen leaves, JOHN SMITH says that they are "obliquely attached by a slender footstalk, which allows them to move freely in a quivering manner, even when the air is still. Their rustling against one-another gives an audible sound which may be heard at some distance, and explains the going 'at the top of the mulberry tree' " (291).

Although no one seems ever to have made the suggestion before, it seems to us that the reference to the noise made by a "shaken leaf" in Leviticus 26: 36 also applies clearly to the aspen. The Authorized Version includes a marginal note giving "driven" as an alternate translation for "shaken", and this is adopted by JASTROW, MOFFATT, and GOODSPEED. The Douay version employs the different term "flying".

In early Christian legendry it is stated that the cross on which JESUS was crucified was made of aspen wood, and that aspen trees everywhere shuddered when the nails were driven into the wood and the sacred blood gushed upon the cross, and have shuddered or trembled ever since. Another legend states that JESUS himself was forced to fashion the cross out of poplar wood, and for this reason poplars are considered sacred in some lands and some French-Canadian lumberjacks even today refuse to cut poplar wood (298). Another legend tells us that when the Holy Family was fleeing from HEROD it passed through a woods. All the other trees bent their heads in adoration as the procession went by except the poplar, which held itself aloof and would not bend its head. The infant JESUS looked upon the stubborn tree reprovingly, whereupon it was seized with uncontrollable mortification and began to tremble, and it has not ceased trembling since (298). The poplar is another of the trees on which ancient legends say that JUDAS ISCARIOT hanged himself (see under *Cercis siliquastrum*, *Ficus carica*, and *Pistacia terebinthus* var. *palaestina*).

167. Prunus armeniaca L.
(FIGURE 77)

GENESIS 2: 9 & 17— . . . and the *tree* of knowledge of good and evil . . . But of the *tree* of knowledge of good and evil, thou shalt not eat of it . . .

---

[106]LEESSER translates this phrase as "the valley of weeping."

GENESIS 3: 6—And when the woman saw that the *tree* was good for food, and that it was pleasant to the eyes, and a *tree* to be desired to make one wise, she took of the fruit thereof, and did eat, and gave also unto her husband with her; and he did eat.
JOSHUA 15: 53—And Janum, and Beth-*Tappuah*, and Aphekah.
JOSHUA 17: 8—Now Manasseh had the land of *Tappuah*: but *Tappuah* on the border of Manasseh belonged to the children of Ephraim.
I CHRONICLES 2: 43—And the sons of Hebron; Korah, and *Tappuah* . . .
PROVERBS 25: 11—A word fitly spoken is like *apples* of gold in pictures of silver.
SONG 2: 3 & 5—As the *apple tree* among the trees of the wood, so is my beloved among the sons. I sat down under his shadow with great delight, and his fruit was sweet to my taste . . . Comfort me with *apples* . . .
SONG 7: 8— . . . the smell of thy nose like *apples*.
SONG 8: 5— . . . I raised thee up under the *apple tree* . . .
JOEL 1: 12—The vine is dried up, and the fig tree languisheth; the pomegranate tree, the palm tree also, and the *apple tree*, even all the trees of the field, are withered . . .

The "apple" of the Bible constitutes another of the most perplexing problems of Biblical botany. There has been a vast amount of discussion and argument about its identity (299). The Hebrew word used is "tappûach" (Greek, μῆλον). From the texts quoted above we learn that the "apple" tree of the Scriptures was a tree which afforded a pleasant shade. Its fruit was enticing to the sight, sweet to the taste, imparting fragrance, with restorative properties, and of a golden color, borne amid silvery leaves (184, 306). It was a tree or fruit which also gave its name to two towns or villages in Palestine, one in the highlands of Judah, the other in the territory of Ephraim, as we see from the Joshua references. A descendant of CALEB also bore the name, according to I Chronicles 2: 43.

Many writers have argued in favor of our common apple, *Malus pumila* Mill.[107]. Dr. THOMSON, for instance, remarks that the entire area about the once famous Philistine city of Ascalon is now filled with apple orchards, and that these apples are extremely large and delicious. POST, however (265, 266), states that while the borders of the native area of the apple—the Caucasus region—touch on the area of the Levant he has never found any wild apples in all of Palestine, Syria, or Sinai. Most botanists are agreed that the common apple is not a native of Palestine, but is a comparatively recent introduction there. Anyhow, it is only comparatively recently that the poor wild fruits of the common apple have been so improved by selection and cultivation as to bring them to a form which would fit the description in the Biblical quotations (299). In the wild state its fruit is small and acid—"wretched woody fruit" according to TRISTRAM—and it is certainly very improbable that it could be the fruit referred to by the Old Testament writers in such glowing terms. HORNER disagrees with THOMSON and states very definitely that modern commentators agree that the Hebrew word "tappûach" should not have been translated "apple" (178). She goes so far as to assert that the common apple is not found at all in Palestine, or, if it does occur there, it is just barely able to survive the excessive heat. Numerous other writers state the same things. This direct contradiction between the statements of HORNER and THOMSON is just one example of the hundreds of similar contradictions in the literature on this subject. It is now believed that THOMSON was referring to quinces, not apples (299).

Some other writers have supposed that the "apples of gold" were oranges, *Citrus sinensis* (L.) Osbeck[108], but that fruit is not known to have been cultivated in Palestine in SOLOMON's time (299). It seems, rather, to be a very recent importation (146). BORAH is one of the very recent writers on this subject who still holds out for the orange or the apple (71). The orange, however, while now one of the principal exports of Palestine, is not indigenous there. It is a native of China and was not known in the Holy Land in Biblical days. The Seville or bitter orange, *C. vulgaris* Risso [109], also suggested, is a native of eastern India, not introduced into the Holy Land until 1000 A.D. (266). It is now widely employed there as a grafting-stock for the common orange.

---

[107]Also referred to as *Pyrus malus* L., *Malus communis* Poir., *M. malus* (L.) Britton, and, incorrectly, *M. sylvestris* Mill.
[108]Also known as *Citrus aurantium* var. *sinensis* L.
[109]Also known as *Citrus aurantium* L., *C. bigaradia* Loisel., and *C. amara* Link.

Many writers identify the "apples" of Scripture with the citron, *C. medica* L. Dr. ROYLE says "The rich color, fragrant odor, and handsome appearance of the citron, whether in flower or in fruit, are particularly suited to the passages of Scripture mentioned above" (306). But here, again, there are the usual contradictions. SMITH (299) says that the citron gives a delightful shade and so might well fit the description in Song 2: 3, while THOMSON says: "The tree is small, slender, and must be propped up. Nobody ever thinks of 'sitting under' its 'shadow', for it is too small and straggling to make a shade" (319). Be that as it may, modern writers do not seriously consider the citron as a Biblical plant. While its fruit is pleasant-scented, it is very acid to the taste and is hard and indigestible, so that it cannot be eaten except when made into preserves. It would therefore not meet the requirements of the sweet fruit as described in the Biblical texts, although THEOPHRASTUS called citrons "Median and Persian apples" and reported that they were used to sweeten the breath in the fashion indicated in the Song of Solomon (299). PRATT is convinced that the "apples" of the Bible were citrons (268, 299). He also reports the popular legend that it was the citron, or, at least, some species of *Citrus*, which was gathered and eaten by EVE in the garden of Eden (Genesis 3: 1-6). The fruit of the "tree of knowledge" is by us in America usually referred to as an apple, due, no doubt, to the influence of Medieval and Renaissance artists who so depicted it. The shaddock, grapefruit, or pomelo, *C. grandis* (L.) Osbeck[110], is still sold in London stores under the name of "forbidden fruit". The citron, however, is a native of India, and the shaddock is from the Fiji and other Pacific islands. The latter was certainly not known in Biblical days, and it is doubtful if the former was either (266).

As has been pointed out by previous writers (172c), the appearance of the "apple" in the famous story of the so-called "Fall of Man" in the Garden of Eden is an example of poetic license. The word does not appear in the Mosaic record, but as the poet, JOHN MILTON, had to sing of

> "The fruit
> Of that forbidden tree whose mortal taste
> Brought death into the world, and all our woe,"

he found himself forced to be more explicit in his description and more specific in his nomenclature. In 1665, in his "Paradise Lost", the identification of the apple with this "forbidden fruit" apparently begins. In the address of the serpent to the "Empress of the World", resplendent EVE, we read:

> "On a day, roving the field, I climbed
> A goodly tree far distant to behold,
> Loaden with fruit of fairest colours mixed,
> Ruddy and gold. I nearer drew to gaze,
> When from the boughs a savoury odour blown,
> Grateful to appetite, more pleased my sense
> Than smell of sweetest fennel, or the teats
> Of ewe or goat dropping with milk at even,
> Unsucked of lamb or kid, that tend their play,
> To satisfy the sharp desire I had
> Of tasting those fair Apples, I resolved
> Not to defer; hunger and thirst at once—
> Powerful persuaders — quickened at the scent
> Of that alluring fruit, urged me so keen."

Other authorities have believed (299) that the "apples" of the Old Testament were quinces, *Cydonia oblonga* Mill.[111] This tree is, indeed, common in Palestine, although chiefly in the cultivated state, perhaps wild in the northern parts of Syria. It is a native of northern Persia and Asia Minor (266). The fruit is yellowish and highly fragrant. The fragrance of the quince caused it to be held in high regard by the ancients. "Its scent," says an Arabic author, "cheers my soul, renews my strength, and restores my breath" (306). It was

---

[110] Also known as *Citrus aurantium* var. *grandis* L., *C. aurantium* var. *decumana* L., *C. decumana* (L.) Murr., and *C. maxima* (Burm.) Merr.

[111] Also known as *Pyrus cydonia* L., *Cydonia vulgaris* Pers., and *C. cydonia* (L.) Karst.

regarded as sacred to VENUS in classical times. However, it is not sweet to the taste, but exceedingly acrid, and so does not meet the requirements of the texts (27, 184, 299).

ETOC claims (115a) that the most ancient of trees of Biblical mention is the one in the Garden of Eden. It cannot, he says, readily be identified from ancient bas-reliefs and similar records because all of these designs are too highly conventionalized. He suggests *Asclepias acida* Roxb., which was commonly called "soma" by the ancient Aryas. This plant is nowadays recognized under the name of *Sarcostemma brevistigma* Wright & Arn. It is native to India and Burma and may therefore safely be eliminated as a possibility.

The descendants of the Portuguese in Ceylon consider the "divi ladner", *Tabernaemontana alternifolia* L., to have been EVE's "apple". This tree is a native of Ceylon, which island they regarded as the site of the Garden of Eden. The fruit is very tempting in appearance, with more or less the shape of an apple minus one bite! It is highly poisonous (25a).

The only fruit which seems to meet all the requirements is the apricot, *Prunus armeniaca*[112] (27, 184, 299). TRISTRAM says that "the apricot is most abundant in the Holy Land, and meets all the requirements of the context, and is the only tree that does so" and "everywhere it is common, and perhaps, with the single exception of the fig, the most abundant fruit of the country. In highlands and lowlands alike, by the shores of the Mediterranean, and on the banks of the Jordan, in the nooks of Judaea, under the heights of Lebanon, in the recesses of Galilee, and in the glades of Gilead, the apricot flourishes and yields a crop of prodigious abundance. Many times have we pitched our tents in their shade. 'I sat down under his shadow with great delight, and his fruit was sweet to my taste.' 'The smell of thy nose (shall be) like apples.' There can scarcely be a more deliciously perfumed fruit than the apricot, and its branches laden with its golden fruit may well be compared to 'apples of gold', and its pale leaves to 'pictures of silver'" (299). Although POST maintains that the apricot is a native of China (266), TRISTRAM asserts that it was originally from Armenia and was introduced into Palestine at about the same time as the grapevine, which was before the time of NOAH (about 2950 B. C.). Certainly it is now widely cultivated there and probably has been so ever since Biblical days. Apricots in Cyprus are still known as "golden apples", which is the literal translation of their modern Greek name. GROSER sums up the situation by saying: "The balance of probability therefore seems to favour the apricot as the 'apple' of the Old Testament" (146).

Some writers cite Leviticus 23: 40 as one of the verses referring to the apricot, probably because the phrase "boughs of goodly trees" is rendered "fruits of goodly trees" in marginal Bibles. MOFFATT, however, does not so translate it and the context of the verse seems to call for the foliage of certain trees to be cut, not the fruit, so it does not seem at all probable that the apricot is involved in this reference. According to modern Jewish authorities the phrase is left "hadar" in Jewish Old Testaments and this is the Hebrew word for the Jewish "Holy Citron" for which the Hebrew word now is "ethrog" or "etrog". This is *Citrus medica* var. *lageriformis* Roem.[113] "Hadar" (Hebrew, "chaydayd", a name meaning "beautiful or splendid") is recorded in Genesis 25: 15 as one of the sons of ISHMAEL. SWINGLE states that "etrog" is the "name applied by the Jews to a citron ... which is imported and used by them for religious ceremonies connected with the Feast of the Tabernacles. The 'etrog' and the 'lulav' (palm leaf with myrtle and willow branches) are carried and waved during the services, especially those of thanksgiving. Since the time of the anti-Jewish demonstrations in Corfu in 1891, the etrog is imported more largely from Palestine than from that island. In addition to the use of the etrog by orthodox Jews for religious ceremonials, the natives of Palestine make salads of the fruit" (312).

The apricot is a round-headed, reddish-barked tree growing to 30 feet tall.

---

[112]Also called *Armeniaca vulgaris* Lam.
[113]Also called *Citrus medica* var. *cucurbitina* Risso & Poit. and var. *cylindrica* Hort.

Its leaves are ovate, acuminate, somewhat heart-shaped at the base, pale and pubescent on the veins beneath, appearing after the pinkish or nearly white flowers. The well-known oval orange-colored fruit is somewhat intermediate in flavor between a peach and a plum (156).

168. Pterocarpus santalinus L. f.

I KINGS 10: 11-12—And the navy also of HIRAM, that brought gold from Ophir, brought in from Ophir great plenty of *almug trees*, and precious stones. And the king made of the *almug trees* pillars for the house of the Lord, and for the king's house, harps also and psalteries for singers: there came no such *almug trees*, nor were seen unto this day.

II CHRONICLES 9: 10-11—And the servants also of HURAM, and the servants of SOLOMON, which brought gold from Ophir, brought *algum trees* and precious stones. And the king made of the *algum trees* terraces to the house of the Lord, and to the king's palace, and harps and psalteries for singers: and there were none such seen before in the land of Judah.

The use of the word "algum" in the second of these references seems to be an error for "almug", as is suggested in a marginal note in the Authorized Version, for the "algum" tree is said in II Chronicles 2: 8 to be a native of Lebanon, not of Ophir. It is not definitely known just where the Biblical "Ophir" was located geographically, but it seems to have been the name of some country or countries lying beyond the Straits of Babelmandeb (299). It has been variously identified as lying in Africa, India, Arabia, Ceylon, the Malay Peninsula (156), Armenia, the Molucca Islands, the Crimea, and even Peru (306)! All the merchandise coming from Ophir was transported by ships up the Red Sea and landed at SOLOMON's port of Ezion-geber (near Elath), at the head of the Gulf of Akaba, and from there was conveyed direct to Jerusalem (299). Modern authorities are divided into three groups in regard to the location of Ophir: those who place it in Arabia, those who place it in India, and those who place it in western Africa, near Mozambique. The evidence and arguments in favor of each theory are well summarized in SMITH's "Dictionary of the Bible" (306).

The identification of the "almug" and "algum" trees of the two references cited above bears a definite relation to the problem of the location of Ophir, since the wood of "almug trees" is stated to have been one of the main imports from that mysterious land. Some commentators have thought (299) that the "almug" tree was the white sandalwood, *Santalum album* (see under *Aquilaria agallocha*). This tree has also been suggested for the "aloes" of Psalms 45: 8, Song 4: 14, and Proverbs 7: 17. It is a tree growing to 20 or 30 feet tall, native to the mountains of the Malabar coast and the islands of the Indian Archipelago. Mostly, however, it does not grow to its maximum height, but remains a low-growing tree resembling a privet. The flowers open pale-yellow, fading to rusty-purple, and are not fragrant, but the wood has been valued highly in the Orient since earliest times on account of its delightful odor. It is employed in India and China for making burnt sacrifices to idols, and it is made into necklaces, fans, boxes, and many other articles of domestic use. Insects are repelled by its strong odor and will not attack it. Some writers suppose that the intense liking for fragrant odors possessed by the Hebrews and other peoples of the Orient would naturally cause this wood to be much in demand for the making of the lyres and lutes on which were to be sounded the high notes which should lead the voices of Israel in praise of JEHOVAH. In China it is still used for the making of musical instruments. In some parts of the East it is powdered, mixed with water, and sprinkled over visitors.

The normally small size of the white sandalwood tree, however, makes it difficult to believe that this was the "almug" of which the "pillars" and "terraces" ["pilasters" and "balustrades" (MOFFATT) or "raised platforms" (GOODSPEED) according to modern versions] of Solomon's palace and temple were made. Its chief value was its odoriferous property, not its lumber. Other writers have suggested the deodar, *Cedrus deodara* Loud., the Sanskrit and Hindu "timber of the gods", but this is a Himalayan tree and probably out of the question. The Septuagint translates the word "almug" as "wrought-wood"; the Greek versions say ξύλα πελεκητά (literally "a hewer of wood")

(184) or ξύλα πεύκινα (literally "made of fir wood") (27). The MOFFATT, JASTROW and GOODSPEED versions all say "sandalwood" or "sandal-wood", without specifying which type. TRISTRAM definitely identifies it as the red sandalwood, *Pterocarpus santalinus*[114] (184), and in this he is followed by JONES, SMITH, and other modern authorities (27, 306). Most of these commentators, however, along with GROSER (146) and HURLBUT and McCLURE (179), regard the "almug" tree as identical with the "algum" of II Chronicles 2: 8 (306). This is not at all probable, even if the words do resemble each other, as MOFFATT has pointed out, for the "algum"[115] is said to have come from Lebanon. WILLIAM SMITH believes that it is still possible for the "almug"[116] and "algum" to have been the same tree if the mention in II Chronicles be regarded as a later and erroneous interpolation by some scribe or on the assumption that King HIRAM had imported the trees from Ophir and later re-shipped them to Solomon from Tyre along with the native Lebanese cedars and firs (306). This reasoning seems a bit far-fetched and we prefer to follow JOHN SMITH in regarding the "algum" as the eastern savin, *Sabina excelsa* (which see). MOFFATT regards it as the cypress, *Cupressus sempervirens* var. *horizontalis* (which see). Some commentators have also suggested that the "almug" was the same tree as the one that produced the "thyine-wood" of Revelation 18: 12, but the latter is by us regarded as *Tetraclinis articulata* (which see). JOSEPHUS gives an elaborate description of the wood imported by SOLOMON from Ophir, but calls it "pine" (see under *Pinus brutia*). HENSLOW (172a) regarded the "almug" as yew.

The red sandalwood or "red saunders-wood" is a large leguminous tree indigenous to India and Ceylon. Its wood is hard and very heavy, of a red or garnet color, takes on a fine polish, and would be very well suited to the purposes for which SOLOMON required it (299, 306). It is still used today for the manufacture of lyres and other musical instruments (184, 299).

169. Punica granatum L.
(PAGE 199; FIGURE 38; FIGURE 76)

EXODUS 28: 33-34—And beneath upon the hem of it thou shalt make *pomegranates* of blue, and of purple, and of scarlet, round about the hem thereof; and *bells* of gold between them round about: A golden *bell* and a *pomegranate*, a golden *bell* and a *pomegranate*, upon the hem of the robe round about.
EXODUS 39: 24-26—And they made upon the hems of the robe *pomegranates* of blue, and purple, and scarlet, and twined linen. And they made *bells* of pure gold, and put the *bells* between the *pomegranates* upon the hem of the robe, round about between the *pomegranates*; a *bell* and a *pomegranate*, a *bell* and a *pomegranate*, round about the hem of the robe to minister in; as the Lord commanded MOSES.
NUMBERS 13: 23—... and they brought of the *pomegranates*, and of the figs.
NUMBERS 20: 5—And wherefore have ye made us to come up out of Egypt, to bring us in unto this evil place? It is no place of seed, or of figs, or of vines, or of *pomegranates*; neither is there any water to drink.
NUMBERS 33: 19-20—And they departed from Rithmah, and pitched at *Rimmon*-parez. And they departed from *Rimmon*-parez...
DEUTERONOMY 8: 8—A land of wheat, and barley, and vines, and fig trees, and *pomegranates*...
JOSHUA 15: 32—And Lebaoth, and Shilhim, and Ain, and *Rimmon*...
JOSHUA 19: 7 & 13—Ain, *Remmon*, and Ether, and Ashan; four cities and their villages... And from thence... goeth out to *Remmon*-methoar to Neah.
JUDGES 20: 45—And they turned and fled toward the wilderness unto the rock of *Rimmon*...
I SAMUEL 14: 2—And SAUL tarried in the uttermost part of Gibeah under a *pomegranate tree* which is in Migron...
II SAMUEL 4: 2—... the sons of *Rimmon* a Beerothite...
I KINGS 6: 32—... and he carved upon them carvings of cherubims and palm trees and *open flowers*.
I KINGS 7: 18 & 20—And he made the pillars, and two rows round about upon the one network, to cover the chapiters that were upon the top, with *pomegranates*... And the chapiters upon the two pillars had *pomegranates* also above, over against the belly which was by the network: and the *pomegranates* were two hundred in rows round about the other chapiter.
II KINGS 5: 18—In this thing the Lord pardon thy servant, that when my master goeth into

---

[114]Also called *Lingoum santalinum* (L. f.) Kuntze.
[115]The Hebrew is "algoomim": LEESSER interprets this as "sandal-wood" or "sandel-wood", with both spellings used in the text.
[116]The Hebrew is "almoogim" and LEESSER translates this as "sandel-trees."

the house of *Rimmon* to worship there, and he leaneth on my hand, and I bow myself in the house of *Rimmon*: when I bow down myself in the house of *Rimmon* . . .
II KINGS 25: 17—. . . and the wreathen work, and *pomegranates* upon the chapiter round about, all of brass . . .
II CHRONICLES 3: 16—. . . and he made an hundred *pomegranates*, and put them on the chains.
II CHRONICLES 4: 13—And four hundred *pomegranates* on the two wreaths; two rows of *pomegranates* on each wreath . . .
SONG 4: 3 & 13—Thy lips are like a thread of scarlet, and thy speech is comely: thy temples are like a piece of *pomegranate* within thy locks . . . Thy plants are an orchard of *pomegranates*, with pleasant fruits . . .
SONG 6: 7 & 11—As a piece of a *pomegranate* are thy temples within thy locks . . . I went down into the garden of nuts to see the fruits of the valley, and to see whether the vine flourished, and the *pomegranates* budded.
SONG 7: 12—Let us get up early to the vineyards; let us see if the vine flourish, whether the tender grape appear, and the *pomegranates* bud forth: there will I give thee my loves.
SONG 8: 2— . . . I would cause thee to drink of spiced wine of the juice of my *pomegranate*.
JOEL 1: 12—The vine is dried up, and the fig tree languisheth; the *pomegranate tree*, the palm tree also, and the apple tree, even all the trees of the field, are withered . . .
HAGGAI 2: 19—Is the seed yet in the barn? yea, as yet the vine, and the fig tree, and the *pomegranate*, and the olive tree, hath not brought forth: from this day will I bless you.

The Hebrew word translated "pomegranate" in all the above-mentioned passages is "rimmôn" (Greek, ῥοά, ῥοιά, or κώδων, meaning "a bell"), and there is no doubt whatever concerning the correctness of the identification. It is probable that the proper names of the towns Remmon and Rimmon in the books of Numbers, Joshua, and Judges refer to the abundance of pomegranates at those localities (27, 184). "Rimmon-parez" ("Remmomphares" in Douay, "Rimmon-perez" in GOODSPEED and MOFFATT) of Numbers 33: 19-20 means literally "pomegranate of the breach" (27). The "Remmon" and "Remmon-methoar" of Joshua 19: 7 & 13 are better rendered "En-Rimmon" and "Rimmon" by MOFFATT and "En-rimmon" and "Rimmonah" by the GOOD-SPEED version, since "methoar" is not part of the place name. They doubtless have a similar derivation. In II Samuel 4: 2 we see that it was also used as a personal name. The Rimmon of II Kings 5: 18, however, was an Assyrian deity worshipped at Damascus, identified by modern authorities with RAMMAN (the Babylonian ADAD and Syrian HADAD), god of thunder, wind, and storms, symbolizing in speculative theology retributive justice (158). Many commentators feel that the pomegranate must have been connected with this worship, perhaps as a symbol of the deity. The Arabic name for the pomegranate is still "rummân" (266, 306).

The pomegranate tree, *Punica granatum*[117], is usually small and bush-like, but may occasionally become a large branching shrub or a small tree reaching a height of 20 or 30 feet (293, 299). The opposite or alternate branches are often thorny. The oblong-lanceolate leaves are entire-margined and are borne in either opposite, alternate, or whorled arrangement on the stems and are quickly shed (266). The showy bell-like flowers are red, yellow, or white in color, usually scarlet. The globular fruit is as large as an orange or a medium-sized apple, has a hard rind of a bright-red or yellowish color when ripe, and is surmounted by the persistent lobes of the calyx, which resemble a rosette or crown. Beneath the rind is a crimson, agreeably acid-tasting, juicy pulp in which are embedded many red seeds (158). The flowers of the pomegranate undoubtedly served as patterns for the "golden bells" (Hebrew, "pa'amon") and "open flowers" (Hebrew, "tzitzim p'turay") embroidered on the temple robes, and the fruit served as model for other ornaments (178). The erect calyx-lobes on the fruit served as inspiration for SOLOMON'S crown and, incidentally, for all crowns from that time on.

The pomegranate is a native of Asia, probably from northern India to the Levant, but has been cultivated since prehistoric times and is now common in the Holy Land, Egypt, and along both shores of the Mediterranean. It is abundantly cultivated in Palestine and occurs wild in Syria, Lebanon, and Gilead (266). In the Bible it is listed as one of the pleasant fruits of Egypt (Numbers 20: 5) and one of the promised blessings of Palestine (Deuteronomy 8: 8). The pulp of the fruit has been used extensively since the days of

---

[117]Misspelled *"Pumica granatum"* by Mrs. BARNEVELD (47).

SOLOMON for making cooling drinks and sherbets, and is also eaten raw. The astringent rind of the unripened fruit yields a red dye and has been used in medicine and for tanning red Morocco leather (101, 158). The flowers also yield a red dye. Pomegranate fruits in their native haunts attain a fine sweetness which makes them highly valued in those hot climates. A spiced wine is made from pomegranate juice (Song 8: 2) and many Moslem sherbets even today owe their flavor to this juice. The soft seeds are also eaten, sprinkled with sugar, or, when dried, as a confectionery. It was the Moors who introduced into Spain from Africa the method of tanning leather with pomegranate rinds, and made Cordova famous for its fine leather.

In very early times the pomegranate came to be regarded as a sacred plant (268). Because of its large number of seeds, it came to be regarded as a symbol of fertility. In Egypt it was held sacred, and its characteristic fruit is easily recognized in Egyptian inscriptions and sculpture (306). In Persia it adorned the head of the royal sceptre. In Rhodes its blossoms formed part of the royal coat of arms. An ancient representation of JUPITER shows him bearing a pomegranate in his hand. In view of all this, it is not at all surprising to find that in the temple of SOLOMON it adorned the trellis-work at the top of the pilasters, and, in blue, purple, and scarlet embroidery, the skirts of the priestly robes or ephods (126). Dr. EVENARI describes a "wonderful pomegranate ornament on the walls of an old Jewish temple at Capernaum in which JESUS is said to have preached." The beauty of the delicately crimson fruits caused them to be used in song and poetry in comparison with the beauty of young blushing cheeks much as we use the peach today when we say that someone's cheeks "had the bloom of ripe peaches".

The statement made by JONES (184) that blood oranges are produced by grafting the branch of an ordinary orange on to a pomegranate tree is false.

Ancient legends say that the pomegranate was the "tree of life" in the garden of Eden, and from this belief it became the symbol of hope of eternal life in early Christian art. In Turkey a bride casts a ripe pomegranate to the ground and the number of seeds that drop out indicates the number of children she will have (298).

In Greek mythology CERES, goddess of the earth, became enraged when ZEUS gave her daughter, PROSERPINE, to PLUTO, god of the underworld, as his wife. CERES left heaven in her rage and came down to the earth, blessing all men who were and cursing all who were not kind to her. But so much did she curse men that ZEUS soon realized his mistake and commanded PLUTO to relinquish PROSERPINE. This he did, but first he asked her to eat of a pomegranate. Doing so, she remained in his power and had to return to Hades with him for six months of every year. So for half of every year CERES is happy in the company of her daughter and the earth is fruitful and green and warm; for the other six months CERES is lonely and angry and the earth is bare and barren and cold (298). The pomegranate thus became, to the Greeks and Romans, the symbol of the nether world and its power and typified all seeds that must be placed underground to germinate, then emerge into the light for a season, only to have their seeds return, in due time, to the darkness beneath the surface of the earth. The original pomegranate, according to Greek mythology, was a beautiful nymph who had been told by a soothsayer that she would one day wear a crown. She was transformed into a pomegranate tree by BACCHUS, god of wine, and a crown was placed at the top of her fruit (298). In China the pomegranate also symbolizes fertility and women offer pomegranates to the goddess of mercy in the hope of being blessed with children. Chinese temple porcelains are decorated with representations of this fruit (298).

FIGURE 28. — View in a garden at Haifa showing a "sakiyeh" used to raise water to the adjacent tank, with date palms (*Phoenix dactylifera*) in the background. This tree was one of almost innumerable uses in Bible times. It was used as a landmark, for shade and lumber, to furnish thatch and insulating material for houses, to cover reed fences, to make cord and rope, dusters and brooms, to furnish edible fruit for man and feed for livestock, to make necklaces and bracelets, and even to furnish "strong drink" and "honey"! (Wood engraving from C. W. Wilson's Picturesque Palestine, 1883; see also fig. 9, fig. 70 and fig. 92).

170. Quercus aegilops L.   171. Quercus coccifera L.
172. Quercus coccifera var. pseudococcifera (Desf.) Boiss.
173. Quercus ilex L.   174. Quercus lusitanica Lam.
(FIGURE 29; FIGURE 84)

GENESIS 12: 6—And ABRAM passed through the land unto the place of Sichem, unto the *plain* of Moreh . . .

GENESIS 13: 18—Then ABRAM removed his tent, and came and dwelt in the *plain* of Mamre, which is in Hebron . . .

GENESIS 14: 13—And there came one that had escaped, and told ABRAM the Hebrew; for he dwelt in the *plain* of Mamre the Amorite . . .

GENESIS 18: 1 & 8—And the Lord appeared unto him in the *plains* of Mamre . . . and he stood by them under the *tree*, and they did eat.

GENESIS 35: 4 & 8— . . . and JACOB hid them under the *oak* which was by Shechem . . . and she was buried beneath Beth-el under an *oak*: and the name of it was called Allon-bachuth.

GENESIS 38: 28 & 30— . . . and the midwife took and bound upon his hand a *scarlet* thread . . . And afterward came out his brother, that had the *scarlet* thread upon his hand . . .

EXODUS 25: 4—And blue, and purple, and *scarlet* . . .

EXODUS 26: 1—Moreover thou shalt make the tabernacle with ten curtains of fine twined linen, and blue, and purple, and *scarlet* . . .

EXODUS 28: 33—And beneath upon the hem of it thou shalt make pomegranates of blue, and of purple, and of *scarlet* . . .

EXODUS 34: 13—But ye shall destroy their altars, break their images, and cut down their *groves*.

EXODUS 35: 23—And every man, with whom was found blue, and purple, and *scarlet* . . . brought them.

EXODUS 39: 24—And they made upon the hems of the robe pomegranates of blue, and purple, and *scarlet* . . .

LEVITICUS 14: 4, 6, & 51-52—Then shall the priest command to take for him that is to be cleansed two birds alive and clean, and cedar wood, and *scarlet*, and hyssop . . . As for the living bird, he shall take it, and the cedar wood, and the *scarlet*, and the hyssop, and shall dip them and the living bird in the blood of the bird that was killed over the running water . . . And he shall take the cedar wood, and the hyssop, and the *scarlet*, and the living bird, and dip them in the blood of the slain bird, and in the running water, and sprinkle the house seven times. And shall cleanse the house . . . with the *scarlet*.

NUMBERS 19: 6—And the priest shall take cedar wood, and hyssop, and *scarlet*, and cast it into the midst of the burning of the heifer.

NUMBERS 24: 6— . . . as the trees of *lign aloes* which the Lord hath planted . . .

DEUTERONOMY 11: 30—Are they not on the other side Jordan, by the way where the sun goeth down, in the land of the Canaanites, which dwell in the champaign over against Gilgal, beside the *plains* of Moreh . . .

DEUTERONOMY 12: 2-3—Ye shall utterly destroy all the places, wherein the nations which ye shall possess served their gods, upon the high mountains, and upon the hills, and under every *green tree*: And ye shall overthrow their altars, and break their pillars, and burn their *groves* with fire . . .

DEUTERONOMY 16: 21—Thou shalt not plant thee a *grove* of any trees near unto the altar of the Lord thy God . . .

JOSHUA 24: 26—And JOSHUA wrote these words in the book of the law of God, and took a great stone, and set it up there under an *oak*, that was by the sanctuary of the Lord.

JUDGES 3: 7—And the children of Israel . . . served BAALIM and the *groves*.

JUDGES 6: 11—And there came an angel of the Lord, and sat under an *oak* which was in Ophrah, that pertained unto JOASH the Abi-ezrite . . .

JUDGES 9: 6—And all the men of Shechem gathered together, and all the house of Millo, and went, and made ABIMELECH king, by the *plain* of the pillar that was in Shechem.

II SAMUEL 18: 9-10— . . . And ABSALOM rode upon a mule, and the mule went under the thick boughs of a great *oak*, and his head caught hold of the *oak*, and he was taken up between the heaven and the earth; and the mule that was under him went away. And a certain man saw it, and told JOAB, and said, Behold, I saw ABSALOM hanged in an *oak*.

I KINGS 13: 14— . . . and found him sitting under an *oak* . . .

I KINGS 14: 23—For they built them high places, and images, and *groves*, on every high hill, and under every green tree.

I KINGS 18: 19— . . . and the prophets of BAAL four hundred and fifty, and the prophets of the *groves* four hundred . . .

II KINGS 17: 10 & 16—And they set them up images and *groves* in every high hill, and under every green tree . . . and made a *grove*, and worshipped all the host of heaven, and served BAAL.

II KINGS 18: 4—He removed the high places, and brake the images, and cut down the *groves* . . .

II KINGS 21: 7—And he set a graven image of the *grove* that he had made in the house . . .
II KINGS 23: 6—And he brought out the *grove* from the house of the Lord, without Jerusalem, unto the brook Kidron, and burned it at the brook Kidron, and stamped it small to powder...
II KINGS 23: 14—And he brake in pieces the images, and cut down the *groves* . . .
I CHRONICLES 10: 12— . . . and buried their bones under the *oak* in Jabesh . . .
II CHRONICLES 2: 7 & 14—Send me now therefore a man cunning to work in gold, and in silver, and in brass . . . and *crimson* . . . skillful to work in gold, and in silver . . . and in *crimson* . . .
II CHRONICLES 3: 14—And he made the vail of blue, and purple, and *crimson* . . .
II CHRONICLES 28: 4—He sacrificed also and burnt incense in the high places, and on the hills, and under every *green tree*.
PSALMS 1: 3—And he shall be like a *tree* planted by the rivers of water, that bringeth forth his fruit in his season; his leaf also shall not wither . . .
ISAIAH 1: 18 & 29-30—Come now, and let us reason together, saith the Lord: though your sins be as *scarlet*, they shall be as white as snow; though they be red like *crimson*, they shall be as wool . . . For they shall be ashamed of the *oaks* which ye have desired, and ye shall be confounded for the gardens that ye have chosen. For ye shall be as an *oak* whose leaf fadeth, and as a garden that hath no water.
ISAIAH 2: 13—And upon all the cedars of Lebanon, that are high and lifted up, and upon all the *oaks* of Bashan.
ISAIAH 6: 13—But yet in it shall be a tenth, and it shall return, and shall be eaten: as a teil tree, and as an *oak*, whose substance is in them, when they cast their leaves . . .
ISAIAH 27: 9— . . . the *groves* and images shall not stand up . . .
ISAIAH 44: 14—He heweth him down cedars, and taketh the cypress and the *oak* . . .
ISAIAH 61: 3— . . . that they might be called *trees* of righteousness, the planting of the Lord . . .
JEREMIAH 4: 30— . . . though thou clothest thyself with *crimson* . . .
JEREMIAH 10: 3— . . . one cutteth a *tree* out of the forest . . .
JEREMIAH 17: 8—For he shall be as a *tree* planted by the waters, and that spreadeth out her roots by the river, and shall not see when heat cometh, but her leaf shall be green; and shall not be careful in the year of drought, neither shall cease from yielding fruit.
EZEKIEL 6: 13— . . . and under every *green tree*, and under every thick *oak*, the place where they did offer sweet savour to all their idols.
EZEKIEL 20: 28— . . . then they saw every high hill, and all the *thick trees*, and they offered there their sacrifices . . .
EZEKIEL 27: 6—Of the *oaks* of Bashan have they made thine oars . . .
EZEKIEL 31: 14—To the end that none of all the *trees* by the waters exalt themselves for their height, neither shoot up their top among the thick boughs, neither their *trees* stand up in their height, all that drink water . . .
DANIEL 4: 10-12— . . . I saw, and behold a *tree* in the midst of the earth, and the height thereof was great. The *tree* grew, and was strong, and the height thereof reached unto heaven, and the sight thereof to the end of all the earth: The leaves thereof were fair, and the fruit thereof much, and in it was meat for all: the beasts of the field had shadow under it, and the fowls of the heaven dwelt in the boughs thereof, and all flesh was fed of it.
HOSEA 4: 13—They sacrifice upon the tops of the mountains, and burn incense upon the hills, under *oaks* and poplars and elms, because the shadow thereof is good . . .
AMOS 2: 9— . . . and he was strong as the *oaks* . . .
MICAH 7: 14—Feed thy people with thy rod, the flock of thine heritage, which dwell solitarily in the *wood*, in the midst of Carmel . . .
ZECHARIAH 11: 2— . . . howl, O ye *oaks* of Bashan; for the forest of the vintage is come down.
SUSANNAH 58—Now therefore tell me, Under what tree didst thou take them companying together? Who answered, Under a *holm-tree*.
HEBREWS 9: 19— . . . he took the blood of calves and of goats, with water, and *scarlet* wool, and hyssop, and sprinkled both the book and all the people.
REVELATION 18: 12— . . . and purple, and silk, and *scarlet* . . .

The Hebrew words translated "scarlet" in the references cited above are "tola", "karmīl", "tola'at" or "tola'ath", "tola'ath shânî" or "tola'at shani", and "shânî" (306). The scarlet dye used in coloring the linen thread and wool in Biblical days was derived from an insect, *Coccus ilicis*, often referred to as "kermes" in older works, which infests the kermes oak, *Quercus coccifera*[118] (299). This oak grows from 6 to 35 or more feet tall and inhabits mountainous regions in Syria, Lebanon, Hauran, and Palestine, covering the rocky hills of Palestine with a dense brushwood of trees usually only 8 to 12 feet tall (293). It branches from the base and is abundantly leafy. Its small, smooth, evergreen leaves are short-petioled, firm-textured, ovate, undulate, and strongly spinous-margined. It yields acorns copiously. The insect which gives out the scarlet dye is a scale-insect which covers the young branchlets with white fluffy masses very much as the closely related cochineal insect, *Coccus cacti*, infests the pricklypear cactus, *Nopalea coccinellifera* Salm-Dyck.

The art of extracting dyes and applying them to various textiles was learned

---

[118]Also known as *Quercus coccifera* var. *genuina* Boiss

by man at a very early date (306). Scarlet thread was employed at the time of ZARAH's birth in about 1727 B.C., and blue and purple are mentioned at the time of the Exodus in about 1491 B.C. It is believed, however, that the Hebrews themselves were not acquainted with these arts. They were probably dependent upon the Phoenicians for the dyes and upon the Egyptians for the art of applying them (306). The purple dye ("argâmân") of the ancients— actually a light-reddish color—was obtained from the shells of certain marine snails of the genus *Purpura* and a few related mollusks, particularly *Murex trunculus* L., found in the Mediterranean and Red Seas. The dye was chiefly produced in and exported from the city of Tyre, and therefore it soon became known as Tyrian purple. It is said that the secret of preparing this dye was known only to the inhabitants of that city and that it was forever lost when Tyre was destroyed. Some of the blue dye ("tecêleth" or "t'chaylet")—really violet in color—was secured from another species of shellfish, *Helix ianthina*, found along the coast of Phoenicia. Other blue dye came from the lichen, *Roccella tinctoria* (which see). Vermillion ("shâshar") was a pigment derived from the mineral cinnabar (158). Yellow dye was obtained from the henna plant (see under *Lawsonia inermis*). Some authorities assert that blue dye was also extracted at that time from the indigo plant (*Indigofera tinctoria* L. and related species) and dyer's woad (*Isatis tinctoria* L.), and red from the madder plant (*Rubia tinctoria* L.) (299). Most commentators, however, are of the opinion that the scarlet and crimson ("carmîl" or "karmīl" were derived exclusively from the scale-insect infesting the kermes oak, particularly abundant in Armenia and other eastern lands. The Arabic name for the insect is still "kermez" (306). The tint produced was crimson rather than scarlet.

Some commentators are certain that the "oak" of Isaiah 2: 13 and 44: 14 was the Valonea oak, *Quercus aegilops*, common to the middle mountain zones and probably as abundant about Bashan in Biblical days as it is now, if not more so. The "oak" of Genesis 35: 4 & 8 is thought to have been the holm oak, *Q. ilex*, [119] common from the coasts of Syria to Judea. Other writers state that three species of oak are now common about Bashan—the Valonea oak, the kermes oak, and the Cyprus oak, *Q. lusitanica*[120] (124d), so the references in Isaiah 2: 13, Ezekiel 27: 6, and Zechariah 11: 2 probably alluded to all three. The "oak" of Joshua 24: 26 and I Kings 13: 14 is thought to have been *Q. coccifera* var. *pseudococcifera*[121]. EVENARI regards all the oaks of Palestine as *Q. calliprinos* Webb, *Q. ithaburensis* (Decne.) Boiss., or *Q. infectoria* Oliv.

Some authorities feel that the "oaks" mentioned in Judges 6: 11 and Amos 2: 9 and the "tree "of Genesis 18: 8 actually were terebinths, *Pistacia terebinthus* var. *palaestina* (which see), but this does not seem nearly as reasonable to most commentators as that they were oaks. Similarly, most modern authorities feel that it was really an oak on which ABSALOM was hanged, not a terebinth as some early writers claimed.

The various Hebrew words used for "oak" in the Old Testament are "êl", "êlon", "îlan", "allah", "allôn", and "êlâh" or "âlâhim". JOHN SMITH gives an excellent summary of the complicated situation here, pointing out, in his "Bible Plants", that some commentators believe that these six words, all translated "oak" in our Authorized Version, are Hebrew designations not only for oak trees, but also for any "large or thick trees" (299). CELSIUS thought that five of the words denote the terebinth and that "allôn" alone denoted true oaks (299). DINSMORE believes that "allôn" applies only to *Q. coccifera* (267). TRISTRAM was of the opinion that "allah" and "allôn" (Greek, βάλανος, meaning "acorn") referred to the evergreen oaks, *Q. ilex, Q. palaestina* Ky., and *Q. coccifera*, while "el", "elôn", and "îlan" (Greek, τερέβινθος, δένδρον, meaning "a tree") referred to the deciduous oaks, *Q. sessiliflora* Salisb.

---

[119] Considered by EVENARI as unknown from Palestine and doubtfully from Lebanon, but reported from Palestine by Löw (218) and POST (267).

[120] Also known as *Quercus infectoria* Oliv., which is the name used by Dr. EVENARI, who considers this species very scarce in Galilee.

[121] Also known as *Quercus pseudococcifera* Desf.; considered a "most doubtful variety" by EVENARI.

[122], *Q. lusitanica*, *Q. aegilops*, *Q. cerris* L., *Q. ehrenbergii* Kotschy, *Q. look* Kotschy, *Q. libani* Oliv., and *Q. syriaca* Ky.[123] (184), all common in the region (299). With 24 kinds of oaks in the Holy Land it seems most probable that the Hebrews and other early inhabitants should have had several words applying to them. Such diverse names as "ballut", "abbas", "sindyân", "likk", "mall", and "mallûl" are recorded by POST for oaks in the region today (267). ROSENMÜLLER awards the terebinth to "el" and "êlâh", and the oak to "allâh", "allôn," and "êlôn" (306). Other commentators are of the opinion that only "allâh" and "allôn" refer to oaks, and "el", "êlah", "êlôn", and "îlan"—all derivatives of the root "el", implying strength—refer to the terebinth. FRASER mentions that "the Hebrew words commonly rendered 'oak' and 'terebinth' are very similar; the difference between them being in part merely a difference in the vowel points which were added to the text by the Massoretic scribes in the Middle Ages. Scholars are not agreed as to the correct equivalents of the words . . . in the warm and dry climate of Moab the terebinth is the principal tree, while the oak flourishes more in the cooler and rainier districts of Gilead and Galilee in the north. It is, therefore, natural that the terebinth should be predominantly the sacred tree of the south and the oak of the north; but throughout Palestine as a whole, if we may judge by the accounts of travellers, the oak appears to be the commoner tree, and, consequently, perhaps, the more frequently revered" (124d). "El" is in most places where it occurs in the Bible rendered "mighty men" or some similar term (27). It seems to us that this word and its derivatives, where they apply to trees, would much more logically refer to the oak—symbol of strength since time immemorial (298)—than to the unrelated terebinth.

In Genesis 12: 6, 13: 18, & 18: 1, Deuteronomy 11: 30, and Judges 9: 6 the Authorized Version translates "êlon" or "aylon" as "plain" or "plains", but modern authorities seem to agree that it should rather have been rendered as a tree of some sort. MOFFATT says "oracular oak", "oaks", "circular oak", and "sacred tree" in these passages. The GOODSPEED and JASTROW versions say "terebinth" for all of them. The Douay version uses "oak" only in Judges 9: 6, "vale" and "valley" in the other passages. In the valleys of Lebanon and where it grows by itself in isolated positions, as at Mamre, *Q. coccifera* forms a large and massive tree (299). According to POST (265) its variety *pseudococcifera* is the one regularly planted by tombs in the East (27). THOMSON, in 1860, said: "I do not believe that ABRAHAM's celebrated tree at Hebron was a terebinth. It is now a venerable oak, and I saw no terebinth in the neighborhood" (299, 319). The oak was always respected and even venerated in Biblical times for its large size and strength, and great men were usually buried under its shade (124d, 299). ABRAHAM's oak was described in 1856 by BONER as a magnificent prickly oak, somewhat isolated, but with other trees not far off. "The protruding knots of root at its base looked almost like pieces of dark brown rock. The stem is enormous, and as rough and shapeless as can be fancied. The branches, spreading widely in several detachments, and with their extremities drooping to the sward, throw their shade over a vast circle" (299). It was of moderate height. Its trunk measured 23 feet in circumference, the spread of its branches 90 feet (184). It was held in high veneration by the Moslems and it was firmly believed that if any person cut or maimed it, he would lose his firstborn son. In the winter of 1850 it suffered the loss of a large limb through a heavy fall of snow, and the fear of touching it made it almost impossible to obtain any workmen to assist in removing the fallen limb (299). Dr. CHARLES EDWARD MOLDENKE, father of the senior author, visited this locality on January 2, 1897, and, at the peril of his life at the hands of Turkish guards, collected a specimen from this tree. This specimen is now on deposit in the Britton Herbarium at the New York Botanical Garden and proves to be *Q. coccifera* var. *pseudococcifera* (233). POST applies the name "Abraham's oak" to *Q. palaestina*, recording both it

---

[122]Also known as *Quercus sessilis* Ehrh., *Q. robur* var. *sessilis* Ait., and *Q. robur* var. *nigra* Lam.
[123]Also known as *Quercus lusitanica* var. *syriaca* A. DC., *Q. lusitanica* var. *latifolia* Boiss., and *Q. infectoria* var. *latifolia* (Boiss.) Nabelek.

and *Q. coccifera* var. *pseudococcifera* from Hebron (267).

TRISTRAM believed that the Hebrew word "elâh" (Greek, τερέβινθos, δρύs) used in Genesis 35: 4 and II Samuel 18: 9 referred to the terebinth, instead of to oaks, while the GOODSPEED version, following the Septuagint, says "terebinth" for the first of these two references, but not for the second, and identifies the "êlôn" of Genesis 12: 6, 13: 18, & 18: 1, Deuteronomy 11: 30, and Judges 9: 6 as the terebinth. If the word "elâh" signifies terebinth, as some writers claim, then it must have been a terebinth in which ABSALOM caught his hair, but THOMSON says: "the tree in which ABSALOM was caught was the *alah*, not the *allon*, and I am persuaded that it was an oak. That battlefield was on the mountains east of the Jordan, always celebrated for great oaks—not for terebinths; and this is true to this day. There is no such thing in this country as a terebinth wood. It was an oak I firmly believe. There are thousands of such trees still in the same locality, admirably suited to catch the long-haired rebels; but no terebinths. I see it asserted by the advocates of this translation that the oak is not a common nor a very striking tree in this country, implying that the terebinth is. A greater mistake could scarcely be made. It is 'simply ridiculous' to compare its strength and size with that of an oak" (299). With regard to the oak forests he says: "Besides the vast grove around us, at the north of Tabor, and in Lebanon and Hermon, in Gilead and Bashan, think of the vast forests extending thirty miles at least, along the hills west of Nazareth, over Carmel, and down south beyond Caesarea Palestina. The terebinth is deciduous, and therefore not a favorite shade tree. It is very rarely planted in the courts of houses, or over tombs, or in the places of resort in villages. It is the beautiful evergreen oak which you find there (*Quercus pseudococcifera*). Beyond a doubt the idolatrous groves so often mentioned in Hebrew history were of oak" (299).

Quercus aegilops is a deciduous species abundant in Syria and northern Palestine in the middle mountain zones. It attains a height of 15 to 50 feet, forming dense forests. In Bashan it grows to especially large size and is undoubtedly the "oak of Bashan" spoken of in Isaiah and Ezekiel and by THOMSON (124d, 299). It bears very large acorns which the natives use as food, and the cups are extensively employed in tanning leather, dyeing, and making ink, and form an important article of commerce. It is supposed that this was the tree referred to by DANIEL. It may also be the "tree planted by the rivers of water" in Psalms 1: 3, for while its leaves are not evergreen, they are leathery, glossy, and very tardy in falling (267). *Q. lusitanica* is a small deciduous tree, seldom more than 20 feet tall, and not as common as the previously discussed species, but it is important because of the galls produced on its leaves and twigs through the stings of small insects. These galls or "oak apples" are an important article of trade, used for tanning leather, dyeing, and ink-making (299). FRASER claims that the acorn cups are used by dyers under the name of "Valonia" (124d). Its very large acorns are eaten by natives (124d).

The climate of Palestine is mostly too warm for the oak to really flourish in the low-lying valleys (268), so that the oaks of the Holy Land are to be found mostly on the hills and mountainsides. On Mount Carmel *Quercus coccifera* var. *pseudococcifera* forms 9/10ths of the arborescent vegetation, and it is almost equally abundant on the western flanks of the Antilebanon and other mountain ranges.

The many references to "groves" in the Old Testament, usually in connection with the worship of BAAL or other heathen gods—Exodus 34: 13, Deuteronomy 16: 21, Judges 3: 7, I Kings 14: 23 & 18: 19, II Kings 17: 16, etc.—have been considered by many writers to refer to groves of sacred oak trees. The worship of idols on "high places" was practiced up to the very end of Bible history and relics of it can still be seen in Palestine today (124d, 299). Dr. THOMSON, in his famous history of Palestine entitled "The Land and the Book" (319), says: "Every conspicuous hilltop has a *willey* or *mazar*, beneath a spreading oak, to which people pay religious visits, and thither they go up to worship and discharge vows. All sects in the country, without exception, have a predilection for these 'high places', as strong as that of the Jews in ancient times... There is one of these high places, with its grove of venerable oaks, on the very summit of Lebanon, east of Jezzin. It is of an oval shape, corresponding to the top of the mountain, and the grove was planted regularly round its outer edge." He mentions another grove in the Lebanon mountains as being "resorted to by Jews, wild Arabs of the desert, Moslems, Metewelies, and Christians" (124d, 299). The word employed in these many verses for "groves" is "asherah" or "asherim". GESENIUS and other writers have identified this with ASHTARTE or ASHTEROTH, the Phoenician and Canaanitish goddess of fertility and reproduction, sometimes described as the wife of BAAL. ASHTEROTH was symbolically represented in the form of a tree, as we know from recovered sculptures in ruined Nineveh. It is probable that such sculptured representations of ASHTARTE were common at her altars on all the "high places", and it appears from II Kings 21:7 and 23:6 that the word "asherah" was applied also to them, for we find it stated that "he set a graven image of the *grove*... in the house" and "he brought out the *grove* from the house of the Lord... and burned it... and stamped it small to powder". An inscription on a sarcophagus found in the tomb of King ASHMUNAZER at Sidon indicates that ASHTARTE was also a goddess of the Sidonians (299). MOFFATT and the GOODSPEED version adopt the term "sacred pole" for some of the Authorized and Douay versions' references to "groves" and "Ashtarte" to others. JASTROW says "Asherah", "Asherim", or "Asheroth" in all these passages—in Deuteronomy 16:21 saying "Asherah of any kind of tree." The reference to "groves" in Genesis 21:33 is different from the ones just discussed. The word there employed is "eshel" and is now regarded as referring to the tamarisk tree (see under *Tamarix articulata*). "Orchard" is the translation given this word in LEESSER.

The "holm tree" of Susannah was supposed by early writers to have been the true holm or holly, *Ilex aquifolium* L. This tree, however, is not definitely known from Palestine, being recorded by POST only from a single collection in the mountains of the northern district of Lebanon. GOODSPEED translates the word "liveoak", and most authorities now agree that it was the evergreen holm oak, *Q. ilex* (299).

WILLIAM SMITH cites Genesis 14:6 as a reference to oaks, but admits that the "El-paran" of that passage may be only a single proper place name (as the Authorized, JASTROW, MOFFATT, and GOODSPEED versions regard it), rather than taken as two words meaning "the oak (terebinth, or grove) of Paran". The Douay version considers it "the plains of Pharam". The word "elim" used in Exodus 15:27 is a plural form of "el" and literally signifies "oaks" (27). In this passage, however, where it is the name of the second station where the Israelites halted after crossing the Red Sea, it is thought to mean "grove" and to refer to the seventy palm trees growing there (306). The other plural forms "eloth" and "elath" are similarly regarded by STANLEY, SMITH, and others as referring to the palm groves at Akaba (306). "Elim", however, occurs also in Isaiah 1:29 & 61:3 and Ezekiel 31:14. In Isaiah 61:3 the Douay version is *"mighty ones* of justice", JASTROW says *"terebinths* of righteousness", MOFFATT says *"sturdy oaks* of goodness", and the GOODSPEED version is *"oak trees* of righteousness". In Isaiah 1:29 these versions say "idols", "terebinths", "sacred trees", and "terebinths", respectively! In the Ezekiel verse all versions use the non-committal term "trees". WILLIAM SMITH feels that in the two latter verses "any strong flourishing trees may be denoted" (306). The JASTROW version uses the word "terebinth" in Genesis 12:6, 13:18, 14:13, 18:1, and 35:4, Deuteronomy 11:30, Judges 6:11 and 9:6, II Samuel 18:9-10, I Kings 13:14, I Chronicles 10:12, and Isaiah 1:29-30 and 6:13, instead of the words "plain", "plains", "oak," and "oaks" used by the King James Version, and the word "aloes" for the "lign aloes" of Numbers 24:6.

In all religions and mythologies in the northern hemisphere the oak has figured as the symbol of strength and sturdiness. It was sacred both to JOVE and THOR (126). The medieval prophet and magician, MERLIN, worked his enchantments beneath an oak. The Druids performed their mystic rites beneath oaks (126). ERISICHTHON was condemned to perpetual hunger for felling an oak sacred to CERES. Oaks were believed to attract lightning or "Jove's thunderbolts" more often than any other tree because their superior strength made them more worthy to receive these bolts (125). Farmers planted oaks to "attract" lightning away from their houses. Ravens habitually perched in the holm oak and croaked their dismal foreboding warnings, hence it became a funereal tree. In Christian legendry all the trees of the forest rebelled at giving their wood for JESUS' cross, except the holm oak (126). Yet JESUS forgave it because it was willing to die with him, and under the shade of a holm oak, after his resurrection, he appeared to the saints (298). Oaks were the traditional homes of dryads. The oak figures largely in the folklore of Germany, Lithuania, Ireland, Greece, Finland, Austria, England (125), and the American Indians. In Greek mythology its roots extended all the way down to Hades; in Christian lore its branches were uplifted to heaven in prayer. It became the tree of MARY, mother of Christ. Oaks were supposed to cure all manner of diseases (125), not only of human beings, but also of horses, and afforded rest to the miserable Wandering Jew. JOAN OF ARC was accused of worshipping a bewitched oak (298).

FIGURE 29. — Abraham's Oak (*Quercus coccifera* var. *pseudococcifera*) near Hebron, in the 1870's. When one of the lower branches was broken down by a heavy fall of snow in the winter of 1850, it was cut up into logs and conveyed to Jerusalem; there were seven camel loads! At present the tree has been nearly destroyed. (Wood engraving from C. W. Wilson's Picturesque Palestine, 1883; see also fig. 84).

175. Retama raetam (Forsk.) Webb. & Berth.
(FIGURE 78)

NUMBERS 33: 18-19—And they departed from Hazeroth, and pitched in *Rithmah*. And they departed from *Rithmah* . . .

I KINGS 19: 4-5—But he himself went a day's journey into the wilderness, and came and sat down under a *juniper tree*: and he requested for himself that he might die; and said, It is enough; now, O Lord, take away my life; for I am not better than my fathers. And as he lay down and slept under a *juniper tree*, behold, then an angel touched him, and said unto him, Arise and eat.

PSALMS 120: 4—Sharp arrows of the mighty, with coals of *juniper*.

The Hebrew words translated "juniper" in the Authorized and Douay versions are "rotem" or "rothem" and "r'tamim", and "ritmah" or "rithmah" (Greek, ῥαθμέν); and it has nothing whatever to do with the true junipers, *Juniperus*. Actually the Scriptural "juniper" was a species of broom known as the white broom, *Retama raetam*[124] (266, 268, 299). Its habit of growth is similar to that of the Scotch broom, *Cytisus scoparius* (L.) Link, but its branches are longer and more flexible, forming an erect dense bush 3 to 12 feet tall. Although its leaves are very small and sparse, simple, and linear, it nevertheless forms a very agreeable shade in desert regions. The white pea-like flowers are sweetly fragrant and are borne in subsessile clusters along the twigs (266, 268). The white broom is a beautiful shrub and is abundant in the Palestinian desert regions, growing on hills, in rocky places, ravines, and sandy situations. It is common around the Dead Sea, in Gilead, in the Jordan valley, on the Syrian desert, in Lebanon, on Mount Carmel, and on all the deserts southward to Arabia Petraea, Sinai, and Egypt (266)[125]. It is also said to be found on the Philistine and Phoenician coasts (326). In the "wilderness" (deserts) it is in many places the only bush that affords any shade (299). TRISTRAM describes this plant very effectively: "This is one of the exquisitely beautiful plants of the country. The gauzy delicate pink-and-white hues of a whole hill-side covered with shrub in blossom, as I have seen it in Gilead, is unsurpassed even by the apple blossom of an English orchard" (326). The "Rithmah" of Numbers 33: 18-19—"Rethma" in the Douay version—is a variant of the same Hebrew word (27), meaning "place of broom" (184), and probably refers to the abundance of these plants at that locality. The GOODSPEED version of the I Kings reference cited above renders the Authorized and Douay versions' "juniper tree" more correctly as "broom tree", JASTROW says "broom-tree," LEESSER gives "broom-bush", while MOFFATT says "brook-bush" and "bush". The Revised Version suggests "broom" in a marginal note (27). The Arabic name for the tree is "retem" (184) or "rethem" (306).

The expression "coals of juniper" used in Psalms 120: 4—"burning coals" in MOFFATT, "live broom coals" in GOODSPEED, "coals of broom" in JASTROW, and "coals that lay waste" in Douay—refers to the fact that the wood of the white broom is used extensively for making charcoal (27). The Royal Psalmist avows that the coals of this plant afford the fiercest fire of any combustible matter that was to be found in the desert, and on this account would provide the most fitting punishment for deceitful tongues (262). This charcoal is said to be of especially fine quality and forms an important article of trade between the Bedouins and the Egyptians (299). The wood of the white broom is said to burn extremely well even when not made into charcoal. JEROME and the writers of the Talmud agree in the belief that this was the wood to which

---

[124] Also known as *Genista monosperma* (L.) Lam., *G. raetam* Forsk., and *Spartium monospermum* L.

[125] EVENARI considers the range of this plant more limited, excluding such places as Lebanon and Mount Carmel from which it has been recorded by POST and other workers.

DANIEL referred when he spoke of "coals of juniper". Some commentators have even suggested that the peculiarly loud and crackling noise made by burning broom wood served DANIEL also as a symbol of the loud and unjust assertions of the calumniators of the righteous man (268).

The "juniper roots" of Job 30: 4 are not the roots of either a juniper or the white broom. The roots of the white broom, like those of the Scotch broom, are extremely nauseous and even somewhat poisonous if eaten (299). They could not be eaten in the manner described by JOB. It is now supposed that JOB'S "juniper roots" were the edible parasitic plant, *Cynomorium coccineum* (which see).

An old Christian legend states that when JESUS was praying on that fateful night in the garden of Gethsemane he was continually disturbed by the noisy crackling and "sawing" of a broom plant. When he was finally led off by the soldiers he said to the broom: "May you always burn with as much noise as you are making now" (298). Another legend says that the crackling of the broom plants among which they were hiding almost revealed the whereabouts of MARY and the infant JESUS to the soldiers of HEROD (298).

176. Rhamnus palaestina Boiss.

GENESIS 3: 18—*Thorns* also and thistles shall it bring forth to thee . . .
EXODUS 22: 6—If fire break out, and catch in *thorns,* so that the stacks of corn, or the standing corn, or the field, be consumed therewith; he that kindled the fire shall surely make restitution.
JOB 41: 2—Canst thou put an hook into his nose? or bore his jaw through with a *thorn?*
PSALMS 58: 9—Before your pots can feel the *thorns,* he shall take them away as with a whirlwind . . .
PROVERBS 15: 19—The way of the slothful man is as an *hedge of thorns* . . .
ECCLESIASTES 10: 8— . . . whoso breaketh an *hedge,* a serpent shall bite him.
ISAIAH 5: 5— . . . I will take away the *hedge* thereof . . .
ISAIAH 7: 23-25—And it shall come to pass in that day, that every place shall be, where there were a thousand vines . . . it shall even be for briers and *thorns* . . . all the land shall become briers and *thorns.* And on all hills that shall be digged with the mattock, there shall not come thither the fear of briers and *thorns* . . .
ISAIAH 10: 17— . . . and it shall burn and devour his *thorns* and briers in one day.
ISAIAH 33: 12— . . . as *thorns* cut up shall they be burned in the fire.
EZEKIEL 2: 6—And thou, son of man, be not afraid of them, neither be afraid of their words, though briers and *thorns* be with thee . . .
EZEKIEL 13: 5—Ye have not gone up into the gaps, neither made up the *hedge* for the house of Israel to stand in the battle in the day of the Lord.
EZEKIEL 22: 30—And I sought for a man among them, that should make up the *hedge* . . .
HOSEA 2: 6—Therefore, behold, I will hedge up thy way with *thorns* . . .
NAHUM 3: 17— . . . thy captains as the great grasshoppers, which camp in the *hedges* in the cold day, but when the sun ariseth they flee away . . .
MARK 12: 1— . . . A certain man planted a vineyard, and set an *hedge* about it . . .
LUKE 14: 23— . . . go out into the highways and *hedges* . . .

Over twenty different words are used in the Bible to refer to thorny or spiny plants. Of these, three have been rendered "thorns" in the Authorized Version—"chadek", "kêtz", and "kimmeshonim". Just as in the case of the "briers", "brambles", "nettles", and "thistles", there has been much difference of opinion about the identity of the "thorns" of the Bible. Under these three words the Hebrews apparently included all the thorny shrubs of the region which did not have well-known names of their own. The "thorns" of Numbers 33: 55 are herein regarded as referring to *Rubus sanctus* (which see), those of Matthew 27: 29 and John 19: 2 as referring to *Paliurus spina-christi* (which see), those of Isaiah 7: 19 and 55: 13, Judges 8: 7, and Matthew 7: 16 as referring to *Zizyphus spina-christi* (which see), those of Matthew 13: 7 and Hebrews 6: 8 to *Centaurea calcitrapa* (which see), and those of Isaiah 34: 13 and Hosea 9: 6 to *Xanthium spinosum* (which see).

The Palestine buckthorn, *Rhamnus palaestina*[126], is a shrub or small tree, attaining a height of 3 to 6 feet, with velvety thorny branches and oblong-cuneate, almost spatulate, obtuse, remotely crenate, evergreen leaves, and clusters of small imperfect flowers blooming in March and April. It grows in

---

[126]Also called *Rhamnus punctata* var. *palaestina* (Boiss.) Post.

thickets and on hillsides from Syria and Lebanon, through Palestine, to Arabia Petraea and Sinai (266). It is known as "suwwayd" in the southern and "ajrayna" in the northern part of its range.

The Hebrew word translated "hedge" in Ecclesiastes 10: 8 is "m'soochah". LEESSER translates it "fence". The "hedge" of the book of Ezekiel, on the other hand, is represented by the Hebrew word "gader".

The Jericho balsam, *Balanites aegyptiaca* (which see), and the European boxthorn, *Lycium europaeum* L., are also prickly shrubs which are widely used as hedges in Palestine and so may be involved in the Proverbs and Hosea references quoted above. *Rhamnus cathartica* L., sometimes suggested, is out of the question since it is not known from that part of the world. Species of hawthorn, *Crataegus*, are possible, because there are five species native to the region (266). The "thorns of the wilderness" of Judges 8: 7 are regarded by some writers as several desert plants like *Capparis sicula*, *Nitraria schoberi* L., *Ononis spinosa* L., and several spiny species of *Acacia* and *Astragalus*. By us they are discussed under *Zizyphus*.

The "thorn" of Job 41: 2 is not regarded as a botanical reference by modern translators. The word used in this verse is translated "hook" by the GOODSPEED and JASTROW versions, "gaff" by MOFFATT, and "buckle" by the Douay version. MOFFATT's rendition of this verse regarding the hunting of a crocodile is: "Can you ... run a cord right through his gills, or carry him with a *gaff* between his jaws?" Zoologists will be interested to note this allusion to "gills" in a crocodile in this 1925 translation.

The phrase "though briers and *thorns* be with thee" of Ezekiel 2: 6 is rendered "although they cut and *wound* you" by MOFFATT, "even when thistles and *thorns* are round about you" by GOODSPEED, "for thou art among unbelievers and *destroyers*" by Douay, and "though defiers and *despisers* be with thee" by JASTROW!

177. Ricinus communis L.
(FIGURE 39)

JONAH 4: 6-7—And the Lord God prepared a *gourd*, and made it to come up over JONAH, that it might be a shadow over his head, to deliver him from his grief. So JONAH was exceeding glad of the *gourd*. But God prepared a worm when the morning rose the next day, and it smote the *gourd* that it withered.

The Hebrew word translated "gourd" by the Authorized version is "kikayon" or "kikajon" (Greek, κολοκύνθη), and there has been considerable argument as to the plant referred to by it. A marginal note in both the Authorized and Revised versions gives the alternative translation of "palmchrist" or "Palma Christi" (27). The Douay version says "ivy".

The "palm-christ" or "Palma Christi" is the ordinary castor-bean, *Ricinus communis* (101), and many commentators have supposed that it was this plant which provided JONAH with shade and was later attacked by a "worm" so that it withered and died. JEROME and other writers have supposed this to be the plant partly because of the similarity of the Egyptian name of the castor-bean plant, "kibil", to the Hebrew word used in Jonah (299). The Arabic name recorded by POST, however, is "khurwa'" (267). The castor-bean is a tender shrub, growing 3 to 12 or more feet tall, the new shoots, petioles, and peduncles more or less glaucous and scurfy. The leaves are often huge, peltate, palmately 7-11-lobed, with lanceolate dentate lobes. The young leaves are usually deep-purple in color, the older ones bright-green. The small imperfect flowers are borne in axillary and terminal racemes. The fruits are oblong-ellipsoid spiny capsules, containing beautifully marked and mottled seeds (267). The huge leaves of this plant are excellently adapted for producing ample shade when growing alongside a bower, booth, or hut or overhanging a bench. The species is said by POST to be found in waste places, especially near water, in both Lebanon and Palestine, and often to be cultivated (265, 267). In hot climates it grows tree-like and affords dense shade by the abundance of its huge umbrella-like leaves. It is remarkable in the Orient for the rapidity of its growth, according to PRATT (268). The oil extracted from the seeds of the castor-bean was used extensively by the Hebrews in their ceremonial rites, and it is mentioned among the five kinds of oil which rabbinical tradition sanctions for such use. It is remarkable, however, that neither the ancient Hebrews nor the modern inhabitants of Palestine and Syria used it for medicine. PRATT is quite convinced that the castor-bean was JONAH'S "gourd" (268), but the early church fathers disputed heatedly over this question. JEROME at one time even decided that it must have been the English ivy, *Hedera helix* (which see), and in this he is followed by the present Douay version.

ST. AUGUSTINE disagreed with the advocates of the castor-bean and the ivy. He decided that JONAH'S "gourd" was indeed a true gourd, which he says grows even faster in the Orient than the castor-bean and wilts just as completely and suddenly when attacked by a subterranean borer. JOHN SMITH agrees with him in this determination (299). Many true gourds have broad leaves which might serve just as well to shade an arbor, and, in fact, are now widely used for just such a purpose in the Holy Land (299). STAUNTON and STURTEVANT (299) assert that JONAH'S "gourd" was *Cucurbita pepo* var. *ovifera* (L.) L. H. Bailey. They say: "This cucurbita is a small pear shaped gourd with wide leaves. It is commonly used for arbors in the East. It often grows a foot or so a day and withers as quickly." Other commentators suggest the pumpkin, *C. pepo* L. The only flaw in this theory is that *Cucurbita pepo* and all its varieties are indigenous to tropical America and so could not have been known in Palestine in Biblical times, even though they — like the pricklypear cactus and black locust — are there now. If JONAH'S plant was a true gourd, it is far more probable that it was the crookneck squash, *C. moschata* Duch., a native of southern Asia, or the white-flowered gourd, *Lagenaria leucantha* (Duch.) Rusby, from Abyssinia. The winter squash, *Cucurbita maxima* Duch., is also possible, although it originated in central Africa and it is doubtful if it could have been introduced into Palestine as early as JONAH'S day, 862 B.C. All three are recorded by POST as being much cultivated in Palestine today (266).

TRISTRAM favored a true gourd, rejecting the castor-bean because it "is a shrub, and unsuitable" (184). The Authorized, Douay, JASTROW, and GOODSPEED versions all unite in stating that it was a "booth" which Jonah made and which was shaded by the gourd. MOFFATT calls it a "hut". In none of these versions is it stated or implied that it was an arbor or trellis. Therefore the castor-bean is just as suitable to the context as a gourd or squash would be. WILLIAM SMITH says: "there can be no reasonable doubt that the *kikâyôn* which afforded shade to the prophet JONAH before Nineveh is the *Ricinus communis*, or castor-oil plant, which, formerly a native of Asia, is now naturalized in America, Africa, and the south of Europe. This plant varies considerably in size, being in India a tree, but in England seldom attaining a greater height than three or four feet. The leaves are large and palmate, with serrated lobes, and would form an excellent shelter for the sun-stricken prophet" (306). Still further evidence in support of the castor-bean plant is the fact that where the "kikayon" is cultivated in Egypt it produces the commercially important "kiki oil", which is certainly not an extract from any gourd (98)!

The "wild gourds" of II Kings 4: 39 are by us discussed under *Citrullus colocynthis* (which see).

178. Roccella tinctoria Lam. & P. DC.

EXODUS 25: 4—And *blue*, and purple, and scarlet . . .
EXODUS 26: 1—Moreover thou shalt make the tabernacle with ten curtains of fine twined linen, and *blue*, and purple, and scarlet . . .
EXODUS 28: 33—And beneath upon the hem of it thou shalt make pomegranates of *blue*, and of purple, and of scarlet . . .
EXODUS 35: 23—And every man, with whom was found *blue*, and purple, and scarlet . . . brought them.
EXODUS 39: 1 & 24—And of the *blue*, and purple, and scarlet, they made cloths of service . . . And they made upon the hems of the robe pomegranates of *blue*, and purple, and scarlet . . .
II CHRONICLES 3: 14—And he made the vail of *blue*, and purple, and crimson . . .
ESTHER 8: 15—And MORDECAI went out from the presence of the king in royal apparel of *blue* and white, and with a great crown of gold, and with a garment of fine linen and purple.
EZEKIEL 27: 7— . . . *blue* and purple from the isles of Elishah was that which covered thee.

According to several recent authorities — FELIX HENNEGUEY, Les Lichens Utiles 57-65 (1883), BRUCE FINK in Contributions from the United States National Herbarium 14: 36 (1910), and LOUIS C. WHEELER in The Bryologist 41: 109 (1938) — at least some of the blue (Hebrew, "t'chaylet") dye of Old Testament days was extracted from the lowly lichen, *Roccella tinctoria*. Even in recent times this lichen has been used in France and elsewhere for obtaining the dye to color silks. Some writers maintain that the purple dye of the Old Testament, at least, also came from this source. However, most commentators feel that the purple color of the ancient Hebrews was the same thing as Tyrian purple mentioned in Mark 15: 17 and that it definitely came from a marine shellfish (see under *Quercus aegilops*). It is not impossible, however, that various tints and shades were obtained by blending the lichen and the shellfish dyes.

179. Rosa phoenicia Boiss.

SONG 4: 13-14—Thy plants are an orchard of pomegranates, with pleasant fruits; camphire, with *spikenard*. Spikenard and saffron . . .
II ESDRAS 2: 19— . . . and seven mighty mountains, whereupon there grew *roses* and lilies, whereby I will fill thy children with joy.
WISDOM OF SOLOMON 2: 8—Let us crown ourselves with *rose-buds*, before they be withered.

The "roses" of the Scriptures, as has been mentioned previously, are among the most controversial of all Biblical plants. There is no unanimity among commentators regarding any of them. It is quite apparent that several different plants were referred to under this name in the Authorized Version. The "rose" of Ecclesiasticus 24: 14 and 39: 13 is now thought to have been the oleander, *Nerium oleander* (which see). The "rose" of Isaiah 35: 1 is by us regarded as *Narcissus tazetta* (which see). The famous "rose of Sharon" mentioned in Song 2: 1 and the "rose" of Ecclesiasticus 50: 8 are discussed under *Tulipa montana* (which see).

Various commentators have suggested different plants for the "roses" and "rose buds" of II Esdras 12: 19 and Wisdom of Solomon 2: 8, including the dog rose (*Rosa canina* L.), species of rockrose (*Cistus*), the oleander (*Nerium oleander*), and even the shrubby-althea or Syrian-rose (*Hibiscus syriacus* L.) — the last-mentioned is now popularly known as "rose-of-Sharon", but, in spite of its name, is a native of China, not of the Holy Land (229). It seems most likely, however, that these "roses" of II Esdras and Wisdom of Solomon were Phoenician roses, *Rosa phoenicia* (306).

"The Rose Annual for 1946" carries the following lukewarm recommendation for these Apocryphal plants as roses: "Here we are on different ground. The books are not only later, but of an entirely different atmosphere and tone of allusion, that of the Graecised Alexandrian, not the Palestinian Jew. Thus the Rose might have been used for illustration, although Jericho was famous for balsam, not for roses. (The so-called 'Rose of Jericho', bought in the Holy Land, which expands and revives when soaked in water after any length of time, is a cruciferous plant.) But the plant spoken of seems to be one that must be grown with artificial irrigation, which roses do not need, even if we had any evidence of their being grown at Jericho. Canon TRISTRAM and Mr. McLEAN (*Encyclopaedia Biblica*) think that the rhododendron is intended (the name, be it noted, means *rose-tree*), but I confess that I am not satisfied of this. If it be not really a rose, why not the balsam (*balsamodendron*), which STRABO (1763), says was grown in a 'paradise' or botanical garden at Jericho, and nowhere else? Anyhow the crowning oneself with rose-buds at a feast is a purely Greek custom borrowed by the Romans; and as we find from MARTIAL vi. 80, that Egypt a century or so later was even growing roses under glass to send to Rome for banquets, we may be pretty sure that the allusion in the Book of Wisdom at least is really to a Rose" (181). In reference to all the Old Testament "roses" this same source feels that they must be given other non-rosaceous identifications.

The Douay and GOODSPEED versions continue to refer to them as they are referred to in the Authorized Version. The Phoenician rose is a bush 3 to

9 feet tall, with long climbing branches that are armed with scattered hooked prickles. The lower leaves have five and the upper have three ovate-elliptic coarsely serrate leaflets, dark-green on the upper and light-green on the lower surface. The white flowers are numerous, borne in corymbose panicles, followed, in due time, by ovate red fruits called "hips". It is common all through Syria, Lebanon, and Palestine (266). The statement made by some authorities (27) that no true roses are found in Palestine is erroneous. At least four kinds are wild in Palestine, although most of them are not common; more kinds grow on the mountains of Syria and Lebanon (266, 306). Five exotic species are also now cultivated in the area.

The Song 4: 13-14 reference is peculiar. The Authorized, Douay, JASTROW, and GOODSPEED versions all use the word "spikenard" or "nard" twice here, but MOFFATT renders the passage: "Your charms are a pomegranate paradise — with henna and *roses*, and spikenard and saffron." Thus we see that he regards the first "spikenard" as actually roses.

The rose figures largely in the legendry, mythology, and folklore of almost every country and people in the northern hemisphere. Representations of roses appear on coins, coats-of-arms, flags, seals, paintings, and objects of art of all sorts. England even had its Wars of the Roses. The rose is connected in story with BRAHMA, VISHNU, MAHOMET, BUDDHA, CONFUCIUS, ZOROASTER, the Pope, the Crusaders, VENUS, CUPID, ZEPHYR, NERO, CLEOPATRA, ALEXANDER THE GREAT, MARY Queen of Scots, MARY Mother of JESUS, ST. VINCENT, ST. FRANCIS OF ASSISI, ELIZABETH of Hungary, and even Lady GODIVA (125, 126, 298)! Even to briefly summarize the many stories about roses in history and mythology would require more pages than are available here.

180. **Rubus sanctus** Schreb.   181. **Rubus ulmifolius** Schott

NUMBERS 33: 55—... those which ye let remain of them shall be pricks in your eyes, and *thorns* in your sides ...
JUDGES 8: 7 & 16—... then I will tear your flesh with the thorns of the wilderness and with *briers* ... And he took the elders of the city, and thorns of the wilderness, and *briers*, and with them he taught the men of Succoth.
ISAIAH 7: 23-25—And it shall come to pass in that day, that every place shall be, where there were a thousand vines ... it shall even be for *briers* and thorns ... all the land shall become *briers* and thorns. And on all hills that shall be digged with mattock, there shall not come thither the fear of *briers* and thorns ...
ISAIAH 9: 18—For wickedness burneth as the fire: it shall devour the *briers* and thorns ...
LUKE 6: 44—... nor of a *bramble bush* gather they grapes.

The Hebrew words rendered "brambles" in the Authorized Version are "atâb" or "atad" (Greek, ῥάμνος) and "choach" (Greek, κνίδη; dictionary translation, "a nettle"), and those rendered "briers" are "sillon" or "silonim", "sirpad", "shamîr", and "chedek" (Greek for the last two, ἄκανθαι; dictionary translation, "a prickly plant that can be painful") and "barkanim" (Greek, ἄκανθαι). In Numbers 33: 35 the word rendered "thorns" is "sikim". Many commentators have blithely supposed that all the passages in which these words occur referred to true brambles of the genus *Rubus*. However, modern authorities believe that this is not the case. For instance, the "bramble" of Judges 9: 14-15 is now regarded as the European boxthorn, *Lycium europaeum* (which see), while that of Isaiah 34: 13 is the Syrian thistle (*Notobasis syriaca*) and the spotted golden-thistle (*Scolymus maculatus*), which see. The "brier" of Micah 7: 4 is herein regarded as *Solanum sodomeum* (which see), and the "briers" of Ezekiel 2: 6 and 28: 24 are prickly butchers-broom, *Ruscus aculeatus* (which see).

The MOFFATT and GOODSPEED versions regard the "brier" of Ezekiel 28: 24 and the "bramble" of Luke 6: 44 as true briers and brambles respectively. JOHN SMITH was of the opinion (299) that the "brambles" of Judges 9: 14-15 and Isaiah 34: 13 referred to *Rubus sanctus*. The "briers" of Judges 8: 7 and Ezekiel 28: 24 he regarded as *Rosa rubiginosa* L., a species not known from the Holy Land area. PRATT considered the "bramble" of Judges 9: 14-15 to be *Zizyphus vulgaris*. Others have suggested *Lycium afra* L. for all the "brambles" of the Bible (299), but this species is unknown in Palestine. Nettles have been suggested by some writers for the "briers". *Rubus fruticosus* L. has also been nominated (184), but that species is likewise not known from Palestine (266).

The Palestine bramble, *Rubus sanctus*[127], and the closely related elmleaf bramble, *R. ulmifolius*[128], are prickly evergreen shrubs, spreading by means of suckers, the stems and young shoots being covered with an intense bloom and short hair. The prickles are strong, erect, and hairy. The leaves are made up of 3 to 5 leathery leaflets, which are unequally sharp-serrate, glabrous or more or less stellate-pubescent and bristly-hirsute on the upper surface, and white-woolly beneath. The elongated racemes are appressed-tomentose and armed with strong hooked prickles. The flowers are white, pink, rose, or purple in color, and the fruit is round or ovoid and black. The Palestine bramble grows mostly near water, while the elmleaf bramble is common in thickets in Lebanon and Palestine (266). The "common bramble" said by TRISTRAM to be "abundant in Palestine, especially about Bethlehem" (184) is doubtless *R. sanctus*, not *R. fruticosus* as stated by JONES.

Some writers have thought that the "thorns" of Psalms 58: 9 were *Rubus sanctus*, but *Rhamnus palaestina* seems more probable to us. In DALMAN's work (98) the Isaiah plant is interpreted as being *Daucus aureus* Desf. Some commentators maintain that the "briers" of Judges 8: 7 & 16 (Hebrew "barkonim") were "a sharp-toothed threshing instrument, or a plant used as such" (27). JONES cites these two verses definitely for *Rubus fruticosus* (184). JASTROW replaces the "thorns" of the King James Version of Numbers 33: 55 with "pricks" and uses "thorns" where the King James uses "pricks".

182. Ruscus aculeatus L.

EZEKIEL 2: 6—And thou, son of man, be not afraid of them, neither be afraid of their words, though *briers* and thorns be with thee . . .
EZEKIEL 28: 24—And there shall be no more a *pricking brier* unto the house of Israel . . .

A glance at the discussions under *Lycium*, *Notobasis*, *Rubus*, and *Solanum* will be sufficient to show how much diversity of opinion there is regarding the identity of the "briers" and "brambles" of the Bible. The reference in Ezekiel 28: 24 was considered to apply to *Rosa rubiginosa* L. by no less an authority than the famous Kew botanist, JOHN SMITH (299), but this species is not recorded by POST from the area (266). Others regarded the reference as being to *Rubus discolor* (233).

The Hebrew word used in both these Ezekiel passages is "sillon" (Greek, 'ἄκανθα ὀδύνης—dictionary translation, "a prickly plant that is painful"). This word is almost identical with the Arabic "shallûm" (267) or "sullaon" (184), which is the present-day name for the prickly butchers-broom or knee-holly, *Ruscus aculeatus*, and so most probably refers to this plant (184, 299). The prickly butchers-broom grows from 1½ to 3 feet tall, with ascending branches. The branchlets and twigs (here called phyllocladia) are flattened and leaf-like, ovate-lanceolate in shape, ½ to 1 inch long, tapering to a spiny tip. The small, imperfect, greenish flowers are borne singly or in pairs, on short pedicels, inserted on the midrib below the middle of the phyllocladium, and are subtended by a firm bract. The fruit is a red berry about ½ inch in diameter. The plant is common in rocky woods in the northern regions, Lebanon, and Palestine, especially about Mount Tabor and Mount Carmel (267).

There is apparently considerable doubt concerning the actual meaning of "sillon" in Ezekiel 2: 6. The Authorized Version's "though *briers* and thorns be with thee" is rendered "although they *cut* and wound you" by MOFFATT. The GOODSPEED version is "even when *thistles* and thorns are round about you", JASTROW says: "though *defiers* and despisers be with thee", and the Douay version says: "for thou art among *unbelievers* and destroyers." Marginal Bibles give the alternative translation of "rebels" for the word "briers".

---

[127] Also called *Rubus anatolicus* Focke, *R. ulmifolius* var. *anatolicus* (Focke) Focke, and (erroneously) *R. discolor* Weihe & Nees.
[128] Also called *Rubus rusticanus* Merc. and *R. discolor* Weihe & Nees.

183. **Ruta chalepensis** var. **latifolia** (Salisb.) Fiori 184. **Ruta graveolens** L.

LUKE 11: 42—But woe unto you, Pharisees! for ye tithe mint and *rue* and all manner of herbs.

The Greek word here translated "rue" is "peganon" (πήγανον) and there is no question as to the general correctness of the translation. The only doubtful point is which species of rue is involved. Most writers assert that it was the common rue, *Ruta graveolens* (98, 306). This is a perennial somewhat shrubby plant with several erect stems 2 to 3 feet tall and deeply cut leaves of a pleasing gray-green color (101, 293, 299). The yellow flowers, borne in terminal cymose inflorescences, are not exactly showy, yet give a welcome dash of color to the plant. A very strong odor emanates from the foliage. This species is a native of the Mediterranean region, is common in many parts of southern Europe, and is said by some writers to grow wild in Palestine and to be especially abundant on Mount Tabor, where it was first reported by the ill-starred HASSELQUIST (299, 306). POST, however, records it only from rocky places about Antioch (266). He asserts that the more common Holy Land plant is a variety of the African rue, *R. chalepensis* var. *latifolia*[129], very similar in appearance to the common rue, from which it is distinguished by its less deeply divided leaves, sharply acute capsules, and deeply fringed petals. It is common on hillsides and in thickets in the northern areas of Syria, Lebanon, the Antilebanons, Palestine, and south to Sinai (266). It is possible that both species were used, and probably also cultivated, by the Hebrews in Biblical days. The fact that rue was not tithed in Old Testament days indicates that it was then a wild plant that was used, probably *R. chalepensis* var. *latifolia*, while the fact that it was tithed in New Testament days implies that it was then a cultivated plant. The latter may have been *R. graveolens* or the more common species which may then not have been as common in the wild state as before.

Rue was very highly thought of by the ancients as medicinal (126, 299). For many centuries it was regarded as a preventive of contagion and an antiseptic. Among its supposed virtues we read that "It drives away the plague if you merely smell of it; it keeps maids from going wrong in affairs of love if only they will pause to eat it when tempted" (298), and also, by carrying a "bundle of rue, broom, maidenhair, agrimony, and ground-ivy, you may know every woman for a witch who is one, no matter how plain or otherwise she may appear to you" (298). Its reputation in medicine was probably due more to its strong odor than to any active medicinal principles (299).

The Talmudical law respecting plants was that "everything eatable, and which is taken care of, cultivated, and nursed in gardens, or in ploughed fields, and which has its growth in the earth, is subject to tithes" (268). ROSENMÜLLER says that the rue was declared free of tithe in olden days because, although it was a kitchen herb, it was wild and not originally a cultivated garden plant. Apparently in JESUS' day, however, it had become an object of cultivation and so was made subject to tithe (306). Rue was used for seasoning dishes. Its fresh leaves applied to the skin will produce a blister. Its pollen will cause an inflammation of the eyes of some people who have an allergy to it (178). Because of the mental association called forth by its English name "rue" with the verb "to rue" (of entirely different etymology), this plant came to be associated with the idea of repentance (158) and sprigs of rue used to be seen frequently on judges' benches in English courts of justice. SHAKESPEARE calls rue the "herb of grace" (158, 299).

Rue was supposed to heal the stings of bees, wasps, and scorpions, and the bites of snakes (125), and to prevent dizziness, dumbness, epilepsy, inflammation of the eyes (!), insanity and the "evil eye" (298). MITHRIDATES VI, king of Pontus (died B.C. 63) used rue as one of the chief constituents of the electuary or antidote by means of which he produced what he regarded as a physiological immunity against internal poisons which his disgruntled subjects were continually administering to him in efforts to murder him (158, 298). Such a potion is now called a mithridate. If gun-flints were boiled with rue and vervain it was asserted that the shot must surely reach the intended victim, no matter how poor the aim of the shooter (298).

[129]Also known as *Ruta latifolia* Salisb. and *R. bracteosa* P. DC.

185. Sabina excelsa (Bieb.) Antoine

II CHRONICLES 2: 8—Send me also cedar trees, fir trees, and *algum trees*, out of Lebanon: for I know that thy servants can skill to cut timber in Lebanon . . .

There has been a large amount of discussion as to the identity of the "algum trees" (Hebrew, "algomim") mentioned in this reference. Most writers have assumed that the "algum trees" of this reference and the "almug trees" of I Kings 10: 11-12 were the same, and that they were the red sandalwood, *Pterocarpus santalinus* (which see). This argument is based chiefly on the striking similarity of the words and the fact that the Authorized Version uses "almug" in I Kings 10: 11-12 and "algum" in II Chronicles 9: 10-11, in both cases referring to a tree imported from the mysterious Ophir. A marginal note in the King James version, however, suggests that the "algum" of the second of these references is an error for "almug" (Hebrew, "almoogim"), and this seems plausible, probably as an error of transcription by some scribe somewhere in the history of the original manuscript. The "almug" was a tree native to Ophir, while the "algum", from the entire context of the verse, was native to Lebanon. MOFFATT apparently concurs in this view for he uses "sandalwood" (an Indian tree) only in the I Kings 10: 11-12 and II Chronicles 9: 10-11 references. In II Chronicles 2: 8 he says "cypress logs", apparently alluding to *Cupressus sempervirens* var. *horizontalis* (which see), which is a Lebanese tree. The GOODSPEED and JASTROW versions regard all three passages as referring to the sandalwood, and so does Webster's dictionary (158). WILLIAM SMITH maintains the same viewpoint, suggesting that King HIRAM may have imported sandalwood logs from Ophir to Tyre and then re-shipped them from Tyre to King SOLOMON along with native Lebanese timber, so that the wood might easily have been referred to by non-botanically trained historians as both from Ophir and from Lebanon (306). We prefer to follow JOHN SMITH in regarding the "algum" as the native Lebanese tree known now as the eastern savin or Grecian juniper, *Sabina excelsa*[130], not the cypress, for which entirely different words are used in the Bible (299).

The eastern savin is a tree from 15 to 65 feet tall, with a pyramidal habit of growth and erect or spreading branches and twigs. The ovate-triangular acute needles of the larger branches are arranged in 3's, but those of the twigs are closely appressed, overlapping, scale-like, and arranged in 4 rows, with an oblong gland on the back of each (267). The staminate flowers are borne in erect or somewhat nodding catkins, and the globular blackish fruits (known technically as galbules) are crowded on short somewhat incurved twigs. The tree is abundant in mountainous woods of the alpine and subalpine zones in Syria, Lebanon, the Antilebanons, and Gilead (267), ascending to 2700 meters elevation.

The Persian walnut, *Juglans regia* (which see), has also been suggested for the "algum" (299), but this view is not held by any present-day authority.

186. Sabina phoenicia (L.) Antoine

LEVITICUS 14: 4-6 & 49-52—Then shall the priest command to take for him that is to be cleansed two birds alive and clean, and *cedar wood*, and scarlet, and hyssop: And the priest shall command that one of the birds be killed in an earthen vessel over running water: As for the living bird, he shall take it, and the *cedar wood*, and the scarlet, and the hyssop, and shall dip them and the living bird in the blood of the bird that was killed over running water . . . And he shall take to cleanse the house two birds, and *cedar wood* . . . And he

---

[130]Also referred to as *Juniperus excelsa* Bieb.

shall take the *cedar wood*, and the hyssop . . . and dip them in the blood . . . And he shall cleanse the house with . . . the *cedar wood*.
LEVITICUS 23: 40—And ye shall take you on the first day . . . the boughs of *thick trees* . . .
NUMBERS 19: 6—And the priest shall take *cedar wood*, and hyssop, and scarlet, and cast it into the midst of the burning of the heifer.

The word "cedar", represented by the Hebrew "erez" (Greek, κέδρος), occurs some 51 times in the Old Testament and in all but a few cases it plainly refers to the cedar-of-Lebanon, *Cedrus libani* (which see). In the above instances from the Pentateuch (184), however, the passages refer to events that took place while the Jews were in the "Wilderness" — *i.e.*, the desert of Sinai — and far away from Lebanon (299). The wood of the cedar-of-Lebanon was at that time unknown and certainly unavailable to them. They must have used, instead, the wood of some tree or shrub which was abundant on the desert. It is supposed that this wood was that of the brown-berried cedar, *Juniperus oxycedrus*, or the Phoenician juniper, *Sabina phoenicia*[131], or both. These are bushes or small trees, common in the desert regions of western Asia, the former being the more common, occupying the lower and middle mountain and subalpine zones throughout Syria, Lebanon, and Palestine, while the latter is indigenous to Sinai. Two other species of *Juniperus* and four of *Sabina* have been recorded from the Holy Land and adjacent regions (265, 267), although not particularly from the Sinai area. All of these species, as well as the two herein discussed, have more or less odoriferous wood. The burning of this cedar wood, or of brushes made of the branches and twigs, probably was for the purpose of destroying the unpleasant odors which would naturally result from the burning of the bodies of the bird and of the cow in the sacrificial offerings. The "thick trees" of Leviticus 23: 40 are thought also to have been *Sabina phoenicia*.

Some writers have supposed that the "box tree" of Isaiah 41: 19 and 60: 13 was actually a species of cedar called "ch'roovim", "scherbin", or "sherbin" (299, 306), perhaps *Sabina phoenicia* (184), but we are inclined to agree with the majority of botanical opinion that *Buxus longifolia* (which see) was actually the plant there referred to.

The brown-berried cedar sometimes grows to become an aromatic tree 18 or more feet tall, but more often is merely a hemispheric or even prostrate shrub. The twigs are triquetrous and the leaves are all needle-like, ternately whorled, and prickly. Its fruits are red or brown. The Phoenician juniper is also a shrub or tree, with erect or ascending terete branches and twigs. The leaves are minute, appressed, scale-like, and overlapping. The globular fruits are tawny-red and glossy, at length fleshy (267).

187. **Saccharomyces cerevisiae** Meyen

GENESIS 45: 23— . . . ten she asses laden with corn and *bread* . . .
EXODUS 12: 15 & 19—Seven days shall ye eat *unleavened bread;* even the first day ye shall put away *leaven* out of your houses: for whosoever eateth *leavened bread* from the first day until the seventh day, that soul shall be cut off from Israel . . . Seven days shall there be no *leaven* found in your houses: for whosoever eateth that which is *leavened*, even that soul shall be cut off from the congregation of Israel . . .
JOSHUA 9: 12—This our *bread* we took hot for our provision out of our houses on the day we came forth to go unto you; but now, behold, it is dry, and it is mouldy.
JUDGES 7: 13— . . . a cake of barley *bread* . . .
JUDGES 8: 5— . . . Give, I pray you, loaves of *bread* unto the people that follow me; for they be faint . . .
I SAMUEL 10: 3— . . . and another carrying three loaves of *bread* . . .
II SAMUEL 6: 19— . . . to every one a cake of *bread* . . .
II SAMUEL 13: 8 & 10— . . . And she took flour, and kneaded it, and made *cakes* . . . and did bake the *cakes* . . . And TAMAR took the *cakes* which she had made, and brought them . . . to AMNON . . .
I KINGS 17: 11-13— . . . Bring me, I pray thee, a morsel of *bread* . . . I have not a *cake* . . . make me thereof a little *cake* . . .
I KINGS 19: 6— . . . behold, there was a *cake* baken on the coals . . .
II KINGS 4: 42—And there came a man . . . and brought . . . *bread* of the firstfruits, twenty *loaves* of barley . . .

[131]Also referred to as *Juniperus lycia* L. and *J. phoenicia* L.

PROVERBS 6: 26—... a man is brought to a piece of *bread* ...
JEREMIAH 37: 21—Then ZEDEKIAH the king commanded that they should commit JEREMIAH into the court of the prison, and that they should give him daily a piece of *bread* out of the *bakers'* street until all the *bread* in the city were spent ...
EZEKIEL 4: 9, 12-13, & 15-17—Take thou also unto thee wheat, and barley, and beans, and lentiles, and millet, and fitches, and put them in one vessel, and make thee *bread* thereof ... And thou shalt eat it as barley *cakes*, and thou shalt bake it ... Even thus shall the children of Israel eat their defiled *bread* ... and thou shalt prepare thy *bread* therewith ... I will break the staff of *bread* in Jerusalem: and they shall eat *bread* by weight ... That they may want *bread* and water ...
HOSEA 7: 4 & 8—... as an oven heated by the baker, who ceaseth from raising after he hath kneaded the dough, until it be *leavened* ... a *cake* not turned.
AMOS 4: 5—And offer a sacrifice of thanksgiving with *leaven* ...
MATTHEW 4: 3-4—... command that these stones be made *bread* ... man shall not live by *bread* alone ...
MATTHEW 7: 9—Or what man is there of you, whom if his son ask *bread*, will he give him a stone?
MATTHEW 13: 33—... The kingdom of heaven is like unto *leaven*, which a woman took, and hid in three measures of meal, till the whole was *leavened*.
MATTHEW 16: 6—... beware of the *leaven* of the Pharisees and of the Sadducees.
LUKE 13: 21—It is like *leaven*, which a woman took and hid in three measures of meal, till the whole was *leavened*.
JOHN 6: 9, 11, & 13—There is a lad here, which hath five barley *loaves* ... And JESUS took the *loaves* ... and filled twelve baskets with the fragments of the five barley *loaves* ...
I CORINTHIANS 5: 6-8—... Know ye not that a little *leaven leaveneth* the whole lump? Purge out therefore the old *leaven*, that ye may be a new lump, as ye are *unleavened* ... Therefore let us keep the feast, not with old *leaven*, neither with the *leaven* of malice and wickedness; but with the *unleavened bread* of sincerity and truth

Bread was used as a food for man since very early times, although the use of the word "lechem" in Genesis 3: 19 (about 4004 B.C.) is not now regarded as referring to bread, but to food in general (306). The earliest reference to actual bread in the Bible is in Genesis 18: 6 (about 1898 B.C.). This bread, offered by ABRAHAM to the visiting angels, is regarded by authorities as probably having been unleavened bread (Hebrew, "matzot" or "matza") as was all the bread used in the temple as a sacred offering — Genesis 19: 3, Exodus 12: 15, 17-20, 34 & 39 and 29: 2, 23, & 40, Leviticus 2: 1-11, 6: 14-18, 7: 9-10, 8: 26, and 24: 5-9, Numbers 6: 15-21 and 15: 19-21, Judges 6: 19-21, I Samuel 1: 24 and 28: 24, I Chronicles 23: 29, Isaiah 44: 15, Jeremiah 7: 18, and Ezekiel 46: 14. This unleavened bread does not concern us in this chapter, but leavened bread does because the "leaven" (Hebrew, "chametz") used by the Hebrews consisted most often of a lump of old dough which had been brought to a high degree of fermentation through the action of certain one-celled plants known as yeasts, of the genus *Saccharomyces*. This mass of "leaven" was inserted into the larger mass of new dough prepared for the baking of the bread or cakes to be eaten by the ordinary people, and, on some occasions, by the priests as well. The prohibition of the use of leaven in sacred offerings was due to an idea that the leaven produced a "corruption" in the dough. This same idea is the basis for the figurative use of the word by JESUS in Matthew 16: 6 and by PAUL in I Corinthians 5: 6-8.

The grain employed in making bread was usually referred to as "corn", and consisted of wheat, spelt, and barley. "Fine flour" was flour of wheat; barley flour was used by the poorest in the land. Sometimes poor people mixed beans, lentils, millet, and "fitches" with the barley, and the bread so made is referred to as "barley cakes". In private homes the baking was usually done by the wife or one of the daughters or, in wealthy homes, by female servants. Professional bakers were men, and in Jerusalem occupied a definite sector known as "bakers' street" (Genesis 40: 16, I Samuel 8: 13, Jeremiah 37: 21). The "furnaces" spoken of in the Bible were bakers' ovens (Nehemiah 3: 11 and 12: 38), and the fuel used to feed them was wood, charcoal, or dried grass and flower-stalks (Matthew 6: 30). The more pastoral Jews, like the Bedouins today, baked bread upon heated stones, among hot embers, or even between layers of smouldering dung (306). Usually this was unleavened bread. The unleavened bread used for sacred offerings was often baked in a pan.

Bread was usually made by mixing the flour with water, or milk, if available, and then kneading it with the bare hands (and feet, in Egypt) in a small wooden bowl or "kneading-trough" until it became dough. Leaven was then added. When time was short, leaven was omitted, and only unleavened cakes were made, as is still done by the Bedouins today. The leavened dough was allowed to stand for a time, during which the yeast or leaven within it began active fermentation, giving off carbon dioxide gas, which inflated or "raised" the dough (158). The dough was then divided into round cakes (Hebrew, "oogah") and mixed with oil or rolled into thin wafers and merely coated with oil. After this the cakes or wafers were placed into the oven (306).

The "mould" spoken of as occurring on old bread in Joshua 9: 12 was probably the universal common gray bread-mold, *Mucor mucedo* L. (which see) (158). The common brewers' and distillers' yeast, *Saccharomyces cerevisiae*, a highly variable species, is the common bakers' yeast throughout the world.

188. Saccharomyces ellipsoideus Reess

GENESIS 9: 21—And he drank of the *wine*, and was drunken . . .
GENESIS 14: 18—And MELCHIZEDEK king of Salem brought forth bread and *wine* . . .
GENESIS 27: 28—Therefore God give thee of the dew of heaven, and the fatness of the earth, and plenty of corn and *wine*.
GENESIS 49: 11-12— . . . he washed his garments in *wine* . . . His eyes shall be red with *wine*.
EXODUS 29: 40— . . . and the fourth part of an hin of *wine* for a drink offering.
LEVITICUS 23: 13— . . . and the drink offering thereof shall be of *wine*, the fourth part of an hin.
NUMBERS 6: 3—He shall separate himself from *wine* and *strong drink*, and shall drink no vinegar of *wine*, or vinegar of *strong drink* . . .
NUMBERS 15: 5, 7, & 10—And the fourth part of an hin of *wine* for a drink offering . . . And for a drink offering thou shalt offer the third part of an hin of *wine* . . . And thou shalt bring for a drink offering half an hin of *wine* . . .
DEUTERONOMY 18: 4—The firstfruit also . . . of thy *wine* . . . shalt thou give him.
DEUTERONOMY 33: 28— . . . the fountain of JACOB shall be upon a land of corn and *wine* . . .
JUDGES 6: 11— . . . his son GIDEON threshed wheat by the *winepress* . . .
JUDGES 9: 13—And the vine said unto them, Should I leave my *wine*, which cheereth God and man . . .
JUDGES 13: 14—She may not eat of any thing that cometh of the vine, neither let her drink *wine* or *strong drink* . . .
JUDGES 19: 19— . . . and there is bread and *wine* also for me . . .
I SAMUEL 25: 18—Then ABIGAIL made haste, and took . . . two bottles of *wine* . . .
II CHRONICLES 2: 10—And, behold, I will give to thy servants . . . twenty thousand baths of *wine* . . .
NEHEMIAH 13: 15—In those days I saw in Judah some treading *wine presses* . . . and bringing in . . . *wine*, grapes, and figs . . .
JOB 24: 11—Which make oil within their walls, and tread their *winepresses*, and suffer thirst.
PSALMS 4: 7—Thou hast put gladness in my heart, more than in the time that their corn and their *wine* increased.
PSALMS 75: 8— . . . the *wine* is red . . .
PSALMS 104: 15—And *wine* that maketh glad the heart of man . . .
PROVERBS 9: 2 & 5— . . . she hath mingled her *wine* . . . Come . . . and drink of the *wine* I have mingled.
PROVERBS 20: 1—*Wine* is a mocker, *strong drink* is raging . . .
PROVERBS 21: 17— . . . he that loveth *wine* and oil shall not be rich.
PROVERBS 23: 20 & 29-32—Be not among *winebibbers* . . . Who hath woe? who hath sorrow? who hath contentions? who hath babbling? who hath wounds without cause? who hath redness of eyes? They that tarry long at the *wine*; they that go to seek mixed *wine*. Look not thou upon the *wine* when it is red, when it giveth his colour in the cup, when it moveth itself aright. At the last it biteth like a serpent, and stingeth like an adder.
PROVERBS 31: 4-6— . . . it is not for kings to drink *wine*; nor for princes *strong drink*: Lest they drink, and forget the law, and pervert the judgment . . . Give *strong drink* unto him that is ready to perish, and *wine* unto those that be of heavy hearts.
ECCLESIASTES 10: 19— . . . and *wine* maketh merry . . .
SONG 8: 2— . . . I would cause thee to drink of *spiced wine* of the juice of my pomegranate.
ISAIAH 5: 2, 11-12, & 22— . . . and also made a *winepress* therein . . . Woe unto them that rise up early in the morning, that they may follow *strong drink*; that continue until night, till *wine* inflame them! . . . and *wine*, are in their feasts . . . Woe unto them that are mighty to drink *wine*, and men of strength to mingle *strong drink*.
ISAIAH 16: 10— . . . the treaders shall tread out no *wine* in their presses; I have made their vintage shouting to cease.
ISAIAH 25: 6— . . . a feast of *wines* on the lees . . . of *wines* on the lees well refined.
ISAIAH 28: 7—But they also have erred through *wine*, and through *strong drink* are out of the way; the priest and the prophet have erred through *strong drink*, they are swallowed up of *wine*; they are out of the way through *strong drink* . . .
ISAIAH 63: 2-3—Wherefore art thou red in thine apparel, and thy garments like him that

treadeth in the *winefat*? I have trodden the *winepress* alone...
JEREMIAH 35: 2, 5-6, 8, & 14—...give them *wine* to drink...and set before the sons... pots full of *wine*, and cups, and I said unto them, Drink ye *wine*. But they said, We will drink no *wine*...Ye shall drink no *wine*...to drink no *wine* all our days...he commanded his sons not to drink *wine*...
JEREMIAH 48: 33—... I have caused *wine* to fail from the *winepresses* ...
LAMENTATIONS 1: 15—... the Lord hath trodden the virgin ... as in a *winepress*.
LAMENTATIONS 2: 12—They say to their mothers, Where is corn and *wine*? ...
EZEKIEL 27: 18—... in the *wine* of Helbon ...
HOSEA 4: 11—... *wine* and *new wine* take away the heart.
HOSEA 7: 5—In the day of our king the princes have made him sick with bottles of *wine* ...
HOSEA 14: 7—... the scent thereof shall be as the *wine* of Lebanon.
AMOS 9: 13—... and the mountains shall drop *sweet wine* ...
HABAKKUK 2: 5 & 15—Yea also, because he transgresseth by *wine*, he is a proud man ... Woe unto him that giveth his neighbour *drink*, that puttest thy bottle to him, and makest him *drunken* ...
ZECHARIAH 9: 15 & 17—... and they shall drink, and make a noise as through *wine* ... corn shall make the young men cheerful, and new *wine* the maids.
ZECHARIAH 10: 7—... and their heart shall rejoice as through *wine* ...
MATTHEW 21: 33—... and digged a *winepress* in it ...
MARK 15: 23—And they gave him to drink *wine* mingled with myrrh ...
JOHN 2: 3 & 9-10—And when they wanted *wine*, the mother of JESUS saith unto him, They have no *wine* ... When the ruler of the feast had tasted the water that was made *wine* ... Every man at the beginning doth set forth good *wine* ... but thou hast kept the good *wine* until now.
ACTS 2: 13—Others mocking said, These men are full of *new wine*.
EPHESIANS 5: 18—And be not drunk with *wine*, wherein is excess ...
I TIMOTHY 5: 23—Drink no longer water, but use a little *wine* for thy stomach's sake ...
REVELATION 14: 19-20—... and cast it into the great *winepress* ... And the *winepress* was trodden without the city, and blood came out of the *winepress* ...
REVELATION 19: 15—... and he treadeth the *winepress* of the fierceness and wrath of Almighty God.

The making of wine is one of the most ancient arts of mankind. In the Bible it is first mentioned in the history of NOAH (about 2347 B.C.), to whom its invention is therefore often attributed. Almost all of the "wine" (Hebrew, "teerosh" and "yayin") of the Bible was most certainly made from the expressed juice of the vine, *Vitis vinifera* (which see), although in Song 8: 2 we are told that wine was also made from the juice of the pomegranate, *Punica granatum* (which see). It is assumed that pomegranate wine was much less potent than that of the vine (306). The "strong drink" (Hebrew, "shaychar") of the Bible, where it is mentioned along with or in contrast to wine, is thought by some authorities (299) to have been the product of the date palm, *Phoenix dactylifera* (which see).

The harvesting of grapes in Palestine in Biblical days took place in September and was accompanied by great rejoicing and festivities. The large baskets of juicy fruit were carried to special winepresses (Hebrew, "yekev" and "gitot") (Jeremiah 6: 9). These presses usually consisted of two receptacles or vats, set one above the other, each hewn out of a piece of solid rock (Joel 3: 13, Isaiah 5: 2, margin, Matthew 21: 33). Such ancient winepresses may still be seen in parts of Palestine today (306). The fresh grapes were dumped into the upper vat (Hebrew for "winevat" is "begat"). A certain amount of juice exuded naturally into the lower vat just as a result of the weight of the grapes upon each other. This was the juice of the ripest and softest of the grapes, and was carefully gathered and kept separate from the juice later pressed out. It was the "sweet wine" (Hebrew, "tirosh"), "new wine" (Hebrew, "ahsis"), or "first wine" of Hosea 4: 11, Amos 9: 13, and Acts 2: 13. After this had been gathered, the pressing of the grapes began through the age-old process of "treading". Depending upon the size of the vat, one or more men would tramp on the grapes with their bare feet, encouraging each other with loud shouting (Isaiah 16: 9-10, Jeremiah 25: 30 & 48: 33). In this process their feet, legs, and loincloths became dyed red with the juice — a fact which we find alluded to in Genesis 49: 11 and Isaiah 63: 2-3. The juice thus expressed, of course, flowed freely into the lower vat, whence it was collected in various vessels. Sometimes this juice was preserved in the unfermented state and consumed as "must", or, as we would now say, as grapejuice, but more usually it was allowed to ferment through the enzymatic action of the wine-ferment yeast, *Saccharomyces ellipsoideus.*

Certain groups of commentators have attempted, from time to time, to prove that the wine of Biblical days was not fermented and thus was non-alcoholic, but the general consensus of impartial students of the Bible today is that most of it, at least, must have been potently alcoholic (306). Wine that was to be stored for any considerable length of time usually had added to it a certain amount of "lees" or dregs from old wine (Isaiah 25: 6). Because of the presence of these lees, old wine had to be strained or "refined" before being served (Isaiah 25: 6).

Unfermented "wine", or "must", was preserved in jars or bottles buried in the earth (306). Wine that was "mingled" had various spices added to it and was used at great festivals, feasts, and drinking parties (Proverbs 9: 2 & 5 and 23: 30, Isaiah 5: 22). The "wine mingled with myrrh", however, was apparently offered to JESUS just before his crucifixion to deaden his pain, much as the unfortunate patients used to be fed intoxicants preceding the butchery that accompanied major operations before the days of anaesthetics. In Biblical days wine was employed as an article of ordinary household hospitality (Genesis 14: 18), as well as at marriage feasts (John 2: 1-11). Under the Levitical ordinances wine was to accompany the daily sacrifices (Exodus 29: 40), the presentation of the firstfruits of the land (Leviticus 23: 13), and numerous other temple offerings (Numbers 15: 5). Tithing was required on wine as it was on all other cultivated products of the land, and a portion of the first vintage had to be offered to the Lord in the temple with all other "first fruits" (Deuteronomy 18: 4, Exodus 22: 29).

189. Saccharum officinarum L.

JOSHUA 16: 8—The border went out from Tappuah westward unto the river *Kanah* ...
JOSHUA 19: 28—And Hebron, and Rehob, and Hammon, and *Kanah* ...
ISAIAH 43: 24—Thou hast bought me no *sweet cane* with money ...

The Hebrew words rendered "sweet cane" in the Old Testament are "keneh" or "kaneh" and "keneh hattôv" or "kaneh hattov" (Greek, κιννάμωμον). There is some doubt as to the identity of the plant referred to by these words. POST lists two species of sugar-cane as indigenous and wild in the Holy Land, but one of these, *Saccharum sara* Roxb., is known only from Lebanon and probably is not involved in these references. The other native species is *S. biflorum* Forsk.[132], inhabiting the banks of ditches and streams from Syria and Lebanon, through Palestine, south to Arabia Petraea and Sinai (95, 267). Possibly this wild cane was employed by the ancient Jews, but most authorities feel that ISAIAH's "sweet cane" was the true sugar-cane, *S. officinarum*. This plant is thought to be native somewhere in the tropics of the Eastern Hemisphere, but the precise land of its origin is not known, as it has been cultivated by man since time immemorial, and has now spread over all the tropical parts of Asia, Africa, and America (299), although it is not now known in the wild state anywhere (158). POST thinks that it came originally from the East Indies, to the Levant through India (267). It is a tall stout perennial grass, of maize-like aspect, with flat distichous leaves, many-jointed stems, and a large plume-like terminal panicle of flowers. It does not bloom in Palestine, although it is widely cultivated there. It is believed to have been introduced into Europe by the Venetians at about the middle of the 12th century and into America by the Spaniards in the 16th century (299).

Although the art of making sugar from this plant was probably unknown to the Hebrews, the canes would nevertheless be highly esteemed — as they are today among primitive peoples — for sweetening food and drinks and for chewing as a confection (299). DALMAN claims that the art of sugar-making was not learned by the ancients until the seventh century A.D.; the sweetness available to the people of Bible times was mainly from honey (98).

Some writers have regarded the "sweet calamus" and "sweet cane" of the Bible as one and the same plant (184), but it seems rather obvious that one was

---

[132]Also called *Saccharum aegyptiacum* Willd. and *S. spontaneum* var. *aegyptiacum* (Willd.) Hack.

a sweet-smelling and the other a sweet-tasting plant. JOHN SMITH, for instance, thought that the "sweet cane" of Jeremiah 6: 20 was also sugar-cane (299), but most commentators, including MOFFATT, think otherwise. SMITH thought that both ISAIAH's and JEREMIAH's plants were sweet-tasting, but the context of the Jeremiah passage plainly indicates a sweet-smelling, not sweet-tasting, plant, and MOFFATT goes so far as to translate the word there "perfume". The Douay version says "sweet smelling cane" in this verse. JEREMIAH's "sweet cane" should have been rendered "sweet calamus" in the King James version. It was probably *Andropogon aromaticus* (which see). The Revised Version, in marginal notes, suggests "calamus" for "sweet cane" (27). The JASTROW version agrees with the Authorized in each of the three passages cited at the head of this chapter.

Because of the similarity between the words "Kanah", used as the name of a brook in Joshua 16: 8 and 19: 28, and the Hebrew word "keneh" here regarded as referring to the sugar-cane, it is supposed that this plant, or the wild variety mentioned above, was especially abundant along that stream. The name's literal translation is given by concordances as "a place of reeds" (27).

JONES and others have suggested the sweet-rush or camels-hay, *Andropogon schoenanthus* L., for the "sweet cane" (27, 184). TRISTRAM has reported finding this species at Gennesaret, but POST does not regard it as native (267). The sugar-cane seems far more likely.

According to WILLIAM SMITH (306) the very similar Hebrew word "kaneh" was a generic term applied to reeds and reed-like plants of various sorts, and was also applied to the "stalk" of wheat (Genesis 41: 5 & 22), the "branches" of a candlestick (Exodus 25 and 37), and the humerus bone of a man's upper arm (Job 31: 22).

The Hindu farmer believes that if his sugar-cane produces flowers it is a sign of impending death in his family. To avoid this catastrophe he burns the unused canes at the end of the season as a votive offering to NAGBELE, the spirit of the plant (298).

190. Salicornia fruticosa (L.) L.   191. Salicornia herbacea (L.) L. 192. Salsola inermis Forsk.   193. Salsola kali L.
(FIGURE 80)

JEREMIAH 2: 22—For though thou wash thee with nitre, and take thee much *sope*, yet thine iniquity is marked before me . . .
MALACHI 3: 2— . . . who shall stand when he appeareth? for he is like a refiner's fire, and like fuller's *sope*.
SUSANNAH 17— . . . Bring me oil and *washing-balls*, and shut the garden-doors, that I may wash me.

The Hebrew word translated "sope" in the Authorized Version is "bôrîth" or "borit" (Greek, πόα, meaning "grass"). It is a general term for any substance with cleansing qualities; but, as we can see from the Jeremiah reference, it was used in contradistinction to "nether", which undoubtedly signifies natron or mineral alkali. Thus, we may assume that "bôrîth" refers to vegetable alkali or potash (306). MOFFATT's translation of the Jeremiah verse is: "You may wash yourself with lye and plenty *soap*, but I see your sin of the deepest dye, says the Lord Eternal". JASTROW merely substitutes the modern spelling "soap" for the Authorized Version's "sope". GOODSPEED changes "washing-balls" in Susannah to "soap".

The alkaline salts that were used in the manufacture of soap in Biblical days were obtained by burning plants of saltwort, *Salsola*, and jointed-glasswort, *Salicornia*, and perhaps related plants. There are twenty kinds of saltwort in the Holy Land area, of which the prickly saltwort, *Salsola kali*, and the unarmed saltwort, *S. inermis*, are the two commonest. The only two species of jointed-glasswort in the area are the shrubby jointed-glasswort, *Salicornia fruticosa*[133], and the slender, marsh, or herbaceous jointed-glasswort, *S. herbacea*[134].

---
[133] Also known as *Salicornia europaea* var. *fruticosa* L. and *S. arborea* L.
[134] Also known as *Salicornia europaea* var. *herbacea* L.

The jointed-glassworts are herbaceous or erect and twiggy, shrubby plants, 4 inches to 3 feet tall, inhabiting salt marshes along the seacoasts. The stems and branches are fleshy and jointed, often rooting where they come in contact with the ground. The leaves are opposite, semi-cylindric, not more than an inch long, with rudimentary blades, and there are opposite, connate, persistent bracts. The tiny flowers are borne in cylindric or club-shaped spikes, are ternate, connate, and immersed in a peculiar structure called the floral-cup. The saltworts are similar. The ones here listed are annual plants, of a bluish-green hue, 6 inches to 3 feet tall, inhabiting sandy and salty maritime situations. The numerous prostrate or ascending succulent branches are brittle, but not jointed, and are usually more or less hairy or papillose-mealy. The leaves are small, linear, lanceolate, or awl-shaped, half-clasping at the base, often spine-tipped, sometimes fleshy and almost terete. The tiny flowers are solitary or in clusters of 2 or 3, loosely spiked along the branches (267).

Soap was made by mixing the ashes (rich in potash) of these plants with olive oil, instead of animal fat. The making of soap was an extensive occupation along the Mediterranean coast of Palestine from remote antiquity. Potash and soap still form a considerable article of trade from Joppa and other Mediterranean ports of Palestine even at the present day (184, 299). Some writers have suggested that the Egyptian fig-marigold, *Mesembryanthemum nodiflorum* L., the bouncing-bet, *Saponaria officinalis*, and some species of goosefoot, *Chenopodium*, may also have been used in soap-making (299, 306). The bouncing-bet is now widely cultivated in the region for its saponaceous roots, which are used for washing woollens ("and which do not shrink when washed with this soap" according to POST), and also in the manufacture of the native "halâwi" (266). It is, however, a native of Asia Minor and Greece and was not present in Palestine in Old Testament days. The fig-marigold and goosefoots, on the other hand, are natives and may well have been so used.

It is of interest to note, in passing, that the Arabic name for the saltwort, "kali" or "el kali", is the source of our word alkali (299). The use of potash is very ancient, and to its discovery on the Belus we owe the invention of glass by the Phoenicians (184).

194. Salix acmophylla Boiss.   195. Salix alba L.
196. Salix fragilis L.   197. Salix safsaf Forsk.

JUDGES 16: 7-9—And SAMSON said unto her, If they bind me with seven *green withs* that were never dried, then shall I be weak, and be as another man. Then the lords of the Philistines brought up to her seven *green withs* which had not been dried, and she bound him with them ... And he brake the *withs*.
JOB 40: 22—The shady trees cover him with their shadow; the *willows* of the brook compass him about.
ISAIAH 15: 7—Therefore the abundance they have gotten, and that which they have laid up, shall they carry away to the brook of the *willows*.
ISAIAH 44: 4—And they shall spring up as among the grass, as *willows* by the water courses.
EZEKIEL 17: 5—He took also of the seed of the land, and planted it in a fruitful field; he placed it by great waters, and set it as a *willow tree*.
AMOS 6: 14— ... and they shall afflict you from the entering in of Hemath unto the river of the *wilderness*.

Two Hebrew words have been rendered "willows" in the King James version — in the Ezekiel passage quoted above the word is "tzaphtzaphah" (omitted by the Septuagint), while in all the other passages it is "aravah" and "aravim" or "'arâbim" (Greek, ἰτέα literally "a willow"). The "willows" of Leviticus 23: 40 and Psalms 137: 2 are usually regarded (299) as having been *Populus euphratica* (which see), although both the GOODSPEED and MOFFATT versions still regard those of the Leviticus verse as "water-willows" ("willows of the brook" in the Douay and JASTROW versions). The LEESSER, Douay, and JASTROW versions also still say "willows" in Psalms 137: 2, but MOFFATT and the GOODSPEED version replace this with "poplars". An ancient legend states that the harps mentioned in this psalm were hung on the branches of the weeping willow, *Salix babylonica* L., whose previously erect branches were weighted down thereby and have remained drooping or pendent ever since (299). Even LINNAEUS seems to have been misled by this legend and named the willow "Babylonian". The species, however, is Chinese (Japanese, according to POST), and trade with China is not thought to have been established yet at that early date (570 B.C.). GROSER, WILLIAM SMITH, and Mrs. BARNEVELD all state that *S. babylonica* was the "willow" on which the captive Israelites are said to have hung their harps (47, 146, 306.) Approximately 21 kinds of willow grow in Palestine and there are few places in that country where one or more may not be found growing along the watercourses (178, 267, 268, 293). The weeping willow, however, is not native to the region, although it is now widely cultivated there and has become naturalized in some places (267).

Some commentators believe that the oleander, *Nerium oleander* (which see), may have been one of the plants referred to as "willow". TRISTRAM inclines to this view (184, 299), especially for the Ezekiel reference. MOFFATT and the GOODSPEED version both say merely "slip" in this passage without any hint as to what kind of tree it was a slip of — MOFFATT's version is: "He also took some seed from the land of Judah and planted it in fertile soil to sprout like a *slip* beside brimming streams." The JASTROW version differs in saying "as a *slip*, and it grew and became a spreading vine." It is, of course, well known that the branchlets and twigs of the brittle or crack willow, *Salix fragilis*, break off very easily and will sprout and grow into new trees readily if they fall into a moist muddy spot along the stream bank or lake shore. Probably this is the type of "slip" referred to by Ezekiel.

The willows of Palestine listed at the head of this chapter are shrubs or trees, usually not attaining the height of willows in Europe or eastern North America. The branches are often reddish in color. The leaves are oblong-elliptic or lanceolate, subtended either by minute or by large lanceolate or heart-shaped stipules. The flowers are borne in lateral catkins and appear in March or April, at the same time as the leaves. The fruiting capsules are glabrous. The oleander, suggested by so many writers, also grows along watercourses. In some parts of the Jordan valley it forms extensive thickets and "in many of the streams of Moab it forms a complete screen, which the sun's rays can never penetrate" (299). Its leaves, also, are elongate, narrow, and willow-like.

Salix aegyptiaca L. and *S. viminalis* L. are suggested by some writers (184, 306), but the first of these species is not common in the region, being recorded by POST only from Tyre (267), and the second is not known at all from the area. It is more probable that the Palestine willow, *S. safsaf*[135], is the willow of most of these references, for it is the commonest in the region. In fact its present-day vernacular names in that area are "safsaf" and "tsaftsâfâh" (267) — practically identical with the Hebrew word used in Ezekiel. Many writers assert that willow branches were used in the construction of the booths for the Feast of the Tabernacles (27, 184), but it is now believed that these "willows" were the Euphrates poplar, *Populus euphratica*.

The "brook of the willows" — "Wady of Willows" according to MOFFATT, "torrent of the willows" in the Douay version — mentioned in Isaiah 15: 7 was one of the boundaries of the land of Moab and is probably identical with the wady mentioned in Amos 6: 18 as the then recognized southern limit of the northern kingdom (306). The latter is rendered "the river of the wilderness" in the King James version. Widely as the two names differ in this version, they are almost identical in the original (306). The Douay version of Amos 6: 18 says "even to the torrent of the *desert*", the GOODSPEED and JASTROW versions render it "unto the brook of the Arabah", while MOFFATT says "to the wady of the Arâbah". The plural form, "'arâbim", is the word translated "willows" in the other passages quoted above.

The Hebrew words rendered "green withs" in the Authorized Version are "yether lach" or "yehter lach" and "y'tarim lachim" (Greek, νευρά ὑγρά, meaning "a moist cord of sinew"). In spite of the fact that the Douay version

---

[135] Also called *Salix subserrata* Willd. and *S. octandra* Sieb.

renders these words "sinews" and the MOFFATT, JASTROW, and GOODSPEED versions say "bowstrings" — in both cases implying an animal origin for them — it seems probable to us that the soft, pliable, green twigs of willow constituted these "green withs". LEESSER gives the more general translation of "moist cords".

WYCLIFFE'S Bible, however, uses a word meaning "willow" in this passage (27, 184). Withes are slender flexible twigs or branches, usually of willow or osier, used as bands or in wicker-work (158).

198. Salvia judaica Boiss.
(FIGURE 81)

EXODUS 37: 17-18—And he made the *candlestick* of pure gold: of beaten work made he the *candlestick*; his shaft, and his branch, his bowls, his knops, and his flowers, were the same: And six branches going out of the sides thereof; three branches of the *candlestick* out of the one side thereof, and three branches of the *candlestick* out of the other side thereof.

The ancient Hebrews, like the Egyptians, Greeks, and Romans, and, in fact, all tribes of mankind, borrowed many of their art forms from plants (276). The acanthus is the source of the decorative scrolls so abundant in ancient, medieval, and modern art. The waterlily was the inspiration for the lily work in SOLOMON's temple and the capitals of Egyptian and Phoenician columns. The pomegranate flower was the inspiration for ornamental bells, and its fruit suggested the form of royal crowns from the time of SOLOMON to the present day. Similarly, the seven-branched candlestick, which has come down to us through the ages as a traditional Jewish symbol, had its origin in the branched inflorescence of a lowly sage plant growing on the mountains and hills of Palestine. This plant's inflorescence (when pressed flat) has almost exactly the shape and form of the seven-branched candlestick (Hebrew, "menorah"), with its central spike and 3 pairs of lateral branches, each bending upwards and inwards in so symmetrical a fashion. On each branch of the plant's inflorescence are whorls of buds, which, again, probably gave the idea to the artist for the "knops" or "knobs" on the Biblical golden candlestick.

The Judean sage, *Salvia judaica*, grows to 3 feet tall. Its stems are 4-angled, stiff, and roughish, panicled above. Its paired leaves are bullate-wrinkled, crenate or dentate-lobed; the basal and sometimes the lower stem-leaves are lyrate-pinnatipartite with ovate obtuse leaflets, the lateral one or two pairs sometimes confluent, the terminal much larger. The basal leaves are petiolate, the stem leaves sessile. The flowers are borne in distant whorls of 6 to 12 each, and the pedicels are shorter than the decidedly red calyx. The corollas are violet, about ¾ of an inch long, strongly 2-lipped (267). The plant now grows from Latakia, Beirut, and Tripoli, south through Nazareth, Hebron, Tiberias, Samaria, Judea, and other parts of Palestine. It has not been reported from Egypt or Sinai, but its distribution may have been more extensive in Biblical days, and there are many closely related species in adjacent areas.

199. Saussurea lappa (Decaisne) C. B. Clarke
(FIGURE 33)

PSALMS 45: 8—All thy garments smell of myrrh, and aloes, and *cassia*...

The Hebrew words rendered "cassia" in the above passage by the King James, JASTROW, Douay, and GOODSPEED versions are "ketzioth" or "k'tziot" and "ketziah" or "k'tziah" (Greek, κασία). Some early writers supposed that the sweet-smelling European orris-root or Florentine iris, *Iris florentina* L., was the plant referred to in this passage (299), and that it was the "cassia" of Exodus 30: 24 as well (27). This species, however, is native only to southern Europe and it is very doubtful whether it would have been available in Palestine at the time of DAVID. Most modern commentators are of the opinion that it was the Indian orris, *Saussurea lappa*[136]. This is a composite from the Himalaya Mountains of Kashmir, whose fragrant roots are gathered in great quantities and shipped to Bombay and thence to ports on the Persian Gulf

---

[136] Also known as *Aucklandia costus* Falc., *Theodorea costus* (Falc.) Kuntze, and *Aplotaxis lappa* Decaisne. The Oxford Bible calls it "*Aplotaxis luffa*" (27).

and Red Sea, as well as eastward to China. It is a strong-rooted perennial plant, with large incised leaves borne in a basal rosette. The thistle-like stem grows 5 to 6 feet tall and bears heads of purple thistle-like flowers (299). It is employed medicinally and as an aphrodisiac (158), but its chief use is as a perfume. In China and India it serves as incense in temples (158). It is quite possible that this plant was known to DAVID and SOLOMON (299). MOFFATT renders the Psalms passage referred to above: "Fragrant are your robes with *orris*, myrrh, and aloes." JONES believed that "ketzioth" referred either to the Indian orris or to "a cassia-bearing tree" (184). The Hebrew word for the cassia, *Cinnamomum cassia* (which see), however, is "kiddah", and LINNAEUS' genus *Cassia*, embracing the sennas, has nothing whatever to do with the "cassia" of the Scriptures.

Some writers think that the "calamus" and "sweet calamus" of Exodus 30: 23, Song 4: 14, and Ezekiel 27: 19 may refer, in part, at least, to this Indian orris, but these are regarded by us as *Andropogon aromaticus* (which see). The Revised Version says "costus" for the "cassia" of Exodus 30: 24, now usually regarded as having been *Cinnamomum cassia*, and other commentators have also regarded them as identical (27, 306).

200. **Shigella ambigua** (Andrews) Weldin
201. **Shigella dysenteriae** (Shiga) Castel. & Chalm.
202. **Shigella paradysenteriae** (Collins) Weldin

II CHRONICLES 21: 18-19—And after all this the Lord smote him in the bowels with an *incurable disease*. And it came to pass, that in the process of time, after the end of two years, his bowels fell out by reason of his sickness: so he died of sore diseases . . .
MATTHEW 9: 20—And, behold, a woman which was diseased with an *issue of blood* twelve years, came behind him, and touched the hem of his garment.
MARK 5: 25-29—And a certain woman, had an *issue of blood* twelve years, And had suffered many things of many physicians, and had spent all that she had, and was nothing bettered, but rather grew worse . . . came . . . and touched his garment . . . And straightway the fountain of her blood was dried up; and she felt in her body that she was healed of her *plague*.
LUKE 8: 43-44—And a woman having an *issue of blood* twelve years, which had spent all her living upon physicians, neither could be healed of any, Came behind him, and touched the border of his garment: and immediately her *issue of blood* stanched.
ACTS 28: 8—And it came to pass, that the father of PUBLIUS lay sick of a *fever* and of a *bloody flux* . . .

Medical literature contains many references to diseases mentioned in the Bible and there is as much uncertainty and argument about the identification of these Scriptural diseases as there is about the flowering plants. The general consensus, however, seems to be that the "incurable disease" of II Chronicles 21: 18-19 was dysentery; and it is very probable that the "issue of blood" of Matthew, Mark, and Luke and the "bloody flux" of Acts 28: 8 also were this disease, although the evidence is not so conclusive in these two latter cases. The "issue of blood" in the Gospel references is regarded by some medical writers as having been caused by a cancer of the urethra, but dysentery is far more probable. The "bloody flux" of Acts is not accompanied by sufficient details to render its identification certain. It might even have been a bloody flux from the mouth and thus symptomatic of acute tuberculosis. In the case of the "incurable disease" of JEHORAM in 889 B.C., however, it seems quite definitely to have been dysentery.

Dysentery is a disease accompanied by fever, inflammation, and ulceration of the large intestine. It is characterized by griping pains, constant urge to evacuate the bowels, and the discharge of mucous and blood (158). It may be acute or chronic, sporadic or epidemic, and has always been very common in Arabia and adjacent lands. When it is epidemic it is often also malignant, and in Arabia and the Holy Land is still described as being "often fatal" (306). So-called tropical or amoebic dysentery is caused by the presence of amoebae in the intestinal canal and therefore does not enter into the scope of this book, but bacillary or Japanese dysentery is due to the presence of certain bacilli in the intestinal canal. Chief of these are *Shigella dysenteriae*[137], which is a

---

[137] Also called *Bacillus dysenteriae* Shiga, *B. shigae* Chester, *B. dysentericus* Ruffer & Willmore,

cause of the disease in man and monkeys, *S. paradysenteriae*[138], which causes it in man and is also a cause of summer diarrhea or "summer complaint" in children, and *S. ambigua*[139], a less-important cause of dysentery in man (52). So-called diphtheritic dysentery is characterized by the formation of a false membrane (158). Other species of *Shigella* are found in the feces in cases of mild dysentery, but are not the causal agents (52).

The "issue of blood" of Leviticus 15: 19-30 refers merely to normal, menstrual discharge causing a temporary uncleanness for a period of usually 7 days, after which the woman was to be "purified" by the usual temple offering (306).

203. Sinapis arvensis L.

PROVERBS 24: 31—And, lo, it was all grown over with thorns, and *nettles* had covered the face thereof . . .

The Hebrew word rendered "nettles" in most translations of this passage is "chârûl" or "charulim" (Greek, φρύγανα ἄγρια, similar to a term meaning "a rough dry stick"). Most commentators think that the translation is accurate, and even JASTROW, MOFFATT, and the GOODSPEED version continue to use the word "nettles" in this passage (306). The Douay version transposes the two botanical terms, saying: "And behold it was all filled with nettles, and *thorns* had covered the face thereof." ROYLE, however, believes that the "nettles" here referred to were not true nettles, but were charlock, *Sinapis arvensis*[140] (299, 306). This plant, rather than nettles, is notorious for its habit of invading the grain fields of the slothful farmer in precisely the manner described in this proverb. Furthermore, among the Arabic names now applied to the charlock are "khardul" and "khardal barri" (266), which are strikingly similar to the word used in Proverbs (299).

The charlock is an annual mustard-like crucifer, 1 to 3 feet tall, more or less hirsute with scattered stiff hairs, or almost smooth, branching above. The leaves are ovate and variously irregularly toothed or lobed, the basal ones sometimes lyrate. The large yellow flowers are clustered in showy terminal racemes. The fruits are spreading, ascending, or erect smooth or slightly bristly pods, slightly constricted between the seeds, tipped with a flattened, elongate-conic beak often almost as long as the rest of the pod (266).

TRISTRAM has suggested the prickly acanthus, *Acanthus spinosus* L., for "chârûl", which he says is "a common troublesome weed in the plains of Palestine" (184). POST does not record the species from the Holy Land at all, but says that the related Syrian acanthus, *A. syriacus* (which see), is common in grainfields and waste places. Its habit of growth and choice of habitat would, indeed, make it as suited to the context of our passage as the charlock, but the latter has the considerable weight of similarity in common names in its favor, which the acanthus does not have. The Arabic names for the acanthus are "kaff-ud-dibb", "mar 'awîla", and "hîs" (267). The acanthus is herein regarded by us as the "nettle" of Job 30: 7 and Zepheniah 2: 9.

Other writers have suggested brambles (*Rubus*), sea-orache (*Atriplex littoralis* L.), butchers-broom (*Ruscus aculeatus*), and thistles for the "chârûl" (306). CELSIUS was of the opinion that it was the Christ-thorn, *Paliurus spina-christi* (which see), an opinion which, as WILLIAM SMITH comments, "is by no means well founded" (306). Of considerable interest is the fact that

---

Bacterium dysenteriae (Shiga) Lehm. & Neum., *B. shigae* (Chester) Topley & Wils., and *Eberthella dysenteriae* (Shiga) Bergey.

[138]Also called *Bacillus dysenteriae* Flexner, *B. dysenteriae* Strong, *B. paradysenteriae* Collins, *B. flexneri* (Castel. & Chalm.) Levine, *Shigella flexneri* Castel. & Chalm., *Bacterium flexneri* (Castel. & Chalm.) Levine, *B. paradysenteriae* (Collins) Holland, *Eberthella flexneri* (Castel. & Chalm.) Weldin & Levine, and *E. paradysenteriae* (Collins) Bergey.

[139]Also called *Bacillus ambiguus* Andrews, *B. dysenteriae* Schmitz, *Bacterium ambiguum* (Andrews) Levine, *B. schmitzii* Weldin & Levine, *Eberthella ambigua* (Andrews) Bergey, and *Shigella schmitzii* (Weldin & Levine) Hauduroy.

[140]Also called *Brassica arvensis* (L.) Rebenh., *B. sinapistrum* Boiss., *B. sinapis* Vis., and *B. arvensis* (L.) B.S.P.

the Revised Version of Proverbs 24: 31 has a marginal note suggesting the alternative translation of "wild vetches" (27). Of the 55 kinds of vetch in the Holy Land not one has a recorded vernacular name remotely resembling "chârûl" (266). The Oxford Bible says categorically that "there is no ground for accepting any one of the different plants which have been suggested for *chârûl;* it seems to be a general term for weeds" (27).

204. Solanum incanum L.
(FIGURE 79)

ISAIAH 10: 17—... and it shall burn and devour his thorns and *briers* in one day.
ISAIAH 55: 13—Instead of the thorn shall come up the fir tree, and instead of the *brier* shall come up the myrtle tree ...
MICAH 7: 4—The best of them is as a *brier*: the most upright is sharper than a thorn hedge ...
HEBREWS 6: 8—But that which beareth thorns and *briers* is rejected, and is nigh unto cursing...

There has been a very great amount of discussion and argument about the identity of the various "thorns", "briers", "brambles", "thistles", and "nettles" of the Bible. There is little unanimity of opinion regarding any of them, since most of the words so translated seem to be generic terms referring to prickly or spiny plants in general. Such plants are characteristic of arid and desert regions and make up a considerable portion of the flora of the Holy Land (27). The consensus, however, seems to be that the "briers" in the four passages cited above refer mostly, at least, to the Palestine nightshade, *Solanum incanum*[141] (184). MOFFATT, on the other hand, believes that the "brier" (Hebrew, "shamir") of Isaiah 55: 13 refers to true nettles (see under *Urtica*). The GOODSPEED version regards the "briers" of Hebrews 6: 8 as "thistles" and completely omits the latter part of Isaiah 10: 17. MOFFATT, WEIGLE, WEYMOUTH, and O'HARA also use the term "thistles" in the Hebrews passage and the Basic English version says "evil plants". The JASTROW rendition is the same as the Authorized in each case. JOHN SMITH was of the opinion that the Palestine nightshade, instead of being the plant referred to as a "brier", was the one referred to as "the vine of Sodom" in Deuteronomy 32: 32 (299), but the latter is now regarded as the colocynth. Some writers have thought that this "apple of Sodom", *Citrullus colocynthis* (which see), is really involved in the "brier" references, but this does not seem very probable to us since the plant is not spiny.

The Palestine nightshade or Jericho potato is a coarse stiff-branched shrubby plant, growing 1½ to 4 or 5 feet tall, and is appressed-woolly throughout, the branches and leaves armed with reddish recurved spines like a true brier. The leaves are ovate, entire-margined or more or less repand or sinuate-lobed. The purple flowers are shaped like those of the potato and are borne in short-stalked bifid cymes, the lower pedicel bearing a fertile flower, the others sterile. The handsome yellow fruit is a berry, about an inch in diameter (267). It is at first pulpy inside, but as it ripens this pulp dries up and, on being pressed, the ripe fruit bursts and emits a cloud of what appears to be "dust and ashes", or, as JOSEPHUS says, "smoke and ashes", the supposed ashes being its seeds (299). This character of the fruit has caused the plant to be known as the apple of Sodom, in allusion to the destruction of Sodom and Gomorrah. These fruits comply, therefore, very well to JOSEPHUS' description of "those Dead Sea fruits that tempt the eye, but turn to ashes on the lips". It is abundant in the lower Jordan valley and in the region about the Dead Sea (299), especially near the remains of what JOSEPHUS calls "the cities of Sodom". POST describes it as a common weed of roadsides and waste places, especially in the lower Jordan valley and about Engedi and Jericho (267). TRISTRAM says "It grows in all the hot valleys" (184). DALMAN writes that it is commonly found on old walls (98).

The Hebrew word used for "brier" in Micah 7: 4 (and in Proverbs 15: 19) is "chedek" (Greek, ἄκανθα), remarkably similar to the present-day Arabic names of the Palestine nightshade, "khâdak" and "hedek" (98, 267). TRISTRAM

---

[141]Also called *Solanum sanctum* L., *S. coagulans* Forsk., *S. hierochunticum* Dun., and *S. sodomeum* L.

believed that the "brier" of Proverbs 15: 19, described there as forming hedges, also referred to this plant, which he says "is used for hedges" (184). It does not seem as plausible to us that a plant of the type of this nightshade would be planted deliberately as a hedge when a woody plant like *Rhamnus palaestina* (which see) was so abundantly available and is described by many observers as being so employed. That the nightshade occurs in such hedges is very probable, but surely only as a weed.

205. Sorghum vulgare var. durra (Forsk.) Dinsm.
(FIGURE 82)

MATTHEW 27: 48—And straightway one of them ran, and took a spunge, and filled it with vinegar, and put it on a *reed*, and gave him to drink.

MARK 15: 36—And one ran and filled a spunge full of vinegar, and put it on a *reed*, and gave him to drink . . .

JOHN 19: 29—Now there was set a vessel full of vinegar: and they filled a spunge with vinegar, and put it upon *hyssop*, and put it to his mouth.

There has been a terrific amount of argument over the passages cited above. The Authorized, Douay, O'HARA, and WEIGLE versions all state in Matthew and Mark that the sponge full of vinegar was placed on a "reed" and in John that it was placed on "hyssop". MOFFATT and GOODSPEED change this to "stick" in Matthew and Mark, the former using "spear" and the latter "pike" in John; LAMSA says "reed" in all three passages; WEYMOUTH uses the terms "cane" in Matthew and Mark and "stalk of hyssop" in John; while the Basic English version is "rod" in the two former and "stick" in the latter passage.

Of all the words in the Bible referring to plants "hyssop" is undoubtedly the most controversial (see under *Origanum*). Many authorities feel that the "reed", "hyssop", "spear", "stick", "rod", "cane", or "pike" on which the sponge of vinegar was placed in the crucifixion story was a dhura stem, *Sorghum vulgare* var. *durra*[142]. Others maintain that it was the stem of the gaint reed, *Arundo donax* (which see), an opinion apparently concurred in, at least in part, by O'HARA, WEIGLE, LAMSA, and the editors of the King James and Douay versions. That the hyssop, *Origanum*, was involved seems to be utterly out of the question because of the low stature of this plant and the weak character of its stem. Some writers suggest, however, that hyssop may have been mixed with the vinegar, but this is pure conjecture in an effort to explain the appearance of the word in John 19: 29 and is not evident from the wording of the texts involved.

The common variety of sorghum in Palestine and Egypt is the one called dhura, durra, Indian millet, yellow milo, or Jersualem corn. This variety is a medium to stout annual maize-like grass, commonly growing to at least 6 feet tall in Egypt and Palestine, but said to attain a height of 16 to 20 feet in the lush river valleys of Nubia. The thick pith is dry and not sweet as in the sugar-cane. The long leaves are often 1½ inches broad, and the flowers are aggregated in compact usually dense terminal panicles, blooming in June and July. The grains are quite large and are often collected, roasted, and eaten and are also used for making a coarse bread (267). According to PRATT (268) large fields of dhura are to be seen in Egypt and Palestine today. POST calls it "very common in cultivation" in the Holy Land because it matures its fruit without irrigation or rainfall (267). Vernacular names for the plant include the Arabic "dhurah-bayda" and the Hebrew "dokhan". The panicles of fruit are often of tremendous size, one stalk sometimes furnishing a meal for an entire Palestinian family. It is thought by some commentators that this was the "parched corn" which BOAZ gave to RUTH (Ruth 2: 14) and the "corn" which JOSEPH sent to his brethren. Some other commentators are of the opinion (299) that the dhura may possibly be involved in the "millet" of Ezekiel 4: 9 or the "pannag" of Ezekiel 27: 17, but these two references are herein discussed under *Panicum miliaceum* (which see). JOHN SMITH thought that

---
[142] Also called *Holcus durra* Forsk., *H. sorghum* var. *durra* (Forsk.) L. H. Bailey, and *Sorghum durra* (Forsk.) Stapf.

since the dhura and the true sorghum, *Sorghum vulgare* Pers.[143] have been cultivated in Egypt since prehistoric times, their broom-like panicles, after the grains were removed, might have formed the famous bunches of "hyssop" which the Hebrews used to paint the lintels of their doorways on the first Passover (Exodus 12: 22) (299). By us this "hyssop" is regarded as a marjoram (see under *Origanum*).

206. Staphylococcus albus Rosenb.
207. Staphylococcus aureus Rosenb.
208. Streptococcus pyogenes Rosenb.

EXODUS 9: 9-11— . . . a *boil* breaking forth with blains upon man . . . And they took ashes of the furnace, and stood before PHARAOH; and MOSES sprinkled it up toward heaven; and it became a *boil* breaking forth with blains upon man, and upon beast. And the magicians could not stand before MOSES because of the *boils*; for the *boil* was upon the magicians.

LEVITICUS 13: 18-23—The flesh also, in which, even in the skin thereof, was a *boil*, and is healed, And in the place of the *boil* there be a white rising . . . it is a plague of leprosy broken out of the *boil* . . . But if the bright spot stay in his place, and spread not, it is a burning *boil;* and the priest shall pronounce him clean.

II KINGS 20: 7—And ISAIAH said, Take a lump of figs. And they took and laid it on the *boil*, and he recovered.

JOB 2: 7—So went SATAN forth from the presence of the Lord, and smote JOB with sore *boils* from the sole of his foot unto his crown.

REVELATION 16: 2— . . . and there fell a noisome and grievous *sore* upon the men which had the mark of the beast . . .

Boils (Hebrew, "ahvaboo'ot") are hard, painful, superficial, inflamed tumors, which on suppurating discharge pus mixed with blood and disclose a small fibrous mass of dead tissue, called the core (158). Single boils are usually instigated by local irritation. The chief causal organisms of boils are *Staphylococcus aureus*[144] and *S. albus*[145], which occur in the skin and mucous membranes and cause not only boils, but also abscesses, furuncles, suppuration in wounds, some types of pimples, etc. (52). *Streptococcus pyogenes*[146] occurs in human infections of very varied types and is also occasionally present in udder infections of cattle (52). It was probably this or a similar organism which caused JOB's systemic affliction of boils. The "burning boil" of Leviticus 13: 23 indicates a sensation of fire or heat and may be the counterpart of our present-day Damascus boil (306). This organism is also one of the likely causal agents of gangrenous infections, which, according to WILLIAM SMITH, "were common in all the countries familiar to scriptural writers" (306). Some writers have supposed that the plague (Hebrew, "gehgah") of boils called down by MOSES upon the Egyptians consisted not of boils as we know them today, but rather of bubonic plague, which is always accompanied by some swellings and which spreads very rapidly (as would be required by the context) (306). However, since it affected both man and "beast", boils would seem the more likely interpretation. The JASTROW version renders "burning boil" as "scar of the boil."

209. Styrax benzoin Dryand.

EXODUS 30: 34— . . . Take unto thee sweet spices, stacte, and *onycha*, and galbanum . . .
ECCLESIASTICUS 24: 15—I gave a sweet smell like cinnamon and aspalathus, and I yielded a pleasant odour like the best myrrh, as galbanum, and *onyx*, and sweet storax . . .

The Hebrew word rendered "onycha" in the Bible is "shechelet" or

---

[143] Also known as *Holcus sorghum* L., *Andropogon sorghum* (L.) Brot., and *A. sorghum* subsp. *sativum* Hack.

[144] Also known as *Staphylococcus pyogenes aureus* Rosenb., *Micrococcus pyogenes aureus* (Rosenb.) Migula, *M. aureus* (Rosenb.) Migula, and *Aurococcus aureus* (Rosenb.) Winslow & Winslow.

[145] Also known as *Staphylococcus pyogenes albus* Rosenb., *Micrococcus pyogenes* Migula, *M. pyogenes albus* (Rosenb.) Migula, *M. albus* (Rosenb.) Buchanan, and *Albococcus pyogenes* (Migula) Winslow & Winslow.

[146] Also known as *Micrococcus scarlatinae* Klein, *Streptococcus erysipelatos* Rosenb., *S. erysipelatis* Zopf, *S. erysipelatosus* Klebs, *S. scarlatinae* Klein, *S. conglomeratus* Kurth, *S. puerperalis* Arloing, *S. longus hemolyticus* Sachs, *S. longissimus* Thalm., *S. hemolyticus* (Sachs) Rolly, *S. hemolysans* Blake, and *S. pyogenes haemolyticus* Weisenb.

"shecheleth" (Greek, ὄνυξ, meaning "onyx"; the Hebrew term for onyx is "tarshish"). It seems most probable that this name was applied to two substances, one a plant product, the other an animal product. In the Arabic versions of the Bible the word "onycha" is replaced by "ladana" (299), thus implying that the substance was identical with ladanum, a gum-resin derived from species of rockrose, especially *Cistus creticus, C. salvifolius*, and *C. villosus* (which see). But some commentators are of the opinion that the Biblical "onycha" may possibly have been the gum-resin known today as "benzoin", derived from *Styrax benzoin*. This tree, however, is a native of Malacca, Sumatra, and Java, and it is very doubtful whether Hebrew trade extended that far east in those early Biblical days (about 1491 B.C.) (299). BOCHART believed that some kind of bdellium was more probable — perhaps the African bdellium, *Commiphora africana* (which see), or the Indian bdellium, *C. roxburghii* (306). Evidently this matter needs more clarifying study.

The animal source of "onycha" is a mollusk known as wing-shell, of the genus *Strombus*, probably *S. lentiginosus* (306). The word "onyx" means a claw or nail, and from this the small horny shield or operculum on the foot of many mollusks, with which the larger shell is closed, gets its name. From this smaller shell or valve a deliciously fragrant essence was obtained which is believed to have been a part of the ingredients of the holy perfume or "frankincense" used in the temple (184). Onyx, as a precious stone, is mentioned in Exodus 28: 20, but the Hebrew word there used for it is "tarshish" (27). The GOODSPEED version replaces the word "onyx" in the Ecclesiasticus passage with the word "onycha."

210. Styrax officinalis L.
(PAGE 225)

EXODUS 30: 34—And the Lord said unto MOSES, Take unto thee spices, *stacte*, and onycha, and galbanum.
ECCLESIASTICUS 24: 15—I gave a sweet smell like cinnamon and aspalathus, and I yielded a pleasant odour like the best myrrh, as galbanum, and onyx, and *sweet storax*, and as the fume of frankincense in the tabernacle.

The Hebrew word involved here is "nataf" or "nâtâph" (Greek, στακτή). The JASTROW, Authorized, Douay, and GOODSPEED versions render it "stacte" in the Exodus passage. MOFFATT translates it "myrrh-oil" and LEESSER "balm". In the Ecclesiasticus verse the Authorized Version says "sweet storax", the Douay version says "storax", and the GOODSPEED version has "stacte". It was one of the ingredients of the holy oil (27, 184). Mrs. BARNEVELD identifies the Biblical "stacte" as a "gum from *Liquidambar styracifolia*". She points out that the name "stacte" is merely the Greek transliteration of a Hebrew word meaning, literally, "a drop", and that "a plant known as *Liquidambar orientale*, found in Cyprus and Anatolia, yields the official storax; this species grows in Palestine, but is considered not to be native there" (47). POST says that *Liquidambar orientale* Mill. has been reported only once, by TRISTRAM, from Lebanon (266). The Revised Version in a marginal note suggests "opobalsamum" (see under *Commiphora opobalsamum*) (27). In Job 36: 27 the same word "nataf" is translated as "drops of water" ("drops of rain" in the Douay version). CELSIUS gives a good review of early opinions about the identity of stacte (86, 87).

Many authorities now hold to ROSENMÜLLER's opinion that the "stacte" and "sweet storax" of the two references given at the head of this chapter refer to the plant now known as the storax-tree, *Styrax officinalis*[147] (27, 184, 299). This is an irregularly stiff-branched shrub or small tree, 9 to 20 feet tall. Its young branches are woolly. The ovate leaves are stalked, smoothish on the upper and whitish-tomentellous on the lower surface. The handsome white flowers are borne in few-flowered cymes and resemble the flowers of the orange or lemon in appearance and fragrance, causing the shrub to be very showy when in bloom. The fruit is a green, hairy, one-seeded drupe (267).

---

[147]Originally published by LINNAEUS as *Styrax officinale* and often so cited.

The storax-tree is abundant on the lower hills and in rocky places from the northern regions, Lebanon, Coelesyria, and Antilebanon, through Palestine, especially about Mount Carmel, Mount Tabor, Gilead, and Judea (267). Its gum, obtained by making incisions in the stems and branches, is highly perfumed and is prized today, as it doubtless was in ancient days, as a perfume (299). The official storax, according to POST, is the inspissated juice of the bark (267), but according to HANBURY, the resin obtained from this tree in olden times, as described, for instance, by DIOSCORIDES and THEOPHRASTUS, has "entirely disappeared from modern commerce". The resin now known as liquid storax is a product of the unrelated *Liquidambar orientale*, a tree native to southwestern Asia Minor (27, 299).

Some writers think that the "spicery" of Genesis 37: 25 and 43: 11 and of Song 5: 1 & 13 and 6: 2 was also the product of the storax-tree, but this is not very probable. "Spicery" is discussed by us under *Astragalus gummifer* (which see). Other commentators have regarded the "rods" of JACOB as described in Genesis 30: 37 as having been made of storax-tree wood, because the Hebrew word there employed — "libneh" — means "white", and the flowers and under surface of the leaves of this tree are white. These "rods", however, are now generally regarded as having been made of white poplar, *Populus alba* (which see), even though Dr. EVENARI reminds us that this species appears only rarely today in cultivation in Palestine proper, the locale of the JACOB story. So much of this whole country has changed from wooded areas to barren sandy or rocky deserts through the passing of almost four thousand years.

FIGURE 30. — *Tamarix* sp. Gnarled old trees of tamarisk growing at the Wady es Sheihk in Sinai. Such trees are common in the deserts of the Near East. It is supposed that the "grove" planted by ABRAHAM, as well as the "shrub" under which the despairing HAGAR threw her child, were tamarisks. A related species yields the "manna" of Baruch. (Wood engraving from C. W. Wilson's Picturesque Palestine, 1883).

211. Tamarix articulata Vahl   212. Tamarix pentandra Pall.
213. Tamarix tetragyna Ehrenb.
(FIGURE 30)

GENESIS 21: 15 & 33—... and she cast the child under one of the *shrubs* ... And ABRAHAM planted a *grove* in Beer-sheba, and called there on the name of the Lord ...
I SAMUEL 22: 6—... now SAUL abode in Gibeah under a *tree* in Ramah ...
I SAMUEL 31: 13—And they took their bones and buried them under a *tree* at Jabesh ...

The Hebrew word translated "grove" and "tree" in the three passages cited above is "eshel". The two words used in these passages in Greek versions are ἄρουρα and ἄλσος. The Douay version uses "grove" and "wood", but the JASTROW, LEESSER, MOFFATT, and GOODSPEED versions replace these non-committal words with "tamarisk" and "tamarisk-tree", and there is little doubt that this is correct. The Hebrew term for the "shrubs" of Genesis 21: 15 is "seechim". To the traveler in Palestine the tamarisk trees, especially the three species listed at the head of this chapter, often provide a soothing touch of green foliage and a promise of cooling shade which are most welcome (268). Beersheba, where ABRAHAM planted the tamarisk, is in a region much troubled with droughts, which would render the cultivation of most other kinds of trees very impracticable.

It appears, however, that in most of the places in Scripture where the word "grove" is employed in the King James and Douay versions, it refers not to the tamarisk, but to the oak (299). In Exodus 34: 13, Deuteronomy 16: 21, Judges 3: 7, I Kings 14: 23 and 18: 19, and II Kings 17: 16 and 23: 6, for instance, the Hebrew word so translated is "asherah". As has been pointed out in our chapter on the oak (see under *Quercus*), this word is now thought to refer to ASTEROTH or ASTARTE, a goddess of the Sidonians, who was symbolically represented in the form of a tree.

In the extremely desolate portions of the desert of Shur — the scene of HAGAR's wandering with her outcast child — the stunted bushes of desert species of tamarisk, like *Tamarix articulata*[148] and *T. tetragyna*[149], still grow in abundance, and it was probably under one such as these that the despairing mother cast the child of her blighted hopes. BURCKHARDT observes that tamarisk trees are always particularly fond of sandy environments such as maritime sands, saline flats, and deserts, and in the driest seasons, when all the vegetation around them is withered, they never lose their verdure. This persistence is due to their extremely small scale-like closely appressed leaves, which lose little moisture by transpiration. The larger of the tamarisks are also valued for their wood in a region where wood is very scarce. This wood is used for building purposes and, partially burned, as an excellent type of charcoal.

The distinguished botanist, JOHN SMITH, states (299) his opinion that the word "eshel" possibly "comprehends all plants having single hard-wooded stems, growing from a few to 100 or more feet in height". He estimates that there are only about 50 kinds of such trees in Palestine, "of which thirty are mentioned by special names in the Bible". TRISTRAM found the lower banks of the Jordan river fringed with a dense mass of tamarisks. "In the hilly country, he says, they form graceful trees, with long, feathery branches and tufts, closely clad with the minutest of leaves, and surmounted in spring with spikes of beautiful pink blossoms, which seem to envelop the whole tree in one gauzy sheet of color; the blossoms have the appearance of catkins, and the growth of the tree is something like that of the Arbor Vitae of our shrubberies" (299). He observed masses of tamarisk below the site of Jabesh-Gilead, where the bodies of SAUL and JONATHAN were laid. This evidence lends

---
[148]Also called *Thuja aphylla* L. and *Tamarix orientalis* Forsk.
[149]Also called *Tamarix deserti* Boiss. and *T. noëana* Boiss.

credence to the supposition that it was under a tamarisk tree that they were buried, yet in I Chronicles 10: 12 it is said to have been under an oak. However, as SMITH himself points out, "if the word *eshel* does not mean tamarisk, there is no allusion to these beautiful trees in the Bible" (299) and that would be a most curious circumstance. Modern commentators are fairly sure that the tamarisk is the tree referred to by this word.

214. Tetraclinis articulata (Vahl) Masters

REVELATION 18: 12—The merchandise of gold, and silver, and precious stones, and of pearls, and fine linen, and purple, and silk, and scarlet, and all *thyine wood*, and all manner vessels of ivory ...

The "thyine wood" of this verse takes its name quite literally from the Greek text where it appears as ξύλον θύϊνον (dictionary translation, "sweet or citrus wood") (184). This has been rendered "etz aboth" or "etz avot" in Hebrew versions (299). There seems to be no doubt that the lamentation of the figurative Babylon for its lost trade in "thyine wood" referred to the wood of the sandarac-tree, *Tetraclinis articulata*[150]. This is a coniferous tree closely related to the arbor-vitae, and is a native of Morocco, Algeria, and other parts of the Barbary Coast and Atlas Mountains of northern Africa (27, 158, 184, 267) and Mauritania (306). It seldom exceeds a height of 30 feet and has hard, dark-colored, durable, fragrant wood which takes on a fine polish (299). It was the most highly prized wood of the ancients (158), and TRISTRAM refers to it as "priceless" (184). The Romans called the tree "citrus" and the wood "citrum" or "citron wood" (27, 306), and it is still so called, although it has nothing whatever to do with the true citron, *Citrus medica*. Both MOFFATT and GOODSPEED use "citron wood" in place of "thyine wood" in the Revelation passage. POST records it as cultivated in Lebanon and Palestine (267). Marginal Bibles refer to it as "sweet wood" (27).

Thyine or citron wood was extensively employed by the ancients for cabinet work (158). CICERO had a table made of it which is said to have cost the equivalent of $45,000, and PLINY states that even much higher prices were paid for articles made of this wood (299). It was commonly referred to as being worth its weight in gold. Doubtless the Phoenician traders carried thyine wood from Carthage, on the north African coast, to their great port of Tyre (in what is now Lebanon) and from there it was transported overland to Babylon, where it apparently was very highly prized. It yields a brittle, faintly aromatic, more or less transparent resin known as "sandarach" or "sandarac", which is still highly valued — as it was by the Greeks and Romans — for making varnish and as an incense (158). It is gathered and sold in the form of pale yellow grains or "tears", which have been burned on Greek altars since time immemorial.

The sandarac-tree furnishes fine timber, which, owing to its resinous properties, is slow to decay and remains practically uninjured by insects. It is the very last vegetable product to be mentioned by name in the Bible (the fig is the first — Genesis 3: 7).

Some commentators have supposed that the thyine wood tree was the same as the "almug" of SOLOMON, but the latter is herein regarded as *Pterocarpus santalinus* (which see).

215. Triticum aestivum L.   216. Triticum compositum L.
(FIGURE 41)

GENESIS 12: 10—And there was a *famine* in the land: and ABRAM went down into Egypt to sojourn there; for the *famine* was grievous in the land.
GENESIS 26: 12—Then ISAAC *sowed* in that land, and received in the same year an hundredfold.
GENESIS 27: 28—Therefore, God give thee of the dew of heaven, and the fatness of the earth, and plenty of *corn* and wine.

---

[150]Also known as *Thuja articulata* Vahl, *Thuya articulata* Desf., and *Callitris quadrivalvis* Vent.

Genesis 30: 14—And Reuben went in the days of *wheat* harvest, and found mandrakes in the field ...

Genesis 41: 5-57—And he slept and dreamed the second time: and, behold, seven *ears of corn* came up upon one stalk, rank and good. And, behold, seven thin *ears* and blasted with the east wind sprung up after them. And the seven thin *ears* devoured the seven rank and full *ears* ... And I saw in my dream, and, behold, seven *ears* came up in one stalk, full and good: And, behold, seven *ears*, withered, thin, and blasted with the east wind, sprung up after them: And the thin *ears* devoured the seven good *ears*: and I told this unto the magicians; but there was none that could declare it unto me. And Joseph said unto Pharaoh, The dream of Pharaoh is one: God hath shewed Pharaoh what he is about to do ... the seven good *ears* are seven years ... and the seven empty *ears* blasted with the east wind shall be seven years of *famine* ... seven years of *famine* ... and the *famine* shall consume the land ... *famine* following ... lay up *corn* ... seven years of *famine* ... through the *famine* ... food of the field ... gathered *corn* as the sand of the sea ... years of *famine* ... the *famine* was over all the face of the earth ... the *famine* waxed sore in the land of Egypt. And all countries came into Egypt to Joseph for to buy *corn*; because that the *famine* was so sore in all lands.

Genesis 42: 1-33—Now when Jacob saw that there was *corn* in Egypt ... Behold, I have heard that there is *corn* in Egypt ... And Joseph's ten brothers went down to buy *corn* in Egypt ... came to buy *corn* ... the *famine* was in the land of Canaan ... go ye, carry *corn* for the *famine* of your houses ... fill their sacks with *corn* ... laded their asses with the *corn* ... take food for the *famine*.

Genesis 43: 1-2—And the *famine* was sore in the land ... when they had eaten up the *corn* ...

Genesis 45: 6 & 11—For these two years hath the *famine* been in the land ... for yet there are five years of *famine* ...

Genesis 47: 13-14 & 19-20— ... for the *famine* was very sore, so that the land of Egypt and all the land of Canaan fainted by reason of the *famine* ... for the *corn* which they had bought ... give us *seed*, that we may live ... because the *famine* prevailed.

Exodus 9: 32—But the *wheat* and the rie were not smitten: for they were not grown up.

Exodus 11: 5—And all the firstborn in the land of Egypt shall die, from the firstborn of Pharaoh that sitteth upon the throne, even unto the firstborn of the maidservant that is behind the *mill*; and all the firstborn of beasts.

Exodus 22: 6—If fire break out ... so that the stacks of *corn*, or the standing *corn*, or the field, be consumed therewith; he that kindled the fire shall surely make restitution.

Exodus 29: 2—And unleavened *bread*, and *cakes* unleavened tempered with oil, and *wafers* unleavened anointed with oil: of *wheaten flour* shalt thou make them.

Leviticus 2: 1-16—And when any will offer a *meat offering* unto the Lord, his offering shall be of *fine flour* ... and he shall take thereout his handful of *flour* thereof... And the remnant of the *meat offering* shall be Aaron's ... *meat offering* baken in the oven ... *cakes* of fine *flour* ... *wafers* ... *meat offering* baken in a pan ... fine *flour* ... it is a *meat offering* ... *meat offering* baken in a frying pan ... fine *flour* ... bring the *meat offering* ... *meat offering* ... that which is left of the *meat offering* shall be Aaron's ... no *meat offering* ... thy *meat offering* shalt thou season with salt. And if thou offer a *meat offering* of thy firstfruits unto the Lord, thou shalt offer for the *meat offering* of thy firstfruits green *ears of corn* dried by the fire, even *corn* beaten out of full *ears* ... it is a *meat offering*. And the priest shall burn the memorial of it, part of the beaten *corn* thereof.

Leviticus 23: 14—And ye shall eat neither *bread*, nor *parched corn*, nor green *ears*, until the selfsame day that ye have brought an offering unto your God ...

Leviticus 24: 5 & 7—And thou shalt take *fine flour*, and bake twelve *cakes* thereof: two tenth deals shall be in one *cake* ... on the *bread* for a memorial ...

Deuteronomy 8: 8—A land of *wheat*, and barley, and vines, and fig trees, and pomegranates ...

Deuteronomy 25: 4—Thou shalt not muzzle the ox when he treadeth out the *corn*.

Deuteronomy 32:14— ... with the fat *kidneys of wheat*; and thou didst drink of the pure blood of the grape.

Deuteronomy 33: 28— ... the fountain of Jacob shall be upon a land of *corn* and wine ...

Judges 6: 11 & 19-21— ... and his son Gideon threshed *wheat* by the winepress, to hide it from the Midianites ... and unleavened *cakes* of an ephah of *flour* ... unleavened *cake* ... the unleavened *cakes* ... and consumed the flesh and the unleavened *cakes*.

Judges 15: 5—And when he had set the brands on fire, he let them go into the standing *corn* of the Philistines, and burnt up both *shocks*, and also the standing *corn* ...

Ruth 2: 14—And she sat beside the reapers: and he reached her *parched corn*, and she did eat, and was sufficed, and left.

I Samuel 17: 17— ... Take now for thy brethren an ephah of this *parched corn*, and these ten *loaves*, and run to the camp to thy brethren.

I Samuel 25: 18—Then Abigail made haste, and took two hundred *loaves*, and two bottles of wine, and five sheep ready dressed, and five measures of *parched corn* ... and laid them on asses.

II Samuel 4: 6—And they came thither into the midst of the house, as though they would have fetched *wheat* ...

II Samuel 17: 19 & 27-28— ... and spread *ground corn* thereon ... and Barzilai the Gileadite of Rogelim, Brought beds, and basons, and earthen vessels, and *wheat*, and barley, and *flour*, and *parched corn*, and beans, and lentiles, and parched pulse.

I Kings 5: 11—And Solomon gave Hiram twenty thousand measures of *wheat* for food to his household ...

I Chronicles 21: 12-28—Either three years' *famine* ... And the angel of the Lord stood by

the *threshingfloor* of ORNAN ... commanded ... that DAVID should go up, and set up an altar unto the Lord in the *threshingfloor* of ORNAN ... Now ORNAN was *threshing wheat*, And as DAVID came to ORNAN, ORNAN looked and saw DAVID, and went out of the *threshingfloor*, and bowed himself to DAVID with his face to the ground. Then DAVID said to ORNAN, Grant me the place of this *threshingfloor* ... and the *threshing instruments* for wood, and the *wheat* for the *meat offering* ... the Lord had answered him in the *threshingfloor* of ORNAN.
II CHRONICLES 2: 10 & 15—And, behold, I will give to thy servants ... twenty thousand measures of *beaten wheat* ... Now therefore the *wheat* ... let him send unto his servants.
II CHRONICLES 3: 1— ... the Lord appeared unto DAVID his father, in the place that DAVID had prepared in the *threshingfloor* ...
JOB 5: 26—Thou shalt come to thy grave in a full age, like as a *shock of corn* cometh in in his season.
JOB 31: 40—Let thistles grow instead of *wheat* ...
PSALMS 4: 7— ... in the time that their *corn* and their wine increased.
PSALMS 65: 9 & 13— ... thou preparest them *corn* ... the valleys also are covered over with *corn* ...
PSALMS 72: 16—There shall be an handful of *corn* in the earth upon the top of the mountains; the fruit thereof shall shake like Lebanon ...
PSALMS 81: 16—He should have fed them also with the finest of the *wheat* ...
PSALMS 147: 14—He ... filleth thee with the finest of the *wheat*.
PROVERBS 11: 26—He that withholdeth *corn*, the people shall curse him: but blessing shall be upon the head of him that selleth it.
ISAIAH 28: 25-28—When he hath made plain the face thereof, doth he not ... cast in the principal *wheat* ... ? ... *Bread corn* is bruised; because he will not ever be *threshing* it, nor break it with the wheel of his cart, nor bruise it with his horsemen.
ISAIAH 41: 15-16—Behold, I will make thee a new sharp *threshing instrument* having teeth: thou shalt *thresh* the mountains, and beat them small, and thou shalt make the hills as *chaff*. Thou shalt fan them, and the wind shall carry them away, and the whirlwind shall scatter them ...
ISAIAH 47: 2—Take the *millstones* and grind *meal* ...
JEREMIAH 50: 11— ... ye are grown fat as the heifer at *grass* ...
LAMENTATIONS 2: 12—They say to their mothers, Where is *corn* and wine? ...
EZEKIEL 27: 17— ... they traded in the market *wheat* of Minnith ...
HOSEA 14: 7— ... they shall revive as the *corn*, and grow as the vine ...
AMOS 8: 5-6— ... When will the new moon be gone, that we may sell *corn*? and the sabbath, that we may set forth *wheat* ... Yea, and sell the refuse of the *wheat*?
ZECHARIAH 9: 17— ... *corn* shall make the young men cheerful ...
MATTHEW 3: 12—Whose fan is in his hand, and he will thoroughly purge his floor, and gather his *wheat* into the garner; but he will burn up the *chaff* with unquenchable fire.
MATTHEW 12: 1—At that time JESUS went on the sabbath day through the *corn*; and his disciples were an hungred, and began to pluck the *ears of corn*, and to eat.
MATTHEW 13: 3-8 & 24-30— ... Behold a sower went forth to sow; And when he sowed, some *seeds* fell by the way side, and the fowls came and devoured them up: Some fell upon stony places, where they had not much earth: and forthwith they sprung up, because they had no deepness of earth: And when the sun was up, they were scorched; and because they had no root, they withered away. And some fell among thorns; and the thorns sprung up and choked them: But other fell into good ground, and brought forth fruit, some an hundredfold, some sixtyfold, some thirtyfold ... The kingdom of heaven is likened unto a man which sowed *good seed* in his field: But while men slept, his enemy came and sowed tares among the *wheat*, and went his way. But when the blade was sprung up, and brought forth fruit, then appeared the tares also ... didst thou not sow *good seed* in the field? ... lest while ye gather up the tares, ye root up also the *wheat* with them. Let both grow together until the harvest ... gather the *wheat* into my barn.
MATTHEW 18: 6—But whoso shall offend one of these little ones which believe in me, it were better for him that a *millstone* were hanged about his neck, and that he were drowned in the depth of the sea.
MARK 2: 23—And it came to pass, that he went through the *corn fields* on the sabbath day; and his disciples began, as they went, to pluck the *ears of corn*.
MARK 4: 28—For the earth bringeth forth fruit of herself; first the blade, then the *ear*, after that the full *corn in the ear*.
MARK 14: 12—And the first day of unleavened *bread* ...
LUKE 6: 1— ... he went through the *corn fields*; and his disciples plucked the *ears of corn*, and did eat, rubbing them in their hands.
JOHN 12: 24—Verily, verily, I say unto you, Except a *corn of wheat* fall into the ground and die, it abideth alone: but if it die, it bringeth forth much fruit.
ACTS 7: 12—But when JACOB heard that there was *corn* in Egypt, he sent out our fathers first.
I CORINTHIANS 9: 9— ... Thou shalt not muzzle the mouth of the ox that treadeth out the *corn* ...
I TIMOTHY 5: 18— ... Thou shalt not muzzle the ox that treadeth out the *corn* ...
REVELATION 18: 13— ... and oil, and *fine flour*, and *wheat* ...

The Hebrew word translated "wheat" is "chittah" or "chitim" (Greek, $\pi\upsilon\rho\delta\varsigma$), but for "corn" eleven different Hebrew words are employed — in Numbers 18: 27, for instance, "dagan" (Greek, $\sigma\tilde{\iota}\tau o\varsigma$, meaning, according to the dictionary, "food made of corn or wheat, bread"); in Judges 15: 5 "kâmah"

(Greek, στάχυς, meaning "standing corn or ears of corn"); in Genesis 41: 49 "bar" (Greek, σῖτος); in Ruth 2: 2 "shibbôleth" or "shibalim" (Greek, στάχυς); in Leviticus 2: 14 "karmel" (Greek, χῖδρον, meaning "corn in the ear or a dish of unripe, toasted, wheaten groats"); in Joshua 5: 11 "'abûr" or "avur" (Greek, σῖτος) and "kâli" (omitted in the Septuagint); in Leviticus 2: 16 "girsah" or "géres" (Greek, χῖδρον); in Ruth 2: 7 "'amâr" or "'omer" (Greek, δράγμα, meaning "sheaves"); in Job 24: 6 "belîl"; and in Genesis 42: 1-3 "shéber" or "shever" (Greek, πράως, σῖτος, meaning "mild or gentle wheat, corn, or grain") (27, 184).

There is no doubt that the "wheat" of the Bible is the commonly cultivated and well-known summer and winter wheat of the present day, *Triticum aestivum*[151] (47). This is an abundant annual grass cultivated in Egypt and other Eastern lands since the earliest recorded times. The exact place of its origin is not known. It is supposed by some writers that many of the plants which have been cultivated since time immemorial in the Mediterranean area and which have never been found native anywhere, may have originated in the area now covered by the Mediterranean Sea — a body of water which geologists are convinced broke through from the Atlantic in only comparatively recent geologic times. Other writers suppose that they came originally from the region about Mesopotamia and the highlands of western Asia, supposed by many to be the cradle of civilization. Grains of wheat have been found in the most ancient Egyptian tombs, and have also been found, along with flint instruments, in the remains of the prehistoric lake-dwellings of Switzerland, indicating, apparently, that this cereal was cultivated even by prehistoric peoples (299). It was certainly the chief grain of Mesopotamia in JACOB'S time, about 1753 B.C. (Genesis 30: 14). Even today it forms the "staff of life" for countless millions of people all over the world.

Five kinds of wheat are native to and still wild today in Palestine and neighboring lands and at least 8 others are cultivated there. It is very probable that most or perhaps even all of these were also used in Biblical days. The native kinds were undoubtedly much more abundant there then than they are now. These native kinds are the so-called one-grained wheat, einkorn, or little spelt, *T. monococcum* L., the thaoudar, *T. thaoudar* Reut., and the wild emmer or wild wheat, *T. dicoccoides* (Koern.) A. Schulz[152] (98), the last-mentioned occurring there in two forms, f. *straussianum* A. Schulz and f. *kotschyanum* A. Schulz, and carefully studied by AARONSOHN to determine its range and distribution (15a, 18, 223a, 223b, 256, 257). Non-native wheats are the spelt, *T. aestivum* var. *spelta* (which see), the Polish wheat *T. polonicum* L., the Egyptian or Poulard wheat, *T. turgidum* L., the hard wheat, *T. durum* Desf., and its f. *depauperatum* Oppenh., the club wheat, *T. compactum* Host, the composite wheat, *T. compositum*, and *T. aegilopoides* (Link) Bal.[153] (267). The composite wheat, with its branched spikes (often as many as seven ears or heads per stalk), is definitely referred to in Genesis 41: 5-27. Depicted on numerous Egyptian monuments and inscriptions, it is still commonly seen in the Nile delta, where it is known as "mummy wheat" (184), and also occurs in Palestine (306).

Three varieties of wheat are now very commonly cultivated in Palestine: on the maritime plains there is a white short-bearded form, while inland there is a short-stemmed, long-bearded, thick-set, coarse-grained form and also a form with longer stems and coarse black beards and husks (184). Mrs. BARNEVELD tells us that "Heshbon wheat" is a bearded kind, bearing several ears on one stalk (probably *T. compositum*), and "Egyptian wheat" was also bearded (*T. turgidum*) (47, 306). Babylonia, Syria, and Palestine all were known in ancient days for the excellent quality of their wheat, but they were all frequently subject to droughts resulting in widespread famine (Psalms 81: 16

---

[151] Also referred to as *Triticum sativum* Lam., *T. vulgare* Vill., *T. hybernum* L. (the awnless form), and *T. tenax* Hausskn.
[152] Also called *Triticum vulgare* var. *dicoccoides* Koern., *T. dicoccum* var. *dicoccoides* (Koern.) Asch. & Graebn., and *T. hermonis* Cook.
[153] Also called *Crithodium aegilopoides* Link and *Triticum baeoticum* Boiss.

and 147: 14, Genesis 12: 10 and 41: 57). After the government of the Hebrews became more or less settled and they gave up their nomadic existence, *i.e.*, from SOLOMON'S reign (about 1015 B.C.) onward, agriculture became more developed among them, and Palestine became a grain-exporting country. Her surplus grain was absorbed in large part by her powerful commercial neighbor to the northwest, Tyre (Ezekiel 27: 17, Amos 8: 5), whose ships sailed all the then-known seas (306).

Wheat is still "trodden out" by oxen (Deuteronomy 25: 4), pressed out by a wooden wheel (Isaiah 28: 28), or threshed with a flail (I Chronicles 21: 20-23, Isaiah 41: 15-16), and then winnowed with a fan and sifted. The time of wheat harvest — from the end of April to well into June, depending on the location, soil, and season — still marks a definite division of the year in Oriental lands. Dates are often reckoned as so many days or weeks before or after the "wheat harvest" (27, 184). In Genesis 30: 14 we read that REUBEN went out "in the days of wheat harvest".

Wheat was also the main constituent of the "corn" of the Bible. Corn is mentioned no less than 71 times in the Scriptures (299). Wheat fields are known even today in the Old World as "corn fields". This "corn" of the Old World must not be confused with our American corn, *Zea mays* L., which should more properly be called Indian-corn or maize. "Parched corn" forms an article of food in Palestine even today. THOMSON describes its preparation as follows: "a quantity of the best ears (of wheat), not too ripe, are plucked with the stalks attached; these are tied into small parcels, a blazing fire is kindled with dry grass and thorn bushes, and the corn heads are held in it until the chaff is mostly burned off" (299). When sufficiently roasted thus, the grain is rubbed out between the hands, and is still a favorite food all over the country. SMITH says that "corn is also prepared by being first boiled, then bruised in a mill to take the husks off, afterwards dried in the sun, and then stored for use" (299).

According to modern commentators, the Hebrew words "dâgân" and "shéver" are general terms for corn in the generic sense, as compared with any other general commodity such as wine; "belîl" is a term signifying provender, as the Revised Version correctly renders it; "kamah" refers to "standing corn" as it grows in the fields; "bar" is applied to winnowed corn or the actual grain itself; "shibboleth" refers to an ear or head of corn individually; "karmel" is applied to the early green sprouts of germinating corn; "'avûr" refers to corn a year old or "old corn"; "kâli" is the term for "parched corn", dried or baked in fire as described above; "girsah" indicates corn beaten out of the ear but not yet winnowed; and "'omer" refers to a handful or sheaf of cornstalks (27, 184). The dish of food sent by JOSEPH in Egypt from his table to BENJAMIN and his brothers is thought to have been "frumenty" or "firmity", which is described as a dish made of hulled wheat, boiled in milk (158, 184). "Corn" in Biblical days often included as a mixture peas, beans, lentils, cummin, barley, millet, and spelt (156, 231), but wheat was always its main and most prized constituent, even as today. Egypt was a great grain-producing country in JACOB'S time and has, in fact, always been famous for its wheat. ABRAM (Genesis 12: 10) and, later, JOSEPH'S brothers (Genesis 42) naturally turned to Egypt for wheat when famine (Hebrew, "ra'av") devastated Canaan, and we read that people from many other lands also appealed to Egypt for food in those times of crisis. Rome and Constantinople regarded Egypt as an inexhaustible granary. It was the chief granary of the Roman Empire (184). Even today Arabia imports all its wheat from Egypt, and the caravans which leave upper Egypt for the Red Sea are laden with precious wheat.

The "meat offerings" (Hebrew, "mincha") so minutely described in Leviticus 2: 1-6 and 6: 15, Jeremiah 17: 26, and elsewhere, were actually not meat, but cereal offerings. MOFFATT and the GOODSPEED versions commendably substitute "cereal offering" for the misleading "meat offerings" of the King James Version. The JASTROW and LEESSER versions use the term "meal-offering" which is likewise accurate. JASTROW further substitutes "kidney-fat of wheat" for "fat kidneys of wheat" in Deuteronomy, changes "spread ground corn" to "strewed groats" in I Samuel 17: 19, and along with the Douay version replaces "finest of the wheat" with "fat of the wheat" in the Psalms references, and in Isaiah uses the more descriptive term "threshing-sledge" in place of "threshing instrument."

Crops of wheat in the Levant are currently expected to yield about 20-fold what was sown. In ancient times, when the land was so much more fertile, the yield was far greater. The parable of JESUS recorded in Matthew 13: 3-8 speaks of wheat yielding 30, 60, and 100-fold. In good soil certain strains of *T. aestivum* will sometimes even today produce ears or heads with 60 or even 100 grains each (306). Wheat seeds were planted in the winter by the Hebrews, and were either sown broadcast and then ploughed in or trampled in by cattle (Isaiah 32: 20), or, more rarely, were painstakingly planted in rows to insure healthier and huskier plants (Isaiah 28: 25). Wheat and spelt in Biblical days — and even today in those lands — are not planted until well after the barley is planted. This fact accounts for the story of the hail destroying PHARAOH'S barley, but not his wheat and spelt, since only the former had grown up (Exodus 9: 32). The mills, millstones, granaries, and threshing-floors mentioned in so many places in the Scriptures all refer to equipment employed in the processing of grain, mostly wheat and spelt, to produce flour. Hulled wheat intended for home consumption was often stored in the central portion of the house. This fact explains the story told in II Samuel 4: 6. It was also sometimes stored in dry wells, even as today (II Samuel 17: 19). The "fine flour" of which the shewbread cakes were made (Leviticus 24: 5) was unquestionably wheat flour (299).

Oats, *Avena sativa* L., are mentioned by the rabbinical writers (306), but most commentators are agreed that this modern cereal was unknown in Biblical times. POST states that it is but little cultivated in the Levant today (267).

217. Triticum aestivum var. spelta (L.) L. H. Bailey

EXODUS 9: 32—But the wheat and the *rie* were not smitten: for they were not grown up.
ISAIAH 28: 25—When he hath made plain the face thereof, doth he not cast abroad the fitches, and scatter the cummin, and cast in the principal wheat and the appointed barley and the *rie* in their place?
EZEKIEL 4: 9—Take thou also unto thee wheat, and barley, and beans, and lentiles, and millet, and *fitches*, and put them in one vessel, and make thee bread thereof.

It is now generally agreed that CELSIUS was correct in his contention (306) that the "rie" mentioned in the Bible was actually spelt, *Triticum aestivum* var. *spelta*[154]. The Hebrew words rendered "rie" in Exodus and Isaiah and "fitches" in Ezekiel are "koosemet" or "cussémoth" and "koos'mim" (Greek, ξειά, dictionary translation "a coarse barley or rye", and ὄλυρα, dictionary translation "a kind of spelt"). The Hebrew name recorded by POST for spelt is "kussemeth" (267), and its Arabic names, according to TRISTRAM, are "chirsanat" (27, 184) and "kirsenni" (170a). Spelt is a hard-grained race of wheat with loose ears and the grains triangular in cross-section, somewhat adherent to the chaff (158, 299). It was the most common form of wheat in early times (158), especially in southern Europe, and is still popular in Germany and Switzerland, especially in poor soils. It was cultivated in Egypt since earliest times and was certainly common in that land (but probably less common in Palestine) at the time of MOSES (27). POST says that it is not now known from Syria, Palestine, or Sinai. He believes that it was originally native to Mesopotamia.

True rye, *Secale cereale* L., is a cereal grain adapted to cultivation in colder and more northern countries only, and therefore it is almost unknown in Egypt and the Levant (184, 268). Even if, as some writers claim, it actually is now "sparingly" cultivated in Egypt and Syria (299), it seems fairly certain that it was not known there or in the Holy Land in Biblical times.

Spelt has a stouter stem than wheat, and strong spikes of grain. Bread

---
[154]Also known as *Triticum spelta* L. and *T. sativum* var. *spelta* (L.) Richt.

made of its flour is very much inferior to that made of wheat, but the chief recommendation of spelt is that it will thrive well in almost any kind of soil and will yield a crop on land which is totally unfit for wheat. The ancients preferred it to barley for bread.

Commentators have suggested rye, "fitches" (probably meaning *Vicia*), and even oats (*Avena sativa* L.) for some or all of the references given at the head of this chapter (306). The einkorn (*Triticum monococcum* L.) [154a] and the rice-wheat (*T. dicoccum* Schrank), have also been suggested (306). MOFFATT, for some reason not clear to us, thinks that the "rie" of Isaiah was really vetches (*Vicia*) [154b], although he admits that the "rie" of Exodus and the "fitches" of Ezekiel were spelt. The GOODSPEED, LEESSER, and JASTROW versions regard all three passages as referring to spelt. The Douay version uses "winter corn" in Exodus, "vetches" in Isaiah, and "fitches" in Ezekiel. The King James Version suggests "spelt" in a marginal note for the "rie" of Isaiah 28: 25 and the "fitches" of Ezekiel 4: 9. The Revised Version uses "spelt" in all three passages (27).

The "fitches" of Isaiah 28: 25 (King James Version) are now regarded by all commentators and translators as "cummin" (see under *Nigella sativa*).

218. Tulipa montana Lindl.   219. Tulipa sharonensis Dinsm.
(FIGURE 83)

SONG 2: 1—I am the *rose* of Sharon and the lily of the valleys.
ECCLESIASTICUS 50: 8—And as the flower of *roses* in the spring of the year, as lilies by the rivers of waters, and as the branches of the frankincense-tree in the time of summer.

POST, SMITH, and others are of the opinion that the "rose" of Song 2: 1 is *Narcissus tazetta*, even as that of Isaiah 35: 1 is conceded to be by MOFFATT (95, 184, 267, 299). Other writers have suggested species of *Crocus* and *Lilium* (181), as well as *Anemone coronaria*, *A. fulgens* (181), *Colchicum autumnale* L., and even *Corchorus olitorius* L., but none of these is very probable (27). The Malta jute, *Corchorus olitorius*, is wild in Syria, Lebanon, and Palestine today, but only as an adventive from India. It is most doubtful whether this plant existed in Palestine in Biblical days, at least in quantities sufficient to justify its being used in figures of speech (266). *Colchicum autumnale* is not recorded from the region at all by POST, although he lists 19 related species and varieties from there (267). The Revised Version suggests "autumn crocus" in marginal notes for Isaiah 35: 1 and Song 2: 1 (27). The plant which we today call the "rose-of-Sharon", *Hibiscus syriacus* L., is a native of eastern Asia, not of Syria as its name would lead us to assume, although POST states that it is now cultivated there (266). Since trade had certainly not been established yet with China in Biblical days, this plant cannot possibly have had any connection with the "rose of Sharon" or any other "rose" of the Scriptures.

DONEY believes that the "rose of Sharon" was *Narcissus tazetta*, while the "rose" of Isaiah 35: 1 was a true rose of the genus *Rosa* (101). In defense of this view it must in fairness be pointed out that the oft-repeated statement that there are no true roses in Biblical lands save in the Lebanon mountains (27, 184) is false. POST records several from other areas in our region, including Palestine (266). PRATT regards all the New Testament references to the "lily" as applying to *Tulipa montana*. He thinks that the "rose of Sharon" may have been a species of rockrose, *Cistus*, whose beautiful single flowers indeed greatly resemble those of the brier rose (268).

MOFFATT translates the Song of Solomon passage: "I am only a *blossom* of the plain, a mere lily of the dale", while the GOODSPEED version says: "I am a *saffron* of the plain, a hyacinth of the valleys." GOODSPEED continues to use the

---

[154a] DE CANDOLLE in his "Origin of Cultivated Plants", p. 363, says "I imagined it was perhaps the allied form, *T. monococcum*, not now in Egypt" (170a).

[154b] DINSMORE claims that *T. spelta* is not grown in Palestine at all and suggests that "*Vicia ervillea* appears more likely to be the plant intended by 'rie'". HENSLOW's criticisms are that no one else has made this suggestion, that such a plant product would more likely have been included in pulse, and that it has only lately been introduced into Egypt (99b, 170a).

word "roses" in the Ecclesiasticus passage, which he renders: "Like roses in the days of first fruits, Like lilies by a spring of water, Like a sprig of frankincense, on summer days."

The most recent work on the subject, done by Dr. EPHRAIM HA-REUBENI, professor of Biblical botany at the Hebrew University in Jerusalem, concludes that the "rose of Sharon" was the mountain tulip, *Tulipa montana*. This is a handsome bulbous plant with oblong-lanceolate to linear-lanceolate, often falcate, wavy-margined leaves. The flowers are 2-4 inches wide, rather pale outside, but crimson within, the tepals being ovate to ovate-oblong or obovate, acute or obtuse at the apex, sometimes cuspidate, with an oblong, blackish, yellow-margined spot at the base. The species is common in the mountainous regions of Syria, Coelesyria, Lebanon, Antilebanon, and Hauran, recorded questionably from Palestine, the Damascus area, and Arabia Petraea (267). Since this is primarily a mountain species, it is more probable that the species involved in our Song of Solomon reference is the closely related Sharon tulip, *T. sharonensis*, which is found in sandy places on the Sharon coastal plain (267). GERARDE, FOLKARD, and some other authors have considered these tulips to be the "lilies of the field" (see under *Anemone coronaria*) (95, 298).

The "rose" of Ecclesiasticus was a spring-blooming flower according to the text, so summer- and autumn-blooming species may well be automatically eliminated. The two tulips mentioned above bloom only in March. The Hebrew word rendered "rose" both in Isaiah 35: 1 and Song 2: 1 is "chavatzelet" or "chăbatzeleth" (Greek, κρίνον, dictionary translation "a lily of any kind", and ἄνθος, dictionary translation "a young bud, sprout, or flower") (27). The word is derived from a root meaning "a bulb", so there seems to be no logical doubt that a bulbous plant was referred to in both places. We regard the Isaiah reference as applying to *Narcissus tazetta* (which see).

The so-called "rose of Jericho" is *Anastatica hierochuntica* (which see).

Tulips were early cultivated in Turkey, where an annual Feast of Tulips was celebrated in the Sultan's seraglio. "The feast and show of tulips lasted all day, and at night the gardens were brilliantly illuminated" (95). The name of this flower is derived from the Persian word "thuliban" or "dulband", meaning "a turban". This evolved into the Turkish "dulbend" and "tulbend", the Old French "tulipan", Italian "tulipano", Latin "tulipa", and the modern Spanish "tulipán", French "tulipe", German "Tulpe", and English "tulip" (158). As Mrs. COTES remarks, the turbaned heads of a seated row or circle of Turks are strikingly suggestive of a tulip bed, and "the name seems particularly suitable to these brilliant striped, flame-petalled, orange and crimson flowers, which make bright dots of colour in the fields of Palestine and in the Levant" (95).

In the folklore of Devonshire tulip flowers form cradles for pixie babies. Parsley leaves were torn ragged, as we know them now, by the enraged pixies as punishment to a scoffer who tore out the tulips and planted the more utilitarian parsley in his garden (125, 298). In Persia the ardent lover gives a crimson tulip to his beloved as a symbol of the burning flame of his love for her and to further indicate that "his heart is charred to a coal by its ferocity, just as the flower's base shows black" (126, 298). Holland has long been famous for its tulips. In 1634 a "tulip mania" gripped the land and fantastic prices were paid for new strains. A single bulb was sold at one time for the equivalent of $1,725 (158, 298).

220. Typha angustata Bory & Chaub.

MATTHEW 27: 29—And when they had platted a crown of thorns, they put it upon his head, and a *reed* in his right hand . . .
MARK 15: 19—And they smote him on the head with a *reed* . . .

There is no unanimity of opinion regarding this "reed" of the New Testament. Many old paintings of the masters, depicting the mock-trial of JESUS, picture him with the inflorescence-stalk of *Typha* in his hand as a sceptre. For this reason many writers include the great reedmace or cattail, *Typha latifolia* L., among the plants of the Bible and consider that it was the "reed" referred to in Matthew and Mark. SMITH, however, seriously doubts whether this "reed" could have been the cattail, for while cattails are said to be common along streams throughout the region, they would hardly have been in flower or fruit in March or April, when the mock-trial of JESUS took place (299). POST lists two species of cattail or reedmace from Palestine and a third from Lebanon, and states that they all bloom in July and August (267). The common reedmace (or "bulrush" according to TRISTRAM), *T. angustata*, is very similar to our American narrow-leaved cattail, *T. angustifolia* L., differing only in its leaves being convex on the outer and flat on the inner face, the pistillate spikes being only pale-brown, with grayish dots, and the bractlets being much longer than the perigonial bristles (267).

It is of interest to note that the Douay, LAMSA, O'HARA, and WEIGLE versions all continue to use the word "reed" in both the Matthew and Mark passages, but the Basic English uses "rod" in Matthew and "stick" in Mark, MOFFATT and GOODSPEED use "stick" in both places, while WEYMOUTH says "cane".

221. Urtica caudata Vahl   222. Urtica dioica L.
223. Urtica pilulifera L.   224. Urtica urens L.

ISAIAH 34: 13—And thorns shall come up in her palaces, *nettles* and brambles in the fortresses thereof: and it shall be an habitation of dragons, and a court for owls.

HOSEA 9: 6— . . . the pleasant places for their silver, *nettles* shall possess them: thorns shall be in their tabernacles.

Because of the fact that two different words are employed in the Hebrew for what the King James version renders "nettles", there is some doubt as to the identity of the plants involved. The word "chârûl" (Greek, θρύγανα ἀγρία, dictionary translation "a wild or rough dry stick") occurs in Job 30: 7, Proverbs 24: 31, and Zephaniah 2: 9, while "kîmmôsh" (Greek, ἄκανθα, "a prickly plant", and ὄλεθρος, "ruin or destruction") is the word used in Isaiah 34: 13 and Hosea 9: 6.

The "nettles" of JOB and ZEPHANIAH were identified as *Acanthus* by TRISTRAM, and in this identification we concur. *A. syriacus* (which see) is a strong-growing spiny-leaved perennial plant, and a common and troublesome weed on the plains of Palestine. We also follow ROYLE in the opinion that the "nettles" of Proverbs were charlock, *Sinapis arvensis* (which see), which quickly invades and infests the fields of the slothful farmer, rendering them very gay with its yellow blossoms. PRATT (268) was impressed with the obvious fact that neither man nor beast would willingly loll under true nettles, and therefore suggested charlock also for the "nettles" of JOB. Those of ISAIAH, however, have been regarded by various writers as camomile, thistle, thorn, caper-plant, and, by PRATT, true nettles (268). SHAW considered all the "nettle" references of Job, Proverbs, Isaiah, Hosea, and Zephaniah as applying to *Urtica* (293). JOHN SMITH, on the other hand, seems to think that only those of Hosea, Isaiah, and Proverbs were true nettles (299). R. LEVI, MARTIN LUTHER, and the Swedish and old Danish versions all regarded the Hebrew word "mallûach" of Job 30: 4 as referring to nettles. By us this passage is regarded as applying to species of *Atriplex* (which see).

The four species of true nettle found in Palestine are the common or great nettle, *Urtica dioica*, the Roman nettle, *U. pilulifera*, the small nettle, *U. urens*, and the related *U. caudata*[155]. Some of these plants often attain a height of 5 or 6 feet. They are all common pests of waste places and fields. It is quite usual to see them occupying ground that was once cultivated but has since been neglected, as well as ground around ruined buildings — even as described by ISAIAH and HOSEA. The sting of the Roman nettle is said to be far more irritating than that of any of our American species (178). MOFFATT and both the Douay and GOODSPEED versions continue to regard the "nettles" of Isaiah and Hosea as *Urtica*.

Nettles were formerly employed medicinally among country folk as a stew to cure almost any ailment (125, 126). The young shoots are still often eaten (126). It is stated that the Roman nettle was introduced into Great Britain and perhaps various other parts of northern Europe by CAESAR's soldiers, "who, not having breeches thick enough to enable them to withstand the climate, suffered much in the cold, raw fogs; so, when their legs were numb they plucked nettles and gave those members such a scouring that they burned and smarted gloriously for the rest of the day" (298).

---

[155]Also called *Urtica membranacea* Poir.

FIGURES 31a, 31b, 32. — *Vitis vinifera*, the common grape-vine of the Old World used as a printer's mark in many PLANTIN publications (left, top) and (right, top) by a modern calligrapher (R. KOCH, with F. KREDEL, Christliche Symbole, Kassel, ca. 1935). Many early manuscript and printed Bibles, in various languages were beautifully illuminated or illustrated with title pages, frontispieces, pictorial initials, headpieces, and marginal decorations, often hand-painted. As pointed out previously (p. xiv) the plants depicted in these illustrations were often supposed, rather than actual, Bible plants. In the present cases the plant is obviously the common grape of the Old World and the reference is to its use by JESUS in the famous Last Supper prototype of Christian Communion.—MOSES sent spies into the land of Canaan to report on its condition; at the brook Eshcol they found grapes whose clusters were so large that they had to be carried on a pole between two men, as shown below, in a drawing made for the Swedish Bible (1927, cf. sub fig. 9; see also fig. 71). The vines of Palestine have always been renowned for the luxuriance of their growth and the large size of the fruit clusters they produce, but the artist's conception here is obviously a symbolic exaggeration.

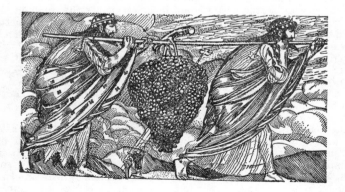

225. Vitis orientalis (Lam.) Boiss.

ISAIAH 5: 2-4—And he fenced it, and gathered out the stones thereof, and planted it with the choicest vine, and built a tower in the midst of it, and also made a winepress therein: and he looked that it should bring forth grapes, and it brought forth *wild grapes* ... What could have been done more to my vineyard, that I have not done in it? wherefore, when I looked that it should bring forth grapes, brought it forth *wild grapes*?
JEREMIAH 2: 21—Yet I had planted thee a noble vine, wholly a right seed: how then art thou turned into a degenerate plant of a *strange vine* to me?
EZEKIEL 15: 2-6— ... What is the *vine* tree more than any tree, or than a branch which is among the trees of the forest? Shall wood be taken thereof to do any work? or will men take a pin of it to hang any vessel thereon? Behold, it is cast into the fire for fuel; the fire devoureth both the ends of it, and the midst of it is burned. Is it meet for any work? Behold, when it was whole, it was meet for no work: how much less shall it be meet yet for any work, when the fire hath devoured it, and it is burned? ... As the *vine* tree among the trees of the forest, which I have given to the fire for fuel ...

The Hebrew word involved in the first two references is "boser" or "beooshim". Because the singular form of this word is rendered "cockle" by the King James Version in Job 31: 40, there has been some argument as to whether or not wild grapes are really intended here. Some have supposed that the hoary Palestine nightshade, *Solanum incanum* L., might fit both the "cockle" reference in Job and the "wild grapes" and "strange vine" (Hebrew, "gefen nachriyah") references here. This nightshade is a common weed along roadsides and in waste places in Palestine and Egypt, and its berries resemble grapes in form (267), although they are narcotic and poisonous. For more information about this plant, see under *Agrostemma githago*.

Because of the context of the verses cited at the head of this chapter, wherein the rebellious people of Israel are compared to "wild grapes" and "the degenerate plant of a strange vine", it seems clear to us that nothing but some close relative of the vine could have been intended (306). TRISTRAM supposed that the plant referred to was the Oriental vine, *Vitis orientalis*[156]. This has been described as "the native wild fox grape of Palestine and Syria" (184, 233, 299). The LEESSER translation affords an excellent additional support for this contention because it speaks of "worthless fruit" from the "ignoble vine". This wild grape has small, black, very acid berries about the size of a currant, with very little juice (266). It has bipinnate leaves with 7-11 leaflets, or the uppermost sometimes 3-foliolate. The leaflets are ovate-oblong, rounded or wedge-shaped at the base, and coarsely serrate or incised. There are no tendrils, and the greenish flowers are borne in dichotomously cymose inflorescences. The plant grows wild in rocky places and hedges in the Mediterranean region (299). POST records it from Syria and Hauran, but not from Palestine. EVENARI also doubts whether it was in Palestine then since it is not there now, but is only in northern Syria. The Jastrow and King James versions both give the same translations in these verses.

JOHN SMITH advances the theory (299) that the plant of these two references may have been nothing more than an inferior or "wild" strain of the ordinary grape vine, *Vitis vinifera* (which see). Indeed, the context of the passages would seem to call for a plant which could not be differentiated from the ordinary vine in its growth and foliage and hence was cultivated for some years in the hope that it would eventually yield the expected fruit. A plant even as closely related as the Oriental vine would certainly not thus be mistaken by the owner of the vineyard, because its leaves and habit of growth are entirely different.

Obviously this "wild grape" of ISAIAH and "strange vine" of JEREMIAH have

---
[156]Also called *Cissus orientalis* Lam. and *C. pinnata* Russ.

nothing whatever to do with the "wild vine" of ELISHA (II Kings 4: 39), which is discussed by us under *Citrullus colocynthis*.

226. Vitis vinifera L.
(FIGURE 31; FIGURE 32; FIGURE 71)

GENESIS 9: 20-21 & 24—And NOAH began to be an husbandman, and he planted a *vineyard*: and he drank of the *wine*, and was *drunken* . . . And NOAH awoke from his *wine* . . .

GENESIS 14: 18—And MELCHIZEDEK, king of Salem brought forth bread and *wine*. . .

GENESIS 40: 9-11— . . . In my dream, behold, a *vine* was before me; and in the *vine* were three branches: and it was as though it budded, and her blossoms shot forth; and the clusters thereof brought forth ripe *grapes*: and PHARAOH's cup was in my hand: and I took the *grapes* . . .

GENESIS 49: 11-12—Binding his foal unto the *vine*, and his ass's colt unto the choice *vine*; he washed his garments in *wine*, and his clothes in the blood of *grapes*: His eyes shall be red with *wine*, and his teeth white with milk.

EXODUS 22: 5—If a man shall cause a field or *vineyard* to be eaten, and shall put in his beast, and shall feed in another man's field; of the best of his own field, and of the best of his own *vineyard*, shall he make restitution.

LEVITICUS 19: 10—And thou shalt not glean thy *vineyard*, neither shalt thou gather every grape of thy *vineyard*; thou shalt leave them for the poor and stranger . . . .

NUMBERS 13: 20 & 23-24— . . . Now the time was the time of the firstripe *grapes* . . . And they came unto the brook of *Eshcol*, and cut down from thence a branch with one cluster of *grapes*, and they bare it between two upon a staff; and they brought of the pomegranates and of the figs. The place was called the brook *Eshcol*, because of the cluster of *grapes* which the children of Israel cut down from thence.

NUMBERS 20: 5—And wherefore have ye made us to come up out of Egypt, to bring us in unto this evil place: it is no place of seed, or of figs, or of *vines* . . .

NUMBERS 22: 24—But the angel of the Lord stood in a path of the *vineyards*, a wall being on this side, and a wall on that side.

NUMBERS 32: 9— . . . the valley of *Eshcol* . . .

DEUTERONOMY 6: 11—And houses full of all good things, which thou filledest not, and wells digged, which thou diggedest not, *vineyards* and olive trees which thou plantedst not.

DEUTERONOMY 8: 8—A land of wheat, and barley, and *vines*, and fig trees, and pomegranates . . .

DEUTERONOMY 24: 21—When thou gatherest the *grapes* of thy *vineyard*, thou shalt not glean it afterward . . .

DEUTERONOMY 32: 14, 32-33, & 38— . . . and thou didst drink the pure blood of the *grape* . . . For their *vine* is the vine of Sodom, and of the fields of Gomorrah: their *grapes* are grapes of gall, their clusters are bitter. Their *wine* is the poison of dragons . . . drank the *wine* of their drink offerings . . .

JUDGES 9: 12-13 & 27—Then said the trees unto the *vine*, Come thou, and reign over us. And the *vine* said unto them, Should I leave my *wine*, which cheereth God and man, and go to be promoted over the trees? . . . And they went out into the fields, and gathered their *vineyards*, and trode the *grapes*, and made merry . . .

JUDGES 11: 33—And he smote them from Aroer . . . unto the plain of the *vineyards* . . .

JUDGES 14: 5—Then went SAMSON down . . . and came to the *vineyards* of Timnath.

JUDGES 15: 5— . . . and burnt up . . . the *vineyards* and olives.

JUDGES 16: 4— . . . in the valley of *Sorek* . . .

RUTH 2: 14—At meal time come thou hither, and eat of the bread, and dip thy morsel in the *vinegar* . . .

I SAMUEL 25: 18—Then ABIGAIL made haste, and took . . . two bottles of *wine* . . . an hundred clusters of *raisins* . . .

I KINGS 4: 25—And Judah and Israel dwelt safely, every man under his *vine* and under his fig tree . . .

II KINGS 18: 31-32— . . . eat ye every man of his own *vine* . . . Until I come and take you away to a land like your own land, a land of corn and *wine*, a land of bread and *vineyards*, a land of oil olive and of honey . . .

I CHRONICLES 12: 40— . . . brought . . . cakes of figs, and bunches of *raisins*, and *wine*, and oil . . . for there was joy in Israel.

NEHEMIAH 1: 11— . . . For I was the king's *cupbearer*.

NEHEMIAH 2: 1—And it came to pass . . . that *wine* was before him: and I took up the *wine* and gave it unto the king . . .

PSALMS 78: 47—He destroyed their *vines* with hail, and their sycomore trees with frost.

PSALMS 80: 8-16—Thou hast brought a *vine* out of Egypt: thou has cast out the heathen, and planted it. Thou preparedest room before it, and didst cause it to take deep root, and it filled the land. The hills were covered with the shadow of it, and the boughs thereof were like the goodly cedars. She sent out her boughs unto the sea, and her branches unto the river. Why hast thou then broken down her hedges, so that all they which pass by do pluck her? The boar out of the wood doth waste it, and the wild beast of the field doth devour it. Return . . . and visit this *vine*; And the *vineyard* which thy right hand hath planted, and the branch that thou madest strong for thyself . . .

PSALMS 105: 33—He smote their *vines* also . . .

PSALMS 107: 37—And sow the fields, and plant *vineyards*, which may yield fruits of increase.

PSALMS 128: 3—Thy wife shall be as a fruitful *vine* by the sides of thine house; thy children like olive plants round about thy table.
ECCLESIASTES 2: 4— ... I planted me *vineyards*.
SONG 1: 14—My beloved is unto me as a cluster of camphire in the *vineyards* of Engedi.
SONG 2: 13 & 15—The fig tree putteth forth her green figs, and the *vines* with the tender *grapes* give a good smell ... Take us the foxes, the little foxes, that spoil the *vines*: for our *vines* have tender *grapes*.
SONG 7: 7-8—This thy stature is like to a palm tree, and thy breasts to clusters of *grapes* ... now also thy breasts shall be as clusters of the *vine* ...
ISAIAH 1: 8—And the daughter of Zion is left as a cottage in a *vineyard* ...
ISAIAH 5: 1-7 & 10—Now will I sing to my wellbeloved a song of my beloved touching his *vineyard*. My wellbeloved hath a *vineyard* in a very fruitful hill. And he fenced it, and gathered out the stones thereof, and planted it with the choicest *vine*, and built a tower in the midst of it, and also made a *winepress* therein: and he looked that it should bring forth *grapes*, and it brought forth wild grapes. And now ... judge ... betwixt me and my *vineyard*. What could have been done more to my *vineyard*, that I have not done in it? Wherefore, when I looked that it should bring forth *grapes*, brought it forth wild grapes? And now go to; I will tell you what I will do to my *vineyard*: I will take away the hedge thereof, and it shall be eaten up; and break down the wall thereof, and it shall be trodden down: And I will lay it waste: it shall not be pruned, nor digged; but there shall come up briers and thorns ... For the *vineyard* of the Lord of hosts is the house of Israel ... Yea, ten acres of *vineyard* shall yield one bath ...
ISAIAH 16: 8-10—For the fields of Heshbon languish, and the *vine* of Sibmah ... Therefore I will bewail with the weeping of Jazer the *vine* of Sibmah ... and in the *vineyards* there shall be no singing, neither shall there be shouting: the treaders shall tread out no *wine* in their *presses*; I have made their *vintage* shouting to cease.
ISAIAH 24: 7, 9, 11, & 13—The new *wine* mourneth, the *vine* languisheth ... They shall not drink *wine* with a song ... There is a crying for *wine* in the streets ... and as the gleaning *grapes* when the *vintage* is done.
ISAIAH 27: 2-11— ... A *vineyard* of red *wine* ... I will water it every moment: lest any hurt it ... when it shooteth forth, thou wilt debate with it ... there shall the calf feed ... and consume the *branches* thereof. When the *boughs* thereof are withered, they shall be broken off: the women come, and set them on fire ...
ISAIAH 29: 9— ... but not with *wine* ...
ISAIAH 55: 1— ... come, buy *wine* and milk without money and without price.
ISAIAH 61: 5— ... the sons of the alien shall be your plowmen and your *vinedressers*.
ISAIAH 62: 8— ... the stranger shall not drink thy *wine*, for the which thou hast laboured.
ISAIAH 63: 2-3—Wherefore art thou red in thine apparel, and thy garments like him that treadeth in the *winefat*? I have trodden the *winepress* alone ...
JEREMIAH 2: 21—Yet I had planted thee a noble *vine*, wholly a right seed: how then art thou turned into the degenerate plant of a strange vine unto me?
JEREMIAH 6: 9— ... They shall thoroughly glean the remnant of Isarel as a *vine*: turn back thine hand as a *grapegatherer* into the baskets.
JEREMIAH 25: 15— ... Take the *wine* cup of this fury at my hand ...
JEREMIAH 25: 30— ... he shall give a shout, as they that tread the *grapes* ...
JEREMIAH 31: 5 & 29-30—Thou shalt yet plant *vines* upon the mountains of Samaria: the planters shall plant, and shall eat them as common things ... the fathers have eaten a sour *grape* ... every man that eateth the sour *grape* ...
JEREMIAH 48: 32-33—O *vine* of Sibmah, I will weep for thee with the weeping of Jazer: thy plants are gone over the sea, they reach even to the sea of Jazer: the spoiler is fallen upon thy summer fruits and upon thy *vintage*. And joy and gladness is taken from the plentiful field, and from the land of Moab; and I have caused *wine* to fail from the *winepresses*: none shall tread with shouting; their shouting shall be no shouting.
JEREMIAH 49: 9—If *grapegatherers* come to thee, would they not leave some gleaning *grapes*? ...
LAMENTATIONS 1: 15— ... the Lord hath trodden the virgin, the daughter of Judah, as in a *winepress*.
EZEKIEL 17: 5-10—He took also of the seed of the land, and planted it in a fruitful field; he placed it by great waters, and set it as a willow tree. And it grew, and became a spreading *vine* of low stature, whose branches turned toward him, and the roots thereof were under him: so it became a *vine*, and brought forth branches, and shot forth sprigs. There was also another great eagle with great wings and many feathers: and, behold, this *vine* did bend her roots toward him, and shot forth her branches toward him, that he might water it by the furrows of her plantation. It was planted in a good soil by great waters, that it might bring forth branches, and that it might bear fruit, that it might be a goodly *vine*. Say thou, Thus saith the Lord God; Shall it prosper? shall he not pull up the roots thereof, and cut off the fruit thereof, that it wither? It shall wither in all the leaves of her spring, even without great power or many people to pluck it up by the roots thereof. Yea, behold, being planted, shall it prosper? shall it not utterly wither, when the east wind toucheth it? It shall wither in the furrows where it grew.
EZEKIEL 18: 2—The fathers have eaten sour *grapes* ...
EZEKIEL 19: 10-11—Thy mother is like a *vine* in thy blood, planted by the waters: she was fruitful and full of branches by reason of many waters. And she had strong rods ... and her stature was exalted among the thick branches, and she appeared in her height with the multitude of her branches.

EZEKIEL 27: 18—... in the *wine* of Helbon...
DANIEL 5: 1-2 & 4—BELSHAZZAR the king made a great feast to a thousand of his lords, and drank *wine* before the thousand. BELSHAZZAR, whiles he tasted the *wine*, commanded to bring the gold and silver vessels... They drank *wine* and praised the gods of gold...
HOSEA 9: 2 & 4—The floor and the *winepress* shall not feed them, and the new *wine* shall fail in her... They shall not offer *wine* offerings...
HOSEA 10: 1—Israel is an empty *vine*, he bringeth forth fruit unto himself...
HOSEA 14: 7—They that dwell under his shadow shall... grow as the *vine*: the scent thereof shall be as the *wine* of Lebanon.
JOEL 1: 5, 7, & 10-12—... all ye drinkers of *wine*, because of the new *wine*... He hath laid my *vine* waste... the new *wine* is dried up... O ye *winedressers*... The *vine* is dried up...
JOEL 2: 22 & 24—... the *vine* do yield their strength... the fats overflow with *wine* and oil.
JOEL 3: 3 & 13—... and sold a girl for *wine*, that they might drink... for the *press* is full...
AMOS 9: 13-14—Behold, the days come... that the plowman shall overtake the reaper, and the treader of *grapes* him that soweth seed; and the mountains shall drop sweet *wine*... and they shall plant *vineyards*, and drink the *wine* thereof...
MICAH 4: 4—But they shall sit every man under his *vine* and under his fig tree; and none shall make them afraid...
MICAH 6: 15—... and sweet *wine*, but shall not drink *wine*.
NAHUM 3: 11—Thou shalt also be *drunken*.
HABAKKUK 2: 15-16—Woe unto him that giveth his neighbour *drink*, that puttest thy bottle to him, and makest him *drunken* also... *drink* thou also...
HABAKKUK 3: 17—... neither shall fruit be in the *vines*...
ZECHARIAH 3: 10—In that day, saith the Lord of hosts, shall ye call every man his neighbour under the *vine* and under the fig tree.
I MACCABEES 6: 34—And to the end they might provoke the elephants to fight, they showed them the blood of *grapes* and mulberries.
MATTHEW 21: 33—... There was a certain householder, which planted a *vineyard*, and hedged it round about, and digged a *winepress* in it, and built a tower..
MATTHEW 26: 27-29—And he took the cup, and gave thanks, and gave it to them, saying, Drink ye all of it... But I say unto you, I will not drink henceforth of this fruit of the *vine*, until that day when I drink it new with you in my Father's kingdom.
MARK 12: 1—... A certain man planted a *vineyard*, and set an hedge about it, and digged a place for the *winefat*, and built a tower, and let it out to husbandmen, and went into a far country.
LUKE 20: 9-16—... A certain man planted a *vineyard*, and let it forth to husbandmen... that they should give him of the fruit of the *vineyard*... Then said the lord of the *vineyard*... I will send my beloved son... So they cast him out of the *vineyard*... What therefore shall the lord of the *vineyard* do unto them?... He... shall give the *vineyard* to others...
JOHN 15: 1-6—I am the true *vine*, and my Father is the husbandman. Every branch in me that beareth not fruit he taketh away: and every branch that beareth fruit, he purgeth it, that it may bring forth more fruit... As the branch cannot bear fruit in itself, except it abide in the *vine*; no more can ye, except ye abide in me. I am the *vine*, ye are the branches: He that abideth in me and I in him, the same bringeth forth much fruit: for without me ye can do nothing. If a man abide not in me, he is cast forth as a branch, and is withered; and men gather them and cast them into the fire, and they are burned.
REVELATION 14: 18-20—... Thrust in thy sharp sickle, and gather the clusters of the *vine* of the earth; for her *grapes* are fully ripe. And the angel thrust in his sickle into the earth, and gathered the *vine* of the earth, and cast it into the great *winepress* of the wrath of God. And the *winepress* was trodden without the city, and blood came out of the *winepress*, even unto the horse bridles, by the space of a thousand and six hundred furlongs.

The common grape-vine, *Vitis vinifera*, is mentioned throughout the Bible, from the days of NOAH (about 2347 B.C.) to those of JESUS. Prophets, patriarchs, psalmists, and apostles all spoke of it, often employing it in a symbolic sense. "The fruitful vine" and "the vine brought out of Egypt" were symbolic of the Jewish people; while JESUS compared himself to that "true vine" of which his disciples were the branches. Its use in this sense is an indication of the very high regard in which it was held by the Israelites. The vine was cultivated by the ancient Egyptians, as is abundantly proved by the paintings and representations of it in the ancient Egyptian tombs, where the various processes of wine-making are fully portrayed (268). It is the very first plant to be recorded in the Bible as cultivated, and it has been cultivated so long by the human race, following the course of civilization from region to region, that its exact origin is now shrouded in mystery (184). It is considered to be a native of the hilly regions of Armenia and the countries bordering on the Caspian Sea, especially on the southern side of this body of water (Azerbaijan and northern Persia), according to JOHN SMITH (299). POST says that "its home is between the southern shores of the Caspian Sea and the Taurus" (95). Because it is first mentioned in the Bible in connection with NOAH, he is usually credited in Jewish and Christian legendry with its introduction into cultivation. From the story of the dream of PHARAOH's butler and from many paintings, inscriptions, and sculptures on ancient Assyrian and Egyptian monuments and records, we know without doubt that it was early cultivated in those lands. Palestine — the "Promised Land" — was described as a "land of wheat and barley, of *vines* and fig-trees and pomegranates". That the vine was early and commonly cultivated in Palestine is indicated by the very numerous references to it in Scripture, and also by the numerous remains of old winepresses found cut in the rocks of that land (299). The quotations given by us at the head of this chapter and the chapter on wine (see under *Saccharomyces ellipsoideus*) represent only a fraction of the Biblical references to this plant and its products.

Dr. ROBERT F. GRIGGS, of George Washington University, in a letter dated March 7, 1940, states that he believes that the "grapes" brought back by JOSHUA and CALEB were bananas, *Musa paradisiaca* subsp. *sapientum* (L.) Kuntze[157]. He says: "It impressed me because bananas are obviously the only fruit a bunch of which could constitute a man's burden as there described, and because interpretation of the text in that way made a rational account of what on the face of it appeared a statement which could not possibly be true. Furthermore, my slight acquaintance with higher criticism has indicated that the translators of the English text were very much at a loss to find English equivalents for the words of the original and upon discovering mention of a fruit in bunches, grapes would be the most natural way to construe it."

The grape-vine of the Old World sometimes assumes the habit of a tree, with a stem up to 1½ feet in diameter (299), the branches then being trained on a trellis, and bearing bunches of grapes 10 or 12 pounds in weight, the individual berries the size of small plums (306). Bunches have even been produced weighing as much as 26 pounds (299). The vines of Palestine were always renowned both for the luxuriance of their growth and for the immense clusters of grapes which they produced. Travelers have again and again commented on the large size of the grape-clusters of Palestine (306). SCHULZ has recorded a vine which he observed at Beitshin, near Ptolemais, whose stem attained a height of 30 feet and whose branches formed a "hut" more than 30 feet in diameter (306). Therefore, it does not seem so improbable to us that the spies sent to the Promised Land should employ a pole or stretcher, carried between them, to transport home some of the clusters, although one would hardly except to find clusters of this size on a wild plant. It is possible that the clusters came from a plant cultivated by the native inhabitants of the land.

MASTERS, *in* LINDLEY's "The Treasury of Botany", reminds us that the specific names applied to the banana, namely *sapientum* and *paradisiaca*, were applied in the belief that bananas constituted the "forbidden fruit" of the "tree of knowledge of good and evil" in the garden of Eden, and that according to classical tradition it was bananas that the spies brought back. However, if bananas really were cultivated in Palestine in those early days it is most remarkable that they would not have been mentioned more often in the Scriptures and that there would not have been a special Hebrew word applying to them. POST regards the banana as a native of India, and states that it is now widely cultivated in our area (267).

The Hebrew word usually translated "vine" in the many Biblical passages quoted above, and elsewhere, is "gefen" or "géphen" (Greek, $ἄμπελος$, meaning "vine"). The Hebrew name still used today for the grape-vine in Palestine is "gefen", according to POST (266). Other words are rendered "grape" (Hebrew, "enav" or "a'navim"), "wine" (Hebrew, "yayin"), "vineyard" (Hebrew, "kerem"), "grape gatherers", "cupbearer" (Hebrew, "mashkeh"), "winefat", "winepress", and "drunken" (Hebrew, "shikor"); and even "vinegar" is considered to have been made from wine on which acetic-acid forming bacteria have been allowed to work (see under *Acetobacter*). The proper place names "Abel-Kerâmim" (Hebrew, "avale keramim", literally "abundance of grapes") in Judges 11: 33 and "Sorek" in Judges 16: 4 are thought to refer to the abundance of grapes and vineyards at those localities. "Sorek" or "shorek" literally means "choice vine" (27).

[157] Also called *Musa sapientum* L.

The grape-vine is at present cultivated everywhere in the Holy Land in numerous varieties, but is nowhere strictly spontaneous (266). In the days when the Hebrews were in Egypt the culture of the vine was of great importance. Its care and culture were well understood by them (179). The "raisins" of I Samuel 25: 18 and I Chronicles 12: 40 were merely dried grapes, as they are today. Palestine has always been renowned for the quantity, quality, and productiveness of its grape-vines, especially in the famous valley of Eshcol (literally "grapes"), and no climate or soil is better adapted for them (27). Special mention is also made in the Bible of the grapes of Sibmah, Heshbon, Elealeh, and Engedi. The vine is the emblem of the Jewish nation and later was adopted as that of the Christian church (184).

The leaves of the common grape-vine differ from those of most of our cultivated American grapes in being rounded heart-shaped, 5-lobed, and coarsely toothed. The branches bear tendrils, and the small, inconspicuous, greenish flowers are panicled (266). According to Egyptian legends the great god of the underworld and judge of the dead, OSIRIS, first taught mankind the art of grape culture (306). Biblical writers employed the vine frequently in their metaphors. To dwell under one's vine and fig-tree was a symbol of domestic happiness, peace, and plenty. The rebellious Israelites were compared to "wild grapes", an "empty vine", and "the degenerate plant of a strange vine".

The ancient Hebrew probably allowed his vines to trail merely upon the ground or over rocks and walls, later using supports and finally trellises (306). The time of vintage was a season of general festivity and usually commenced in September. The towns were then practically deserted and the people lived in temporary tents and "lodges" among the vineyards. The grapes were gathered with great and joyous shouting, and were carried in baskets on the head and shoulders of the grape-gatherers or else were slung upon a yoke. The finest ones were saved for eating and kept in flat open wicker baskets. From these finest ones were also made the dried raisins (Hebrew, "tz'mookim") (306). The rest of the harvest was carried to the big stone winepresses, dug or hewn out of the rocky soil. There the treaders pressed out the juice (see under *Saccharomyces ellipsoideus*). This juice, according to Dr. ROBINSON, "is boiled down to a sirup, which, under the name of *dibs*, is much used by all classes, wherever vineyards are found, as a condiment with their food" (306). Vineyards in Biblical days were usually planted on a hill and were surrounded by walls or hedges to keep out the wild boars, jackals, and foxes. One or more towers, usually of stone, were built within the vineyard, in which the vinedressers, whose function was to prune and cultivate the vines and keep out thieves, lived (306).

The Parable of the Vineyard in Luke 20: 9-16 was entitled "Parable of the Vinegar" in the text of the Clarendon Press at Oxford in 1717. Because of this fact that Bible became known as the Vinegar Bible (158).

The grape-vine is said to have been introduced into France in 540 B.C., and soon thereafter into the southern parts of Europe by the Romans (299). Its successful cultivation is more or less limited to the zone between 36° and 48° in both the Northern and Southern Hemispheres. The heat or cold beyond these limits hinders the full perfection of its fruit. The cooking "plums" and the "currants" of English shops are grapes imported from Spain and Greece — the "currants" coming chiefly from the neighborhood of Corinth, from which they derive their name (158, 299).

The "vine of Sodom" (Hebrew, "gefen s'dom"), mentioned in Deuteronomy 32: 32, is now regarded as being identical with the "apple of Sodom", which Webster's Dictionary and numerous other authorities identify as *Solanum sodomeum* (158), but which we regard as *Citrullus colocynthis* (which see).

227. Xanthium spinosum L.

ISAIAH 34: 13—And *thorns* shall come up in her palaces . . .
HOSEA 9: 6— . . . the pleasant places for their silver, nettles shall possess them: *thorns* shall be in their tabernacles.

There has been much argument over the identity of the "thorns" of the Bible. Spiny plants are very common now in the Levant, as they are in all dry, barren, and desert areas of the world, and as was forecast by the prophets of old. Fully 22 Hebrew and Greek words are used in the Bible to refer to these spiny, thorny, and prickly plants. Hardly any two authorities agree on the identity of any one of them. After reviewing the involved literature on the subject and weighing the evidence presented, it is our conclusion that the "thorns" of Isaiah 7: 19 and 55: 13, Judges 8: 7, and Matthew 7: 16 are probably *Zizyphus spina-christi* (which see); those of Numbers 33:55 are *Rubus sanctus* (which see); those of Genesis 3: 18, Proverbs 15: 19, Isaiah 7: 23-25, 10: 17, and 33: 12, Ezekiel 2: 6, Micah 7: 4, and Psalms 58: 9 are *Rhamnus palaestina* (which see); those of Matthew 13: 7 and Hebrews 6: 8 are *Centaurea calcitrapa* (which see); and those of Matthew 27:29 and John 19: 2 are probably *Paliurus spina-christi* (which see).

There remain, then, the passages about "thorns" in Isaiah 34: 13 and Hosea 9: 6 (Hebrew, "kimosh" or "kimmesonim"). The MOFFATT and GOODSPEED versions render the Hosea verse: "nettles covering the rare silver idols, *thorns* springing in your shrines" and "their desirable places nettles shall possess; *Thorns* shall be in their tents" respectively. JASTROW also uses the word "thorns" in these two verses. The Douay version, however, advances a new concept for the word which these and the King James versions render "thorns". The editors of the Douay translation believe that HOSEA's plant was a "bur". That the burs of Palestine should not have been referred to by Biblical writers seems difficult to believe. Therefore it seems quite probable to us that the Douay version is here the most nearly correct. From the context it would appear that the Isaiah plant was similar to that of Hosea.

Among the bur-producing plants of the region are the branching bur-reed, *Sparganium erectum* L. (which is not annoying in any way and therefore probably not involved in this reference); six species of bur-parsley, *Caucalis* L. (whose burs are too small to be really annoying); the great burdock, *Arctium vulgare* (Hill) Druce[158], an unarmed weedy herb with prickly bur-like fruits; and 2 kinds of clotbur or burweed, *Xanthium strumarium* var. *antiquorum* (Wallr.) Boiss.[159] and *X. spinosum*. Of these the spiny clotbur is perhaps the most annoying and painfully prickly, and for this reason is the most logical plant for the Hosea reference. In addition to its very prickly fruits, characteristic of the genus, it possesses vicious yellow spines at the base of each usually 3-lobed or tripartite leaf. The plant grows to 3 feet tall. Its leaves are canescent beneath, short-petiolate, wedge-shaped at the base, oblong-lanceolate in outline, with the middle lobe much longer than the two lateral ones. The staminate heads of tiny greenish flowers are borne at the apex of the stems; the pistillate are usually solitary, axillary, and nodding (267).

DALMAN offers the names of two other plants as candidates for the Hosea reference: first choice, *Urtica urens* L., second choice, *Ochradenus baccatus* Del. (98).

---

[158]Also referred to as *Lappa vulgaris* Hill, *L. major* Gaertn., *Arctium majus* (Gaertn.) Thuill., and *A. lappa* Willd.
[159]Also known as *Xanthium brasilicum* Vell. and *X. antiquorum* Wallr.

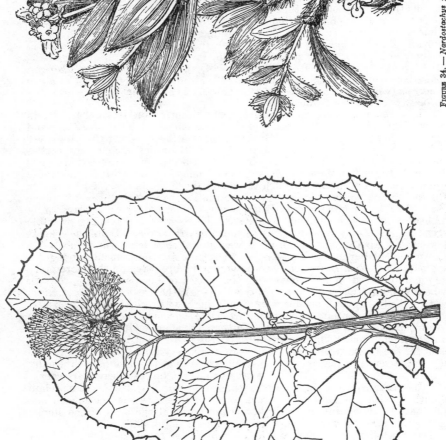

FIGURE 34. — *Nardostachys jatamansi*, the Himalayan spikenard. Its fragrant roots were used as a perfume and stimulant, commonly imported from India as an ointment in sealed alabaster boxes. (Callcott, Scripture Herbal, 1842).

FIGURE 33. — *Saussurea lappa*, the Indian orris, now generally regarded as the "cassia" of Psalms, but not that of Exodus or Ezekiel. Its fragrant roots were used as a medicine, as an aphrodisiac, and especially as a perfume and incense. (After Jacquemont, Voyage dans l'Inde, 1844).

228. Zizyphus lotus (L.) Lam.

JOB 40: 21-22—He lieth under the *shady trees*, in the covert of the reed, and fens. The *shady trees* cover him with their shadow; the willows of the brook compass him about.

The GOODSPEED version renders the above verse as follows: "Beneath the *lotus bushes* he lies down, In the depths of reed and swamp. The *lotus bushes* screen him as his shade; The willows of the brook surround him." Moffatt uses "lotus-trees" in the first part of the quotation, but "thorny thickets" in the second part. The Douay version says "shadow" and "shades", perhaps wisely ignoring the botanical connotations. The Jastrow version uses "lotus-tree" in both cases. The Hebrew for these "shady trees" is "tzeh'ehlim".

The lotus bush of the Levant, *Zizyphus lotus*[160], is a shrub or low tree to about 5 feet tall, with smooth zigzag whitish branches, bearing a pair of slender recurved thorns at the base of each leaf, one thorn of each pair being shorter than the other. The leaves are small, nearly 2-ranked, alternate, leathery, ovate-oblong, obscurely crenate, 3-nerved, and only about $\frac{1}{2}$ inch long. The small greenish flowers are borne in axillary clusters, and are followed by nearly globular drupes about the size of a large pea (266). According to POST, this species inhabits dry places in Antilebanon, the Syrian coast, and Palestine.

The reasons for modern translators identifying the plant of Job's reference with this lotus bush are not apparent to us. MOFFATT and the GOODSPEED version identify the animal which is so glowingly described in this portion of the 40th chapter of Job as the hippopotamus, a more or less aquatic animal of rivers and swamps, lying mostly submerged in the water. It is not at all likely that a hippopotamus would ever lie under a lotus bush or even be found in places where these bushes grow — nor is it certain that hippopotami ever lived where they would likely be known to JOB. Nor would whales or crocodiles — the other suggested identifications of the animal here described — ever lie under lotus bushes. That the animal "munches grass like an ox" would seem to eliminate both the whale and the crocodile; that it has "a tail stiff as any cedar" does not suggest a hippopotamus! It would seem to us that one or the other — or both — of the organisms mentioned in this chapter is misidentified. If the animal is a hippopotamus, then it is difficult to believe that the "shady trees" under which he lay were the low-growing small-leaved lotus bushes of dry ground. Some large-leaved tree habitually inhabiting wet places, like the planetree, *Platanus orientalis* (which see), or the oleander, *Nerium oleander* (which see), would seem far more likely, although neither is spiny. However, the animal may still be grossly misidentified, and in that case the final version may be such as to render the lotus bush more plausible.

For the benefit of readers who are zoologically inclined and who may be interested in attempting to solve this problem, the following description of the animal in the MOFFATT version is given in full: "Look at the hippopotamus there; munching grass like an ox. Look at the strength of his thighs, and the stout muscles of his belly. His tail is stiff as any cedar, the sinews of his thighs are closely knit; his bones are tubes of bronze, his ribs like iron bars. He is God's masterpiece, made to be lord of his fellows. The rivers furnish him with food; wild animals are all amazed at him, as there he lies beneath the *lotus-trees*, in covert of the reed and fen, in the shade of *thorny thickets*, surrounded by the water-willows. He never trembles, though the torrent rages; he is unmoved amid the swollen streams. Who catches him with any barb? Who runs a rope through his nose?"

---

[160]Also called *Rhamnus lotus* L.

229. Zizyphus spina-christi (L.) Willd.

JUDGES 8: 7— . . . then I will tear your flesh with the *thorns* of the wilderness and with briers.
ISAIAH 7: 19—And they shall come, and shall rest all of them in the desolate valleys, and in the holes of the rocks, and upon all *thorns*, and upon all bushes.
ISAIAH 9: 18—For wickedness burneth as the fire: it shall devour the briers and *thorns* . . .
ISAIAH 55: 13—Instead of the *thorn* shall come up the fir tree, and instead of the brier shall come up the myrtle tree . . .
MATTHEW 7: 16—Ye shall know them by their fruits. Do men gather grapes of *thorns*, or figs of thistles?

As has been stated in the previous chapters dealing with spiny plants, there is much uncertainty regarding the identity of the various "thorns" of the Bible. The term was apparently applied to many kinds of spinose bushes and shrubs, and perhaps also to some spiny herbs. The "thorns" of Numbers 33: 55 are considered by us to be *Rubus sanctus* (which see); those of Genesis 3: 18, Proverbs 15: 19, Isaiah 7: 23-25, 10: 17, & 33: 12, Ezekiel 2: 6, Micah 7: 4, and Psalms 58: 9 as *Rhamnus palaestina* (which see); those of Matthew 27: 29 and John 19: 2 as *Paliurus spina-christi* (which see); those of Isaiah 34: 13 and Hosea 9: 6 as *Xanthium spinosum* (which see); and those of Hebrews 6: 8 as *Centaurea calcitrapa* (which see). This leaves the "thorns" of Judges 8: 7, Isaiah 7: 19 & 55: 13, and Matthew 7: 16 to be accounted for. The Hebrew words for these "thorns" are "kotz", "shazit", "naasus" or "nasasu-sim" and "nahatzootz" or "n'ahtzootzim". It is supposed that in these passages the Syrian Christ-thorn, *Zizyphus spina-christi*[161], is the thorny plant to which reference is made (98).

The Syrian Christ-thorn[161a] is a shrub or small tree, 9 to 15 feet tall, with smooth white branches, bearing a pair of stout, unequal, recurved spines at the base of each leaf. The alternate, nearly 2-ranked, leathery leaves are elliptic to ovate or oblong, 1-1½ inches long, rounded or subcordate at the base, obscurely crenate along the margins, 3-nerved, and smooth or puberulent on the nerves beneath. The small greenish flowers are borne on woolly pedicels in axillary clusters, followed by ovate-globular drupes about the size of a hazel nut, with rather dry astringent pulp (266). It grows abundantly on the plains from Syria and Lebanon, through Palestine, to Arabia Petraea and Sinai.

Many commentators have suggested the related jujube, *Zizyphus officinarum* Medic.[162], for the "thorns" of Isaiah and Matthew. This is a spiny middle-sized tree with edible berry-like fruit with a two-seeded nut in its center. The fleshy exterior of the fruit is saffron-colored when ripe and has the form and size of an olive. The tree bears a great abundance of these fruits, which are said to be much relished by the natives (268). According to POST it is cultivated in the region, but as an introduction from China or India, only "doubtfully native" in Palestine and Syria (266). For this and other reasons we are not regarding it as involved in the references cited at the head of this chapter. The context of these references, furthermore, does not call for a plant with edible fruit, and would, in fact, seem to preclude one whose fruit was edible and highly relished.

Some commentators have even suggested *Zizyphus officinarum* for the "bramble" of Judges 9: 14-15, which, they say, is singularly fitted to be classed with the fig and grape-vine as eligible for the title of "king of the trees" (233). It is also suggested by these writers that the statement "let fire come out of the bramble, and devour the cedars of Lebanon" in this same story applies very well to *Zizyphus*, for its wood is singularly combustible (98, 268). DALMAN gives *Prosopis stephaniana* (Willd.) Kunth as a second choice (98). However, most authorities now believe that the reference there was to *Rubus sanctus* (which see). The "brambles" of Isaiah 34: 13 are also often considered to have been *Zizyphus spina-christi* or *Z. officinarum*, but according to MOFFATT

---

[161]Also known as *Rhamnus spina-christi* L. and *R. nabeca* Forsk.
[161a]Also known by the following common names: "döm" (by the Europeans), "sidr", and "nulok" (124d).
[162]Also known as *Zizyphus jujuba* Mill., *Z. vulgaris* Lam., and *Z. sativa* Gaertn.

should be rendered "thistles", and are by us regarded as *Notobasis syriaca* (which see).

230. Zostera marina L.

JONAH 2: 5—The waters compassed me about, even to the soul: the depth closed me round about, the *weeds* were wrapped about my head.

The Hebrew word here rendered "weeds" in the King James and Jastrow versions, "seaweed" in the GOODSPEED, "sea-weeds" by MOFFATT, and "sea" by the Douay version is "sûph" (Greek, τὸ ἕλος, dictionary translation "a general term for water weeds"). LEESSER says "sea weeds" in Jonah, but "flags" in Exodus. In spite of the Douay version's rendition, this term is a general one for water weeds, including sea-weeds as well as rank marsh vegetation growing on rivers' edges (184). Although the same Hebrew word is used in Exodus 2: 3 & 5, the "flags" and "bulrushes" in which MOSES' ark was concealed are now generally regarded as having been papyrus. The "weeds" of JONAH are obviously a very different type of plant. The context here is such that it points directly and unequivocally to the eelgrass, sea-wrack, or grass-wrack, *Zostera marina* (78). This is a plant which greatly resembles grass, growing in sandy or muddy water at the mouths of tidal rivers and even in the sea itself to a depth of 35 feet, with ribbon-like leaves 3 or 4 feet long, floating in submerged masses (299). Swimmers and bathers often have the unpleasant experience of coming suddenly in contact with masses of this slimy eelgrass under water, and know that it wraps itself around one's body exactly as JONAH describes, although certain large leafy algae like *Ulva* and *Fucus* will do the same thing. Another species of eelgrass, *Zostera nana*[163] also occurs in the region, but has very short leaves and grows on the sea-bottom in much shallower water (to 6 feet deep) (267) and therefore is not very likely as JONAH's plant.

Some commentators have supposed that the "flags" of Exodus 2: 3 & 5, Isaiah 19: 6, and Job 8: 11 were also *Zostera marina*, but this is not at all probable for the context is entirely different. Those of Exodus are regarded by us as *Cyperus papyrus* (which see), and those of Isaiah and Job as *Juncus effusus* (which see). Certain writers have based their contention in this matter on the claim that the Hebrew word "sûph" is used in all these references. Actually it occurs only in the Exodus, Isaiah, and Jonah passages (27, 184, 299). In Job 8: 11 the word employed is "achú" (184).

---

[163]Also known as *Fucagrostis minor* Cavol. and *Zostera minor* (Cavol.) Nolte.

FIGURE 35.—*Tabernaemontana alternifolia*, the divi-ladner, a non-European and non-Biblical plant of interest for having been considered in S. E. Asiatic folklore as a Bible plant. The Portuguese in Ceylon long considered that island to have been the site of the Garden of Eden and this curious tree to have been the tree of Knowledge of Good and Evil. The fruit is said to resemble somewhat an apple with one bite removed, but is poisonous. (La Belgique Horticole, 1854, after Paxton).

Unidentified Plant References

GENESIS 1: 11-12 & 29—Let the earth bring forth *grass*, and *herb* yielding *seed*, and the *fruit tree* yielding *fruit* after his kind . . . And the earth brought forth *grass*, and *herb* yielding *seed* after his kind, and the *tree* yielding *fruit* whose *seed* was in itself . . . Behold, I have given you every *herb* bearing *seed* which is upon the face of the earth, and every *tree*, in which is the *fruit* of a *tree* yielding *seed* . . .

GENESIS 22: 7 & 13— . . . Behold the fire and the *wood*: but where is the lamb for a burnt offering? . . . a ram caught in a *thicket* by his horns . . .

GENESIS 24: 25 & 32— . . . We have both *straw* and *provender* enough, and room to lodge in . . . and gave *straw* and *provender* for the camels . . .

EXODUS 5: 7-18—Ye shall no more give the people *straw* to make brick, as heretofore: let them go and gather *straw* for themselves . . . Thus saith PHARAOH, I will not give you *straw* . . . get your *straw* where ye can find it . . . to gather *stubble* instead of *straw* . . . Fulfil your works, your daily tasks, as when there was *straw* . . . There is no *straw* given unto thy servants . . . there shall no *straw* be given you.

EXODUS 15: 7— . . . by wrath which consumed them as *stubble*.

EXODUS 15: 25— . . . and the Lord shewed him a *tree*, which when he had cast into the waters, the waters were made sweet . . .

LEVITICUS 19: 19— . . . thou shalt not sow thy field with mingled *seed* . . .

NUMBERS 13: 26— . . . shewed them the *fruit* of the land.

DEUTERONOMY 14: 22—Thou shalt truly tithe all the increase of thy *seed* that the field bringeth forth year by year.

DEUTERONOMY 20: 19-20—When thou shalt besiege a city . . . thou shalt not destroy the *trees* thereof by forcing an ax against them: for thou mayest eat of them, and, thou shalt not cut them down (for the *tree* of the field is man's life) to employ them in the siege. Only the *trees* which thou knowest that they be not *trees* for meat, thou shalt destroy and cut them down: and thou shalt build bulwarks against the city . . .

DEUTERONOMY 21: 22-23— . . . and thou hang him on a *tree*: His body shall not remain all night upon the *tree* . . .

DEUTERONOMY 26: 2—That thou shalt take of the first of the *fruit* of the earth . . .

DEUTERONOMY 28: 38—Thou shalt carry much *seed* out into the field, and shalt gather but little in, for the locust shall consume it.

DEUTERONOMY 28: 42—All thy *trees* and *fruit* of thy land shall the locust consume.

DEUTERONOMY 33: 14— . . . precious *fruits* brought forth by the sun . . .

I SAMUEL 22: 5— . . . Then DAVID departed, and came into the *forest* of Hareth.

II SAMUEL 17: 28— . . . and parched *pulse*.

II KINGS 19: 30— . . . shall again take *root* downward.

JOB 14: 2 & 7-9—He cometh forth like a *flower*, and is cut down . . . For there is hope of a *tree*, if it be cut down, that it will *sprout* again, and that the tender *branch* thereof will not cease. Though the *root* thereof wax old in the earth, and the *stock* thereof die in the ground; Yet through the scent of water it will *bud*, and bring forth *boughs* like a *plant*.

JOB 15: 32— . . . and his *branch* shall not be green.

JOB 21: 18—They are as *stubble* before the wind, and as *chaff* that the storm carrieth away·

JOB 24: 20— . . . wickedness shall be broken as a *tree*.

JOB 30: 7—Among the *bushes* they brayed . . .

JOB 41: 27-29—He esteemeth iron as *straw*, and brass as rotten *wood* . . . slingstones are turned with him into *stubble*. Darts are counted as *stubble* . . .

PSALMS 1: 4—The ungodly are not so: but are like the *chaff* which the wind driveth away.

PSALMS 37: 2— . . . and wither as the *green herb*.

PSALMS 50: 10—For every beast of the *forest* is mine . . .

PSALMS 83: 13— . . . as the *stubble* before the wind.

PSALMS 103: 15—As for man, his days are as grass: as a *flower* of the field, so he flourisheth.

PSALMS 104: 12-16— . . . fowls . . . which sing among the *branches*, *grass* to grow for the cattle, and *herb* for the service of man . . . the *trees* of the Lord are full of *sap*.

PSALMS 105: 33—He smote their vines also and their fig trees; and brake the *trees* of their coasts.

PSALMS 126: 6—He that goeth forth and weepeth, bearing precious *seed*, shall doubtless come home again rejoicing, bringing his sheaves with him.

PSALMS 141: 7— . . . as when one cutteth and cleaveth *wood* upon the earth.

PROVERBS 3: 18—She is a *tree* of life to them that lay hold upon her . . .

PROVERBS 8: 19—My *fruit* is better than gold . . .

PROVERBS 11: 30—The *fruit* of the righteous is a *tree* of life . . .

PROVERBS 12: 14—A man shall be satisfied with good by the *fruit* of his mouth . . .
PROVERBS 18: 20—A man's belly shall be satisfied with the *fruit* of his mouth . . .
PROVERBS 26: 20—Where no *wood* is, there the fire goeth out . . .
PROVERBS 27: 25— . . . And the *herbs* of the mountains are gathered in.
ECCLESIASTES 2: 5-6—I made me *gardens* and *orchards*, and I planted *trees* in them of all kind of *fruits*: I made me pools of water, to water therewith the *wood* that bringeth forth *trees*.
ECCLESIASTES 11: 3 & 6— . . . if the *tree* fall toward the south . . . in the place where the *tree* falleth, there it shall be . . . In the morning sow thy *seed* . . .
SONG 4: 16— . . . Let my beloved come into his garden, and eat his pleasant *fruits*.
ISAIAH 3: 10—Say ye to the righteous . . . they shall eat the *fruit* of their own doings.
ISAIAH 5: 10 & 24— . . . *seed* of an homer shall yield an ephah . . . Therefore as the fire devoureth the *stubble*, and the flame consumeth the *chaff*, so their *root* shall be as rottenness, and their *blossom* shall go up as dust . . .
ISAIAH 7: 11 & 19— . . . and the lion shall eat *straw* like the ox . . . and upon all *bushes*.
ISAIAH 9: 18— . . . For wickedness burneth as the fire: it shall devour the briers and thorns, and shall kindle in the *thickets* of the *forest* . . .
ISAIAH 15: 6— . . . There is no *green thing*.
ISAIAH 17: 11— . . . in the morning shalt thou make thy *seed* to flourish . . .
ISAIAH 21: 13— . . . In the *forest* of Arabia shall ye lodge . . .
ISAIAH 25: 10— . . . and Moab shall be trodden down under him, even as *straw* is trodden down for the dunghill.
ISAIAH 28: 4—And the glorious beauty . . . shall be a fading *flower*, and as the hasty fruit . . .
ISAIAH 29: 17— . . . and the fruitful *field* shall be esteemed as a *forest*?
ISAIAH 32: 15— . . . and the fruitful *field* be counted for a *forest*.
ISAIAH 33: 11—Ye shall conceive *chaff*, ye shall bring forth *stubble* . . .
ISAIAH 41: 2— . . . as driven *stubble* to his bow.
ISAIAH 53: 2— . . . as a *tender plant*, and as a *root* out of a dry ground . . .
ISAIAH 55: 10— . . . that it may give *seed* to the sower, and bread to the eater . . .
ISAIAH 56: 3— . . . I am a dry *tree*.
ISAIAH 57: 19—I create the *fruit* of the lips . . .
ISAIAH 61: 3 & 11— . . . they might be called *trees* of righteousness . . . For as the earth bringeth forth her *bud* . . .
ISAIAH 64: 6— . . . we all do fade as a *leaf*; and our iniquities, like the wind, have taken us away.
ISAIAH 65: 22— . . . for as the days of a *tree* are the days of my people . . .
JEREMIAH 4: 7 & 29—The lion is come up from his *thicket* . . . The whole city shall flee for the noise of the horsemen and bowmen; they shall go into the *thickets* . . .
JEREMIAH 5: 6—Wherefore a lion out of the *forest* shall slay them . . .
JEREMIAH 13: 24—Therefore will I scatter them as the *stubble* that passeth away by the wind of the wilderness.
JEREMIAH 17: 10— . . . give every man according . . . to the *fruit* of his doings.
JEREMIAH 21: 14—But I will punish you according to the *fruit* of your doings . . . and I will kindle a fire in the *forest* thereof . . .
JEREMIAH 26: 18— . . . and Jerusalem shall become heaps, and the mountain of the house as the high places of a *forest*.
JEREMIAH 32: 19— . . . according to the *fruit* of his doings.
JEREMIAH 46: 22-23— . . . as hewers of *wood*. They shall cut down her *forest* . . .
EZEKIEL 1: 4 & 27— . . . and a brightness was about it, and out of the midst thereof as the colour of *amber* . . . And I saw as the colour of *amber*, as the appearance of fire . . .
EZEKIEL 8: 2— . . . as the appearance of brightness, as the colour of *amber*.
EZEKIEL 15: 2-3 & 6— . . . among the *trees* of the *forest*? Shall *wood* be taken thereof to do any work? . . . As the vine tree among the *trees* of the *forest* . . .
EZEKIEL 20: 46-47— . . . and prophesy against the *forest* of the south field. And say to the *forest* of the south, Hear the word of the Lord . . . I will kindle a fire in thee, and it shall devour every *green tree* in thee, and every *dry tree* . . .
EZEKIEL 47: 12—And by the river upon the bank thereof, on this side and on that side, shall grow all *trees* for meat, whose *leaf* shall not fade, neither shall the *fruit* thereof be consumed: it shall bring forth new *fruit* according to his months, because their waters they issued out of the sanctuary: and the *fruit* thereof shall be for meat, and the *leaf* thereof for medicine
DANIEL 1: 12 & 16— . . . let us give them *pulse* to eat . . . and gave them *pulse*.
HOSEA 2: 12— . . . I will make them a *forest* . . .
HOSEA 10: 13— . . . ye have eaten the *fruit* of lies . . .
JOEL 1: 17—The *seed* is rotten under their clods . . .
AMOS 3: 4—Will a lion roar in the *forest*, when he hath no prey?
AMOS 4: 9—I have smitten you with *blasting* and *mildew* . . .
MICAH 3: 12— . . . Jerusalem shall become heaps, and the mountain of the house as the high places of the *forest*.
MICAH 7: 13— . . . for the *fruit* of their doings.
HAGGAI 1: 10— . . . the earth is stayed from her *fruit*.
HAGGAI 2: 17—I smote you with *blasting* and with *mildew* . . .
HAGGAI 2: 19—Is the *seed* yet in the barn? . . .
ZECHARIAH 8: 12—For the *seed* shall be prosperous . . .
MALACHI 4: 1— . . . and all that do wickedly, shall be *stubble*: and the day that cometh shall burn them up, saith the Lord of hosts, that it shall leave them neither *root* nor *branch*.

MATTHEW 3: 8 & 10—Bring forth therefore *fruits* meet for repentance . . . And now also the ax is laid unto the *root* of the *trees*, therefore every *tree* which bringeth not forth good *fruit* is hewn down, and cast into the fire.

MATTHEW 12: 33—Either make the *tree* good, and his *fruit* good; or else make the *tree* corrupt, and his *fruit* corrupt: for the *tree* is known by his *fruit*.

MARK 8: 24— . . . . . I see men as *trees*, walking.

LUKE 3: 8-9—Bring forth therefore *fruits* worthy of repentance . . . And now also the axe is laid unto the *root* of the *trees*: every *tree* therefore which bringeth not forth good *fruit* is hewn down, and cast into the fire.

LUKE 11: 42— . . . and all manner of *herbs*.

LUKE 21: 29-30— . . . Behold the fig tree, and all the *trees;* When they now shoot forth, ye see and know of your own selves that summer is now nigh at hand.

ROMANS 1: 13— . . . that I might have some *fruit* among you also.

ROMANS 6: 21—What *fruit* had ye then in those things whereof ye are now ashamed?

ROMANS 7: 4— . . . we should bring forth *fruit* unto God.

I CORINTHIANS 3: 12—Now if any man build upon this foundation gold, silver, precious stones, *wood, hay, stubble*.

II CORINTHIANS 9: 10—Now he that ministereth *seed* to the *sower* both minister bread for your food, and multiply your *seed* sown, and increase the *fruits* of your righteousness.

GALATIANS 3: 13— . . . Cursed is every one that hangeth on a *tree*.

GALATIANS 5: 22—But the *fruit* of the Spirit is love, joy . . .

EPHESIANS 5: 9—For the *fruit* of the Spirit is in all goodness . . .

PHILIPPIANS 4: 17— . . . I desire *fruit* that may abound to your account.

II TIMOTHY 2: 6—The husbandman that laboureth must be the first partaker of the *fruits*.

JAMES 5: 7— . . . the husbandman waiteth for the precious *fruit* of the earth . . .

I PETER 1: 23—Being born again, not of corruptible *seed*, but of incorruptible . . .

I PETER 2: 24—Who his own self bare our sins . . . on the *tree* . . .

JUDE 12— . . . *trees* whose *fruit* withereth, without *fruit*, twice dead, plucked up by the *roots*.

REVELATION 2: 7— . . . To him that overcometh will I give to eat of the *tree* of life, which is in the midst of the paradise of God.

REVELATION 7: 3—Saying, Hurt not . . . the *trees* . . .

REVELATION 22: 2 & 14—In the midst of the street of it, and on either side of the river, was there the *tree* of life, which bare twelve manner of *fruits*, and yielded her *fruit* every month: and the leaves of the *tree* were for the healing of the nations . . . they may have right to the *tree* of life . . .

Extensive and intensive botanical, bibliographic, linguistic, and Biblical studies over a period of many years have enabled the authors to recognize with varying degrees of assurance many of the plants and plant products mentioned in the books of the Bible. The main method of carrying out this research has been careful study and comparison of the various English-language translations — both old and new — Protestant, Catholic, and Jewish — of the various books of the Old and New Testaments and Apocrypha, supplemented by constant study of standard Biblical dictionaries, encyclopedias, and concordances, always in conjunction with studies of the actual recorded flora of Bible-lands, past and present. The present volume lists 231 different plants to which we are fairly certain that reference is made; but there are doubtless many others still unrecognized by us. It lists most — but not always all — of the Bible verses containing direct or indirect references to plants and plant products. The indexes at the close of this chapter will help the reader to find the passages or the topics to which references are made in this or other books on the subject. This work, however, cannot yet answer all of the problems of plant identification involved in the Bible. It is hoped that this volume will stimulate others to think along these lines and to make original investigations, preferably in the Levant and Egypt. Further studies in the science of medicine will doubtless bring to light more Biblical references to diseases caused by plant parasites. Some diseases whose causal organisms are not yet definitely known may some day be found to be caused by plant parasites. It is confidently expected that Dr. HA-REUBENI and his associates at the Hebrew University in Jerusalem will be able eventually to settle many still disputed identifications and perhaps add many more plant names to our census of Bible plants.

In numerous Scriptural passages words are employed which seem to be very general in their meaning. Among these are the "grass" (Hebrew, "desheh") and "herb" (Hebrew, "aysehv") of Genesis 1: 11, the "green herb" of Psalms 37: 2, the "herbs" (Hebrew, "achu") of Proverbs 27: 25, the "green thing" (Hebrew, "desheh yerek") of Isaiah 15: 6, and the "herbs" of Luke 11:

42 — the last-mentioned being rendered "vegetables" by MOFFATT. Some authors have attempted to limit these terms to definite kinds of plants, but the evidence in most of these cases does not seem yet to be conclusive. We have therefore preferred to withhold any definite determination in these cases. Some of the many passages in which such as yet unidentified plant references occur are given at the head of this final chapter. In some cases, however, we are able to give some hints as to possible identifications.

The "pulse" of II Samuel 17: 28 and of Daniel 1: 12 and 16 probably was comprised of the dried leguminous seeds of both beans and lentils, or perhaps of only one or the other. The Hebrew word used in the II Samuel passage is "kali", while that used in Daniel is "zeronim". The latter is rendered "vegetables" by LEESSER. The Septuagint uses ὄσπρια in both books. The "chaff" (Hebrew, "motz"), "straw" (Hebrew, "tehven"), and "stubble" (Hebrew, "kahsh") mentioned so many times in the Old Testament consisted of the dried stalks of the various grains cultivated at that time — spelt, barley, millet, and the various types of wheat — separately or in various combinations, as well as stalks of wild grasses. In some cases a study of the context of the passage may reveal which of these grains would have been the more likely to have furnished the straw or stubble at that time and place, but in most cases this seems to be a hopeless task. Similarly, the "provender" (Hebrew, "mispo") of Genesis 24: 25 and 32 consisted of food for camels. What grains and other plants were included in this food is not certain. What kind of grain stems were employed for the straw used in brick-making as recounted in Exodus 5: 7-18 might eventually be ascertained, but is not definitely known as yet to us.

The "blasting" (Hebrew, "shidaphon") and "mildew" (Hebrew, "yea' rakon" or "yirakon") of Amos 4: 9 and Haggai 2: 17 were most certainly caused by some parasitic fungi such as cause blights and mildews on our garden crops today. If the Biblical writers had specified which of the valuable plants of the Jews were so smitten, it would be an easy matter to hazard an identification of the causal fungus. But in both cases the "blasting" and "mildew" are described in such a way as to apply to no specific plants, but to all the cultivated plants of the Israelites in general. The "leprosy" (Hebrew, "nehga") described in Leviticus 14: 33-57 as attacking houses was probably a form of mold or mildew growing on damp walls. Perhaps even forms of lichens, algae, and higher fungi are involved here. Although the Bible country is in general more arid than moist, there would always be enough damp areas or objects under certain conditions that would favor the rapid growth of various mildews and molds. This applies also to the "leprosy" (Hebrew, "nehga") of garments described in Leviticus 13: 47-59, which was probably a mold.

The words "bush", "bushes" (Hebrew "nahalolim"), "tree" (Hebrew, "etz") and "trees" occur in many passages in such a general way that identification to a particular species seems impossible at this time. Various translators have rendered these terms in different ways, but mostly have not attempted to be more specific. The "bushes" of Isaiah 7: 19, for instance, are called "branches" by JASTROW and "shrubs" by the Douay version. In many verses the word "wood" (Hebrew, "etz") is mentioned in reference to fires. Obviously the fire-makers would use whatever wood was accessible and available to them. The identity of the wood or woods would depend upon where in Bible lands the fire-makers were at each time. In some cases this might be worked out with a fair degree of probable accuracy, but in most cases it is a hopeless task. Some of such "wood" references occur in Genesis 22: 7, Psalms 141: 7, Proverbs 26: 20, and Jeremiah 46: 22.

The word "amber" appears three times in Ezekiel 1: 4 and 27 and 8: 2, in each case spoken of as a color, which is a familiar clear bright yellow. Amber — or electrum, as it is often called and as some translations render the word — is a fossilized resin from various coniferous trees. It is usually found in association with lignite (158). The Hebrew word involved here is "chashmal". It is not certain that true plant amber was known in EZEKIEL's time (about 595 B.C.). The King James, MOFFATT, and Douay versions all use the term "amber", while JASTROW employs the synonymous term "electrum". Some modern authorities are of the opinion that the Hebrew word involved here refers to some shining metal such as brass, and not to a plant product. The matter is still uncertain.

There are many references to "forest" (Hebrew, "ya'ar" and "ya'rak") and "forests" in the Bible. Some of these are quoted at the head of this chapter. Where these terms are used in connection with a definitely and specifically identified area or locality, or where the component members of the forest are described in some detail, the botanical character of the forest may be inferred. But often the terms are used very generally, as, for instance, in Psalms 50: 10, where it says "For every beast of the *forest* is mine". The concept imparted by the original relaters and writers in such cases is merely that of a wooded land area, and not of any particular kind of tree. In Jeremiah 46: 23, however, the "forests" to be cut down are those of Egypt, and, furthermore, of the very limited part of that country with which the Jews were acquainted. In such cases the term could probably be assigned to one or a very few species. All the translations employ the same or a similar indefinite word in these passages.

The term "tree" (Hebrew, "etz") is in many passages employed in an allegorical sense; as, for instance, in Isaiah 56: 3 ("I am a dry tree") and Isaiah 61: 3 ("they might be called trees of righteousness"). But even in these cases it is very possible that the writers had a definite kind of tree in mind. Some one kind of tree might have suggested the concept of a "dry tree" — possibly species of tamarisk, with their scale-like leaves — and of a "tree of righteousness" — possibly the evergreen laurel or myrtle. More study is required of these allegorical uses of the term before definite identifications are suggested. In Mark 8: 24 the blind man who was just beginning to regain his sight said "I see men as trees, walking". Here the reference is doubtless not to any particular kind of tree, but to the general outline of an erect slender tree, seen indistinctly and dimly as through a fog. The columnar cypress so common in the Levant would come to mind at once as a tree to which walking men might thus be compared. In Revelation 7: 3 and in Deuteronomy 20: 19-20 the use of the term "trees" is generic. In both cases it applies only to trees for which the Israelites had a definite use, either as food or lumber. Trees which bore edible fruit were not to be destroyed in warfare, but trees with no direct value to the Israelites could be destroyed and could be used for lumber. The value of all trees to the ecology of a region was obviously not recognized. In Job 14: 7-9 a tree is referred to which, when cut down, was able to sprout from adventitious buds around the margins of the stump. Olive trees and willows are noted for this property.

"Thickets" (Hebrew, "sovech") are mentioned a number of times in the Scriptures and like "forests" cannot yet be definitely identified in most cases as to their dominant floristic element. Usually these "thickets" are not described in sufficient detail or their location noted in sufficiently precise words. Possibly some may be identified after more study of the context. Oleanders, for instance, are well known to form very dense and characteristic thickets along the banks of streams. In some places it seems hopeless to identify the plant or plants comprising the thicket, as, for instance, in Isaiah 9: 18 where it is said that fire "shall kindle in the thickets of the forest."

In many passages separate plant parts are referred to — parts such as leaves, (Hebrew, "aleh"), roots, branches, stems, fruit, seed, etc. For example, in Psalms 104: 2 it is stated that fowls "sing among the branches". In some cases the word "branches" is used incorrectly, as in the passages which speak of "branches of palm trees" — referring obviously to the large pinnate leaves of the date-palm. In Isaiah 64: 6 man is warned that he will "fade like a leaf", referring, apparently, to the leaf of almost any deciduous tree. "Root" (Hebrew, "shoresh") is used in a general sense in Isaiah 53: 2 ("as a tender plant, and as a root out of a dry ground") or even in a probably inaccurate sense referring to the base of the trunk of a tree (as in Matthew 3: 10 — "the ax is laid unto the root of the trees"). In some cases it is used without any definite botanical connotation — as in I Timothy 6: 10, "the love of money is the root of all evil", where the obvious meaning is that evil practices issue from an inordinate love of money much as a tree's trunk and ramifying branches issue from its underground roots. A similar usage of botanical terms is seen in Isaiah 11: 1 — "And there shall come forth a *rod* out of the *stem* of Jesse, and a *Branch* shall grow out of his *roots*."

The term "flower" (Hebrew, "tzeetz") is used several times, and often with enough descriptive context to make an identification possible; but in Isaiah 28: 4 the glorious beauty of the fertile valley country of Ephraim is described as nothing but a "fading flower", and in Job 14: 2 man's life span on this earth is allegorically described in these words: "He cometh forth like a *flower*, and is cut down." The word "bud" (Hebrew, "tzimcha" or "tzehmach") is employed in several passages to refer to the general springtime germination of all plant life, as, for instance, in Isaiah 61: 11, where it is stated: "For the earth bringeth forth her *bud*." Elsewhere it is used more specifically and can be identified with reasonable accuracy. Since the Israelites were in such large degree a pastoral nation in Biblical days, well acquainted with the soil and its products and phenomena, it is natural that preachers, prophets, and the tellers of folk tales should often draw upon plants and plant parts to illustrate their parables and stories. That only the most common plants would be employed in such cases is a logical corollary, for the commoner the plant the more likely would it be that all members of the audience would be acquainted with it and would therefore appreciate the allegorical reference. Thus, it seems safe to assign most of these words to the commoner plants of the area, and to the ones most intimately connected with the daily lives of the people.

The terms "fruit" (Hebrew, "p'ri" and "t'noovah") and "fruits" are used very many times all through the Bible. In many cases the reference is very definitely and obviously to the fig or grape or olive, the three most important "fruits" to the Hebrews after they left Egypt. The "fruit" of the legendary "tree of knowledge of good and evil" (124d) (Genesis 2: 17), of course, has no botanical counterpart, although even here a thorough study of the folklore of the Middle East area might bring to light some tree to whose fruit such a power is ascribed by simple natives. Medieval artists usually represented the Tree of Life (Revelation 2: 7) as the date palm (101, 102). In Song 4: 16 there is the invitation "Let my beloved come into his garden, and eat his pleasant *fruits*". The reference here is obviously to fruits cultivated by SOLOMON in his garden. It is generally believed that his gardens contained plants from the far corners of the then-known world, so the word here probably covers more kinds of fruit than in most of the other places where it is employed. In Deuteronomy 26: 2 is the law concerning the offering of a part of the "firstfruits" of the land to the Lord ("then thou shalt take of the first of the *fruit* of the earth"). This practice led directly into the tithing of later days, and it is generally agreed that only the products of cultivated plants were to be tithed. The "firstfruits" of the land therefore included the products of all the plants cultivated by a man, not only his cereal grains, figs, olives, and the like, but also the wine made from his grapes. The "precious *fruits* brought forth by the sun", alluded to in Deuteronomy 33: 14, probably included the same items. In many passages the word is used in a figurative sense, as in Galatians 5: 22 where love and joy are described as the "fruit" of the spirit. In Isaiah 27: 6 several botanical terms are applied to man and his generations: "He shall cause them that come of Jacob to take *root*: Israel shall *blossom* and *bud*, and fill the face of the world with *fruit*."

In various New Testament passages a tree is spoken of which may bring forth "good" or "corrupt" fruit (Matthew 3: 10 & 12: 23, Luke 3: 8-9, Romans 1: 13, etc.). The tree with the good fruit is to be preserved and cherished, that with the corrupt fruit is to be cut down and destroyed. The reference here seems to be very plainly to the fig, which, under favorable conditions will produce delicious fruit and in other circumstances will yield only hard and inedible fruit. The references in Amos 8: 1-2 to "a basket of summer *fruit*" and in Isaiah 28: 4 to "hasty fruit" are also plainly to figs (see under *Ficus carica*).

As with the word "fruit", the related word "seed" (Hebrew, "zehra") has been given several meanings in relatively uniform pattern throughout the various Bible translations. Where the seeds are definitely identified, as the "coriander seed" of Exodus 16: 31, or where it is stated that "wheat" seed was "sown", as in Jeremiah 12: 13, the major part of the identification problem is solved. But in many cases "seed" is mentioned in reference to some grain crop, and it is not certain whether wheat, spelt, barley, or millet was the seed used. Wheat, being the most common, most highly esteemed, and most valuable, it may in many cases be assumed that the seed sown by the sower was wheat. The Hebrews were specifically forbidden to mix seeds of various grains (Leviticus 19: 19) in their fields so in some cases it may be assumed with a fair degree of probable accuracy what seed is the one involved in a particular passage. In other passages it seems hopeless to attempt any such identification — for example, Deuteronomy 28: 38 ("thou shalt carry much *seed* out into the field"), Ecclesiastes 11: 6 ("in the morning sow thy *seed*"), and Deuteronomy 14: 22 ("thou shalt truly tithe all the increase of thy *seed*"). The term is also used figuratively, as in "Being born again, not of corruptible *seed*, but of incorruptible" (I Peter 1: 23), and non-botanically to refer to future generations of men much as the word "fruit" is used. A few instances of such a usage are seen in Malachi 2: 15 ("he might seek a godly *seed*"), Deuteronomy 11: 9 ("which the Lord sware unto your fathers to give unto them and their *seed*"), and Isaiah 61: 9 ("And their *seed* shall be known among the Gentiles"). It is also used to refer to semen, as in Genesis 38: 8-10.

In the final identification of the many still unidentified and the many more still disputed botanical references in the Bible, the help of King SOLOMON, who was reputedly "wiser than all men" and who was able to speak, presumably intelligently, "of trees, from the cedar tree that is in all Lebanon even unto the hyssop that springeth out of the wall", would be of inestimable service! Such accurate and definite identification would add to our concepts of various phases of Bible history and would greatly increase our appreciation of the Bible as a great collection of literature and as an invaluable source of religious inspiration.

Rosa Hierichuntica. Rosen von Jericho.
18

Mandelbaum. Amygdalus.
97

Mit dir kan ich Kriegsvolck zerschmeissen,
und mit meinem Gott über die mauren springen. 130.
Die Ros erinnert uns der Hierichunter Statt
die Israël mit dem gebet bezwungen hat:
Also wer Gott vertraut vil grosses kan verbringe
und mit sighaffter fäust die mauren überspringn.

Ihr gerechten freuet euch des Herzen. 32. 12.
Der süsse mandelbaum früh blühet in dem Merz
und schenckt uns seine frucht gestaltet als ein herz;
So soll von kindheit an das herz zu Gott gericht
Ihm treulich danckbar seyn, stets leben Ihm ver-
pflichtet.

Granatenblüh. Malus punica flore pleno.
33

Wunderbaum. Ricinus.
104

Unser herz freüet sich sein. v. 21.
Der Granatapffelbaum der augenlust erfüllt
die schwachen labt u. stärckt herz, leber, magen kühlt:
Auch also Gottes trost und güte stets ist offen
denselben die darnach ohn tück u. schalkheit hoffen.

Herr wie sind deine wercke so groß u. vil. v. 24.
Ist nicht ein wunderbaum die wundersame welt?
worinn so ordentlich und schön sich alles helt.
der gärten wunderbaum ist zwar schön anzusehen
doch dises weltgebäu müß weit darüber gehen.

Bibliography

I. Bibles — entire or partial translations:—

1. The Holy Bible, translated from the Latin Vulgate, diligently compared with the Hebrew, Greek, and other editions in divers languages; the Old Testament first published by the English College at Douay, A.D. 1609, and the New Testament first published by the English College at Rheims, A.D. 1582, with annotations, references, and an historical and chronological table. 1406 pages. 1914.
2. The precious promise Holy Bible: the Authorized Version of the Scriptures. 1149 pages. 1915.
2A. Nowy Testament, pana naszego Jezusa Chrystusa. B Greckiego jezyka na Polski pilnie i wiernie przetlumnaczony. 434 pages. 1855.
2B. Ksiega Psalmow. 117 pages. 1855.
2C. The Holy Bible, containing the Old and New Testaments: translated out of the original tongues; and with the former translations diligently compared and revised. American Bible Society. 43rd edition. 951 pages. 1879.
2D. The New Testament of our Lord and Saviour Jesus Christ translated out of the Greek: being the version set forth A.D. 1611 compared with the most ancient authorities and revised A.D. 1881. Printed for the Universities of Oxford and Cambridge. 350 pages. 1881.
2E. Bibeln eller den Heliga Skrift, innehållande Gamla och Nya Testamentets Kanoniska Böcker. Ofwersättning utgifwen 1878 af Kgl. Bibelkommissionen. 1279 pages. 1889.
2F. The Holy Bible, containing the Old and New Testaments, translated out of the original tongues: and with the former translations diligently compared and revised: containing sixty thousand original and selected parallel references and marginal readings. International series, self-pronouncing edition. 929 pages. n.d.—The Bible readers' aids: being brief treatises upon and outlines of topics related to the study and understanding of the Holy Scriptures, edited by C. H. H. WRIGHT. American edition, rearranged, with an extended word book, including index, concordance, etc., etc. 268 pages. 1895.
3. The self-pronouncing Sunday-school teachers' Bible, containing the Old and New Testaments: translated out of the original tongues: and with the former translations diligently compared and revised; the text comparable to that of the Oxford Bible. 1000 pages. 1895.
3A. The New Covenant commonly called the New Testament of our Lord and Saviour Jesus Christ translated out of the Greek: being the version set forth A.D. 1611 compared with the most ancient authorities and revised A.D. 1881: newly edited by the New Testament members of the American Revision Committee A.D. 1900. Standard edition. 527 pages. 1901.
3B. Bibeln eller den Helige Skrift. 1238 pages. 1927.
3C. BUGENHAGEN, J., De Psalmen na de plattdütsche Oewersettung. 110 pages. 1885.
3D. BUGENHAGEN, J., Dat Nie Testament vun unser Herrn un Heiland Jesus Christus na de plattdütsche Oewersettung. 446 pages. 1885.
3E. BUNSEN, C. C. J., Die Bibel oder die Schriften des Alten und Neuen Bundes nach die überlieferten Grundtexten übersetzt und für die Gemeinde erklärt. II. Die Propheten. 826 pages. 1860.
3F. CHAPMAN, J. W., and W. E. BIEDERWOLF, R. A. WALTON, and H. OSTROM, Personal Worker's New Testament. 516 pages. 1905.
3G. DELITZSCH's Hebrew New Testament. Fifth and electrotype edition. Berlin. 471 pages. 1883.
4. GOOD, J. M., The book of Job literally translated from the original Hebrew. 1812.
5. GOODSPEED, E. J., et al., The complete Bible: an American translation; the Old Testament translated by J. M. POWIS SMITH and a group of scholars; the Apocrypha and the New Testament translated by EDGAR J. GOODSPEED. 1357 pages. 1939.
6. HARKAVY, A., The Pentateuch with Haftaroth and five Megiloth — revised English translation. 550 pages. 1928.
7. HOOKE, S. H., The New Testament in basic English. 556 pages. Edition 13. 1946.
7A. KNOX, R. A., The New Testament. 576 pages. 1944.
7B. KNOX, R. A., —— Edition 2. 512 pages. 1946.
8. LAMSA, G. M., The New Testament according to the Eastern text, translated from original Aramaic sources. 555 pages. 1940.
9. LEESSER, I., Twenty-four books of the Holy Scripture carefully translated after the

---

FIGURES 36-39. — Four "true Bible plants" as depicted in HOHBERG's "Psalter Davids" (1680), cf. figs. 2-4. Illustrated are *Anastatica hierochuntica*, the Palestinian tumbleweed, referred to as a "wheel", a "rolling thing", and the "rose plant in Jericho", in the Old Testament (see also fig. 22); *Amygdalus communis*, the almond (see also p. 21 and p. 274); *Punica granatum*, the pomegranate (see also p. 199 and fig. 76); and *Ricinus communis*, the castorbean, referred to as a "gourd" in the book of Jonah.

best Jewish authorities. 1240 pages. 1913.
9A. LUTHER, M., Die Bibel oder die ganze Heilige Schrift des Alten und Neuen Testaments nach der deutschen Uebersetzung D. MARTIN LUTHER'S. Durchgesehene Ausgabe mit dem von der Deutschen Evangelischen Kirchenkonferenz genehmigten Text. 256 pages. 1906.
9B. LUTHER, M., Die Bibel oder die ganze Heilige Schrift des Alten und Neuen Testaments nach der deutschen Uebersetzung Dr. MARTIN LUTHER'S. 1435 pages, n.d. [includes the Apocrypha].
10. MOFFATT, J., The New Testament: a new translation, together with the Authorized Version. Parallel Edition with introduction. 676 pages. 1922.
11. MOFFATT, J., The Old Testament: a new translation. Volume I, Genesis — Esther. 571 pages. 1924.
12. MOFFATT, J., The Old Testament: a new translation. Volume II, Job — Malachi. 482 pages. 1925.
13. [O'HARA, E. V.], The New Testament of our Lord and Savior Jesus Christ, translated from the Latin Vulgate; a revision of the Challoner-Rheims version, edited by Catholic scholars under the patronage of the Episcopal Committee of the Confraternity of Christian Doctrine. 780 pages. 1941.
13A. SCOFIELD, C. I., The Scofield Reference Bible. The Holy Bible containing the Old and New Testaments, Authorized Version, with a new system of connected topical references to all the greater themes of Scripture, with annotations, revised marginal renderings, summaries, definitions, and index, to which are added helps at hard places, explanations of seeming discrepancies, and a new system of paragraphs. 1375 pages. 1909.
14. [WEIGLE, L. A.], The New Covenant commonly called the New Testament of our Lord and Savior Jesus Christ. Revised Standard Edition. Translated from the Greek, being the version set forth A.D. 1611, revised A.D. 1881 and A.D. 1901; compared with the most ancient authorities and revised A.D. 1946. 560 pages. 1946.
15. WEYMOUTH, R. F., The New Testament in modern speech: an idiomatic translation into every day English from the text of the resultant Greek testament. Newly revised by J. A. ROBERTSON. Edition 5. 734 pages. 1943.
16. MARGOLIES, M. [editor], The Holy Scriptures according to the Masoretic text; a new translation with the aid of previous versions and with constant consultation of Jewish authorities. 1152 pages. Jewish Publication Society. 1917 & 1946.
16A. The Holy Bible, containing the Old and New Testaments: translated out of the original tongues: and with the former translations diligently compared and revised. A. J. HOLMAN edition. 804 pages. n.d.
16B. The Holy Bible, containing the Old and New Testaments: translated out of the original tongues: and with the former translations diligently compared and revised, by His Majesty's special command. The S. S. Teacher's edition. Appointed to be read in churches. Oxford University Press. 1056 pages. n.d.—The Oxford Cyclopedic Concordance containing new and selected helps to the study of the Bible arranged in one alphabetical order with illustrations and a new series of maps. 332 pages. n.d.
17. The Holy Bible: the great light in Masonry: containing the Old and New Testaments according to the Authorized or King James version together with illuminated frontispiece, presentation and record pages and helps to the Masonic student. 1249 pages. 1940.

II. Secular literature:—

18. AARONSOHN, A., Über die in Palästina und Syrien wildwachsend aufgefundenen Getreidearten, *in* Verhandlungen Zoologisch-botanischen Gesellschaft in Wien, vol. 59, pp. 485-509. 1909.
19. ALEXANDER, W. L. [editor], Kitto's Cyclopaedia of Biblical Literature. Edition 3, 3 volumes. 1866.
20. ALLEN, D. B., Diseases mentioned in the Bible, *in* Dietetic and Hygienic Gazette, volume 24, pages 285-291 (in reprint form, pages 1-24). 1908.
21. ANDRIEUX, F. G. J. S., L'olivier, le figuier, la vigne et le buisson; fable de Joatham, tirée de la Bible, Juges, c. ix, v. 8. 1793.
22. ANOMOEUS, C., Sacrorum arborum, fruticum et herbarum. 2 vols. 223 pages. 1609.
23. Anonymous; Christian florist. 1832.
24. Anonymous; —— Edition 2. 1835.
25. Anonymous; Scripture garden walk. 1832. (Professes to notice every Biblical plant!)
25A. Anonymous; The divi ladner, *in* Horticultural Register, vol. 1, p. 762. 1832.
26. Anonymous; Juvenile conversations on the botany of the Bible. 1851.
27. Anonymous; Helps to the study of the Bible, comprising summaries of the several books, with copious explanatory notes and tables illustrative of Scripture history and the characteristics of Bible lands, embodying the results of the most recent researches together with a dictionary of proper names, a Biblical index, concordance, and revised series of maps. 476 pages. 1895.
27A. Anonymous; Ergebnisse der Sinai-Expedition 1927 der Hebräischen Universität, Jerusalem. 143 pages, 24 plates. 1929.
27B. Anonymous; *in* Nature, vol. 124, pp. 1003-1004. 1929.
27C. Anonymous; *in* Agricultural History, vol. 6, p. 44. 1932.
27D. Anonymous; *in* Biological Abstracts, vol. 10, p. 546. 1936.
28. Anonymous; Carmel Biblical garden, *in* Sunset, January 1941 issue, pp. 23-24 & 43.

29. Anonymous; What flowers did Jesus refer to when He spoke of "lilies of the field?", *in* Plainfield (N. J.) Courier News, February 18, 1945.
30. Anonymous; Land of Yemen, *in* Life, vol. 21, no. 21, pp. 59-62. 1946.
30A. Anonymous; Christ's crown of thorns came from Paluirus bush, *in* The American Eagle and Horticultural Review [Estero, Florida], vol. 42, p. 8. 1947.
30B. Anonymous; Flowers from the Holy Land. Jerusalem. 13 pages. n.d.
31. ASCHERSON, P.F.A., Die Herkunft des Namens "*Lilium convallium*", *in* Naturwissenschaftliche Wochenschrift, vol. 9, pp. 17-22. 1894.
32. ASCHERSON, P.F.A., Nachschrift zu meinen Aufsatze: Die Herkunft des Namens "*Lilium convallium*", *in* Naturwissenschaftliche Wochenschrift, vol. 9, p. 310. 1894.
33. ASCHERSON, P.F.A., & GRAEBNER, R.O.R.P.P., Synopsis der mitteleuropäischen Flora. vol. 3, pp. 172-174. 1905.
34. BAILEY, L. H., Manual of cultivated plants. 851 pages. 1924.
35. BAILEY, L. H., The standard cyclopedia of horticulture. 3 vols. 3663 pages. 1935.
36. BALFOUR, J. H., Lessons from Bible plants. 1851.
37. BALFOUR, J. H., —— [Another edition]. 1870.
38. BALFOUR, J. H., Phyto-theology; or, Botanical sketches, intended to illustrate the works of God in the structure, functions, and general distribution of plants. 258 pages. 1851.
39. BALFOUR, J. H., —— Edition 2. 1863.
40. BALFOUR, J. H., The plants of the Bible; trees and shrubs. London. 58 pages. 1857.
41. BALFOUR, J. H., —— Edinburgh. 200 pages. 1857.
42. BALFOUR, J. H., —— [Another edition]. 192 pages. 1866.
43. BALFOUR, J. H., —— New and enlarged edition. 249 pages. 1885.
44. BALFOUR, J. H., Botany and religion; or, Illustrations of the works of God, in the structure, functions, arrangements, and distributions of plants. Edition 3. 476 pages. 1859.
45. BALFOUR, J. H., —— Edition "2". 476 pages. 1863.
46. BANG, C. F., De plantis quibusdam sacrae botanicae cujus particulam primam edidit. 26 pages. 1767.
47. BARNEVELD, M. VAN, Plants and flowers found in the Bible and at home in the Southwest. 16 pages. 1935.
48. BARREIRA, F. I. DE, Tractado das significacoens das plantas, flores e fructos que se referem na Sagrada Escriptura. 582 pages. 1622.
48A. BARRETT, M. F., The sycomore fig of ancient lineage, *in* Journal of the New York Botanical Garden, vol. 48, pp. 254-262. 1947.
49. BATCHELDER, A., Is there no balm in Gilead?, *in* Ladies Home Journal, vol. 58, no. 7, p. 123. 1941.
50. BEKE, C. T., Remarks on Mr. CARTER's paper on the gopherwood, *in* London, Edinburgh and Dublin Philosophical Magazine, ser. 3, vol. 4, pp. 280-282. 1834.
51. BERCHTOLD, F., Graf VON, Das Todte Meer, insbesondere der Sodomsapfel, *in* Lotos, vol. 1, pp. 17-20. 1851.
52. BERGEY, D. H., BREED, R. S., MURRAY, E. G. D., and HITCHINS, A. P., Bergey's Manual of determinative bacteriology. 1032 pages. 1939.
53. BERKELEY, M. J., [Manna], *in* Gardeners' Chronicle & Agricultural Gazette for 1849, pp. 611-612. 1849.
54. BERKELEY, M. J., [Manna], *in* Gardeners' Chronicle & Agricultural Gazette for 1856, p. 84. 1856.
55. BERKELEY, M. J., [Manna], *in* Gardeners' Chronicle & Agricultural Gazette for 1864, pp. 769-770 & 794. 1864. (The former with R. MURCHISON).
56. [CELSIUS, O.], [Beroth, berosch] 2 Sam. vi. 5, *in* Acta Literaria et Scientiarum Sueciae, vol. 4, pp. 139-165. 1742.
56A. BLANCHAN, N., Wild flowers, *in* The New Nature Library, vol. 6, 415 pages. 1916.
57. BLESSNER, G., Flora sacra. 208 pages. 1864.
58. BLOOD, Mrs. S. W., The flowers and trees of the Bible. 8 pages. 1934.
59. BLUM, J. L., The identification of Bible plants, *in* The Science Counselor, volume 7, pages 104-106 & 128. 1941.
60. [BOISSIER, E.], Botanique biblique, ou courtes notices sur les végétaux mentionnés dans les Saintes Ecritures. 195 pages. 1861-1862.
61. BOISSIER, E., Flora orientalis sive enumeratio plantarum in oriente a Graecia et Aegypto ad Indiae fines hucusque observatarum, vol. 1. 1051 pages. 1867.
62. BOISSIER, E., Flora orientalis sive enumeratio plantarum in oriente a Graecia et Aegypto ad Indiae fines hucusque observatarum, vol. 2. 1159 pages. 1872.
63. BOISSIER, E., Flora orientalis sive enumeratio plantarum in oriente a Graecia et Aegypto ad Indiae fines hucusque observatarum, vol. 3. 1033 pages. 1875.
64. BOISSIER, E., Flora orientalis sive enumeratio plantarum in oriente a Graecia et Aegypto ad Indiae fines hucusque observatarum, vol. 4. 1276 pages. 1879.
65. BOISSIER, E., Flora orientalis sive enumeratio plantarum in oriente a Graecia et Aegypto ad Indiae fines hucusque observatarum, vol. 5. 868 pages. 1884.
66. BOISSIER, E., Flora orientalis sive enumeratio plantarum in oriente a Graecia et Aegypto ad Indiae fines hucusque observatarum supplementum, edited by R. BUSER. 466 pages. 1888.
67. BORAH, A., Trees of the Bible. I. Evergreens of the Old Testament, *in* American Forests & Forest Life, vol. 34, pp. 715-717 & 765. 1928.

68. BORAH, A., —— II. The oaks of Palestine, *in* American Forests & Forest Life, vol. 35, pp. 13-15 & 60. 1929.
69. BORAH, A., —— III. The date palm and the pomegranate, *in* American Forests & Forest Life, volume 35, pp. 89-92 & 128. 1929.
70. BORAH, A., —— IV. The olive tree and the fig, *in* American Forests & Forest Life, vol. 35, pp. 155-157 & 190. 1929.
71. BORAH, A., —— V. The tamarisk and the sycamore, *in* American Forests & Forest Life, vol. 35, pp. 231-233. 1929.
72. BORAH, A., —— VI. The sycamine and the almond, *in* American Forests & Forest Life, vol. 35, pp. 293-295. 1929.
73. BRIM, C. J., Medicine in the Bible — The Pentateuch (Torah). 384 pages. 1936.
74. BROWNE, T., Observations upon several plants mentioned in Scripture, *in* his Works, edited by S. WILKIN, vol. 4, pp. 121-173. 1835.
75. BROWNE, T., —— [Another edition], vol. 4, pp. 121-173. 1846.
76. BROWNE, T., —— [Another edition], vol. 3, pp. 151-203. 1852.
*172B. BREWER, E. W., and M. R. HICKERNELL, Adam's herbs — their story from Eden on. 77 pages. 1947.
76A. BRUNTON, T. L., The Bible and science. 416 pages. 1881.
77. BUCKHAM, P. W., Remarks on *Phytolacca dodecandra* or the mustard tree of Scriptures. 1829.
77A. BUSCHAN, G., Vorgeschichtliche Botanik der Cultur- und Nutzpflanzen der Alten Welt auf Grund prähistorischer Funde. 280 pages. 1895.
78. CALLCOTT, M., A Scripture herbal. 568 pages. 1842.
79. CALLCOTT, M.; A Scripture herbal, by Lady MARIA CALCOTT, [book review] *in* Church of England Quarterly Review, vol. 13, no. 1, pp. 113-136. 1843.
80. CARPENTER, W., Scripture natural history. 608 pages. 1828.
81. CASTELLI, E., Oratio in scholis theologicis habita cum in Academia cantabrigiensi praelectiones suas in secundum Canonis Avicennae librum auspicaretur, quibus via praestruitur ex scriptoribus orientalibus ad clarius enarrandum botanologicam S. Scripturae partem. 1667.
82. CELSIUS, O., Botanici sacri exercitatio prima, qua [oren] ex Arabum scriptis illustratur. 29 pages. 1702.
83. CELSIUS, O., De arbore scientiae boni et mali. 22 pages. 1715.
84. CELSIUS, O., [Aylon, allon] Gen. xiv.6. Terebinthus Judaica. Diosc. i.91. 1738. (Reprinted in Acta Literaria et Scientiarum Sueciae, vol. 4, pp. 389-409. 1742.)
85. [CELSIUS, O.] [Ahalot, ahalim] Num. xxiv. 6, Agalochum praestantissimum. Arbor Aloes. C. Bauh. Pin. 393, *in* Acta Literaria et Scientiarum Sueciae, vol. 4, pp· 280-311. 1742.
86. CELSIUS, O., Hierobotanicon; sive, De plantis Sacrae Scripturae. 2 vols. 1188 pages. Upsala, 1745 & 1747.
87. CELSIUS, O., Hierobotanicon; sive, De plantis Sacrae Scripturae dissertationes breves, 2 vols. Amsterdam, 1748.
88. CHARLES, R. H. [editor], The Apocrypha and Pseudepigrapha of the Old Testament, vol. 1, p. 169. 1913.
88A. CHARTON, M. E. DE, Le tour du monde, pp. 391-416. 1860.
89. CHASE, F., Christmas cities, *in* Holiday, vol. 1, no. 10, pp. 97-102, 118-119, & 151-153. 1946.
90. CHRIST, K.H.H.,Nochmals die Lilie der Bibel, *in* Zeitschrift der Deutschen Palästina Verein, vol. 22, pp. 65-80. 1899.
91. CLEWBERG, C., De [rotem] arbore sub qua Elias profugus recubuisse legitur I Reg. xix. 4.5. 19 pages. 1758.
92. COCQUIUS, A., Historia ac contemplatio sacra plantarum arborum & herbarum quarum fit mentio in Sacra Scriptura. 277 pages. 1664.
93. COCQUIUS, A., Observationes et exercitationes philologico-physiologicae ad vetus testamentum, in quibus contextus sacer illustratur, imprimis ubi agitur de rebus naturalibus. 263 pages. 1671.
93A. CONDIT, I. J., The fig. 240 pages. 1947.
94. [CORNELL, R. D.], John the Baptist ate carob bean, not locust, *in* New York Herald Tribune, July 4, 1943.
95. COTES, R. A., Bible flowers. 288 pages. 1904.
95A. CROWFOOT, G. M., and L. BALDENSPERGER, From cedar to hyssop: a study in the folklore of plants in Palestine. 204 pages. 1932.
96. CULTRERA, P., Flora biblica; ovvero spiegazione delle piante menzionate nella Sacra Scrittura. 486 pages. 1861.
97. DALMAN, G., Arbeit und Sitte in Palästina, vol. 1, parts 1 and 2. 697 pages. 1928.
98. DALMAN, G., Arbeit und Sitte in Palästina, vol. 2. 384 pages. 1928.
99. DALMAN, G., Orte und Weg Jesus, *in* Beiträge zur Forderung Christlicher Theologie, ser. 2, vol. 1, pp. 169-237. 1924.
99A. DE SIMINI, N. [*editor*], Album souvenir — Flowers from the Holy Land. 12 pages. n. d.
99B. DINSMORE, J. E., Plants of the Bible, *in* The Gardeners' Chronicle, series 3, vol. 43, p. 179. 1908.
100. DINSMORE, J. E., The Jerusalem catalogue of Palestine plants. Jerusalem 1912.
101. DONEY, C., Flowers of the Bible. 5 pages. 1939.
102. DRIVER, S. R., TURNER, C. H., PAUES, A. C., HENSON, H. H., *et al.*, Bible, *in* the

	Encyclopaedia Brittanica, edition 11, vol. 3, pp. 849-905. 1910.
103.	DUNCAN, J. S., Botano-theology, an arranged compendium. 1825.
104.	DUNCAN, J. S., Botanical theology, or evidences of the existence and attributes of the deity. Edition 2. 90 pages. 1826.
105.	DUNS, J., Biblical natural science; being an explanation of all references in Holy Scripture to geology, botany, zoology and physical geography. Volume 1. 581 pages. 1863.
106.	DUNS, J., —— Volume 2. 632 pages. 1868.
107.	DU PAS, C., Cognoscite lilia agri quomodo crescant, non laborant neque nent; attamen dico vobis ne Solomonem quidem in universa gloria sua sic amictum fuisse, ut unum ex his: Matth. 6. cap. 48 pages. 16——.
108.	DUSCHAK, M., Zur Botanik des Talmud. 1870.
108A.	EBSTEIN, W., Die Medezin im Alten Testament. 192 pages. 1901.
109.	EIG, A., A contribution to the knowledge of the flora of Palestine, *in* Agricultural Experiment Station (Tel Aviv) Bulletin 4. 1926.
110.	EIG, A., Les éléments et les groupes phytogéographiques auxiliaires dans la flore palestinienne, *in* FEDDE, Repertorium Specierum Novarum Regni Vegetabilis, vol. 63, pp. 1-201. 1931.
111.	EIG, A., On the phytogeographical subdivision of Palestine, *in* Palestine Journal of Botany, J series, vol. 1, pp. 4-12. 1938.
112.	EIG, A., On the vegetation of Palestine, *in* Agricultural Experiment Station (Tel Aviv) Bulletin 7. 1927.
113.	ELLACOMBE, H. N., Thrum-eyed polyanthus, *in* The Gardeners' Chronicle, new series, vol. 5, page 85. 1876.
114.	ERICSON, A. L., Plants of the Holy Scriptures expanded into a book, [review] *in* Journal of the New York Botanical Garden, vol. 42, p. 273. 1941.
115.	ESSEN, J. G., Spring shows. The International Flower Show, *in* The Gardeners' Chronicle of America, vol. 45, p. 112. 1941.
115A.	ETOC, G., Les plantes dans la Bible, *in* the Bulletin de l'Académie Internationale de Géographie Botanique, vol. 9, pp. 74-77. 1900.
*258.	EVENARI, M., and OPPENHEIMER, H. R., Reliquiae Aaronsohnianae II, Florula cisiordanica — Révision critique des plantes récoltées et partiellement déterminées par AARON AARONSOHN ou cours de ses voyages (1904-1916) en Cisjordanie, en Syrie et au Liban, *in* Bulletin de la Société Botanique de Genève, ser. 2, vol. 31, pp. 1-423. 1940.
115B.	ETT, B. and R., Dell'agricoltura presso gli antichi ebrei: monografia. 82 pages. 1892.
116.	FEINBRUN, N., Materials for a revised flora of Palestine, I, *in* Proceedings of the Linnean Society of London, vol. 157, part I, pp. 46-54. 1945.
117.	FERGUSON, W., Scripture botany of Ceylon; being familiar illustrations of those plants mentioned in the Bible, and which are indigenous to or are related to genera and species growing or known in Ceylon. 48 pages. 1859.
118.	FERNALD, M. L., & KINSEY, A. C., Edible wild plants of eastern North America 452 pages. 1943.
119.	FLETCHER, A., Scripture natural history. 2 vols. 700 pages. [no date].
120.	FONDA, M. E., The Lord's land. 28 pages. 1947.
121.	FORSKÅL, P., Descriptiones animalium, avium, amphibiorum, piscium, insectorum, vermium; quae in itinere Orientali observavit ... Adjuncta est materia medica Kahirina atque tabula Maris Rubri geographica. Edited by C. NIEBUHR. 164 pages. 1775.
122.	FORSKÅL, P., Flora aegyptiaco-arabica, sive descriptiones plantarum, quas per Aegyptum inferiorem et Arabiam Felicem detexit. Edited by C. NIEBUHR. 377 pages. 1775.
123.	FORSKÅL, P., Icones rerum naturalium quas in itinere Orientali depingi curavit ... Edited by C. NIEBUHR. 15 pages, 43 plates. 1776.
124.	FORSKÅL, P., —— [Another edition] 12 pages, 43 colored plates. 1776.
124A.	FRAZER, J. G., Folk-lore in the Old Testament. Volume 1. 569 pages. 1918.
124B.	FRAZER, J. G., —— Volume 2. 571 pages. 1918.
124C.	FRAZER, J. G., —— Volume 3. 566 pages. 1918.
124D.	FRAZER, J. G., —— Abridged edition. 476 pages. 1923; also 1927 reprinting.
125.	FRIEND, H., Flowers and flower lore. 2 vols. 720 pages. 1883.
126.	FRIEND, H., —— Edition 3. 720 pages. 1886.
127.	FROST, J., Remarks on *Phytolacca dodecandra*, or the mustard tree of the Scriptures, *in* Quarterly Journal of Science, vol. 20, pp. 57-59. 1826.
128.	FROST, J., Remarks on the mustard tree mentioned in the New Testament (*Phytolacca dodecandra*). 22 pages, colored plates. 1827.
128A.	GEIKIE, J. C., Life and words of Christ. Edition 3, 838 pages. 1891.
129.	GESSNER, J., Phytographia sacra generalis. 28 pages. 1759.
130.	GESSNER, J., Phytographiae sacrae generalis pars practica. Volume 1. 56 pages. 1760.
131.	GESSNER, J., —— Volume 2. 54 pages. 1762.
132.	GESSNER, J., —— Volume 3. 30 pages. 1763.
133.	GESSNER, J., —— Volume 4. 31 pages. 1764.
134.	GESSNER, J., —— Volume 5. 35 pages. 1765.
135.	GESSNER, J., —— Volume 6. 34 pages. 1766.
136.	GESSNER, J., —— Volume 7. 33 pages. 1767.
137.	GESSNER, J., Phytographiae sacrae specialis. Volume 1. 27 pages. 1768.
138.	GESSNER, J., —— Volume 2. 25 pages. 1769.

139. GESSNER, J., —— Volume 3. 32 pages. 1773.
140. GORDON, B. L., Ophthalmology in the Bible and in the Talmud, *in* the Archives of Ophthalmology, new series, vol. 9, pp. 751-788. 1933.
141. GORRIE, D., Illustrations of Scripture from botanical science. 160 pages. 1853.
142. GORTER, D. DE, Kruidkundige verhandeling over Jerem. xvii. vs. 6, *in* Hollandsche Maatschappij der Wetenschappen Verhandelingen, vol. 15, pp. 126-147. 1774.
143. GREENE, H. B., Pressed flowers from the Holy Land. 1896.
144. GRINDON, L., Scripture botany, *in* The Gardeners' Chronicle, new series, vol. 2, pp. 240-241. 1874.
145. GROSER, W. H., Scripture natural history. I. The trees and plants mentioned in the Bible. 1888.
146. GROSER, W. H., —— Edition 2. 243 pages. 1895.
147. [GROSVENOR, G.], The Society's new map of Bible lands, *in* the National Geographic Magazine, vol. 90, pp. 815-816. 1946.
148. GUIBBORY, M., The Bible in the hands of its creators: Biblical facts as they are: a Biblical research, 2010 pages. 1943.
149. HALLMAN, D. Z., Dissertatio theologica inauguralis de στεφάνω ἐξ ἀκανθων, corona de spinis. 97 pages. 1757.
150. HAMILTON, F., La botanique de la Bible, étude scientifique, historique, littéraire et exégétique des plantes mentionnées dans la Sainte-Ecriture. 216 pages. 1871.
151. HAMILTON, F., —— [Another edition]. 2 vols. 20 illust. 1872.
151A. HANFORD, M. A. C., Bible plants from Bible lands; collected in Syria 1881-1886. 50 pages. 1894.
152. HA REUBENI, E.; The lilies of the field [extract from a lecture; translated by D. DE SOLA POOL, *in* Torreya, vol. 25, pp. 35-38. 1925.
153. [HA REUBENI, E.], Garden of prophets and sages gets under way in Jerusalem, *in* New York Times, August 5, 1944.
154. HARRIS, T. M., The natural history of the Bible. 297 pages. 1793.
155. HARRIS, T. M., —— [Edition 2]. 1820.
156. HARRIS, T. M., The natural history of the Bible; or, a description of all the quadrupeds, birds, fishes, reptiles and insects, trees, plants, flowers, gums and precious stones mentioned in the Sacred Scriptures, collected from the best authorities and alphabetically arranged. 462 pages. 1824.
157. HARRIS, T. M., A dictionary of the natural history of the Bible. 1833.
158. HARRIS, W. T., & ALLEN, F. S., [editors], Webster's new international dictionary of the English language. 2620 pages. 1917.
159. HART, H. C., Some account of the fauna and flora of Sinai, Petra, and Wady 'Arabah. 255 pages. 1891.
160. HASSELQUIST, F., Iter palaestinum; eller resa till Heliga Landet förrättad ifrån år 1749 til 1752. Edited by C. LINNAEUS. 619 pages. 1757.
161. HASSELQUIST, F., Reise nach Palästina in den Jahren von 1749 bis 1752. Edited by C. LINNAEUS. 621 pages. 1762.
162. HASSELQUIST, F., Voyages and travels in the Levant in the years 1749, 50, 51, 52: containing observations in natural history, physick, agriculture, and commerce: particularly in the Holy Land and the natural history of the Scriptures. 464 pages. 1766.
163. HASSELQUIST, F., Voyages dans le Levant dans les années 1749, 50, 51 & 52. 203 pages. 1769.
164. HASTINGS, G. T., Plants of the Bible, [review] *in* Torreya, vol. 42, pp. 114-115. 1942.
165. HAUPT, P., The burning bush and the origin of Judaism, *in* Proceedings of the American Philosophical Society, vol. 48, pp. 354-369. 1909.
166. HAUPT, P., Manna, nectar, and ambrosia, *in* Proceedings of the American Philosophical Society, vol. 61, pp. 227-236. 1922.
167. HAYNALD, L., A szentirási mézgák és gyanták termönövényei [The plants which produce the gums and resins mentioned in the Bible], *in* Magyar Növénytani Lapok, vol. 3, pp. 177-222. 1879.
168. HAYNALD, L., Des plantes qui fournissent les gommes et les résins mentionnés dans les livres saints. 13 plates. 1894.
169. HEGI, G., Illustrierte Flora von Mittel-Europa, vol. 5, part 1, p. 65. 1919.
170. HEHN, V., Kulturpflanzen und Hausthiere in ihren Übergang aus Asien nach Italien und Griechenland, ed. 3, p. 216. 1877.
171. HENSLOW, G., The plants of the Bible. 128 pages. 1895.
172. HENSLOW, G., The plants of the Bible: their ancient and mediaeval history popularly described. 309 pages. 1906.
172A. HENSLOW, G., The plants of the Bible, review *in* The Gardeners' Chronicle, series 3, vol. 43, p. 53. 1908.
172B. HICKERNELL, M. R., and E. W. BREWER, Adam's herbs — their story from Eden on. 77 pages. 1947.
172C. HIBBERB, S., An alphabet of appledom, *in* the Gardeners' Chronicle, series 2, vol. 20, p. 523. 1883.
173. HILDEBRAND, J. R., Cotton: foremost fiber of the world, *in* the National Geographic Magazine, vol. 79, pp. 137-184. 1941.
174. HILLER, M., De plantis in Scriptura Sacra memoratis. 2 vols. 40 pages. 1716.
175. HILLER, M., Hierophyticon; sive, commentarius in loca Scripturae Sacrae quae plantarum faciunt mentionem, distinctus in duas partes, quarum prior de arboribus, posterior

de herbis dicta complectitur; cui accedit praefatio Salomonis Pfisteri. 2 vols. 833 pages. 1725.
176. HOHBERG, W. H. VON, Verfassung des gantzen Psalter Davids in teutsche Reim-gebande, 526 pages. 1680.
177. HOLMES, E. M., The manna of the Scripture, in the American Journal of Pharmacy, vol. 92, pp. 174-179. 1920.
178. "HORNER, N. C.", Plants of the Bible — see under MOORE, G. T., 1931.
179. HURLBUT, J. L., & MCCLURE, A. J. P. [editors], The international teacher's handy Bible encyclopedia and concordance. 390 pages. 1908.
180. HURST, E., Poisonous plants of New South Wales. 1942.
181. J., G. E., The Rose of the Bible, in The Rose Annual for 1946, pp. 78-80.
182. JACKSON, B. D., Guide to the literature of botany; being a classified selection of botanical works, including nearly 6000 titles not given in PRITZEL's "Thesaurus". 626 pages. 1881.
183. JOHNS, C. A., Flora sacra; or, the knowledge of the works of nature conducive to the knowledge of the God of nature. 48 pages. 1840.
184. JONES, J. R., [copyright holder], Helps to the study of the Bible: containing copious analytical and explanatory notes, and summaries of the several books; historical, chronological and geographical tables, lists of animals, birds, reptiles, plants, precious stones, etc. found in Scripture; geology of Bible lands; table of weights, measures, time and money; words obsolete or ambiguous; together with a new index to the Bible; a new and complete concordance; a dictionary of Scripture proper names, with their pronunciation and meaning, and a series of maps. 352 pages. 1895.
185. KAUTZSCH, E. F., Die Heilige Schrift, ed. 3, vol. 2, p. 369. 1910.
186. KESSEL, D. [photographer], The Holy Land: it echoes the Bible's story, in Life, vol. 21, April 1, 1946, pp. 53-63. 1946.
186A. KILLERMANN, S., Die Blumen des Heiligen Landes. Botanische Auslese einer Frühlingsfahrt durch Syrien und Palästina. 170 pages. 1917.
187. KING, E. A., Bible plants for American gardens. 203 pages. 1941.
188. KING, E. A.; Bible plants for American gardens, [announcement] in New Books, Fall, 1941, Final Announcement, The Macmillan Company. 1941.
*164. KING, E. A.; Bible plants for American gardens; Plants of the Bible, [review] by G. T. HASTINGS (see under HASTINGS, G. T., 1942).
189. KING, E. A., Bible plants for American gardens, [advertising leaflet and letter from] The Macmillan Company, A-151, A-41. 1941.
190. KING, E. A.; —— [review] in Gardeners' Chronicle of America, vol. 45, pp. 352 & 358. 1941.
191. KING, E. A.; —— [announcement] in New York Times Book Section, June 15, 1941.
*114. KING, E. A.; —— Plants of the Holy Scriptures expanded into a book, review by A. L. ERICSON (see under ERICSON, A. L., 1941).
192. KING, E. A., Plants of the Holy Scriptures, in Journal of the New York Botanical Garden, vol. 42, pp. 50-65. 1941.
192A. KING, E. A., —— New and revised edition. 1948.
193. KING, E. A., —— [review] in Zions Herald, vol. 119, pp. 398-399. 1941.
194. KING, E. A.; —— Bible plants [review] in the Earnest Worker, p. 394. 1941.
195. KING, E. A., Plants of the Holy Scriptures, in Biblical Digest, vol. 5, no. 4, pp. 37-39, and no. 5, pp. 39-43. 1941.
196. KING, E. A., Plants of the Holy Scriptures, in the Evangelical Messenger, vol. 94, no. 31, pp. 4-11. 1941.
197. KING, E. A.; —— Bible plants, [review] in the Sabbath Recorder, vol. 130, no. 17, p. 271. 1941.
198. KING, E. A.; —— In a nutshell, [review] in the Christian Leader, vol. 123, p. 412. 1941.
199. KING, E. A.; —— [review] in Religion in Life, summer number. 1941.
200. KING, E. A.; —— [review] in the Presbyterian Record, vol. 66, p. 177. 1941.
*311. KING, E. A.; —— [review] by R. S. STEIN in The American Biology Teacher (see under STEIN, R. S., 1941).
201. KING, E. A., Plants of the Holy Scriptures, in the Christian Observer, vol. 130, no. 46, pp. 5-9. 1942.
202. KITTO, J., Palestine: the Bible history (the physical geography and natural history) of the Holy Land. 2 vols. 1841.
203. KRAASBÖL, H. C., De arboribus sodemaeis. 12 pages. 1705.
*253. LANDER, G. D., Veterinary toxicology, ed. 3, edited by J. A. NICHOLSON. 328 pages. 1945.
203A. KRAUSE, K., Ueber Allium Kurrat Schweinf., in Notizblatt des Botanischen Gartens und Museums zu Berlin-Dahlem, vol. 9, pp. 523-524. 1926.
203B. KÜKENTHAL, G., in ENGLER, Das Pflanzenreich, vol. 4, part 20, pp. 44-48. 1935.
204. LEDWARD, W. J., The trees of the Bible and their spiritual lessons. 80 pages. 1895.
205. LEEMANN, A. C., A short summary of our botanical knowledge of Lolium temulentum, in Onderstepoort Journal of Veterinary Science and Animal Industry, vol. 1, p. 213. 1933.
206. LEMMENS [LEMNIUS], L., Herbarum atque arborum quae in Bibliis passim obviae sunt et ex quibus sacri vates similitudines desumunt, ac collationes rebus accommodant, dilucida explicatio; in qua narratione singula loca explanantur quibus Prophetae observata stirpium natura, conciones suas illustrant, divina oracula fulciunt.

161 pages. 1566.
207. LEMMENS [LEMNIUS], L., Similitudinum ac parabolarum quae in Bibliis ex herbis atque arboribus desumuntur dilucida explicatio; in qua narratione singula loca explanantur, quibus Prophetae, obseruata stirpium natura, conciones suas illustrant, diuinaque oracula fulciunt. 163 pages. 1568.
208. LEMMENS [LEMNIUS], L., —— [Another edition]. 150 pages. 1581.
209. LEMMENS [LEMNIUS], L., —— [Another edition]. 304 pages. 1591.
210. LEMMENS [LEMNIUS], L., —— [Another edition]. 304 pages. 1652.
211. LEMMENS [LEMNIUS], L., An herbal for the Bible; containing a plaine . . . exposition of such similitudes, parables and metaphors . . . as are . . . taken from herbs, plants, trees, fruits and simples, by observation of their vertues . . . drawen into English (with alteration) by T. NEWTON. 302 pages. 1587.
212. LENZ, H. O., Botanik der alten Griechen und Römer. 776 pages. 1859.
213. LIDDELL, H. G., and SCOTT, R., A Greek-English lexicon based on the German work of FRANCIS PASSOW, ed. 1. 1740 pages. 1846.
214. LIDDELL, H. G., and SCOTT, R., A Greek-English lexicon based on the German work of FRANCIS PASSOW, ed. 2. 1734 pages. 1858.
*302. [LINDLEY, J.], Books received: a dictionary of the Bible, [reviews] (*see under* SMITH, W., 1864).
215. LINDLEY, J., The treasury of botany. 2 vols. 1352 pages. 1884.
216. Löw, I., Aramaeische Pflanzennamen. 498 pages. 1881.
217. Löw, I., Die Flora der Juden. Vols. 2 and 3. 1054 pages. 1924.
218. Löw, I., —— Vol. 1. 807 pages. 1928.
218A. Löw, I., Jardin et parc, *in* Revue des Etudes Juives, January-March no. 1931.
219. MACMILLAN, H., The poetry of plants. 386 pages. 1902. (Includes a chapter on Bible trees).
220. MAURILLE DE SAINT MICHEL, Pére, Phytologie sacrée; ou discours moral sur les plantes de la Sainte Ecriture; symboles des mystères de la foy des verités chrestiennes; divisée en six parterres. 787 pages. 1664.
221. MAYER, L., Views in Egypt. 102 pages. 1801.
222. MAYER, L., —— [Another edition]. 102 pages. 1804.
223. MAYER, L., Reise nach Aegypten, p. 226.
223A. MAW, G., A monograph of the genus *Crocus*. 356 pages. 1886.
224. MCCLINTOCK, J., & STRONG, J., Cyclopedia of Biblical, theological, and ecclesiastical literature, vol. 10, pp. 316-317. 1881.
225. MEURS, J., Arboretum sacrum; sive, de arborum, fruticum & herbarum, consecratione, proprietate, usu ac qualitate. 159 pages. 1642.
226. MEURS, J., —— [Another edition]. 132 pages (appended to RAPIN, R., Hortorum libri IV). 1672.
227. MEYER, G. A., De sycomoro, quam Zachaeus publicanorum magister ascenderat (Luc. xix. 1-4). 20 pages. 1694.
228. MITFORD, J., Remarks on the Sinapi or mustard tree of Scripture. 1853.
229. MOGHADAM, S., Étude historique, botanique et biochimique des mannes de Perse, *in* Travaux Laboratoires de Matière Médicale et de Pharmacie Galénique (Paris), vol. 20, part 6, I-X, pp. 1-145. 1930.
230. MOLDENKE, C. E., Ueber die in altägyptischen Texten erwähnten Bäume und deren Verwerthung. 149 pages. 1886.
231. MOLDENKE, C. E., The trees of ancient Egypt. Being additional notes on the dissertation entitled "Ueber die in altägyptischen Texten erwähnten Bäume und deren Verwerthung". (mss.). 150 pages. 1887.
232. MOLDENKE, C. E., In alten Städten und an heiligen Stätten. Beschreibung einer Orientreise in Briefen, veröffentlicht in "Lutherischen Herold" (New York) vom 28. November 1896 bis 18. September 1897. 200 pages. 1897.
233. MOLDENKE, H. N., Plants of the Bible. 135 pages. 1940.
234. MOLDENKE, H. N., —— (Revised Edition) [typescript]. 190 pages. 1940.
235. MOLDENKE, H. N., Check-list of plants that are mentioned in the Bible, *in* Biblical Digest, vol. 5, no. 4, pp. 40-41, and no. 5, pp. 44-45. 1941.
236. MOLDENKE, H. N., Some flowering plants of the Bible, *in* Wild Flower, vol. 22, pp. 39-66. 1946.
237. MONTCLAIR, P., Consider the lilies of the field, *in* King Features Syndicate release for April 22, 1941.
*178. MOORE, G. T., Plants of the Bible, *in* the Bulletin of the Missouri Botanical Garden, vol. 19, pp. 149-159. 1931. (Erroneously attributed to N. C. HORNER in some bibliographies).
238. M[OORE, T.], Sacred botany: on hyssop, *in* Gardeners Magazine of Botany for 1850, part 1, pp. 37-38. 1850.
239. M[OORE, T.], —— the tamarisk—manna, *in* Gardeners Magazine of Botany for 1850, part 1, pp. 77-78. 1850.
240. M[OORE, T.], —— the lentil, *in* Gardeners Magazine of Botany for 1850, part 1, pp. 141-142. 1850.
241. M[OORE, T.], —— the terebinth—nuts, *in* Gardeners Magazine of Botany for 1850, part 1, pp. 162-163. 1850.
242. M[OORE, T.], —— the plane tree, *in* Gardeners Magazine of Botany for 1850, part 1, pp. 239-241. 1850.

243. M[OORE, T.], —— flax—linen, *in* Gardeners Magazine of Botany for 1850, part 2, pp. 79-80. 1850.
244. M[OORE, T.], —— lign aloes, *in* Gardeners Magazine of Botany for 1850, part 2, pp. 212-214. 1850.
245. M[OORE, T.], —— balm tree, myrrh, bdellium, *in* Gardeners Magazine of Botany for 1851, pp. 10-14. 1851.
246. M[OORE, T.], —— the cereals, *in* Gardeners Magazine of Botany for 1851, pp. 135-136. 1851.
247. MÜCKE, F., Wald und Wild in der Bibel. 133 pages. 1896.
248. MUNDELSTRUP, J. N., De pomis sodomiticis. 2 pl. 1683.
249. MURCHISON, R., & BERKELEY, M. J., [Manna], *in* the Gardeners' Chronicle & Agricultural Gazette for 1864, pp. 769-770. 1864.
250. NEUSTÄTTER, O., Where did the identification of the Philistine plague (I Samuel, 5 and 6) as bubonic plague originate?, *in* the Bulletin of the History of Medicine, vol. 11, no. 1, pp. 36-47. 1942.
251. NEVIN, J. W., A summary of Biblical antiquities. 447 pages. 1849.
*211. NEWTON, T. [*translator*], An herbal for the Bible (*see under* LEMMENS, L., 1587).
252. NICHOLS, W. A., New Bible version upsets old, *in* New York World-Telegram, August 5, 1939.
253. NICHOLSON, J. A. [*editor*], LANDER's Veterinary toxicology, ed. 3. 328 pages. 1945.
254. OEDMAN, S., Vermischte Sammlungen aus der Naturkunde zur Erklarung der heiligen Schrift. 1786.
255. OLDS, N. S., [review] *in* The Villager, [Greenwich Village, New York City], June 12, 1941.
255A. OPPENHEIMER, H. R., An account of the vegetation of the Huleh swamps, *in* Palestine Journal of Botany, Rehovot series, vol. 2, pp. 34-40. 1938.
256. OPPENHEIMER, H. R., Esquisse de la Géographie botanique de Transjordanie, *in* Bulletin de la Société Botanique de Genève, ser. 2, vol. 22, pp. 410-438. 1930.
257. OPPENHEIMER, H. R., Reliquiae Aaronsohnianae: Florula transiordanica, *in* Bulletin de la Société Botanique de Genève, ser. 2, vol. 22, pp. 126-410. 1930.
258. OPPENHEIMER, H. R., & EVENARI, M., Reliquiae Aaronsohnianae: II, Florula Cisiordanica — Révision critique des plantes rècoltèes et partiellement dèterminèes par AARON AARONSOHN au cours de ses voyages (1904-1916) en Cisjordanie, en Syrie et au Liban, *in* Bulletin de la Société Botanique de Genève, ser. 2, vol. 31, pp. 1-423. 1940.
258A. OPPENHEIMER, H. R., and M. EVENARI, Reliquiae Aaronsohnianae III: Une contribution à la connaissance de la flore du Bosphore, *in* Palestine Journal of Botany, Rehovot series, vol. 2, pp. 17-34. 1938.
259. [CELSIUS, O.], [Oren] Esai xliv. 14 an [aran], arbor arabica, spinosa, baccifera, Abi lfadli?, *in* Acta Literaria et Scientiarum Sueciae, vol. 4, pt. 1732, pp. 101-109. 1738.
260. OSBORN, H. S., Plants of the Holy Land, with their fruits and flowers. 174 pages. 1860.
261. OSBORN, H. S., —— [Another edition]. 174 pages. 1861.
261A. PATAI, R., A survey of Near-Eastern anthropology, *in* Transactions of the New York Academy of Sciences, series 2, vol. 12, pp. 200-209. 1948.
261B. PETERS, E. J., Weihrauch und Myrrhe, *in* Wiener Illustrierte Gartenzeitung, vol. 30, pp. 34-36. 1905.
262. PAXTON, G., Illustrations of the Holy Scriptures; in three parts. I from the geography of the East, II from the natural history of the East, III from the customs of ancient and modern nations, vol. 1, pp. 1-608. 1822.
263. PICKERING, C., Chronological history of plants: man's record of his own existence illustrated through their names, uses, and companionship. 1222 pages. 1879.
264. PIEL, G., BW, Biological warfare, *in* Life, vol. 21, no. 21, pp. 118-130. 1946.
265. POST, G. E., Flora of Syria, Palestine, and Sinai. 919 pages. 1883-1896.
266. POST, G. E., —— Edition 2. Volume I. 683 pages. 1932.
267. POST, G. E., —— —— Volume II. 946 pages. 1933.
268. [PRATT, A.], Plants and trees of Scripture. 192 pages. 1851.
269. PRITZEL, G. A., Thesaurus literaturae botanicae omnium gentium inde a rerum botanicarum initiis ad nostra usque tempora, quindecim millia operum recenseus. Edition 2. 579 pages. 1872.
270. RAVIUS, C., De dudaim Rubenis. 1656.
271. REDGROVE, H. S., The plants of the Bible, *in* the Gardeners' Chronicle, series 3, vol. 93, p. 100. 1933.
272. REDGROVE, H. S., An old Bible herbal, *in* the Gardeners' Chronicle, series 3, vol. 98, pp. 9-10. 1935.
273. REHDER, A., The Bradley bibliography. A guide to the literature of the woody plants of the world published before the beginning of the twentieth century, vol. 5, 1040 pages. 1918.
273A. REID, C., *in* A. BULLEID and H. ST. G. GRAY, The Glastonbury lake dwellings, vol. 2, p. 628. 1917.
273B. RICKETT, H. W., How the American sycamore acquired its name, *in* Journal of the New York Botanical Garden, vol. 48, p. 262. 1947.
273C. ROCHEBRUNE, A.-T. DE, Toxicologie Africaine: Etude botanique, historique, ethnographique, chimique, physiologique, thérapeutique, pharmacologique, posologique etc. Volume 1. 917 pages. 1897.
273D. ROCHEBRUNE, A.-T. DE, —— Volume 2, 500 pages. 1898.

274. ROHR, J. B., Phytotheologia, oder vernunst- und schriftmässiger Versuch, wie aus dem Reiche der Gewächse die Allmacht, Weisheit, Güte und Gerechtigkeit das grossen Schöpfers erkannt werden möge. 590 pages. 1740.
275. ROHR, J. B., —— Edition 2. 450 pages. 1745.
275A. ROMANES, G. J., The Bible and science, [review] *in* Nature, vol. 24, pp. 332-335. 1881.
276. ROSENBERG, P., Flowers of the Bible, *in* Nature Magazine, vol. 24, pp. 254-256. 1934.
277. ROSECRANS, J. H., Historical evidence of the origin of syphilis, *in* Journal of the Medical Society of New Jersey for 1917, pp. 1-11. 1917.
278. ROSENFELD, D., De [chavatzelet hasharon]; sive, Rosa saronitica. D. philologica ad illustrandum Com. I. cap. II. Cantic. 24 pages. 1715.
279. ROSENMÜLLER, E. F. K., Biblisch Naturgeschichte. Erster Theil. Das biblische Mineral- und Pflanzenreich. 348 pages. 1830.
280. ROSENMÜLLER, E. F. K., Handbuch der biblischen Alterthumskunde, IV — The mineralogy and botany of the Bible; translated from the German with additional notes by T. REPP & N. MORREN. 352 pages. 1840.
280A. ROSNER, J., A Palestine picture book. 141 pages. 1947.
281. ROYLE, J. F., On the hyssop of Scripture, *in* Proceedings of the Royal Society, vol. 5, pp. 510-520. 1844.
282. ROYLE, J. F., On the hyssop of Scripture, *in* the Journal of the Royal Asiatic Society, vol. 7, pp. 193-221. 1846.
283. ROYLE, J. F., Identification of the mustard tree and the hyssop of Scripture, *in* the Journal of the Royal Asiatic Society, vol. 8, pp. 113-137. 1846.
284. RUDBECK, O., Jr., Dudaim Rubenis, quos neutiquam Mandragorae fructus fuisse, aut flores amabiles, Lilia, Violas, Narcissos, Leucoia, species Melonis, Vaccinia, Chamaebatum, Rosam, Solanum, Halicacabum, certas uvas, tubera, Maiisch, Circaeam, Hordeum, philtra spinosi, allatae hic rationes satis videntur evincere. 18 pages. 1733.
285. RUMETIUS, L., Sacrorum Bibliorum arboretum morale. 118 pages. 1606.
286. RUMETIUS, L., Scripturae Sacrae viridarium literale et mysticum, in tres libros et sexaginta arboreta digestum. 901 pages. 1626.
287. RUMETIUS, L., Scripturae Sacrae . . . de frugiferis arboribus, infrugiferis & aromaticis. 920 pages. 1718.
287A. SANTINI DE RIOLS, Les plantes dans l'antiquité: légendes, poésie, histoire, etc. etc. l'amandier, *in* Le Naturaliste, ser. 2, vol. 19, pp. 178-179. 1897.
288. SARGEANT, A. M., The vegetable kingdom, *in* YOUNG, J., Scripture natural history, pp. 207-262. 1849.
288A. SCHEUCHZER, J. J., Kupfer Bibel, in welcher die Physica Sacra, oder geheiligte Naturwissenschaft derer in heil. Schrifft vorkommenden natürlichen Sachen, deutlich erklärt und bewahrt . . . etc. Illustrated by J. A. PFEFFEL and others. 761 plates. 1731-1735.
289. SCHEUCHZER, J. J., Physica sacra iconibus oeneis illustrata, in qua de variis Scripturae Sacrae plantis, procurante ANDREA PFEFFEL, chalcographo Augusteno. 5 volumes. 650 plates. 1732-1735.
290. SCHEUCHZER, J. J., Physique sacrée; ou, histoire naturelle de la Bible, traduite du latin. 8 volumes. 750 plates. 1732-1737.
291. SCOT, D., Contributiones ad historiam naturalem biblicam. 4 parts. 1830.
292. SCOTT, D., On the mustard plant mentioned in the Gospel, *in* Memoirs of the Wernerian Society, Edinburgh, vol. 6, pp. 430-442. 1826-1831.
293. [SHAW, H.], Plants of the Bible at the Missouri Botanical Garden. 16 pages. 1884.
294. SHAW, T., Travels or observations relating to several parts of Barbary and the Levant. 717 pages. 1738.
295. SHAW, T., Travels or observations relating to several parts of Barbary and the Levant. A supplement. 112 pages. 1746.
296. SIEBER, F. W., Reise von Cairo nach Jerusalem und wieder zurück, nebst Beleuchtung einiger heiligen Orte. 167 pages. 1823.
297. SIEBER, F. W., —— Edition 2. 168 pages. 1826.
298. SKINNER, C. M., Myths and legends of flowers, trees, fruits, and plants in all ages and in all climes. 302 pages. 1911.
298A. SKINNER, C. M., Myths and legends of flowers, trees, fruits, and plants in all ages and in all climes, reviewed *in* The Journal of Botany (British and Foreign), vol. 64, p. 196. 1926.
299. SMITH, J., Bible plants, their history, with a review of the opinions of various writers regarding their identification. 265 pages. 1878.
300. SMITH, W. [editor], A dictionary of the Bible, comprising its antiquities, biography, geography, and natural history. Volume I. 1860.
301. SMITH, W. [editor], Natural history of the Bible, [review] *in* the London Quarterly Review, vol. 114, pp. 22-39. 1863.
302. SMITH, W. [editor], Books received: a dictionary of the Bible. Parts XII and XIII, [review by J. LINDLEY] *in* the Gardeners' Chronicle & Agricultural Gazette for 1864, p. 343. 1864.
303. SMITH, W. [editor]; Notices of books, [review by J. LINDLEY] *in* the Gardeners' Chronicle & Agricultural Gazette for 1864, p. 726. 1864.
304. SMITH, W. [editor], Hyssop, *in* the Gardeners' Chronicle & Agricultural Gazette for 1864, pp. 895-896. 1864.
305. SMITH, W. [editor], Algum or almug trees, *in* the Gardeners' Chronicle & Agricultural Gazette for 1864, p. 895. 1864.
306. SMITH, W. [editor], A dictionary of the Bible, comprising its antiquities, biography,

geography, and natural history. 1033 pages. 1876.
*152. SOLA POOL, D. de [*translator*], The lilies of the field [extract from a lecture by Dr. E. HA REUBENI] (*see under* HA REUBENI, E., 1925).
307. SONNINI DE MANONCOUR, C. N. S., Voyage dans la haute et basse Egypte. 3 volumes. 1266 pages. 1799.
308. SONNTAG, C., Paralipomena quibus ligna Sittim explicata et applicata sistuntur. 28 pages. 1710.
309. STAUNTON, R., & STURTEVANT, B., The Church of the Wayfarer and its Biblical garden. 22 pages. 1940.
310. STEIN, D.G., Poisoning of human beings by weeds contained in cereals, *in* Onderstepoort Journal of Veterinary Science and Animal Industry, vol. 1, pp. 219-266. 1933.
311. [STEIN, R. S.]; King, Eleanor. Plants of the Holy Scriptures, [review] *in* the American Biology Teacher, vol. 4, no. 3, p. 100. 1941.
311A. STEUER, R. O., Myrrhe und Stakte, *in* Verlag der Arbeitsgemeinschaft der Ägyptologen und Afrikannten. 48 pages. 1933.
312. SWINGLE, W. T., Etrog, *in* BAILEY, L. H., Standard cyclopedia of horticulture, vol. 1, p. 1148. 1935.
313. TALEGON, J. G., Flora bíblico-poética; o, historia de las principales plantas elogiadas en la Sagrada Escritura. 519 pages. 1871.
314. TAYLOR, C., AUG. CALMET's great dictionary of the Bible, with continuation and Scripture illustrated by means of natural science in botany. 4 volumes. 1797-1803.
315. TAYLOR, J., Bible garden. 1836.
316. TAYLOR, J., —— Edition 2. 1839.
317. TEMPLE, A. A., Flowers and trees of Palestine. 184 pages. 1907.
317A. TEMPLE, A. A., —— New edition, illustrated in color by K. M. REYNOLDS. 160 pages. 1929.
318. THOMAS, M. P., [Biblical plants], *in* Flower Grower, vol. 28, p. 243. 1941.
318A. THOMPSON, C. J. S., The mystic mandrake. 253 pages. 1934.
319. THOMSON, W. M., The land and the Book. Volume I. 1860.
320. THOMSON, W. M., —— Volume II. 1871.
321. THUNBERG, C. P., Afhandling om de wäxter, som i Bibelon omtales. 9 parts. 128 pages. 1828.
322. TRISTRAM, H. B., The natural history of the Bible; being a review of the physical geography, geology, and meteorology of the Holy Land; with a description of every animal and plant mentioned in the Holy Scripture. 524 pages. 1867.
323. TRISTRAM, H. B., —— Edition 3. 526 pages. 1873.
324. TRISTRAM, H. B., —— Edition 6. 528 pages. 1880.
325. TRISTRAM, H. B., —— Edition 7. 520 pages. 1883.
326. TRISTRAM, H. B., The survey of western Palestine. The fauna and flora of Palestine. 477 pages. 1884.
327. TYAS, R., Flowers from the Holy Land; being an account of the chief plants named in Scripture; with historical, geographical, and poetical illustrations. 207 pages. 1851.
328. URSINUS, J. H., Arboretum biblicum in quo arbores & frutices passim in S. Literis occurrentes, notis philologicis, philosophicis, theologicis exponuntur & illustrantur. 638 pages. 1663.
329. URSINUS, J. H., —— Edition 2. 638 pages. 1672.
330. URSINUS, J. H., —— Edition 3. 638 pages. 1685.
331. URSINUS, J. H., Continuatio historiae plantarum biblicae; sive, de sacra phytologia, herbarius sacer, & hortus aromaticus, cum sylva theologiae symbolicae recusa. 502 pages. 1665.
332. URSINUS, J. H., —— Edition 2. 559 pages. 1672.
333. URSINUS, J. H., —— [Edition 3]. 559 pages. 1685.
334. URSINUS, J. H., Arboretum biblicum, in quo arbores & frutices passim in S. Literis occurrentes, ut & plantae . . . exponuntur & illustrantur. 638 pages. 1699.
335. URSINUS, J. H., Arboreti biblici continuatio; sive, historiae plantarum Biblicae libri tres . . . una cum sylva theologiae symbolicae; cui nunc accessere indices necessarii. New edition. 534 pages. 1699.
336. URSINUS, J. H., Arboretum biblicum et continuatio historiae plantarum Biblicae. 2 volumes. 832 pages. 1699.
336A USTERI, A., Die Hölzer des Kreuzes und ihre Beziehung zur Flora der Mittelmeerländer und benachbarter Gebiete. 47 pages. 1942.
337. VALLES, F., De iis, quae scripta sunt physice in Libris Sacris; sive, de sacre philosophia liber singularis. 978 pages. 1588.
338. VALLES, F., De sacra philosophia; sive, de iis, quae in Libris Sacris physice scripta sunt liber singularis. Edition 6. 440 pages. 1652.
339. VIREY, J. J., Pharmacon hieron; ou, botanique sacrée; notice des plantes saintes, servant au culte de diversés religions, et á des pratiques d'exorcisme, *in* Journal de Pharmacie et de Chimie, Paris, vol. 18, pp. 188-194. 1832.
340. WALDSTEIN, A. S., Hebrew-English dictionary. 1081 pages. 1938.
341. WARBURG, O., Heimat und Geschichte der Lilie, *in* FEDDE, Repertorium Specierum Novarum Regni Vegetabilis Beihefte 56, pp. 167-204. 1929.
342. WEDEL, G. W., Do corona Christi spinea. 2 parts. 16 pages. 1696.
343. WEDEL, G. W., De lignis thyinis Apocalypseos in genere. 8 pages. 1707.
344. WEEKS, M. E., An exhibit of chemical substances mentioned in the Bible, *in* Journal of

Chemical Education, vol. 20, pp. 63-76. 1943.
345. WESTMACOTT, W., θεολοβοτονολογια; sive, historia vegetabilium sacra; or, a Scripture herbal; wherein all the trees, shrubs, herbs, plants, flowers, fruits, &c. both foreign and native, that are mentioned in the Holy Bible . . . are in an alphabetical order, rationally discoursed. 260 pages. 1694.
345A. WHITE, W. E., Trees and the Bible. Bulletin 37, Texas Forest Service, Agricultural & Mechanical College. 16 pages. 1947.
346. WILLIAMS, M. O., Syria and Lebanon taste freedom, in the National Geographic Magazine, vol. 90, pp. 729-766. 1946.
347. [WOODWARD, C. H.], Check-list of plants that are mentioned in the Bible, in Journal of the New York Botanical Garden, vol. 42, pp. 67-74. 1941.
347A. WOODWARD, C. H., —— New revised edition. 1948.
348. [WOODWARD, C. H.], Biblical plants at flower show, in Journal of the New York Botanical Garden, vol. 42, p. 49. 1941.
349. [WOODWARD, C. H.], Biblical botany, in Time, vol. 37, no. 13, p. 53. 1941.
*235. [WOODWARD, C. H.] Check-list of plants that are mentioned in the Bible, in the Biblical Digest (see under MOLDENKE, H. N., 1941).
350. [WOODWARD, C. H.], Flower Show, in Metropolitan New York Section of the Parents' Magazine for March, 1941.
351. [WOODWARD, C. H.], WPA helped arrange Biblical flower exhibit, in The (Bronx, New York) Home News, March—, 1941.
352. [WOODWARD, C. H.], Bible flowers in coming show, in the New York Sun, March 8, 1941.
353. [WOODWARD, C. H.], Garden of Biblical days duplicated for exhibit, in the New York Herald-Tribune, March 9, 1941.
354. [WOODWARD, C. H.], Flower Show opens at Grand Central Palace on March 17, in the Westchester County (New York) Times, March 14, 1941.
355. [WOODWARD, C. H.], Bible plants, in the Scarsdale (New York) Inquirer, March 14, 1941.
356. [WOODWARD, C. H.], Bible flowers at the show, in the New York Sun, March 15, 1941.
357. [WOODWARD, C. H.], Flowers of Biblical days, in the New York World-Telegram, March 15, 1941.
358. [WOODWARD, C. H.], Botanical Gardens' exhibits, in the New York Times, Annual Garden Section, March 16, 1941.
359. [WOODWARD, C. H.], Bronx Park botanical garden is awarded Flower Show medal for "Biblical" exhibit, in The (Bronx, New York) Home News, March 18, 1941.
360. [WOODWARD, C. H.], [Lilies of the field], in the Rochester Times-Union, March 18, 1941.
361. [WOODWARD, C. H.], [Bible plants], in the New York Times, March 20, 1941.
362. [WOODWARD, C. H.], Flowers of the Bible, in the New York Sun, March 22, 1941.
363. [WOODWARD, C. H.], Bible garden, in the New York Sun, March 22, 1941.
364. [WOODWARD, C. H.], Flowers that Christ knew shown in Biblical garden exhibit, in The Newark (New Jersey) Evening News, March 22, 1941.
365. [WOODWARD, C. H.], Work of Watchung resident inspires Biblical garden exhibit, in the Plainfield (New Jersey) Courier News, March 27, 1941.
366. [WOODWARD, C. H.], Nativity scene, in the New York Sun, December 19, 1941.
367. [WOODWARD, C. H.], Botanical Garden shows inn yard at Bethlehem, in the New York Herald-Tribune, December 21, 1941.
368. [WOODWARD, C. H.], New York Botanical Garden display depicts nativity, in the New York Herald-Tribune, December 22, 1941.
369. [WOODWARD, C. H.], Nativity reproduced with trees of year 1, in the New York Times, December 22, 1941.
370. [WOODWARD, C. H.], Bethlehem scene at Botanical Garden, in The Tablet, Brooklyn, N. Y., December 27, 1941.
371. [WOODWARD, C. H.], Replica of old Bethlehem frames nativity scene, in The Catholic News, December 27, 1941.
372. [WOODWARD, C. H.], Holy Land plants on view, in the New York Sunday Times, April 18, 1943.
373. WRIGHT, A. E. [copyright owner], Twentieth century encyclopedia and dictionary. 5 volumes. 3261 pages. 1907.
374. WRIGHT, W. [editor], Bible helps. The illustrated Bible treasury by Prof. SAYCE [and others]. 726 pages. 1896.
375. WRIGHT, W. [editor], —— New edition. 739 pages. 1899.
376. YONGE, C.D., An English-Greek lexicon, edited by H. DRISLER, ed. 1. 778 pages. 1870.
377. YONGE, C.D., An English-Greek lexicon, edited by H. DRISLER, ed. 2. 893 pages. 1871.
378. YOUNG, J., Trees and flowers in Scripture. 1848.
379. YOUNG, J., Scriptural natural history. 1849. (The vine and palm tree are treated in the zoological section of the work).
380. YOUNG, J., Scripture natural history. Edition 3. 1859.
381. YOUNG, R., Analytical concordance to the Bible. 1889.
382. ZELLER, H. G., Feldblumen aus dem Heiligen Land. 54 plates. 1875.
383. ZELLER, H. G., Wild flowers of the Holy Land. 54 plates. 1880.

— *Additions to the bibliography** —

457. Liber psalmorum cum canticis breviarii Romani. Nova e textibus premigeniis interpretatio latina cum notis criticis et exegeticis, cura professorum ponteficii I. B edita. Edit. Americana a P. I. approbata. Benziger, New York. 1945.
458. ABBOTT, L. [*reviser*], The pictorial New Testament, with notes by the ABBOTTS. 998 pp. 1881.
459. Anonymous; Forests and forest devastation in the Bible. United States Department of Agriculture Forest Service [unnumbered; U. S. Government Printing Office M-5204, 8-9820]. 8 pp. 1938.
460. Anonymous; The legend of the dogwood, *in* Royle Forum, March 1947, p. 11. 1947.
461. Anonymous; Cradle of Christianity, *in* Sunday News [New York City], December 19, 1948, pp. 25 & 26. 1948.
462. Anonymous; Many secrets discovered by explorers this year, *in* New York Herald Tribune, December 26, 1948, section 2, p. 1. 1948.
463. ASCH, J., Some sacred plants of our neighbors' faiths, *in* The Garden, vol. 1, no. 2, pp. 14-16. 1948.
464. BALL, C. R., *in* Scientific Monthly, vol. 69, p. 49. 1949.
465. BATES, E. S., The Bible designed to be read as living literature. The Old and the New Testaments in the King James Version. 1296 pp. 1936.
466. BENGTSON, B., The trees of the Bible, *in* Frontiers, vol. 7, no. 1, p. 13. 1942.
467. BENGTSON, B., The birds of the Bible, *in* Frontiers, vol. 7, no. 4, p. 100. 1943.
468. BENGTSON, B., Flowers of the Bible, *in* Frontiers, vol. 9, no. 1, p. 18. 1944.
469. BERNSTEIN, P. S., What the Jews believe, *in* Life, volume 29, no. 11, pp. 161, 162, 164, 167-172, 174, & 179. 1950.
470. BODENHEIMER, F. S., Animal life in Palestine. 515 pp. 1935.
471. BOYKS, E., A glimpse at Israel's flora and fauna, *in* The Jewish Herald, December 31, 1949, p. 8. 1949.
472. BRAGA, E., Notas sobre a flora e fauna de Ophir, *in* Revista da Sociedade Scientifica de São Paulo, no. 2, pp. 90-98. 1905.
473. CALDER, R., Need one-third of the world be desert?, *in* The New York Times Magazine, July 9, 1950, pp. 16-17, 34, & 35. 1950.
474. ESTERNAUX, G., Evander Childs High School Biology Department Meeting, March 22, 1948.
475. FOLKARD, R., Plant lore, legends, and lyrics. 610 pp. 1884.
476. FRITSCH, F. E., Structure and reproduction of the algae, vol. 2, p. 784. 1945.
477. HARDY, E., A handbook of the birds of Palestine. 54 pp. 1946.
478. HOEHNE, F. C., Plantas e substancias vegetais toxicas e medicinais. 355 pp. 1939.
479. JENNISON, G., Animals for show and pleasure in ancient Rome. 223 pp. 1937.
480. KOLB, G., Manna der Natur und der Bibel, *in* Natur und Offenbarung, vol. 38, pp. 1-13. 1892.
481. LEMMENS [Lemnius], L., Similitudinum ac parabolarum quae in Bibliis ex herbis atque, Arboribus desumuntur, dilucida explicatio: In qua narratione singula loca explanantur, quibus Prophetae, obseruata stirpium natura, conciones sisas illustrant, diuina oracula fulciunt. Frankfurt. 166 pp. 1596.
482. LUNDGREEN, F., Heilige Bäume in Alten Testament. 43 pp. 1908.
483. LUNDGREEN, F., Die Benutzung der Pflanzenwelt in der alttestamentlichen Religion, *in* Zeitschrift für die Alttestamentliche Wissenschaft, vol. 14, pp. i-xxiii, 1-191. 1908.
484. MATSUMURA, J., [On the fig], in Report of the Commercial Department of Mabie Memorial School, Yokohama, vol. 51. 1939.
485. MATSUMURA, J., [On the olive and its translation into Japanese], *in* Baptist News, no. 1102. 1936.
486. MATSUMURA, J., [Wormwood and its translation into Japanese], *in* Baptist News, no. 1127. 1938.
487. MATSUMURA, J., [Mandrake is not love apple], *in* Baptist News, no. 1114. 1938.
488. MATSUMURA, J. [Two plants which originate in southern islands], *in* Amatores Herbarii, vol. 10, nos. 1-2. 1942.
489. MATSUMURA, J., [Trees and herbs of Palestine. 1], *in* Baptist News, no. 1136. 1939.
490. MATSUMURA, J., [Trees and herbs of Palestine. 2], *in* Baptist News, no. 1137. 1939.
491. MATSUMURA, J., [Trees and herbs of Palestine. 3], *in* Baptist News, no. 1141. 1939.
492. MATSUMURA, J., [Trees and herbs of Palestine. 4], *in* Baptist News, no. 1142. 1939.
493. MATSUMURA, J., [On the carob bean], *in* Amatores Herbarii, vol. 6, no. 4. 1938.
494. MATSUMURA, J., [Plants and the Bible: on the carob bean], *in* Kohan No Koe [Voice of Lake "Biwa"], vol. 27, no. 8. 1939.
495. MATSUMURA, J., [On the wormwood and its translation "Inchin"], *in* Kohan No Koe [Voice of Lake "Biwa"], vol. 27, nos. 9 & 11. 1939.
496. MATSUMURA, J., [On flax], *in* Kohan No Koe [Voice of Lake "Biwa"], vol. 28, no. 4, pp. 24-27, & no. 5. 1940.
497. MATSUMURA, J., [On nard], *in* Kohan No Koe [Voice of Lake "Biwa"], vol. 28, no. 12. 1940.

---

*Inasmuch as in the original bibliography the items run from no. 1 through no. 383, with 73 interpolated lettered items, these additions are made to begin with no. 457.

498. MATSUMURA, J., [On the love apple], *in* Kohan No Koe [Voice of Lake "Biwa"], vol. 29, nos. 3 & 4. 1941.
499. MATSUMURA, J., [On the olive], *in* Kohan No Koe [Voice of Lake "Biwa"], vol. 29, no. 6, pp. 10-12; no. 7, pp. 14-16; no. 8, pp. 14-16; and no. 9, pp. 17-18. 1941.
500. MATSUMURA, J., [On the fig in the autumn], *in* Kohan No Koe [Voice of Lake "Biwa"] vol. 29, no. 12, pp. 9-13 (1941) & vol. 30, no. 1, pp. 10-15. 1942.
501. MATSUMURA, J., [On the pomegranate], *in* Kohan No Koe [Voice of Lake "Biwa"], vol. 30, no. 5, pp. 17-20, and no. 6, pp. 14-16. 1942.
502. MATSUMURA, J., [On the grape], *in* Kohan No Koe [Voice of Lake "Biwa"], vol. 30, no. 11, pp. 10-15; no. 12, pp. 10-12 (1942) and vol. 31, no. 3, pp. 12-15; no. 8; and no. 9. 1943.
503. MILLER, M. S. and J. L., Encyclopedia of Bible life. 493 pp. 1944.
504. MOLDENKE, H. N., Consider the lilies . . ., *in* The Garden, vol. 2, no. 2, pp. 11-13. 1949.
505. MURRAY, J. J., Wild wings. 123 pp. 1947 [chapter entitled "Behold the fowls of the air", pp. 117-123, contains information on the birds of the Bible].
506. NELSON, T., and son [*publishers*], Bible plants, or botanical allusions and illustrations in Holy Writ explained. 64 pp. 1870.
507. ROTHERY, A. The joyful gardener. 274 pp. 1949.
508. SCHUMANN, K., *in* Engler & Prantl, Die natürlichen Pflanzenfamilien, part 12-125. 1895.
509. STELLFELD, C., Mandragora a droga de quarenta seculos, *in* Tribuna Farmaceutica, vol. 3, no. 10, pp. 135-138. 1935.
510. STOW, M. A., and TRIMBLE, W., *in* Newsweek, vol. 35, no. 15, p. 8. 1950.
511. TÄCKHOLM, V., and DRAR, M., Flora of Egypt, vol. 2, 568 pp. 1950.
512. TEESDALE, M. J., The manna of the Israelites, *in* Science Gossip, new series, vol. 3, pp. 229-232, 5 text figs. 1897.
513. TEESDALE, J., Manna, *in* Nature, vol. 55, p. 349. 1897.
514. THOMPSON, F. C., [*editor*], The new chain-reference Bible. Second revised edition, containing THOMPSON'S chain-references and text cyclopedic to which has been added a new and complete system of Bible study. 740 pp. 1917.
515. TIMOTHY, B., The origin of manna, *in* Nature, vol. 55, p. 440. 1897.
516. FISHER, L. J., Trees and the Bible. Texas Forest Service, Agricultural & Mechanical College of Texas Bulletin 37. 14 pp. 1946.
517. FISHER, L. J., —— Second printing. 1947.
518. WOOD, J. G., Bible animals. 681 pp. 1870.
519. WOOD, J. G., The City and the Land, *in* TRISTRAM, H. B., The natural history of Palestine. 1892.
520. ZOHARY, M., A vegetation map of western Palestine, *in* Journal of Ecology, vol. 34, pp. 1-19. 1947.
521. ZOHARY, M., Phytogeographical problems of the Near East countries. VII International Botanical Congress, Stockholm 1950, Section PHG, Abstract of Communication. 1950.
522. Just's Botanischer Jahresbericht, vol. 20, part 2, p. 51. 1895.
523. Just's Botanischer Jahresbericht, vol. 23, part 2, p. 56. 1898.
524. Just's Botanischer Jahresbericht, vol. 25, part 1, p. 318. 1900.
525. Just's Botanischer Jahresbericht, vol. 25, part 2, p. 175. 1900.
526. Just's Botanischer Jahresbericht, vol. 33, p. 374. 1908.
527. FITZPATRICK, H. M., History of mycology. Outline of a one hour lecture. 21 pp. 1929.
528. KELLEY, A. P., Mycotrophy in plants. 240 pp. 1950.
529. NEILL, J. C., The endophyte of rye-grass (*Lolium perenne*), *in* New Zealand Journ. Sci. & Techn., vol. 21, pp. 280A-291A. 1940.
530. SMITH, C. R., The physician examines the Bible.
531. MARQUAND, Mrs. A., The history of plant illustration, *in* Bulletin of the Garden Club of America, series 7, no. 14, pp. 18-23. 1941.
532. Anonymous; Forest and flame in the Bible. The advertising Council, State Foresters, and the U. S. Department of Agriculture, Forest Service, PA-93 [U. S. Government Printing Office 875962]. 16 pp. 1950.
533. MACKAY, A. I., Farming and gardening in the Bible. 280 pp. 1950.
534. BOYKO, H., *Lilium candidum* L. wild in the Carmel area, *in* Palestine Journal of Botany, Rehovot Series, vol. 5, no. 1, pp. 124-125. 1945.
535. WHITE, W. L., Deterioration of Quartermaster fabrics in the tropics, *in* The Quartermaster Review for November-December 1946.
536. MOLDENKE, H. N., Plants and "The Man of Peace", *in* The Garden Journal, vol. 1, pp. 71-72. 1951.
537. FRIES, T. M., Bref och skrifvelser af och till Carl von Linné, volume 1, pp. 273-277. 1907.
538. MEYERHOFF, M., The earliest mention of a manniparous insect, *in* Isis, vol. 37, pp. 31-36. 1947.
539. HOWES, F. N., Vegetable gums and resins. 210 pp. 1949.
540. THIÉBAUT, J., Flore libano-syrienne, part 1, 198 pp. 1936.
541. THIÉBAUT, J., Flore libano-syrienne, part 2. 372 pp. 1940.
542. KELLERMAN, S., Die Blumen des Heiligen Landes: botanische Auslese einer Frühlingsfahrt durch Syrien und Palästina. 170 pp. 1947.
543. FERRY DE LA BELLONE, C. DE, La truffe, p. 11. 1888.

544. GLUECK, N., The river Jordan. 268 pp. 1946.
545. CHAMBERLIN, R. B., & FELDMAN, H., The Dartmouth Bible, an abridgement of the King James Version, with aids to its understanding and literature, and as a source of religious experience. 1295 pp. 1950.
546. O'CONNELL, J. P. [editor], The Holy Bible, illustrated by J. J. J. TISSOT. 288 pp. 1951.
547. BUHL, F. [editor], WILHELM GESENIUS' Hebräisches und Aramäisches Handwörterbuch über das Alte Testament. 1032 pp. 1949.
548. VIGOUROUX, F., Dictionaire de la Bible, contenant tous les noms de personnes, de lieux, de plantes, d'animaux mentionnés dans les saintes écritures, les questions théologiques, archeologiques, scientifiques, critiques relative à l'ancien et au nouveau testament et des notices sur les commentateurs anciens et modernes avec de nombr. renseignements bibliographiques. 40 vols. 1891-1912.
549. EMERSON, E. R., A lay thesis on Bible wines. 63 pp. 1902.
550. SALAMASIUS, C., De manna et saccharo commentarius. 95 pp. 1663.
551. SCHEUCHZER, J. J., Bibliotheca scriptor, historiae naturali. 250 pp. 1716.
552. ASCHERSON. P., *Cephalaria syriaca*, ein für Menschen schädliches Getreideunkraut Palästinas und die biblischen zizania (Matth. 13, 25-30), *in* Zeitschrift der Deutschen Palästinavereins, vol. 12, pp. 152-156. 1889.
553. SCHWERIN, F., Graf VON, Kreuzeholz und Dornenkrone, *in* Mitteilungen der Deutsche Dendrologische Gesellschaft, vol. 45, pp. 155-157. 1933.
554. THOMASSON, R. R., Beauty survives mandrake's legends, *in* New York Herald Tribune, Sunday, July 8, 1951, section 4, page 15. 1951.
555. TRISTRAM, H. B., The natural history of the Bible: being a review of the physical geography, geology, and meteorology of the Holy Land; with a description of every animal and plant mentioned in Holy Scripture. Second edition, revised and corrected. 526 pp. 1868.
556. Anonymous; A few observations on Miss CALLCOTT's "Scripture Herbal". mss. 4 pp. inscribed as "Extracted from The Nephelococcygia Quarterly Review, Jan. 1843". Chronica Botanica Archives, acq. 1950, Ins. sub. CAL.
557. Picture Bible, published by the Bible Corporation of America, New York City. Editions of 1897 and 1926.
558. Anonymous; Die Bäume, Pflanzen, Blumen und Früchte der Heiligen Schrift, on pp. 49-56 of the supplementary portion of "Die neue Illustrierte Familien Bibel für häusliche Erbauung und Belehrung . . . ", published by A. J. Holman & Company, Philadelphia. 1891.
559. Anonymous; Scriptural natural history—botany, one unnumbered text page and one colored plate *in* "The Illustrated Family Bible", published by A. J. Holman & Company, Philadelphia. 1882.
560. BIRDWOOD, G. C. M., The perfumes of the Bible: aloes, balm, cassia, frankincense, myrrh, saffron, on page 31 of the supplementary portion of "The Illustrated Family Bible", published by A. J. Holman & Company, Philadelphia. 1882.
561. HONORATI, B., La Saincte Bible en François. Lyon, 1582.
562. SOUTHWELL, H., The universal family Bible or Christian's divine library, published by J. Cooke, London. 1775.
563. FEINBRUN, N., and ZOHARY, M., Flora of the land of Israel: iconography, 12 pages and 50 plates. 1949.
564. ROSEN, F., Die Natur in der Kunst. 355 pp. 1903.
565. KENNEDY, R. W., The Renaissance painter's garden. 60 plates, 30 unnumbered pages of text and index. 1948.
566. WILSON, A. M., The wines of the Bible: an examination and refutation of the unfermented wine theory. 387 pp. 1877.
567. KERR, N. S., Unfermented wine a fact: a review of the latest attempt to show that the existence of unfermented wine among the ancients was impossible. 21 pp. 1878.
568. LEES, F. R., The wine question. 14 pp. 187-.
569. FISCHEL, O., The works of TITIAN reproduced in 284 illustrations. 243 pp. 1921.
570. MAYER, A., Meisterwerke des Prado in Madrid. 1922.
571. STEVENSON, R. A. M., RUBENS paintings and drawings. 302 pp. 1939.
572. TIETZE, H., TINTORETTO, the paintings and drawings, with three hundred illustrations. 383 pp.
573. Löw, I., Die flora der Juden, vol. 4, 740 pp. 1934.
574. WILSON, C. W., Picturesque Palestine, Sinai and Egypt. New York edition, vol. 1. 480 pp. 1881.
575. WILSON, C. W., ——— vol. 2, 476 pp. 1883.
576. WILSON, C. W., ——— London edition, vol. 3, 240 pp. 1884.
577. WILSON, C. W., ——— London edition, vol. 4, 236 pp. 1884.
578. FERGUSON, J., Symbolism of the vine, *in* Gardeners' Chronicle and Agricultural Gazette for 1873, pp. 1048-1050. 1873.
579. ZOHARY, M., The arboreal flora of Israel and Transjordan and its ecological and phytogeographical significance, *in* Imperial Forestry Institute, University of Oxford, Institute Paper 26. 50 pp., 2 maps. 1951.
580. HAUSMAN, E. H., A Christmas herb, *in* Horticulture, vol. 28, pp. 437 & 458. 1950.

581. PALLAS, P. S., Reise durch verschiedene Provinzen des Russischen Reichs, 3 vols., 592 pp., 61 pl. 1778.
582. DORÉ, G., The Bible gallery. 1880.
583. DALZIEL, G. & E., Bible gallery. 1881.
584. BROWNE, L., The graphic Bible. 159 pp. 1928.
585. BENTWICH, N., A wanderer in the Promised Land. 263 pp. 1933.
586. Anonymous; Sheba lived like a queen, in New York World-Telegram and Sun, p. 1. August 17, 1951.
587. WOODCOCK, H. B. D., and STEARN, W. T., Lilies of the world, 431 pp. 1950.
588. SPRUNT, A., JR., Animals of the Bible, in American Forests, vol. 37, pp. 717-720. 1931.
589. DAY, G. E., and THAYER, J. H. [editors], The Holy Bible: containing the Old and New Testaments translated out of the original tongues: being the version set forth A.D. 1611 compared with the most ancient authorities and revised 1881, 1885: newly edited by the American Revision Committee 1901. Thomas Nelson & Son. 1435 pp. 1901.
590. FREEDMAN, H., and SIMON, M. [editors], Midrash Rabbah, vol. 1. Genesis I, by H. FREEDMAN. 537 pp. 1939.
591. FREEDMAN, H., and SIMON, M. ——— vol. 2. Genesis II, by H. FREEDMAN. 500 pp. 1939.
592. FREEDMAN, H., and SIMON, M. ——— vol. 3. Exodus, by S. M. LEHRMAN. 589 pp. 1939.
593. FREEDMAN, H., and SIMON, M. ——— vol. 4. Leviticus, by J. ISRAELSTAM and J. J. SLOTKI. 479 pp. 1939.
594. FREEDMAN, H., and SIMON, M. ——— vol. 5. Numbers I, by J. J. SLOTKI. 492 pp. 1939.
595. FREEDMAN, H., and SIMON, M. ——— vol. 6. Numbers II, by J. J. SLOTKI. 403 pp. 1939.
596. FREEDMAN, H., and SIMON, M. ——— vol. 7. Deuteronomy, Lamentations, by J. RABBINOWITZ and A. COHEN. 253 pp. 1939.
597. FREEDMAN, H., and SIMON, M. ——— vol. 8. Ester, Song of Songs, by M. SIMON. 1939.
598. FREEDMAN, H., and SIMON, M. ——— vol. 9. Ruth, Ecclesiastes, by L. RABINOWITZ and A. COHEN. 334 pp. 1939.
599. FREEDMAN, H., and SIMON, M. ——— vol. 10. Index volume, by J. J. SLOTKI, 344 pp. 1939.
600. CHRISTENSEN, C., Index to PEHR FORSSKÅL, Flora Aegyptiaco-arabica 1775 with a revision of Herbarium Forskalii contained in the Botanical Museum of the University of Copenhagen, in Dansk Botanisk Arkiv, vol. 4, no. 3, 54 pp. 1922.
601. CHIZIK, B., Cucurbits in Palestine in past and present. 1937.
602. HA-REUBENI, E. & H., Thesaurus plantarum terrae Erez-Israel: de Urginea maritima L. et Asphodelus microcarpa Viv. 1941.
603. HALPER, D., Post Biblical Hebrew literature. Jewish Publication Society. 1921.
604. HEGI, G., Illustrierte Flora von Mittel-Europa, vol. 5, part 1, pp. 327-329. 1925.
605. ROSENBAUM, I. J., and TARCOV, O., Your neighbor celebrates. 31 pp. 1951.

Supplementary Notes

During recent months considerable additional relevant information has come to light. It has been thought advisable to include some of this in the present Supplementary Notes. Further supplements will be issued from time to time in CHRONICA BOTANICA.

Historical Sketch (p. 1).— It is of considerable interest to all botanists to note that the Royal Swedish Bible Commission in 1773 appointed LINNAEUS as one of the commissioners, and he went through the Swedish version of the Bible very carefully to see where it needed correction in regard to botanical and zoological nomenclature. The few emendations, however, that he induced his fellow commissioners to adopt did not compensate for the time he devoted to the work (537).

To the list of important recent Bible versions should be added the English Revised Version (1885), the American Standard Version (1901), the Revised Standard Version (New Testament, 1946; Old Testament, probably 1952), the Dartmouth Bible (an abridgement of the King James Version, with aids to its understanding as history and literature, and as a source of religious experience) edited by R. B. CHAMBERLIN and H. FELDMAN (1950), and the new O'CONNELL edition of the Catholic Bible (1951).

Dr. TREVOR has recently informed us that we now have a Hebrew text of two books of the Old Testament which go back to the first century B.C. at least. He states, also, that the Samaritan Pentateuch actually goes back to the fourth century B.C., rather than to the second century A.D. as stated on page 7 of our text, and that the Latin translation of JEROME was not completed until 405 A.D. The Authorized Version of King James I, referred to in many places in our text, is sometimes cited by other authors as the "Authorized Version of King James VI". King JAMES VI of Scotland was King JAMES I of England when the version was authorized. The "version of CRANMER" referred to by us on page 9 of our text is often spoken of as the Great Bible, since CRANMER's preface did not appear until 1541. The version which we refer to in our text as the "GOODSPEED" version is often called the Chicago Version.

Description of the Land (p. 13).—ELIZABETH BOYKS has recently written a very succinct resumé of the prominent aspects of the flora and fauna of Palestine (471) which we regard as so good that we would like to quote the major part of it here, with her kind permission. "It takes one's breath away, these changes of landscape, climate and aspects in so tiny a country as Israel. ... In the course of two hours' motoring one can pass from the soft Mediterranean area across the semi-desert and desert areas of the Judean mountains to the subtropical valleys of the Jordan and to the deep depression of the Dead Sea. Another ride from there up north leads to the Galilean Mountains and up to the foot of Mount Hermon and Lebanon, passing by the Hule-lake district. And of course each one of these areas has a very specific flora and fauna, too. The Mediterranean area, for instance —that is, roughly the coastal plains and mountain ranges as far as Mount Carmel—are at their best in early spring. In February and March the meadows and hills, the roadsides and every stretch of soil are covered with a multitude of blossoms; thousands of anemonies, mostly bright scarlet, but in some places of every shade from white to pink, mauve, purple and red. There are tulips and hyacinths that could proudly be displayed in every florist's shop. Cyclamen, bachelor's buttons, gladiolus, and many, many others, are all lovely, multi-coloured and abundant in this luxurious spring season. This is the effect of the soil's moisture accumulated during the winter rains and combined with the soft spring sunshine which pulls them up. Later on, in May-June, the soil becomes drier and drier, the heat and burning sun increase in potency and put a stop to the gay display. Bulb [ous plants], like tulips, hyacinths, gladiolus, lilies, withdraw into their bulbs and remain there undisturbed by drought and blazing sun. Annuals by then have completed their short life-cycle and have fruit, and the lush green of meadows and hills changes into [the] yellow and brown of burnt weeds.

"It is quite different in an oasis like Jericho, where underground springs feed the roots of slender-stemmed date-palms, of papaya trees and grapes, and where spring flowers build only the top-decoration between stretches of date-palms, oranges, and banana-groves. In the Jordan valley and up [to] the Hule-lake spring comes early and summer is now being defeated by intensive irrigation with Jordan-water and the waters of Lake Tiberias. . . . The hills, of course, come latest with blooms of crops and fruit. But late grapes and pomegranates, apples and figs are grown there in the small cultivated areas and could perhaps later be grown all over the hill area, barren and eroded now after centuries of ill-treatment, wood-vandalism and ignorance and lack of an enterprising population, ready and willing to up-build.

"Figs and olives are the fruit trees best adapted to the soil and weather-conditions in the hills and the remains of gardens and terraces show age-old and gnarled specimens of both these trees. Calculations . . . . about the age of olive trees have proved that some of these reach the age of 3,000 years, hollowed and dead inside they still sprout and bear fruit and may still continue to do so for some time. . . . .

"In devastated and poor areas, desert-land and poor mountain-soil, animal life is poor as well, for lack of feed and shelter. Still, bird-lovers find their compensation when in spring and autumn they watch the migrating birds fly over Palestine, flocks and flocks of them; they rest here for days and weeks and continue their ways in due course; storks and starlings, blackbirds and quails come and pass. But the most thrilling area for an ornithologist is the Huleh-Swamp. This area is in fact like a mid-African island . . . . and paddling by boat through dense papyrus forests on the border of the lake gives one the illusion of being right in the heart of Africa. And what a collection of birds! aigrettes and flamingoes, ibis, king-fishers and honey-birds, pelicans and many, many more, nesting and multiplying in this ideal natural zoological garden. Water buffalos take the place of ordinary cattle in this area, wild boars are not infrequent here although they were to be found 30 years back in several places in Palestine.

"Leopards are rare but may be encountered near the frontier in Galilee and Gepaids in the Negev; porcupines hide in the Judean hills. The hyena and jackal are prominent features of nearly all the areas of Palestine. In the plains, as soon as dusk falls the howling of the jackals is a non-stop concert in all the villages and colonies and even at the border of the towns. But their activities are by far not limited to the musical. Many a farmer tells the tale of broken-in chicken pens, stolen poultry, destroyed eggs, uprooted carrots, spoiled strawberry-beds, spoiled grapes and unearthed peanuts. Nothing tasty or sweet is safe from them and these destroyers always spoil far more than they can eat." Insects and spiders are very abundant. "Scorpions, yellow or black, on stony grounds are frequent, but if left in peace they do not attack. Locusts are still one of the plagues of these areas adjoining the big deserts."

FISHER has recently brought together (516, 517, 532) a very splendid bulletin on conservation of our forests wherein all the text is Bible references. Bible lands certainly can teach us a powerful lesson on the folly of non-conservation of the natural forests of a land! This bulletin is apparently based on an earlier publication (459) of the United States Department of Agriculture Forest Service. For the benefit of others who might wish to do similar things with this most laudatory theme, we give the Bible references here in the sequence in which he uses them: Genesis 1: 11-12 & 2: 9; Psalms 29: 9; Isaiah 60: 13; II Corinthians 3: 2; Psalms 93: 12-13, 104: 12 & 16-17; Isaiah 41: 19-20 & 44: 4; Psalms 40: 5 & 148: 7-8; Isaiah 14: 7-8 & 55: 12; I Chronicles 16: 33; Numbers 24: 5-6; Job 40: 21-22 & 14: 7-9; Revelation 7: 1-2 & 9: 4; Psalms 74: 7-8 & 83: 14; Isaiah 10: 18-19; James 3: 5-6; Joel 1: 18-20 & 2: 3; Exodus 21: 1 & 22: 6; Psalms 68: 11; Acts 8: 6; and Hebrews 8: 6.

The impression that passionflowers (*Passiflora*) are Biblical plants is very widely held, but is entirely erroneous. There are about 300 species of passionflowers, most of which are natives of America. There are a few in Asia and one in Madagascar. They are not referred to in the Bible. Their history in connection with Christian symbolism begins in 1610 when a Mexican Augustinian friar, EMANUEL DE VILLEGAS, brought a drawing of a passionflower to JACOMO BOSIO, who was preparing a monumental work on the Cross of Calvary (34, 475). The bud of the flower was taken to be symbolic of the Eucharist; the half-opened flower suggested the Star of the East that guided the Wise Men (463). The normally ten sepals and petals were regarded as representing the ten apostles present at the crucifixion (PETER and JUDAS being absent). The corona which is usually present as an outgrowth of the receptacle inwardly from the corolla, is taken to be symbolic of the crown of thorns placed on JESUS' head (or by some it is regarded as emblematic of the halo). The usually 5 stamens suggest the five wounds inflicted in JESUS' body on the cross. The three central styles, with their capitate stigmas, represent the three nails used to fasten him to the cross. In species with only 3 stamens these are regarded as representing the hammers used to drive in the nails. The long axillary coiling tendrils are thought to be symbolic of the cord-like strands of the scourges used to beat him. The often digitately lobed leaves represent the hands of the persecutors. In some species the corona is tinged with red, representing to some the bloody thorns of the crown, to others the bloody scourges. In some species there are 5 red spots on the corolla segments, suggesting blood from the 5 wounds. In some species there are said to be 72 filamentous divisions to the corona, suggesting the 72 thorns said by tradition to have pierced JESUS' brow. In some species the leaves are shaped like the head of a lance or pike, symbolic of the spear used to pierce his side. Some species have the lower surface of the leaves marked with round spots, suggesting the 30 pieces of silver for which he was betrayed.

As far back as 1680 HOHBERG (176) refers to *Passiflora* in his "Psalter Davids" and uses it in his illuminated borders.

Another plant often cited as a Bible plant, although entirely on assumption, is the so-called "sainfoin", "saintfoin", "esparcet","holy clover", or "holy hay", *Onobrychis viciaefolia* Scop. [also called *O. vulgaris* Hill, *O. sativa* Lam., and *Hedysarum onobrychis* L.]. According to legend it was this forage plant which formed the bed for the infant JESUS in the manger at Bethlehem. ROTHERY (507) claims that this plant is common in the region. She calls it "alfalfa or lucerne", but in error, for those names apply to *Medicago sativa* L., common in fields and on hills throughout the area (266). POST suggests that the sainfoin was "perhaps introduced from south of the Caucasus" (266).

HAUSMAN (580) calls attention to a legend that *Galium verum* L., known popularly as "our Lady's bedstraw", grew abundantly in the fields about Bethlehem and was mowed down with bracken, *Pteridium aquilinum* (L.) Kuhn, to make the "straw" used as cattle-bedding in the stables. "It was on this rather harsh yet fragrant bed that the Christ child lay. It is said that upon his birth, the withered hay burst into fresh bloom, the flowers of the bedstraw changing from white to gold. The painter NICHOLAS POUSSIN (1594-1665) is said to have painted a Nativity showing the bedstraw receiving its gilding from the celestial rays streaming from the Holy Child."

Another plant very commonly supposed to be Biblical in the eastern flowering-dogwood, *Benthamidia florida* (L.) Spach [also and more commonly known as *Cornus florida* L. or *Cynoxylon floridum* (L.) Raf.]. Although this plant is native only to eastern North America and was completely unknown in Bible days, it is widely believed by the general public to have been the tree of which the cross on Calvary was made. The legend is that the flowering-dogwood in those days was as large and strong a tree as the oak and for this reason its wood was used to make the cross. "To be thus used for such a cruel purpose greatly distressed the tree, and JESUS, nailed upon it, sensed this and in His gentle pity for all sorrow, said to it: 'Because of your regret and pity for my suffering, I make you this promise: Never again shall the dogwood tree grow large enough to be used for a cross. Henceforth it shall be slender, bent, and twisted, and its blossoms shall be in the form of a cross. And in the center of the outer edge of each petal there will be nail prints, brown with rust and stained with red, and in the center of the flower will be a crown of thorns, and all who see it will remember; and this tree shall not be mutilated or destroyed, but cherished and protected as a reminder of my agony upon the cross," (460). Of course, the "petals" in this story are actually bracts, while the actual flowers of the tree comprise the central "crown of thorns".

Still another plant associated with the life of JESUS in legendry is the christmas-rose, *Helleborus niger* L. According to tradition, a little shepherdess, seeing the Wise Men kneeling before the manger and offering their precious gifts to the infant JESUS, wept bitterly because she had no gift to offer. And lo, where every tear fell to the ground there blossomed up a white flower with a crown of gold. Such a flower had never bloomed before. It was the christmas-rose. Actually, *Helleborus niger* is native only to Europe, but there are related species in Asia Minor, *H. orientalis* Lam. and *H. vesicarius* Auch.

2. **Loranthus acaciae** (p. 23). — A recently issued news release tells of the paintings of Bible plants made by Miss WINIFRED WALKER, F.L.S. An exhibition of these paintings won her a Gold on Silver Grenfell medal from the Royal Horticultural Society in London in January, 1950. The collection illustrates 65 Bible plants and took Miss WALKER over 4 years to complete. Included are frankincense, myrrh, and spikenard. "At Westwood University, Los Angeles, she asked to see the specimen of the 'burning bush' mentioned by ISAIAH, and found that it was a crimson-flowered mistletoe, a parasitic plant bearing red berries. This explains why the bush appeared to be burning when the light of Heaven descended upon it." The reference here is doubtless to MOSES in Exodus 3: 2-4, not to ISAIAH.

3. **Acacia seyal** (p. 24). — The Midrash (592) uses the correctly translated term "acacia" for the wood employed in the building of the Tabernacle, in place of the untranslated term "shittim" used in most of the Christian versions.

4. **Acacia tortilis** (p. 24). — A synonymous binomial often employed for this species, under which illustrations of it have been published in widely read books, is *A. raddiana* Savi.

19. **Alhagi maurorum** (p. 31). — MEYERHOFF (538) points out that the great east Persian scholar and scientist, ABU'RRAYHAN MUHAMMAD IBN AHMAD AL BERUNI (born 973, died 1050), was apparently the first person to publish the observation that the "manna" (taranjubin) obtained from the leaves of *Alhagi maurorum* and its var. *camelorum* is actually produced by a tiny insect. As MEYERHOFF says, "manna is a saccharine exudation from a number of different plants belonging to various natural orders. In the present case it is the manna produced on the camel-thorn (*Alhagi*), called alhagi-manna and in Persian tar-angubin ('honey-dew'). Manna was considered, indeed, in the Orient as well as in the Occident, as a kind of sweet dew falling from the heaven on certain plants. It was only in 1822 that J. L. BURCKHARDT's observations in his 'Travels in Syria and the Holy Land' were published, in which he discusses the observation of Capt. FREDERICK from India, concerning three kinds of insects which the latter held responsible for the production of manna on certain plants. Then came the report by EHRENBERG and HEMPRICH (1829) about the *Coccus manniparus* Ehrenb., thought to provoke the exudation of manna on the Sinai tamarisks. This cochineal is now called *Gossyparia mannifera* (Hardw.) Signoret." He quotes IBN SARABIYUN, a Syrian Christian physician of the 9th century, as follows: "When the vapour rising from fruit trees and water and earth is refined by the action of the sun in the higher spheres and is cooked, it acquires a sweet and thick quality, and when it is congealed by the cold of the night it becomes thick, solid and heavy and descends upon the earth on to trees like dew; this is the honey-dew (in Arabic: 'asal at-tall') and that is manna."

MEYERHOFF states that in India about 14 plants are known to yield manna due to the action of parasitic insects upon them. "This," he says, "is regularly collected and, like honey, enters more largely than sugar into the pharmaceutical preparations of the Hindus. ... The question of the manna-production by the alhagi plant is still not settled, as some scholars (BURCKHARDT, FREDERICK) think that it must be caused by a manniparous insect, while others believe that the exudation is provoked by lesions of the plants." MEYERHOFF is of the opinion that AL BERUNI confused the spiny camel-thorn with spiny *Echinops globethistles*, and that the insect to which he refers was the beetle *Larinus maculatus* Fald., which lives on these plants and manufactures a sweet-tasting cocoon (called "trehala-manna") used in Persia as an expectorant to relieve obstinate coughs. He states that on the Sinai desert manna is produced by (1) the manna tamarisk (*Tamarix nilotica* var. *mannifera*, called in Arabic "tarfa") and is called "menn et-tarfa", (2) a species of wormwood, *Artemisia herba-alba* (in Arabic "shih"), and called "menn esh-shih", and (3) a saltwort, *Haloxylon articulatum* Biss. (in Arabic "rimt"), producing the "menn er-rimt". The first of these, however, is the only one now gathered commercially by the Bedouins who supply it to the monks of the Convent of Santa Catherine on Mount Sinai, who, in turn, use it in place of sugar. "The whole yearly quantity of this manna collected in the Sinai region does not, according to estimates by BURCKHARDT and WELSTED, exceed 600 pounds. It is collected from the shrubs and does not fall in sufficient quantity to cover the earth 'like rime', as described in the Bible. It is not probable that the Sinai desert has ever produced this manna in such quantity to enable a whole population to live on it."

21. **Tamarix mannifera** (p. 31). — TIMOTHY (515) was of the opinion that all the manna of the Bible came from the resinous exudations of species of *Tamarix* and most probably from *T. mannifera*. TEESDALE (513) suggested *T. gallica* L., a species native from western Europe to the Himalaya Mountains (34) and certainly not as likely as the source of Biblical manna as the Arabian *T. mannifera*. In all this discussion the reader should remember that we regard only the manna of Baruch 1: 10 as being derived from this source.

27. **Aloë succotrina** (p. 35). — DR. TREVOR tells us that the correct transliteration of the Hebrew words cited by us would be, according to the system followed by the authorities we are quoting, " 'ahalim" and " 'ahaloth" — the apostrophe coming before rather than after the "a", since it represents a consonant which precedes the vowel. "The 'oth' and 'ot' ending is highly unnecessary, since it is purely a matter of pronunciation, upon which the Jews differ. The 'oth' and 'im' endings do distinguish words, since one is masculine and the other is feminine, so you are probably right in including the two forms."

28. **Amygdalus communis** (p. 35). — Dr. TREVOR has written to us that "it is now fairly well established that the Exodus occurred at about 1290 B.C.", not two hundred years earlier as claimed by Archbishop USSHER. He states also that the discussion of AARON's "rod" is irrelevant because this story "was written down long after the Hebrews had entered Palestine" and therefore the budding almond would have been a natural and well-known occurrence to the writer and its presence in Sinai would have been taken for granted by him. "The Pentateuch is full of anachronisms of this type."

The Midrash (594) uses the term "almond" in the Jeremiah 1: 11 reference, just as does the King James version.

29. **Anastatica hierochuntica** (p. 38). — The Arabic names of "kaf maryam" (hand of Mary) and "kaf fatima" (hand of Fatima) are recorded for this species. Dr. TREVOR reports the species as common. In a letter to us, dated August 6, 1951, he points out that the correct transliteration of the word involved here is "galgal". He says that it is quite possible that some Jewish readers pronounce it "gulgal", but technically that word does not exist. There is a gulgeleth, but that does not refer to a plant at all. In fact anything round or circular could be meant by the word, and it is only by implication that a plant is suggested. Thus the difference in translation found among scholars. It is not certain whether the writer means a weed in the cases you cite, but parallelism strongly suggests it, and I agree that probably a tumbleweed is meant in these cases. In Psalms 78: 18 the same word must mean 'whirlwind'; while in Isaiah 5: 28, Jeremiah 47: 3, and perhaps in Ezekiel 10: 2, 6, & 13 the same word means 'wheel'". The spelling "gulgal" for the Hebrew word used in the Bible references is given by JOHN SMITH, distinguished ex-curator at the Royal Botanic Gardens, Kew, in his "Bible Plants", page 71 (1878).

32. **Anemone coronaria** (p. 41). — Löw states that the anemonies bloom in Palestine from about the middle of December to the middle of March. In regard to the recent discovery of *Lilium candidum* as a valid member of the indigenous flora of Bible lands (534) BOYKO has pointed out that N. NAFTOLSKY, a well-known Palestinian plant-collector, found plants of this species growing completely wild and apparently indigenous in the mountains of upper Galilee near the frontier between Palestine and Lebanon. Here the lily was growing on the walls of a deep pit of karstic origin, between shrubs of maqui. The details of this discovery were published by EIG (109). Following this discovery WARBURG (341) elaborated a comprehensive study on the origin and history of the madonna lily, based on all available botanical and historical evidence. In this he came to the conclusion that this lily, or one of its closely related varieties, originated in the area of Lebanon and upper Galilee.

Some years later TUVIA KUSHNIR found the lily on the slopes of Mount Carmel, but this discovery was not published. On May 15, 1945, BOYKO made the second discovery of wild specimens of *Lilium candidum* in this area. This was at a spot on Mount Carmel far away from any recent, and probably from any ancient, settlement. "In this setting three brilliantly white blooms of this majestic forest plant, on a 121 cm. long stem, shone forth among the limestone rocks and the shrubs of the moderately dense maqui." The specimens on Mount Carmel, on tall stems and in full bloom in mid-May, apparently were growing in a habitat quite favorable to the species. "By contrast, the specimens collected by NAFTOLSKY in the Upper Galilee were all very low, and there was no indication of the formation of inflorescences even at the end of June; out of a number of bulbs taken to the Hebrew University, only one could be induced to blossom after two years' careful tending. We may thus assume that the locality in which *L. candidum* was then found approaches the limit of the ecological amplitude of the plant."

39. **Artemisia judaica** (p. 48). — Dr. TREVOR writes: "I do not see why you include Hosea 10: 4 in section 38, when its Hebrew is *rôsh* and not *la 'anah* and would thus have the meaning 'gall' and belong with section 73, where you do not list it. The King James version translates this 'hemlock', but that was their mistake, and I think the passage belongs with your listing under 73. I am recommending for the Revised Standard Version that they use 'gall' in the text of Hosea 10: 4 with a footnote referring to Deuteronomy 29: 18 where the footnote says 'poisonous herb.' As an alternative I suggest that they might put in 'Or, *colocynth*' but I doubt they will want to be this specific. King James version uses 'hemlock' for *la 'anah* in Amos 6: 12, but 'wormwood' is used in American Standard Version and will probably be retained in Revised Standard Version."

We appreciate the cogency of Dr. TREVOR's argument on philologic grounds, but it still seems to us botanically more probable that the wormwood was the plant referred to as growing up in the furrows of the cultivated fields. The colocynth is a creeping and climbing vine and is not a typical erect-growing weed of cultivated fields.

Dr. TREVOR also points out that the Job 30: 4 reference which we include in this chapter should not be included here. The word "bushes" which we italicized in our text, he says, is represented by "siach" in the Hebrew and means "bush". The meaning of the passage is that the entire bush of saltwort is pulled up at once, and does not imply that there was an additional bush, such as wormwood, which was also taken at the same time. In this we now agree with him, and therefore the Job 30: 4 reference should be deleted from this chapter.

40. **Arundo donax** (p. 50). — The NELSONS (506) identify the reed of Isaiah 42: 3 as *Arundo tenax* Vahl, a related species common in thickets throughout Lebanon and Palestine (266), now usually called *A. plinia* Turra or *Ampelodesmos tenax* (Vahl) Link [also called *Donax tenax* Beauv., *Arundo ampelodesmos* Cyrillo, *A. mauritanica* Poir., and *Ampelodesmos mauritanicus* (Poir.) Dur. & Schinz].

44. **Atriplex halimus** (p. 53). — The American Standard Version uses "salt-wort" and "broom" in the Job 30: 4 reference. The expression "by the bushes" in this verse implies that the entire bush of saltwort was gathered; it does not imply that another kind of plant, or "bush", was gathered along with it, as some writers have supposed.

47. **Balanites aegyptiaca** (p. 55). — Dr. TREVOR, in a letter to us dated June 13, 1951, states his belief that the "balm" of Genesis 43: 11 was also derived from this plant. However, inasmuch as *Balanites* must have been a common plant already in Egypt at that time it seems hardly plausible that it would have been carried there as a gift. We therefore regard the use of this word in this passage as an error and are of the opinion that the "balm" of Genesis 43: 11 was the product of *Pistacia lentiscus*.

LINNAEUS, in his letter to the Royal Swedish Bible Commission (537), suggested *Amyris gileadensis* [now known as *Commiphora opobalsamum*, which see] as the balm of Gilead ["balsam i Gilead" in the Swedish text].

48. **Boswellia carterii** (p. 56). — Recent archeologic explorations in southern Arabia (586) indicate that the Queen of Sheba lived and ruled "in a virtual paradise in southern Arabia about 950 years before the birth of Christ." The excavations at Mareb, now a wild desert fortress of a Bedouin tribe, indicate that this city is built largely of stones from the ruins of previous cities at the same site. Ruins extend to at least 70 feet beneath the surface and some of these "may possibly go back to ABRAHAM." In this area is the once great Mareb dam, considered the finest ancient achievement of its kind, built in the 8th century B.C. This, it is claimed, "must have made the area a virtual paradise . . . . Arab writers . . . . attribute the downfall of southern Arabia to the destruction of the dam."

51. **Brassica nigra** (p. 59). — ZOHARY (579) says that *Salvadora persica* is a typical Sudano-Deccanian tree, now limited in Palestine to the lower Jordan valley where it is "a remnant saved from devastation by being an arboretum sanctum of the local Arabs."

53. **Buxus longifolia** (p. 62). — The NELSONS (506) agree with SHAW in referring the references cited at the head of our chapter to *B. sempervirens*, but probably only due to the fact that they were not aware of modern botanical opinion regarding the specific identity of the Palestinian plant.

60. **Ceratonia siliqua** (p. 72). — STOW and TRIMBLE (510) agree with HENSLOW that the "locusts" eaten by JOHN the Baptist were carob beans. They note that "for 2000 years, at least, the carob bean, still called 'locusts', has been a standard article of food in many Mediterranean countries." In answer to this the editor of "Newsweek" states "Whether ST. JOHN ate insects or fruit has been a matter of dispute for years. There is just as good a case for one as for the other. Eating the grasshopper-like locusts has been a standard practice in certain parts of the Near East and Africa."

61. **Cercis siliquastrum** (p. 73). — HOHBERG (176) as far back as 1680 gives *Cercis* as the plant on which JUDAS hanged himself. He refers to it as the "Judasbaum" or "arbor Judae".

62. **Cichorium endivia** (p. 74). — LINNAEUS, in his letter to the Royal Swedish Bible Commission (537), suggested *Centaurea calcitrapa* for the "bitter herbs" ["bitter salso" in the Swedish version] of Exodus 12: 8.

68. **Cinnamomum cassia** (p. 75). — The NELSONS (506) discuss this plant under the name of *Cassia lignea* on page 26 of their work. This is a binomial apparently overlooked by the editors of the "Index Kewensis".

71. **Cistus salvifolius** (p. 77). — HOWES (539) states that labdanum "is the nearest approximation to ambergris in the Vegetable Kingdom and is a valuable fixative in perfumery." He states that it is also a constituent of most artificial ambers, is used for scenting certain classes of tobacco and soaps, and is used in making fumigating pastilles. E. R. VAN LIEW, in a letter to us dated April 17, 1951, states that labdanum is employed today in perfumery as a binder, a fixative, and to give the perfume a residual character in the dried-out stage. "Percentagewise", however, he says, "even when compared to natural oils, including the florals, such as rose and jasmin, it is not used in what might be called an appreciable quantity. On the other hand when compared to the tonnage of synthetic aromatics, isolates and derivatives of such products as Oil of Citronella and Oil

of Lemongrass that is sold and used in perfumery, Labdanum itself is used in a proportion of what might be compared to a very small drop in a very sizable barrel."

73. **Citrullus colocynthis** (p. 78). — Dr. HOEHNE (478) agrees with the identification of the "wild vine" and "wild gourds" of II Kings 4: 39-40 as this species. SCHUMANN (508), however, maintains that the "vine of Sodom", popularly referred to as "apple of Sodom", was *Calotropis procera*.

74. **Citrullus vulgaris** (p. 80). — Dr. HOEHNE (478) believes that the "melons" of Numbers 11: 5 were a species of *Lagenaria*. He doubtless refers to *L. leucantha* (Duch.) Rusby, a native of Abyssinia, discussed by us under species no. 177 (p. 203) as a possibility for the "gourd" of Jonah. The Hebrew word used in the Numbers reference, however, is "avatiach", while that used in Jonah is "kikayon". It does not seem likely to us that the same plant is intended in these two references.

79. **Commiphora opobalsamum** (p. 84). — LINNAEUS (537) has suggested this plant as the "balm" of Gilead in Genesis 37: 25, Jeremiah 8: 22, and Jeremiah 46: 11.

89. **Cupressus sempervirens** var. **horizontalis** (p. 89). — Dr. TREVOR, in a letter to us dated June 13, 1951, states that in his opinion the "tirzah" in Isaiah 44: 14, rendered "cypress" in the King James Version, refers not to the cypress nor to the planetree, but to the holm oak, *Quercus ilex*.

91. **Cyperus papyrus** (p. 92). — TÄCKHOLM and DRAR (511) have pointed out that papyrus boats were frequently employed in the Egyptian Dynastic Period. The Isaiah 18: 2 reference indicates that as late as the 8th century B. C. the sea passage of about 100 miles from Pelusium to Gaza was negotiated by government officials in boats made of papyrus. Generally these papyrus boats were compact floats, not navicular and watertight as wooden ships have always been. "They consisted of three bundles of papyrus stems strung together and narrowly tied at the ends. These were broad enough to allow persons to stand on them. They were also so tight that they could be easily carried on the shoulder." These distinguished writers on the flora of Egypt also agree with us that the ark of "bulrushes" made by the mother of MOSES was constructed of papyrus, and that the word "papyrus" should have been used by the translators for the Hebrew word "gōme" wherever that appears.

98. **Elaeagnus angustifolia** (p. 97). — Dr. TREVOR, in a letter to us dated June 13, 1951, states that "etz shamen" is a pausal form only and should not be used in this connection. He states, further, that commentators probably based their use of "pine branches" on the Septuagint, rather than on the Authorized Version of Nehemiah 8: 15 as we have stated, because the Septuagint uses "juniper, pine, and cypress" wherever it translates the phrase at all.

ZOHARY (579) questions if *Elaeagnus angustifolia* is really native in Palestine. It is certainly native in Syria and probably also in northern Palestine. At present the Acre Plain marks its southernmost limit. In view of the destruction of the native flora all through the Middle East during the course of its long history and pre-history, it is always dangerous to assert that the absence of a species from a particular area there now, when it does occur relatively near-by, indicates that it was always absent.

104. **Ficus carica** (p. 103). — LINNAEUS, in his letter to the Royal Swedish Bible Commission (537), makes the extremely interesting suggestion that the leaves of the plantain, *Musa paradisiaca* L., were the leaves used to make the aprons in Genesis 3: 7 rather than the much smaller (and possibly inadequate?) fig leaves ["fikone löf" in Swedish].

Dr. TREVOR writes us that "no scholar would put David as early as 1060 B. C." However, as has been stated by us in our Preface, we are following the Archbishop USSHER chronology, realizing full well its many defects, since it is still rather widely used by lay readers of the Bible, and this is the date assigned to the event in that chronology. The "Twentieth Century Encylopaedia and Dictionary", published by the Scientific Research Association in 1907, gives DAVID's birth date as 1085 B.C. and death date as 1015.

Dr. TREVOR also states that "pag" in our text should be "paggāh", and that "pageha" is an entirely incorrect rendering of the form with the feminine suffix, used only in Song 2: 13 and not "elsewhere". He says that "bikkûrāh" is the correct form — that "there is no such form as 'bi'kurah'". Our reply is that these various forms of transliteration occur in the literature on Biblical botany cited in our bibliography and are included in our text for the same reason that various vernacular names and various scientific names for the same plant, with all variant spellings of both, are listed.

105. **Ficus sycomorus** (p. 106). — Dr. TREVOR, in a letter to us dated June 13, 1951, has pointed out that the "California sycamore" — more accurately called the California

planetree — *Platanus racemosa* Nutt., has been widely confused with the "sycomore" of the Bible, and, of course, has no more connection with the Biblical plant than the eastern North American or Old World species do.

ZOHARY (579) has come to the conclusion that *Ficus sycomorus* is not indigenous to Palestine, but occurs there only as a remnant "of abandoned culture" even when found far from present habitations. Seedlings of the tree are never found in Palestine, indicating that it is sterile there, as it is also in Egypt.

111. **Hyacinthus orientalis** (p. 114). — LINNAEUS, in his letter to the Royal Swedish Bible Commission (537), suggested *Pancratium illyricum* Forsk. (now called *P. aegyptiacum* Roem.) as the "lily" of the Song ["Salomons liljor" in Swedish]. This species, according to POST (267) inhabits only the desert sands near the coast. Our opinion remains unchanged that the hyacinth is far more likely as the plant intended.

119. **Scirpus maritimus** (p. 120). — A report recently published by the National Geographic Society in Washington (462) indicates findings which purport to show that MOSES led the Exodus not across the Red Sea, as commonly supposed, but across the Reed Sea, ten miles south of where Port Said now stands. This is a brackish body of water taking its name from the abundance of reeds on its shores.

124. **Lecanora esculenta** (p. 125). — TEESDALE (512) gives a popular discussion of the manna produced by the lichens of the genus *Lecanora*. In addition to the three species which we name in this work, he also lists *L. tartarea* as edible. He seems to believe that the manna of Numbers 11: 6-9 came chiefly from *L. esculenta* and *L. affinis*. In this connection the reader should see also the excellent work by KOLB (480).

A quotation from MEYERHOFF (538) is well worth repeating here: "There is another desert plant, not producing a sweet manna, but growing so rapidly that it allows, in certain conditions, entire tribes to live on it. This is a lichen, *Lecanora* or *Sphaerothallia esculenta* Nees, which has been called by REICHARDT and HAUSSKNECHT 'Mannaflechte' (manna lichen). Growing on stones and rocks in desert regions, it becomes detached from its base by heat and drought, and is rolled into small balls of the size of peas or hazelnuts which are carried by the wind over long distances, sometimes forming high heaps and sometimes raining from the sky. It contains no sugar, but is largely composed of oxylate of lime and a lichen jelly; this latter may be the substance which gives it value as food for nomadic tribes in times of dearth. Kurdic nomads call it 'bread from heaven' or 'bread from the earth', and the Persians call it shir-zad ('milk-producer') because they think it to be a galactagogue for women in confinement. The Persian name of the plant is gawz-gundum ('wheat-nut') on account of its form and taste. This lichen is found over a wide area, stretching from the Iranian deserts and the Kirghiz steppes through South Russia and Anatolia to Northern Syria. Then, after a lacuna through the Sinai Desert, Egypt and Eastern North Africa, the manna lichen is found again in Algeria and Morocco at the southern slopes of the Atlas Mountains. Everywhere the nomadic tribes know it well and occasionally make use of it, grinding the lumps to a meal for baking. We remind the reader that MAHMUD AS SUQABADI has mentioned that some old authors have compared the biblical manna with bread. Though this manna lichen is now absent from the Sinai Peninsula it is not impossible that it existed there 3,000 years ago. Therefore REICHARDT, HAUSSKNECHT and KAISER believe that, if one is disinclined to consider the biblical report as a mere legend, the only vegetable matter which could have been responsible for the subsistence of the Jewish people in the desert can hardly have been anything else but the manna lichen. Since, however, this lichen grows quickly after rains but then disappears for a considerable period, continuous subsistence on it 'during forty years' is out of the question."

126. **Nostoc** (p. 125). — JOHN SMITH reports (299) that these gelatinous algae are known as "star-jelly" in England. Growths such as described in the Bible have been observed in various parts of the world; the British species acting thus is *N. commune* Vaucher and is often found on walks and paths. *N. collinum* Kütz. in 1855 covered an area of several square miles in the Presidency of Bombay. It is called "meat" by the Scindians, who firmly believe that it falls from heaven. It is apparently well known in western Pakistan. In China *N. edule* Berk. & Mont. is dried and forms an ingredient of soups.

130. **Lolium temulentum** (p. 133). — KELLEY (528) reports that the poisonous nature of darnel seeds being due to the presence of fungal infection was first discovered by VOGL in 1897. He says that "the constant presence of hyphae in the hyaline layer of the caryopsis and typical stages of digestion, are curious phenomena that have nevertheless been established by repeated researches, especially at the hands of McLENNAN in Australia. The endophyte also occurs, according to NEILL" (529) "in the leaves of *Lolium* but not in those of other pasture grasses." LINDAU in 1904 reported that not only does this fungus

exist in present-day darnel, but it has been found in grains recovered from an Egyptian tomb about 4000 years old.

ASCHERSON (552) has suggested the Syrian scabious, *Cephalaria syriaca*, for the "tares" of Matthew 13: 24-30, and POST also claims (266) that this is "one of the tares or darnel spoken of in the N.T." It is a very injurious weed in grainfields. Its minute black seeds must be picked out by hand from the threshed grain or else the bread made therefrom becomes unwholesome. Because the vegetative aspect of the scabious is so very different from that of grain plants it does not fit the context of the story any better than true tares do. *Lolium* is the only suggested plant which fits the context perfectly.

132. **Mandragora officinarum** (p. 137). — The NELSONS agree (506) in the identification of the mandrakes of Genesis 30: 14-16, although they use for the plant the synonymous designation "*Atropa mandregora*" on page 63 of their work.

As recently as July 8, 1951, one of our leading newspapers carried a story by R. R. THOMASSON, entitled "Beauty survives mandrake's legends" (554) in which it is asserted that the mandrakes of the Bible were the North American *Podophyllum peltatum!* This plant was unknown in Bible times.

Recently there has come to our attention a theory that the word "dudaim" in the Bible refers not to the mandrake, but to truffles (*Tuber* spp.) FERRY DE LA BELLONE, for instance, says (543) "Je ne m'occuperai donc pas des *Dudaims* de la Bible que Lia paya si cher à Rachel et qui feraient remonter à 1620 et avant JESUS-CHRIST la gourmandise de l'homme pour la Truffe, si les Dudaims sont des Truffes . . ." It has been claimed that the Hebrew word in its etymology indicates that it applies to something produced underground. Mr. BENJAMIN D'ARLON, head of the French department at Evander Childs High School, New York City, who is well acquainted with truffles in France states categorically that he has never heard the name "dudaim" applied to them. ALEXANDER MARX, Director of Libraries at the Jewish Theological Seminary of America, in a letter to us dated June 8, 1951, states that "there does not seem to be any etymology on the 'dudaim'. In the latest most scholarly Hebrew Lexicon which is being printed in The Hague now there is no etymological note added to it, nor is there one found in the treatment of the plant in LOEW's Flora der Juden, Vol. III."

JOSHUA BLOCH, Chief of the Jewish Division of the New York Public Library, in a letter to us dated May 24, 1951, states that the standard Hebrew dictionary of GESENIUS (A Hebrew and English lexicon of the Old Testament, ed. *Brown, Driver,* and *Briggs*) derives "dudaim" from the Hebrew root "dud", meaning "to fondle, to love". The last German edition (16th ed. by F. BUHL) says that it is "probably connected with Dod (love) in popular etymology". CHEYNE, in the Encyclopaedia Biblica, volume 3 (1913), under "Mandrakes", says that the Hebrew name, "dudaim", was no doubt popularly associated with "dodim" (love), "but its real etymology is obscure." S. R. DRIVER, in his "The Book of Genesis", says in commenting on Genesis 30: 14 "The Hebrew name is akin to the Hebrew word for (sexual) love" used in Ezekiel 16: 8. A. DILLMAN's "Commentary on Genesis" says in speaking of Genesis 30: 14 that the word "is to be rendered *amatoria*, love-apples, from . . . . Dudai". As to mandrakes as an aphrodisiac and as possessing magic qualities, BLOCH points out that there are many studies. CROWFOOT and BALDENSPERGER (95a) on pages 114-119 discuss dudaism. A particularly valuable treatment is given by LOEW (217) under "Alraune" and "Mandragora", especially in volume 3, pages 363-368, where the legendary lore in Greek, Jewish, and Arabic literature and popular beliefs are fully discussed with a fine bibliography appended.

134. **Morus nigra** (p. 140). — Dr. TREVOR has informed us that the transliteration "shikmah" which is given for the word appearing in some Hebrew translations of the New Testament for the "sycamine" of Luke 17: 6, is an error of translation, as this word is merely a form of the word for the sycomore fig, *Ficus sycomorus* (which see).

136. **Mycobacterium leprae** (p. 142). — The Midrash (594) identifies not only the common form of leprosy and of gonorrhea, but states that there were actually twenty-two different kinds of "leprosy" mentioned in Holy Writ!

151. **Origanum maru** (p. 160). — It is of considerable interest to note that LINNAEUS in his letter to the Royal Swedish Bible Commission (537) suggested a moss, *Bryum truncatulum* L., for the "hyssop" ["isop" in Swedish versions] not only of Exodus, Leviticus, and Psalms, but also of John 19: 29.

The Midrash (593) states that at the Feast of the Tabernacles MOSES exhorted the people to take not only the fruit of the "goodly trees" and the branches of 4 or 5 other trees, but also a bunch of hyssop. The latter seems to have been something quite scarce, because the price that had to be paid for a bunch of it by those who could not collect their own was "four maneh". The same, however, is true of the "palm branches", which were likewise quite expensive. The Midrash in another place (598) gives a valuation for the hyssop used at the time of the first Passover through the interpretation of Rabbi JUDAH, who quotes MOSES as follows: "If you want to be redeemed, you can obtain your redemption with a very simple thing; *viz.* 'and ye shall take a bunch of hyssop and dip'. They said to him, 'Our Master MOSES, how much is this bundle of hyssop worth? Four farthings or five!' He replied to them: "Were it worth only one, it will [would] enable you to acquire the spoil of Egypt and the spoil of the Red Sea." This seems to indicate plainly, again, that the "hyssop" was something that was not very common and had to be purchased.

153. Ornithogalum umbellatum (p. 162). — LINNAEUS, in his letter to the Royal Swedish Bible Commission (537), suggested *O. luteum* L. for the "dove's dung" ["dufwoträck" in the Swedish version] of II Kings 6: 25. This plant is now generally known as *Gagea arvensis* (Pers.) Dum. We do not concur in this determination.

154. Paliurus spina-christi (p. 165). — GEORGE HENSLOW in the Gardeners Chronicle, series 3, 43: 53 (1908) says "The Christ-thorn was more probably *Paliurus*, as it is a very common shrub, with very flexible and spiny branches, and therefore suitable for twisting into a crown." HEGI (604) quotes WARBURG as the authority for the statement that *P. spina-christi* is extremely plentiful ["massenhaft wächst"] in the immediate vicinity of Jerusalem, especially in the valley of Kidron. ZOHARY, however (579), states that the species is now very rare in Palestine, confined to lowlands, despressions, and otherwise moist protected places. He disagrees with HEGI and maintains that it is not a Mediterranean plant, but is "either a relic of a northern element which once occurred in certain parts of the Mediterranean and has since retreated from them leaving some remnants behind, or is a recent immigrant which has infiltrated into secondary habitats after the local vegetation has been destroyed". He says "As seen from its distribution in Palestine, it does not occur at all in the Judean mountains and therefore the discussion of HEGI . . . . . about the 'thorn of Christ' as referring to this plant is groundless. *P. spina-christi* is an intricate shrub never used here for hedges or any other purpose."

ZOHARY also states — and this is of special interest in view of the statements made above — that *Zizyphus spina-christi* is now widespread in Palestine, but is confined in general to alluvial plains only. He says that "until recently it has been limited to the Lower Jordan Valley. But with the devastation of vegetation in the plains and valleys. . . . *Z. spina-christi* conquered new areas in the Coastal Plain, in the Upper Jordan Valley and in the adjacent plains." He has seen a tree of it over 1000 years old.

LINNAEUS, in his letter to the Royal Swedish Bible Commission (537), lent his authority to the theory that it was *Zizyphus spina-christi* out of which the crown of thorns ["Christi törnekrona" in Swedish] was made. SCHWERIN (553) also maintains this view. He states, as a matter of more than passing interest, that when he visited the Garden of Gethsemane before the First World War he found growing there a splendid tree of *Gleditsia triacanthos* L. about 2m. tall, which was pointed out to him as the species from which the crown of thorns had been made. When he pointed out to the monk in charge that this was an American species unknown at the time of JESUS which could not possibly have been used for that purpose, the monk replied that two seedlings of this species had been brought over to him from America in flower pots by two extremely nice old lady pilgrims who asserted that this was the plant used to make the crown of thorns, and that in view of the devotion which caused them to make this arduous trip to bring him this offering he would be far more inclined to believe them than he would the unsupported word of a mere botanist who had given no evidence of religious zeal! This story is quite typical of the confused thinking which surrounds the subject of Biblical botany in the minds of a considerable part of the general public even today.

SCHWERIN also gives a very interesting resumé of some of the ideas current in Europe as to the wood from which the cross on which JESUS was crucified was made. Pieces of the "original cross" are said to be extant and from these small sections have been taken and set into other wood to make small crosses which are found on the altars of thousands of Catholic churches all over the world. It would be of great interest if an anatomic study could some day be made of some of this wood, for its identity could probably then be established. It is, however, improbable that this has ever or will ever be done. In parts of Germany it is believed that the pear, *Pyrus communis* L., was the tree that furnished the wood for the cross. The legend states that the wood took root and produced blood-red flowers after the death of JESUS; and that the fruits it produced were blood-red, and even the leaves were red-veined. In Pomerania it is believed that a pine (*Pinus*) native to that part of central Europe was the tree involved and that since this event the pine has always produced its branches in cross-like whorls. In Poland it is believed that an aspen (*Populus tremula* L.) was the tree of the cross and that its leaves have trembled in fear of God's wrath ever since. It is further believed that this tree will afford protection against lightning while the related poplars will attract lightning. In England there is a legend that the mistletoe (*Viscum album* L.) was a tree originally and its wood was used to manufacture the cross; after which it was condemned to be an humble and despised parasite until the end of time. In northern Italy it is believed that the *Clematis* native there was originally a tree and furnished the wood of the cross, after which it was condemned to be an humble low-growing vine hiding its face near the ground far beneath the crowns of ordinary trees. In other parts of Europe it is believed that the alder (*Alnus*) was the tree involved — its wood even now "turns red" when it is cut. In Greece *Quercus ilex* is the local candidate for this distinction.

157. **Phoenix dactylifera** (p. 169). — ASCH takes the date palm to be symbolic (463) of Islam or Mohammedanism ("the religion of beauty and temperance"), as he takes the passionflower as symbolic of Christianity, the etrog or citron (*Citrus medica*) as symbolic of Judaism (the "mother of religions"), and the sacred bo tree (*Ficus religiosa* L.) as symbolic of Hinduism ("the world's oldest religion") and Buddhism ("the religion of the middle way"). He points out that the date is the sacred tree of the Arabs and "has been revered by all peoples since earliest antiquity. In all ancient classical writings the date palm was compared to the sun, and even today it is said that it grows best when its roots are in water and its head is in the fiery heat of the desert sun. . . . An ancient legend relates that soon after Creation a magnificent bird of enormous dimensions was on its way from earth to the sun. It was flying too fast and lost one of its feathers. This feather, after a long voyage downward, eventually landed on the ground. Finding there a fertile soil it formed roots and developed into a magnificent palm, whose fronds resemble the feathers of the large mythical bird. In Islamic tradition it was MOHAMMED who created the palm, causing it to spring from the earth at his command."

The Midrash (593) states that "palm branches" were quite expensive when they had to be purchased for the Feast of the Tabernacles by those of the people who could not collect their own.

LINNAEUS, in his letter to the Royal Swedish Bible Commission (537), suggested *Nerium oleander* for the "tree" ["arbor justi" according to LINNAEUS] of Psalms 1: 3.

158. **Phragmites communis** (p. 172). — POST refers (266) to the plant with hard, dark-brown stems, of which reed pens so much used in the Orient are made, as *Arundo scriptoria* L., and states that it has been reported from the Holy Land area. Mr. JASON R. SWALLEN, Curator of the Division of Grasses, United States National Museum, in a letter to us dated November 3, 1950, states "Apparently there has been a great confusion about the identity of the plant from which 'reed-pens' were made in the Orient in the early days. The best information I have on this is that given by TÄCKHOLM in the 'Flora of Egypt' (1: 213. 1941). In this work it is given as *Phragmites communis* var. *isiacus* (Del.) Coss. & Dur. For your purpose I think it would be better to use *Phragmites communis* Trin., as the varietal name is invalid, and there is some question whether this form is worthy of varietal rank. C. E. HUBBARD, in the 'Flora of Tropical Africa', does not recognize any varieties. *Arundo scriptoria* L., as originally published, is a nomen nudum, and I think it is safe to assume that it is a synonym of *Phragmites communis*." The "Index Kewensis", however, reduces it to *Arundo donax* L.!

159. **Pinus brutia** (p. 173). — The Midrash identifies the "thick trees" of Leviticus 23: 40 as myrtles. This does not seem correct because in the Nehemiah passage "thick trees" and myrtle trees are both listed in a way that indicates plainly that they are distinct from each other. Furthermore, the myrtle is normally only a straggling bush or shrub 3 to 10 feet tall (34), and even when it grows into the form of a tree is not noted for its massive trunk. The terebinth, olive, elm, planetree, and some of the oaks, on the other hand, are noted for their huge gnarled trunks; yet these trees either did not occur in the region or are referred to by other words in the Bible. It still seems most plausible to us that the Brutian pine was the plant referred to as "thick trees".

160. **Pinus halepensis** (p. 175). — The following Bible references should be added:

I KINGS 6: 15—. . . . and he covered them on the inside with wood, and covered the floor of the house with planks of *fir*.

I KINGS 9: 11—. . . . Now HIRAM the king of Tyre had furnished SOLOMON with cedar trees and *fir* trees. . . .

II CHRONICLES 3: 5—And the greater house he cieled with *fir tree*. . . .

Dr. TREVOR, in a letter to us dated June 1, 1951, states it was TRISTRAM who suggested that there was an Arabic word "aran" for a kind of ash tree found near Petra. He says also that an Assyriologist has recently stated that a word with a similar root origin is found in cuneiform meaning "cedar". Dr. TREVOR is of the opinion that the reason MOFFATT and GOODSPEED omit in their translations the reference to "fir wood" in II Samuel 6: 5 is that they have adopted the parallel reading in I Chronicles 13: 8 where the similarity of text indicates that the interpolation of "fir wood" in the II Samuel passage is a textual error.

He says that the Hebrew word "ôran" in Isaiah 44: 14 is rendered πιγυς in the two Greek versions in which the word is used (not πιγυς as stated in our text). In all other Greek versions κύριος is used alone. He states, further, that the GOODSPEED version (the book of Isaiah having been done by A. R. GORDON) follows the Hebrew for only the first half of this verse and then shifts to the Septuagint for the rest, and therefore does not "play safe with 'some other tree of the forest'" as we have stated, but rather emends the text with the Septuagint.

It is worthy of note that the JASTROW version uses "cedar" instead of "fir tree" in Isaiah 41: 19 and "cypress" in Isaiah 60: 13.

161. **Pistacia lentiscus** (p. 177). — Dr. TREVOR, in a letter to us dated June 13, 1951, states that in his opinion the "balm" of Genesis 43: 11 was the same as that of Genesis 37: 25, Jeremiah 8: 22, 46: 11, and 51: 8, which we discuss under *Balanites aegyptiaca*. Because the latter was without doubt well known in Egypt at this time, we cannot concur in his view.

162. **Pistacia terebinthus** var. **palaestina** (p. 178). — The Midrash (592) lends its support to the theory that the "oaks" of Judges 6: 11 were really terebinths.

164. **Platanus orientalis** (p. 180). — The following Biblical reference should be added:
ISAIAH 44: 14—He heweth him down cedars, and taketh the *cypress* and the oak. . . .

There has been considerable argument concerning the identity of the tree rendered "cypress" in this passage by the King James version. We follow MOFFATT and GOODSPEED in regarding it as the planetree. Dr. TREVOR, in a letter to us dated June 13, 1951, states his opinion that the "cypress" in this passage ("tirzāh" in the original) is *Quercus ilex*, the holm oak, and that the "oak" ("'allôn" in the Hebrew) which follows is *Quercus aegilops*, the Valonea oak. He says "I suppose *Quercus coccifera* could be understood here, but the picture of the forest here seems to me to argue against it, since these oaks are more often in groves by themselves. . . . I noted that you follow the Chicago version (GOODSPEED) in emending the word *tirzāh* to *tidhār*, first suggested by G. A. SMITH, and thus reading 'plane-tree'. I do not agree with this, for reasons too complicated to state here." He points out that the Septuagint uses "'elatē"— 'ελατή — for which the dictionary translation is "silver fir or pine" in Ezekiel 31: 8.

165. **Populus alba** (p. 181). — The Midrash in certain of its explanations, as was characteristic of the times when it was composed, confuses the processes of grafting and pollination in respect to their effects or supposed effects on the production and characters of fruit in the date (593) and fig (591). It likewise holds to the idea of prenatal influences on the physical appearance and characteristics of offspring, as is evidenced by the story told by Rabbi AKIBA of the white child born to the black king of Arabia and his black wife, and the story of JACOB's speckled flock.

167. **Prunus armeniaca** (p. 184). — In support of our contention that the mis-identification of the fruit of the "tree of knowledge" in the Garden of Eden as the apple was popularized, if not originated, by the Renaissance artists of western Europe, we have a statement from MARGARETTA M. SALINGER, Research Fellow in the Department of Paintings at the Metropolitan Museum of Art in New York City. This authority points out that the temptation of ADAM has been depicted by many artists. The fruit pictured is an apple in the painting of "Adam and Eve" by RUBENS (1577-1640) in the Mauritshuis at The Hague, in "Adam and Eve" by TITIAN (1477?-1576) in the Prado at Madrid, in "Adam and Eve" by TINTORETTO (1518-1594) in the Academia at Venice, and in "Temptation" by HUGO VAN DER GOES (active from 1467, died 1482) in the Museum at Vienna. An Italian painting at Duveen's, New York, ascribed to PIERO DI COSIMO (1462-1521?) shows ADAM and EVE with the forbidden fruit pictured as a fig. RAPHAEL (1483-1520), too, in the Biblical scenes in the Loggi of the Vatican at Rome shows them with a fig. She goes on to say "One of the most famous representations of ADAM and EVE is that by the brothers VAN EYCK, in the wings of the great altarpiece of the Mystic Lamb, in the church of St. Bavon in Ghent. Here EVE is shown holding in her hand an undetermined fruit, which bears some resemblance to a partially peeled lemon [more probably the etrog or citron — H.N.M.]. Another very famous and problematic statement of the theme is ALBRECHT DÜRER's 1504 engraving of the "Fall of Man". Here the fruit seems to be an apple but the leaves of the tree on which it grew are surely not apple leaves." Mrs. MARQUAND (531) thinks that the "tree of life" of DÜRER was a mountain-ash (*Sorbus*).

In this connection it must be pointed out that many Renaissance painters depicted Bible characters in settings of Italian, French, Spanish, Portuguese, Belgian, German, or Dutch gardens. It is generally recognized by students of art that such use of European plant material in these paintings was not intended to imply that the artist believed these plants to have grown in Bible lands in Bible times. Many of the plants used in these paintings were inserted with a very definite symbolic meaning, often actually prescribed by the Church and required to be placed in every painting of that particular scene to be used in chapels, on altars, or for frescoed walls of religious edifices. The European rose, for instance, was to be inserted as symbolic of love, the lily of sinlessness and virginity, the tomato as poison, the gourd as symbolic of salvation, the olive of peace, etc. The rod sprouting lilies, often seen in JOSEPH's hand, was to symbolize marriage and legal but not physical paternity; the garland of peaches, cucumbers, and pears on the Virgin MARY's throne in many paintings was to symbolize good works (565).

HOHBERG (176) in 1680 implied that *"Malus assyrica"* was the fruit of the tree of knowledge. He calls this species the "Adamsöpffel." This binomial is not listed in the "Index Kewensis" and obviously does not represent an apple. It seems to us most probable that it is the artist's attempt to depict *Ficus sycomorus*.

The studies in the Midrash (591) in reference to the "ethrog" of Genesis 3: 6 indicate the belief of its authors that the citron (*Citrus medica*) was probably the tree which attracted EVE's eye and tongue, but this same Old Testament commentary also indicates that the fig (*Ficus carica*) was the choice of Rabbi JOSÉ and that the grape (*Vitis vinifera*) was the choice of Rabbi AIBU. Rabbi LEVI sidestepped the issue with the statement "The Holy One, blessed be He, did not and will not reveal to man what the tree was." The Midrash adds the complicating statement that the rest of the tree was edible, as well as the fruit! We do not know of any Holy Land fruit tree whose wood is also edible! The original idea might have been that the wood, leaves, roots, etc., merely conveyed some of the same flavor (acid, in the case of the citron) as the fruit. In fact, the Midrash in its original Hebrew uses a word meaning "taste"; the English translation substitutes the word "eat".

168. **Pterocarpus santalinus** (p. 188). — BRAGA (472) has discussed the plants and animals said to have been introduced from Ophir. He reaches no definite conclusion as to the identity of the "almug" trees.

173. **Quercus ilex** (p. 193). — Dr. TREVOR has pointed out to us that Dr. WILLIAM REED has recently published a treatment on the "Asherah" which is widely accepted by scholars today. In this he confirms our conclusion that this word does not refer to a tree at all, but to a goddess image. The King James Version has long been considered in error in its use of "grove" for this word.

175. **Retama raetam** (p. 201). — The NELSONS (506) correctly identify the plant in I Kings, although they use for it the synonymous designation *Genista monosperma*. The recent BENZIGER translation of the Psalms into Latin (457) reads for "coals of juniper" *"carbones genistarum."* The English version of this translation says simply "burning coals". Our good friend, DON BENTO PICKEL, of São Paulo, Brazil, has suggested whin or gorse because of the dictionary definitions of *Genista* available to him. The name whin, furze, and gorse, however, apply to the genus *Ulex* (not to *Genista*), more especially to the European species, *U. europaeus* L., not known from Old Testament lands. LINNAEUS, in his letter to the Royal Swedish Bible Commission (537), suggested a species of *Chamaerops*, undoubtedly *C. humilis* L., the dwarf fan-palm, for the "juniper" ["enträd" in the Swedish version] of I Kings 19: 4-5 and Psalms 120: 4 and also for that of Job 30: 4. This stemless palm is a native of the Mediterranean region (34), not recently seen in Bible lands (267).

176. **Rhamnus palaestina** (p. 202). — The following Biblical reference should be added:
PSALMS 118: 12—They compassed me about like bees; they are quenched as the fire of *thorns*. . . .

The Douay version of this verse is "they burned like fire among *thorns"*; GOODSPEED says "though they burn like a fire of *thorns"*. The Jastrow version is identical with that of the King James. MOFFATT says "they blazed like a fire among thorns."

Dr. OPPENHEIMER has called our attention to an article by FEINBRUN in the "Palestine Journal of Botany, Jerusalem Series", in which it is pointed out that *R. palaestina* is actually deciduous, while it is *R. punctata* Boiss., previously overlooked, which is evergreen.

The Midrash commentary (591) adds some additional thought to the problem of the many prickly plants of the Bible. The "thorns" (boz) of Genesis 3: 18 are identified as artichokes. This translation is based on the expression "shall it bring forth to thee" in this verse, implying, it is claimed, some edible growths. The "thistles" (dardar) of this verse are rendered "cardoon". This study also admits that these identifications are often interchanged. We are not convinced that either of these words referred to plants that were really important or occurred at all in the food economy of the people of Bible times, but believe, rather, that the references are to plants like *Rhamnus palaestina* and *Centaurea calcitrapa* (which see).

The cardoon is *Cynara cardunculus* L., a spiny, robust, and large-leaved plant of southern Europe which resembles the artichoke and which is cultivated for its edible root and thickened young inner leaf-stalks, which are rendered crisp, white, and tender by blanching. Its flowers are used to coagulate milk. POST states that it is not native to

Palestine. The true artichoke (*C. scolymus* L.) is a less stout plant which is probably a derivative of the cardoon.

180. **Rubus sanctus** (p. 206). — Dr. OPPENHEIMER, in a letter to us, questions whether this species is truly evergreen. He thinks that the plant has some deciduous period during the winter months.

188. **Saccharomyces ellipsoideus** (p. 212). — There has been considerable argument among writers on the wines of the Bible as to whether they were fermented or nonfermented (566, 567, 568). Löw describes (217) an "Alcohol-free Bible" or "Dry Bible" published by the American Prohibitionists, in which the word "wine" is replaced by "raisin-cakes", "grapejuice", or "juice" in all places where it is used in a non-condemnatory sense and from which the entire story of the wedding at Cana is omitted. He states that this Bible was produced by Prof. CHARLES FOSTER KENT, professor of Biblical Literature at Yale University, and Prof. C. T. TORREY, also of Yale. The "flagon of wine" of II Sam. 6: 19 is replaced by raisin-cakes ("Rosinenkuchen"). The Midrash (594) states that the "mingled wine" of the Old Testament consisted of one part wine and two parts of water.

197. **Salix safsaf** (p. 216). — BALL has pointed out (464) that "of the six references to willows in our Old Testament Scriptures, four make them the symbols of fertility, vigor, and rejoicing; only one associates them with mourning (Psalms 137: 1-2)". As has been pointed out by us, however, under *Populus euphratica*, the "willows" of the Psalms reference are now usually regarded as aspens.

The NELSONS (506) fell into the common error of referring the Biblical willow references to the Chinese *S. babylonica*.

215. **Triticum aestivum** (p. 228). — According to FOLKARD (475), pages 16 and 456, there is a well-known tradition among the Arabs, now taken over by Christians, that when ADAM and EVE were driven out of the Garden of Eden ADAM took with him only three things: "an ear of Wheat, which is the chief of all kinds of food; Dates, which are the chief of fruits; and the Myrtle, which is the chief of sweet-scented flowers."

It is interesting to note that the Midrash (593) uses the term "meal-offering" in place of the very misleading "meat-offering" of the King James version in the Leviticus, Jeremiah, and other references where this term occurs. This is in accord with the more modern Bible translations.

219. **Tulipa sharonensis** (p. 234). — Dr. URIAH FELDMAN has recently suggested (in a letter to Mr. HAROLD WESLER of Paterson, N. J.) that "the true Rose of Sharon is *Pancratium maritimum* L." However, the Hebrew word used for this plant in Song 2: 1 is the same word used in the book of Ecclesiasticus, chapter 50, verse 8, where it is definitely stated that the "rose" referred to blooms in the spring of the year. *Pancratium*, according to POST, blooms in September and October; *Tulipa* blooms in March.

A writer in "Hatteva' Vehaaretz (Nature and Country)" for 1932, page 338, describes *Pancratium maritimum* as starting to bloom on the plain of Sharon in the month of August. "Out of a large bulb hidden in the sand peers forth a fleshy stem that carries at its head a bunch of flowers. The flowers are from 10 to 15 centimeters long, are white, and give forth a sharp odor. They are numbered among the largest and most beautiful flowers of the land. Afternoon, when the flowers are open on their tall stems, they live to their greatest strength on the desolate sandy hills along the length of the shore. . . . . During August and September these lovely flowers are sold in the streets of Tel-Aviv. At this time there are no leaves on the plant. They appear only at the beginning of the fall. . . . . This lovely plant is gradually disappearing from the coastal plain with the development of building and landscaping. The Germans are also removing bulbs to Germany as the rose-of-Sharon that is mentioned in the Song of Songs, and from year to year it is becoming more and more scarce."

The pictorial imagery of the Ecclesiasticus (or Wisdom of BEN SIRA, as the book is often called) verses is exquisite and full of beautiful floral forms, and many listeners and readers over the centuries have conjured up many different flowers to fit into these passages. The most widely known translations of this book are from an earlier Greek one, and it was on the basis of these and statements made in the literature cited in our bibliography that we decided that the "flower of roses in the spring of the year" was either *Tulipa montana* or *T. sharonensis*.

In 1896 the late Prof. SOLOMON SCHECHTER discovered in a Cairo synagogue a Hebrew text of parts of this book. This text may be an earlier and more authentic one than the Greek; or, conversely, it may be just another divergent translation. At all events, in it there are used different words for the passage that interests us in Ecclesiasticus 50: 8, and the rendition is (after being put into English) "flower (or blossom) of Lebanon in the days of the summer". If this be the more accurate translation, then it is quite possible that some attractive, summer-blooming flower common in Lebanon is here being referred to, rather than the bulbous spring-blooming "rose" of Sharon, *Tulipa sharonensis*. The possibility that the Sharon plant was *T. sharonensis* and the Lebanon plant *T. montana* should not be overlooked. *T. montana* is common in the Lebanon and Antilebanon region. It should be emphasized here that while at first scholars hailed the SCHECHTER text as unquestionably the more authentic, they are by no means agreed today on this point. The original text, composed in 180 B.C. by JESUS BEN SIRA, was known in the 10th century, but has since been lost. According to information kindly supplied us by HAROLD WESLER and Rabbi OPHER, the Hebrew words used in the SCHECHTER text are "perah lebanon", while the word used for the "rose" of Sharon is "habatzeleth", in both cases the "b" being rendered soft, like a "v".

226. **Vitis vinifera** (p. 240). — Dr. OPPENHEIMER writes us that in Jerusalem now the vintage begins "at the end of July or early in August. September might be just for Hebron or Es-salt."

The Midrash (594) considers the fruit carried out of the Promised Land of Canaan, from the valley of Eshcol, at the instigation of CALEB, to have been the grape, as we do also. MOSES sent out these spies at a time of the year which corresponded to that of the first-ripe grapes (595).

229. **Zizyphus spina-christi** (p. 248). — Dr. OPPENHEIMER, in a letter to us, says that he believes that this plant attains to heights greater than 15 feet.

LINNAEUS was of the opinion (537) that the crown of thorns mentioned in Matthew 27: 29 and John 19: 2 was made of this species, rather than from *Paliurus* as suggested by us. In this he is supported also by SCHWERIN (553).

232. **Trichodesmium erythraeum** Ehrenb.

EXODUS 4: 9—. . . . thou shalt take of the water of the river, and pour it upon the dry land: and the water which thou takest out of the river shall become blood upon the dry land.
EXODUS 7: 17-19—. . . . I will smite with the rod that is in mine hand upon the waters which are in the river, and they shall be turned to blood. And the fish that is in the river shall die, and the river shall stink; and the Egyptians shall loathe to drink of the water of the river . . . . and upon all their pools of water, that they may become blood.
REVELATION 8: 8—. . . . and the third part of the sea became blood.
REVELATION 11: 6—. . . . and have power over waters to turn them to blood. . . .
REVELATION 16: 4—And the third angel poured out his vial upon the rivers and fountains of waters; and they became blood.

The periodic reddening of the water which has given the Red Sea its name is generally assumed to be caused by the alga, *Trichodesmium erythraeum*, although Dr. WILLIAM RANDOLPH TAYLOR, in a letter to us dated October 12, 1950, says that it may be due also to members of the genus *Peridinium* and perhaps other genera. He states that there is quite some literature of a superficial kind relating to *Trichodesmium* "but nothing very substantial beyond enough to certify its name." It is his opinion that not much has been done with the *Peridinium* flora of that area. FRITSCH (476) states that the *Trichodesmium* is normally a bottom-living species consisting of several color forms or races. It sometimes floats to the surface and occasionally becomes so abundant in the surface-floating plankton that it imparts a reddish color to the water. He says that the red-colored forms of the species are commonly found in deep water and are associated with green-colored forms, but that they will retain their color for several months even at the surface. It has furthermore been reported that the green form of the species can be transformed into the red form through exposure to monochromatic light and that after the change has been brought about the individuals will retain their red color even in white light through many generations.

It has been suggested by some students, and, it seems to us, with good reason, that the Bible references given above may well be connected in some way with this plant. The sudden appearance of these plants in large quantities on the surface of a body of water where previously they had been invisible at the bottom, or their sudden rapid reproduction, might cause observers to describe the water as bloody, especially when we take into account the natural exaggeration to which all such stories are always subjected.

233. **Erysiphe graminis** P.DC.   234. **Tilletia caries** (P. DC.) Tul.
235. **Tilletia foetens** (B. & C.) Trel.   236. **Ustilago tritici** (P.) Rostr.

GENESIS 41: 23—And, behold, seven ears, withered, thin, and *blasted* with the east wind. . . .
DEUTERONOMY 28: 22—The Lord shall smite thee with . . . . *blasting*, and with *mildew*. . . .
I KINGS 8: 37—If there be in the land famine, if there be pestilence, *blasting*, *mildew*. . . .
II CHRONICLES 6: 28—If there be dearth in the land, if there be pestilence, if there be *blasting*, or *mildew*.
AMOS 4: 9—I have smitten you with *blasting* and *mildew*. . . .
HAGGAI 2: 17—I smote you with *blasting* and with *mildew*. . . .

While it is impossible to be certain what plant diseases are referred to here without knowing the crops which were thus smitten, it seems logical to suppose that the crops were wheat, barley, and spelt, the most important and essential crops of the Israelites. Dr. B.

O. DODGE, of the New York Botanical Garden, has suggested, in this case, that the "mildew" probably was *Erysiphe graminis*, the powdery mildew of grains, and that the "blasting" was a smut, probably *Tilletia caries* [also called *T. tritici* (Bjerk.) Wint.] or *T. foetens*, the bunt or stinking smut, and *Ustilago tritici*, the loose smut of wheat (527).

237. **Microsporium gyseum** (Badin) Guiart & Grigorakis.   238. **Oidium** sp.
239. **Aspergillus niger** Van Tieghem.   240. **Actinomyces albus** (Rossi-Doña) Krainsky.
241. **Cladosporum herbarum** (Pers.) Link.

LEVITICUS 13: 47-59—The garment also that the plague of *leprosy* is in, whether it be a woollen garment, or a linen garment; Whether it be in the warp, or woof; of linen, or of woollen; whether in a skin, or in any thing made of skin; And if the plague be greenish or reddish in the garment, or in the skin, either in the warp, or in the woof, or in any thing of skin; it is a plague of *leprosy*, and shall be shewed unto the priest: And the priest shall look upon the plague; and shut up it that hath the plague seven days: And he shall look on the plague on the seventh day: if the plague be spread in the garment, either in the warp, or in the woof, or in a skin, or in any work that is made of skin; the plague is a fretting *leprosy*; it is unclean. He shall therefore burn that garment, whether warp or woof, in woollen or in linen, or any thing of skin, wherein the plague is: for it is a fretting *leprosy*; it shall be burnt in the fire. And if the priest shall look, and, behold, the plague be not spread in the garment, either in the warp, or in the woof, or in any thing of skin; Then the priest shall command that they wash the thing wherein the plague is, and he shall shut it up seven days more. And the priest shall look on the plague, after that it is washed: and behold, if the plague have not changed his colour, and the plague be not spread; it is unclean; thou shalt burn it in the fire; it is fret inward, whether it be bare within or without. And if the priest look, and, behold, the plague be somewhat dark after the washing of it; then he shall rend it out of the garment, or out of the skin, or out of the warp, or out of the woof: And if it appear still in the garment, either in the warp, or in the woof, or in any thing of skin; it is a spreading plague: thou shalt burn that wherein the plague is with fire. And the garment, either warp, or woof, or whatsoever thing of skin it be, which thou shalt wash, if the plague be departed from them, then it shall be washed the second time, and shall be clean. This is the law of the *plague of leprosy* in a garment of woollen or linen, either in the warp, or woof, or any thing of skin, to pronounce it clean, or to pronounce it unclean.

It seems quite certain that the plague of "leprosy" attacking garments, as described in such detail in the quotation given above, was caused by the action of certain of the lower plants. Some botanists have suggested species of *Actinomyces* (now often called *Streptomyces*), but Dr. RALPH G. H. SIU, Research Director of the General Laboratories of the Philadelphia Quartermaster Depot, United States Army, in a letter to us dated October 24, 1950, states that in his researches he has found so far only *Microsporum gyseum* attacking woolen clothes. This is a white organism. Linen goods he has found attacked by a species of *Oidium*, also a white organism; while leather is attacked by *Aspergillus niger*, a black organism. It is interesting to note that two green organisms, according to Dr. SIU, have been found to attack cotton goods. These are *Penicillium glaucum* Link and *Chaetomium globosum* Kunze — the former green and the latter dark-olive. A red organism, *Cytophaga rubra* Winogradsky, has also been found by him to attack cotton clothes. Whether these organisms also attack woolen and linen garments and thus might be the ones involved in the "greenish" and "reddish" leprosy in the Bible reference, is not yet known, but is possible.

W. L. WHITE (535) states that *Memnoniella echinata* (Rivolta) Galloway is the most common and destructive species of fungus to attack fabrics, especially cotton ones, in the tropics, occurring sporadically also in the temperate regions. Others he lists are *Metarrhizium glutinosum* Pope, *Aspergillus fumigatus* Fresenius, *A. unguis* Emile-Weil & Gaudin, and a species of *Stachybotrys*. Of these *A. unguis* attacks silk and woolen fabrics, the others attack cotton and linen. In a letter to us dated November 10, 1950, he states that only a few organisms are capable of attacking woolen fabrics unless such fabrics are buried in the ground. One of the worst is *Actinomyces albus*. "The molds attacking leather are usually green and nearly always are species of *Penicillium* or *Aspergillus*, of which there are many species that may be involved. Black molds on leather or other skins may be *Aspergillus niger* . . . . or *Cladosporum herbarum*. . . . . Linen and cotton fabrics under moist conditions are susceptible to attack by a great array of common molds and I believe it would be totally useless to even take a guess as to what might be involved on the basis of color alone."

242. **Citrus medica** var. **lageriformis** Roem.

LEVITICUS 23:'40—And ye shall take you on the first day the boughs of *goodly trees*. . . .

A preliminary discussion of this plant was given by us under *Prunus armeniaca* (which see), but it seems to us now that those who have advocated the etrog for the "goodly trees" of the above reference are probably correct. ASCH (463) has summed up the case well when he says "There are several reasons why the etrog was chosen as the 'Goodly tree' by the Jews during the time of MOSES. The small tree is beautiful to behold during its time of bloom, and when covered with golden fruit. In addition, the fruit has a very pleasing penetrating odor when ripe, and the rind is delicious when candied. Furthermore, this was probably the only fruit they knew which retains all parts of the female organ or pistil of the flower when ripe. Perfectly shaped, unblemished fruits with persistent styles only can be brought to the Temple during the Feast of the Tabernacles. These ceremonial fruits are grown mainly on the Island of Corfu, Greece, and to a limited degree in Israel. They are carefully plucked with scissors, packed in cotton, and often shipped in individual boxes."

In reference to the Feast of the Tabernacles, the Midrash (593) discusses the "goodly trees" of Leviticus 23: 40, considering them to be four (unnamed) species of "lulab". It is interesting to note that this version speaks of the *fruit* of these goodly trees, while the King James and GOODSPEED versions indicate that the *boughs* were to be collected. The Douay version of this reference is of interest: "And you shall take you on the first day the fruits of the fairest tree, and branches of palm trees, and boughs of thick trees, and willows of the brook. . . ." Here the implication of four different species appears just as in the Midrash, and the admonition that the fruit was to be taken of the first, while the branches were to be the parts used of the other three. In Nehemiah 8: 15 the branches of five different trees are listed for gathering. If these "goodly trees" or "fairest tree" really differed from all the others in having their fruits used, rather than the branches, it is very possible that they were the etrog or citron (*Citrus medica* var. *lageriformis*).

In Esther 5: 14 there is a reference to the proposed hanging of MORDECAI. HAMAN, to relieve himself of his wrath against MORDECAI, quickly accepts his wife's suggestion of death by hanging, and causes a 50-cubit-high gallows to be constructed. The Midrashian writers relate (598) an interesting tale as to the choice of timber for these gallows: "God called all the trees of creation and said: 'Which will offer itself for this man to be hanged on?' The fig tree said: 'I will offer myself, for from me Israel bring first-fruits, and what is more, Israel are also compared to a first-ripe fruit, as it says, I saw your fathers as the first-ripe in the fig-tree at her first season' (Hos. ix, 10). The vine said: 'I offer myself since Israel are compared to me, as it says, Thou didst pluck up a vine out of Egypt' (Ps. lxxx, 9). The pomegranate said: 'I offer myself, since Israel are compared to me, as it says: Thy temples are like a pomegranate split open' (S.S. iv, 3). The nut-tree said: 'I offer myself, because Israel are likened to me, as it says, I went down into the garden of nuts' (ib. vi. 11) [footnote: "God is supposed to be speaking to Israel"]. The citron said: 'I offer myself, because Israel take my fruit for a religious ceremony, as it says, And ye shall take you on the first day the fruit of goodly trees' (Lev. xxiii, 40) [footnote: "Which according to the Rabbis is the ethrog (citron)"]. The myrtle said: 'I offer myself, because Israel are compared to me, as it says, And he stood among the myrtles that were in the bottom' (Zech. 1, 8). The olive said. 'I offer myself, since Israel are compared to me, as it says, The Lord called thy name a leafy olive tree, fair with goodly fruit' (Jer. xi, 16). The apple tree said. 'I offer myself, because Israel are compared to me, as it says. As an apple tree among the trees of the wood, so is my beloved among the sons' (S.S. ii, 3) [footnote: "This citation seems inappropriate, as it applied to God and not to Israel"], and it also says, And the smell of thy countenance like apples' (ib., vii, 9). The palm-tree said: 'I offer myself, because Israel are likened to me, as it says, This thy stature is like a palm tree' (ib. 8). The acacia trees and fir trees said: 'We offer ourselves, because from us the Tabernacle was made and the Temple was built.' The cedar and the palm-tree said: 'We offer ourselves, because we are likened to the righteous, as it says, The righteous shall flourish like the palm-tree; he shall grow like a cedar in Lebanon' (Ps. xcii, 13). The willow said: 'I offer myself because Israel are compared to me, as it says, They shall spring up . . . . as willows by the watercourses (Isa. xliv, 4), and they also use me for the ceremony of the four species of the lulab'. Thereupon the thorn said to the Holy One, blessed be He: 'Sovereign of the Universe, I who have no claim to make offer myself, that this unclean one may be hanged on me, because my name is thorn and he is a pricking thorn [footnote: "V. Ezek. xxviii, 24"], and it is fitting that a thorn should be hanged on a thorn.' So they found one of these and they made the gallows."

It is apparent from this tale that the Rabbinical writers definitely considered the "goodly trees" to be citrons. At the present time (605) the Jews celebrate the ancient Feast of the Tabernacles during a holiday which they call Succoth. It comes five days after Yom Kippur and lasts for nine days. During these days many orthodox Jews eat their meals and spend part of their time in a "succah" — a little booth or hut covered with leafy branches and twigs and decorated with the fruits of the season and assorted green vegetables as a harvest thanksgiving. In olden times almost every family had its own succah, but now it is more common for the synagogue to build one large succah for the entire congregation and to hold joyous gatherings during the holiday week in it. The cantor during these celebrations holds a palm branch (lulav) with a few sprigs of myrtle and willow in one hand and a bright yellow citron (ethrog) in the other and waves these back and forth to indicate symbolically that God, who is being thanked for his gifts of the harvest, is everywhere. All members of the congregation who are able to secure these articles, do the same with them.

Unidentified plant references (p. 251). — The reference in Exodus 15: 25 is most interesting, for here is mentioned a tree which God pointed out to MOSES at Marah. Pieces of this tree thrown into "bitter" water had the property of making this water "sweet". What tree's sap in that area has the power to transform alkali water into potable water? According to the Midrash (592), Rabbi NATHAN has identified the plant as "a kind of creeper", whose berries were known to be injurious to animals. This authority also states that "it was also the name for the trunk of ivy". Rabbi JOSHUA has identified it as a willow tree and Rabbi ELIEZER as the olive.

FIGURE 40. — One of many medieval and early Renaissance representations of the Garden of Eden, with the Lord who points out the Tree of Knowledge of Good and Evil to ADAM and EVE. Various birds and beasts and many trees and flowers, most of which have a special symbolic meaning, are depicted, and leading from the Garden are the rivers Euphrates, Tigris, Gihon, and Pison. (Wood cut from a Flemish incunable, Ludolphus's Leven ons Heeren Jesu Christi, Antwerp, 1487).

Index to Bible Verses

This index includes all Bible verses quoted or referred to on the preceding pages. *Those verses that are quoted in part or in full at the head of a chapter are italicized.* The books of the Bible, including the Apocrypha, are arranged in one alphabetic sequence. The abbreviated names of the books are mostly written as they appear in the Authorized Version of King James I, but included also are the several references in the text to the titles as they appear in other versions.

ACTS OF THE APOSTLES
1: 12—*158*
1: 25—73
2: 13—*213*
7: 12—*230*
8: 6—276
8: 23—*78*, 79, 80
11: 19—*170*
15: 3—*170*
27: 12—*170*, 171
28: 8—*97*, *219*
ADDITIONS TO ESTHER
11: 3—*41*
16: 18—*41*
AMOS
2: 9—*68*, 179, *194*, 195
3: 4—*252*
4: 5—*211*
4: 9—*103*, 142, 143, *157*, *252*, 254, *289*
5: 7—*48*, 49
6: 12—*48*, 49, *78*, 79, 279
6: 14—*20*, *216*
6: 15—*20*
6: 18—*217*
7: 14—*106*
8: 1-2—*103*, 256
8: 5—*232*
8: 5-6—*230*
9: 13—*213*
9: 13-14—*242*
BARUCH
1: 10—*31*, 126, *278*
CANTICLE OF CANTICLES
6: 1-2—20
6: 6—20
6: 10—20
7: 8—20
7: 12-13—20
I CHRONICLES
2: 43—*185*
3: 9—*169*
9: 29—*56*, 58
10: 12—*194*, 198, 228
11: 13—*113*, *128*
12: 40—19, *103*, *240*, 244
12: 41—19
13: 8—*285*
14: 1—*66*
14: 14-15—*141*, *183*

15: 27—*130*, 131, 132
16: 33—276
17: 1—*66*
21: 12-28—*229*
21: 20-23—*232*
22: 3-4—*66*
23: 29—211
27: 28—*97*, *106*, 108, *157*
II CHRONICLES
1: 15—*67*, 106
1: 16—*130*, 132
2: 3—*67*
2: 6-7—19
2: 7—*194*
2: 7-8—19
2: 8—*67*, 89, 90, 119, *175*, 176, 188, 189, *209*
2: 9—19
2: 10—19, *157*, *212*, *230*
2: 13-14—19
2: 14—*194*
2: 14-15—19
2: 15—*230*
3: 1—*230*
3: 5—*285*
3: 14—*130*, 132, *194*, *204*
3: 16—*190*
4: 5—42, *154*
4: 13—*190*
5: 12—*130*, 132
6: 28—*289*
8: 4—*169*
8: 17—*169*, 172
9: 10-11—90, *188*, *209*
9: 27—*67*, 106
20: 2—*169*
21: 18-19—*219*
22: 14—70
25: 18—*67*, 71
26: 2—*169*, 172
26: 19-21—*142*
26: 23—*142*
28: 2—151
28: 4—*56*, *194*
28: 15—*169*, 171
31: 5—*169*
34: 25—*56*, 59
I CORINTHIANS
3: 12—*253*
5: 6-8—*211*

9: 9—*230*
10: 10—*167*, *253*
II CORINTHIANS
3: 2—276
9: 10—*253*
DANIEL
1: 12—*252*, 254
1: 16—*252*, 254
4: 7-9—20
4: 10-12—*20*, *194*
5: 1-2—*242*
5: 4—*242*
10: 5—*131*
12: 4—*173*
DEUTERONOMY
2: 8—*169*, 172
4: 3—*149*, 150
6: 11—*157*, *240*
7: 13—*157*
8: 3—*126*
8: 7-9—14
8: 8—*103*, 104, *111*, *157*, *189*, 190, *229*, *240*
8: 16—*126*
10: 3—*24*
11: 9—*257*
11: 15—*28*, 29
11: 30—*193*, 196, 197, 198
12: 2-3—*193*
14: 22—*251*, 257
14: 26—*170*
16: 21—*193*, 197, 198, 227
18: 4—*212*, 214
20: 19-20—*251*, 254
21: 11-12—*125*
21: 22-23—*251*
22: 11—*130*, 131
24: 8—*142*
24: 20—*157*
24: 21—*240*
25: 4—*229*, 232
26: 2—*251*, 256
28: 22—*97*, *143*, *289*
28: 27—*167*
28: 38—*251*, 257
28: 40—*157*
28: 42—*251*
29: 6—*170*
29: 17—19
29: 18—19, *48*, 49, *78*, 79, 279

32: 2—*28*
32: 13—*157*
32: 14—*229, 240*
32: 32—*78*, 79, 221, 244
32: 32-33—*240*
32: 38—*240*
33: 14—*251*, 256
33: 24—*157*
33: 28—*212, 229*
34: 3—*169*
ECCLESIASTES
2: 4—*241*
2: 5-6—*252*
10: 1—*157*
10: 8—*202*, 203
10: 19—*212*
11: 1—*112*, 113
11: 3—*252*
11: 6—*252*, 257
12: 5—*36, 65*
ECCLESIASTICUS
1: 8—43
24: 13—*89*
24: 14—*38, 151, 180,* 205
24: 15—20, *30*, 31, *102, 223,*
  224
24: 16—*178*
24: 21—20
39: 13—*151*, 205
39: 14—*117*
40: 4—*131*
50: 8—*42*, 58, *84*, 85, *117*, 205,
  234, 288
50: 10—*89*
EPHESIANS
5: 9—*253*
5: 18—*213*
II ESDRAS
2: 19—*205*
14: 24—*62*, 63, 173
16: 29—*158*
ESTHER
1: 2—*41*, 117
1: 5-6—*109*
1: 6—*130*
2: 7—*143*
2: 12—*82*, 84
5: 14—291
8: 15—*130*, 131, 132, 204
11: 3—*117*
16: 18—*117*
EXODUS
2: 3—*92*, 93, 249
2: 5—*92*, 249
3: 2—25
3: 2-4—*23,* 277
4: 9—*289*
5: 7-18—*251*, 254
7: 17-19—*289*
9: 9-11—*223*
9: 22—29
9: 22-35—112
9: 31—*111, 129,* 131, 133
9: 32—*229, 232*
11: 5—*229*
12: 8—*74*, 140, 280
12: 22—*65, 160,* 162, 223
12: 15—*210*, 211
12: 17-20—211
12: 19—*210*
12: 22—*65*, 162
12: 34—211
12: 39—211
15: 7—*251*
15: 25—*251*, 292
15: 27—*169*, 172, 198
16: 4—*125*
16: 12-13—*127*
16: 13-15—*32*
16: 13-21—*126*
16: 13-35—*125, 126*
16: 31—*86*, 257
20: 5—*149*, 150
20: 14—10
21: 1—276

22: 5—*240*
22: 6—*202, 229,* 276
22: 29—214
23: 11—*157*
25—215
25: 4—*129*, 132, *193, 204*
25: 5—*24*
25: 10—*24*
25: 13—*24*
25: 23—*24*
25: 28—*24*
25: 33-36—*36*
26: 1—*129*, 131, *193, 204*
26: 15-16—*24*
26: 26—*24*
26: 31—*129*
26: 32—*24*
26: 36—*129*
26: 37—*24*
27: 1—*24*
27: 6—*24*
27: 16—*129*
27: 18—*129*
27: 20—*157*
28—110
28: 5-6—*129*
28: 8—*129*
28: 15—*129*
28: 20—224
28: 33—*193, 204*
28: 33-34—*189*
28: 39—*129*
28: 42—*129*, 132
29: 2—211, *229*
29: 23—211
29: 40—211, *212*, 214
30: 1—*24*, 56
30: 5—*24*
30: 7—59
30: 7-9—*56*
30: 23—*40, 76, 82,* 219
30: 23-24—*39, 75*
30: 24—218, 219
30: 34—*56, 102, 223, 224*
34: 13—*193*, 197, 227
35: 6—*130*, 132
35: 7—*24*
35: 23—*130, 193, 204*
35: 24—*24*
35: 25—*130*
35: 35—*130*
36: 8—*130*
36: 20—*24*
36: 31—*24*
36: 35—*130*
36: 36—*24*
36: 37—*130*
37—215
37: 1—*24*
37: 4—*24*
37: 10—*24*
37: 15—*24*
37: 17-18—*213*
37: 19-20—*36*
37: 25—*24*
37: 28—*24*
38: 1—*24*
38: 6—*24*
38: 9—*130*
38: 18—*130*
38: 23—*130*
39: 1—*204*
39: 2—*130*
39: 3—*130*
39: 5—*130*
39: 8—*130*
39: 24—*130, 193, 204*
39: 24-26—*189*
39: 27—*131*
39: 27-29—*130*
EZEKIEL
1: 4—*252*, 254
1: 27—*252*, 254
2: 6—*202*, 203, 206, *207*, 245,
  249

2: 9-10—173
4: 8-12—112
4: 9—*101, 112, 128,* 152,
  166, 211, 233, 234
4: 12—*112*
4: 12-13—*211*
4: 15-17—*211*
6: 13—*194*
8: 2—*252,* 254
9: 2—131
9: 2-3—*130,* 173
10: 2—*130,* 279
10: 6—279
10: 6-7—*130*
10: 13—39, 279
13: 5—*202*
13: 19—*112*, 113
15: 2-3—*252*
15: 2-6—*239*
15: 6—*252*
16: 8—283
16: 10—*130, 140*
16: 13—*130, 140*
17: 3—*67*
17: 5—*216*
17: 5-10—*241*
17: 22-24—*67*
18: 2—*241*
19: 10-11—*241*
20: 28—*194*
20: 46-47—*252*
22: 30—*202*
27: 5—*68, 176*
27: 6—*62, 194,* 195
27: 7—*130,* 132, *204*
27: 15—*95*
27: 16—*130*
27: 17—*84,* 85, *157, 166, 178,*
  222, *230,* 232
27: 18—*213, 242*
27: 19—*39,* 40, *75,* 76, 149,
  219
27: 24—*67*
28: 24—206, *207,* 291
29: 6-7—*50,* 51
31: 3-18—*67*
31: 8—68, 70, *176, 180,* 286
31: 14—*194,* 198
40: 3—*50,* 51, *131*
40: 31—*169,* 170
44: 17-18—*131*
46: 14—*211*
47: 10—10
47: 12—*252*
47: 19—*169*
48: 28—*169*
EZRA
3: 7—*67*
6: 4—*67,* 70
7: 6—173
GALATIANS
3: 13—*253*
4: 29—10
5: 22—*253,* 256
GENESIS
1: 11—*29,* 253
1: 11-12—*28, 251,* 276
1: 11-13—20
1: 29—*251*
1: 30—29
2: 5—29
2: 8-9—20
2: 9—*184,* 276
2: 12—*81*
2: 15-17—20
2: 17—*184,* 256
3: 1-6—*20,* 186
3: 6—*185,* 287
3: 7—10, *103,* 104, 228, 281
3: 17-18—*70*
3: 18—*71,* 165, *202,* 245, 248,
  287
3: 19—211
6: 14—*89,* 90, 152

8: 11—*157*, 159
9: 20-21—*240*
9: 21—*212*
9: 24—*240*
10: 29—82
12: 6—*193*, 196, 197, 198
12: 10—*228*, 232
13: 18—*193*, 196, 197, 198
14: 6—198, 262
14: 7—*169*
14: 13—*193*, 198
14: 18—*212*, 214, *240*
18: 1—*193*, 196, 197, 198
18: 6—211
18: 8—*178*, *193*, 195
19: 3—211
21: 15—*227*
21: 33—198, *227*
22: 7—*251*, 254
22: 13—*251*
24: 25—*251*, 254
24: 32—*251*, 254
24: 61—10
25: 15—187
25: 18—82
25: 29-34—*128*
26: 12—*228*
27: 28—*157*, *212*, *228*
28: 18—*157*
28: 19—*35*, 37
30: 14—*229*, 231, 232, 283
30: 14-16—*137*
30: 37—*35*, 37, *180*, *181*, 225
30: 37-39—182
35: 4—*193*, 195, 197, 198
35: 6—*35*, 37
35: 8—*193*, 195
36: 41—*178*, 179
37: 25—*51*, 52, 55, 77, 83, 85, 178, 225, 281, 286
38: 6—*169*
38: 8-10—257
38: 28—*193*
38: 30—*193*
40: 9-11—*240*
40: 16—211
41: 2—*51*, *62*, *120*, *172*
41: 5—215
41: 5-27—231
41: 5-57—*229*
41: 22—215
41: 23—*289*
41: 42—*129*, 132, 140
41: 49—231
41: 57—*232*
42—232
42: 1-3—231
42: 1-33—*229*
43: 1-2—*229*
43: 11—*35*, *51*, 77, 83, 84, 119, 170, *177*, *179*, 225, 280, 286
45: 6—*229*
45: 11—*229*
45: 23—*210*
47: 13-14—*229*
47: 19-20—*229*
49: 11—213
49: 11-12—*212*, *240*

Habakkuk
2: 5—*213*
2: 15—*213*
2: 15-16—*242*
3: 17—*103*, *157*, *242*

Haggai
1: 10—*252*
2: 17—*142*, *143*, *252*, 254, *289*
2: 19—*103*, *190*, *252*

Hebrews
3:17—*167*
6: 8—*71*, 72, 165, 202, *221*, 245, 248
8: 6—276
9: 4—*126*
9: 19—*65*, *160*, 162, *194*

Hosea
2: 5—*131*, *157*
2: 5-6—20
2: 6—*202*
2: 7-8—20
2: 8—*157*
2: 8-9—20
2: 9—*131*, 133
2: 9-10—20
2: 12—20, *103*, *252*
2: 14—20
3: 2—*112*, 113
4: 11—*213*
4: 13—*175*, *178*, *181*, 182, *194*
7: 4—*211*
7: 5—*213*
7: 8—*211*
9: 2—*242*
9: 4—*242*
9: 6—165, 202, *237*, *245*, 248
9: 10—*103*, 104, 291
10: 1—*242*
10: 4—*48*, 279
10: 8—*15*, 71
10: 13—*252*
12: 1—20, *157*
12: 2—20
14: 5—*43*, *117*, 181, *182*
14: 6—*157*
14: 7—*213*, *230*, *242*
14: 7-8—20
14: 8—*46*, 177
14: 8-9—20
14: 9—70

Isaiah
1: 6—*157*
1: 8—*88*, *241*
1: 18—*194*
1: 29—198
1: 29-30—*194*, 198
2: 13—*67*, *194*, 195
3: 10—*252*
3: 23—*130*, 131, 132
5: 1-7—*241*
5: 2—*212*, 213
5: 2-4—*239*
5: 5—*202*
5: 10—*241*, *252*
5: 11—170
5: 11-12—*212*
5: 22—*212*, 214
5: 24—*252*
5: 28—279
6: 13—*178*, 179, *194*, 198
7: 2—*183*
7: 11—*252*
7: 15—170
7: 19—165, 202, 245, *248*, *252*, 254
7: 23-25—*202*, *206*, 245, 248
9: 9—20
9: 10—20, *67*, 106
9: 14—51, *120*
9: 18—*206*, *248*, *252*, 255
10: 17—*202*, *221*, 245, 248
10: 18-19—276
11: 1—256
14: 7-8—276
14: 8—*67*, 176
15: 6—*28*, 29, 34, *252*, 253
15: 7—*216*, 217
16: 8-10—*241*
16: 9-10—213
16: 10—*212*
17: 6—*157*
17: 11—*252*
17: 13—*38*
18: 2—*92*, 93, 94, 281
19: 6—*50*, *120*, 249
19: 6-7—*92*, 93, 94
19: 9—*130*, 131, 132
19: 15—*51*, *120*
21: 13—*252*
24: 7—*241*
24: 9—170, *241*
24: 11—*241*
24: 13—*157*, *241*
25: 6—*212*, 214
25: 10—*252*
27: 2-11—*241*
27: 6—195
27: 9—*194*
28: 4—*103*, 105, *252*, 256
28: 7—170, *212*
28: 25—*89*, *152*, *233*, 234
28: 25-28—*230*
28: 27—*89*, *152*
28: 28—232
29: 9—*241*
29: 11—173
29: 17—*252*
32: 13—15
32: 15—*252*
32: 20—233
33: 9—15
33: 11—*252*
33: 12—165, *202*, 245, 248
34: 4—173
34: 13—*34*, *153*, 165, 202, 206, *237*, *245*, 248
35: 1—*41*, *147*, 205, 234, 235
35: 7—*28*, 29, *50*, 51, *92*, 93
36: 16—*103*
37: 5—93
37: 24—*67*, 176
37: 27—34
38: 21—*103*
39: 2—*51*, 52, *84*, 85
40: 6-8—*28*, 34
41: 2—*252*
41: 15-16—*230*, 232
41: 19—*24*, *62*, 63, 67, *89*, 90, 91, *97*, 98, *144*, *173*, 174, *176*, 180, 210, 286
41: 19-20—*276*
42: 3—*50*, 51, *130*, 132, 133, 280
42: 15—29
43: 23—*56*, 58, 59
43: 24—*40*, *214*
44: 4—*34*, *216*, 276, 291
44: 14—32, *67*, 90, *176*, 180, *194*, 195, 267, 281, *286*
44: 15—211
47: 2—*230*
51: 12—34
53: 2—*252*, 255
55: 1—*241*
55: 10—*252*
55: 12—276
55: 13—*144*, 165, *176*, 202, *221*, 245, *248*
56: 3—*252*, 255
57: 19—*252*
58: 5—*51*, *92*, 93
60: 6—*56*, 58, 59
60: 13—*62*, *89*, 90, 91, *173*, 174, *176*, 180, 210, 286
61: 3—*194*, 198, *252*, 255
61: 5—*241*
61: 9—*257*
61: 11—*252*, 256
62: 8—*241*
63: 2-3—*212*, 213, *241*
64: 6—*252*, 255
65: 3—*56*, 58, *181*
65: 22—*252*

Isaias
9: 10—20

James
1: 10—*28*
3: 5-6—276
3: 12—*104*, 158
5: 7—*253*
5: 14—*158*

Jeremiah
1: 11—*29*, *36*
2: 21—*239*, *241*

2: 22—*215*
4: 7—*252*
4: 29—*252*
4: 30—*194*
5: 6—*252*
5: 17—*103*
6: 9—213, *241*
6: 20—*39*, 40, *56*, 58, 149, 215
7: 18—211
8: 13—*103*, 104
8: 14—*78*, 79
8: 22—10, *55*, 85, 178, 281, 286
9: 14—20
9: 15—20, *48*, *78*, 79
10: 3—*194*
10: 5—*169*
11: 12—*56*, 59
11: 16—291
11: 17—*56*, 59
12: 10—14
12: 13—*257*
13: 1—*130*
13: 24—*252*
14: 6—*28*, 29
17: 6—*32*, *121*, 122, 264
17: 8—*194*
17: 10—*252*
17: 26—*56*, 58, 59, 232
19: 5—151
21: 9—*167*
21: 14—*252*
22: 7—*67*
22: 14-15—*67*
22: 23—*67*
23: 15—*48*, *78*, **79**
24: 1-8—*103*
25: 15—*241*
25: 30—213, *241*
26: 18—*252*
31: 5—*241*
31: 29-30—*241*
32: 19—*252*
35: 2—*213*
35: 5-6—*213*
35: 8—*213*
35: 14—*213*
36: 14—173
37: 21—*211*
41: 5—*56*, 58, 59
46: 11—*55*, 178, 281, 286
46: 22—*254*
46: 22-23—*252*
46: 23—255
47: 3—*279*
48: 6—*32*, *121*, 122
48: 32-33—*241*
48: 33—*213*
48: 35—*56*, 59
49: 9—*241*
50: 11—*230*
51: 8—*55*, 178, 286
51: 32—*50*, 51
JEREMIAS
 9: 15—20
JOB
 2: 7—*223*
 5: 26—*230*
 8: 11—*50*, *62*, *92*, 93, *120*, 249
 8: 12—34
13: 26—80
14: 2—*251*, 256
14: 7-9—*251*, 255, 276
15: 32—*251*
15: 33—*157*
16: 13—80
19: 24—173
20: 14—80
20: 16—80
20: 25—80
21: 18—*251*
24: 6—231
24: 11—*212*
24: 20—*251*

29: 6—*157*
30: 4—*48*, *53*, *91*, 202, 237, 279, 280, 287
30: 7—*26*, 220, 237, *251*
31: 22—215
31: 40—*29*, 71, *112*, *153*, *230*, 239
36: 27—224
40: 15—*28*, 29, 34
40: 16-17—19
40: 17—*67*
40: 21—19, *50*
40: 21-22—19, *247*, 276
40: 22—*216*
41: 2—19, 51, *202*, 203
41: 20—51
41: 27-29—*251*
JOEL
 1: 5—*242*
 1: 7—*103*, *242*
 1: 10—*157*
 1: 10-12—*242*
 1: 12—*103*, *185*, *190*
 1: 17—*252*
 1: 18-20—276
 2: 3—276
 2: 22—*103*, *242*
 2: 24—*157*, *242*
 3: 3—*242*
 3: 13—*25*, 213, *242*
 3: 18—*20*, *24*
 4: 18—20
JOHN
 1: 48—*104*
 1: 50—*104*
 2: 1-11—214
 2: 3—*213*
 2: 9-10—*213*
 4: 52—*97*
 6: 9—*112*, *211*
 6: 10—*28*
 6: 11—*211*
 6: 13—*112*, *211*
 6: 31—*126*
 9: 1-3—*149*, 150
11: 18—*169*
11: 44—131
12: 3—*148*, 149
12: 13—*169*, 170
12: 24—*230*
13: 4-5—131
15: 1-6—*242*
19: 2—*165*, 202, 245, 248, 289
19: 29—65, 161, *222*, 283
19: 29-30—*27*
19: 39—*35*, 47, *82*
19: 40—*131*, 132
II JOHN
 12—173
III JOHN
 13—*172*, 173
JONAH
 2: 5—20, *249*
 2: 6—20
 4: 6-7—*203*
JONAS
 2: 6—20
JOSHUA
 2: 1—*24*
 2: 6—*130*, 131, **133**
 3: 1—*24*
 5: 11—231
 5: 12—*126*
 9: 12—*141*, *210*, **212**
13: 26—*179*
15: 8—*183*
15: 32—*189*
15: 38—88
15: 53—*185*
16: 2—*36*, 37
16: 8—*51*, *214*, 215
17: 8—*185*
18: 13—*36*, 37
19: 7—*189*, 190

19: 13—*189*, 190
19: 28—*214*, 215
24: 13—*157*
24: 26—*193*, 195
JUDE
 5—*167*
 12—*253*
 16—10
JUDGES
 1: 16—*169*, 172
 1: 23—37
 2: 10-13—150
 3: 7—*193*, 197, 227
 3: 13—*169*, 172
 4: 5—*169*
 6: 11—*179*, *193*, 195, 198, *212*, *229*, 286
 6: 19-21—211, *229*
 6: 25—150
 7: 13—*111*, *210*
 7: 13-15—113
 8: 5—*210*
 8: 7—*165*, 202, 203, *206*, 207, 245, *248*
 8: 16—*206*, 207
 8: 33—150
 9: 6—*193*, 196, 197, 198
 9: 8—260
 9: 8-9—*157*
 9: 8-15—*135*, 260
 9: 10-11—*103*
 9: 12-13—*240*
 9: 13—*212*
 9: 14—135
 9: 14-15—*134*, 153, 206, 248
 9: 15—*66*
 9: 27—*240*
10: 10—150
11: 33—*240*, 243
13: 3—*170*
13: 4—*170*
13: 14—*212*
14: 5—*240*
14: 12—132
14: 12-13—*130*
15: 5—*157*, *229*, 230, 240
15: 14—*130*
16: 4—*240*, 243
16: 7-9—*216*
19: 19—*212*
20: 33—*169*
20: 45—*189*
I KINGS
 4: 25—19, *103*, *240*
 4: 28—19, *112*
 4: 33—19, 65, *66*, *160*, 161
 5: 5—19
 5: 6-10—*66*
 5: 8—19, *175*, 176
 5: 10—19, *175*, 176
 5: 11—19, *157*, *229*
 5: 13—19
 5: 22—19
 5: 24—19
 5: 25—19
 6: 9—*66*
 6: 15—*285*
 6: 15-16—*66*
 6: 18—*66*, *78*, 79
 6: 23—*97*, 98, 159
 6: 29—*169*, 170
 6: 31-33—*97*, 159
 6: 32—*169*, 170, *189*
 6: 34—*175*, 176
 6: 36—*66*
 7: 2-3—*66*
 7: 7—*66*
 7: 11-12—*66*
 7: 18—*189*
 7: 19—*42*, *154*
 7: 20—*189*
 7: 22—*42*, *154*
 7: 24—*78*, 79
 7: 26—*42*, *154*

8: 37—*289*
9: 11—*66, 285*
9: 26—*169*, 172
10: 10—*39, 84*
10: 11-12—*90*, *188*, *209*
10: 27—*106*
10: 28—*130*, 132
13: 14—*193*, 195, 198
14: 23—*193*, 197, 227
16: 8—*178*, 179
16: 31-33—150
17: 11-13—*210*
18: 5—*28*, 29
18: 15—*34*
18: 19—*193*, 197, 227
18: 19-22—150
19: 4-5—*201*, 262, *287*
19: 6—*210*

II KINGS
4: 39—88, 204, 240
4: 39-40—*78*, 281
4: 42—*112, 210*
5: 1—*142*
5: 3—*142*
5: 6—*142*
5: 11—*142*
5: 18—*189*, 190
5: 27—*142*
6: 25—*162*, 284
7: 3—*142*
7: 8—*142*
8: 27—151
10: 27—*66*
11: 18—151
14: 9—*66*, 69, *71*
14: 22—*169*, 172
16: 3—151
16: 6—*169*, 172
17: 10—*193*
17: 10-11—*56*, 58, 59
17: 16—150, *193*, 197, 227
18: 4—*56, 193*
18: 21—*50*, 51
18: 31—*103*
18: 31-32—*240*
19: 23—*66*, *175*, 176
19: 26—*28*
19: 30—*251*
20: 7—*103*, 104, 105, *223*
20: 13—*51*, 52, *84*, 85
21: 3—151
21: 7—*194*, 198
23: 5—*56*, 58, 59
23: 6—*194*, 198, 227
23: 14—*194*
25: 5—59
25: 17—*190*

III KINGS
4: 25—19
4: 28—19
4: 33—19
5: 8—19
5: 10—19
5: 11—19

LAMENTATIONS
1: 15—*213, 241*
2: 12—*213, 230*
3: 5—*78*
3: 15—*48*, 49
3: 19—*48*, 49, *78*, **79**

LEVITICUS
2: 1-2—*56*, 57
2: 1-6—*232*
2: 1-7—*157*
2: 1-11—211
2: 1-16—*229*
2: 14—231
2: 15—*157*
2: 15-16—*56*, **57**
2: 16—231
5: 11—*56, 157*
6: 8—19
6: 10—*130*, 131, 132
6: 14—19

6: 14-18—211
6: 15—19, *56*, 57, 232
6: 21—19, 157
7: 9-10—211
8: 26—211
10: 1—*56*
10: 9—170
13—*142*, 143, 150
13: 2—*149*
13: 18-23—*223*
13: 23—*223*
13: 30-37—*98*, *99*, **143**
13: 47-48—*130*
13: 47-59—254, *290*
13: 52—*130*
13: 59—*130*
14—*142*, 143
14: 4—68, *160*, *193*
14: 4-6—*209*
14: 4-52—162
14: 6—65, *160*, *193*
14: 6-8—68
14: 33-57—254
14: 49-52—68, *209*, *210*
14: 51-52—*193*
14: 52—*160*
14: 54—*98*, 143
15—150
15: 2-7—*149*
15: 19-30—220
16: 4—*130*
16: 12-13—*56*
16: 23—*130*
16: 24—132
16: 32—*130*
19: 10—*240*
19: 19—*130*, 131, *251*, **257**
22: 4—*142*
22: 4-5—*149*, 150
23: 13—*212*, 214
23: 14—*229*
23: 40—*169*, *173*, *183*, 187, *210*, 216, 285, *290*, 291
24: 5—*229*, 233
24: 5-9—211
24: 7—*56*, *229*
26: 16—*97*, *143*
26: 36—*183*, 184
27: 16—*111*, 113

LUKE
1: 9-10—*57*, 59
1: 15—170
1: 63—173
3: 8-9—*253*, 256
5: 12-13—*142*
6: 1—*230*
6: 44—*103*, *206*
8: 43-44—*219*
10: 34—*158*
11: 42—*139*, *208*, *253*, **254**
12: 27-28—*41*
12: 28—*28*
13—105
13: 6-9—*104*
13: 19—*59*
13: 21—*211*
14: 23—*202*
14: 26—10
15: 16—*72*
16: 19—*131*, 132
17: 6—*59*, *140*, 141
17: 12—*142*
19: 1-4—*266*
19: 4—*106*
19: 29—*104*, *158*, *169*
20—10, *104*
20: 9-16—*242*, 244
21: 29-30—*104*, *253*
21: 37—*158*
23: 36—*27*
23: 53—*131*

I MACCABEES
6: 34—*140*, 141, *242*
13: 51—*169*

II MACCABEES
6: 7—*111*
10: 7—*169*
14: 4—*169*

III MACCABEES
4: 20—*172*, 173

MALACHI
2: 15—257
3: 2—*215*
3: 12—14
4: 1—*252*

MARK
1: 40—*142*
2: 23—*230*
4: 28—*230*
4: 31-32—*59*
4: 32—61
5: 25-29—*219*
6: 13—*158*
8: 24—*253*, 255
11: 1—*169*
11: 13—*103*, 105
11: 20—*103*
12: 1—*202*, *242*
14: 3—*142*, *148*, 149
14: 12—*230*
14: 51—131
14: 51-52—*131*
15: 17—205
15: 19—*235*
15: 23—*80*, *82*, 84, *213*
15: 36—*27*, 50, 161, 173, *222*
15: 46—*131*

MATTHEW
2: 11—*56*, 58, *82*, 84
3: 4—73
3: 8—*253*
3: 10—*253*, 255, 256
3: 12—*230*, 256
4: 3-4—*211*
5: 9—10
6—263
6: 17—*158*
6: 28-30—*41*
6: 30—*28*, 211
7: 9—*211*
7: 16—*71*, *103*, 165, 202, 245, *248*
7: 17-20—*103*
8: 2—*142*
8: 14-15—*97*
9: 20—*219*
11: 7—*50*, 51, 173
12: 1—*230*
12: 33—*253*
13: 3-8—*230*, 233
13: 7—*71*, 165, 202, 245
13: 24-30—*133*, *134*, *230*
13: 31-32—*59*
13: 33—*211*
13: 43—10
16: 6—*211*
17: 19—20, 61
17: 20—20, *59*, 61
18: 6—*230*
21: 1—*103*
21: 19—106
21: 19-21—*103*
21: 33—*213*, *242*
23: 23—*46*, *89*, *139*
24: 32—*103*
25: 3-4—*158*
25: 8—*158*
26: 27-29—*242*
27: 5—*73*
27: 28-29—166
27: 29—*50*, *165*, 202, *235*, 245, *248*, 289
27: 34—*27*, *78*, 79, 80, 83
27: 48—*27*, 50, 161, 173, *222*
27: 59—*131*

MICAH
2: 11—170
3: 12—*252*

4: 4—*103*, *242*
6: 5—*24*
6: 7—*97*, 98
6: 15—*157*, *242*
7: 4—206, *221*, 245, 248
7: 13—*252*
7: 14—*194*

NAHUM
2: 3—11, 20, *176*, 177
2: 4—20
3: 11—*242*
3: 12—*103*
3: 17—*202*

NEHEMIAH
1: 1—*41*, 117
1: 11—*240*
2: 1—*240*
2: 8—*67*
3: 11—211
5: 11—*157*
8: 15—*97*, *143*, 159, *169*, *173*, 174, 281, 291
9: 20—*126*
9: 25—*157*
12: 38—211
13: 5—*56*
13: 9—*56*
13: 15—*212*

NUMBERS
5: 2—*142*
5: 2-3—*149*, 150
5: 15—*56*, *111*, 113
6: 3—170, *212*
6: 15-21—211
7: 14—*56*
7: 20—*56*
7: 26—*56*
7: 32—*56*
7: 38—*56*
7: 44—*56*
7: 50—*56*
7: 56—*56*
7: 68—*56*
7: 74—*56*
7: 80—*56*
7: 86—*56*
9: 11—*74*, 140
11: 5—29, *32*, *33*, *34*, 80, 88, 281
11: 6-9—*126*, 282
11: 7—*81*, *86*
11: 9—126
12: 10—*142*, 167
13: 20—*240*
13: 23—*103*, *189*
13: 23-24—*240*
13: 26—*251*
13: 32—*167*
14: 37—*167*
15: 5—*212*, 214
15: 7—*212*
15: 10—*212*
15: 19-21—211
16—168
16: 46—*56*, 59
16: 46-50—19, *167*
17: 1-8—19, *36*
17: 11-15—19
17: 16-23—19
18: 27—230
19: 6—*68*, *160*, *193*, *210*
19: 6-18—162
19: 18—*65*, *160*
20: 5—*189*, 190, *240*
22: 4—*28*
22: 24—*240*
22: 41—150
24: 5-6—276
24: 6—*47*, *66*, 68, *193*, 198, 262
25: 1—*24*
25: 8-9—*149*, 150
26: 9-11—168
27: 3—168

32: 9—*240*
33: 9—*169*, 172
33: 18-19—*201*
33: 19-20—*189*, 190
33: 49—*24*, 25
33: 55—165, 202, *206*, 207, 245, 248

OSEE
2: 5-6—20
2: 8-9—20
2: 12—20
12: 1—20
14: 8-9—20

I PARALIPOMENON
12: 40—19

II PARALIPOMENON
2: 7-8—19
2: 10—19
2: 14-15—19

I PETER
1: 23—*253*, 257
1: 24—*28*
2: 24—*253*

PHILIPPIANS
4: 17—*253*

PROVERBS
3: 18—*251*
5: 4—*48*
6: 26—*211*
7: 16—*130*, 132
7: 17—*47*, *76*, *82*, 188
8: 15—*251*
9: 2—*212*, 214
9: 5—*212*, 214
10: 26—*27*
11: 26—*230*
11: 30—*251*
12: 14—*252*
15: 19—165, *202*, 221, 222, 245, 248
18: 20—*252*
20: 1—170, *212*
21: 17—*212*
23: 20—*212*
23: 29-32—*212*
23: 30—214
24: 13—170
24: 31—*27*, *220*, 221, 237
25: 11—*185*
25: 16—170
25: 20—*27*
26: 20—*252*, 254
27: 7—170
27: 9—*157*
27: 18—*103*
27: 25—*28*, 29, 34, *252*, 253
31: 4—170
31: 4-6—*212*
31: 6—170
31: 13—*130*, 133
31: 19—133
31: 22—*130*, 140
31: 24—*130*

PSALMS
1: 3—*169*, *194*, 197, 285
1: 4—*251*
4: 7—19, *212*, *230*
4: 8—19
19: 10—170
23: 5—*157*
24—15
29: 5—*67*
29: 9—276
36: 35—19
37: 2—*28*, 34, *251*, 253
37: 12—19
37: 35—19, *123*, 152
37: 36—19
38: 3-7—*149*
38: 11—19, *149*, 151
38: 12—19
40: 5—276

40: 7-8—173
44—19
44:1,9—19
45—19
45, title—*41*, 43
45: 1—173
45: 8—19, *47*, 48; 75, 76, *82*, 188, *218*
45: 9—19
50: 9—19
50: 10—*251*, 255
51: 7—19, 65, *160*, 162
51: 9—19
51: 10—19
52: 8—19, *157*
52: 10—19
57: 10—19
58: 9—19, 165, *202*, 207, 245, 248
58: 10—19
59—19
60—19
60, title—*41*, 43, 117
64: 10—19
64: 12—19
65: 9—19, *230*
65: 10—19
65: 13—19, *230*
65: 14—19
68: 11—276
68: 22—19
69: 21—19, *27*, *78*, 79, 80
69: 22—19
71: 6—19
71: 16—19
72: 6—19, *28*, 29
72: 16—19, *230*
73: 18—39
74: 5-6—*173*
74: 7-8—276
74: 9—19
75: 8—19, *212*
75: 9—19
77: 24—20
77: 47—20
78: 18—279
78: 24—20
78: 24—*126*
78: 47—20, *106*, 107, 108, *240*
78: 66—168
79: 9-17—20
80: 8-16—20, *240*
80: 9-17—20
80: 10—*67*
80: 17—20
81: 16—20, *230*, 231
81: 17—20
82: 14—20
83: 6-7—20
83: 13—20, *38*, *251*
83: 14—20, 276
84: 6—20, *183*, 184
84: 7—20
90: 5—28
90: 5-6—*28*
90: 15—34
91: 5—9, 10
91: 13—20
92: 12—20
92: 12-14—*67*, *169*
92: 13—20, 291
93: 12-13—276
101: 4—20
101: 12—20
102: 4—20, *28*
102: 5—20
102: 11—20, *28*
102: 12—20
102: 15—20
103: 15—20, *28*, 34, *251*
104: 2—255
104: 12—276
104: 12-16—*251*
104: 14—*28*, 29, 34

104: 15—*212*
104: 16—*67*
104: 16-17—*276*
104: 17—*175*
105: 33—*103*, *240*, *251*
106: 28-30—*149*, 150
107: 37—*240*
118: 12—*287*
119: 4—20
119: 161—10
120: 4—20, *201*, 287
126: 6—*251*
127: 3—20
128: 3—20, *157*, *241*
129: 6—*28*, 34
136: 2—20
137: 1-2—*288*
137: 2—20, *183*, 216
141: 7—*251*, 254
147: 8—34
147: 14—*230*, 232
148: 7-8—*276*
148: 9—*67*

REVELATION
2: 7—*253*, 256
2: 17—*126*
5: 1—173
5: 8—*57*, 59
6: 6—*112*
6: 13—*104*
7: 1-2—*276*
7: 3—*253*, 255
7: 9—*170*
8: 3-4—*57*
8: 7—*28*
8: 8—*289*
8: 10-11—*48*
9: 4—*28*, 276
10: 9—170
11: 4—*158*
11: 6—*289*
14: 18-20—*242*
14: 19-20—*213*
15: 6—*131*
16: 2—*223*
16: 4—*289*
18: 12—*131*, *140*, 189, *194*, *228*
18: 13—*57*, 59, *76*, *82*, 84, *158*, *230*
18: 16—*131*
19: 8—*131*, 132
19: 14—*131*
19: 15—*213*
22: 2—*253*
22: 14—*253*

ROMANS
1: 13—*253*, 256
6: 21—*253*
7: 4—*253*
11: 17—*158*
11: 17-24—*159*
11: 24—*158*

RUTH
1: 22—*112*
2: 2—231
2: 7—231
2: 14—*27*, 222, *229*, *240*
2: 17—*112*
2: 23—*112*
3: 2—*112*
3: 15—10
3: 15-17—*112*

I SAMUEL
1: 24—211
2: 18—*130*
5—168, 267
5: 6—*167*, 267
5: 9—*167*
5: 11—*167*
5: 12—*167*
6—168
6: 4—*167*
6: 11—*167*
6: 17—*167*
6: 19—*167*
7: 4—150
8: 13—211
8: 14—*157*
10: 1—*157*
10: 3—*210*
12: 3—*157*
12: 5—*157*
14: 2—*189*
14: 25—170
15: 7—82
17: 2—*178*
17: 2-49—179
17: 17—*229*
17: 19—*178*, 233
22: 5—*251*
22: 6—*227*
22: 18—*130*
25: 18—*103*, 104, *212*, *229*, *240*, 244
28: 24—211
30: 12—*103*
31: 13—*227*

II SAMUEL
3: 29—*149*, 150
4: 2—*189*, 190
4: 6—*229*, 233
5: 11—*66*
5: 23-24—141, *183*
6: 5—*175*, 176, 261, 285
6: 14—*130*
6: 19—*210*, 288
7: 2—*66*
13: 1—*169*
13: 8—*210*
13: 10—*210*
14: 27—*169*
14: 30—*112*
15: 30—*157*
17: 19—*229*, 233
17: 27-28—*101*, *229*
17: 27-29—*123*
17: 28—*251*, 254
18: 9—179, 197
18: 9-10—*193*, 198
21: 9—*112*
23: 11—*123*

SONG OF SOLOMON
1: 12—*148*, 149
1: 13—*82*, 83
1: 14—90, *124*, *241*
1: 17—*67*, *176*
2: 1—*42*, 77, 116, 147, 205, *234*, 235, 268, 288
2: 1-2—*114*
2: 3—*185*, 186, 291
2: 5—*185*
2: 13—*103*, 104, *241*, 281
2: 15—*231*
2: 16—*114*

3: 6—*56*, *82*, *84*, 85
3: 9—*67*
4: 3—*190*, 291
4: 5—*114*
4: 6—*56*, 58
4: 10—*51*, 52
4: 11—170
4: 13—90, *124*, *190*
4: 13-14—*148*, 149, *205*, 206
4: 14—*39*, 40, *47*, *51*, 52, *56*, 58, *76*, 86, *87*, 149, 188, 219
4: 16—*252*, 256
5: 1—*51*, 52, 85, 225
5: 5—*82*
5: 13—*42*, *51*, 52, *82*, 117, *129*, 225
5: 15—*67*
6: 2—*51*, 52, 116, 225
6: 2-3—20
6: 2-4—*114*
6: 4—*86*, 114
6: 7—20, *190*
6: 11—20, *119*, *190*, 291
7: 3—117
7: 7-8—*169*, *185*, *241*
7: 8—20, 291
7: 9—291
7: 12—*190*
7: 12-13—20
7: 13—*137*, 138
8: 2—*51*, *190*, 191, *212*, 213
8: 5—*185*
8: 9—*67*
8: 14—*51*, 52

SONG OF SONGS
6: 2-3—20
6: 7—20
6: 11—20
7: 9—20
7: 13-14—20

SUSANNAH
17—*215*
54—*177*, 178
58—*178*, *194*

I TIMOTHY
5: 18—*230*
5: 21—10
5: 23—*213*
6: 10—255

II TIMOTHY
2: 6—*253*
4: 13—173

WISDOM OF SOLOMON
2: 8—*205*
16: 20-21—*126*

ZECHARIAH
1: 8—*144*, 291
1: 10-11—*144*
3: 10—*103*, *242*
4: 3—*158*
4: 11-14—*158*
5: 1—173
8: 12—*252*
9: 15—*213*
9: 17—*213*, *230*
10: 7—*213*
11: 1-2—*68*, *176*
11: 2—*175*, *194*, 195
14: 4—*158*

ZEPHANIAH
2: 9—*26*, 220, 237
2: 14—*68*

FIGURE 41. — The Parable of the Tares as recounted in the 13th chapter of Matthew. Depicted are the sowing of the seed; then the sowing of the "tares" by the farmer's enemy; followed by the report of the servant's to their master concerning the mixed crop; then the separation of the "tares" (darnel-grass, *Lolium temulentum*, see also fig. 72); and finally, the sheaves of grain preserved while the "tares" are consumed by fire. Around the upper pictures are stalks of wheat (*Triticum aestivum*). (Mid-19th Century print, courtesy New York Public Library).

GENERAL INDEX

Plants are listed under their Latin or vernacular names, as these occur in the text, with cross references for the more common plants. — Figure numbers of plant illustrations will be found under the Latin plant names (for a complete numerical index of all illustrations see p. xv-xix of the front matter of this book.)

AARON, xvii, 21, 36, 56, 126, 130, 142, 167, 278
AARONSOHN, A., 38, 71, 231, 260, 263, 267
abattichim, 80, 81
abbas, 196
ABBOTT, L., 271
ABDIAS — See OBADIAH
Abel-has Shittim, 25
Abel-Kerâmim, 243
Abel-shittim, 25
Abies cilicica, 174; *also see* pines
Abi-ezrite, 193
ABIGAIL, 105, 212, 229, 240
ABIHU, 56
ABIMELECH, 193
ABRAHAM, xvii, 62, 178, 196, 211, 280; fig. 30
Abraham's oak — *see* oak, Abraham's; *also see Quercus* spp.
ABRAM, 193, 228, 232
AUBRIET, CLAUDE, fig. 64
ABSALOM, 169, 179, 193, 195, 197
abscesses, 223
absinthe, 48
Absinthus, 49
abur, 231
Abyssinia (-n, -ns), 57, 61, 82, 93, 94, 133, 161, 204, 281
Acacia; acacia (-s), 6, 14, 24, 25, 26, 203, 277, 291
Acacia arabica, 25
Acacia arabica var. *nilotica*, 23; *also see* burning bush
Acacia catechu, 125
Acacia farnesiana, 23
Acacia nilotica, 23, 25
Acacia raddiana, 277
Acacia seyal, xvii, 23, 24, 25, 26, 63, 277; illus.: p. 110; *also see* shittah & shittim
Acacia spirocarpa, 26
acacia strap-flower, xix, 23
Acacia suma, 125
acacia, thorny, 23
Acacia tortilis, xvii, 24, 26, 277; illus.: fig. 15; *also see* shittah & shittim
Academia, 286
Academus, 182
Acanthus, acanthus, xvii, 8, 26, 27, 218, 237
acanthus, prickly, 220
Acanthus spinosus, xix, 27, 220; illus.: fig. 48; *also see* nettles
Acanthus syriacus, 26, 27, 220, 237; *also see* nettles
acanthus, Syrian, 220

Acer pseudoplatanus, 108, 181; *also see* maple
acetic acid, 243
acetic acid-forming bacteria, 27; *also see Acetobacter* spp.
Acetobacter, 243; *also see* vinegar
Acetobacter acetigenum, 27
Acetobacter acetum, 27
Acetobacter plicatum, 27
Acetobacter xylinum, 27
acetum, 27, 83
achbar, 167
achú, 62, 120, 249, 253
acid (-forming), 27, 143, 186, 190, 287
aconite, 29
Aconitum, 29
Aconitum napellus, 30
acorn, 195
Acorus calamus, 40; *also see* sweet-flag
Acre Plain, 281
Actinomyces, 290
Actinomyces albus, 290
Acts of the Apostles, 19
ADAD, 190
ADAM, iii, xv, xix, 15, 44, 70, 105, 144, 160, 286, 288, 292; figs. 40, 95, p. 329
Adamsäpffel, xiv, 287
Adam's apple, xiv, xv
adas (-ha, -him), 128
adder, 212
Additions to Esther, 18
ADONIS, 45, 91
Adonis palestina, 43, 117; *also see* lily of the field
adulterate; adulterating, 76, 87
Adulterous Bible, 10
adultery, 10, 24
adventitious, 255
Aegilops variabilis, 28; *also see* grass
AESON, 87
Africa (-n), 14, 35, 37, 48, 55, 58, 63, 75, 81, 82, 83, 92, 102, 110, 121, 127, 128, 152, 154, 155, 168, 170, 171, 188, 191, 204, 214, 228, 276, 280, 282
African bdellium, 81, 82; *also see Commiphora africana*
African rue, 208
agallochum, 47; *also see Aquilaria agallocha* & *Santalum album*
agam (-im, -in), 50, 51

Agaricus campestris, 138; *also see* mushroom & mandrake
agate, 131
Agave americana, 35; *also see* aloe
AGGEUS — *see* HAGGAI
agmōn, 50, 51, 120
agriculture; agriculturists, 2, 15, 171, 232
agrimony, 208
Agrostemma githago, 29, 30, 239; *also see* cockle
ague, 72, 97, 143
ahâlim; 'ahalim, 35, 47, 278
a'hâlot; ahâlōth; 'aholoth, 35, 47, 278
AHASUERUS, 41, 109
ahgmon, 120
AHOLIAB, 130
ahrahzim, 68
ahsis, 213
ahvaboo'ot, 223
Ai, 114
AIBU, Rabbi, 287
aigrettes, 276
Ain, 189
Ain Gidy, 171
Aintab, 175
Air, 148
ajrayna, 203
Akaba, Gulf of, 53, 198
AKIBA, Rabbi, 286
Akkar, 60
alabaster, 148
alah (-im), 195, 197
Albococcus pyogenes, 223; *also see* boils
alcohol, 49
alder, 6, 285
ale, 48, 113
aleh, 255
Aleppo, 53, 135, 139, 147, 175, 180
Aleppo pine, *see* pine, Aleppo & *Pinus halepensis*
ALEXANDER THE GREAT, 90, 109, 206
ALEXANDER, W. L., 260
Alexandretta, 165
Alexandria (-n), 74, 109, 205
Alexandrian translators, 8
alfalfa, 277
alga (-e, -l), 126, 127, 143, 254, 282, 289
algaroba (bean), 73
Algeria, 228, 282
algomin, 209
Algonquin, 75

algoomim, 189
algum (trees), 67, 89, 90, 119, 175, 176, 188, 189, 209; also see *Pterocarus santalinus*
Alhagi; alhagi, 31, 278
Alhagi camelorum var. *turcorum,* xix, 30, 31; illus.: fig. 42; *also see* aspalathus
alhagi-manna, 278
Alhagi mannifera, 31; *also see* manna
Alhagi maurorum, 30, 31, 126, 278
Alhagi maurorum var. *camelorum,* 278
alhenna, 124; also see *Lawsonia inermis*
alkali, 215, 216, 292
allah, 179, 195, 196
ALLEN, D. B., 168, 260
ALLEN, F. S., 264
Allium, 32
Allium ascalonicum, 32; *also see* garlic
Allium cepa, 32, 33; *also see* onion
Allium escallonicum, 32
Allium kurrat, 35
Allium porrum, 29, 34; *also see* leek
Allium sativum, 32; *also see* garlic
Allium schuberti, 39; *also see* rolling thing
alloeh, 35
allon; 'allôn, 179, 195, 196, 197, 286
Allon-bachuth, 193
All Souls Day, 172
allspice, 52
almond (-s), xv, xvii, 6, 8, 11, 14, 35, 36, 37, 38, 46, 51, 65, 77, 159, 177, 179, 182, 278, 279; also see *Amygdalus communis*
almoogin, 189, 209
almug (trees), 90, 188, 189, 209, 228, 287; also see *Santalum album* & *Aquilaria agallocha*
Alnus, 285
aloe (-s), 8, 35, 47, 48, 51, 75, 76, 82, 188, 198, 218, 219; also see *Aquilaria agallocha* & *Santalum album* & *Aloë succotrina*
aloes, lign, 198; also see *Quercus* spp.
Aloë socotrina, 35
Aloë succotrina, xix, 35, 47, 278; illus.: fig. 49
aloes wood, 47
Aloë vulgaris, 48
Aloëxylum agallochum, 47
Alopecurus anthoxanthoides, 28; *also see* grass
alpine, 86, 139, 148
Alraune, 283
alruna; alrunen, 139
altar; altars, 24, 31, 44, 45, 56, 71, 77, 84, 151, 181, 182, 193, 198, 228, 284, 287
altarpiece, 286
Althaea frutex, 53; *also see* treemallow
Althaea hirsuta, fig. 87
Amanus, 147
amâr, 231
amaryllid, 116
amatoria, 283
amber, 58, 182, 252, 254, 280
ambergris, 280
amebae; amebic, 97
America (-n), xiv, 4, 5, 6, 14, 23, 33, 39, 41, 49, 52, 57, 60, 61, 64, 69, 76, 101, 134, 144, 161, 165, 173, 180, 186, 204, 214, 237, 276, 284, 288
American aloe, 35
American Bible Society, viii, 8
American Geographical Society, vii
American Museum of Natural History, vii
American Standard version, 275, 279, 280
Amerinds, 109
Amiens, xv
Ammon (-ian), 8, 26
AMNON, 210
amomum, 59
Amorite, 68, 69, 193

AMOS, 18, 49, 69, 103, 107, 108; fig. 62
Ampelodesmos mauritanicus, 280
Ampelodesmos tenax, 280
amphibians, 3
Amygdalus communis, xvii, 35, 278; illus.: p. 21, fig. 37, p. 274; *also see* almond
Amyris gileadensis, 58, 84, 280; *also see* incense & frankincense
Amyris kataf, 82; *also see* myrrh
Amyris opobalsamum, 84; *also see* spices
anaesthetic (-s), 139, 143, 214
ANAK, 69
Anastatica, 39
Anastatica hierochuntica, xvii, 38, 39, 235, 279; illus.: figs. 22, 36; *also see* wheel
Anastatica hiero-chuntina, 38
Anatolia, 224, 282
a'navim, 243
ANDRIEUX, F. G. J. S., 260
Andropogon, 40
Andropogon aromaticus, 39, 40, 41, 85, 149, 219; *also see* calamus
Andropogon muricatus, 40
Andropogon schoenanthus, 40, 215
Andropogon sorghum, 223
Andropogon sorghum ssp. *sativum,* 223; *also see* reed
Anemone, 45
Anemone coronaria, xix, 28, 41, 42, 114, 117, 129, 147, 234, 235, 279; illus.: fig. 46; *also see* lily of the field
Anemone fulgens, 147, 234; *also see* rose
Anemone nemorosa, 45
anemonies, anemony, xix, 6, 28, 42, 43, 45, 129, 275
Anethum graveolens, 46; *also see* anise
angel (-s), xv, xix, 23, 43, 44, 48, 49, 70, 72, 131, 144, 160, 162, 172, 178, 193, 201, 211, 229, 240, 242, 289
angel, destroying, 66
angels' food, 126, 127
animal (-s), 3, 11, 13, 42, 68, 73, 80, 97, 116, 132, 138, 150, 163, 216, 218, 224, 247, 276, 287, 292; figs. 94, 95
animal-parasites, 143
animals, domestic, 99
anise, 8, 46, 89, 139
Annunciation, xv, xix, 41; fig. 65
anodyne, 80
anoint (-ed, -ing), 76, 157, 158; fig. 25
ANOMOEUS, C., 1, 260
antelope, 9
anthem, 69
Anthemis palaestina, xix, 41, 73, 117; illus.: fig. 50; *also see* lily of the field
ANTHONY, SAINT, *see* SAINT ANTHONY
anthos, 147
anthropologists, 13
Antilebanon, 13, 14, 117, 121, 122, 180, 197, 208, 209, 225, 235, 247, 289
Antioch, 141, 165
antispasmodic, 32, 87, 102
ants, 31
Antwerp, xx
Aphekah, 185
aphrodisiac, 65, 66, 75, 137, 219, 283; fig. 50
APHRODITE, 150
APICIUS, 140
Apinus pinea, xvii, 46, 174, 177; illus.: p. 236; *also see* fir tree & evergreen
Aplotaxis lappa, 218; *also see* cassia
Aplotaxis luffa, 218
Apocalypse — *see* Revelation
Apocrypha (-l), vii, xvii, 1, 7, 9, 30, 31, 39, 44, 151, 178, 205, 253
APOLLO, 91, 95, 114, 124, 172
apostles, 276
apple (-s), xiv, xvi, xix, 6, 9, 105, 138, 172, 185, 186, 187, 190, 201, 276, 286, 287, 291; figs. 91,

95; also see *Prunus armeniaca* & *Tabernaemontana alternifolia*
apple of Sodom, 1, 221, 244, 281; also see *Citrullus colocynthis* & *Solanum incanum*
apples, devil's, 137; also see *Mandragora officinarum*
apples, golden, 187; also see *Prunus armeniaca*
apples, love's, 138; also see *Mandragora officinarum*
apples, Median, 186
apples of gold, 185, 187
apples of Jan, 138
apples, Persian, 186
apricot, xix, 8, 14, 105, 187; also see *Prunus armeniaca*
aprons, 10, 103, 104, 105, 281
APULEIUS BARBARUS, fig. 66
AQUILA translation, 7
Aquilaria agallocha, xix, 35, 47, 57, 188, 300; illus.: fig. 44; *also see* aloes
Aquilaria ovata, 47
Arab (-ia, -ian, -ic, -s), 2, 3, 8, 13, 14, 23, 24, 25, 35, 37, 40, 42, 43, 52, 57, 58, 60, 62, 68, 69, 71, 72, 74, 75, 76, 77, 78, 81, 82, 83, 84, 85, 86, 87, 93, 97, 98, 104, 105, 109, 115, 117, 121, 124, 128, 134, 135, 137, 138, 149, 152, 153, 161, 163, 166, 167, 170, 176, 177, 179, 181, 184, 188, 190, 195, 198, 201, 203, 207, 216, 219, 220, 221, 232, 233, 252, 278, 279, 280, 283, 285, 286, 288
Arabah, Wady, 217
Arabia Felix, 83, 166
Arabia Petraea, 31, 121, 171, 173, 176, 178, 183, 201, 203, 214, 235, 248
Arabian Sea, 95
Arabic versions, 8, 63, 137, 224
arabim, 216, 217
Arabs, Bedouin, 132
Araceae, 29
Aralia racemosa, 149; *also see* spikenard
Aramaic, 10
aran, 176, 285
arar, 'ar-âr, 121, 122
aravah, aravim, 183, 216
arboretum sanctum, 280
arbor Judae, xiv, 280
arbor justi, 285
arbors, 204
abor-vitae, 57, 227, 228
arbutus, 14
architect (-s, -ure), 154, 170
architrave, 69
ARCHIUS, 72
arctic, 13, 14, 127
Arctium lappa, 245; *also see* burdock
Arctium majus, 245
Arctium vulgare, 245
argâmân, 195
ARISTOTLE, 1
ark (-s), Ark, 15, 24, 89, 90, 91, 92, 120, 152, 159, 167, 168, 249, 281; figs. 15, 19, 26
Armenia (-n), 8, 90, 187, 188, 195, 242
Armeniaca vulgaris, 187; *also see* apricot
Armenian version, 8
armôn, 180, 181
Armoracia lapathifolia, 74; *also see* horseradish
Arnold Arboretum, vii, xiv, xx
aro 'air, 121, 122
Aroer, aro'er, 121, 240
aromatic (-al), 34, 40, 46, 48, 60, 76, 83, 86, 89, 152, 153, 162, 177, 210, 228, 281
arot (-h), 92, 94
Artemisia absinthium, 48, 49; *also see* wormwood
Artemisia arborescens, 48
Artemisia arenaria, 48
Artemisia camphorata, 48
Artemisia cinerea, 48
Artemisia fruticosa, 48
Artemisia herba-alba, xix, 48, 278; illus.: fig. 43
Artemisia judaica, 48, 279
Artemisia nilotica, 49
Artemisia vulgaris, 49

artichoke (-s), 39, 287, 288; also see herbs & rolling thing
Arundo, 29, 51
Arundo ampelodesmos, 280
Arundo donax, xix, 50, 51, 120, 172, 173, 280, 285; illus.: fig. 55; also see bulrush & reed
Arundo mauritanica, 280
Arundo maximus, 172; also see pens
Arundo phragmites, 172
Arundo plinia, 280
Arundo scriptoria, 285
Arundo tenax, 280
Arundo vulgaris, 172
arvay nachal, 183
Aryas, 187
ASA, 178
asaf, 161
asafetida, 102
asal at-tall, 278
ASAPH, 67
Ascalon (-ia), 32, 185
Ascension, xv, 41
ASCH, J., 271, 285, 290
ASCHERSON, P. F. A., 116, 261, 273, 283
Asclepias acida, 187; also see apple
ash (-es), 6, 31, 32, 72, 92, 93, 176, 177, 216, 221, 223, 285
ash, flowering, 31, 32, 176; also see *Fraxinus ornus* & manna
ash, manna, 31
ash, mountain, 176
Ashan, 189
Asherah, asherim, Asheroth, 198, 227, 287
Ashkenazi, viii
ASHMUNAZAR, 198
Ashod, 167
ASHTARTE, ASTARTE, 150, 198, 227
ASHTEROTH, ASTEROTH, 198, 227
Ashur (-ites), 62
ashur wood, 63
Asia (-n, -tic), 6, 13, 14, 32, 40, 50, 63, 73, 74, 78, 81, 88, 101, 105, 110, 115, 124, 127, 128, 129, 144, 148, 155, 159, 166, 168, 171, 173, 179, 180, 190, 204, 214, 231, 234, 276
Asia Minor, 13, 43, 46, 58, 87, 102, 168, 179, 180, 186, 216, 225, 277
aspalathus, 30, 31, 223, 224; also see *Alhagi camelorum* var. *turcorum*
asparagus, 53
aspen (-s), 8, 141, 183, 184, 284, 288
aspen, Euphrates, 183
Aspergillus, 290; also see leprosy
Aspergillus fumigatus, 290
Aspergillus niger, 290
Aspergillus unguis, 290
Asphodelus, 43; also see lily
Asphodelus microcarpa, 274
Aspicilia alpinodesertorum f. *affinis*, 126; also see manna
Aspicilia alpinodesertorum f. *fruticulosa*, 127
Asplenium ruta-muraria, 161; also see rue
Asplenium trichomanes, 161
ass (-es), 28, 112, 122, 162, 210, 229, 240
Asshur (-im), 62
Assur, 62
ass, wild, 122
Assyria (-n, -ns), 13, 62, 67, 69, 132, 181, 190, 243, 285
Astragalus, 31, 85, 203; also see spice
Astragalus gummifer, 51, 52, 85, 225
Astragalus tragacantha, xvii, 51, 52; illus.: fig. 16
astringent, 83, 177
atab; atad, 135, 206
ATHENE, 160
Athens, 159, 160
Atlas Mountains, 228, 282
Atriplex, 53, 54, 237; also see mallows
Atriplex dimorphostegia, 53
Atriplex halimus, 53, 280
Atriplex littoralis, 220
Atriplex rosea, 53
Atriplex tatarica, 53

Atropa mandragora, 137, 283; also see mandrakes
atzai-shemen, 97
Aucklandia costus, 218; also see cassia
Augusteno, 268
Augustinian, 276
Aurea Chersonesus, 175
Aurococcus aureus, 223
Authorized Version of King JAMES I — see version, Authorized
autumn-crocus, 147
Avale keramim, 243
AVALLO, A. DE, 59, 61
avatiach, 80, 281
avatichim, 80
Avena sativa, 233, 234; also see oats
Avena sterilis, 28; also see grass
avkat, 84
avkat rochel, 85
avur, 231, 232
aylôn, 196
aysehv, 253
Azerbaijan, 242
AZZ EDDIN, 145

BAAL (-im), 56, 150, 151, 193, 197, 198
BAAL-HANAN, 97, 106, 157
BAAL-PEOR, 149, 150, 151, 168
BAAL-TAMAR, 169, 172
BAASHA, 178
babeer, 93
Babel, tower of, 15
Bab-el-mandeb; Babelmandeb, 75
Babylon (-ia, -ian, -ians), 55, 61, 131, 140, 170, 171, 173, 190, 217, 228, 231
Baca, bâcâ, 183, 184
baccalaureate, 124
Bacchanalian feasts, 106
BACCHUS, 106, 111, 191
bachelor's buttons, 275
bacillary; bacillus, 97, 143
Bacillus ambiguus, 220
Bacillus dysenteriae, 219, 220
Bacillus dysentericus, 219
Bacillus flexneri, 220
Bacillus paradysenteriae, 220
Bacillus pestis, 168
Bacillus pestis bubonicae, 168
Bacillus shigae, 219
BACON, Lord, 139
Bacopa monnieria, 162
bacteria; bacterium, 17, 27, 97, 141, 243
bacteria, acetic-acid forming, 27
Bacterium ambiguum, 220
Bacterium dysenteriae, 220
Bacterium flexneri, 220
Bacterium paradysenteriae, 220
Bacterium pestis, 168
Bacterium schmitzii, 220
Bacterium shigae, 220
bâd, 132
badger, 9
badgers' skins, 24
baharas, 138
baharet, 151
BAILEY, L. H., 60, 81, 87, 166, 204, 261, 269
baka, 184
bake (-d, -s), 126, 210, 211, 229
bakers, professional, 211
BALAAM, 24, 47
BALAK, 24
Balanites, 85, 98, 280
Balanites aegyptiaca, 55, 84, 85, 98, 178, 203, 280, 286; also see balm
balasan, 85
BALDENSPERGER, L., 262, 283; figs. 21, 43
BALFOUR, J. H., 4, 261
BALL, C. R., 271, 288
ballut, 196
balm, 10, 35, 51, 55, 77, 84, 85, 86, 157, 166, 177, 178, 179, 224, 280, 281, 286
balm, Jericho, 98; also see *Balanites aegyptiaca*
balm-of-Gilead, 84, 85, 86, 280, 281; also see *Commiphora opobalsamum*
balsam (-ic), 52, 57, 84, 85, 102, 166, 182, 205

Balsamea africana, 81; also see bdellium
Balsamea mukul, 81
Balsamea myrrha, 82
balsam-flower, 52, 129
balsam i Gilead, 280
balsam, Jericho, 55, 203
Balsamodendron, 205
Balsamodendron africanum, 81; also see bdellium
Balsamodendron gileadense, 84
Balsamodendron kataf, 82
Balsamodendron mukul, 81
Balsamodendron myrrha, 82
Balsamodendron opobalsamum, 84
Balsamodendron roxburghii, 81
Balsamodendrum kataf, 82
Balsamodendrum myrrha, 82
balsam-of-Gilead, 84
balsam-tree, 183; also see *Populus euphratica*
Baluchistan, 81
bamboo, 50
banana (-s), vii, 6, 243, 275; also see *Musa paradisiaca*
BANDMANN, viii, fig. 51, 53, 55, 66
BANG, C. F., 261
banyan, 104; see *Ficus*
bapolim, 168
bar (-s), 231, 232
bara, 138
Barbary Coast, 228
bark (-s), 25, 49, 54, 57, 69, 75, 76, 83, 91, 225
barkanim, barkonim, 206, 207
barley, xv, 6, 27, 29, 71, 88, 101, 103, 111, 112, 113, 128, 129, 153, 166, 170, 189, 210, 211, 229, 232, 233, 234, 240, 243, 254, 257, 289; also see *Hordeum*
barley cakes, 113
barley, common, 112
barley-corns, 113
barley, spring, 112
barley, winter, 112
barn, 134, 252
BARNEVELD, M. VAN, 50, 74, 113, 128, 160, 217, 224, 231, 261
BARNHART, J. H., ix
BARRAS, M. & C., vii
BARREIRA, F. I. DE, 1, 261
barren, 15, 127
BARRETT, M. F., 261
BARRIE, T. K., vii
BARUCH, 19, 31, 126; fig. 30
BARZELLAI, BARZILAI, 101, 128, 229
basam, 52, 85
Basel, 10
basham; Bashan, 6, 15, 62, 68, 85, 122, 158, 194, 195, 197
Bashmuric versions, 8
Basic English version — see version, Basic English
bastard cinnamon, 76; also see *Cinnamomum cassia*
bastard saffron, 87; also see *Carthamus tinctorius*
bastions, 51
bat (-s), 13
batam, 179
BATCHELDER, A., 261
batékh, 81
BATES, E. S., 271
bath, 125
Batna, 180
battîkh-akhar, 81
battîkh-asfar, 81
battle, 118, 183, 202
battlefield, 144
battle shield, 118
bay-laurel, 123; also see *Laurus nobilis*
bay-rum, 124
bay-tree (green), xix, 64, 123, 124, 152, 176; also see *Laurus nobilis*
bazaars, 129, 144
bdellium, 81, 82, 86, 126, 224; also see *Commiphora africana*
bdellium, African, 81, 82, 224
bdellium, Indian, 81, 224
b'dolach, 81
beads, iris root, 118
beams, 66, 67
bean (-s), 101, 112, 128, 166, 211, 229, 232, 233, 254

bean, algaroba, 73; also see *Ceratonia siliqua*
bean, broad, 102; also see *Faba vulgaris*
bean, carob, 280
bean, string, 128
bean, Windsor, 101
beard (-s), 77
beast (-s), 20, 28, 45, 49, 66, 68, 71, 159, 194, 223, 229, 237, 240, 251, 255, 292
beautiful wood, 98
becaim, 139, 183
BECK, V. E., viii
BECKET, 150
bedôloch, 81
Bedouin (-s), 113, 201, 211, 212, 278, 280
bed (-s) of spices, 82, 114
bedstraw, Our Lady's, 277
beer, 113
Beerothite, 189
Beersheba, 78, 227; fig. 78
bees, 150, 170, 208, 287
beet, 137
beetle, 278
begat, 213
behemoth, 9
Beirut, 51, 121, 125, 147, 165, 171, 218
Beitshin, 243
BEKE, C. T., 261
Bel and the Dragon, 19
Belgian, 286
belil, 231, 232
bell (-s), 189, 190, 218
bells, golden, 190; also see *Punica granatum*
belly, 72
belsal, 33
BELSHAZZAR, 242
Belus, 216
benches, 62, 63
BENGTSON, B., 271
Benjamin, 56, 158, 232
BENSON, 43
Benthamidia florida, 277; also see flowering dogwood
BENTLEY and TRIMEN, figs. 49, 54, 61
BENTWICH, N., 274
BENZIGER, 287
benzoin, 224
beooshim, 239
BEOR, 24
BERCHTOLD, F., 261
BERGEY, D. H., 17, 261
BERKELEY, M. J., 261, 267
BERNSTEIN, P. S., 271
berosh, beroth, 176, 177
berosh raanan, 46
berries, berry, 30, 60, 65, 207, 221, 239, 248, 292
BERUNI, A., 278
beryls, 82
besem, 85
Bethany, 142, 148, 169, 172
Bethel, Beth-el, 35, 36, 193
Beth-el-luz, 37
Beth-jesimoth, 24
Bethlehem, xvii, 112, 144, 158, 207, 277; fig. 14
Bethphage, 103, 104, 169
Beth-shean, 101
Beth-shemesh, 167
Beth-tappuah, 185
Bethulia, 44
Betonim, 179, 180
betsâlîm, 33
bet teainah, 172
BEZALEEL, 24
Bhutan, 148
Bible (-s), xvii, xix, 1, 50, 276, 281, 282, 286, 287, 288, 289, 290
Bible, Adulterous, 10
Bible, Alcohol-free, 288
Bible, Bamberg, 9
Bible, Berleberg, 9
Bible, Bishops, 9, 10, 55
Bible, Breeches, 10, 104
Bible, Brothers, 9
Bible, Bug, 9
Bible, Catholic, 275
Bible, Chained, 9
Bible Concordance, 50
Bible, Dartmouth, 275
Bible, Discharge, 10
Bible, Douay — *see* version, Douay

Bible, Dry, 288
Bible, Ears-to-ear, 10
Bible Encyclopedia and Concordance, 29
Bible, Flemish, xv
Bible, Geneva, 9, 10, 104, 105
Bible, Great, 9
Bible, Greek — *see* version, Greek
Bible, GUTENBERG, 9
Bible, He, 10
Bible, Hebrew — *see* version Hebrew
Bible, Kralitz, 9
Bible, Leopolita, 9
Bible, LUTHER's, 9
Bible, MAZARIN, 9
Bible, Murderers', 10
Bible of CRANMER, 9
Bible of Ferrara, 9
Bible of Forty-two Lines, 9
Bible of the Bear, 10
Bible of Thirty-six Lines, 9
Bible, Ostrog, 9
Bible, Oxford, 120, 218, 221
Bible, PFISTER, 9
Bible, Placemakers, 10
Bible, Printer's, 10
Bible, Proof, 9
Bible, REBEKAH's Camels, 10
Bible, Rheims, 9
Bible, Rosin, 10, 55
Bible, SAINT WENCESLAUS, 9
Bible, SCHELHORN's, 9
Bible, She, 10
Bible, Swedish, xv, xvii, xviii, xix, 275
Bibles, Marginal, 29, 38, 39, 52, 59, 79, 85, 90, 98, 124, 152, 178, 187, 228
Bible Society of Cambridge, 10
Bible, Standing Fishes, 10
Bible, Thumb, 10
Bible, To-remain, 10
Bible, Treacle, 10, 55
Bible, Vinegar, 10, 28, 244
Bible, WENCESLAUS, 9
Bible, WENZEL, 9
Bible, Wertheim, 9
Bible, Wicked, 10
Bible, Wifehaters, 10
Bible, WUJEK's, 9
Bible, WUYECK's, 9
Bible, WYCLIFFE's, 218
Bible, Zurich, 9
Biblia del Oso, 10
Biblia Latina, 9
Biblical diseases, 97, 98, 99, 142, 143, 149, 150, 151, 167, 168, 169
BIDDULPH, W., 53
BIEDERWOLF, W. E., 259
bikkûrah, bi'kurah, 104, 281
bilberry, 145
bile, 80
binder, 280
BION, 45
bird (-s), 3, 37, 59, 61, 111, 128, 137, 160, 166, 173, 193, 209, 276, 285, 292
bird-milk, 162; also see *Ornithogalum umbellatum*
BIRDWOOD, G. C. M., 58, 273
bisabol; bissabol, 83
bitter (-ness), 48, 49, 78, 79, 83, 98, 240, 292
bitter drink, 28, 83
bitter herbs, 74, 75, 280; also see *Cichorium*
bitter salso, 280
bitter wine, 28
bitumen, 93
blackberry, 30
blackbirds, 276
black cummin, 153; also see *Nigella sativa*
Black Death, 168, 169
black elder, 73
black locust, 26, 204
black mulberry, 141; also see *Morus nigra*
black mustard, 59, 60; also see *Brassica nigra*
black poplar, 182
Black Sea, 175
black seed, 153
BLANCHAN, N., 42, 60, 261
blanching, 131
BLANCO, M., fig. 20
BLANE, Sir G., 40, 149

blasted, blasting, 28, 142, 229, 252, 254, 289, 290
Blastophaga grossorum, 106
blessed seed, 153; also see *Nigella sativa*
blessed thistle, 72; also see *Cnicus benedictus*
BLESSNER, G., 261
blight (-s), 142, 143, 254
blind (-ness), 134, 149, 150, 255
BLOCH, J., viii, 283
blood, 45, 66, 72, 91, 140, 160, 161, 162, 168, 184, 193, 194, 209, 210, 219, 220, 223, 229, 240, 242, 277, 289
BLOOD, S. W., 261
bloody flux, 97
blue, 204, 205
blue lotus, xvii, 154; also see *Nymphaea caerulea*
BLUM, J. L., 261; fig. 27
boar (-s), 240, 244, 276
boards, 66
BOAZ, 27, 112, 222
BOCHART, 132, 161, 163, 224
bod, 132
BODENHEIMER, F. S., 271
bog, 92
bog rush, 120
Bohemian Version, 9
boil (-s), 103, 105, 142, 168, 223
boil, Damascus, 223
BOISSIER, E., 115, 261
boll (-ed, -s), 110, 129, 133
Bombay, 58, 85, 218, 282
BONER, 119, 196
BONFILS, fig. 85
Books of the Bible, 18
booth (-s), 97, 143, 144, 173, 174, 203, 204, 271, 291
bootz, 110, 132
BORAH, A., 4, 171, 185, 261
Borassus flabellifer, 81; also see bdellium
Borassus flabelliformis, 81
borit (-h), 215
bosam; bosem, 52, 85
boser, 239
boshâh, 29
BOSIO, J., 276
Boswellia, 57
Boswellia carterii, 56, 57, 58, 280; also see frankincense
Boswellia papyrifera, 56, 57
Boswellia serrata, 57
Boswellia thurifera, 56, 57, 85, 182
botch of Egypt, 167
botnim, 119, 179
bo tree, sacred, 285
BOTTICELLI, 44
bouncing-bet, 216
bough (-s), 64, 67, 70, 173, 291
bowels, 219
bowls, 36
bowstrings, 218
box (-es), 9, 26, 63, 64, 91, 174, 176; also see *Buxus longifolia*
box, long-leaved, 63
boxthorn, European, 135, 203, 206
box-tree (-s), 24, 26, 62, 63, 89, 173, 210
boxwood, 62, 63, 173
BOYKO, H., 272, 279
BOYKS, E., 271, 275
boz, 237
bracken, 277
BRAGA, E., 271, 287
BRAHMA, 154, 206
bramble (-s, bush), 6, 66, 103, 134, 135, 153, 165, 202, 206, 207, 220, 221, 237, 248
bramble, common, 207; also see *Rubus sanctus*
bramble, elmleaf, 207; also see *Rubus ulmifolius*
bramble, Palestine, 207; also see *Rubus sanctus*
branch (-es), xv, xvii, 36, 52, 55, 59, 63, 64, 65, 67, 68, 69, 70, 72, 73, 75, 76, 84, 85, 97, 98, 251, 255, 256, 281, 285
brass, 24, 50, 66, 131, 190, 194, 251, 255
Brassica, 60, 61; also see mustard
Brassica alba, 59

Brassica arvensis, 220; also see nettles
Brassica nigra, 59, 61, 280
Brassica sinapis, 220
Brassica sinapistrum, 220
Brazil, 110, 287
bread, 14, 27, 31, 32, 74, 88, 89, 91, 98, 101, 112, 113, 125, 126, 127, 128, 134, 141, 142, 152, 154, 163, 166, 170, 210, 211, 212, 222, 233, 234, 240, 252, 253, 282, 283
bread, barley, 112, 113, 210
bread from heaven, 282
bread from the earth, 282
bread, leavened, 210
bread-mold, 141, 212
bread, unleavened, 210, 229, 230
breeches, 10, 104, 105, 129, 130, 131
Breeches Bible, 10, 104
BREED, R. S., 261
brew (-ers, -ing), 39, 48
BREWER, E. W., 262, 264
brick (-s), 173, 251
brick-making, 254
bread of the Nile, 154
brier (-s), 15, 27, 71, 72, 135, 165, 202, 206, 207, 221, 222, 241, 248, 252; also see *Solanum incanum*
BRIM, C. J., 138, 262
British Brewers Society, xvi
British Museum, xix, 3
BRITTON Herbarium, 196
brook (-s), 14, 28, 50, 51, 92, 120, 151, 152, 173, 183, 194, 215, 216, 217, 240, 247, 291
broom, 54, 91, 161, 201, 202, 208, 223, 280
broom-bush, 91, 201; also see *Cynomorium coccineum*
broom roots, 91
broom, Scotch, 201, 202; also see *Cytisus scoparius*
broom-tree, 201
broom, white, xix, 201, 202; see also *Retama raetam*
b'rosh (-im), b'rotim, 176
BROWNE, L., 274
BROWNE, T., 262
BRUCE, 94
BRUEGHEL, J., fig. 94
BRUNTON, T. L., 262
brush, 161
BRYANT, W. C., 69
Bryonia alba, 138
bryony, white, 138
Bryum truncatulum, 283
bubo (-s), 168, 169
Bubon galbanum, 102
bubonic plague, 72, 167
BUCKHAM, P. W., 262
buckle, 203
buckthorn, Palestine, 202; also see *Rhamnus palaestina*
bud (-ding, -s), xvii, 252, 256
BUDDHA, GAUTAMA, 61, 125, 154, 155, 206
Buddhist, 88
Bug Bible, 9
BUGENHAGEN, J., 259
bugges, 9
BUHL, F., 273, 283
building (-s), built, 69, 89, 90, 227, 237, 277
bulb (-iferous, -ous, -s), 32, 33, 34, 114, 147, 162, 235, 275, 279, 288, 289
Bulboschoenus maritimus, 121
bull, 9
BULLEID, A., 267
bulrush (-es), xix, 50, 51, 92, 93, 94, 120, 236, 249, 281; also see *Arundo donax*
bulrush tall, 121
Bumelia, 26
BUNSEN, C. C. J., 259
bunt, 289
bur (-s), 245
BURCKHARDT, J. L., 69, 101, 227, 278
burdock, great, 245
burial; buried; bury; xvii, 47, 64, 93, 131, 149, 228
Burma, 104, 141, 187
burn (-ed, -eth, -ing, -t), xv, 26, 39, 44, 59, 83, 84, 102, 130, 134, 143, 151, 157, 182, 188, 193, 194, 202, 206, 210, 215, 221, 223, 227, 230, 237, 239, 242, 248, 251, 287
burning ague, 97
burning bush, 23, 25, 277; also see *Loranthus acaciae*
burning coal, 287
burning heat, 97
burnt offering (-s), 24, 31
bur-nut, 71
bur-parsley, 245; also see *Caucalis* spp.
bur-reed, branching, 245; also see *Sparganium erectum*
burweed, 245
BUSCHAN, G., 262
bush (-es), 23, 26, 48, 49, 53, 66, 77, 91, 111, 201, 232, 251, 252, 254, 279, 280
bush, burning, 23, 25, 277
bushes, lotus, 247; also see *Zizyphus lotus*
bush, prickly, 23
butchers-broom, 206, 220; also see *Ruscus aculeatus*
butchers-broom, prickly, 207
butm, 178
Butomus umbellatus, 62, 173; also see flowering-rush
buts; butz, 110, 132
butter, 40, 178
buttercup, 114, 152
Buxus, 63; also see box
Buxus longifolia, 26, 62, 63, 91, 177, 210, 280
Buxus sempervirens, 63, 280
buzzard, 9
Byzantine, 118

cabbage, 6
CABEL, 243
cabinet-work, 47, 63, 95, 158, 228; also see *Nopalea coccinellifera* & *Opuntia ficus-indica*
CACUS, 182
CAESAR, JULIUS, 90, 237
Caesarea Palestina, 147, 197
Cairo, 34, 88, 107, 125, 288
cakes (-s), 31, 87, 88, 89, 103, 104, 105, 112, 126, 152, 153, 157, 166, 210, 212, 229
cakes, barley, 112, 113, 211
cakes, shewbread, 233
cakes, unleavened, 229
calamus, 40, 56, 75, 76, 82, 87, 149, 215, 219
Calamus aromaticus, 40
calamus, sweet, 75, 85, 149, 214, 215, 219
Calcitropa, 71; also see thistles
CALDER, R., 271
caldron, 51
CALEB, 15, 185, 289
calendar, natural, 15
calf; calves, 160, 178, 194, 241
California, 106, 119, 281
CALLCOTT, M., 4, 160, 262, 273; figs. 34, 42
Callitris quadrivalvis, 228; also see thyine wood
CALMET, A., 269
Calotropis procera, 79, 281; also see vine of Sodom
caltrop, 71
Calvary, 276, 277
Calycotome villosa, 71; also see thistles
camah, 230
Cambridge University Press, 17
camel (-s), 10, 51, 55, 77, 78, 95, 132, 170, 251, 254
camels hay, 40, 215; also see *Andropogon schoenanthus*
camel-thorn; camel's thorn, xix, 30, 31, 278; also see *Alhagi camelorum* var. *turcorum*
cammoin, 89
camomile, 43, 237
camp (-s), 38, 126, 142, 149; also see *Lawsonia inermis*
camphor, 48, 124
Camphora officinarum, 124
campion, 30
Cana, 288
Canaan (-ite, -itish), 15, 35, 107, 126, 131, 150, 168, 177, 193, 198, 229, 232, 289

Canary Islands, 30, 48, 92
cancer, 143, 219
candies, 72, 291
candles, devil's, 138; also see *Mandragora officinarum*
candlestick (-s), 36, 215, 218; fig. 81
CANDOLLE, P. DE, 234
cane (-s), 40, 51, 166, 222, 236
cane, scented, 40; also see *Cyperus pertenuis*
cane, spiced, 40
cane, sugar — see sugarcane & *Saccharum officinarum*
cane, sweet, 149, 214; also see *Andropogon aromaticus* & *Saccharum officinarum*
cane, wild, 214
Canticle of Canticles — see Song of Solomon
Canticles, 147
caoshah, 29
Cape of Good Hope, 102
caper, xix, 65, 66, 160, 161, 162; also see *Capparis sicula*
caper-berry, 37, 65
caper buds, 66
caper-fruit, 65
Capernaum, 191
caper-plant, 237; also see *Urtica* spp.
caper, thorny, 161; also see *Capparis spinosa*
caper tree, 65
capitalization, 17
Capparis, 65
Capparis sicula, xix, 65, 160, 203; illus.: fig 51
Capparis spinosa, 65, 161
Capparis spinosa var. *canescens*, 65
caprification; caprifig, 106
captivity, 109, 140
carat, 73
caravan, 95
caraway, 46, 86, 89, 153; also see *Cuminum cyminum*
carbasini coloris, 110
carbon dioxide, 212
carbones genistarum, 287
carcom, 87
Cardamine fontanum, 74; also see water-cress
cardiac glucosides, 152
cardoon, 287, 288
Carduus marianus, 71; also see lady's-thistle
Carduus syriacus, 153; also see Syrian thistle
Carmel (Mount), 14, 67, 77, 194, 197
carmîl, 195
carminative, 46, 86, 139, 153
carob, 6, 14, 72, 73, 280; also see *Ceratonia siliqua*
carob bean, 73
carob-tree, xix, 72
carpas, 110
carpenters, 50, 66
CARPENTER, W., 262
carpets, 53
carrot (-s), 6, 83, 89, 102, 276
CARRUTHERS, 53, 71, 79, 90, 102, 153
CARTER, 261
Carthage, 228
carthamine, 87
Carthamus glaucus, 71; also see thistles
Carthamus oxyacantha, 153; also see brambles
Carthamus tinctorius, 87; also see bastard saffron
cart wheel, 89
carved; carving (-s), 41, 42, 47, 63, 66, 78, 98, 106, 169, 173, 189
caskets, 145
Cassia; cassia, 30, 39, 40, 41, 47, 52, 58, 75, 76, 82, 166, 218, 219; also see *Saussurea lappa*
cassia-bark, 75
Cassia cinnamomum, 75
Cassia lignea, 76, 280
cassia, Meuzal, 75
Castanea castanea, 180
Castanea sativa, 180; also see chestnut
Castanea vesca, 180
Castanea vulgaris, 180

CASTELLI, E., 1, 262
castles, 51
castor-bean; castor-oil, xvii, 203, 204; also see *Ricinus communis*
Carania, 120
cataracts, 128
catechu, 125
cathartic, 78
Catholic version, 9, 275
cattail (-s), 50, 93, 235, 236; also see *Typha angustata*
cattail, American narrow-leaf, 236
cattle, 24, 28, 29, 40, 51, 62, 72, 101, 112, 128, 133, 223, 233, 251, 276, 277
Caucalis, 245; also see bur-parsley
Caucasus (Mountains), 45, 180, 185, 277
cedar (-s), 24, 26, 57, 62, 63, 66, 67, 68, 69, 70, 71, 78, 89, 90, 97, 106, 123, 144, 160, 175, 176, 177, 180, 189, 194, 209, 210, 247, 257, 285, 286, 291; also see *Cedrus libani*
cedar, brown-berried, 121; also see *Juniperus oxycedrus*
cedar-of-Lebanon; cedar of Lebanon, iv, xv, xvii, xix, 14, 47, 68, 69, 70, 115, 123, 134, 177, 181, 210, 248; also see *Cedrus libani*
cedar, sharp, 121; also see *Juniperus oxycedrus*
cedar-trees, 90, 285
Cedrus deodara, 188; also see deodar
Cedrus libani, iv, xv, xvii, xix, 66, 68, 123, 210; illus.: fig. 1, p. 64, fig. 52; also see cedar of Lebanon
Cedrus libanotica, 68
cellulose, 110
CELSIUS, O., 2, 3, 25, 27, 30, 51, 53, 65, 79, 90, 109, 115, 121, 160, 163, 177, 179, 182, 184, 195, 220, 224, 233, 261, 262, 267
Celtic, 101
cemeteries, 64, 90
censer, 56
Centaurea calcitrapa, 153, 165, 202, 245, 248, 280, 287; also see star-thistle
Centaurea iberica, 71; also see thistle
Centaurea microcephala, 39; also see galgal
Centaurea pallescens, 71; also see durdar
Centaurea venustum, 71; also see dwarf centaury
Centaurea verutum, 71; also see dwarf centaury
centuryplant, 35; also see *Agave americana*
Ceos, 91
Cephalaria syriaca, 283
Ceratonia siliqua, xix, 72, 280; illus.: fig. 56; also see husks & carob
Cercis, 179, 280
Cercis siliquastrum, xiv, xv, 73, 106, 184, 280; illus.: fig. 2; also see Judas-tree
cereal (-s), 13, 112, 134, 231, 233
ceremonial; ceremonies; ceremony, 47, 64, 66, 87, 137, 151, 187, 291
CERES, 91, 101, 191, 198
Ceylon, 13, 40, 75, 76, 95, 187, 188, 189
chabatzeleth, 235
chablat zeleth, 147
chadek, 202
Chaetochloa italica, 166; also see millet
Chaetomium globosum, 290
chaff, 38, 39, 251, 252, 254
Chaldee Targums, 7, 42
Chaldee versions, 7, 124
CHALLONER, Bishop, 9
Challenor version, 9
Challenor-Rheims version, 10, 17
Challenor-Rheims-Douay version, 17
Chamaerops, 287
Chamaerops humilis, 287
CHAMBERLIN, R. B., 273, 275

chametz, 211
chamomile, xix
chapels, 84, 287
chapiters, 154, 189, 190
CHAPMAN, J. W., 259
charchur, 97
charcoal, 26, 201, 211, 227
charcoal burners, 26
chargers, 11
chariot (-s), 66, 67, 69, 175
CHARLEMAGNE, 38, 72
CHARLES, R. H., 262
charlock, 27, 220, 237; also see *Sinapis arvensis*
CHARTON, M. E., 252
chârûl, 26, 27, 220, 221, 237
charûlîm, 220
CHASE, F., 262
chasmal, 254
châtsîr; châtzîr, 29, 34
chatz'tzon tamar, 172
chavatzelet, 147, 235
chaydayd, 187
chazit, 248
chedek, 165, 206, 221
chelbenah, 102
chenar, 181
Chenopodium, 53, 216; also see mallows
cherev, 73
CHERSON, 175
cherubims, 97, 159, 169, 189
chestnut (tree), 9, 35, 67, 180; also see *Castanea sativa*
chestnut, common, 180
chewing-gum, 177
CHEYNE, 283
Chian, 178
Chicago version, see version, Goodspeed
chicken, 61, 134
chicken pens, 276
chick-pea, 163
chicory, 74, 75, 140; also see *Cichorium* spp.
child (-ren), 49, 66
China; Chinese, 5, 45, 53, 57, 70, 76, 110, 119, 124, 140, 141, 166, 167, 185, 187, 188, 191, 205, 217, 219, 234, 248, 282, 288
Chinese cinnamon, 76; also see *Cinnamomum cassia*
Chio (-s), 178
chirsanat, 233
chiryonim, 162
chitim; chittah; chittim, 62, 230
CHIZIK, B., 274
CHLORIS, 45
ch'metz, 27
choach, 71, 153, 165, 206
chômets, 27
CHRIST (JESUS), xix, 6, 49, 70, 101, 155, 158, 277, 280, 283; also see JESUS
CHRIST, K. N. H., 116, 117, 262
CHRISTENSEN, C., 274
Christian (-s), 44, 50, 115, 160, 172, 180, 184, 198, 202, 276, 277, 278, 288
Christianity, 118
Christi törnekrona, 284
Christmas, 4, 33, 45, 111
Christmas-rose, 277
Christ-thorn, 27, 165, 220, 284; also see *Paliurus spina-christi*
Christ-thorn, Syrian, 248; also see *Zizyphus spina-christi*
Chronicles, I & II, 18, 69, 132
chronology, viii, 13
ch'roovim, 210
church (-es), 4, 9, 111, 161, 244, 287
Cicer arietinum, 163; also see chick-pea
Cicer pinnatifidum, 163
CICERO, 102, 228
Cichorium endivia, 74, 140, 280; also see herbs
Cichorium intybus, 74
Cichorium pumilum, 74
Cicuta maculata, 49; also see hemlock
cinnabar, 195
Cinnamomum camphora, 124
Cinnamomum cassia, 75, 219, 280
Cinnamomum zeylanicum, xvii, 76; illus.: p. 257
cinnamon, xvii, 30, 39, 41, 47, 52, 56, 57, 58, 75, 76, 77, 82, 87, 223, 224

cinnamon, bastard, 76
cinnamon, Chinese, 76
Cirsium syriacus, 153; also see Syrian thistle
Cissus orientalis, 239; also see Oriental vine
Cissus pinnata, 239
Cistus, 77, 147, 151, 205, 234
Cistus creticus, 77, 224; also see myrrh
Cistus ladaniferus, 77
Cistus salviaefolius, 77
Cistus salvifolius, xvii, xix, 77, 225, 280; illus.: p. 122, fig. 53
Cistus villosus, 77, 224
Cistus villosus var. *creticus*, 77
citron (holy), xvi, 14, 138, 171, 186, 187, 285, 286, 287, 291; also see *Citrus medica*
citronella, 280
citron, true, 228
Citrullus citrullus, 80
Citrullus colocynthis, xix, 49, 78, 79, 81, 88, 204, 221, 240, 244, 281; illus.: fig. 54; also see gourd
Citrullus vulgaris, 80, 281
citrum, 228
Citrus; citrus, 5, 186, 228; also see apple
Citrus amara, 185
Citrus aurantium, 185
Citrus aurantium var. *decumana*, 186
Citrus aurantium var. *grandis*, 186
Citrus aurantium var. *sinensis*, 185
Citrus bigaradia, 185
Citrus decumana, 186
Citrus grandis, 186
Citrus maxima, 186
Citrus medica, 138, 186, 228, 285, 287
Citrus medica var. *cucurbitina*, 187
Citrus medica var. *cylindrica*, 187
Citrus medica var. *lageriformis*, 187, 290, 291
Citrus sinensis, 185
Citrus vulgaris, 185
Cladonia rangiferina, 127; also see reindeer-moss
Cladosporium herbarum, 290
Cladrastis lutea, 91; also see gopher-wood
Clarendon Press, 10, 28, 244
CLARKE, 71
clean (-sing), 66, 98, 99, 142, 160, 173, 193, 209, 223
Clematis, 285
CLEOPATRA, 206
clergymen, xx, 108
CLEWBERG, C., 262
climate (-s), 5, 6, 13, 31, 60, 66, 141, 149, 275
clone, 87
clotbur, 245
cloth (-e, -es, -ing), 41, 43, 44, 53, 78, 87, 105, 109, 132, 142, 290
clover, 34, 54, 277
cloves, 32
CLOVIS, King, 118
clubrush, 120; also see *Scirpus* spp.
clubrush, cluster-headed, 121
clubrush, lake, 121
clubrush, saltmarsh, 121
clubrush, sea, 121
Cnicus benedictus, 72; also see Our Lady's Thistle
Cnicus syriacus, 153
Coa, 132
coals, 56, 201, 210, 287
coals of juniper, 201, 202, 287; also see *Retama raetam*
coash, 29
coat-of-arms, 118, 172, 206
Coccus cacti, 194; also see cochineal insect
Coccus ilicis, 194
Coccus manniparus, 31, 278
Cochin-china, 47
cochineal, 277
cock (-s), 32
cockle, 29, 71, 112, 134, 153, 239; also see *Vitis orientalis*
cocoon, 278

COCQUIUS, A., 1, 262
Code, Levitical, 168
Coelosyria, 121, 225, 235
coffin (-s), 64, 90, 101
COHEN, A., 274
coins, 206
Colchicum, 116
Colchicum autumnale, 147, 234; also see autumn-crocus
College of Physicians & Surgeons, vii
colocynth, xix, 78, 79, 80, 81, 221, 279; also see Citrullus colocynthis
Colossians, 19
Columbia University, ii, vii
COLUMBUS, C., 109
COLUMELLA, 32
columns, 218
combs, 63
commandments, xv
commerce; commercial (-ly), 2, 48, 59, 62, 76, 77, 82, 85, 87, 95, 102, 109, 110, 119, 124, 149, 158, 178, 197, 204, 225, 232
Commiphora abyssinica, 82
Commiphora africana, 81, 82, 224; also see bdellium
Commiphora kataf, xvii, 82, 83; illus.: fig. 27; also see myrrh
Commiphora molmol, 84
Commiphora mukul, 81
Commiphora myrrha, xvii, 77, 82, 83, 177; illus.: p. 99
Commiphora myrrha var. molmol, 84
Commiphora opobalsamum, 52, 58, 82, 84, 85, 177, 178, 184, 224, 280, 281; also see myrrh
Commiphora playfairii, 84
Commiphora roxburghii, 81, 224
Community Church, New York, vii
Community Church, Unity, vii
condiment, 86, 89, 124, 139
CONDIT, I. J., 104, 105, 262
cone (-s), 46, 63
confection (-ery), 46, 86, 87, 166, 180, 191, 214
confinement, 282
Confraternity Edition — see version, O'HARA
CONFUCIUS, 206
congregation, 125, 130, 167, 292
conifer (-ous, -s), 46, 174, 175, 176, 177, 181, 228, 254
Conium maculatum, 49; also see poison-hemlock
Constantinople, 169, 232
consumption, 49, 97, 143
contour-plowing, 14
Convallaria majalis, 114, 117; also see lily-of-the-valley
Convent of Saint Catherine, 278
Convolvulus floridus, 30; also see lignum Rhodium
Convolvulus scoparius, 30
cook (-ed, -ery, -ing, -s), 32, 46, 65, 74, 87, 89, 139, 140, 180
copher, 124
Coptic, 8, 42
Coptic Version, 8
coral, 131
Corchorus olitorius, 53, 147, 234; also see mallows
cord (-age, -s), 53, 132, 217, 218
Cordova, 191
Corfu, 187, 279
coriander (seed), xvii, xix, 81, 82, 86, 126, 257; also see Coriandrum sativum
Coriandrum sativum, xvii, xix, 86; illus.: fig. 18, 57
Corinth, 145, 244
Corinthians, I & II, 19
cormorant, 9
corn, xv, 28, 29, 36, 56, 112, 126, 157, 169, 202, 210, 211, 212, 213, 228, 229, 230, 231, 232, 233; also see Triticum spp.
corn, American, 232
corn-cockle, 30
corn, ground, 229
corn, Jerusalem, 222
corn of heaven, 126, 127
corn, old, 232
corn, parched, 101, 128, 222, 229, 232
corn, standing, 232
corn, winnowed, 232

corn, winter, 234
CORNELL, R. D., 262
Cornus florida, 277; also see flowering dogwood
Corregio, 44
CORTEZ, 109
Corylus avellana, 37; also see hazel
COSIMO P. DI, 286
cosmetic, 84, 125
cosmogeny, 33
costus, 58, 76, 219
COTES, R. A., 18, 44, 45, 138, 235, 262
cotton, 6, 41, 93, 109, 110, 131, 132, 133, 290, 291; also see Gossypium herbaceum
cotton, Levant, 109
cotton, sea-island, 110
cotton, tree, 110
Cotula anthemoides, 49; also see hemlock
Coturnix vulgaris, 127; also see quail
coughs, 118
court, 34, 66, 129, 130, 131
courtyards, 105
COUTANCE, A., fig. 26
covenant, 48
Coverdale version, 9, 10
cow, 170, 210
cradle, 106; fig. 63
cramps, 139
CRANACH, LUCAS, xix, fig. 95
crane, 9
Cranmer version, 9, 275
Crataegus, 203
crater, 91
creation; Creator, 20, 69, 285, 291
creel, 93
creeper, 292
Cretans; Crete, 90, 102, 171
cricket, 72
Crimea, 175, 188
crimson, 194
Crithodium aegilopoides, 231
crocodile (-s), xix, 9, 203, 247
Crocus; crocus, 6, 86, 87, 116, 147, 234
crocus, autumn, 234
Crocus cancellatus var. damascenus, 86, 87
crocus, Damascus, 87
Crocus hyemalis, 86, 87
crocus, ringed, 87
crocus, saffron, xix, 86, 87
Crocus sativus, xix, 86, 87; illus.: fig. 58; also see saffron
crocus, Syrian, 87
Crocus vitellinus, 86, 87
crocus, winter, 87
Crocus·zonatus, 86, 87
cross, xix, 5, 45, 49, 66, 70, 160, 172, 184, 198, 276, 277, 284, 285
cross-pollination, 106
CROWFOOT, G. M., 262, 283; figs. 21, 43
crown (-ing, -s), 84, 93, 118, 131, 144, 148, 155, 165, 170, 181, 190, 191, 204, 205, 218, 223, 235, 284, 289
crown-imperial, 43; also see Fritillaria imperialis
crown-of-thorns, 165, 289; also see Paliurus spina-christi & Zizyphus spina-christi
crucifer (-ous), 205, 220
crucified; crucifixion; crucify, xiv, xv, 5, 45, 70, 80, 160, 161, 172, 184, 214, 222, 276, 284
CRUDEN, 50
Crusade (-rs, -s), 32, 85, 118, 206
CTESIAS, 109
cubit (-s), 50, 66, 78, 79, 97, 113, 129, 154
cucumber (-s), 32, 33, 34, 78, 79, 80, 88, 287; also see Cucumis spp.
cucumber, common, 88
cucumber field, 88
cucumber, hairy, 88
cucumber, squirting, 78
Cucumis anguria, 88; also see gourd
Cucumis chate, 88
Cucumis citrullus, 80; also see melon
Cucumis colocynthis, 78
Cucumis colocynthus, 78

Cucumis dudaim, 137; also see Mandragora officinarum
Cucumis melo, 80; also see melon
Cucumis prophetarum, 78
Cucumis sativa var. chaete, 88
Cucumis sativus, 88
cucurbita, 204
Cucurbita bettich, 80; also see melon
Cucurbita citrullus, 80
Cucurbita maxima, 204; also see squash
Cucurbita moschata, 204
Cucurbita pepo, 204
Cucurbita pepo var. ovifera, 204
culms, 50
cultivate (-d); cultivation; cultivators, 6, 14, 32, 33, 37, 40, 46, 52, 53, 59, 61, 63, 73, 74, 81, 86, 87, 88, 89, 101, 105, 106, 107, 110, 112, 113, 114, 116, 119, 121, 124, 128, 131, 133, 139, 140, 141, 142, 151, 152, 159, 166, 167, 171, 180, 185, 186, 187, 190, 203, 208, 214, 223, 227, 228, 231, 233, 234, 237, 242, 243, 244, 248, 254, 256
CULTRERA, P., 262
cumin, 89
Cuminum, 153
Cuminum cyminum, xvii, 89; illus.: fig. 17; also see cummin
Cuminum cyminum var. hirsutum, 89
Cuminum sativum, 89
cummin, xvii, 46, 89, 139, 152, 232, 233, 234
cummin, black, 153; also see Nigella sativa
cup (-s), 36
cupar, 91
cupbearer, 240, 243
cuper, 91
CUPID, 145, 154, 206
Cupressus, 124, 152
Cupressus sempervirens var. horizontalis, 89, 119, 174, 176, 189, 209, 281; also see cypress
Cupressus sempervirens var. pyramidalis, 90
Cupressus sempervirens var. stricta, xvii, illus.: fig. 19
cure, 66, 72
currants, 60, 244
curries, 86, 87, 152
curtain (-s), 129, 130, 193, 204
cussemoth, 152, 233
Cyclamer; cyclamens, 6, 43, 275
Cyclamen persicum, 43, 117
Cydonia cydonia, 186; also see quinces
Cydonia oblonga, 186
Cydonia vulgaris, 186
Cymric, 35
Cynara cardunculus, 287
Cynara scolymus, 288
Cynara syriaca 27; also see thistles
Cynomorium coccineum, xix, 91, 202; illus.: fig. 60; also see juniper roots
cynomorium, scarlet, 91
Cynoxylon floridum, 277; also see flowering dogwood
CYPARISSOS, 91
Cyperus, 40, 51; also see sedge
Cyperus esculentus, 62; also see flag
Cyperus papyrus, xix, 92, 120, 249, 281; illus.: figs. 59, 63
Cyperus papyrus var. antiquorum, 92; also see bulrushes & paper reeds
Cyperus papyrus var. palaestinae, 92
Cyperus pertenuis, 40
cypress (-es), 9, 11, 14, 25, 26, 46, 63, 70, 89, 90, 91, 98, 124, 152, 160, 173, 174, 175, 176, 177, 180, 181, 189, 194, 209, 281, 286
cypress, columnar, xvii, 255
cypress, evergreen, 89; also see Cupressus sempervirens var. horizontalis
cypress tree (-s), 46, 89, 90, 91
Cyprus, 30, 62, 84, 91, 124, 178, 187, 224
CYRUS (King), 67, 69, 91
Cytisus scoparius, 201; also see broom
Cytophaga rubra, 290

dâgân, 230, 232
daisy, 6, 43
daleket; dalleketh, 97
DALMAN, G., 15, 27, 39, 43, 60, 71, 74, 109, 134, 207, 214, 221, 245, 248, 262
DALZIEL, E., 274
DALZIEL, G., 274
dam, 280
Damascus, 30, 117, 121, 144, 147, 180, 182, 190, 235
Damietta, 88
Dan, 39, 75
dandanah, 139
dandelion, 6, 74, 75, 140; also see *Taraxacum officinale*
DANIEL, 18, 197, 202, 254
Daniell, W., fig. 59
Danish, 54, 72, 237
Danish version, 237
DAPHNE, 124
DAPHNIS, 71
dardar, 71, 153, 287
Dar el Kodîb, iv
D'ARLON, B., viii, 283
darnel (-grass), xvii, xix, 134, 282, 283, 300; also see *Lolium temulentum*
darnel, drunken, 134
Dartmouth Bible, 275
date (-palm, -s, tree), xv, xvii, xviii, xix, 6, 14, 95, 120, 144, 170, 172, 255, 275, 285, 286, 288; also see *Phoenix dactylifera*
date-plum, 95; also see *Diospyros* spp.
dates, Trebizond, 98; also see *Elaeagnus angustifolia*
Daucus aureus, 207; also see thorns
DAVID, 38, 41, 43, 66, 67, 69, 83, 105, 108, 123, 130, 157, 175, 176, 179, 218, 219, 230, 251, 281
DAVID, Père, xix, fig. 91
DAY, G. E., 43, 51, 274
dead; death, xv, xvii, 28, 55, 61, 68, 78, 83, 88, 102, 131, 134, 138, 142, 148, 149, 155, 157, 162, 167, 168, 169, 291
Dead Sea, 13, 23, 25, 38, 50, 53, 55, 60, 71, 78, 79, 91, 121, 135, 152, 171, 201, 275; fig. 79, 86
debelah, 104
DEBORAH, 169
Deccan, 104
DEDAN, 62
deer, xv
deficiency, vitamin, 143
dehsheh, 29
Delean, 88
delirium, 48
delta, 13, 231
Deluge, 120
demulcent, 32
deodar, 188; also see *Cedrus deodara*
desert (-s), xvii, 3, 12, 13, 15, 24, 25, 26, 29, 31, 33, 34, 36, 39, 41, 48, 55, 62, 63, 66, 69, 78, 80, 84, 89, 117, 120, 121, 122, 126, 127, 147, 161, 171, 173, 176, 198, 201, 203, 210, 217, 221, 227, 245, 275, 276, 278, 280, 282
desert-thorn, 135; also see *Lycium europaeum*
déshe, 29
deshseh, 253
deshseh yerek, 253
Deuteronomy, xv, xvii, 18; fig. 9
devil (-s), 119, 141, 160
dew, 117, 125, 126, 278
dhura, xix, 222, 223; also see *Sorghum vulgare* var. *durra*
dhura-bayda, 222
diabolical, 139
DIANA, 44, 119
diarrhea, summer, 220
dibs, 244
Dictamnus albus, 23; also see gasplant
Dictamnus fraxinella, 23
Dilan, 88
dilb, 181
Dilean, 88
dill, 8, 46, 86, 89, 152
DILLMAN, A., 283
dill-water, 46
DINSMORE, J. E., 71, 77, 173, 174, 195, 234, 262
DIODORUS SICULUS, 83

DIONYSUS, 111, 150
DIOSCORIDES, 1, 2, 30, 41, 43, 75, 83, 140, 149, 162, 178, 225
Diospyros ebenaster, 95; also see ebony
Diospyros ebenum, 95
Diospyros melanoxylon, 95
Diplococcus gonorrhoeae, 150
DIS, 148
disease (-s), 33, 45, 66, 72, 97, 98, 99, 124, 141 142, 143, 149, 150, 151, 167, 168, 169, 198, 219, 220, 253
disease, incurable, 219
disease, inflammatory, 169
diseases, plant, 289
Dispensation, New, xv
Dispensation, Old, xv
distillation, 46, 113
diuretic, 32
DIVES, 132
divi ladner, xvii, 187; also see *Tabernaemontana alternifolia*
dochan, 166, 167
doctrine, 28
dod; dodim, 283
DODGE, B. O., 289, 290
DODOENS, xx
dog (-s), 32, 137
dog-rose, 6
dokhan, 22
doleful creatures, 9
döm, 248
domestic (animals), 13, 73, 93, 133
DON, Prof., 31
Donax tenax, 280
DONEY, C., 41, 144, 147, 234, 262
door (-s), 67, 68, 90, 97
door-posts, 66
DORÉ, G., 274; fig. 52
Douay version, see version, Douay
dough, 210, 211, 212
DOUGHTY, C. M., 105
dove (-s), xvii, 157, 159
dove's dung, 162, 163, 284; also see *Ornithogalum umbellatum*
Dracaena ombet, fig. 92
Dracocephalum canariense, 86; also see balm-of-Gilead
dragon (-s), 9, 28, 50, 92, 237, 240
DRAR, M., 272, 281
dream (-ed), 111, 113, 138, 172, 229, 240, 243
drink (-s), 27, 48, 49, 51, 55, 78, 79, 80, 81, 86, 191, 222, 289
drink, bitter, 28
drink, strong, 170, 212, 213
DRISLER, H., 270
DRIVER, S. R., 262, 283
dromedaries, 112
dropsy, 118
drought (-s), 112, 127, 227, 231
drug (-ed, -s), 35, 79, 80, 83, 87
Druids, 107, 198
drunk (-en, -enness), 48, 49, 134, 212, 213, 240, 242, 243
drunken darnel, 134
Druses, 141
Dryads, 198
duckweed, 6
dud, 283
Dudai, 283
dudâim, 137, 283
dudaism, 283
dufwoträck, 284
dukhn; dukhun, 166, 167
dulband; dulbend, 235
DUNCAN, J. S., 263
dung, 57, 104, 127, 128, 211
dung, dove's, 162, 163; also see *Ornithogalum umbellatum*
dung, quail's, 127
dung, sparrow's, 163
DUNS, J., 263
DU PAS, C., 1; see also PAS, C. DU
DÜRER, A., 286; fig. 92
DUSCHAK, M., 263
dust, 38, 39, 252
dust-bowls (-storms), 69
Dutch, xx
DUVEEN'S, 286
d'vash, 170
d'velet; d'velim, 104
dwarf centaury, 71; also see *Centaurea verutum*
dwarf elder, 30

dwarf mallow, 53; also see *Malva rotundifolia*
dye (-ing, -s), vii, 11, 87, 125, 191, 194, 195, 197, 205
dye, blue, 195
dye, red, 195
dye, yellow, 195
DYER, R. A., 121
dysentery, 92, 97, 168, 219
dysentery, amebic, 219
dysentery, bacillary, 219
dysentery, diphtheritic, 220
dysentery, Japanese, 219
dysentery, tropical, 219

eagle, 67, 241
eagle-tree, 47; also see *Aquilaria agallocha*
eaglewood, xix, 8, 47, 48, 52, 58, 300; also see *Aquilaria agallocha* & *Santalum album*
early fig, 104; also see *Ficus carica*
ears (of corn), 229, 230, 231, 232, 288, 289; also see *Triticum* spp.
Earth; earth, xv, 20, 28, 33, 37, 45, 48, 63, 68, 71, 91, 104, 120, 133, 138, 144, 148, 154, 191, 194, 229, 230, 242, 251, 256, 278, 285
Easter, 41, 171
East Indies, 47, 50, 57, 75, 76, 102, 151, 155, 166, 214
eat (-en, -ing), 46, 53, 72, 73, 74, 78, 80, 86, 88, 92, 93, 101, 113, 128, 134, 162, 163, 237, 251, 252, 287
Eberthella, 141
Eberthella ambigua, 220
Eberthella dysenteriae, 220
Eberthella flexneri, 220
Eberthella paradysenteriae, 220
Eberthella typhi, 97
ebony, 41, 95; also see *Diospyros* spp.
ebony, Ceylonese, 95
ebony, Ethiopian, 95
ebony, Indian, 95
EBSTEIN, W., 263
Ecbalium elaterium, 78; also see squirting-cucumber
Ecballium elaterium, 78
Ecclesiastes, xxiv, 18, 37; figs. 50, 83, p. 293
Ecclesiasticus, 19, 30, 39, 147, 178, 179, 224, 235, 288
Echinops, 278
ECHO, 148
EDDIM, AZZ, 145
Eden, xv, xvi, xvii, 15, 67, 68, 70, 81, 82, 104, 105, 172, 186, 187, 191, 243, 286, 288, 292; *see also* Garden of Eden
edible, 46, 59, 62, 106, 155, 163, 180, 248, 287
EDOM, 128
eelgrass, 249; also see *Zostera marina*
eggs, 137, 276
egöz, 119
Egypt (-ian, -ians), xix, 3, 5, 6, 13, 14, 23, 24, 25, 30, 32, 33, 34, 35, 36, 41, 45, 46, 50, 51, 53, 54, 55, 58, 59, 62, 65, 70, 72, 74, 75, 77, 78, 80, 81, 83, 84, 86, 87, 88, 89, 90, 92, 93, 101, 107, 109, 110, 112, 113, 120, 121, 124, 125, 126, 128, 130, 131, 132, 133, 141, 152, 153, 154, 155, 157, 160, 161, 162, 166, 167, 168, 169, 170, 171, 172, 177, 178, 189, 190, 191, 195, 201, 203, 204, 205, 218, 222, 223, 228, 229, 231, 232, 233, 234, 239, 240, 242, 243, 244, 253, 255, 280, 281, 282, 283, 284, 286, 289, 291
Egyptian Dynastic period, 281
Egyptian mimosa, 23; also see *Acacia nilotica*
Egyptian privet, 124; also see *Lawsonia inermis*
Egyptian version, 8
Eh'limah, 172
eh'lit, 172
EHRENBERG, 83, 278
EIG, A., 5, 263, 279
Einkorn, 231, 234
Eisenach German Protestant Church Congress, 9
el, 179, 195, 196, 198

Bible Plants — 309 — General Index

Elaeagnus angustifolia, 97, 98, 159, 174, 281; *also see* wild olive
Elah; elah, 175, 178, 179, 195, 196, 197
'elate, 286
Elath, 169, 172, 188, 198
elder, black, 73
Elealeh, 244
electrum, 254
elephants, 140, 141, 242
elephantiasis graecorum, 143
ELIJAH, fig. 78
ELIEZER, Rabbi, 292
Elim; elim, 169, 172, 179, 198
ELISHA, 78, 79, 80, 81, 204, 240; fig. 54
ELIZABETH, mother of JOHN, xv
ELIZABETH of Hungary, 206
el kali, 216
ELLACOMBE, H. N., 263
elm (-s), 5, 9, 26, 174, 178, 179, 181, 194, 285; *also see Pistacia terebinthus* var. *palaestina*
elm, common, 175
elon, 179, 195, 196, 197
e'lot; elot (-h), 169, 170, 172, 198
El-paran, 198
embalm (-ed, -ing, -s), 35, 47, 58, 83, 131; fig. 49
embroidered; embroiderer, 130, 190
emeralds, 130
emerods, 167, 168, 169
EMERSON, E. R., 273
emetic, 137
emmenagogue, 35, 87
emmer, wild, 231
Empire (Roman), 106, 232
emperor, 72, 123, 124, 158, 171
Empress THEODORA, 118
enav, 243
enchantress, 87
endive (-s), 6, 34, 74, 75, 140; *also see Cichorium* spp.
endive, Indian, 74
Engaddi; Engedi; En-gedi; Engeddi, 78, 124, 125, 151, 169, 171, 180, 221, 241, 244
England; English, 2, 5, 6, 13, 14, 36, 37, 38, 45, 60, 64, 73, 81, 83, 84, 87, 88, 93, 101, 108, 118, 119, 129, 137, 138, 139, 141, 145, 152, 198, 201, 204, 206, 208, 235, 244, 275, 282, 284, 287, 288
ENGLER and DRUDE, fig. 60
ENGLER, A., 265
English version, 8, 89, 105
English Revised version, 275
En-rimmon, 190
enträd, 287
ephah, 111, 112, 229, 252
Ephesians; Ephesus, 19, 168
ephod (-s), 110, 129, 130, 191
EPHRAIM, 158, 182, 185, 256
epidemic (-s), 72, 151, 168, 169, 219
Epidermophyton, 97, 141; *also see* plague
Epidermophyton rubrum, 98, 143
epileptic fits, epilepsy, 120, 143, 208
Episcopal Committee of the Confraternity of Christian Doctrine, 17
ER, 169
Eragrostis megastachya, 28; *also see* grass
erez, 68, 123, 210
Erica orientalis, 121; *also see* heath
Erica vagans, 121
Erica verticillata, 121
ERICSON, A. L., 263, 265
ERISICHTON, 199
ermine, 149
erosion, 15, 69
eruptions, 91
Ervalenta, 129
Ervum lens, 128, 129; *also see* lentils
Ervun lens, 128
Eryngium, 39; *also see* thistle & galgal
erysipelas, 97
Erysiphe graminis, 290
ESAU, xix, 128; fig. 90
Escalom, Escalon, 32
Esdraëlon, 71, 147, 158
ESDRAS I — *see* EZRA

ESDRAS II — *see* NEHEMIAH
eseb, 29
Eshcol, 240, 244, 289
eshel, 198, 227, 228
esparcet, 277
Es-salt, 289
ESSEN, J. G., 263
ESTERNAUX, Mrs. G., vii, 271
ESTHER, 18, 43, 109, 110, 132, 143, 144
Eternal; eternal, 36, 47, 57, 114, 176
Etham, 52, 119
Ether, 189
Ethiopia (-n), 13, 89, 95, 109
Ethiopian version, 8
ethrog, 187, 287, 291; *also see Citrus medica*
etoc, 187
ETOC, G., 263
etrog, 187, 285, 286, 290, 291; *also see Citrus medica*
ets shemen, 97
ETT, B., 263
etun, 132
etz, 178, 254, 255
etz aboth; etz 'aboth, 174, 228
etz avot; etz a'vot, 174, 228
etz shamen; etz shemen, 97, 98, 174, 281
Eucystia pestis, 168
Eugenia pimenta, 52; *also see* spicery
EUMINIDES, 148
Euonymus americanus, 23; *also see* burning-bush
Euonymus atropurpureus, 23
Euphorbia millii, 165; *also see* crown-of-thorns
Euphrates, 53, 78, 292; fig. 40
EURIPEDES, 131
EUROPA, 145
Europe (-an), viii, 6, 13, 14, 29, 30, 31, 32, 34, 37, 40, 46, 49, 52, 57, 63, 73, 76, 77, 89, 90, 92, 101, 109, 119, 123, 127, 128, 131, 134, 138, 141, 144, 152, 159, 160, 167, 168, 173, 177, 180, 204, 208, 214, 217, 218, 233, 244, 248, 277, 278, 284, 285, 286, 287,
Evander Childs High School, New York, ii, iii, vii, viii
EVE, iii, xv, xix, 15, 105, 186, 187, 286, 287, 288, 292; figs. 40, 95, p. 329
EVENARI, M., vii, 26, 30, 37, 38, 39, 46, 48, 63, 65, 66, 71, 74, 77, 79, 98, 102, 111, 117, 134, 162, 165, 171, 174, 181, 191, 195, 201, 225, 239, 263, 267; figs. 51, 53, 55, 63, 66, 70, 81
evergreen (-s), 46, 72, 85, 107, 111, 144, 152, 158, 174, 175, 176, 177, 181, 194, 195, 197, 198, 207, 255
evil, 119, 120, 134, 148, 167, 221, 240
ewe, 186
excavated, 93
exhaustion, 49
Exodus; exodus, xv, 18, 36, 40, 57, 75, 94, 161, 195, 224, 233, 234, 240, 249, 278, 282, 283; figs. 33, 45
expectorant, 32, 278
exotic, 5, 6, 14, 26
extreme burning, 97
exudations, 31, 32
eye (-s), 27, 89, 97
EZECHIEL — *see* EZEKIEL
EZEKIEL, 18, 40, 50, 62, 63, 68, 75, 84, 95, 140, 166, 167, 178, 181, 198, 203, 207, 216, 217, 233, 254; fig. 33
Ezion-geber, 169, 188
ezob; ezov, 160, 161, 162
EZRA, 18
EZRA, ABEN, 74
Ezrach, 123
ezzof, 161

Fagus castanea, 180; *also see* chestnut
fairs, 75
fakirs, Indian, 143
Fall of Man, 186, 286
famine, 127, 162, 169, 228, 229, 231, 232, 289
fan-palm, dwarf, 287
Far East, 77, 109
FARLEIGH, J., xvi, fig. 7
farmer (-s), 60, 134, 198, 215, 237, 276, 300
farthings, 284
Fates, Roman, 148
fat (-s), 157, 216, 242
FATIMA, 279
fatted fowl, 9
fauna (-s), 5, 13, 275
Favotrichophyton violaceum, 98; *also see* scall
fears, 65
feast (-s), 74, 109, 111, 140, 211, 212, 213, 214, 242
Feast of the Tabernacles, 64, 144, 171, 174, 187, 217, 283, 285, 291; fig. 68
Feast of Tulips, 235
feasts, Bacchanalian, 106
feather (-s), 38, 67, 241, 285
fecundity, 112
fed; feed (-ing), 72, 78, 114, 120, 128, 133, 134, 154
FEINBRUN, N., 263, 273, 287; figs. 24, 46
FELDMAN, H., 273, 275
FELDMAN, U., 265
fen (-s), 50, 247
fence (-s), 170, 203, 239, 241
FENN, H., fig. 88
fennel, 46, 89, 102, 152, 153, 186
fennel-flower, yellow, 153
fenugreek; fenu-grec, 34, 54
FERGUSON, J., 273
FERGUSON, W., 263
ferment (-ed, -ation), 27, 55, 113, 211, 212, 213, 214
FERNALD, M. L., 263
FERRY DE LA BELLONE, C. DE, 272, 283
fertile; fertility, 26, 37, 63, 198, 256, 285
fertilization, 104, 106, 171
Ferula, 102; *also see* galbanum
Ferula communis, fig. 87
Ferula erubescens, 102
Ferula ferulago, 102
Ferula galbaniflua, xix, 102; illus.: fig. 61
Ferula gummosa, 102
Ferula persica, 102
Ferula rubricaulis, 102
Ferula schair, 102
festivals, Saturnalian, 111
Feuerlilien, 43
fever (-s), 81, 97, 120, 168, 169, 172, 219
fiber (-s), 53, 110, 131, 133, 149
Ficus bengalensis, 104
Ficus benghalensis, 104
Ficus carica, 73, 103, 104, 107, 184, 256, 281, 287; *also see* fig.
Ficus carica var. *globosa*, 105
Ficus carica var. *silvestris*, 103, 106
Ficus indica, 104
Ficus religiosa, 285
Ficus sycomorus, 107
Ficus sycomora, 107
Ficus sycomorus, xiv, xix, 106, 107, 140, 281, 282, 283, 287; illus.: figs. 4, 62; *also see* sycomore
field (-s), xix, 5, 6, 20, 28, 40, 41, 42, 43, 44, 48, 53, 59, 61, 62, 67, 68, 71, 78, 81, 86, 88, 101, 102, 112, 114, 126, 128, 129, 130, 133, 137, 138, 151, 153, 162, 166, 180, 185, 190, 194, 202, 229, 232, 237, 240, 241, 251, 252, 257, 277, 279
fields, corn, 230
fields, grain, 117
fields of Gomorrah, 240
fig (-s), xv, xvi, 6, 8, 10, 14, 33, 71, 103, 104, 105, 106, 108, 110, 159, 161, 172, 175, 185, 187, 189, 212, 223, 240, 248, 256, 276, 281, 286, 287, 291; *also see Ficus* spp.
fig, common, 107
fig leaves, 103, 281
fig-marigold, Egyptian, 216; *also*

see *Mesembryanthemum nodiflorum*
fig-mulberry, 107; also see *Ficus sycomorus*
figs, dried, 104
figs, green, 241
fig, sycomore — see *Ficus sycomorus* & sycomore
fig tree (-s), 73, 103, 104, 105, 106, 111, 158, 175, 190, 229, 241, 242, 243, 244, 251, 253
fikone löf, 281
finch (-es), 14, 61
fingernails, 50
FINK, B., 205
fir (-s), 25, 46, 63, 66, 67, 68, 173, 174, 175, 176, 177, 180, 189, 209, 221, 248, 285, 286; also see *Pinus halepensis* & fir tree & evergreen
fir, silver, 286
fire (-s), 23, 66, 67, 68, 74, 77, 118, 130, 134, 176, 182, 201, 202, 206, 215, 223, 229, 230, 232, 239, 241, 242, 248, 251, 252, 254, 255, 287, 290, 300
fireplaces, 64
fire-worshippers, 91
fir, Scotch, 177
firstfruit (-s), 112, 169, 210, 212, 214, 256, 291
firstripe figs, 103, 104
fir tree (-s), xvii, 11, 24, 26, 46, 62, 89, 90; *also see* fir
FISCHEL, O., 273
FISCHER, 43
fish (-es), 3, 10, 32, 33, 34, 80, 88, 89, 112, 150, 289
fisher (-s), 10, 66
FISHER, L. J., 272, 276
fishing-rods, 50
FITCH, W. H., fig. 45
fitches, xvii, 89, 101, 112, 128, 152, 153, 166, 211, 233, 234; also see *Nigella sativa*
fits, 118
FITZPATRICK, H. M., 272
fixatives, 280
flag (-s), 50, 62, 76, 92, 94, 120, 121, 131, 206, 249
flag, yellow, 117; also see *Iris pseudacorus*
flagon of wine, 288
flail, 153, 232
flake manna, 32
flame (of fire), 23, 252
flamingoes, 276
flavor (-ing), 32, 34, 35, 46, 89, 139, 287
flax, xvii, 50, 111, 129, 130, 131, 133; also see *Linum usitatissimum*
flaxseed, 133
flax, smoking, 132
fleas, 168, 172
Flemish, 292
Flemish Bible, xv
flesh, 28, 29, 74, 248
FLETCHER, A., 263
fleur-de-lis; fleur de Louis; fleur de luce; fleur-de-lys, 118; also see *Iris germanica*
flies, 157
floats, 281
flocks, 29, 51, 286
Flood; flood, 15, 159
floodplain, 88
flour, 56, 82, 89, 101, 128, 134, 154, 157, 158, 163, 210, 229, 233, 234
flour, fine, 211, 230
flower (-s), xiv, 5, 18, 36, 37, 38, 42, 44, 45, 46, 48, 50, 52, 53, 55, 57, 60, 62, 65, 66, 72, 73, 76, 78, 82, 83, 84, 85, 86, 87, 89, 91, 95, 97, 234, 235, 251, 252, 256, 287, 288, 289, 290, 292
flowering ash, 31, 32; also *Fraxinus ornus*
flowering-dogwood, 277; also see *Benthamidia florida*, *Cornus florida*, *Cynoxylon floridum*
flowering-rush, 62; also see *Butomus umbellatus*
flower of the field, 28; also see *Anemone coronaria*
flowers, open, 189
flux, bloody, 219
foal, 240
fodder, 29, 60, 73, 101

Foeniculum vulgare, 152; also see fennel
FOLKARD, R., 235, 271, 288
folklore, 17, 45, 138, 198, 206, 235
FOLKROD, 91
FONDA, M. E., 15, 263
food (-s), 15, 20, 24, 28, 29, 31, 32, 33, 34, 35, 53, 54, 66, 72, 73, 79, 80, 81, 88, 91, 92, 98, 101, 104, 105, 106, 107, 112, 113, 127, 128, 134, 139, 144, 154, 155, 162, 166, 170, 171, 185, 197, 211, 214, 229, 232, 247, 253, 254, 280, 282, 287, 288
forage, 166, 277
forcing bed for cucumbers, 88
forest (-s), 4, 6, 66, 67, 68, 69, 92, 115, 116, 171, 178, 179, 182, 184, 194, 197, 239, 251, 252, 255, 276, 279, 286
Forest of Lebanon, 69; fig. 52
FORSKÅL, P., 3, 5, 34, 83, 263, 274
fossil (-ized, -s), 6, 254
fountain (-s), 24, 28, 289
fowl (-s), 59, 61, 66, 67, 68, 194, 230, 251, 255
fox (-es), 9, 239, 241, 244
FRA ANGELICA, 44
FRACASTOR, H., 150
fragrance; fragrant, xvii, 47, 57, 58, 68, 69, 70, 77, 83, 86, 98, 114, 119, 123, 124, 125, 140, 144, 149, 177, 182, 185, 188, 219, 228, 277
fragrant gum, 77
France, 2, 5, 45, 72, 118, 128, 141, 205, 244, 283
frankincense, 30, 39, 40, 41, 56, 57, 58, 59, 76, 82, 84, 87, 102, 182, 224, 235, 277; also see *Boswellia* spp.
frankincense-tree, 39, 84, 85, 117, 234
FRASER, 196, 197
fraxinella, 23
Fraxinus ornus, 31, 126, 176; also see flowering ash & manna
Fraxinus parviflora, 176
Fraxinus parvifolia, 176
Fraxinus syriaca, 176
FRAZER, J. G., 263
FREDERICK, Capt., 278
FREEDMAN, H., 274
French, 2, 125, 139, 145, 235, 286
French-Canadian, 184
FRIEND, H., 263
FRIES, T. M., 272
Fritillaria, 116; *also see* fritillary & lily
Fritillaria imperialis, 43, 117
Fritillaria libanotica, 116, 129
fritillary, common, 116
FRITSCH, F. E., 271, 289
frost, 106, 108, 125, 240
FROST, J., 60, 263
fruit (-s, -ful) xvii, xix, 5, 6, 18, 20, 21, 25, 36, 45, 55, 60, 67, 70, 71, 72, 73, 78, 79, 80, 81, 83, 85, 88, 119, 126, 137, 138, 140, 147, 148, 155, 157, 159, 160, 165, 169, 177, 180, 185, 186, 187, 188, 190, 191, 194, 205, 207, 210, 218, 220, 221, 224, 230, 240, 242, 243, 244, 248, 251, 252, 253, 255, 256, 276, 283, 286, 287, 288, 290, 291
fruit, corrupt, 256
fruit, forbidden, 105, 186, 243, 286
fruit, good, 256
fruit, hasty, 252, 256
fruit, summer, 256
fruit-tree, 119, 278
frumenty, 232
Fucagrostis minor, 249; *also see* eelgrass
Fucus, 249
FUDO, 154
fuel, 24, 91, 93, 159, 211, 239
FÜSSLI, J. M., fig. 90
fumigating; fumigation, 47, 58, 280
fungous; fungus; fungi, 99, 134, 141, 142, 254, 282, 290
Fungus melitensis, 91; *also see* scarlet cynomorium
Furies, 148
furnace (-s), 43, 211, 223
furuncles, 223
furze, 287

GABRIEL, xix, 44; fig. 65
gad, 86
gader, 203
gaff, 203
Gagea arvensis, 284
Galaad, 10, 55
galactogogue, 282
Galatians, 19
galbanum, xix, 30, 48, 56, 58, 102, 223, 224
Galbanum officinale, 102
galbules, 121, 209
GALEN, 75
galgal, 39, 279
Galilean; Galilee, 6, 41, 61, 63, 114, 122, 144, 165, 171, 187, 196, 275, 276, 279
Galilee, Sea of, xix, 60; fig. 84
Galium verum, 277
gall, 27, 48, 49, 78, 79, 80, 240, 279; also see *Citrullus colocynthis*
gallows, 291
galls, oak, 197
galleys, 90
Gallipoli, 120
gall-juice, 173
gall of bitterness, 78, 80
gall, water of, 80
gander (-s), 32
Ganges, 82, 154
GANYMEDE, 182, 183
garden (-s), xvi, xvii, 4, 20, 34, 45, 47, 51, 52, 56, 57, 59, 60, 61, 67, 70, 81, 83, 88, 101, 104, 105, 109, 114, 116, 119, 124, 139, 150, 157, 159, 160, 172, 180, 181, 186, 187, 190, 191, 194, 202, 205, 235, 243, 252, 254, 256, 276, 284, 286, 291, 292
garden lettuce, 74
garden, nuts, 119, 291
Garden of Eden, iii, vii, xv, xvi, xvii, 15, 20, 32, 160, 286, 288; figs. 5-7, 35, 40, 91, 95, 292
Garden of Gethsemane, *see* Gethsemane
Garden of Paradise, fig. 94
garland, 111, 287
garlic (-h, -k), 32, 33, 34, 80, 88, 118; also see *Allium ascalonicum*
garment (-s), 27, 47, 48, 75, 82, 110, 130, 131, 142, 143, 212, 218, 219, 240, 241, 254, 290
gasplant, 23; also see *Loranthus acaciae*
gate (-s), 129, 130, 131, 137
gawz-gundum, 282
gayz, 29
Gaza, 125, 135, 138, 179, 281
gazelle, 122
GEDDES, H.
Gederite, 97, 106, 157
gefen, 243
gefen nachriyah, 239
gefen sadeh, 78
gefen s'dom, 244
gehgah, 291
GEIKIE, J. C., 5, 263
Genesis, xv, 7, 18, 52, 55, 62, 71, 77, 81, 82, 84, 90, 105, 137, 138
Geneva version, 9
Genista, 19
Genista monosperma, 201, 287; *also see* broom & juniper
Genista raetam, 201
genital, 150
Gennesaret, 40, 71, 92, 171, 215
Gentiles, 257
GEORGE III, 84
George Washington University, vii
Gepaids, 276
gephen, 243
geranium (-s), 6
GERARDE, 30, 43, 134, 138, 235
geres, 231
German (-y), 2, 5, 8, 37, 133, 138, 139, 145, 147, 198, 233, 235, 283, 284, 286, 288
German version, 9
germinating; germination, 232, 256
germs, 169
Geryon, 182
GESENIUS, W., 79, 80, 132, 147, 198, 273, 283
GESSNER, C., xx
GESSNER, J., 263, 264

Gethsemane, xvii, 44, 158, 160, 202, 284; fig. 25
gez, 29
Ghent, 286
gherkins, 88; also see *Cucumis anguria*
gherkin, West Indian, 88
Ghor, 13
Ghor-es-Sofi, 38
ghosts, 101
Gibeah, 189, 227
Gibraltar, 14
GIDEON, 113, 150, 212, 229
Gihon, 292
Gilead (-ite), 10, 14, 51, 55, 77, 101, 119, 122, 128, 138, 158, 174, 178, 187, 190, 196, 197, 201, 209, 225, 229
Gilgal, 24, 79, 193
gills, 203
ginger, 40
ginger-grass, 40, 41; also see *Andropogon aromaticus*
ginseng, 166; also see *Panax ginseng*
girdles, 105
girsah, 231, 232
gith, 89, 152
Githago segetum, 30; also see corn-cockle
gitot, 213
Gladiolus, 116, 275; also see lily
glean (-ed, -ing), 112, 240, 241
glede (-s), 9
Gleditsia triacanthos, 284
globe-cucumber, 78, also see *Cucumis prophetarum*
globethistles, 278
GLUECK, N., 273
gnat-tree, 184
goat (-s), 77, 98, 129, 160, 173, 186, 194
God; god (-ess, -esses, -s), xv, 20, 21, 23, 28, 31, 33, 38, 43, 44, 45, 56, 57, 64, 67, 68, 69, 70, 71, 72, 77, 78, 84, 91, 93, 95, 101, 106, 111, 112, 114, 120, 122, 124, 133, 139, 140, 142, 145, 148, 149, 150, 154, 155, 157, 159, 160, 172, 175, 180, 182, 190, 191, 193, 197, 198, 203, 212, 213, 227, 228, 229, 240, 241, 242, 244, 247, 253, 284, 287, 291, 292
GOES, H. VAN DER, 286
gofer, 90
gogul, 58
gold (-en), 24, 36, 39, 51, 56, 57, 58, 81, 82, 84, 97, 106, 126, 129, 130, 131, 158, 167, 169, 185, 186, 188, 189, 194, 204, 218, 228, 242, 251, 253, 277
golden bells, 190; also see *Punica granatum*
golden fleece, 87
golden-thistle, spotted, 206; also see *Scolymus maculatus*
Golgotha, 45, 165
GOLIATH, 179
gomê; gome, 92, 281
gômeh, 92
Gomorrah, 26, 78, 221
gonorrhea, 150, 151, 283
GOOD, J. M., 259
goodly tree (-s), 169, 173, 177, 183, 187, 283, 290, 291
GOODSPEED, E. J., 7, 10, 17, 27, 259, 285, 286, 287
GOODSPEED version, see version, GOODSPEED
GOODYEAR, 155
goose, 9
goosefoot, 216
gopher, 90
gopher-wood, 89, 90, 91, 152; also see *Cupressus sempervirens* var. *horizontalis*
GORDON, A. R., 17, 286
GORDON, B. L., 264
GORRIE, D., 264
gorse, 287
GORTER, D. DE, 264
Gospel (-s), 60, 80
Gossiparia mannifera, 278
Gossypium arboreum, 110; also see cotton
Gossypium barbadense, 110
Gossypium herbaceum, 109
Gothic version, 8
Goths; gothic, 8, 139
gourd (-s), 6, 74, 78, 79, 80, 81, 88, 203, 204, 281, 287; also see *Citrullus colocynthis*; fig. 36
gourds, true, 204
gourds, wild, 88, 204, 281
gourd, white-flowered, 204
graft (-ed, -ing), 158, 159, 185, 191, 286
grain (-s), xv, 46, 59, 61, 82, 86, 88, 112, 113, 134, 166, 167, 175, 211, 222, 231, 232, 254, 257, 283, 290, 300
grainfield (-s), 30, 46, 80, 117, 134, 159, 220, 283
grains, cereal, 256, 289
granary, 232
grape (-s, -vine), vii, xvi, xvii, 14, 30, 71, 72, 78, 79, 103, 116, 123, 124, 140, 157, 177, 187, 190, 206, 212, 213, 229, 239, 240, 241, 242, 243, 244, 248, 256, 275, 276, 287, 289; also see *Vitis vinifera*
grapefruit, 186; also see *Citrus grandis*
grape-gatherer (-s), 241, 243, 244
grapejuice, 213, 288
grapes, American, 244
grapes of gall, 78, 79
grapes, sour, 241
grapes, wild, 29, 239, 241, 244
grass (-es, -y), vii, xvii, xix, 6, 20, 28, 29, 34, 40, 41, 43, 50, 51, 54, 87, 91, 92, 94, 116, 134, 166, 211, 215, 216, 230, 231, 232, 247, 249, 251, 253, 254, 282, 300
grasshopper (-s), 36, 65, 202, 280
grass, river, 121
grass, Sorghum, 161; also see *Sorghum vulgare*
grass-wrack, 249; also see *Zostera marina*
grave, 64, 65, 68, 70, 172, 230
gravecloth (-es), 87, 132
GRAY, H. ST. G., 267
graze (-ed, -ing), 51, 91, 116, 163
Great Bible of Cranmer, 9, 275
Great Pyramid of Cheops, 33
Great Pyramid of Khufu, 33
Greece; Greek (-s), vii, 2, 3, 8, 23, 25, 29, 30, 32, 33, 34, 35, 37, 40, 42, 44, 45, 46, 47, 48, 53, 55, 57, 59, 61, 62, 64, 66, 68, 71, 72, 73, 74, 75, 76, 77, 78, 80, 81, 82, 83, 86, 87, 88, 90, 91, 92, 93, 101, 102, 104, 109, 111, 112, 114, 116, 119, 120, 121, 123, 124, 129, 131, 132, 134, 137, 139, 140, 144, 148, 150, 152, 153, 155, 159, 160, 166, 167, 170, 174, 176, 177, 178, 179, 180, 181, 182, 183, 185, 187, 188, 190, 191, 197, 198, 201, 203, 205, 206, 207, 208, 210, 214, 215, 216, 217, 218, 220, 221, 224, 227, 228, 230, 231, 235, 244, 245, 249, 283, 285, 286, 288, 291
Greek version, 7, 71, 286
green bay tree, 123, 152, 176; also see *Laurus nobilis*
GREENE, H. B., 264
green fir tree — see tree, green fir & *Apinus pinea*
green herb (-age), 94, 251, 253
GREENMAN, J. M., vii
greens, 34
green thing, 253
green tree, 193, 194; also see *Quercus* spp.
GREGORY MARTIN's version, 9
GRIGGS, R. F., vii, 243
GRINDON, L., 264
gristah, 131
groats, 231, 233
GROSER, W. H., 4, 26, 34, 35, 57, 58, 60, 90, 98, 187, 189, 217, 264
GROSVENOR, G., 264
ground-ivy, 208
groundsel, 6
ground-squirrels, 168
grove (-s), 56, 59, 68, 69, 145, 151, 158, 159, 170, 171, 181, 182, 193, 194, 197, 198, 227, 286, 287; also see *Quercus* spp.
guardian thistle, 72; also see *Cnicus benedictus*
gugal; gugul, 39, 81
GUIBORY, M., 264
guinea pigs, 168
Gulf of Akaba, 188
Gulf of Oman, 95
gulgal, 38, 279
gum (-my), 25, 31, 47, 52, 55, 57, 58, 81, 82, 83, 85, 86, 93, 102, 126, 127, 177, 184, 224, 225
gum arabic, 25, 26; also see *Acacia* spp.
gum cistus, 77; also see *Cistus ladaniferus*
gum galbanum, 102; also see *Ferula galbaniflua*
gum-resin, 224
gum-tragacanth, xvii, 52
guncotton, 110
Gundelia; gundelia, 39
Gundelia tournefortii, xvii, 38, 39; illus.: fig. 21
gun-flints, 208
Gustavus Adolphus Church, New York, viii
GUTENBERG, 9
Gymnarrhena micrantha, 27; also see thistles
Gypsophila, 39

Habacuc — see Habakkuk
HABAKKUK, 18
habatzeleth, 147, 289
HADAD, 190
hadar, 187
hadas, 144
Hadassah, 143, 144
Hadera, 92
Hades, 182, 191, 198
Hadramaut, 58
HAGAR, 227; fig. 30, 70
HAGGAI, 18
Hague, The, 286; fig. 94
Haifa, xvii, xix, 81; fig. 28
hail (-storm), 106, 112
hair, 40, 49, 65, 75, 98
halawi, 216
halbeh, 34
HALL, Miss E. C., vii
HALLMAN, D. Z., 264
Haloxylon articulatum, 278
HALPER, D., 274
Haman, 291
Hamath, 38
Hamburg, 139
HAMILTON, F., 59, 264
hammers, 173
Hammon, 214
HANBURY, 225
HANFORD, M. A. C., 264
hang, (-ed, -eth, -ing), xiv, 73, 106, 193, 195, 251, 253, 280, 291
HANNAH, xv
HARDY, E., 274
harehkâch, 52
Hareth, 251
HA-REUBENI, E., 4, 43, 235, 253, 264, 269, 274
HA-REUBENI, H., 274
HARKAVY, A., 259
harp (-s), 57, 175, 176, 183, 188, 216, 217
HARPER, H. A., fig. 84
HARRIS, T. M., 2, 3, 264
HARRIS, W. T., 264
hart, 51
HART, H. C., 38, 71, 264; fig. 45
harvest (-ed, -ing), 31, 112, 113, 116, 134, 137, 138, 229, 232, 291, 292
HASSELQUIST, F., 2, 3, 5, 30, 33, 55, 81, 88, 115, 125, 138, 144, 208, 264
HASTINGS, G. T., 264, 265
hasty fruit, 103, 105; also see *Ficus* spp.
HAUPT, P., 264
Hauran, 18, 194, 235, 239
HAUSMAN, E. H., 273, 277
HAUSSKNECHT, 282
Havilah, 82
havnim, 95
hawk, 61
hawthorn, 6, 14, 203
hay, 28, 29, 34, 43, 166, 253, 277
hayfever, 48
HAYNALD, L., 264
hazel, xvii, 8, 35, 37, 180, 282; also see *Amygdalus communis*
Hazeroth, 201
Hazezon-tamar, 169, 171
headache, 168
healing, 55, 85, 86
heath (-s), 8, 32, 121, 122
heathen, 67, 68, 111, 197, 240
heaven, xv, 33, 38, 39, 48, 56, 59,

67, 68, 70, 88, 104, 125, 131, 133, 167, 168, 172, 191, 194, 198, 211, 212, 223, 228, 230, 277, 278, 282
Hebrew (-s), vii, viii, 2, 3, 9, 13, 15, 19, 23, 24, 25, 26, 27, 28, 29, 30, 31, 32, 33, 34, 36, 37, 38, 39, 40, 42, 46, 47, 48, 49, 50, 51, 52, 53, 55, 57, 58, 59, 62, 65, 68, 71, 73, 74, 75, 76, 77, 78, 79, 80, 81, 82, 83, 84, 85, 86, 87, 88, 89, 90, 91, 92, 95, 97, 98, 99, 101, 102, 104, 106, 107, 109, 110, 111, 112, 113, 114, 117, 119, 121, 122, 123, 124, 127, 128, 129, 131, 132, 135, 137, 139, 140, 141, 142, 143, 144, 147, 148, 149, 150, 151, 152, 153, 154, 158, 159, 160, 161, 162, 165, 166, 167, 168, 170, 171, 172, 174, 176, 177, 178, 179, 180, 181, 183, 184, 185, 187, 189, 190, 194, 195, 196, 197, 201, 203, 205, 206, 207, 209, 210, 211, 212, 213, 214, 215, 217, 218, 219, 220, 221, 223, 224, 225, 227, 228, 230, 232, 233, 235, 237, 239, 243, 244, 245, 248, 249, 254, 255, 256, 257, 278, 279, 281, 283, 286, 287, 288
Hebrew University at Jerusalem, vii, 4, 39, 43, 115, 235, 253, 279
Hebrew version, 7
Hebron, xvii, 123, 144, 158, 178, 185, 193, 196, 197, 214, 218, 289; fig. 29
hedek, 221
Hedera helix, 111, 204; also see ivy
hedge (-s), 5, 23, 135, 159, 202, 203, 221, 222, 240, 241, 242, 244, 284; also see *Lycium europeaeum* & *Rhamnus palaestina*
Hedypnois taraxacum, 75; also see bitter herbs
Hedysarum alhagi, 31; also see manna
Hedysarum onobrychis, 277
HEGI, G., 264, 274, 284; fig. 47
HEHN, V., 116, 129, 264
heifer, 162, 193, 210
Helbon, 213, 242
Heliades, 182
Heliopolis, 84
Helix ianthina, 195
Helleborus niger, 277
Helleborus orientalis, 277
Helleborus vesicarius, 277
Hellenistic, 8
Hellespont, 150
Hemath, 216
hemlock, 48, 49, 78, 79, 279; also see *Artemisia* spp.
hemlock tree, Canadian, 49
hemorrhoids, 168
hemp, 131
HEMPRICH, 83, 278
HENGSTENBERG, 80
henna (-blossom, -flowers, -plant), xvii, 11, 90, 124, 125, 149, 195, 206; also see *Lawsonia inermis*
HENNEGUEY, F., 205
HENRY VI, 111
HENRY VIII, 138
HENSLOW, G., 48, 73, 189, 234, 264, 280, 284
HENSON, H. H., 262
herb (-aceous, -age, -s), 20, 28, 29, 34, 39, 52, 59, 61, 72, 78, 79, 86, 135, 139, 152, 153, 160, 208, 251, 252, 253, 279
herbal (-ists), 2, 4, 72, 118, 138
herba-alba, 48; also see *Artemisia* spp.
herb of forgetfulness, 49
herbs, bitter, 140, 280; also see *Cichorium* spp. & *Mentha longifolia*
Herculaneum, 93
HERCULES, 154, 182
herd (-s, -smen), 106, 107, 150
heretic (-s), 1
HERLING, A. K., vii
Hermon, 6, 89, 138, 197
HEROD, 106, 184, 202; fig. 92
HERODOTUS, 1, 33, 75, 83, 89, 109, 170, 173
Heshbon, 179, 241, 244
Hesychius version, 7
Hexateuch, 7
HEZEKIAH, 51, 84, 105, 138; fig. 16

HIBBERB, S., 264
Hibiscus syriacus, 5, 53, 151, 205, 234; also see mallows
HICKERNELL, M. R., 262, 264
highland (-s), 13
HILDA, 133
HILDEBRAND, J. R., 264
hill (-s, -sides), 14, 15, 24, 37, 63, 65, 67, 68, 71, 83, 118, 275, 276, 277, 288
HILLER, M., 2, 61, 264
Himalaya (-n, -s), xvii, 115, 119, 148, 188, 218, 278
hin, 39, 212
Hindu (-s), 149, 152, 154, 188, 215, 278
Hinnom, 141
HIPPOCRATES, 101, 143
hippopotamus, 9, 247
hips, 147, 206
HIRAM, 66, 69, 89, 157, 175, 188, 189, 209, 229, 285
HITCHINS, A. P., 261
Hivites, 141
HJORTZUNG, OLLE, xv, xvii, fig. 9
hodnim, 95
HOEHNE, F. C., 271, 281
HOHBERG, W. H. VON, xiv, 265, 277, 280, 287; figs. 2-4, 36-39
Holcus durra, 222; also see reed & hyssop
Holcus sorghum, 223
Holcus sorghum var. *durra*, 222
holly, 90, 111, 178, 198; also see *Ilex aquifolium*
holm, 14, 90, 178, 198; also see *Ilex aquifolium*
HOLMAN, A. J., 260, 273
HOLMES, E. M., 265
HOLMES, J. H., vii
holm oak, 90, 281, 286
holm-tree, 90, 178, 194, 198
Holy Citron, 187; also see *Citrus medica*
Holy Family, 184
holy hay, 277
Holy Land, 2, 3, 4, 5, 6, 23, 26, 27, 29, 32, 34, 37, 39, 40, 41, 42, 43, 46, 48, 49, 50, 52, 53, 60, 61, 63, 65, 69, 71, 72, 74, 77, 81, 85, 86, 87, 91, 93, 97, 102, 109, 114, 115, 117, 121, 122, 123, 125, 129, 131, 133, 135, 139, 140, 141, 151, 152, 153, 154, 158, 160, 161, 165, 171, 173, 174, 179, 180, 182, 183, 185, 187, 190, 196, 197, 204, 205, 206, 208, 210, 214, 215, 220, 221, 233, 244, 285, 287
holy oil, 83, 224; also see *Olea europaea*
holy thistle, 72; also see *Cnicus benedictus*
HOMER, 131, 155
homer, 111, 112, 252
honey, 15, 35, 51, 69, 77, 84, 86, 126, 157, 166, 169, 170, 177, 179, 214, 240; also see *Phoenix dactylifera*
honeybirds, 276
honeydew, 31, 278
honeysuckle, 14
HONORATI, B., 273
HOOKE, S. H., 259
HOOKER, Sir J., 60, 68, 144, 180; fig. 44
HORACE, 72
Hordeum, 29, 50; also see barley
Hordeum distichon, 111, 112, 113
Hordeum hexastichon, 111, 112, 113
Hordeum murinum, 29
Hordeum sativum, 112
Hordeum spontaneum, 113
Hordeum vulgare, 111, 112, 113
Hordeum vulgare var. *spontaneum*, 113
Horeb, xv
horehound, 6
HORNER, N. C., 40, 41, 42, 65, 82, 92, 158, 185, 265, 266
horns, 95
horse (-s), xvii, 59, 60, 71, 72, 73, 78, 101, 112, 125, 130, 131, 132, 144, 160, 177, 198, 242
horse-killer, 152
horseradish, 74, 75; also see *Armoracia lapathifolia*
HORUS, 155
HOSEA, 18, 26, 46, 49, 71, 117, 179, 182, 203, 236, 237, 245

HOSKINS, Miss B., vii
house (-s), 36, 50, 66, 67, 78, 84, 87, 88, 97, 105, 130, 131, 141, 142, 143, 144, 147, 148, 149, 151, 160, 162, 169, 172, 175, 176, 188, 190, 193, 194, 197, 198, 202, 207, 209, 210, 229, 233, 240, 241, 254, 285
house of green figs, 104
housetops, 28
HOWES, F. N., 102, 272, 280
HUBBARD, C. E., 285
Hûlah; Huleh, 92, 93, 117, 275, 276; fig. 63
Hülch lily, 42, 43
HUME, 31
hunger, 3, 33, 65, 198
Huns, 118
Hunter College, ii
HURAM, 67, 188
HURLBUT, J. L., 189, 265
hurricane, 39
HURST, E., 265
husbandmen, 240, 242, 253
husks, 72, 73, 232; also see *Ceratonia siliqua*
hyacinth (-s), 6, 41, 114, 115, 116, 129, 234, 275, 282; also see *Hyacinthus orientalis*
HYACINTHUS, 114, 154
Hyacinthus orientalis, 41, 114, 117, 129, 282; also see lily & hyacinths
hyena (-s), 9, 29, 51, 276
hyssop, xix, 27, 65, 66, 160, 161, 162, 193, 194, 209, 210, 222, 223, 257, 283, 284; also see *Capparis sicula* & *Origanum maru* & *Sorghum vulgare* var. *durra*
Hyssopus officinalis, 160; cf. fig. 86
hyssop, water, 162

Iberian centaury, 71; also see *Centaurea iberica*
ibis, 276
Ibri text, 7
icons, 70
idols, 90, 176, 188, 194, 197, 198, 245
ilan, 195, 196
ilex, 90
Ilex aquifolium, 90, 111, 178, 198; also see holm & holly
image (-ry, -s), 38, 56, 69, 98, 151, 167, 193, 194, 198, 287
Immaculate Conception, 44
imprisonment, 91, 138
incantations, 137
incense, 24, 31, 39, 40, 47, 56, 57, 58, 59, 84, 85, 151, 178, 181, 182, 194, 219, 228; also see *Boswellia* spp.
incense, holy, 102
Index Kewensis, 89, 280, 287
India (-n), 2, 6, 13, 14, 30, 36, 40, 41, 47, 50, 53, 55, 57, 58, 61, 63, 74, 75, 76, 77, 81, 82, 86, 88, 90, 95, 104, 109, 110, 119, 124, 125, 129, 133, 140, 148, 152, 154, 166, 171, 185, 186, 187, 188, 189, 190, 204, 209, 214, 219, 234, 243, 248, 278
Indian bdellium, 81, 82, 224; also see *Commiphora africana*
Indian-corn, 232; also see *Zea mays*
Indian orris, xvii; also see orris
indigo, 6, 195
Indigofera tinctoria, 195; also see indigo
infection (-s), 99, 118
inflammation, 97, 208, 219
ink, 172, 197
inkwell, 173
Inquisition, 50
insanity, 101, 138, 143, 148, 208
inscriptions, 231
insect (-s), 3, 47, 70, 73, 106, 188, 194, 197, 228, 276, 278, 280
insect, cochineal, 194
instruments, 95, 175, 231
instruments, musical, 177, 188, 189
instruments, threshing, 89, 207, 230
intestinal; intestine, 97, 219
intercourse, 151, 168
intermittent fevers, 97

intoxicants; intoxicating, 55, 113, 214
Ionians; Ionic, 155, 173
Iranian, 282
Irano-Turanian, 112
IRBY, L. I., 60
Ireland, 198
Iris; IRIS; iris, 40, 42, 43, 75, 76, 116, 117, 118
Iris bismarkiana, 117
Iris florentina, 118, 218
iris, Florentine, 218
Iris germanica, 116, 118
Irishman, 35
Iris palaestina, 117
iris, Palestinian, 116
Iris pseudacorus, 43, 117, 129
iron, 39, 66, 75, 251
irrigate (-d); irrigation, 88, 205, 222, 276
IRVING, W., 73
ISAAC, 228
ISAIAH, iii, xv, xvii, 18, 25, 26, 30, 39, 40, 63, 88, 89, 94, 98, 103, 105, 120, 121, 132, 141, 147, 152, 153, 174, 175, 176, 179, 181, 207, 214, 215, 223, 233, 234, 235, 237, 245, 248, 249, 277, 286; figs. 23, 24
ISAIAS — see ISAIAH
Isatis tinctoria, 195; *also see* dyer's woad
ISCARIOT, JUDAS, 184
Ishmael (-ites); Ishmeelites, 51, 55, 77, 178, 187
ISIS, 112
island (-s); isles, 35, 62, 63, 77, 91, 95
isop, 283
Israel (-ites), viii, xvii, 5, 8, 24, 25, 31, 33, 36, 41, 42, 47, 50, 56, 58, 66, 67, 68, 69, 74, 80, 88, 102, 103, 106, 117, 120, 125, 127, 128, 131, 133, 142, 144, 149, 150, 151, 157, 161, 162, 166, 167, 168, 169, 174, 175, 176, 178, 198, 202, 207, 210, 211, 217, 239, 240, 241, 242, 244, 254, 255, 256, 275, 289, 291
ISRAELSTAM, J., 274
issue (-s), 149, 150
issue of blood, 220
issue, running, 149
italicization, 17
Italian; Italy, 63, 93, 141, 150, 152, 160, 163, 235, 285, 286
ivory, 9, 62, 63, 95, 228
ivy, 6, 14, 65, 111, 203, 292; *also see Hedera helix*
ivy, English, 111, 204

Jabesh, 227
Jabesh-Gilead, 227
JABLONSKI, 25
jackal (-s), 9, 29, 51, 244, 276
JACKSON, B. D., 265
JACOB, 35, 36, 77, 84, 128, 137, 177, 179, 180, 181, 182, 193, 212, 225, 229, 230, 231, 256, 286; fig. 67
Jaffa, 78, 81, 125, 135, 165
JAMES, 19
JAMES I, 141
JANSSEN, GERARD, iii, xv
Janum, 185
Japan (-ese), 5, 41, 154, 155, 217, 219
JAQUEMONT, fig. 33
jasmin (-e), 114, 138, 280
JASON, 87
Jastrow version — see Version, Jastrow
jatamansee, 149
jaundice, 72
Java (-n), 39, 75, 224
Jazer, 241
Jebesh, 194
JEHORAM, 219
JEHOVAH, 188
jelly, lichen, 282
JENNISON, G., 271
JEREMIAH, 18, 36, 40, 49, 55, 80, 84, 121, 122, 211, 215, 239, 288
JEREMIAS — see JEREMIAH
Jericho, xvii, 38, 39, 55, 71, 78, 84, 85, 107, 125, 135, 147, 151, 169, 171, 180, 205, 221, 275
Jericho balm, 98
JEROME, 7, 201, 203, 204, 275
Jerome's version, 7, 275

Jerusalem, vii, xix, 4, 53, 56, 61, 65, 66, 67, 86, 97, 103, 105, 106, 114, 115, 119, 135, 141, 143, 144, 147, 158, 161, 165, 169, 171, 172, 179, 188, 194, 211, 252, 284, 287, 289
JESSE, 256
Jesuit (-s), 9
JESUS, iii, xv, xix, 5, 27, 39, 42, 43, 44, 46, 50, 59, 60, 61, 66, 70, 72, 83, 89, 97, 106, 107, 131, 132, 141, 142, 148, 149, 150, 160, 161, 165, 172, 184, 191, 198, 202, 211, 214, 230, 233, 235, 236, 242, 276, 277, 283, 284, 292; figs. 46, 49, 56, 62, 65, 82
jewel caskets, 63
JEWETT, 141
Jewish Publication Society version, 8, 17
Jewish Theological Seminary, vii, 283
Jezzin, 198
Joab, 112, 149
JOAN OF ARC, 198
JOASH, 157, 193
JOB, 13, 18, 26, 27, 29, 30, 49, 53, 54, 62, 71, 91, 92, 94, 120, 121, 153, 202, 223, 237, 247
JOEL, 18, 25
JOHN, 19, 222
JOHN OF LEMBERG, 9
JOHN the Baptist, iii, xv, 73, 280
JOHNS, C. A., 265
JOHNSON, B., 138
jointed-glasswort, 215, 216; also see *Salicornia* spp.
jointed - glasswort, herbaceous, 215
jointed-glasswort, shrubby, 215
JONAH, 15, 18, 88, 203, 204, 249, 281; fig. 36
JONAS — see JONAH
JONATHAN, 227
JONES, J. R., 43, 47, 49, 65, 75, 95, 98, 102, 175, 189, 191, 207, 215, 219, 265
Joppa, 67, 147, 216; fig. 80
Jordan (valley), 12, 13, 14, 24, 25, 29, 30, 50, 58, 60, 93, 98, 113, 121, 135, 138, 151, 152, 171, 180, 183, 187, 193, 197, 201, 217, 221, 227, 275, 276, 280, 284; fig. 79
JOSÉ, Rabbi, 287
JOSEPH, xix, 107, 129, 131, 132, 172, 222, 229, 232, 287; fig. 92
JOSEPHUS, 1, 58, 84, 119, 127, 137, 138, 162, 171, 173, 175, 178, 189, 221
JOSHUA, 7, 15, 18, 24, 25, 87, 141, 174, 185, 190, 193, 243
JOSHUA, Rabbi, 292
JOSUE — see JOSHUA
JOVE, 87, 198
Jove's thunderbolts, 198
Judah, 24, 56, 103, 151, 166, 169, 178, 212, 217, 240, 241, 284
JUDAS, iii, xiv, xv, 73, 106, 179, 276, 280; fig. 2
Judasbaum, xiv, 280
Judas-tree, xv, 73, 106; *also see Cercis siliquastrum*
Jude, 19
Judea (-n), xix, 12, 65, 76, 109, 138, 165, 187, 195, 218, 225, 275, 276, 284; fig. 86
Judges, xiii, xv, 18, 132, 190
JUDITH, 18, 44
Juglans regia, 119, 209; *also see* walnut
juice, 35, 79, 80, 288
jujube, 248
JULIET, 141
Juncus, 93, 120; *also see* rush
Juncus compressus, 121
Juncus effusus, 120, 249
Juncus glaucus, 121
Juncus inflexus, 121
Juncus lamprocarpus, 121
Juncus maritimus, 120
Juncus rigidus, 120
jungle (-s), 50, 92
juniper (-s), 6, 8, 14, 37, 54, 91, 121, 122, 173, 177, 201, 281, 287
juniper, common, 122
juniper, Grecian, 209
juniper, Phoenician, 121, 174, 210
juniper roots, 48, 53; *also see Cynomorium coccineum*

juniper, savin, 121
juniper tree, 201
Juniperus, 201, 210
Juniperus communis, 122
Juniperus drupacea, 174
Juniperus excelsa, 209
Juniperus lycia, 57, 210
Juniperus macrocarpa, 174
Juniperus oxycedrus, 32, 57, 121, 210
Juniperus phoenicia, 57, 210
Juniperus sabina, 121
Juniperus thurifera, 174
JUNIUS, 132
JUNO, 44, 87
JUPITER, 191
JUSTINIAN, 118
jute, 53
jute, Malta, 234
JUVENAL, 72

kadachat; kaddachath, 97
Kaf Fatima, 279
Kaf Maryam, 279
kaff-ud-dibb, 220
kahsh, 254
KAISER, 282
kali, 216, 231, 232, 254
KAMADIVA, 154
kamah, 232
kammon, 89
Kanah, 51, 214, 215
kǎněh, 51, 214
kaneh hattov, 214
KAPLAN, S., vii
karkom, 86, 87
karmel, 231, 232
karmil, 194, 195
karpas, 109
Kashmir, 218
KAUTZSCH, E. F., 116, 117, 265
kazha, 152
Kedisha River, 68
KELLERMAN, S., 272
KELLEY, A. P., 272, 282
keneh, 40, 214, 215
keneh bosem, 40
keneh hattob; keneh hattov, 40, 214
Kenite, 169
KENNEDY, R. W., 273
KENT, C. F., 288
kerem, 243
kermes; kermez, 194, 195
kernel, 37
KERR, N. S., 273
KESSEL, D., 265
ketoret, 57
ketsah, 152
Ketubim, 7
ketyaeh, 152
ketz, 202
ketzach, 152
ketziah; ketzioth, 75, 218, 219
Keve, 132
Kew, 4, 60, 129, 207
khadak, 221
Khân-Minyah, 92
khardal, 60
khardal barri, 220
khardul, 220
Khufu, Great Pyramid of, 33
khurwa, 203
kibil, 203
Kibroth-hattaavah, 33
kiddad, 75
kiddah, 75, 219
kidney-fat, 232
kidneys of wheat, 229
Kidron, 194, 284
kikajon; kikayôn, 203, 204, 281
KILLERMANN, S., 265
KIMCHI, 94
kimmeshonim; kimmesonim; kimsonim, 165, 202, 245
kimmosh; kimosh, 26, 237, 245
kine, 62, 120
king (-dom, -s), 24, 39, 41, 45, 56, 66, 67, 68, 69, 70, 72, 84, 95, 105, 106, 108, 109, 130, 132, 133, 134, 138, 148, 150, 151, 157, 159, 166, 169, 175, 178, 188, 189, 193, 204, 208, 211, 212, 213, 240, 242, 257, 285
KING, E. A., 4, 265, 269
King CLOVIS, 118
king-fishers, 276
King JAMES version — *see* versions, Authorized and King JAMES

King LOUIS IX, 45
King MANASSEH, 141
Kings, I, II, III, IV, 18, 287
kinnamon; kinnemon, 76
Kinnamomum cassia, 75; *also see* cassia
KINSEY, A. C., 263
Kirghiz, 282
kirsenni, 233
kishuim, 88
kitter, 57
kitteroth, 57
Kittites, 62
KITTO, J., 2, 3, 4, 34, 36, 78, 87, 88, 161, 265
Kiveh, 132
k'la'at p'kaim, 78, 80
knapweeds, 153
kneading, 210, 211
knee-holly, 207; *also see Ruscus aculeatus*
Knights of St. Mary of the lily, 45
knobs; knops, 36, 66, 78, 79, 80, 218
KNOX, R. A., 259
KOLB, G., 271, 282
koosemet, 233
koos'min, 233
koost, 76
kopher, 124
Korah (-ite), 41, 43, 168, 185
korunda-guahah, 76
KOTELES, MARTHA, viii, xi, xv
koter, 57
kotz (-im), 165, 248
KRAASBÖL, H. C., 265
KRALITZ, 9
KRAUSE, K., 265
krinon, 147
k'soos, 111
k'tzia; k'tziot, 75, 218
Kuê, 132
KUHN, 162
KÜKENTHAL, G., 265
KUNTH, C. S., 115
Kupferbibel, fig. 90
kupros, 91
Kurdic, 282
kurkum, 87
KUSHNIR, T., 279
kussemeth, 233

la'anah, 48, 279
labdanum, vii, 77, 224, 280, 281
Lactuca sativa, 74; *also see* bitter herbs
Lactuca scariola var. *sativa*, 74
ladan, 77
ladana, 249
ladanum, xvii, 77, 224; *also see Cistus* spp. & *Styrax benzoin*
Lady GODIVA, 206
lady's-thistle, 71; *also see Silybum marianum*
LAGARDE, 129
Lagenaria, 281
Lagenaria leucantha, 204, 281; *also see* gourd
lair, 51
lake (-s), 62, 92, 123, 171, 276
lake-dwellings, 101, 133, 231
Lake of Galilee, 42, 151
Lake of Gennesaret, 119
Lake of Tiberias, 151, 276
Lakshmi, 154
LAMARCK, J. B. A. P. M. DE, 58
lamb, 25, 66, 74, 162, 186, 251
Lamentations, 18, 79
lämmergeier, 9
lamp (-s), xv, 48, 56, 133, 157, 158, 159
lamp-black, 173
Lampsacus, 150
lampstand, 36
LAMSA, G. M., 10, 17
Lamsa version — *see* version, Lamsa
LANCKORONSKA-OEHLER, fig. 90
land (-s), 15, 28, 35, 37, 42, 47, 51, 53, 64, 81, 82, 87, 88, 92, 94, 101, 119, 289
LANDER, G. D., 265, 267
Lapland, 5
Lappa major, 245; *also see* burdock & thorns
Lappa vulgaris, 245
larch (-es), 26, 62, 63, 174; *also see Larix*
larch, European, 177

Larinus maculatus, 278
Larix decidua, 63, 177; *also see* larch
Larix europaea, 177
Larix larix, 177
Laryx Europoea, 177
Latakia, 120, 121, 135, 218
Latin, 7, 8, 82, 147, 287
Latin version, 7, 8
laudanum, 77
laurel, 76, 77, 123, 124, 144, 255; *also see Laurus nobilis*
Laurentium, 124
Laurus camphora, 124; *also see* camphire or camphor
Laurus cassia, 75; *also see* cassia
Laurus cinnamomum, 76; *also see* cinnamon
Laurus kinnamomum, 76
Laurus nobilis, xix, 76, 123, 152; illus.: figs. 47, 95; *also see* green bay tree & laurel
lava, 14
lavender, 52
law (-s), 46, 87, 89, 99, 125, 139, 173, 193, 208, 212, 256
lawns, 132
Lawsonia alba, 124; *also see* camphire & henna
Lawsonia inermis, xix, 90, 124, 195; illus.: fig. 20
Lawsonia spinosa, 124
laxative, 32
lead, 173
leaf; leaves, xv, xvii, 10, 27, 33, 34, 35, 37, 38, 40, 46, 50, 52, 53, 55, 57, 60, 62, 64, 69, 72, 73, 74, 76, 77, 78, 79, 83, 85, 86, 89, 91, 95, 252, 255, 287
leafy tree, 174
LEAH, 137; fig. 67
leather, 144, 191, 197
leaven, 210, 211, 212
Lebanon, iv, xv, 5, 6, 13, 14, 23, 30, 31, 32, 53, 57, 61, 62, 63, 65, 66, 67, 68, 69, 70, 71, 86, 89, 90, 113, 114, 115, 117, 119, 120, 121, 122, 124, 134, 135, 138, 140, 141, 144, 160, 161, 173, 174, 175, 176, 177, 178, 179, 180, 181, 182, 187, 188, 189, 190, 194, 195, 196, 197, 198, 201, 203, 206, 207, 208, 209, 210, 214, 224, 225, 228, 230, 234, 235, 236, 242, 248, 257, 279, 280, 288, 289, 291; figs. 42, 52
Lebaoth, 189
lebonah, 57
Lecanora, 86, 127, 282; *also see* manna
Lecanora affinis, 32, 125, 126, 282
Lecanora desertorum, 126
Lecanora desertorum var. *esculenta*, 126
Lecanora esculenta, xvii, 125, 126, 282; illus.: fig. 27
Lecanora fruticulosa, 125, 127
Lecanora tartarea, 282
lechem, 211
LEDWARD, W. J., 265
leech, 139
leek (-s), 29, 32, 33, 34, 35, 80, 88; *also see Allium porrum*
leek-eater, 35
LEEMANN, A. C., 265
lees, 214
LEES, F. R., 273
LEESSER, I., 17, 259
legend (-ry, -s), 18, 20, 38, 39, 44, 45, 49, 70, 72, 73, 75, 81, 84, 87, 88, 91, 105, 106, 114, 119, 138, 141, 172, 179, 182, 184, 186, 191, 198, 202, 206, 216, 217, 242, 244, 277, 283, 284
Leghorn, 118
legume (-s); leguminous, 25, 30, 34, 37, 72, 254
LEHRMAN, S. M., 274
LEMMENS, L., 1, 3, 265, 266, 267, 271; figs. 10-12
LEMNIUS, L., xv, xx, 265, 266
lemon, 286
lemon-grass, 40, 280, 281; *also see Andropogon schoenanthus*
Lens culinaris, 128; *also see* lentils
Lens esculenta, xix, 113, 128; illus.: fig. 90

lentil (-es, -s), xix, 101, 112, 113, 128, 129, 166, 211, 229, 232, 233, 254; *also see Lens esculenta*
lentisk, vii, xix, 84, 177, 178; *also see Pistacia lentiscus* & mastic
LENZ, H. O., 266
Leontodon taraxacum, 75; *also see* bitter herbs
Leontodon vulgaris, 75; *also see* bitter herbs
leopards, 276
leper (-s); leprosy; leprous, 66, 98, 99, 130, 142, 143, 148, 149, 150, 162, 167, 223, 254, 283, 290
lepidaries, 36
lepra alphoides, 143
lepra nigricans, 143
lepra vulgaris, 143
lettuce, 6, 34, 74, 75, 140; *also see Lactuca sativa* & bitter herbs
Levant (-ine), 2, 5, 31, 52, 81, 90, 92, 129, 138, 168, 170, 171, 174, 177, 185, 190, 214, 233, 235, 245, 247, 253, 255
Levi (-tes), 36, 54, 175, 237
LEVI, Rabbi, 287
leviathan, 9
Levitical Code, 125, 168
Leviticus, 18, 142, 143, 161, 174, 183, 216, 283, 288
levonah, 57
Libanus, 58
libneh, 181, 225
lichen (-s), 126, 127, 128, 143, 195, 205, 254, 282
Lichen esculentus, 126; *also see* manna
lichen jelly, 282
lichen, manna, xvii, 282
licorice, 72, 93
LIDDELL, H. G., 266
life, elixir of, 87
lightning, 120, 124, 172, 198, 284, 285
lightning plant, 72
lign aloes, 47, 48, 193, 198; *also see Aquilaria agallocha*
lignites, 254
lignum Rhodianum; lignum Rhodium, 30; *also see Alhagi camelorum* var. *turcorum*
likk, 196
lilies; lily, vii, xix, 5, 28, 41, 42, 43, 44, 45, 82, 84, 114, 115, 116, 117, 129, 147, 154, 181, 205, 234, 235, 275, 282, 287; *also see Hyacinthus orientalis* & *Anemone coronaria* & *Lilium candidum*
lilies (lily) of the field, xix, 1, 41, 42, 43, 114, 116, 235; *also see Anthemis palaestina*
LILITH, 44
Lilium, 234
Lilium bulbiferum, 116
Lilium candidum, xix, 41, 42, 44, 114, 115, 116, 117, 129, 279; illus.: p. 17, figs. 64, 89
Lilium chalcedonicum, 41, 42, 114, 116, 117, 129
Lilium longiflorum var. *eximium*, 41
lily, Madonna, xix, 115, 116, 129, 279
lily, martagon, 129
Lily of the Law, 43
Lily-of-the-valley; lily of the valleys, 41, 86, 114; *also see Anemone coronaria*
lily, scarlet, 129
lily-work, 42
lime, 6, 179
LINDAU, 282
LINDLEY, J., 266, 268; figs. 17, 18
LINDSAY, 71
linen, xvii, 93, 109, 110, 129, 130, 131, 132, 133, 140, 179, 193, 194, 204, 228, 290; *also see Linum usitatissimum*
linen, fine, 110, 132, 133
linen, white, 110
Lingoum santalinum, 189; *also see* sandelwood & almug & algum
LINK, J. H. F., 71
LINNAEUS, C., 2, 3, 5, 30, 55, 78, 115, 134, 137, 160, 162, 166, 217, 219, 264, 275, 280, 281, 282, 283, 284, 285, 287, 289

linnets, 61
lintel (-s), 66, 97, 160, 223
Linum sativum, 131
Linum usitatissimum, xvii, 129; illus.: p. 135; *also see* flax & linen
lion, 68, 252
LIPPO-LIPPI, 44
Liquidambar orientale, 224, 225; *also see* storax
Liquidambar styracifolia, 224
liquor (-s), 87, 118, 170, 177
Lithuania, 120, 198
livenim, 181
live oak (-s), 75, 198
livneh, 181, 182
loaf; loaves, 112, 153
loaves, barley, 112
loblolly pine, 57
locust (-s), 9, 37, 73, 159, 251, 276, 280
locust, American, 5
locust, black, 26, 204; *also see Robinia pseudo-acacia*
locust pod, 73
locust-tree, 72
lodge (-s), 64, 88, 244
Loggi, 286
Lolium, 29, 282, 283
Lolium temulentum, xvii, xix, 133, 134, 282, 300; illus.: figs. 41, 72; *also see* darnel & tares
Lombardy, 90
London, 169, 181, 186, 277
Loranthus acaciae, xix, 21, 25, 277; illus.: fig. 45; *also see* burning bush
Lord, 42, 47, 50, 56, 57, 67, 68, 78, 80, 97, 103, 120, 125, 126, 130, 142, 143, 144, 149, 167, 169, 175, 176, 188, 189, 193, 194, 198, 201, 202, 203, 213, 214, 215, 219, 223, 224, 227, 229, 230, 240, 241, 242, 247, 251, 292; fig. 40
Los Angeles, vii, 277
lôt, 77
lotophagi, 155
lotus, 77, 154, 155; *also see Nymphaea* spp.
lotus, blue, xvii, 154
lotus bushes, 247; *also see Zizyphus lotus*
lotus-eaters, 155
lotus, golden, 154
lotus-trees, 247
lotus, white, 154
love, 283
love-apple, 137, 138, 283; *also see Mandragora officinarum*
love-in-the-mist, 137
love-potions, 137
Löw, I., 74, 116, 171, 266, 273, 279, 283, 288
LOWDERMILK, W. C., 15
lucerne, 277
Lucian version, 7
LUDOLF, 34
LUDOLPHUS, fig. 40
LUKE, 19, 28, 43, 44, 140
lulab; lulav, 171, 187, 291
lumber (-ing), 68, 188
LUNDGREEN, F., 271
LUNDGRUN, 43
LUTHER, M., 9, 54, 134, 237, 260
Luther's Bible, 9
Luz; luz, xvii, 35, 36, 37, 38
l'vanon, 182
LYCEUS, 106
Lychnis githago, 30; *also see* corn-cockle
Lycium, 207; *also see* brambles
Lycium afra, 266
Lycium europaeum, 134, 135, 153, 203, 206
Lycium mediterraneum, 135
Lycium spinosum, 135
Lycopersicum esculentum, 138; *also see* love-apple
lye, 215

MACCABEES I & II, 19
MACHABEES I & II — *see* MACCABEES
MACCABAEUS, JUDAS, 170, 171
MACKAY, A. I., 147, 272
MACMILLAN, H., 158, 266
Madagascar, 165, 276
madder, 195; *also see Rubia tinctoria*

Madonna, 41
Madonna lily, xix, 41, 44, 279; *also see Anemone coronaria* & *Lilium candidum*
Madrid, 289
magaypha; magepha, 150, 167
magic (-ian, -ians), 32, 33, 49, 198, 223, 229
maglore, 139
magnolias, 75
maidenhair, 208
MAILLET, 34
MAIMONIDES, 42, 58
main-de-gloire, 139
maize, 6, 232; *also see Zea mays*
Malabar, 75, 104, 188
Malacca, 224
MALACHI, 18
MALACHIAS — *see* MALACHI
malaria (-l), 45, 97
Malaya (-n); Malay Peninsula, 47, 75, 76, 104, 125, 188
mall,196
mallow (-s), 6, 9, 48, 53, 54, 91; *also see Atriplex* spp.
mallow, dwarf, 53
mallow, Jews, 53
mallow, marsh, 53
malluach, 53, 54, 237
mallûl, 196
malt, 113
Malta, 73, 89, 91, 92
Malta-cross, 71
Malus assyrica, xiv, 287
Malus communis, 185; *also see* apple & apricot
Malus malus, 185
Malus pumila, 185
Malus sylvestris, 185
Malva, 53
Malva rotundifolia, 53; *also see* mallows
Malva sylvestris, 53
mammals, 3, 173
Mamre, 193, 196
man; mân, 32, 50
MANASSAH; MANASSEH, 19, 158, 185
Mandragora, 283
Mandragora officinalis, 137; *also see* mandrake & dudâim
Mandragora officinarum, xix, 137, 283; illus.: figs. 66, 67
mandrake (-s), xix, 1, 137, 138, 139, 229, 283; *also see Mandragora officinarum*
mandrake, American, 138
maneh, 283
manger, 277
MANGLES, 60
manna (-s), xvii, xix, 31, 32, 72, 81, 82, 86, 125, 126, 127, 128, 278, 282; illus.: figs. 27, 93; *also see Alhagi maurorum & Fraxinus ornus & Tamarix mannifera & Lecanora* spp. & *Nostoc*
manna ash, 31
manna, common, 32
manna, fat, 32
Mannaflechte, 282
manna, flake, 32
Manna hebraica, 31
manna lichen, 282
manna tamarisk, 31
mannin, 32
MANONCOURT, S. DE, 125
maple, 6, 108, 181
maqui, 279
mar, 83
Maracot, xiv
mar 'a wila, 220
marble, 109
Mareb, 280
MARGOLIES, M., 260
marigold, French, 119
marine, 13, 205
maritime, 12, 29, 53, 91, 122
marjoram, 6, 52, 162, 223; *also see Origanum maru*
marjoram, Egyptian, 161
marjoram, Syrian, 161
MARK, 19, 61, 80, 83, 222, 236
market (-s), 39, 73, 75, 84, 95, 230
MARQUAND, Mrs. A., 272, 286
marriage, 33, 45, 214
marsh (-es, -y), 51, 62, 70, 91, 120, 173, 216
Martagon lily, 41; *also see Lilium chalcedonicum*

MARTIAL, 205
MARTIN'S Rheims-Douay Bible, 9
MARX, A., 283
MARY (mother of JESUS), iii, xv, xix, 38, 45, 83, 106, 107, 148, 172, 198, 202, 206, 279, 287; figs. 65, 92
Maryland, 14
MARY, Queen of Scots, 206
Mary's flower, 39; *also see Anastatica hierochuntica*
Maschil, 41
mashkeh, 243
masons, 66
Masoretic text, 7, 196
MASTERS, 243
mastic (-h, -he, -k), 177, 178; *also see Pistacia lentiscus*
mastic-tree, 177
masts, 69
materia medica, 3, 34
matgrass, English, 148
Matharee, 107
mathematical instruments, 63
mats, 93
MATSUMURA, J., 271, 272
MATTHEW, xvii, 19, 27, 28, 43, 44, 61, 80, 83, 89, 222, 236, 248, 300; fig. 41
MATTHEW version, 9
mattock, 202, 206
matza; matzot, 211
Mauritania, 228
Mauritshuis, 286
MAW, G., 87, 266
MAYER, A., 273
MAYER, L., 34, 266
mazar, 197
MCCLINTOCK, J., 266
MCCLURE, A. J. P., 189, 265
MCLEAN, 205
MCLENNAN, 282
meadow (-s), 13, 25, 28, 29, 51, 62, 92, 94, 120, 172, 275
meadow-grass, 62, 93; *also see Butomus umbellatus & Cyperus papyrus*
meal, 88, 128, 129, 163, 211, 230, 240, 288
meal, barley, 112, 113
meal, wheat, 113
measuring-rods, 50
meat (-s), 33, 48, 53, 74, 78, 79, 89, 91, 126, 148, 194, 251, 252, 282
meat offering, 56, 288; *also see* meal
Mecca, 55
MEDEA, 87
Medicago sativa, 277
medical; medicine; medicinal (-ly); medication, 17, 32, 34, 46, 47, 55, 77, 78, 81, 83, 86, 87, 89, 92, 98, 99, 102, 110, 124, 137, 139, 150, 151, 152, 159, 160, 166, 167, 177, 191, 204, 208, 219, 237, 252, 253
medicine, horse, 34, 35
medieval, 115, 138, 139, 160
Mediterranean, 12, 14, 32, 38, 53, 63, 68, 71, 73, 77, 78, 89, 101, 115, 120, 123, 134, 138, 144, 147, 152, 159, 176, 177, 180, 187, 190, 195, 208, 216, 231, 239, 275, 280, 284, 287
MEEK, T. J., 17
Megiddo, 138
MELCHIZEDEK, 212, 240
Melissa officinalis, 86
melon (-s), 6, 32, 33, 34, 79, 80, 81, 88, 281; *also see Citrullus vulgaris & Cucumis melo*
melon, round-leaved Egyptian, 88
MELPOMENE, 91
Memnoniella echinata, 290
Memphitic versions, 8
MENDELSSOHN, 8
menn er-rimt, 278
menn esh-shih, 278
menn et-tarfa, 278
menorah, 218
menstrual, 220
Mentha alba, 140
Mentha arvensis, 139; *also see* mint & bitter herbs
Mentha gentilis, 139
Mentha longifolia, 139
Mentha sativa, 139
Mentha spicata var. *longifolia*, 139

Moldenke — 316 — Bible Plants

Mentha sylvestris, 139
merchant (-s), 56, 67, 84, 85, 95, 130, 166
MERCURY, 145, 150, 154
Merismopedia gonorrhoeae, 150
MERLIN, 198
Merom, 92
merorah; merôrîm, 74, 80
Mesembryanthemum, 54
Mesembryanthemum nodiflorum, 216; *also see* fig-marigold
meshi, 140
Mesopotamia, 13, 15, 133, 231, 233
Metarrhizium glutinosum, 290
Metewelies, 198
Metropolitan Museum of Art, 286
Metz, 9
MEURS, J., 1, 266
Meuzal (cassia), 75
Meuzalites, 75
MEYER, G. A., 1, 266
MEYERHOFF, M., 272, 278, 282
MICAH, 18, 25
mice, 49, 167, 168, 172
MICHAES — *see* MICAH
Michtam, 41
Micrococcus albus, 223
Micrococcus aureus, 223
Micrococcus gonococcus, 150
Micrococcus gonorrheae, 150
Micrococcus gonorrhoeae, 150
Micrococcus pyogenes, 223
Micrococcus pyogenes albus, 223
Micrococcus pyogenes aureus, 223
Micrococcus scarlatinae, 223
Microspironema pallidum, 150
Microsporium gyseum, 290
Midian (-ites), 111, 113, 150, 229
Midrash, 277, 279, 283, 284, 285, 286, 287, 288, 289, 291, 292
midwife, 193
mignonette, 6
migrations, 69
Migron, 189
mikshah, 88
miktar, 57
mikvay; mikveh, 132
mildew, 142, 143, 252, 254, 289, 290
milk (-y), 15, 24, 40, 69, 79, 178, 186, 211, 232, 240, 241, 282, 287
milk-producer, 282
mill (-s), 126, 229
MILLER, J. L., 272
MILLER, M. S., 272
millet, xv, 6, 101, 112, 128, 166, 211, 232, 233, 254, 257; *also see Panicum miliaceum*
millet, German, 166
millet, Indian, 222
millet, Italian, 166
Millo, 193
millstones, 230
milo, yellow, 222
MILTON, J., 104, 186
mimosa, Egyptian, 23
Mimosa farnesiana, 23; *also see Acacia*
Mimosa nilotica, 23
Mimosa scorpioides, 23
Mincha, 232
MINERVA, 145, 160
mingled (wine), 212, 288
Minnith, 84, 157, 166, 230
mint (-s), 6, 46, 74, 75, 89, 139, 140, 162, 208; *also see Mentha longifolia*
mint, horse, 139
MINTHO, 140
miracle (-s); miraculously, 45, 78, 127
mire, 50, 62, 92, 120
MIRIAM, 142, 167; fig. 63
mirre, 82, 83
mispo, 254
Missouri Botanical Garden, vii
mistletoe, 23, 25, 33, 277, 285; *also see Loranthus acaciae*
mistletoe, crimson-flowered, 23
MITFORD, J., 266
mithridate, 208
MITHRIDATES VI, 208
Moab (-ites), 24, 26, 39, 77, 113, 122, 138, 150, 151, 174, 196, 217, 241, 252; fig. 86
mock-sceptre, 50
mock-trial, 50, 235, 236
MOFFATT, J., 10, 26, 27, 38, 39, 43, 152, 260, 285, 286, 287

MOFFATT version — *see* Version, MOFFATT
MOCHADAM, S., 266
Mohammedan (-s); Mohammet, 8, 38, 55, 90, 91, 125, 172, 206, 285
mold (-s); mouldy, 141, 143, 210, 212, 254, 290
MOLDENKE, A. L., ii
MOLDENKE, A. R., viii
MOLDENKE, C. E., ii, 196, 266; fig. 85
MOLDENKE, H. N., ii, 266, 270, 272
mollusk, 224
Molucca Islands, 188
Momordica elaterium, 78; *also see* squirting-cucumber & gourd
monasteries; monastery, 4, 7
money, 31, 33, 67, 87, 241, 255
monk (-s), 7, 39, 278, 284
monkeys, 220
MONTCLAIR, P., 266
MOORE, G. T., 265, 266
MOORE, T., 138, 266
Moors, 191
môr, 77, 82
moraines, 68
Moravia, 9
MORDECAI, 130, 204, 291
Moreh, 158, 193
morning-glory, 30
Morocco, 191, 228, 282
MORREN, N., 268
Morus alba, 5
Morus nigra, xv, xvii, xviii, 108, 140, 141, 183, 283; illus.: fig. 23; *also see* mulberry & sycamine
Mosaic law, 89
MOSCHUS, 114
MOSES, iii, 15, 23, 25, 32, 33, 36, 56, 65, 74, 75, 93, 125, 126, 150, 161, 162, 169, 174, 189, 223, 224, 233, 249, 277, 281, 282, 283, 284, 289, 290, 292; figs. 14, 45, 63
Moslem (-s), 191, 196, 198
MOSS, fig. 80
moss (-es), 94, 283
mosyletis; Mosyllon; mosyllos, 75
moth (-s), 48, 49, 78
Mother of the Gods, 106
mother-of-vinegar, 27
motz, 254
mount (-ain, -ainous, -ains), 6, 13, 14, 23, 24, 25, 35, 38, 39, 59, 63, 65, 66, 67, 68, 69, 71, 86, 89, 90, 91, 97, 117, 119, 121, 122, 123, 127, 143, 147, 169, 173, 174, 175, 177, 178, 179, 180, 181, 193, 194, 195, 197, 198, 205, 209, 210, 213, 218, 230, 234, 235, 242, 252, 275, 276, 279, 284
mountain-ash, xvi, 14, 286
Mount Carmel, 81, 123, 138, 144, 201, 207, 225, 275, 279
Mount Ebal, 147
Mount Etna, 91
Mount Hebron, 98
Mount Hermon, 14, 89, 140, 275
Mount Hor, 121
Mount Ida, 87
Mount Lebanon, 89, 140, 275
Mount of Olives, 42, 158, 160, 171; fig. 64
Mount Olivet, 106
Mount Olympus, 148
Mount Sinai, 24, 278
Mount Tabor, xix, 98, 144, 165, 207, 208, 225; fig. 87
Mount Vesuvius, 93
mourn (-ers, -ing), 65, 68, 91, 288
mowed; mown grass, 29, 101
Mozambique, 188
m'reeroot, 80
m'rorim, 74
m'shoach, 158
m'soochah, 203
m'tzorah, 143
MÜCKE, F., 267
Mucor mucedo, 141, 212
mugwort, 48, 49, 50
mukul, 81
mulberries; mulberry, 5, 8, 14, 140, 141, 183, 242; *also see Morus nigra*
mulberry, black, xvii, 140, 141
mulberry-fig, 107; *also see Ficus sycomorus*

mulberry, white, 141
mule (-s), 28, 193
mummies; mummy, 101, 125, 131
mummy-cases; mummy-coffins, 24, 70, 107; fig. 62
mummy-cloths, 110
MUNDELSTRUP, J. N., 1, 267
MURCHISON, R., 261, 267
Murderers' Bible, 10
Murex trunculus, 195
MURILLO, 44
murr (-a, -ha), 82
MURRAY, E. G. D., 261
MURRAY, J. J., 272
murrophore, 83
Musa paradisiaca, 281
Musa paradisiaca subsp. *sapientum*, 243; *also see* bananas
Musa sapientum, 243
Muse, 91
Museum, 286
mushroom, 138; *also see Agaricus campestris*
musical; musician, 41, 50
muskmelon, 80, 81; *also see Cucumis melo*
Musselmen, 72
must, 213, 214
mustard (-s), 6, 38, 59, 60, 61, 62, 74, 140, 220; *also see Brassica nigra*
mustard, black, 59, 60
mustard-tree, 60
mustard, white, 59
Muzri, 132
Mycobacterium, 97, 141
Mycobacterium leprae, 142, 143, 167, 283
Mycobacterium tuberculosis var. *hominis*, 143
MYRENE, 145
Myrica sapida, 30
myrobalanum, 84
myrrh (-e), 30, 35, 39, 47, 51, 55, 56, 57, 58, 75, 76, 77, 80, 82, 83, 84, 102, 129, 177, 179, 214, 218, 219, 224, 277; *also see Commiphora kataf*, *C. myrrha* & *Cistus* spp.
MYRRHA, 84
myrrh, European, 83
Myrrhis odorata, 83
myrrh-oil, 224
MYRSINE, 145
MYRTILUS, 145
myrtle (-s, trees), xvii, xix, 14, 24, 25, 26, 33, 62, 63, 64, 67, 97, 143, 144, 145, 171, 173, 221, 248, 255, 285, 288, 291; *also see Myrtus communis*
myrtleberry, 145
myrtle, three-leaved, 175
Myrtus communis, xvii, xix, 143, 144; illus.: p. 145, fig. 68; *also see* myrtle
Mystic Lamb, 286
myth (-ology, -s), 17, 18, 39, 45, 71, 87, 91, 95, 105, 106, 124, 133, 140, 141, 145, 148, 160, 182, 191, 198, 206

NAAMAN, 142
naasus, 248
na'atzootz (-im); n'ahtzootz (-im), 165, 248
NABELEK, 38
Nachal-Shittim, 25
nâcôth, 51, 85
NADAB, 56
NAFTOLSKY, N., 42, 115, 279
NAGBELE, 215
nahalolim, 254
NAHUM, 18
naird, 148
NAPOLEON, 90
Narcissus; narcissus, 116, 147; *also see* lily & rose
Narcissus jonquilla, 114, 117
narcissus, polyanthus, xvii, 147
Narcissus tazetta, xvii, 41, 117, 147, 151, 205, 234, 235; illus.: p. 154
narcotic, 30, 80, 137, 239
nard (-as), 148, 149, 206
nard, Ganges, 149
Nardostachys, 148
Nardostachys jatamansi, xvii, 40, 148, 149; illus.: fig. 34; *also see* spikenard
nard, pistic, 149

Nardurus orientalis, 28; also see grass
Nardus stricta, 148; also see spikenard
NARKISSOS, 148
narkos, 147
nasasusim, 248
Nasturtium fontanum, 74; also see bitter herbs
Nasturtium officinale, 74
nataf; nataph, 224
NATHAN, Rabbi, 292
National Council of Churches of Christ, vii
National Geographic Society, 282
Nativity scene, 4
natron, 215
natural histories of the Bible, 2
Nature-worship, 107
nauseous, 79, 91, 202
Nazareth, xix, 65, 71, 138, 165, 197, 218; fig. 88
n'chot, 51
n'coth, 85
Near East, 280
Nebi'im, 7
NEBUCHADNEZZAR, 44
neca'at, 52
Nechôth; nĕcôth, 51, 52
Negev, 276
NEHEMIAH, 18, 43, 97, 98, 144, 174, 175, 285
nehga, 151, 254
NEILL, J. C., 272, 282
Neisseria, 167, 168
Neisseria gonorrhoeae, 149, 150
NELSON, T., 272, 274, 280, 283, 287, 288
Nelumbium speciosum, 155
Nepal, 30, 148
nerd, 148
Nerium oleander, xix, 123, 151, 205, 217, 247, 285; illus.: fig. 69; also see rose
NERO, 35, 124, 206
nest (-s), 59, 61, 67, 137
nĕtêk, 99
nether, 215
nettle (-s), 6, 8, 26, 27, 54, 153, 202, 206, 220, 221, 237, 245; also see *Acanthus spinosus* & *A. syriacus*
nettle, great, 237; also see *Urtica dioica*
nettle, Roman, 237; also see *Urtica pilulifera*
nettle, small, 237; also see *Urtica urens*
nettles, true, 237; also see *Urtica* spp.
NEUSTÄTTER, O., 168, 267
NEVIN, J. W., 267
Newark State Teachers College, ii
NEWTON, T., 1, 161, 266, 267
New York Academy of Medicine, vii
New York Botanical Garden, ii, vii, 4, 196, 290
New York City, 73, 286
New York Public Library, vii, viii, 283, 300
New York State, 14
NICHOLS, W. A., 267
NICHOLSON, J. A., 265, 267
NICODEMUS, 35, 47, 83; fig. 49
nidaf aleh, 183
NIEBUHR, C., 3, 263
Nigella damascena, 152
Nigella orientalis, 153
Nigella sativa, xvii, 152, 234; illus.: p. 249; also see fitches
Nigella sativa var. *brachyloba*, 152
nightshade, 30, 137, 222, 239
nightshade, hoary, 30, 239
nightshade, Palestine, xix, 221, 239; also see *Solanum incanum*
nikoodim, 141
Nile, xix, 13, 51, 62, 88, 92, 93, 94, 101, 128, 154, 231; fig. 59
Nimrim, 28
Nineveh, 15, 173, 198, 204
Nitraria schoberi, 203; also see hedge
nitre, 27, 215
NOAH, xvii, 15, 90, 152, 159, 187, 189, 213, 240, 242; figs. 25, 26, 77
nodules, tubercular, 143

nomads; nomadic, 107, 232, 282
nomenclature, botanical, 6, 275
non-alcoholic, 214
Nopalea coccinellifera, 194; also see pricklypear cactus
Normandy, 45
North America, 26, 29, 40, 107, 217, 277, 282, 283
NORTON, M., vii
NORTON, R., vii
Norway, 57, 169
Nostoc, 125, 282; also see manna
Nostoc collinum, 282
Nostoc commune, 282
Nostoc edule, 282
Notobasis, 207
Notobasis syriaca, 71, 153, 206, 249; also see thistle
Nubia, 92, 124, 222
nulok, 248
Numbers, 18, 25, 82, 88, 161, 190, 281
NUN, 24
nut (-s), 35, 46, 51, 77, 119, 120, 177, 179, 180, 190, 291
nutmeg, 152
nutmeg-flower, xvii, 152, 153; also see *Nigella sativa*
nuts, pistachio, 119, 179; also see *Pistacia vera*
nut-tree, 291
nymph, 45, 124, 140, 145, 148, 154, 191
Nymphaea, 42, 154; also see lily
Nymphaea alba, 129, 154
Nymphaea caerulea, xvii, 154; illus.: p. 155
Nymphaea lotus, 42, 154

oak (-s), 6, 14, 47, 62, 68, 89, 107, 175, 178, 179, 181, 182, 193, 194, 195, 196, 197, 198, 227, 228, 277, 285, 286; also see *Quercus* spp.
oak, Abraham's, xvii, 196; also see *Quercus coccifera* var. *pseudococcifera*
oak apples, 197
oak, circular, 196
oak, Cyprus, 195; also see *Quercus lusitanica*
oak, holly, 178; also see *Quercus ilex*
oak, holm, 195, 198, 281, 286
oak, kermes, 194, 195
oak of Bashan, 197; also see *Quercus aegilops*
oak, oracular, 196
oak, Valonia, 195, 286; also see *Quercus aegilops*
oasis; oases, 13, 275
oats, 112, 233, 234; also see *Avena sativa*
OBADIAH, 18
Occident (-al, -als), 11, 44, 83, 138, 278
Oceanica, 50
Ochradenus baccatus, 245; also see thorns
O'CONNELL, J. P., 273, 275
odor; odour (-s), 30, 32, 33, 47, 57, 59, 76, 77, 82, 86, 102, 118, 124, 138, 148, 182, 188, 208, 210, 223, 224, 291
Odyssey, 155
OEDMAN, S., 2, 267
offering (-s), burnt, cereal, drink, meal, meat, wine, 59, 125, 132, 139, 151, 155, 159, 210, 211, 212, 214, 215, 220, 229, 230, 232, 240, 242, 251, 256, 277, 288
O'HARA, E. V., 10
O'HARA version — see version, O'HARA
Oidium, 290
oil, 23, 30, 37, 39, 40, 46, 55, 56, 57, 59, 61, 82, 84, 97, 98, 119, 124, 126, 133, 139, 144, 149, 153, 158, 159, 166, 169, 171, 203, 204, 212, 229, 230, 240, 242, 280, 281; also see *Olea europaea*
oil, colza, 59
oil, ginger-grass, 40; also see *Andropogon aromaticus*
oil, holy, 76, 83, 224
oil, kiki, 204; also see *Ricinus communis*
oil, linseed, 133; also see *Linum usitatissimum*

oil of bay, 124; also see *Laurus nobilis*
oil of cinnamon, 76; also see *Cinnamomum zeylanicum*
oil of citronella, 280
oil of lemongrass, 280, 281
oil of Rhodium, 30; also see *Convolvulus floridus* & *Convolvulus scoparius*
oil, olive, 39, 119, 216; also see *Olea europaea*
oil-presses, 160
oil (-y) tree, 24, 62, 67, 97, 98, 144, 174
oil, Zackum, 98; also see *Balanites aegyptiaca*
ointment (-s), 30, 51, 57, 76, 82, 84, 148, 149, 157, 158, 159
OLDS, N. S., 267
Old World, xvii, 5, 60, 88, 107, 232, 243, 282
Olea europaea, xvii, 98, 157, 158; illus.: p. 15, figs. 14, 25, 26; also see olive & orchard
oleander, xix, 6, 14, 123, 151, 152, 205, 217, 247; also see *Nerium oleander* & rose
oleander wood, 90; also see gopher wood
oleaster, 98, 159, 174; also see *Elaeagnus angustifolia*
oleoresin, 77
olibanum, 58
olive (-s), xvii, 6, 14, 25, 33, 39, 63, 70, 98, 104, 108, 143, 157, 158, 159, 160, 173, 241, 248, 256, 276, 285, 287, 291, 292; also see *Olea europaea* & *Elaeagnus angustifolia*
Olivet, 157, 158
olive-press (-es), 15
olive tree (-s), 26, 97, 103, 106, 151, 157, 158, 159, 180, 190, 240, 255, 292
olive (-s), wild, 98, 159, 174; also see *Elaeagnus angustifolia*
olive wood, 98
oliveyard (-s), 157, 159
OLOFERNES, 44
Olympian, 123
OM, 154
omelet, 128
omer, 125, 126, 231, 232
onion (-s), vii, 32, 33, 34, 35, 39, 80, 88; also see *Allium cepa*
onion, Bermuda, 119
onion, Egyptian, 33
Onobrychis sativa, 277
Onobrychis viciaefolia, 277
Onobrychis vulgaris, 277
Ononis spinosa, 203
onycha, 56, 58, 102, 223, 224; also see *Styrax benzoin*
onyx, 30, 81, 102, 223, 224
oogah, 212
operculum, 224
Ophel, 141
OPHER, Rabbi, viii, 289
Ophir, 175, 188, 189, 209, 287
Ophrah, 193
opobalsamum, 224; also see *Commiphora opobalsamum*
Opoidia galbanifera, 102; also see galbanum
opopanax, 5, 83; also see *Commiphora kataf*
opopanax, yellow-flowered, 23; also see *Vachellia farnesiana*
Oporanthus luteus, 41, 114; also see lily
OPPENHEIMER, H. R., viii, 92, 263, 267, 287, 288, 289; fig. 73
opthalmia, gonorrheal, 150
Opuntia ficus-indica, xix, 5; illus.: fig. 88; also see cactus
orach, 53
orach, shrubby, 53; also see *Atriplex halimus* & mallows
oracle (-s), 97, 101, 124
ŏran, 286
orange (-s), 6, 11, 14, 78, 185, 190, 191, 275; also see *Citrus sinensis* & apples
orange, bitter, 185
orange, blood, 11, 191
orange, Seville, 185
orchard (-s), 85, 105, 119, 124, 148, 158, 159, 160, 185, 190, 198, 201, 205, 252
orchids, 61

Order of the Thistle, 72
ŏren, 176
orient (-al, -als), 11, 32, 35, 37, 44, 46, 47, 48, 59, 61, 65, 87, 105, 119, 124, 125, 128, 132, 134, 138, 145, 152, 159, 168, 169, 170, 173, 180, 188, 203, 204, 232, 278, 285
Origanum, 222, 223; also see hyssop & reed
Origanum aegyptiacum, 161
Origanum maru, 46, 160, 161, 162, 283
Origanum maru var. *aegyptiacum*, 160, 161, 162
Origanum maru var. *sinaicum*, 161
Origanum nervosum, 161
Origanum smyrnaeum, 162
Origanum syriacum, 161
ORIGEN translation, 7
ornament (-al, ation, -s), 36, 95, 155, 170, 181, 190, 191
ORNAN, 230
Ornithogalum luteum, 284
Ornithogalum umbellatum, 162, 284; also see doves' dung
orris, 41, 76, 219
orris, Indian, xvii, 218, 219
orris-root, 76, 118
orris-root, European, 218
OSEE — see HOSEA
osier (-s), 132, 218
OSIRIS, 155, 244
OSBORN, H. S., 267
osnan, 163
ossifrage, 9
ostrich (-es), 9, 51
Ostrog, 9
OSTROM, H., 259
Our Lady's bedstraw, 277
Our Lady's thistle, 72; also see *Cnicus benedictus*
Ourouparia gambier, 125; also see henna
oven, 41, 43, 44, 211, 212
over-cultivation, 6
OVID, 114
owl (-s), 9, 13, 237
ox (-en), 9, 28, 112, 182, 229, 230, 232, 247, 252
Oxford University Press, 17
ox, wild, 9
oxylate of lime, 282

pa'amon, 190
PACHEL, fig. 65
Pacific, 186
pack-animals, 166
pag, 104, 281
paggah, 281
pagan, 125, 150, 182
pageha, 104, 281
pain (-s), 139, 168, 214, 219
PAINE, 71
painters; painting; paints, 41, 50, 67, 206, 242, 243, 277, 286
paka, 79
Pakistan, 282
pakknoth-sadeh, 78
palace (-s), 41, 67, 69, 109, 153, 170, 175, 237, 245
palanquins, 69
Palestine; Palestinian (-s), viii, xvii, xix, 2, 3, 4, 5, 6, 8, 13, 14, 15, 23, 25, 26, 28, 30, 31, 32, 33, 36, 38, 39, 40, 41, 42, 43, 45, 46, 47, 48, 50, 52, 53, 54, 55, 59, 60, 61, 62, 63, 65, 71, 73, 74, 75, 77, 78, 79, 81, 84, 85, 87, 88, 90, 92, 98, 101, 105, 107, 109, 110, 112, 113, 114, 115, 116, 119, 120, 121, 128, 129, 131, 134, 139, 140, 144, 147, 151, 152, 153, 158, 159, 161, 162, 166, 167, 170, 171, 174, 175, 176, 177, 179, 180, 182, 183, 184, 185, 186, 187, 190, 194, 196, 197, 198, 201, 203, 204, 205, 206, 207, 208, 210, 213, 214, 216, 217, 218, 220, 222, 224, 225, 227, 228, 231, 232, 233, 234, 235, 236, 237, 239, 243, 244, 245, 247, 248, 276, 278, 279, 280, 281, 282, 284, 288; fig. 42
Palestine caper, xix
Palestine chamomile, 43
Palestine nightshade, xix
Paliurus, 284, 289
Paliurus aculeatus, 27, 165
Paliurus australis, 165

Paliuris spina-christi, xvii, 165, 202, 220, 245, 248, 284; illus.: fig. 27; also see crown-of-thorns
PALLAS, P. S., 273
palm (-s), (date), xv, xvii, xviii, xix, 6, 44, 67, 69, 81, 97, 141, 143, 151, 169, 170, 171, 172, 173, 180, 183, 185, 189, 190, 198, 213, 241, 255, 256, 283, 285, 291; also see *Phoenix dactylifera*
Palma Christi; palm-christ, 263
palm, dwarf fan-, 287
palmerworm, 9, 103, 157, 159, 172
Palm Sunday, 171, 172
Palmyra, 172
palsy, 143
pampas-grass, 50
panaxeia, 166
Panax ginseng, 166; also see ginseng
Panax schinseng, 166
Pancratium, 288
Pancratium aegyptiacum, 282
Pancratium illyricum, 282
Pancratium maritimum, 288
Panicum italicum, 166
Panicum miliaceum, 166, 167, 222; also see millet
Pannag, 84, 157, 166, 222
papaya, 275
paper, 93, 172, 173
paper reeds, xix, 92, 94; also see *Cyperus papyrus*
papyrus (reeds), vii, 62, 92, 93, 94, 133, 173, 249, 276, 281
Papyrus antiquorum, 92; also see bulrush & reed
parable, (-s), 10, 28, 59, 60, 61, 72, 103, 113, 300
Parable of the Lilies, xix
Parable of the Tares, 300, fig. 41
Parable of the Vinegar, 28, 244
Parable of the Vineyard, 28, 244
parachutists, 50
Paradise, xvi, 144, 205, 253
Paradise Lost, 104
Paran, 198
parasitic; parasites, 25, 91, 99, 202, 253, 254, 277, 278, 285
parched; parchment, 50, 93, 173
Paris, 48, 118
PARKHURST, 184
Parmelia esculenta var. *affinis*, 126; also see manna & lichen
parsley, 46, 235
partridge, 127
PAS, C. DU (PAS, C. VAN DE), 1, 263; fig. 58
Paschal, 74, 140, 162
Paschal Lamb, 75
Passiflora, xv, 276, 277; illus.: fig. 3
Passiflora incarnata, xiv
Passionblumen, xiv
passionflowers, xiv, 276, 285
Passover, 74, 161, 223, 284
PASSOW, F., 266
Pasteurella, 97, 141
Pasteurella pestis, 167
pastilles, 280
pasture (-s), 29, 51, 282
PATAI, R., 267
PATERSON, 288
PATON, W. R., 104
PAUES, A. C., 262
PAUL, 159
PAUSANIAS, 109
PAUSANIUS, 109
PAXTON, G., 166, 267
pea (-s), 6, 31, 72, 152, 201, 232, 282
peacemakers, 50
peach, 14, 37, 188, 287
peanuts, 276
pear, 204, 284, 287
pear, common, 183
pearls, 82, 131, 228
pear, Syrian, 184
peganon, 208
pelican, 9, 276
Pelusium, 281
Penicillium, 290
Penicillium glaucum, 290
pens (reed), 172, 173, 285
Pentapera sicula var. *libanotica*, 121
Pentateuch, 7, 13, 68, 74, 173, 210, 275, 278
pepper, 153
perah lebanon, 289

perfume (-d, -ry, -s), 30, 40, 47, 52, 58, 59, 76, 77, 82, 83, 87, 101, 144, 148, 149, 157, 159, 187, 215, 219, 224, 225, 280, 281
Peridinium, 289
Persia (-n, -ns), 13, 36, 42, 50, 67, 76, 77, 84, 86, 91, 95, 101, 102, 109, 119, 127, 129, 140, 181, 186, 191, 235, 242, 278, 282
Persian Gulf, 82, 95, 218
Peru, 188
Peshitta version, 10
pestilence, 45, 124, 167
PETER, 97, 132, 276
Peter I & II, 19
PETERS, E. J., 267
Petra, 285
Peucedanum galbanum, 102; also see galbanum
Peucedanum graveolens, 46
PFEFFEL, A., 268
PFISTER, A., 9
PFISTER's Bible, 9
PHAETON, 182
Phakussa, 128
phallic, 88
Pharam, 198
PHARAOH, 41, 62, 69, 92, 112, 129, 172, 223, 229, 240, 243, 251; fig. 52
Pharisees, 46, 89, 139, 208, 211
Phenice, 170, 171
Philadelphia Quartermaster Depot, 290
PHILEMON, 19
PHILIP, 104
Philippians, 19
Philistia, 158
Philistine (-s), xv, 128, 167, 178, 185, 201, 216, 229
PHILO, 8
PHINEHAS, 149
Phoenicia (-n, -ns), 60, 76, 90, 95, 115, 121, 150, 154, 158, 170, 173, 195, 198, 201, 216, 218
Phoenician cedar, 63
Phoenis, 171
Phoenix, 171
Phoenix dactylifera, xv, xvii, xix, 95, 169, 170, 170, 213, 285; illus.: figs. 9, 28, 70, 92; also see date & palm
Phragmites, 29, 173; also see pens
Phragmites communis, 51, 172, 285
Phragmites communis var. *isiacus*, 285
Phragmites maximus, 172
Phragmites phragmites, 172
Phragmites scriptorum, 172
physician (-s), xx, 55, 102, 150, 219, 278
phytogeography, 13
Phytolacca americana, 60; also see mustard & pokeberry
Phytolacca decandra, 60, 161
Phytolacca dodecandra, 61, 161
Phytolacca pruinosa, 161
Phytotheologia, 61
Picea abies, 57; also see spruce & frankincense
Picea orientalis, 174; also see pines
PICKEL, DON B., 287
PICKERING, C., 267
pickle (-s), 46, 88
PIEL, G., 168, 267
pigs, 72
pigeon, 134, 162
pillars, 56, 66, 82, 84, 154, 188, 189, 193
Pimenta dioica, 52; also see spices
Pimenta officinalis, 52
pimpernel, 6
Pimpinella anisum, 46; also see anise
pimples, 223
pine (-s), 6, 14, 24, 62, 89, 90, 97, 98, 143, 174, 175, 176, 177, 180, 181, 189, 281, 284, 286
pine, Aleppo, xix, 14, 46, 175, 176; also see *Pinus halepensis*
pine, Brutian, 174, 175, 177, 285; also see *Pinus brutia*
pine, loblolly, 57; also see *Pinus taeda*
pine, Scotch, 177; also see *Pinus sylvestris*

Bible Plants — 319 — General Index

pine, stone, xvii, 46, 177; also see *Apinus pinea*
Pinus, 284
Pinus brutia, 173, 174, 176, 177, 181, 189, 285; *also see* pine
Pinus carica, 174; *also see* Brutian pine
Pinus cedrus, 68; *also see* cedar
Pinus halepensis, xix, 32, 46, 57, 68, 70, 174, 175, 176, 177, 181, 285; illus.: fig. 73; *also see* pine, Aleppo
Pinus halepensis var. *brutia*, 174
Pinus hierosolymitana, 176
Pinus larix, 177; *also see* larch & pine
Pinus libanotica, 68; *also see* cedar
Pinus maritima, 174, 176; *also see* Aleppo pine
Pinus peuce, 115
Pinus pinea, 46; *also see* stone pine
Pinus pyrenaica, 174; *also see* Aleppo pine
Pinus sylvestris, 177; *also see* Scotch pine & Scotch fir
Pinus taeda, 57; *also see* loblolly pine
pipes, 50
pishtah; pishtim, 131
Pisidia, 43
Pison River, 82, 292; fig. 40
pista, 131
pistachio (-s, trees), 6, 159, 180
Pistacia lentiscus, xix, 55, 84, 177, 280, 286; illus.: fig. 74; *also see* nuts & pistachio
Pistacia terebinthus var. *palaestina*, 73, 178, 184, 195, 286; illus.: fig. 75
Pistacia trifolia, 179
Pistacia vera, 119, 179
pitch, 92, 93
PIVAL, D., 9
pixie, 235
PIZARRO, 109
plague (-s), 33, 98, 99, 130, 131, 142, 149, 151, 167, 168, 169, 219, 223, 290
plague-boils, 167, 168
plague, bubonic, 72, 150, 151, 167, 168, 223
plague, pneumonic, 168
plain (-s), 12, 13, 29, 39, 68, 71, 81, 85, 88, 92, 97, 114, 121, 127, 147, 158, 169, 171, 193, 196, 198, 220, 231, 234, 235, 237, 275, 276, 288
planetree (-s), 9, 14, 25, 63, 67, 90, 107, 151, 174, 175, 180, 181, 182, 247, 281, 285, 286; also see *Platanus orientalis*
planetree, California, 281, 282
planetree, London, 181
planetree, Oriental, 180, 181
planks, 91
plankton, 289
plantain, 281
plantations, 88
PLANTIN, iii, xv, xvii
plant-lice, 31
plants, well-rooted, 94
plant, tender, 252
plasters, 77
platanus, 180
Platanus acerifolia, 181; *also see* chestnut & London planetree
Platanus occidentalis, 107 181; *also see* chestnut & American planetree
Platanus orientalis, 90, 107, 175, 180, 247, 286; *also see* chestnut & Oriental planetree
Platanus racemosa, 282
plateau (-s), 14
PLATO, 1
PLINY, 1, 30, 32, 35, 45, 83, 106, 109, 110, 115, 116, 137, 170, 178, 228
Pliocene, 115
plowman; plowmen, 241, 242
plum (-s), 6, 60, 83, 137, 188, 243
plums, cooking, 244
PLUTARCH, 1, 33
PLUTO, 95, 140, 148, 191
pod (-s), 25, 61, 72, 73

Podophyllum peltatum, 138, 283; *also see* American mandrake
poet (-s), 123, 124, 144, 150
poison (-ing, -ous, -s), 30, 48, 49, 78, 79, 80, 91, 105, 120, 134, 137, 151, 152, 163, 187, 202, 208, 239, 240, 279, 282
poison-hemlock, 49; also see *Conium maculatum*
poison-water, 78
pokeberry, 60; also see *Phytolacca* & mustard
pokeweed, 61, 161; also see *Phytolacca* & mustard
pôl, 101
Poland, 284
Polish, 9
polish, 95, 228
pollination, 286
POLLUX, J., 109
polyanthus narcissus, xvii, 11
Polypogon monspeliensis, 28; *also see* grass
pomegranate (-s), xv, xvii, xix, 6, 11, 14, 51, 103, 111, 124, 130, 148, 149, 185, 189, 190, 191, 193, 204, 205, 206, 212, 213, 218, 229, 240, 243, 276, 291; *also see Punica granatum*
pomelo, xvii, 186; also see *Citrus grandis*
Pomerania, 284
Pompeii, 93
Pontus, 208
pool, 50, 51, 92, 289
POOL, D. DE S., 264, 269
Pope, 9, 206, 214
poplar (-s), 6, 14, 35, 73, 117, 178, 180, 181, 182, 183, 184, 194, 216, 284; also see *Populus alba*
poplar, black, 182
poplar, Euphrates, 183, 217; also see *Populus euphratica*
poplar, silver, 181
poplar, white, 181, 182, 225; also see *Populus alba*
poppies; poppy, 30, 80, 152
Populus alba, 70, 225, 286; *also see* poplar
Populus candicans, 86
Populus euphratica, 73, 141, 152, 183, 216, 217, 288
Populus nigra, 182
Populus tremula, 183, 284
Porch of Judgement, 69; fig. 52
porcupines, 276
Porrophagus, 35
Port Said, 282
Portugal; Portuguese, 106, 187, 286
posca, 27
POSEIDON, 160
POST, G. E., 4, 13, 17, 25, 27, 29, 30, 33, 38, 40, 41, 43, 49, 51, 52, 53, 58, 59, 60, 61, 63, 65, 74, 77, 79, 81, 82, 85, 86, 87, 90, 92, 102, 113, 114, 121, 122, 123, 125, 129, 138, 140, 147, 151, 152, 153, 162, 165, 166, 173, 174, 175, 176, 181, 182, 183, 185, 187, 196, 198, 201, 203, 204, 207, 208, 215, 216, 217, 220, 221, 224, 225, 228, 233, 234, 236, 239, 242, 243, 247, 248, 267, 277, 282, 283, 285, 287, 288
potash, 215
potato (-es), 6, 137, 221
potato, Irish, 119
potato, sweet, 6
pot-herb (-s), 83, 94
pot-pourri, 87
pottage, 78, 128
pottage, red, xix, fig. 90
poultry, 276
POUSSIN, N., 277
powder (-s), 61, 84, 85, 118, 144
POWIS SMITH, J. M., 259
Prado, 286
Prague, 9
PRATT, A., 4, 30, 41, 47, 52, 54, 59, 60, 65, 74, 77, 78, 81, 85, 128, 139, 154, 179, 184, 186, 203, 204, 206, 222, 234, 237, 267
precious things, 51, 52, 84, 85
pregnant, 182
prenatal influences, 182, 286
present, 35, 36, 51, 77, 95
presses, 241
p'ri, 256
PRIAPUS, 106, 150
prickles; prickly; pricks, 29, 39,

71, 78, 79, 135, 206, 207, 221, 237, 246, 287
prickly alhagi, 31; also see *Alhagi maurorum* & manna
prickly bush, 23; also see *Acacia nilotica*
prickly herb, 39; also see *Gundelia tournefortii*
pricklypear cactus, 5, 6; also see *Opuntia ficus-indica*
prickly plant, 29; *also see* cockle
pride of India, 6
priest (-s), 33, 56, 76, 77, 98, 99, 102, 110, 111, 123, 130, 131, 133, 142, 143, 144, 151, 160, 193, 209, 210, 223, 229, 290
primrose, 137
prince (-s), 36, 68, 212, 213
Printer's Bible, 10
prison (-ed, -ers), 70, 80, 125, 211
PRITZEL, G. A., 265, 267
privet, 188
privet, Egyptian, 124
Probe-Bibel, 9
prodigal son, 72
Prohibitionists, 288
Promised Land, xv, xviii, 15, 31, 243, 289
prophecy; prophecies, 63, 71, 124, 252
prophet (-s), 7, 13, 62, 68, 69, 101, 106, 198, 245, 256
Prophets, Books of the, 9
PROSERPINE, 140, 148, 191
Prosopis stephaniana, 248; *also see* brambles
provender, 232, 251, 254
Proverbs, 18, 47, 48, 203, 220, 237
prune (-d, -s); pruning, 46, 241, 244
Prunus armeniaca, xix, 187, 286, 290; illus.: fig. 77; *also see* apricot & apple
psalm; Psalms, xiv, 18, 27, 38, 43, 47, 75, 79, 80, 123, 127, 161, 216, 219, 233, 283, 287, 288; fig. 33
Psalter Davids, xiv, 265, 277
psoriasis, 143
PSYCHE, 145
ptelea, 179
Pteridium aquilinum, 277
Pterocarpus santolinus, 90, 175, 188, 189, 209, 228, 287; *also see* almug & algum
PTOLEMAIS, 243
PUBLIUS, 219
puddings, 86
pulse, 252, 254
pulse, parched, 101, 128, 229, 251
Pumica granatum, 190; *also see* pomegranate
pumpkin (-s), 6, 204; also see *Cucumis pepo* & gourd
punctuation, 17
pungent, 47, 76, 83, 89, 102, 152
Punica granatum, xviii, xix, 189, 190, 213; illus.: p. 199, figs. 38, 76; *also see* pomegranate
punishment (-s), 133, 150
purgative, 35, 78, 137
purification (-s); purify (-ing), 82, 84, 162, 220
Puritan Bible, 9
Puritans, 9
purple, 205
purple, Tyrian, 195
Purpura, 195
purslane, sea, 53; also see *Atriplex halimus* & mallows
pus, 223
putrid fever, 97
pyramid, 33
Pyramid of Cheops, 33
Pyramid of Khufu, 33
PYRAMUS, 141
Pyrus communis, 183, 284; *also see* pear
Pyrus cydonia, 186; *also see* quinces
Pyrus malus, 185; *also see* apple
Pyrus syriaca, 184
PYTHAGORAS, 102
Pythian, 95, 123

quails, 127, 128, 276
quail-dung, 127
Queen; queen, 2, 39, 70

Queen of Sheba, 70, 84, 85, 160, 274, 280
queen of the cucumbers, 88
Quercus, 179, 227; also see oaks
Quercus aegilops, 90, 193, 195, 196, 197, 205, 286; also see Valonea oak
Quercus calliprinos, xix, 195
Quercus cerris, 196
Quercus coccifera, 193, 194, 196, 286
Quercus coccifera var. *genuina*, 194
Quercus coccifera var. *pseudococcifera*, xvii, 193, 195, 196, 197; illus.: figs. 29, 84; also see Abraham's oak
Quercus ehrenbergii, 196
Quercus ilex, 90, 178, 193, 195, 198, 281, 285, 286, 287; also see holm oak
Quercus infectoria, 195; also see Cyprus oak
Quercus infectoria var. *latifolia*, 196
Quercus ithaburensis, 195; fig. 87
Quercus libani, 196
Quercus look, 196
Quercus lusitanica, 193, 195, 196, 197; also see Cyprus oak
Quercus lusitanica var. *latifolia*, 196
Quercus lusitanica var. *syriaca*, 196
Quercus palaestina, 196; also see Abraham's oak
Quercus pseudococcifera, 195, 197
Quercus robur var. *nigra*, 196
Quercus robur var. *sessilis*, 196
Quercus sessiliflora, 195
Quercus sessilis, 196
Quercus syriaca, 196
quills, 76
quinces, 185, 186; also see *Pyrus cydonia*

ra'av, 232
rabbi (-nical, -s), 42, 73, 133, 137, 150, 204, 233, 291
RABBINOWITZ, J., 274
rabbit-foot, 138
RABINOWITZ, L., 274
RACHEL, 137, 283; fig. 67
Radicula nasturtium, 74; also see bitter herbs
radish (-es), 33
raiment, 41, 130, 140
rain (-fall, -s, -ed, -y), 14, 15, 25, 28, 45, 112, 126, 140, 176, 222, 224, 275, 282
raisin-cakes, 288
raisins, 244
ram (-s), 24, 97, 251
Ramah, 165, 227
Ramath-mizpah, 179
Ramman, 190
ranunculus, 6, 42
Ranunculus asiaticus, 43, 117; also see lily
RAPHAEL, 286
RAPIN, R., 45, 266
RASHI, 58, 138
Râs-ul-Ayn, 92
rationalistic version, 9
rats, 167, 168
ravens, 198
RAVIUS, C., 1, 267
REDGROVE, H. S., 161, 267
Red Sea, 35, 78, 95, 188, 195, 198, 219, 232, 282, 284, 289
reed (-s), 6, 28, 29, 40, 50, 51, 62, 92, 93, 94, 120, 130, 132, 161, 170, 172, 173, 215, 222, 235, 236, 247, 280, 282, 285; also see *Arundo donax* & *Sorghum vulgare* & *Phragmites communis*
reed, carpenter's, 50
reed, common, 172; also see *Phragmites communis*
reed, giant, xix, 50, 51; also see *Arundo donax*
reed-grass, 51, 62, 120, 172; also see *Scirpus lacustris*
reedmace, 50, 93, 236; also see *Typha angustifolia*
reedmace, giant, 235
reed, measuring, 51
reed, paper, xix; also see reed & *Cyperus papyrus*
reed-pens, 285

Reed Sea, 282
REED, W., 287
reforestation, 171
Reformers, 7
REHDER, A., vii, 267
Rehob, 214
REICHARDT, 282
REID, C., 267
reindeer-moss, 127; also see *Cladonia rangiferina*
Remmon, 189, 190
Remmon-methor; Remmon-methoar, 189, 190
Remmon-phares, 190
REMUS, 106
Renaissance, xiv, xvi, 186, 286, 292
RAPP, T., 286
resin (-ous, -s), 31, 52, 55, 57, 58, 81, 82, 83, 84, 91, 102, 178, 182, 225, 228, 254
restorative, 137
resurrection, 41, 198
resurrection-plant, 38, 39; also see *Anastatica hierochuntica* & *Gundelia tournefortii*
ret, 133
retam, 14
Retama raetam, xix, 37, 91, 201, 287; illus.: fig. 78
retem; rethem; Rethma, 201
REUBEN, 137, 229, 232; fig. 67
Revalenta, 129
Revelation, 19, 49, 140, 228
Revised standard version — see version, Revised Standard
REYNOLDS, K. M., 269
REZIN, 169
Rhamnus cathartica, 203; also see hedge & thorns
Rhamnus lotus, 247; also see shady trees & lotus bushes
Rhamnus nabeca, 248
Rhamnus palaestina, 165, 202, 207, 222, 245, 248, 287; also see hedges & thorns
Rhamnus paliurus, 165; also see crown of thorns
Rhamnus punctata, 287
Rhamnus punctata var. *palaestina*, 202
Rhamnus spina-christi, 248
RHEA, 106
rheumatic, 47
Rhodes, 191
rhododendron, 205; also see oleander
rice, 6
rice-wheat, 234; also see *Triticum dicoccum*
Ricinus communis, xvii, 88, 203, 204; illus.: fig. 39; also see gourd
RICKETT, H. W., 267
Rickettsia prowazeki, 97
rie, 229, 233, 234; also see rye & *Triticum aestivum* var. *spelta*
RIKLI, M., fig. 78
rime, 278
Rimmon (-ah), 189, 190
Rimmon-parez; Rimmon-peres, 189, 190
rimt, 278
ringworm, 99, 143
RIP VAN WINKLE, 73
Rithmah; ritmath, 189, 201
river (-s), xix, 24, 47, 48, 50, 53, 62, 67, 78, 84, 88, 92, 93, 97, 117, 120, 144, 154, 157, 169, 172, 183, 194, 197, 214, 216, 227, 234, 247, 249, 252, 253, 289, 292
robe (-s), 132, 133, 189, 191, 193, 219
ROBERTSON, J. A., 260
robins, 14
Robinia pseudo-acacia, 5, 26; also see acacia
ROBINSON, 179, 244
Roccella tinctoria, 195, 204, 205; also see lichen
ROCHEBRUNE, A.-T. DE, 267
rock (-s, -y), xv, 43, 65, 86, 173, 225, 243, 244, 248, 282
rockrose (-s), xvii, 77, 205, 224, 234; also see *Cistus* spp.
rod (-s), xv, xvii, 35, 36, 51, 89, 145, 152, 180, 181, 182, 194, 222, 225, 236, 241, 256, 278, 287, 289
roe (-s), 51, 114

Rogelim, 101, 128, 229
ROHR, J. B., 2, 61, 268
rolling thing, xvii, 38, 39; also see *Anastatica hierochuntica*
Roman (-s); Rome, xvii, 2, 3, 19, 27, 30, 32, 35, 43, 44, 45, 63, 64, 77, 80, 83, 85, 90, 93, 101, 102, 106, 119, 123, 124, 137, 139, 140, 144, 148, 149, 158, 159, 160, 170, 171, 181, 191, 205, 218, 228, 232, 244, 256
Roman Catholic, 9
ROMANES, G. J., 268
ROMEO, 141
ROMULUS, 106
roof (-ing, -s), 49, 69, 91, 130, 131, 144, 151
root (-s), 39, 40, 42, 48, 49, 54, 59, 67, 68, 70, 72, 76, 78, 79, 91, 251, 252, 253, 255, 256, 275, 287
roots, juniper, 202, also see *Cynomorium coccineum*
root, sorcerer's, 138; also see *Mandragora officinarum*
rootstock (-s), 40, 62, 93
rope, 132, 133, 170
Roripa nasturtium-aquaticum, 74; also see bitter herbs
Rosa, 147, 234
Rosa canina, 151, 205; also see rose
Rosa phoenicia, 151; also see rose
Rosa rubiginosa, 206, 207; also see rose
rose (-s), xvii, 5, 38, 39, 41, 42, 44, 45, 77, 84, 114, 117, 123, 124, 125, 145, 147, 149, 151, 152, 180, 205, 206, 234, 235, 280, 287, 288, 289; also see *Narcissus tazetta*, *Nerium oleander*, *Rosa phoenicia*, *Tulipa montana*, *Tulipa sharonensis*
rose, brier, 234
rose-buds, 205
rosebushes, 39
ROSECRANS, J. H., 268
rose, dog, 205
rosemary, 52
ROSEN, F., 273
ROSENBAUM, I. J., 274
ROSENBERG, A., viii
ROSENBERG, P., 43, 268
ROSENFELD, D., 268
ROSENMÜLLER, E. F. K., 74, 79, 87, 139, 179, 183, 196, 224, 268
rose of Jericho, 39, 235; also see *Anastatica hierochuntica*
Rose of Phoenicia, 205
rose of Sharon; rose-of-Sharon, 5, 39, 42, 53, 77, 147, 205, 234, 235, 288, 289; also see *Tulipa montana* & *Hibiscus syriacus*
rose-tree, 205
Rosetta Stone, 110
rôsh, 49, 78, 79, 80, 279
rosh'ay besamim, 51
rosin, 10, 55, 85, 166
Rosinen Kuchen, 288
ROSNER, J. A., 268
rot, 25
rotem; rothem, 201
ROTHERY, A., 272, 277
Royal Botanic Garden, 129, 279
Royal Horticultural Society, vii, 277
Royal Swedish Bible Commission, 275, 280, 281, 282, 283, 284, 285, 287
ROYLE, J. F., 27, 30, 41, 60, 61, 123, 160, 167, 184, 186, 220, 237, 268
r'tamim, 201
r'tamim sho'resh, 91
RUBEN, xix
RUBENS, P. P., 273, 286; fig. 94
Rubia tinctoria, 195; also see madder
Rubus, 207, 220; also see bramble
Rubus anatolicus, 207
Rubus discolor, 207
Rubus fruticosus, 207
Rubus rusticanus, 207
Rubus sanctus, 165, 202, 206, 207, 245, 248, 288
Rubus ulmifolius, 206, 207
Rubus ulmifolius var. *anatolicus*, 207
RUDBECK, O., 268

Bible Plants — 321 — General Index

rue, 139, 208; also see *Ruta* spp.
rue, common, 208
rue, wall, 161
RUFUS, 168
RUMETIUS, L., 1, 268
Rumex acetosella var. *multifidus*, 74, 75; also see bitter herbs
Rumex acetoselloides, 75
Rumex multifidus, 75
rumman, 190
Ruscus aculeatus, 206, 207, 220; also see briers
rush (-es), 6, 28, 29, 50, 51, 62, 92, 94, 120, 121, 132; also see *Arundo donax*, *Cyperus papyrus*, *Juncus* spp.
rush, bog, 120; also see *Juncus effusus*
rush, sea hard, 120; also see *Juncus maritimus*
rush, soft, 120; also see *Juncus effusus*
rush, sweet-, 40; also see *Andropogon aromaticus*
Russia (-n), 49, 120, 144, 183, 282
Ruta bracteosa, 208; also see rue
Ruta chalapensis var. *latifolia*, 208
Ruta graveolens, 208
Ruta latifolia, 208
RUTH, 18, 27, 222
rye (true), 134, 233, 234; also see rie & *Triticum aestivum* var. *spelta*

SAADIAS, 63
Saadya, Gaon, 8
Saba, 40, 58,
Sabina, 210
Sabina excelsa, 90, 119, 189, 209; also see algum,
Sabina phoenicia, 57, 63, 68, 121, 174, 209, 210; also see cedar wood
Sabina thurifera, 174
Sabina vulgaris, 121
Saccharomyces, 141, 211
Saccharomyces cerevisiae, 210, 212
Saccharomyces ellipsoideus, 212, 213, 243, 244, 288
Saccharum, 29
Saccharum aegyptiacum, 214
Saccharum biflorum, 214
Saccharum officinarum, 40, 51, 214; also see sweet cane
Saccharum sara, 214
Saccharum spontanaeum var. *aegyptiacum*, 214
sachet, 118
sackcloth, 92
sacred, 33, 70, 76, 91, 107, 145, 155
sacred bo tree, 285
sacrifice (-s); sacrificial, 56, 58, 66, 194, 214
Sadducees, 211
sadin, 132
safflower, 87
saffron, xix, 39, 56, 58, 76, 87, 114, 148, 205, 206, 234; also see *Crocus sativus*
saffron, bastard, 87; also see *Carthamus tinctorius*
safsaf, 217
SAGARA, 88
sage, 6, 52, 218; also see *Salvia judaica*
sagebrush, 48
sage, Judean, xix, 218; also see *Salvia judaica*
Sahara Desert, xix, 127
sail (-s), 130, 131, 132
sainfoin, saintfoin, 277
saint (-s), 57, 69, 131, 198
SAINT AGATHA, 120
SAINT ANTHONY, 44
SAINT AUGUSTINE, 204
SAINT BAVON, 286
SAINT BERNARD, 44
SAINT CALIXTUS, xv
SAINT CHRISTOPHER, 172
SAINT CLARA, 44, 172
Saint David's Day, 35
SAINT DOMINICK, 44
SAINT FRANCIS OF ASSISI, 44, 172, 206
SAINT JAMES — see James

SAINT JEROME, 91
SAINT JOHN, 172 — also see JOHN
Saint John's bread, 73; also see *Ceratonia siliqua* & husks
Saint Johnswort, 43
SAINT JOSEPH, 44, 152
SAINT JUDE — see Jude
SAINT KATHERINE OF SIENA, 44
SAINT LOUIS DE GONZAGUA, 44
SAINT LUKE — see Luke
SAINT MARK — see Mark
SAINT MATTHEW — see Matthew
SAINT MICHEL, M. DE, 266
SAINT PAUL to the Romans — see Romans
SAINT PETER, 90 — also see PETER
SAINT THOMAS, 33, 44
SAINT VINCENT, 206
sakiyeh, fig. 28
salad, 34, 74, 140, 187
salai, 58
SALAMASIUS, C., 273
Salem, 212, 240
Salicornia, 215; also see soap & sope
Salicornia arborea, 215
Salicornia europaea var. *fruticosa*, 215
Salicornia europaea var. *herbacea*, 215
Salicornia fruticosa, 215
Salicornia herbacea, 215
SALINGER, M. M., 286
Salix, 152; also see willow & withs
Salix acmophylla, 216
Salix aegyptiaca, 217
Salix alba, 216
Salix babylonica, 5, 216, 217, 288
Salix fragilis, 216, 217
Salix octandra, 217
Salix safsaf, 216, 217, 288
Salix subserrata, 217
Salix viminalis, 217
Salomons liljor, 282
Salsola, 215; also see soap & sope
Salsola inermis, 215
Salsola kali, xix, 215; illus.: fig. 80
salt (-ed, -iness, -y), 26, 53, 54, 58, 81, 91, 162
saltpetre, 114
saltwort (-s), 9, 49, 53, 54, 91, 215, 278, 279, 280
saltwort, prickly, xix, 215; also see *Salsola kali*
saltwort, unarmed, 215; also see *Salsola inermis*
Salvadora, 60, 61; also see mustard
Salvadora persica, 60, 280
Salvia judaica, xix, 218; illus.: fig. 81; also see candlestick & sage
Samaria, 56, 98, 142, 144, 151, 162, 170, 218, 241
Samaritan translation, 7, 275
Sambucus nigra, 73; also see black elder & Judas' tree
SAMSON, xiii, xv, 216, 240
SAMUEL, xv, 69, 105, 130, 150, 157, 167, 168
Samuel I & II, 18
Samur, 165
sanctuaries; sanctuary, 70, 89
sand (-s, -y), 13, 15, 62, 78, 90, 122, 171, 201, 227, 229, 235, 282, 288
sandals, 154
sandalwood, 41, 47, 52, 57, 90, 189, 209; also see *Santalum album* & *Pterocarpus santalinus*
sandalwood, red, 90, 189, 209
sandalwood, white, 188
sandarac (-h), 228
sandel-trees, 189; also see *Pterocarpus santalinus*
sandel-wood, 189
Sanskrit, 188
sant, 25
Santalum album, 35, 47, 57, 188; also see sandalwood
santh, 25
SANTINI DE RIOLS, 268
São Paulo, 287
sap, 31, 85, 177, 251, 292
sa'pachat, 151

saponaceous, 216
Saponaria officinalis, 5, 216
SARABIYUN, IBN, 278
sarcophagi, 107
Sarcostemma brevistigma, 187; also see apple
SARGEANT, A. M., 268
sassafras, 76
SATAN, 32, 223
Satureia thymbia, 161; also see hyssop & savory
Satureja graeca, 162
Satureja juliana, 162
Satureja thymbra, 161
satyrs, 145
SAUL, 178, 179, 189, 227
saunders-wood, red, 189; also see *Pterocarpus santalinus*
Saussurea lappa, xvii, 75, 76, 218; illus.: fig. 33; also see cassia
savin, 121
savin, eastern, 189, 209; also see *Sabina excelsa*
savory, 52
savory, whorled, 161; also see *Satureja thymbra*
sawdust, 17
SAYCE, 270
sayt, 151
scab, 142, 149, 151
scabious, Syrian, 283
scale-insect, 31, 194, 195
scall, 98, 99, 143
scarlet, 193, 194, 195
scent (-ed, -ing), 32, 34, 40, 145, 148
sceptre, 36, 166
SCHARL, JOSEPH, xvi, fig. 8
SCHECHTER, S., 288, 289
SCHELHORN's Bible, 9
scherbin, 63, 210
SCHEUCHZER, J. J., 2, 168, 268, 273; fig. 90
schizophytes, 17, 141
Schoenoplectus lacustris, 121; also see bulrush & flag
SCHONGAUER, MARTIN, xix, fig. 92
Schôschanna, 116
SCHUMANN, K., 272, 281
Schüschan, 116
SCHWERIN, F., 273, 284, 289
Scindians, 282
Scio, 178
Scirpus, 51, 93, 120; also see flag & meadow
Scirpus corymbosus, 121
Scirpus holoschoenus var. *linnaei*, 120, 121
Scirpus lacustris, 51, 93, 120, 121
Scirpus linnaei, 121
Scirpus maritimus, 120, 121, 282
SCOFIELD, C. I., 260
Scolymus maculatus, 153, 206; also see golden-thistle & briers
scorpions, 208, 276
SCOT, D., 268
Scotland; Scots (-man, -men), 5, 14, 35, 71, 72, 275
SCOTT, D., 268
SCOTT, R. A., 266
scribe (-s), 46, 89, 139, 173, 189, 196, 209
Scriptural; Scripture (-s), xiv, xv, xx, 2, 5, 7, 8, 17, 18, 25, 42, 65, 77, 85, 92, 129, 138, 148, 288
scrofula, 118
scroll (-s), 27, 218
scrub, 27, 122
s'deenim, 132
sea, 53, 59, 66, 76, 78, 79, 92, 93, 113, 140, 145, 154, 241, 249, 289
Sea, Caspian, 242
seacoast (-s), 13, 120, 216
sea cow (-s), 9
Sea, Dead, see Dead Sea
sea-dragons, 49
Sea of Galilee, see Galilee, Sea of
sea-orache, 220; also see *Atriplex littoralis* & nettles
Sea, Red, see Red Sea
season (-s), 25
seasoning, 34, 153
seaweed, 249; also see *Zostera marina*
sea-wrack, 249; also see *Zostera marina*
Secale cereale, 233; also see rye & rie

sedge, 62, 120, 173
sedge-bush, 62; also see *Butomus umbellatus*
sedge-grass, 93; also see *Cyperus papyrus*
seechim, 227
seed (-s), 20, 34, 38, 43, 46, 59, 60, 61, 70, 72, 73, 75, 79, 81, 86, 89, 152, 153, 166, 189, 191, 217, 221, 229, 230, 239, 240, 241, 242, 251, 252, 253, 254, 255, 257, 282, 300
seed, black, 153; also see *Nigella sativa*
seed, blessed, 153; also see *Nigella sativa*
seed, corruptible, 257
seed, incorruptible, 257
semen, 257
Semite (-s); Semitic, 34, 173
seneh, 23, 25
Senir, 176
senna (-s), 25, 219
SENNACHERIB, 14, 51, 68
seorah, 112
Sephardic, viii
Septemberbibel, 9
Septuagint, vii, 7, 8, 26, 51, 63, 65, 71, 76, 85, 90, 95, 98, 108, 123, 127, 132, 137, 147, 153, 166, 176, 177, 179, 180, 188, 197, 216, 254, 281, 286
sepulcher; sepulchre, 83, 131
sequoias, 69
seraglio, 235
Sermon on the Mount, fig. 89; also see *Anemone coronaria* & lily of the field
serpent (-s), xv, xvii, 20, 21, 77, 80, 172, 186, 202, 212; fig. 94
Setaria italica, 166, 167; also see millet
SETH, 70, 160
setim, 24, 25
Seventy, The — see Septuagint
seyal, 25
Seyal, Wady, 25
shaddock, 6, 186; also see *Citrus grandis* & apple
shade; shadow, 59, 70, 105, 107, 119, 124, 134, 141, 160, 178, 181, 185, 186, 187, 194, 197, 198, 201, 203, 204, 216, 227, 240, 247
shajrat-al-bak, 184
shâkêd, 36, 37, 38
SHAKESPEARE, W., 4, 138
shallot, 32; also see *Allium ascalonicum* & garlic
shallum, 207
shamir, 206, 221
shamrock, 35
shanî, 194
Sharon, xix, 15, 46, 71, 114, 147, 158, 234, 288, 289; fig. 83
shâshar, 195
SHAW, H., 2, 3, 35, 40, 41, 47, 52, 60, 63, 65, 124, 179, 237, 268, 280
SHAW, T., 268
Shawondesee, 75
shaychar, 213
sheaves, 231, 251, 300
Sheba (Queen of), 39, 56, 58, 274
sheber, 231
shechelet (-h), 223, 224
Schechem, 56, 158, 171, 193
sheep, xv, 127, 170, 182, 229
shekels, 39, 56, 75, 76, 82, 111
shelav, 127
shellfish, 195, 205
shell, walnut, 120
shemen, 158
shepherd (-ess), xv, 107, 277
sherbets, 72
sherbin, 83, 210
shesh, 110, 132, 140
shever, 231, 232
shewbread cakes, 233
shibalim; shibboleth, 231, 232
shidaphon, 254
Shigella, 97, 141, 220
Shigella ambigua, 219, 220
Shigella dysenteriae, 219
Shigella flexneri, 220
Shigella paradysenteriae, 219, 220
Shigella schmitzii, 220
shih, 278
shikmah, 140, 283

shikmin; shikmoth, 106
shikor, 243
Shilhim, 189
Shiloh, 56
ship (-building, -s), 90, 95, 132, 169, 175, 176, 177, 188, 281
shir-zad, 282
shittah (tree); shittim, xvii, 24, 25, 26, 56, 62, 63, 67, 97, 144, 277; also see *Acacia seyal* & *Acacia tortilis*
Shittim, Abel-, 24
Shittim, Abel-has, 24
Shittim, Nachal, 24
shittim wood, 24, 25, 26; also see *Acacia seyal* & *Acacia tortilis*
SHOHET, D., viii
shoomim, 32
shorek, 243
shoresh, 255
shôsannâh, 42
shoshanah; shoshanim; Shoshannim; shoshanot, 41, 42, 43, 114, 129
shrews, 49
shrouds, 131
shrub (-by, -s), 23, 27, 30, 31, 32, 52, 53, 60, 78, 83, 90, 98, 122, 123, 124, 135, 151, 153, 160, 202, 203, 204, 207, 210, 217, 224, 227, 247, 248, 254
shrubby-althea, 205; also see *Hibiscus syriacus*
Shur, 227
shûsan; shushân; Shushan, 41, 42, 43, 109, 117, 154
Shushan-eduth, 41, 43
Sîach, 279
Siam, 155
Sibmah, 241, 244
SIBTHORP, F., figs. 57, 68, 69, 74, 75, 79
Sichem, 193
Sichron Yaakole, fig. 50
Sicily, 104, 141
Sidon (-ians), 66, 78, 138, 147, 170, 198, 227
sidr, 248
SIEBER, F. W., 268
sikim, 206
silk (-s), 87, 130, 131, 132, 140, 205, 228, 290
silkworms, 140, 141
sillon; silonim, 206, 207
silver, 33, 51, 66, 67, 73, 84, 106, 109, 111, 112, 131, 162, 182, 183, 185, 194, 228, 237, 242, 245, 253, 277
Silybum marianum, xix, 71; illus.: figs. 87, 92; also see thistles & thorns
SIMINI, N. DE, 262
SIMON, 142, 148
SIMON, M., 274
Simplon Pass, 90
Sinai, xvii, 4, 13, 14, 23, 24, 25, 31, 36, 38, 50, 53, 61, 65, 66, 120, 121, 127, 135, 171, 172, 173, 174, 182, 185, 201, 203, 208, 210, 214, 218, 233, 248, 278, 279, 282; fig. 30
Sinai-manna, 31
sinapi, 59
Sinapis, 60; also see mustard
Sinapis alba, 59, 60
Sinapis arvensis, 27, 60, 220, 237
Sinapis arvensis var. *orientalis*, 59
Sinapis nigra, 59
Sinapis orientalis, 59
sindyân, 196
sinew (-s), 217, 218
sin offering, 31
SIRA, J. BEN, 288, 289
Sirach, Wisdom of — see Ecclesiasticus
sirpad, 206
Sisymbrium nasturtium-aquaticum, 74; also see bitter herbs
SIU, R. G. H., 290
SIVA, 154
siwan, 134
skin (-s), 24, 98, 99, 142, 143, 149, 159, 168, 173, 223, 290
SKINNER, C. M., 33, 39, 49, 268
skylark, 14
slaves, 88
Slavonic, 9
Slavonic version, 8

slime, 92
slingstones, 251
slip, 217
SLOTKI, J. J., 274
slough, 92
smell, 30, 35, 47, 137, 138, 148, 218
SMITH, C. R., 272
SMITH, F. W., viii
SMITH, G. A., 286
SMITH, J. M. P., 17
SMITH, JOHN, 4, 23, 25, 30, 32, 40, 42, 49, 52, 57, 60, 63, 71, 74, 76, 77, 81, 82, 83, 84, 88, 98, 102, 107, 109, 121, 129, 134, 139, 147, 161, 166, 175, 179, 181, 184, 186, 195, 204, 206, 207, 209, 215, 221, 222, 228, 232, 234, 236, 237, 239, 242, 268, 279, 282
SMITH, W., 25, 27, 32, 40, 52, 55, 61, 63, 80, 85, 94, 101, 104, 108, 138, 161, 167, 172, 177, 180, 184, 188, 189, 198, 204, 209, 215, 217, 220, 223, 227, 266, 268
smoke (-s); smoking, 27, 56, 57, 82, 84, 182
smoke-raiser, 58
smut, 290
Smyrna, 3, 106
snails, 195
snake (-bite, -s), 118, 182, 208
s'neh, 23
snow, 35, 45, 194
soap (-making), 119, 216, 280; also see sope & *Salsola kali*
soapwort, 5
Socotra, 35; also see aloes
SOCRATES, 49
Sodom, 1, 2, 6, 58, 78, 221, 281
Sodom, apple of, 1, 221, 244, 281; also see *Citrullus colocynthis* & briers
soil (-s), 5, 15, 26, 49, 63, 69, 79, 276
Solanum, 207
Solanum coagulans, 221
Solanum hierochunticum, 221
Solanum incanum, xix, 29, 30, 221, 239; illus.: fig. 79; also see briers & nightshade, Palestine
Solanum sanctum, 221
Solanum sodomeum, 79, 206, 221, 244
soldier (-s), 5, 27, 80, 83, 106, 161, 165, 202, 237
SOLOMON, xvi, xvii, 39, 41, 42, 43, 52, 57, 66, 67, 68, 69, 70, 75, 83, 84, 85, 86, 87, 89, 114, 116, 119, 124, 129, 130, 138, 140, 154, 157, 159, 160, 161, 169, 170, 172, 175, 177, 185, 186, 188, 189, 190, 191, 209, 218, 219, 228, 229, 232, 257, 285; figs. 8, 16, 52, 76, 85, 89, p. 301
soma, 187
Somali (coast, -land), 13, 57, 58, 82, 83, 84
Song (of Solomon, of Songs), 18, 40, 47, 48, 52, 86, 87, 116, 124, 129, 138, 154, 234, 235, 282, 288; figs. 20, 83
Song of the Three Children, 19
SONNINI DE MANONCOUR, C. N. S., 34, 269
SONNTAG, C., 269
sont, 25
sope, 215; also see soap & *Salsola kali*
SOPHONIAS — see ZEPHANIAH
s'o'rah, 112
Sorbus, 176, 286; also see fir trees
Sorek, 240, 243
sorghum; *Sorghum*, 29, 161, 166, 222
Sorghum durra, 222
sorghum, true, 223
Sorghum vulgare, 50, 161, 166, 167, 172, 173, 223
Sorghum vulgare var. *durra*, xix, 222; illus.: fig. 82
s'o'rim; s'orim, 112, 128
sorrel, 74, 75, 140; also see *Rumex acetosella* var. *multifidus* & bitter herbs
SOUCIOT, 49
soup (-s), 86, 128, 282
South Africa, 30
South Carolina, 172

SOUTHWELL, H., 273
sovech, 255
sow (-ed, -er, -eth, -ing, -n), 59, 133, 230, 233, 240, 242, 252, 253, 257, 300
Spain; Spaniards; Spanish, 6, 9, 10, 30, 44, 45, 53, 73, 106, 134, 141, 191, 214, 235, 244, 286
Spanish version, 9
Sparganium erectum, 245; also see bur-reed & thorns
sparrow, 61
Spartium monospermum, 201; also see broom & juniper tree
spear (-s), 11, 38, 50, 222
spelling, 17, 281
spelt, xv, 152, 211, 231, 232, 233, 234, 254, 257, 289; also see *Triticum aestivum* var. *spelta*
spelt, little, 231
SPENSER, 89
Sphaerophorus gelatinosus, 127; also see manna & lichen
Sphaerothallia esculenta, 282
spice (-ry, -s), 35, 39, 40, 47, 51, 52, 55, 58, 59, 75, 76, 77, 82, 83, 84, 85, 87, 89, 102, 129, 131, 152, 159, 177, 178, 179, 223, 224, 225
spiced wine, 51
spiders, 276
spikenard; spike-nard; xvii, 39, 40, 41, 56, 58, 76, 87, 124, 148, 149, 205, 206, 277; also see *Andropogon aromaticus* & *Nardostachys jatamansi*
spikenard, American, 149; also see *Aralia spinosa*
spin, 130
spinach, 53
spine (-s); spiny, 26, 30, 31, 71, 79, 153, 165, 202, 203, 221, 245, 247, 248, 278, 284
spirit (-s), xix, 65, 119, 120, 138, 139, 160
Spirochaeta pallida, 150
Spirochaete pallidum, 150
Spironema pallidum, 150
spleenwort, maidenhair, 161
sponge, 27, 66, 161, 222; fig. 82; also see spunge
spoons, 63
spot, bright, 142, 149, 151
SPRENGEL, C. P., 53, 102, 114, 137, 184
spring (-s), 42, 50, 84, 113, 117, 235, 275, 288
sprinkle (-d, -r); sprinkling, 161, 162, 193, 194
sprout, 251
spruce, 57; also see *Picea abies*
spruce-fir, 57; also see *Picea abies*
SPRUNT, A., 274
spunge, 222; also see sponge
squash, crookneck, 204; also see *Cucurbita moschata* & gourd
squash, winter, 204; also see *Cucurbita maxima* & gourd
squirting-cucumber, 79; also see *Ecballium elaterium*
stables, 277
Stachybotrys, 290
stacte, xvii, 56, 58, 76, 102, 223, 224
staff, xv, 38, 50, 51, 89, 152, 153, 172, 240
stag, 91
stalks, 254
STANLEY, 25, 198
STANTON, 40
Staphylococcus, 97, 141
Staphylococcus albus, 223
Staphylococcus aureus, 223
Staphylococcus pyogenes, 223
Staphylococcus pyogenes albus, 223
Staphylococcus pyogenes aureus, 223
staples, 88
star (-s), 48, 49, 104
star-jelly, 282
starling, 276
star-of-Bethlehem, 162, 163; also see *Ornithogalum umbellatum* & doves' dung
star-thistle, 71; also see *Centaurea calcitrapa*
STAUNTON, R., 204, 269
STEARN, W. T., vii, 32, 87, 93, 101, 129, 274; fig. 64

STEIN, D. G., 134, 269
STEIN, R. S., 265, 269
STELLFELD, C., 272
stem (-s), 29, 31, 50, 51, 59, 63, 69, 71, 77, 79, 85, 86, 89, 93, 255, 256, 281, 288
steppe, 112, 282
sterility, 139
Sternbergia aurantiaca, 41, 114; also see lily
Sternbergia fischeriana, 114
Sternbergia lutea, 41, 114, 117
Sternbergia pulchella, 114
STEUER, R. C., 84, 269
STEVENSON, R. A. M., 273
stew (-s), 87, 89, 128
stimulant, 32, 87, 89, 148
sting, 237
stink, 29, 125, 126, 149, 290
Stipa, 134
stock of Sodom, 79
stomach (-ic), 55, 86, 140
stone (-s), 66, 67, 81, 84, 142, 211, 239, 241, 280, 282
stones, precious, 39, 131, 188, 224, 228, 253
storax, xvii, 52, 181; also see *Styrax officinalis*
storax, liquid, 225
storax, official, 224, 225
storax, sweet, 223, 224
storax-tree, 181, 182, 224, 225
stork, 175, 276
storm (-s), 70, 190
STOW, M. A., 272, 280
STRABO, 83, 170, 205
Straits of Babelmandeb, 188
strap-flower, acacia, xix, 23; also see *Loranthus acaciae* & burning bush
straw, 38, 112, 251, 252, 254, 277
strawberry, 276
stream (-s, -sides), 6, 13, 40, 62, 117, 118, 122, 123, 139, 175, 181, 214, 217
street (bakers', -s), 65, 211
Streptococcus conglomeratus, 223
Streptococcus erysipelatis, 97, 223
Streptococcus erysipelatos, 223
Streptococcus erysipelatosus, 223
Streptococcus hemolysans, 223
Streptococcus hemolyticus, 223
Streptococcus longissimus, 223
Streptococcus longus hemolyticus, 223
Streptococcus puerperalis, 223
Streptococcus pyogenes haemolyticus 223
Streptococcus scarlatinae, 223
Streptomyces, 290
string, 132
Strombus lentiginosus, 224; also see onycha
STRONG, J., 266
strong drink, 27; also see *Phoenix dactylifera*
stubble, 38, 251, 252, 253, 254
STURTEVANT, B., 40, 204, 269
stylus, 173
Styrax benzoin, 223, 224; also see storax
Styrax officinalis, xvii, 52, 181, 224; illus.: p. 225
subalpine, 43, 86, 121
succah, 291
Succoth, 206, 291
succulent, 35
sugar (-making), 81, 113, 214, 278
sugar-cane, 6, 40, 41, 50, 51, 214, 215, 222; also see *Saccharum officinarum*
sullaon, 207
Sultan, 235
Sumatra, 224
summer, 15, 84, 88, 103, 112, 115, 117, 124, 141, 275, 288, 289
summer fruit, 103
sun (-light, -stroke), 15, 28, 56, 77, 84, 125, 126, 172, 276, 285
sunbirds, 13
sunbul, 149
Sunda Islands, 81
sunt, 23, 25
superstitions; superstitious, 49, 88, 119, 137
sûph, 249

SUQABADI, M., 282
surgeons; surgery; surgical, 53, 55, 110, 138, 159
SUSA (-N), 41, 42, 43, 109, 117
SUSANNAH; SUSANNA, 19, 43, 116, 178, 198, 215
Susquehanna University, ii
suwwayd, 203
SWALLEN, J. R., vii, 285
swallows, 9
swamp (-s), 92, 93, 121, 173, 247
swastika, 155
Sweden; Swedish, xvii, 2, 3, 54, 237, 280, 281, 283
Swedish version, xv, xvii, xviii, xix, 237, 275, 280, 284, 287
sweet (-ness), 191, 222, 251, 278, 292
sweet bark, 58
sweet-bay, 123, 124; also see *Laurus nobilis*
sweet calamus, 40, 75, 85, 149, 214, 215, 219; also see *Andropogon* spp.
sweet cane, 39, 40, 215; also see *Saccharum officinarum* & *Andropogon* spp.
sweet-flag, 40; also see *Acorus calamus*
sweet odours, 85
sweet-rush, 40, 215; also see *Andropogon aromaticus*
sweet-scented (-smelling), 30, 40, 147, 215, 218
sweet sedge, 40; also see *Acorus calamus*
sweet storax, xvii, 30, 102
sweet-tasting, 40, 215, 278
swine, 57, 72
SWINGLE, W. T., 187, 269
Switzerland, 101, 133, 231, 233
sword, 48, 68, 137, 183
sycamenea, 140
sycamine; sycaminos, 8, 59, 108, 140, 141, 183, 283; also see *Morus nigra*
sycamore, 6, 107, 108, 281; also see *Acer pseudoplatanus*
sycomore (-fig, -tree), xiv, xix, 1, 5, 8, 24, 66, 67, 97, 106, 107, 108, 140, 157, 240, 281, 283; also see *Ficus sycomorus*
Sycomorus antiquorum, 107
syconium, 105
symbol (-ic, -ism, -s), xv, 88, 93, 218, 242, 276, 286, 292
symbol of beauty, 159
symbol of bitter calamity, 48, 79; fig. 43
symbol of Buddhism, 285
symbol of cheapness, 113
symbol of Christianity, 118, 285
symbol of crown of thorns, 276
symbol of crucifixion, xiv
symbol of cruelty, 48
symbol of death, 45
symbol of dignity, 68
symbol of divine blessing, 159
symbol of divine generosity, 144
symbol of domestic happiness, 244
symbol of drooping head, 65
symbol of elegance, 170
symbol of Eucharist, 276
symbol of everlasting life, 111
symbol of fecundity, 88, 106, 119
symbol of fertile life, 44
symbol of fertility, 191, 288
symbol of friendship, 111, 159
symbol of glory, 68
symbol of good fortune, 70; fig. 47
symbol of good luck, 44
symbol of good works, 287
symbol of grace, 170
symbol of grandeur, 68
symbol of grief, 71
symbol of halo, 276
symbol of Hinduism, 285
symbol of home, 145
symbol of hope of eternal life, 191
symbol of humility, 35, 120
symbol of ill omen, 33
symbol of immortality, 144, 145
symbol of innocence, 44
symbol of Islam, 285
symbol of Jewish people, 242

symbol of joy, 144
symbol of Judaism, 285
symbol of justice, 144
symbol of legal but not physical paternity, 287
symbol of light, 172
symbol of lofty stature, 68
symbol of love, 144, 235, 287
symbol of luxuriance, 159
symbol of majesty, 68
symbol of marriage, 287
symbol of martyrdom, 172
symbol of might, 68
symbol of misfortune, 33; fig. 43
symbol of Mohammedanism, 285
symbol of noble position, 47
symbol of pain, 71
symbol of passion, 145
symbol of peace, 106, 144, 145, 159, 244, 287
symbol of plenty, 244
symbol of poison, 287
symbol of poverty, 113
symbol of power, 68
symbol of pride, 120
symbol of procreation, 172
symbol of prosperity, 106, 123, 159; fig. 47
symbol of rejoicing, 288
symbol of reproductive generation, 62
symbol of restfulness, 145
symbol of resurrection, 39
symbol of retributive justice, 190
symbol of salvation, 172, 287
symbol of sanctity, 172
symbol of scorn, 113
symbol of scourges, 276
symbol of sensual love, 145
symbol of shortness of life, xv
symbol of sinlessness, 287
symbol of sorrow, 48
symbol of sovereignty, 159
symbol of spear, 277
symbol of spring's awakening, 36, 37
symbol of strength, 159, 196, 198
symbol of sturdiness, 198
symbol of the hope of heaven, 38
symbol of the nether world, 191
symbol of thriftlessness, 133
symbol of triumph, 123, 170
symbol of trying hard, 37
symbol of universe, 33
symbol of victory, 172
symbol of vigor, 288
symbol of virginity, 44, 287
symbol of wide expanse, 68
symbol of worldly riches, 172
symbol of worldly strength, 68
symbol of worthlessness, 113
SYMMACHUS translation, 7
synagogue (-s), 4, 140, 291
syphilis, 150
SYPHILUS, 150
Syracuse, 104
Syria (-c, -n, -ns), 4, 5, 8, 13, 14, 37, 38, 42, 43, 50, 53, 58, 61, 63, 65, 72, 78, 87, 89, 90, 92, 93, 101, 102, 105, 113, 115, 119, 120, 121, 122, 124, 129, 130, 132, 133, 134, 140, 152, 162, 169, 177, 178, 179, 180, 181, 182, 183, 185, 186, 190, 194, 195, 197, 201, 203, 204, 206, 208, 209, 210, 214, 231, 233, 234, 235, 239, 247, 248, 278, 281, 282; fig. 42
Syriac version, 8, 137, 166
Syrian Christ-thorn, 248; also see *Paliurus spina-christi* & *Zizyphus spina-christi*
Syrian-rose, 205; also see *Hibiscus syriaeus*
Syrian scabious, 283
syrup, 72

tabernacle (-s), xvii, 24, 30, 36, 47, 57, 58, 76, 129, 130, 132, 167, 193, 204, 224, 237, 245, 277, 291
Tabernacles, Feast of the, 64, 144, 171, 174, 187, 217, 283, 285, 291
Tabernaemontana alternifolia, xvii, 187; illus.: fig. 35; *also see* apple & divi ladner
table, 228
tableland, 13, 14
tablets, writing, 173
Tabor, xix, 197

TÄCKHOLM, V., 272, 281, 285
Tadmor, 169, 172
tail, 67, 247
TALEGON, J. G., 269
talent (-s), 33, 39, 84
Talmud (-ic), 32, 61, 63, 73, 86, 133, 171, 173, 201, 208
Tamar; tamar (-ah), 169, 170, 210
tamaric, 122; also see *Juniperus oxycedrus*
tamarisk, (-tree), xvii, 6, 31, 122, 159, 198, 227, 228, 255, 278; also see *Juniperus oxycedrus & Tamarix* spp.
tamarisk, manna, 31, 278; also see *Tamarix mannifera*
Tamarix, xvii, 122, 278; illus.: fig. 30
Tamarix articulata, 198, 227; *also see* tamarisk tree
Tamarix deserti, 227; *also see* tamarisk & tree
Tamarix gallica, 278
Tamarix gallica var. *mannifera*, 31; *also see* manna
Tamarix mannifera, 31, 32, 122, 126, 159, 278; *also see* manna
Tamarix nilotica var. *mannifera*, 31, 278; *also see* manna
Tamarix noëana, 227; *also see* tamarisk & tree
Tamarix orientalis, 227
Tamarix pentandra, 227
Tamarix tetragyna, 227
t'anim, 104
Tannhäuser, 38
tanning, 25, 191, 197
tantrums, 118
tapher, 65
Tappuah, tappûach, 185, 214
tar, 98
tar-angubin, 278
Taranjubin, 278
Taraxacum dens-leonis, 75; *also see* bitter herbs
Taraxacum officinale, 74, 75
Taraxacum vulgare, 75
TARCOV, O., 274
tares, xvii, 133, 134, 230, 283, 300; also see *Lolium temulentum*
tarfa, 278
Targum (-s), 8, 42, 137, 147, 166
Targums, Chaldee, 42
tarshish, 224
t'ashur, 62
taste, 33, 40, 89, 287
Taurus, 242
tavern, 111
Taverner version, 9, 10
tax, 132
TAYLOR, C., 2, 269
TAYLOR, J., 269
TAYLOR, W. R., 289
t'aynah, 104
t'chaylet, 195, 205
tea, 39
tears, 228
teasshûr, 62, 63
teceleth, 195
teenah; teenim, 104
teerosh, 213
TEESDALE, J., 272
TEESDALE, M. J., 272, 278, 282
teeth (-ing), 27, 118
tehven, 254
teil-tree, 175, 178, 179, 194; *also see Pistacia terebinthus* var. *palaestina*
Tel-Aviv, 288
TELLUS MATER, 71
temarin, 170
tempest, 39
temple (-s), 57, 58, 59, 69, 73, 77, 83, 89, 91, 93, 97, 103, 110, 111, 125, 131, 132, 133, 151, 152, 154, 155, 159, 168, 170, 175, 177, 190, 191, 211, 214, 218, 219, 220, 224, 290, 291
TEMPLE, A. A., 269
Temple of Solomon, 69
Temptation, xv, xvi, xix, 286; figs. 91, 95
tendrils, 78, 79
tenesmus, 167
TENNANT, E., 76
tent (-s), 38, 105, 107, 111, 125, 127, 129

terebinth (-ine, -os), 9, 14, 57, 175, 177, 179, 182, 195, 196, 197, 198, 285, 286
terebinth, Palestine, 178; also see *Pistacia terebinthus* var. *palaestina*
terrace-type cultivation, 14, 276
terror, 97
Testament, New, xv, xvi, 1, 9, 27, 29, 41, 42, 47, 48, 50, 71, 97, 104, 129, 132, 139, 173, 183, 208, 234, 235, 253, 256, 275, 283
Testament, Old, vii, xiv, xv, xvi, 1, 5, 6, 7, 9, 27, 29, 35, 40, 41, 42, 47, 71, 79, 80, 85, 97, 99, 104, 110, 117, 123, 127, 132, 143, 154, 161, 182, 183, 186, 187, 195, 197, 205, 208, 210, 214, 216, 253, 254, 275, 283, 287, 288
Tetraclinis articulata, 189, 228; *also see* thyine wood
Teutonic, 71, 133
textile (-s), 110, 131
Thanksgiving, 64, 211, 291
thaoudar, 231
Thargelia, 104
THAYER, J. H., 274
Thebaic version, 8
THENIUS, 168
THEODORA, Empress, 118
Theodorea costus, 218; *also see* cassia & orris
THEODOTION translation, 7
THEOPHRASTUS, 1, 2, 30, 58, 83, 109, 134, 140, 178, 186, 225
Thessalonians I & II, 19
thicket (-s), 25, 151, 160, 178, 203, 207, 208, 217, 247, 251, 252, 255, 280
thick trees, 97, 285; also see *Elaeagnus angustifolia*
THIEBAUT, J., 272
thirst (-y), 3, 27, 50, 65, 92
THISBE, 141
thistle (-s), 6, 15, 27, 29, 35, 39, 66, 67, 69, 71, 72, 87, 103, 112, 114, 135, 153, 202, 203, 207, 219, 220, 221, 230, 237, 248, 249, 287; also see *Centaurea* spp. & *Silybum marianum & Notobasis syriaca*
thistle-down, 38, 39; also see *Anastatica hierochuntica*
thistle, holy, 72; also see *Cnicus benedictus*
thistle, spotted golden, 153; also see *Scolymus maculatus*
thistle, Syrian, 153, 206; also see *Notobasis syriaca*
THOMAS, M. P., 269
THOMASSON, R. R., 273, 283
THOMPSON, C. J. S., 269
THOMPSON, F. C., 272
THOMSON, W. M., 38, 42, 60, 92, 138, 185, 186, 196, 197, 232, 269
THOR, 72, 198
thorn (-s, -y), xix, 15, 26, 27, 29, 39, 71, 72, 83, 103, 114, 135, 153, 159, 165, 176, 190, 202, 203, 206, 207, 220, 221, 230, 232, 235, 237, 241, 245, 247, 248, 252, 277, 287, 289, 291
thorn-bush (-es), 23, 71; also see *Acacia nilotica*
thorny acacia, 23; also see *Acacia nilotica*
thread, 194, 195
thresh (-ed, -ing), 152, 212, 230, 232
threshingfloor, 112, 116, 230
threshing instrument, 89, 207, 230
threshing-sledge, 233
Thuja aphylla, 227; *also see* tamarisk
Thuja articulata, 228
Thuja occidentalis, 57; *also see* frankincense & arbor-vitae
thuliban, 235
THUNBERG, C. P., 3, 269
thunder, 72, 190
Thuya articulata, 228; *also see* tamarisk
thyine-wood, 189; also see *Tetraclinis articulata*
thyme, 6, 52
Tiberias (Lake of), 60, 71, 92, 218
TIBERIUS, Emperor, 124

Bible Plants — 325 — General Index

tidhar, 174, 175, 286
TIETZE, H., 273
Tigris, 78, 292; fig. 40
Tilia europaea, 179; also see teil
tillage, 88
Tilletia caries, 290
Tilletia foetens, 290
Tilletia tritici, 290
Timath, 240
timber (-s), 14, 24, 66, 67, 70, 90, 91, 93, 119, 170, 175, 177, 209, 228
Timothy I & II, 19
TIMOTHY, B., 272, 278
tinea circinata, 99
tinea tonsurans, 99
TINTORETTO, xix, 273, 286; fig. 93
tirosh, 213
Tirzah; tirzah, 86, 90, 114, 180, 281, 286
TISSOT, J. J. J., 273
Titan, 106
tithe (-d, -s); tithing, 46, 89, 139, 208, 214, 251, 256, 257
TITIAN, 44, 273, 286
TITUS, 19, 171
TITUS VESPASIAN, 85
tizza, 90
t'noovah, 256
tobacco, 6, 118, 280
TOBIAS — see TOBIT
TOBIT, 18
toenails, 50
tola, 194
tola'at (-h), 194
tola'at (-h) shani, 194
tomato, 137, 138, 287; also see *Lycopersicum esculentum* & love-apple
tomb (-s), xvii, 35, 44, 70, 91, 93, 101, 107, 115, 133, 154, 196, 197, 242, 283
tonic, 48, 159
toot, 140
toothache, 139
Torah, 7, 8
torrent of thorns, 25
torrent tree, 25; also see *Acacia seyal*
TORREY, C. T., 288
TOURNEFORT, 71, 178
tower, 159, 239, 241, 242
Tower of Babel, 15
tragacanth — see gum-tragacanth
Transjordania, 13, 14, 53
treacle, 10
Treacle Bible, 10
treader (-s); treading, 213, 241, 242, 244
treasure (-s, -house), 52, 57, 82, 85
Trebizond dates, 98; also see *Elaeagnus angustifolia*
tree (-s), vii, 4, 6, 11, 18, 20, 21, 23, 25, 26, 30, 31, 32, 33, 35, 39, 46, 47, 49, 50, 54, 55, 56, 57, 58, 59, 60, 62, 63, 65, 66, 67, 68, 70, 73, 75, 76, 82, 83, 85, 87, 90, 91, 95, 98, 103, 106, 108, 119, 123, 126, 134, 135, 141, 144, 158, 160, 170, 171, 173, 175, 177, 178, 179, 180, 181, 182, 183, 185, 186, 187, 188, 189, 190, 193, 194, 196, 197, 198, 201, 202, 209, 210, 216, 224, 225, 227, 228, 239, 240, 243, 247, 248, 251, 252, 253, 254, 255, 256, 257, 277, 285, 286, 287, 291, 292
tree, acacia, 24, 25, 291; also see *Acacia* spp.
tree, algum, 89, 90, 119, 175, 176, 188, 209; also see *Cupressus sempervirens* var. *horizontalis* & *Sabina excelsa* & *Pterocarpus santalinus*
tree, almond, 36, 37, 38; also see *Amygdalus communis*
tree, almug, 90, 209; also see *Pterocarpus santalinus* & *Sabina excelsa* & *Cupressus sempervirens* var. *horizontalis*
tree, apple, 185, 190, 291; also see *Prunus armeniaca*
tree, balsam, 84; also see *Commiphora opobalsamum*
tree, bay, xix, 64, 123, 124, 152, 176; also see *Laurus nobilis*
tree, box, 24, 26, 62, 63, 89, 173, 210; also see *Acacia seyal* &

Acacia tortilis & *Buxus longifolia*
tree, broom, 201; also see *Retama raetam*
tree, caper, 65; also see *Capparis sicula*
tree, carob, xix, 72; also see *Ceratonia siliqua*
tree, cassia-bearing, 219
tree, cedar, 90, 175, 209, 285; also see *Cedrus libani* & *Sabina phoenicia*
tree, chestnut, 9, 35, 67, 180; also see *Platanus orientalis*
tree, corrupt, xv
tree, cypress, 46, 89, 90, 91; also see *Apinus pinea* & *Cupressus sempervirens* var. *horizontalis*
tree, dry, 252, 255
tree, fairest, 291
tree, fig, xvi, 73, 103, 104, 105, 106, 111, 158, 175, 190, 229, 241, 242, 243, 244, 251, 253; also see *Ficus carica*
tree, fir, 173, 175, 176, 209, 221, 248, 286; also see *Pinus halepensis*
tree, frankincense, 39, 84, 85, 117, 234; also see *Boswellia thurifera*
tree, fruit, 20, 119, 251, 278
tree, good, xv
tree, goodly, 169, 173, 177, 183, 187, 283, 290, 291
tree, green, 193, 194
tree, green bay, 123, 152, 176; also see *Laurus nobilis*
tree, green fir, xvii, 177; also see *Apinus pinea*
tree, holm, 90, 178, 194, 198; also see *Quercus ilex*
tree, Judas — see *Cercis siliquastrum*
tree, juniper, 201; also see *Retama raetam*
tree, leafy, xix, 174
tree, lentisk, vii, 177; also see *Pistacia lentiscus*
tree, lonely, 122; also see *Juniperus oxycedrus*
tree-mallow, 53, also see *Hibiscus syriacus*
tree, mastic, 177; also see *Pistacia lentiscus*
tree, mulberry, 141, 183; also see *Morus nigra* & *Populus euphratica*
tree, myrrh, 84; also see *Commiphora kataf* & *Commiphora myrrha*
tree, myrtle, 144, 221, 248, 285; also see *Myrtus communis*
tree, nut, 291
tree, oak, 179; also see *Quercus* spp.
tree of knowledge, xiv, 20, 105, 172, 184, 186, 243, 256, 286, 287, 292; figs. 35, 40, 94, 95
tree of life, iii, xv, 20, 70, 191, 256, 286
tree of oil, 98; also see *Elaeagnus angustifolia*
tree of righteousness, 194, 252, 255
tree, oil, 24, 62, 67, 97, 98, 144, 174
tree, olive, xvii, 26, 97, 103, 106, 151, 157, 158, 159, 180, 190, 240, 255; also see *Olea europaea*
tree, palm, 97, 141, 151, 169, 170, 172, 173, 180, 183, 190, 241, 255, 291; also see *Phoenix dactylifera*
tree, papaya, 275
tree, pear, 183
tree, pine, 90, 175, 176; also see *Pinus brutia*
tree, pomegranate, 190; also see *Punica granatum*
tree, rose, 205; also see *Rosa phoenicia*
tree, sacred, 196, 198; also see *Pistacia terebinthus* var. *palaestina* & *Quercus* sp.
tree, sacred bo, 285
tree, sandarac, 228; also see *Tetraclinis articulata*
tree, sandel, 189; also see *Pterocarpus santalinus*

tree, shady, 247
tree, shittah, xvii, 24, 67, 97, 144; also see *Acacia* spp.
tree, solitary, 122; also see *Juniperus oxycedrus*
tree, storax, xvii, 181, 182, 224, 225; also see *Populus alba* & *Styrax officinalis*
tree, sycamine, 140; also see *Morus nigra*
tree, sycomore, xix, 97, 157, 240; also see *Ficus sycomorus*
tree, tamarisk, 198, 227, 228; also see *Tamarix* spp.
tree, teil, 175, 178, 179, 194; also see *Pistacia terebinthus* var. *palaestina*
tree, terebinth — see tree, teil
tree, thick, 143, 144, 173, 174, 183, 195, 210, 285, 291; also see *Myrtus communis* & *Quercus* spp. & *Sabina phoenicia*
tree, timber, 93
tree, torrent, 25; also see *Acacia seyal* & *Acacia tortilis*
tree, turpentine, 57, 178, 179; also see *Pistacia terebinthus* var. *palaestina*
tree, vine, 239, 252; also see *Vitis orientalis*
tree, walnut, 119; also see *Juglans regia*
tree, willow, 216, 241, 292; also see *Salix* spp.
tree-wool, 109
trehala-manna, 278
TREMELLIUS, 132
Treponema pallidum, 149, 150
TREVOR, J. C., vii, viii, 275, 278, 279, 280, 281, 282, 283, 285, 286, 287; fig. 71
tribes (-men), 127, 150
Tribulus, 71; also see thistles
Tribulus terrestris, 71
Trichodesmium, 289
Trichodesmium erythraeum, 289
Trichoon phragmites, 172; also see pens
Trichophyton rosaceum, 98
Trigonella foenum-graecum, 34, 54; also see fenugreek
TRIMBLE, W., 272, 280
Tripoli, 68, 121, 125, 165, 218
trispium, 88
TRISTRAM, H. B., 4, 23, 25, 26, 29, 38, 40, 42, 46, 48, 52, 63, 65, 69, 77, 79, 81, 92, 98, 115, 121, 129, 153, 158, 161, 166, 175, 176, 179, 180, 185, 187, 189, 195, 197, 201, 204, 205, 207, 215, 217, 220, 221, 224, 227, 228, 233, 236, 237, 269, 273, 285; fig. 76
Triticum, 29; also see wheat & grain & corn
Triticum aegilopoides, 231
Triticum aestivum, xvii, 29, 228, 231, 233, 288, 300
Triticum aestivum var. *spelta*, 152, 231, 233; also see spelt
Triticum baeoticum, 231
Triticum compactum, 231
Triticum compositum, 228, 231
Triticum dicoccoides, 231
Triticum dicoccoides var. *kotschyanum*, 231
Triticum dicoccoides var. *straussianum*, 231
Triticum dicoccum, 234
Triticum dicoccum var. *dicoccoides*, 231
Triticum durum f. *depauperatum*, 231
Triticum hermonis, 231
Triticum hybernum, 231
Triticum monococcum, 231, 234
Triticum polonicum, 231
Triticum sativum, 231
Triticum sativum var. *spelta*, 233
Triticum spelta, 233, 234; also see spelt
Triticum thaoudar, 231
Triticum turgidum, 231
Triticum vulgare, 231
Triticum vulgare var. *dicoccoides*, 231
Troglodytic, 77
tropic (-al, -s), 6, 13, 98, 169, 214, 290

truffles, 283
trunks, 69, 70, 72, 73, 292
tryacle, 10, 55
Tryacle Bible, 10
Trypanosoma luis, 150
tsaftsafah, 217
Tsuga canadensis, 49; *also see* hemlock
Tuber, 283
tuberculosis, 143, 219
Tufts College, vii
tulbend, 235
tulip (-s), 6, 42, 116, 147, 235, 275
tulip, mountain, 235; *also see Tulipa montana*
tulip, Sharon, xix, 235; *also see Tulipa sharonensis*
Tulipa, 235, 288; *also see* tulip
Tulipa montana, 42, 77, 117, 147, 151, 205, 234, 235, 288, 289
tulipán (-o), 235
Tulipa sharonensis, xix, 234, 235, 288, 289; illus.: fig. 83; *also see* rose
tulipe, 235
Tulips, Feast of, 235
Tulpe, 235
tumble-weed, xvii, 38, 39, 279; *also see Anastatica hierochuntica* & *Gundelia tournefortii*
tumors, 167, 168, 223
tundra, 127
turban, 235
Turkestan, 89, 102
Turk (-ey, -ish, -s), 34, 55, 58, 85, 144, 191, 235
TURNER, C. H., 262
turnery, 63, 95, 158
turpentine, 98, 178, 179
turpentine-tree, 57, 178, 179; *also see Pistacia terebinthus* var. *palaestina*
TURPIN, 38
turtle-dove, 37
twigs, 31, 67
twigs, thorny, 166
twigs, willow, 132
twine, 133
TWINING, L., xv
TYAS, R., 269
TYNDAL version, 9
Typha, 50, 93, 235; *also see* reed
Typha angustata, 50, 235, 236
Typha angustifolia, 50, 236
Typha latifolia, 50, 51, 235, 236
typhoid fever, 97
typhus, 97, 169
Tyre; Tyrian, 62, 66, 67, 69, 75, 76, 95, 170, 189, 195, 205, 209, 217, 228, 232, 285
tzaphtzaphah, 216
tzaraat, 143
tzari, 55, 84
tzeetz, 256
tzeh'ehlim, 247
tzehmach, 256
tzeri, 55
Tzidon, 170
tzimcha, 256
tzitzim p'turay, 190
tz'mookim, 244
tzori; tzo'ri, 55, 84
tzrai, 177

udder, 223
ulceration (-s), 143, 219
Ulex, 287
Ulex europaeus, 287
ul-Ghawr, 92
Ulmus campestris, 175, 179; *also see* elm & thick trees
Ulva, 249
umbelliferone, 102
underworld, 140, 148, 155, 191, 244
unfermented, 213
unicorn, 9
Union County Park Commission, ii
Union Theological Seminary, vii
United Brethren, 9
United States, 4, 5, 26, 91, 110, 119, 166,
United States Army, 290
United States Department of Agriculture, 276
United States National Herbarium, vii, 205, 285

University of Southern California, vii
unleavened, 74, 157, 212, 229
unripened, 104
urethra, 219
Urginea maritima, 274
urn, xvii
URSINUS, J. H., 1, 3, 269
Urtica, 26, 221, 237; *also see* nettles
Urtica caudata, 237
Urtica dioica, 237
Urtica membranacea, 237
Urtica pilulifera, 237
Urtica urens, 237, 245
USSHER chronology, viii, 7, 278, 281
USTERI, A., 269
Ustilago tritici, 290
Uz, 53
UZZIAH, 142

Vachellia farnesiana, 5, 23; *also see* opopanax
vail (-s), 129, 130, 132
valerian, 148
Valeriana jatamansi, 148; *also see* spikenard
Valeriana officinalis, 148
VALLES, F., 1, 269
valley (-s), 13, 24, 25, 29, 30, 60, 65, 67, 68, 77, 103, 114, 119, 121, 144, 151, 169, 171, 175, 178, 184, 196, 201, 217, 221, 230, 234, 240, 244, 248, 256, 275, 284
Valley of Shittim, 25
Valonia, 197
Valonia oak, 90, 286; *also see Quercus aegilops*
VAN EYCK, 286
VAN LIEW, E. R., vii, 280
varnish (-es), 87, 102, 177, 182, 228
VARRO, 109
vat, 213
Vatican, 286
VAUCHER, 282
vegetable (-s), 38, 61, 72, 74, 101, 254, 291
vegetable kingdom, 280
vegetation, 6, 15, 68, 92
veil, 132; *also see* vail
veneer, 95
venereal, 149, 167, 168
Venetian, 214
Venice, 286
venom (-ous), 77, 80, 182
VENUS, 45, 64, 91, 145, 187, 206
Verbena hastata, 5
VERDOORN, F., ii, vii
vermifuge, 48
vermillion, 195
version, American Standard, 275, 279, 280
version, Ammonian, 8
version, Aquila, 7
version, Arabic, 8, 63, 137, 224
version, Armenian, 8
version, Authorized, 8, 9, 17, 18, 19, 20, 24, 25, 27, 29, 34, 36, 37, 38, 41, 46, 47, 49, 51, 52, 53, 57, 58, 62, 65, 68, 71, 75, 77, 79, 80, 85, 86, 89, 90, 91, 94, 97, 105, 107, 110, 113, 114, 117, 120, 124, 127, 129, 132, 134, 135, 137, 140, 141, 147, 148, 152, 165, 169, 172, 173, 174, 175, 176, 177, 179, 180, 182, 183, 188, 195, 196, 198, 201, 202, 203, 204, 205, 206, 207, 209, 215, 217, 222, 224, 275, 281
version, Bashmuric, 8
version, Basic English, 10, 28, 43, 46, 51, 59, 61, 72, 76, 80, 84, 89, 134, 139, 141, 149, 162, 171, 221, 236
version, Bohemian, 9
version, Catholic, 9, 275
version, Chaldee, 7, 124
version, Challoner, 9
version, Challoner-Rheims (revised at Douay), 10, 17
version, Challoner-Rheims-Douay (revised by O'HARA), 17
version, Chicago — see version, GOODSPEED
version, Coptic, 8
version, Coverdale, 9, 10
version, CRANMER, 9, 275
version, Danish, 237

version, Dartmouth, 275
version, Douay, 9, 10, 11, 18, 19, 23, 24, 25, 26, 27, 37, 38, 40, 43, 46, 47, 49, 51, 52, 55, 58, 61, 62, 63, 65, 71, 76, 79, 80, 81, 85, 86, 88, 89, 90, 91, 93, 97, 98, 105, 107, 109, 110, 114, 122, 123, 124, 128, 132, 134, 137, 140, 141, 142, 147, 148, 149, 152, 153, 161, 162, 166, 167, 170, 173, 174, 175, 176, 177, 178, 179, 180, 183, 184, 196, 198, 201, 203, 204, 205, 215, 216, 217, 218, 220, 222, 224, 227, 233, 234, 236, 247, 249, 254, 287, 291
version, Egyptian, 8
version, English, 8, 89, 105
version, English Revised, 275
version, Ethiopian, 8
version, Geneva, 9
version, German, 9
version, GOODSPEED, viii, 10, 11, 17, 23, 24, 26, 27, 29, 30, 31, 36, 37, 38, 39, 40, 43, 46, 47, 49, 51, 52, 54, 55, 57, 58, 61, 62, 63, 65, 72, 76, 77, 79, 80, 82, 83, 84, 85, 86, 88, 89, 90, 91, 93, 97, 98, 99, 105, 107, 109, 111, 113, 114, 116, 117, 119, 120, 122, 123, 124, 126, 128, 129, 132, 134, 137, 138, 141, 147, 149, 152, 153, 154, 161, 162, 165, 166, 168, 171, 173, 174, 175, 176, 177, 179, 180, 181, 182, 183, 184, 188, 189, 190, 196, 197, 198, 201, 203, 204, 205, 206, 209, 215, 216, 217, 218, 220, 221, 222, 224, 227, 228, 232, 234, 236, 237, 245, 247, 249, 275, 286, 291
version, Gothic, 8
version, Greek, 7, 71, 286
version, GREGORY MARTIN's, 9
version, Hebrew, 7
version, Hesychius, 7
version, Ibri, 7
version, JASTROW, 10, 11, 17, 18, 19, 23, 24, 27, 28, 29, 36, 37, 38, 40, 46, 47, 49, 51, 54, 55, 59, 62, 63, 65, 72, 76, 77, 79, 80, 81, 82, 85, 86, 89, 90, 91, 94, 97, 98, 99, 105, 107, 109, 113, 114, 117, 120, 122, 123, 124, 127, 128, 129, 132, 137, 142, 147, 151, 152, 153, 161, 166, 172, 174, 176, 177, 179, 180, 183, 184, 189, 196, 201, 203, 204, 206, 207, 209, 215, 216, 218, 220, 221, 224, 227, 232, 234, 237, 247, 249, 254, 286, 287
version, JEROME, 7, 275
version, Jewish Publication Society, 8, 17
version, King JAMES, 8, 9, 10, 11, 17, 18, 23, 24, 37, 40, 41, 43, 46, 49, 58, 59, 61, 65, 72, 76, 80, 81, 90, 93, 94, 105, 109, 114, 115, 120, 121, 123, 127, 132, 149, 151, 161, 167, 170, 176, 198, 207, 209, 216, 217, 218, 227, 232, 234, 237, 239, 245, 249, 254, 275, 279, 281, 286, 287, 288, 291; fig. 52
version, LAMSA, 10, 17, 27, 44, 46, 59, 61, 72, 80, 134, 149, 162, 170, 171, 236
version, Latin, 7, 8
version, LEESSER, 10, 17, 62, 71, 77, 80, 82, 91, 94, 98, 109, 122, 123, 124, 128, 132, 147, 152, 153, 166, 169, 175, 179, 180, 189, 198, 201, 203, 216, 218, 224, 227, 232, 234, 239, 249, 254
version, LUCIAN, 7
version, MATTHEW, 9
version, Memphitic, 8
version, MOFFATT, 10, 11, 17, 20, 23, 24, 26, 27, 29, 36, 37, 38, 39, 40, 43, 46, 47, 49, 51, 52, 54, 55, 57, 58, 61, 62, 63, 65, 72, 77, 79, 80, 82, 84, 85, 86, 88, 89, 90, 91, 93, 97, 98, 99, 105, 107, 109, 113, 114, 117, 119, 120, 122, 123, 124, 125, 127, 128, 129, 132, 134, 141, 142, 144, 147, 149, 153, 154, 161, 162, 166, 170, 171, 172, 174, 175, 176, 177, 179, 180, 181, 182, 183, 184, 187, 188, 189, 190, 196, 198, 201, 203, 206, 207, 209, 215, 216, 217, 218, 219, 220, 221, 222, 224, 227, 228, 232, 234, 236, 237, 245, 247, 248, 249, 254
version, O'CONNELL, 275
version, O'HARA, viii, 28, 43, 46,

59, 61, 72, 79, 134, 141, 149, 170, 171, 221, 222, 236,
version, ORIGEN, 7
version, Peshitta, 10
version, Polish, 9
version, Puritan, 9
version, rationalistic, 9
version, Revised Standard, 10, 17, 24, 37, 62, 63, 65, 75, 76, 77, 89, 90, 92, 94, 98, 109, 120, 122, 123, 124, 132, 134, 138, 147, 149, 166, 167, 175, 176, 178, 179, 180, 181, 201, 203, 215, 219, 221, 224, 232, 234, 275, 279
version, Samaritan, 7, 275
version, Septuagint, 7, 8, 26, 51, 63, 65, 71, 76, 85, 90, 95, 98, 108, 123, 127, 132, 137, 147, 153, 166, 176, 177, 179, 180, 188, 197, 216, 254
version, Slavonic, 8, 9
version, Spanish, 9
version, Swedish, 237, 275, 284, 287
version, SYMMACHUS, 7
version, Syriac, 8, 137, 166
version, Targum, 8, 42, 137, 147, 166
version, TAVERNER, 9, 10
version, Thebaic, 8
version, THEODOTION, 7
version, TYNDALL, 9
version, Vulgate, 7, 9, 63, 65, 85, 110, 127, 137, 147, 166, 176
version, WEIGLE, 17, 134, 141, 149, 162, 171, 221, 222, 236
version, WEYMOUTH, 10, 17, 28, 44, 46, 59, 61, 72, 113, 134, 141, 149, 162, 165, 166, 170, 171, 221, 236
version, WYCLIFFE, 9
vervain, 6, 208
vervain, blue, 5
vessels (papyrus), 92, 93
VESPASIAN, 171
VESPATIAN, TITUS, 158, 178
vetch (-es), 128, 152, 221, 234; also see *Nigella sativa* & *Lolium temulentum*
vetches, wild, 221
veterinarian (-s), 34, 35
Vicia, 152, 234
Vicia ervillea, 234
Vicia faba, 101
Vicia sativa, 134
Victorian, 93
Vienna, 286
VIETZ, fig. 62
VIGOUROUX, F., 273
VILLEGAS, E. DE, 276
Vinca minor, 145; *also see* myrtle & periwinkle
vine (-s, -yard, -yards), xv, xix, 5, 6, 10, 28, 33, 65, 78, 79, 80, 88, 103, 104, 105, 106, 111, 124, 150, 157, 159, 185, 189, 190, 202, 206, 212, 213, 217, 229, 230, 239, 240, 241, 242, 243, 244, 251, 291
vine, choice, 243; also see *Vitis vinifera*
vinedressers, 241, 244
vine, empty, 244
vinegar, 10, 27, 28, 65, 66, 78, 79, 80, 161, 222, 243
vinegar of strong drink, 212
vinegar of wine, 212
vine, grape, xvi, xvii, 239; also see *Vitis vinifera*
vine of Sodom, 78, 79, 221, 240, 244, 281; also see *Citrullus colocynthis* & *Solanum sodomeum*
vine, oriental, 239; also see *Vitis orientalis*
vine, strange, 239, 244; also see *Vitis orientalis*
vine tree, 239, 252
vine, wild, 80, 240, 281; also see *Vitis orientalis*
vineyard of Sodom, 79
vintage, 68, 212, 214, 241, 244, 289
violet, 114; *also see* lily
VIREY, J. J., 269
VIRGIL, 38, 150
virgin, 55, 213, 241
Virgin, Blessed, 44
Virgin MARY, 38, 287
virility, 64

virus, 143
Viscum album, 285
VISHNI, 206
VISHNU, 154
Vitis orientalis, 239; *also see* Oriental vine
Vitis vinifera, xvii, xix, 79, 213, 239, 240, 242, 287, 289; illus.: figs. 31, 32, 71, 95; *also see* grape & vine
v'kaneh hatov, 40
VOGL, 282
volcanic cones, 14
VRIES, HILDE DE, fig. 50
VULCAN, 91
Vulgate version — *see* version, Vulgate
Wacholder, 37
wadies; wady, 13, 25, 26, 91, 151, 217
Wady Arabah, 30, 38
Wady es Sheikh, fig. 30
Wady of Willows, 217
Wady Seyal, 25
Wady Sunt, 25
wafers, 86, 126, 157, 212, 229
WAGNER, R., 38
wakeful; wakeful-tree; waketree, 11, 36, 37, 38; also see *Amygdalus communis*
WALDSTEIN, A. S., 269
Wales, 35
WALKER, WINIFRED, vignettes by, viii, xvii, 64, 99, 110, 122, 135, 145, 155, 199, 225, 236, 249, 257, 247, 277
walking-sticks, 50, 51
wall, (-s), xix, 43, 65, 66, 111, 142, 143, 159, 160, 161, 162, 169, 191, 212, 221, 240, 244, 254, 257
walnut (-s), 6, 14, 119, 120, 159; also see *Juglans regia*
walnut-bower, 119
walnut, common, 119
walnut, English, 119
walnut, Persian, 119, 209
walnut shell, 120
WALTON, R. A., 259
Wandering Jew, 198
War of the Roses, 206
WARBURG, O., 42, 115, 116, 269, 279, 285
WARBURTON, 69
wash (-ed, -ing), 215, 216, 240
washing-balls, 215; also see *Salicornia* spp. & *Salsola* spp.
Washington, 282
wasp, fig. 106
wasps, 208
waste (-land), 63, 220, 221, 237, 239
watches, 104
watchman, 88
watchtower, xix
water (-s), xv, 24, 28, 45, 48, 49, 50, 51, 53, 62, 66, 67, 68, 71, 76, 79, 84, 92, 93, 112, 117, 120, 133, 151, 155, 160, 169, 180, 181, 189, 193, 194, 197, 203, 205, 207, 209, 211, 224, 231, 234, 235, 241, 247, 249, 251, 252, 276, 278, 282, 288, 289, 292
water buffaloes, 276
watercourse (-s), 5, 6, 23, 29, 119, 151, 152, 216, 217, 291
water-cress, 6, 74, 140; also see *Nasturtium officinale* & bitter herbs
water gladiole, 62; also see *Butomus umbellatus* & flag
water-hemlock, 49; also see *Cicuta maculata* & *Artemisia* spp.
water, holy, 161
watering-troughs, 182
water-lilies; water-lily, 6, 42, 154, 155, 218; also see *Nymphaea* spp.
water lily, European white, 154
water lily, white, 92
water lily, yellow, 92
WATERMAN, L., 17
watermelon, 80, 81; also see *Citrullus vulgaris*
water of gall, 48, 78, 79, 80; also see *Citrullus colocynthis*
water-willow (-s), 216, 247; also see *Salix* spp.
wax, 166, 173

Weary-glen, 184
weather, 139, 276
weaving, 133
WEBSTER, N., 83
wedding (-s), 34, 87, 119, 288
WEDEL, G. W., 269
weed (-s, -y), 6, 27, 29, 30, 46, 49, 60, 74, 134, 135, 220, 221, 239, 245, 249, 275, 279, 283
weeds, foul, 29; also see *Agrostemma githago* & *Solanum incanum*
weeds, noisome, 29; also see *Agrostemma githago* & *Solanum incanum*
weeds, noxious, 29; also see *Agrostemma githago* & *Solanum incanum*
weeds, poisonous, 49; also see *Artemisia* spp.
weeds, sea, 249; also see *Zostera marina*
weeds, water, 249; also see *Zostera marina*
WEEKS, M. E., 269
WEIGLE, L. A., 10, 260
WEIGLE version — *see* version, WEIGLE
well (-s), 105, 119, 183
Welsh (-men), 35
Wertheim Bible, 9
WESLER, H., viii, 288, 289
WESSELY, 129
West Indies, 110
WESTMACOTT, W., 2, 270
Westwood University, 277
WEYMOUTH, R. F., 10, 260
WEYMOUTH version — *see* version, WEYMOUTH
whale (-s), 9, 247
wheat, xvii, 6, 14, 29, 44, 71, 82, 84, 101, 103, 111, 113, 116, 128, 133, 134, 137, 138, 144, 153, 157, 158, 166, 189, 211, 212, 229, 230, 231, 232, 233, 234, 240, 243, 254, 257, 288, 289, 290, 300; also see *Triticum* spp.
wheat, beaten, 230
wheat, bewitched, 134; also see *Lolium temulentum*
wheat, club, 231
wheat, composite, 231
wheat, Egyptian, 231
wheat, hard, 231
wheat, Heshbon, 231
wheat, hulled, 232, 233
wheat, mummy, 231
wheat-nut, 282
wheat, one-grained, 231
wheat, Polish, 231
wheat, Poulard, 231
wheat, wild, 231
wheel, 38, 39, 232, 279; also see *Anastatica hierochuntica* & *Gundelia tournefortii*
WHEELER, L. C., vii, 205
whin, 287
whirling dust, 38
whirlwind, 38, 39, 202, 279; also see *Anastatica hierochuntica* & *Gundelia tournefortii*
white aconite, 30; also see *Aconitum napellus* & cockle
white broom, xix
white mulberry, 5
White Nile, 92
WHITE, W. E., 269
WHITE, W. L., 272, 290
Whitsuntide, 64
whortleberry, 145; also see *Myrtus communis*
wicker basket, 92
wicker-work; wicks, 133, 218
wilderness, xix, 24, 26, 31, 56, 62, 64, 65, 67, 82, 84, 97, 121, 122, 125, 126, 127, 144, 147, 167, 169, 189, 201, 203, 206, 210, 216, 217, 248, 252
Wilderness of Sin (-ai), 23
wild gourd, 79, 281; also see *Ecballium elaterium*
wild-rice, 134; also see *Zizania aquatica* & tares
WILKIN, S., 262
willey, 197
WILLIAMS, M. O., 270
willow (-s), 6, 14, 152, 173, 180, 183, 184, 216, 217, 218, 241, 247, 255, 288, 291, 292; also see *Salix* spp.

willow, brittle, 217
willow, crack, 217
willow, Palestine, 217
willow, weeping, 5, 217
WILSON, A. M., 273
WILSON, C. W., iv, 273; figs. 1, 14, 15, 19, 23, 25, 28, 29, 30, 84, 86, 87, 88
wind (-s), 15, 38, 39, 50, 51, 75, 104, 114, 127, 128, 173, 190, 229, 241, 251, 252, 282, 289
windflower, 42, 45; also see *Anemone coronaria* & lily of the field
winding-sheets, 131
wine, 24, 27, 28, 49, 51, 57, 72, 79, 80, 82, 83, 86, 87, 106, 111, 141, 157, 158, 169, 170, 191, 212, 213, 214, 228, 229, 230, 232, 240, 241, 242, 243, 256, 288,
wine, bitter, 28
wine, common, 28
wine, mingled, 214, 288
wine, new, 213
wine, old, 213, 214
wine, sour, 28
wine, spiced, 190, 191
wine, sweet, 213
wine, unfermented, 214, 288
winebibbers, 212
winedressers, 242
winefat, 213, 241, 242, 243
wine-making, 242
winepress, 212, 213, 229, 239, 241, 242, 243, 244
WINER, 79
winevat, 213
wing-shell, 224
winnowed, 232
winter, 15, 37, 45, 78, 104, 124, 127, 275, 288
Wisdom of BEN SIRA, 288
Wisdom of SIRACH — see ECCLESIASTICUS
Wisdom of SOLOMON, xvii, 19
Wise Men, 84, 276, 277
witch (-craft, -es), 45, 49, 120, 124, 139, 160, 208
wither (-ed, -eth), 28, 50
withs, green, 216, 217, 218; also see *Salix* spp.
Wittenberg, 9
woad, dyers, 195; also see *Isatis tinctoria*
wold, 122
WOLSTEAD, 31
WOOD, W., fig. 59
wood (-s, -y), 5, 25, 30, 47, 56, 57, 63, 68, 69, 70, 73, 83, 90, 91, 93, 95, 116, 119, 159, 160, 173, 174, 175, 176, 183, 184, 185, 189, 194, 201, 202, 207, 210, 211, 225, 228, 232, 239, 248, 251, 252, 253, 254, 255, 284, 285, 287

wood, acacia, 24; also see *Acacia* spp.
wood, aspen, 184; also see *Populus euphratica*
wood, cedar, 193; also see *Cedrus libani* & *Sabina phoenicia*
wood, citron, 228; also see *Tetraclinis articulata*
wood, citrus, 228
wood, fig, 106; also see *Ficus carica*
Wood, J. G., 272
wood, setim, 24; also see *Acacia* spp.
wood, shittim, xvii, 24; also see *Acacia* spp.
wood sweet, 228; also see *Tetraclinis articulata*
wood, thyine, 228; also see *Tetraclinis articulata*
wood carving, 63
WOODCOCK, H. B. D., 274; fig. 64
WOODWARD, C. H., vii, 4, 270
WOODWARD, J. D., fig. 87
wool (-en), 78, 130, 131, 194, 216, 290
work, lily-, 154
worm (-s), 3, 125, 126, 203
wormwood, xix, 48, 49, 78, 79, 91, 278, 279; also see *Artemisia* spp.
worship (-ped), 33, 91, 111, 150, 151, 190, 197
wounds, 55, 149, 158, 223
wreath (-ed, -s), 77, 111, 145, 148, 165, 166, 190
wren, 14
WRIGHT, A. E., 270
WRIGHT, C. H. H., 259
WRIGHT, W., 270
writers; writing (tablets), 63, 93, 124, 172, 173, 186
wrought-wood, 188; also see *Pterocarpus santalinus*
WYCLIFFE version, 9

Xanthium antiquorum, 245; also see thorns
Xanthium brasilicum, 245
Xanthium spinosum, 165, 202, 245, 248
Xanthium strumarium var. *antiquorum*, 245

ya'ar, 225
Yale University, 288
ya'rak, 255
yard, 50
yarn (linen), 132
yayin, 213, 243
yea'rakon, 254
yeast (-s), 141, 211, 212; also see *Saccharomyces cerevisiae*
yeast, bakers, 212

yeast, brewers, 212
yeast, distillers, 212
yeast, wine-ferment, 213
yehter lach, 217
yekev, 213
Yemen, 13, 82, 84, 184
yêrek, 28
yether lach, 217
yew, 189
yirakon, 142, 254
Yom Kippur, 291
YONGE, C. D., 270
YOUNG, J., 268, 270
YOUNG, R., 270
Y'richo, 171
y'tarim lachim, 217
yule log, 33

ZACCHAEUS, 107; fig. 62
ZACHARIAS — see ZECHARIAH
zackum (oil), 98
za'faran; zafran, 87
zait (-h), 98, 159
ZARAH, 195
za'tim, 98
Zauberwurzel, 138
zawan, 134
zayit, 158, 159
zaytim, 158, 159
Zea mays, 232; also see corn
ZECHARIAH, 18
ZEDEKIAH, 211
zehra, 256
ZELLER, H. G., 270
ZEPHANIAH, 18, 27, 237
ZEPHYR, 206
ZEPHYRUS, 45, 114
zeronim, 254
ZEUS, 182, 183, 191
Zion, 88, 183, 241
zizania; zizanion, 134
Zizania aquatica, 134; also see wild-rice & tares
Zizyphus, 165, 203
Zizyphus jujuba, 248; also see thorns
Zizyphus lotus, 247; also see shady trees & lotus-bushes
Zizyphus officinarum, 248
Zizyphus paliurus, 165
Zizyphus sativa, 248
Zizyphus spina-christi, 165, 202, 245, 248, 284, 289; also see crown of thorns
Zizyphus vulgaris, 206, 248
ZOHARY, M., viii, xv, 272, 273, 280, 281, 282, 284; figs. 24, 46
Zoroaster (-ian), 77, 91, 206
Zostera marina, 249; also see weeds & grass-wrack
Zostera minor, 249
Zostera nana, 249
zov, 150
zukkum, 98
zuwan, 134

PLATES

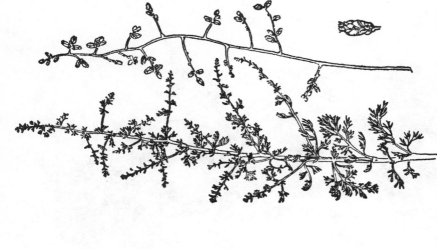

FIGURE 43. — *Artemisia herba alba*, wormwood, a strongly aromatic, woolly plant, which, because of its intensely bitter juice, was considered (with "gall") as symbolic of bitter calamity and misfortune. (Crowfoot & Baldensperger, From Cedar to Hyssop, 1932).

FIGURE 42. — *Alhagi camelorum* var. *turcorum*, the camel's-thorn. This spiny shrub is found in Lebanon, Syria, and Palestine, and yields a sweet fragrance. It is referred to in the Apocrypha as "aspalathus". (Callcott, Scripture Herbal, 1842).

FIGURE 45.—*Loranthus acaciae*, the acacia strap-flower. The crimson flowers of this parasitic mistletoe, growing abundantly on yellow-flowered thorn-bushes, may well have given to Moses the impression that the bush was aflame but "not consumed" by the flames, as reported in Exodus. (After a drawing by W. H. Fitch in H. C. Hart, Flora of Sinai, 1891).

FIGURE 44.—*Aquilaria agallocha*, the eaglewood. This and the sandalwood (*Santalum album*) were probably the source of the Old Testament "aloes", but not those of the New Testament. The eaglewood is a tall tree native of northern India, Cochin-china, and Malaya. (After Hooker, Icones Plantarum, 1837.)

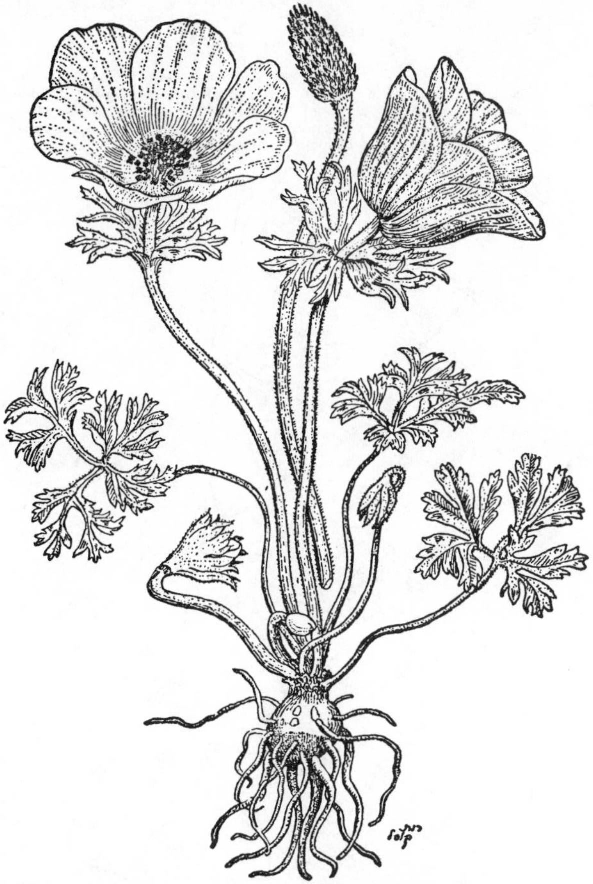

FIGURE 47. — *Laurus nobilis*, the bay-tree. Having evergreen leaves among trees whose leaves were mostly shed in the autumn, the bay-tree and true myrtle became symbolic of the prosperity, good fortune, and wealth, which enables man to flourish even in the time of cold adversity. The Greeks and Romans used the leaves to crown their heroes and immortals to signify that their fame would not fade. (After Hegi, Illustrierte Flora von Mittel-Europa, 1913).

FIGURE 46. — *Anemone coronaria*, the Palestine anemony. These plants are now generally accepted as the "lilies of the field" mentioned in the Sermon on the Mount. One of the most abundant plants of Palestine, the anemony is still common on the Mount of Olives and must have been well known to all members of Jesus' audience. It varies in color from scarlet or rose to blue, purple, and even yellow or white. (Feinbrun and Zohary, Iconographia Florae Terrae Israëlis, 1949).

FIGURE 48.—*Acanthus spinosus*, the leaves of which have furnished the models for the scroll decorations so popular in art. It is a close relative of the Syrian acanthus, incorrectly called "nettles" in the Old Testament. (Botanical Magazine, 1816).

FIGURE 49.—*Aloë succotrina*, source of the "aloes" brought by NICODEMUS to embalm the body of JESUS. The art of embalming with the juice of this plant was learned by the Israelites from the ancient Egyptians. (Bentley and Trimen, Medicinal Plants, 1879).

FIGURE 51.—*Capparis sicula*, the Palestine caper, among rocks near Sichron Ysakob. Being supposedly aphrodisiac, it is referred to in Ecclesiastes in connection with old age. (Photograph by Dr. Bandmann, courtesy M. Evenari, Jerusalem).

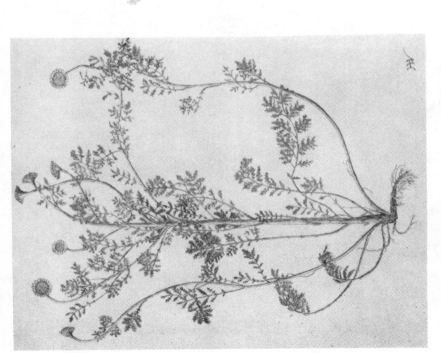

FIGURE 50.—*Anthemis palaestina*, the Palestine chamomile. Some distinguished students of Biblical botany feel that the "lilies of the field" were not the showy anemonies depicted in fig. 46, but flowers of which the innate beauty was not obvious and had to be pointed out carefully on a plant present at the time of hay-gathering. They suggest this native daisy-like chamomile. (Drawing by Hilde de Vries).

FIGURE 52.— *Cedrus libani*. King SOLOMON sent 183,300 men to Lebanon to cut down these mighty cedars for the building of the Temple, the House of the Forest of Lebanon, the Porch of Judgement, and the house for PHARAOH's daughter. (Wood engraving after a drawing by Gustave Doré; see also fig. 1 and p. 64).

FIGURE 54.—*Citrullus colocynthis*, the colocynth. The fruit of this watermelon-relative contains an intensely bitter pulp which is a drastic cathartic. It is referred to as "gall" in the Old and New Testaments. These handsomely colored fruits were the "wild gourds" of ELISHA. (Bentley and Trimen, Medicinal Plants, 1880).

FIGURE 53.—*Cistus salvifolius*, one of the Palestinian rockrose species from which the gum known as "ladanum" is obtained during the hottest hours of the day, often gathered in the beards of browsing goats. It is referred to as "myrrh" in some passages in the King James version of the Bible. (Photograph by Dr. Bandmann, courtesy M. Evenari, Jerusalem; see also p. 122).

FIGURE 55.— *Arundo donax*, the giant reed, growing on the banks of the Crocodile River near Binyamina, much as it must have appeared in Bible times. (Photograph by Dr. Bandmann, courtesy M. Evenari, Jerusalem).

FIGURE 56.— *Ceratonia siliqua*, the carob-tree. In JESUS' parable of the Prodigal Son mention is made of "husks that the swine did eat". These "husks" were the long locust-like pods of the carob-tree, eaten by the poorest people and by livestock in Mediterranean and Near Eastern lands even today. (The Garden, 1878).

FIGURE 57.—*Coriandrum sativum*, the coriander, whose pearl-like seeds are quite aromatic. They were employed in Bible times as a condiment in soups, curries, and wines, and as a medicine. (Sibthorp, Flora Graeca, 1819; see also fig. 18).

FIGURE 58.—*Crocus sativus*, the saffron crocus. It requires at least 4,000 stigmas from this plant to make an ounce of commercial saffron, used for coloring curries and stews, as a yellow dye, in medicine, and as a scent. (Crispyn van de Pas, Hortus Floridus, 1614).

FIGURE 59.—*Cyperus papyrus* on the banks of the Nile. In Old Testament times this paper reed was very common on the banks of the Nile and in Palestine, but it is now very rare in Lower Egypt and most of Palestine. (Lithograph by W. Daniell in W. Wood's Zoography, 1807; see also fig. 63).

FIGURE 61.—*Ferula galbaniflua*, source of the Biblical "galbanum". It is native to Syria and Persia, and yields an ill-scented yellowish or brownish gum-resin used in ancient medicine. (Bentley and Trimen, Medicinal Plants, 1877).

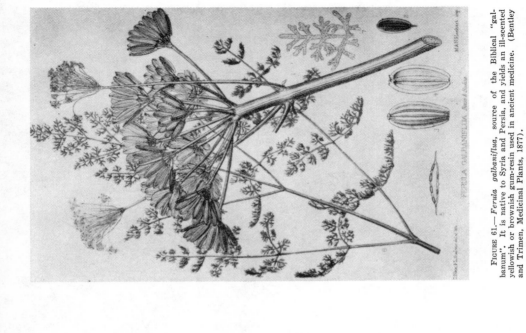

FIGURE 60.—*Cynomorium coccineum*, a scarlet- or crimson-colored parasitic plant which attaches itself to the roots of various woody plants in salt marshes and on maritime sands throughout the Near East and Mediterranean areas. In times of scarcity it is eaten, and probably comprised the "juniper roots" of Job. (Engler and Drude, Vegetation der Erde, 1910, after Hooker, Weddell, and Hofmeister).

FIGURE 62.— *Ficus sycomorus*, the "sycomore" of the Bible. This is the tree which ZACCHAEUS climbed in order to see JESUS pass by. The prophet AMOS tended orchards of these trees. Their fruits were the figs eaten by the poorest of the people; the wood was used to make mummy coffins. (Vietz, Mediz.-ökon.-techn. Pflanzen, 1822).

FIGURE 63.— *Cyperus papyrus* in the Huleh Swamp. This plant is the Egyptian "bulrush" or "paper reed" of the Bible. It furnished the papyrus, or paper, for ancient manuscripts; its stems made the floating cradle into which MIRIAM placed the baby MOSES. (Photograph by M. Evenari, Jerusalem; see also fig. 59).

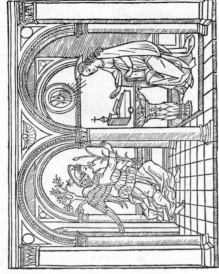

FIGURE 64 (*left*).—*Lilium candidum*, the Madonna lily. Most of the "lilies" of the Old Testament were blue hyacinths, but the Madonna lily has recently been proven native to Palestine and may well have been far more abundant there in ancient times. (Drawing by Claude Aubriet, 1651-1743; from Woodcock and Stearn, *Lilies of the World*, 1950). — FIGURE 65 (*above*).—Wood cut by PACHEL depicting the Annunciation. The angel GABRIEL, bearing a Madonna lily in his hand, tells MARY that she will be the mother of JESUS. The Church instructed artists in medieval times always to employ the Madonna lily when representing this scene. The three blossoms on the stalk suggest the presence of the Trinity. The lily was symbolic of innocence, virginity, and purity, although it is not at all certain that it is a legitimate Bible plant. In this connection see also the initial on p. 17 and fig. 89. (Missale Romanum, 1499; courtesy Metropolitan Museum of Art).

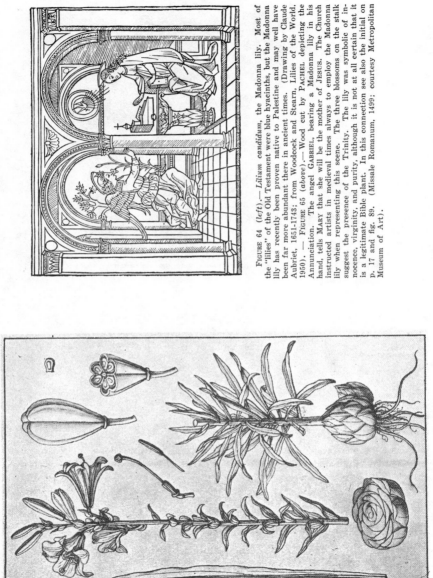

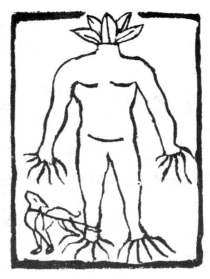

FIGURE 66 (*above*).—Much superstition and folklore are attached to the mandrake. Its roots often have a somewhat human configuration and were considered to be aphrodisiac and to ensure conception. Elaborate ceremony was required to dig them; a black dog had to be tied to the plant in order to pull it from the ground, whereupon it was supposed to shriek in agony and the dog to die! Plants with darker color and more feminine form were called "womandrakes". (Woodcut from Apuleius Barbarus (Pseudo-Apuleius), Herbarium, 1481?, probably the first printed book with a series of plant illustrations; courtesy Metropolitan Museum of Art). — FIGURE 67 (*below*).— *Mandragora officinarum*, the mandrake, the fruits of which were gathered by REUBEN and figure so largely in the story of JACOB, LEAH, and RACHEL. The yellowish fruits are the size of a plum and lie like bird's eggs in the center of a handsome rosette of leaves. (Photograph by Dr. Bandmann, courtesy M. Evenari, Jerusalem).

FIGURE 68.—*Myrtus communis*, the true myrtle. The myrtle of the Bible has nothing to do with the various plants called "myrtle" in America, but is an evergreen tree used especially at the Feast of the Tabernacles. The Greeks and Romans made wreaths of it to crown their heroes. (Sibthorp, Flora Graeca, 1825; see also p. 145).

FIGURE 69.—*Nerium oleander*, the oleander. Like the narcissus, the oleander is also referred to as a "rose" in English translations of the Bible. It forms groves and thickets along Palestinian watercourses. (Sibthorp, Flora Graeca, 1819).

FIGURE 70.— *Phoenix dactylifera*, the date palm, seen in a view near Haifa. In Old Testament days the date palm was as abundant in Palestine as in Egypt. Thoughtless exploitation of the land largely eliminated the trees in later years, except where carefully cultivated. (Photograph courtesy M. Evenari, Jerusalem; see also fig. 9, fig. 28, and fig. 92).

FIGURE 71.— Watchtower in a grape vineyard (*Vitis vinifera*). "I will stand upon my watch, and set me upon the tower, and will watch to see what people say unto me, and what I shall answer when I am reproved". (Habakkuk 2:1). (Recent photograph by John C. Trevor; see also fig. 31 and fig. 32).

FIGURE 73.—*Pinus halepensis*, the Aleppo pine, growing in the yard of the Rockefeller Museum in Jerusalem. This tree is referred to as "fir tree" and "ash" in the Bible. (Photograph by H. R. Oppenheimer).

FIGURE 72.—*Lolium temulentum*, the bearded darnel-grass. This is the plant erroneously referred to as "tares" in most English versions of the New Testament. It is a strong-growing grass greatly resembling a cereal grain in its early stages. A fungus growing in the seeds gives them an ergot-like poisonous property which is transmitted to flour and bread made from grain mixtures containing them. (Courtesy U. S. National Herbarium, Washington; see also fig. 41).

FIGURE 74.—*Pistacia lentiscus*, the lentisk. This tree is referred to as "mastic-tree" in the book of Susannah, and its gum-resin is one of the substances called "balm" in Genesis. (Sibthorp, Flora Graeca, 1840).

FIGURE 75.— *Pistacia terebinthus*, the common terebinth, closely related to the Palestine terebinth referred to in the Bible as "teil tree", "turpentine-tree", and "elm". (Sibthorp, Flora Graeca, 1840).

FIGURE 76.— *Punica granatum*, the pomegranate. Its flowers served as models for the "golden bells" in the Temple decorations and on the robes of the priests. From its fruit were made sherbets and cooling drinks; juice from its rind tanned leather. Its mature calyx served as model for SOLOMON's crown. (Courtesy U. S. Department of Agriculture; see also p. 199 and fig. 38).

FIGURE 77.— *Prunus armeniaca*, the apricot. This is supposed to be the "apple tree" of the Old Testament. Of the many trees suggested, this is the only one which fits both the text and context of all Biblical passages involved. It was introduced into Palestine before the time of NOAH. (Photograph by J. H. Lovell, courtesy Arnold Arboretum of Harvard University).

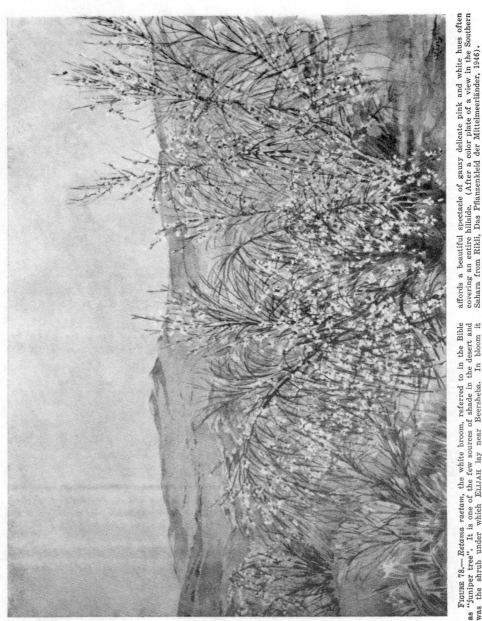

FIGURE 78.—*Retama raetam*, the white broom, referred to in the Bible as "juniper tree". It is one of the few sources of shade in the desert and was the shrub under which ELIJAH lay near Beersheba. In bloom it affords a beautiful spectacle of gauzy delicate pink and white hues often covering an entire hillside. (After a color plate of a view in the Southern Sahara from Rikli, Das Pflanzenkleid der Mittelmeerländer, 1946).

FIGURE 79.—*Solanum incanum*, the Palestine nightshade, supposed to be one of the spiny plants referred to as "briers" in the Bible. It is now frequently known as "apple of Sodom", and is abundant in the lower Jordan Valley and about the Dead Sea. (Sibthorp, Flora Graeca, 1819).

FIGURE 81.—*Salvia judaica*, the Judean sage. It is supposed that the stiffly branched inflorescence of this plant served as model for the seven-branched candlestick of the Temple. (Photograph by M. Evenari, Jerusalem).

FIGURE 80.—*Salsola kali*, the prickly saltwort. This was one of several plants commonly burned in the making of soap. The potash-rich ashes were mixed with olive oil and solidified. Soap made in this way still forms a considerable article of trade from Joppa and other Mediterranean ports. (Moss, Cambridge British Flora, 1914).

FIGURE 82.—*Sorghum vulgare* var. *durra*, the dhura. The 6- to 20-foot long stems of this plant are thought by many authorities to have been the "reed" or "hyssop" used to convey the vinegar-soaked sponge to the lips of JESUS on the cross. (Journal of the New York Botanical Garden, 1941).

FIGURE 83.—*Tulipa sharonensis*, the Sharon tulip, growing in huge numbers in a field on the Plain of Sharon. It is widely held by experts that this is the "rose of Sharon" of the Song of Solomon and perhaps also the "flower of roses in the spring of the year" mentioned in Ecclesiasticus. (Courtesy American Forests, 1929).

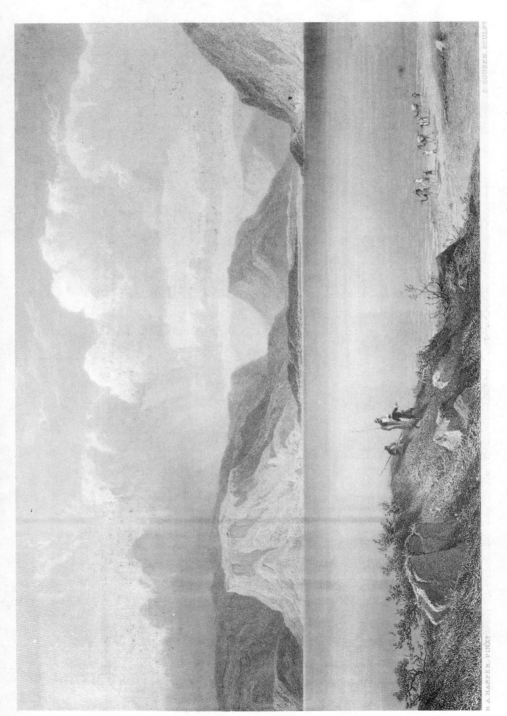

FIGURE 84.—A view of the Sea of Galilee from the heights of Safed. The low trees in the foreground are *Quercus coccifera* var. *pseudococcifera*. (Steel engraving after H. A. Harper from C. W. Wilson's Picturesque Palestine, 1883; see also fig. 29.)

FIGURE 85.— SOLOMON, we are told, "spake of trees, from the cedar tree that is in Lebanon even unto the hyssop that springeth out of the wall." There is much controversy about the identity of this "hyssop". Many plants grow in crevices of walls in Palestine, as is well illustrated in this photograph. Leading contenders for the honor are the caper and the Syrian marjoram. (Photograph by Bonfils; from C. E. Moldenke's Holy Land photograph collection).

FIGURE 86.— A typical scene in the "wilderness" (desert) of Judea, in southern Palestine, showing the Dead Sea and the mountains of Moab at sunrise. (Wood engraving from C. W. Wilson's Picturesque Palestine, 1883).

FIGURE 87.— A view of dome-shaped Mount Tabor in northern Palestine the crown of which is 1800 feet above sea level. The broad-topped trees are *Quercus ithaburensis*; those to the right of the church are columnar cypresses. The parsnip-like plant in the lower left is the common giant-fennel, *Ferula communis*; to its right is a hollyhock-like mallow, *Althaea hirsuta*. In back and to the right of the sheep are holy thistles, *Silybum marianum* (see also fig. 92). (Steel engraving after J. D. Woodward from C. W. Wilson's Picturesque Palestine, 1883).

FIGURE 88.—A view of the valley of Nazareth. The prickly-pear cactus (*Opuntia ficus-indica*), in the foreground, is an American plant, not known in the Old World in Bible times, but now very common there and for this reason often erroneously included in paintings purporting to show Biblical characters in Biblical scenes. (Steel engravings after H. Fenn in C. W. Wilson's Picturesque Palestine, 1883).

FIGURE 89.— Many artists have attempted to depict the Sermon on the Mount and especially the beautiful parable concerning the "lilies of the field" which were more beautiful than SOLOMON "in all his glory." In numerous instances the artists have used the Madonna lily *(Lilium candidum)* to illustrate these "lilies", as is the case in this late 19th century print from Peterson's Monthly. See also the initial on p. 17, and figs. 64-65 in this connection. (Culver Service Photos).

FIGURE 90 (right).— ESAU eating the "red pottage" *(Lens esculenta,* the lentil plant), for which he had sold his birthright. Around the frame twine lentil plants of two varieties. An interesting engraving from JOHANN SCHEUCHZER's (1684-1738) monumental *Kupferbibel* (1731/35), for generations the standard work on the plants and animals of the Bible, the most elaborate natural history of the Bible ever attempted, still praised for its illustrations by CUVIER. The *Kupferbibel* is today chiefly of interest from the point of view of the paleontologist, the anatomist, and the historian of art, as SCHEUCHZER did not have a very direct knowledge of the Flora of the Terra Biblica. Valuable also remains his extensive bibliography. Most of the engravings, made after drawings by J. M. FÜSSLI, incorporated, in an encyclopedic manner, the knowledge of the period concerning botany, zoology, astronomy, geography, architecture, etc., with interesting emblematic and symbolic designs. "Auch die aus dem Ornamentstich bekannten Motive, Phantasieumrahmungen aller Art, sind in vorbildlicher Form hier zu finden. Die von PREISSLER geschaffenen, abwechselnden 'Nebenzierraten' gehören zu den phantasievollsten Leistungen ihrer Gattung. So bildet die Kupferbibel einen Höhepunkt der Augsburger graphischen Kunst des 18. Jahrhunderts, gleichzeitig ein wichtiges Denkmal des Tafelwerkes". (LANCKORONSKA-OEHLER 1:32). (Courtesy Widener Library of Harvard University).

GENESIS Cap. XXV. v. 34.
Esauus ΦΑΚΟΦΑΓΟΣ.

I. Buch Mosis Cap. XXV. v. 34.
Esaü Linsen Speise.

PRIMVS OMNIVM PECCAVIT LVCIFER.   25.

Quis sese primus scelere incestauit iniquo?
Lucifer: æternâ Veri flatione relicta.

Die vuasser vors so sor, die eerst sonde heeft ghedaen,
Lucifer: die in Godt, noost waerheyt, en bleef staen.

Sçait on la matrice, D'ou vient la malice, Du premier peché?
C'est l'orgueil damnable, De ce miserable, Du Ciel trebuche.

FIGURE 91.—The Temptation in the Garden of Eden, in a less usual, interesting representation, reproduced from one of Père DAVID's Emblem Books. The fruit here depicted seems plainly to be the apple. (Veridicus Christianus, Officina Plantiniana, 1601; courtesy Houghton Library of Harvard University).

FIGURE 92.—During the flight of JOSEPH and MARY and the infant JESUS into Egypt to escape the soldiers of HEROD, according to legend, a date palm (*Phoenix dactylifera*) bent down its head as they passed and permitted them to feast on its fruit. Depicted is the famous engraving by MARTIN SCHONGAUER (1445-1491) of this scene. The other trees are *Dracaena ombet* (not a bible plant) and the plant by the donkey's head is the holy thistle or Our Lady's thistle (*Silybum marianum*). Approximately the same plants are shown in DÜRER's well known "Flight into Egypt". (Courtesy Metropolitan Museum of Art).

FIGURE 98.— The Gathering of the Manna, "the bread which the Lord hath given you to eat", (Exodus xvi.15), as painted by TINTORETTO (about 1594). This famous painting, in the San Giorgio Maggiore, Venice, with its companion piece, "The Last Supper", are concerned with the idea of the gift of daily bread and its spiritual connection with heavenly bread. The white pebble-like material on the ground in the foreground represents the artist's conception of the "manna" that "fell from heaven", now thought to have been *Lecanora* lichens (see fig. 27). (Courtesy Fogg Museum of Harvard University).

FIGURE 94.—The Garden of Paradise, a famous oil painting by PETER PAUL RUBENS (1577-1640), in the Hague Gallery, with a singularly complete and accurate assemblage of Bible animals. The fruit on the Tree of Knowledge of Good and Evil, with the serpent coiling among its branches, seems definitely to be apples. This painting is probably one of those to which we owe the presently widely held misconception that the apple is a Bible plant. Though a formidable landscapist and a lover of gardens and animals, RUBENS often had the collaboration of JAN BRUEGHEL, ("the Velvet BRUEGHEL", 1568-1625), originally a flower painter, who painted most of the birds and flowers in this painting. It shows approximately the same avifauna as we meet in ROELANDT SAVERY'S (1576-1639) splendid Landscape with Birds in the Vienna Museum. (Courtesy Fogg Museum of Harvard University).

FIGURE 95.—ADAM and EVE by LUCAS CRANACH the Elder (1472-1553), one of the most beautiful representations of the Temptation in the Garden of Eden, by the German Master, Court Painter to the House of Saxony. Here, again, the apple has been chosen as the Tree of Knowledge of Good and Evil. In the background is a laurel (*Laurus nobilis*) and in the foreground a grape vine (*Vitis vinifera*) with conveniently placed leaves. The lion, king of beasts, lying down peaceably with deer, is symbolic of the peace which was supposed to have existed among all wild animals before the "fall" of man. The Renaissance painters, as KENNEDY has said, transferred every sacred story into terms of the daily life they knew and loved retaining their right to rely on their imagination when they chose. (Courtesy Fogg Museum of Harvard University).